FOR USE IN LIBRARY ONLY
T5-BCJ-274

The Handbook of New Zealand Mammals

The Handbook of New Zealand Mammals

Edited by Carolyn M. King

Auckland
OXFORD UNIVERSITY PRESS
Melbourne Oxford New York
in association with
The Mammal Society,
New Zealand Branch

RANDALL LIBRARY UNC-W

Oxford University Press

Oxford University Press, Walton Street, Oxford OX2 6DP

OXFORD NEW YORK TORONTO
DELHI BOMBAY CALCUTTA MADRAS KARACHI
PETALING JAYA SINGAPORE HONG KONG TOKYO
NAIROBI DAR ES SALAAM CAPE TOWN
MELBOURNE AUCKLAND
and associated companies in
BERLIN IBADAN

Oxford is a trade mark of Oxford University Press

First published 1990
©C. M. King 1990

This book is copyright.
Apart from any fair dealing for the purpose of private study, research, criticism, or review, as permitted under the Copyright Act, no part may be reproduced by any process without the prior permission of the Oxford University Press.

The publisher, editor, and contributors gratefully acknowledge the assistance of:

the New Zealand Lottery Board,

the John Ilott Charitable Trust,

the Ministry of Agriculture and Fisheries,

and the Department of Scientific and Industrial Research.

ISBN 0 19 558177 6

Cover illustrated by Ulco Glimmerveen and designed by Chris O'Brien
Typeset in Bembo by Rennies Illustrations Ltd.,
and printed in Hong Kong
Published by Oxford University Press
1A Matai Road, Greenlane, Auckland 5, New Zealand

Endpapers

IFC (a) Colour variation among species of wallaby now or formerly present in New Zealand (Table 6). A, *Macropus eugenii;* B, *M. rufogriseus;* C, *M. parma;* D, *M. dorsalis;* E, *Petrogale penicillata;* F, *Wallabia bicolor* (D. Watts/NPIAW; I. R. McCann/NPIAW; A. G. Wells/NPIAW; C. Andrew Henley/NPIAW; G. B. Baker/NPIAW, C. Andrew Henley/NPIAW).

(b) Colour variation among fallow deer in New Zealand (C. N. Challies, G. Nugent, G. B. Baker/NPIAW, C. Andrew Henley/NPIAW).

IBC (a) Colour variation among possum skins taken in New Zealand (P. E. Cowan).

(b) The three main colour morphs of the ship rat (see Table 38) (J. G. Innes).

Ref QL 734 .H36 1990

Contents

Foreword, by the late K. A. Wodzicki vii

Preface viii
- Aims and limitations viii
- Coverage ix
- Corrections and additions ix
- Citation ix

Contributors x

Acknowledgements xi

Part I The Mammals of New Zealand: General Information 1

Introduction *C. M. King* 3

Species List 10

Identification 22
- From the whole animal 22
- Key to skulls 24

Preamble to Species Accounts 31
- Taxonomy 31
- Maps 31
- Layout of Species Accounts 32

Part II Species Accounts 33

1. Dama wallaby — *B. Warburton & R. M. F. Sadleir* 35
2. Bennett's wallaby — *B. Warburton & R. M. F. Sadleir* 44
3. Parma wallaby — *B. Warburton & R. M. F. Sadleir* 51
4. Black-striped wallaby — *B. Warburton & R. M. F. Sadleir* 57
5. Brushtailed rock wallaby — *B. Warburton & R. M. F. Sadleir* 58
6. Swamp wallaby — *B. Warburton & R. M. F. Sadleir* 64
7. Brushtail possum — *P. E. Cowan* 68
8. European Hedgehog — *R. E. Brockie* 99
9. NZ long-tailed bat — *M. J. Daniel* 117
10. Lesser short-tailed bat — *M. J. Daniel* 123
11. Greater short-tailed bat — *M. J. Daniel* 131
12. Little red flying fox — *M. J. Daniel* 136
13. European rabbit — *J. A. Gibb & J. M. Williams* 138
14. Brown hare — *J. E. C. Flux* 161
15. Kiore, Polynesian rat — *I. A. E. Atkinson & H. Moller* 175

16. Norway rat — *P. J. Moors* 192
17. Ship rat — *J. G. Innes* 206
18. House mouse — *E. C. Murphy & C. R. Pickard* 225
19. NZ fur seal — *M. C. Crawley* 246
20. NZ sea lion — *M. C. Crawley* 256
21. Southern elephant seal — *M. C. Crawley* 262
22. Weddell seal — *M. C. Crawley* 268
23. Leopard seal — *M. C. Crawley* 271
24. Crabeater seal — *M. C. Crawley* 274
25. Ross seal — *M. C. Crawley* 277
26. Kuri, Maori dog — *A. J. Anderson* 281
27. Stoat — *C. M. King* 288
28. Weasel — *C. M. King* 313
29. Ferret — *R. B. Lavers & B. K. Clapperton* 320
30. House cat — *B. M. Fitzgerald* 330
31. Feral horse — *R. H. Taylor* 349
32. Feral pig — *J. C. McIlroy* 358
33. Feral cattle — *R. H. Taylor* 373
34. Chamois — *C. M. H. Clarke* 380
35. Himalayan tahr — *K. G. Tustin* 392
36. Feral goat — *M. R. Rudge* 406
37. Feral sheep — *M. R. Rudge* 424
38. Red deer — *C. N. Challies* 436
39. Wapiti — *C. N. Challies* 458
40. Sika deer — *M. M. Davidson* 468
41. Sambar deer — *M. J. W. Douglas* 477
42. Rusa deer — *M. J. W. Douglas* 483
43. Axis deer — *C. M. King* 488
44. Fallow deer — *M. M. Davidson & G. Nugent* 490
45. White-tailed deer — *M. M. Davidson & C. N. Challies* 507
46. Moose — *M. M. Davidson & K. G. Tustin* 514

Glossary and Abbreviations 521

References 528

Index 598

Foreword

This book describes all 46 species of land-breeding mammals which are or have been living wild in New Zealand (11 native and 35 introduced). Very little was known about any of them until 1946–1948, when the Department of Scientific and Industrial Research (DSIR) commissioned a survey of introduced mammals, the results of which were published as DSIR Bulletin No. 98, *Introduced Mammals of New Zealand* (Wodzicki, 1950). This first comprehensive account has remained useful for a surprisingly long time. Its most important predecessor (Thomson, 1922) had been written almost 30 years before, and the acceleration of learning would suggest that a new compilation should have appeared about 1970. An updating of the 1950 Bulletin was indeed considered, but G. R. Williams's book *The Natural History of New Zealand* temporarily filled the gap in 1973. A new work has clearly been needed for many years, and it is with great pleasure that I welcome it.

Dr C. M. King has very effectively collated the efforts of 29 authors, all authorities in their fields of research. I note with some pride that no fewer than twelve of them are, or have been, members of DSIR Ecology Division, which I founded and directed from 1947 to 1965. A comparison of the present account of the ecology of introduced mammals with that of 40 years ago will provide information on changes affecting native flora and fauna, agriculture and other economic aspects. It also demonstrates the enormous increase in our knowledge of mammals in New Zealand. Most importantly, however, it should stimulate research by pointing out topics that require further study.

Kazimierz Wodzicki
March 1987

Preface

Aims and Limitations

The Handbook of New Zealand Mammals has two aims. First, to bring together for convenient reference authoritative, fully documented descriptions of all the land-breeding mammals that are or have been resident in New Zealand. The extensive information collected by the authors of each species account includes much material previously unpublished or gathered from inaccessible or widely scattered sources. Second, to stimulate new research. The three successive editions of *The Handbook of British Mammals* (Southern, 1964; Corbet & Southern, 1977; Corbet & Harris, 1990) have documented progressive increases in information about practically every British species, at least partly because each new edition stimulated ideas and pointed out gaps in available information. We hope this book will do the same. For example, many hypotheses suggested in these pages await rigorous testing (pp. 191, 197, 220); some ideas developed overseas could be better tested in New Zealand than almost anywhere else (p. 309). Many questions of taxonomy (pp. 131, 225, 320), or even basic biology (pp. 180, 307), remain to be examined. Readers may also detect differences of opinion between contributors, which had to be allowed to stand because they can be resolved only by new studies.

The Handbook of New Zealand Mammals is not, however, an introduction to the biology of mammals in general, a field guide, or a compendium of techniques for study of mammals. All these subjects are well catered for elsewhere, e.g., in *The Encyclopaedia of Mammals* (Macdonald, 1984); *Mammal Ecology* (Delaney, 1982); *Collins Guide to the Mammals of New Zealand* (Daniel & Baker, 1986); and the nine-part series by various authors, *Techniques of Mammalogy*, published intermittently in *Mammal Review* between 1975 and 1979.

Coverage

All known land-breeding mammals which are or have been established in the wild in the New Zealand region are described, including vagrants (one bat, and four Antarctic seals); introductions which survived for at least 25 years but have since become extinct (Polynesian dog, black-striped wallaby, axis deer); and established feral populations of domestic stock.

All known species landed but never established as wild populations are mentioned but not described. These include (1) exotic mammals confined to zoos, experimental farms or other premises, e.g., Pѐre David's deer (p. 436), alpacas and llamas (p. 358), and chinchillas (p. 173); and (2) failed introductions, except as mentioned under the appropriate family or genus headings and listed in Table 2.

Excluded are (1) the cetaceans (whales and dolphins), except as a species list (Table 2), because this special group is well catered for elsewhere (Baker, 1983); (2) controlled farm stock and pets (sheep, goats, cattle, donkeys, horses, working and town dogs, house cats, guinea pigs, fancy rabbits, etc.); and (3) the mythical "New Zealand otter" (p. 287).

Corrections and Additions

Every effort has been made to check all information given here. The editor and publishers would be grateful to hear of any errors or omissions, which can be corrected in the next edition.

Citation

Individual chapters from this book may be cited in the literature as follows.

Brockie, R. E., 1990. Hedgehog. *In* C. M. King (Ed.): The Handbook of New Zealand Mammals, pp. 99-113. Oxford University Press, Auckland.

Contributors

Anderson, A. J. Department of Anthropology, University of Otago, PO Box 56, Dunedin.

Atkinson, I. A. E. Botany Division, DSIR, Soil Bureau, Private Bag, Lower Hutt.

Brockie, R. E. Ecology Division, DSIR, Private Bag, Lower Hutt.

Challies, C.N. Forestry Research Centre, PO Box 31-011, Christchurch.

Clapperton, B. K. PO Box 772, Whangarei.

Clarke, C. M. H. Forestry Research Centre, PO Box 31-011, Christchurch.

Cowan, P. E. Ecology Division, DSIR, Private Bag, Lower Hutt.

Crawley, M. C. Ecology Division, DSIR, Private Bag, Lower Hutt.

Daniel, M. J. Ecology Division, DSIR, Private Bag, Lower Hutt.

Davidson, M. M. PO Box 146, Leigh, North Auckland.

Douglas, M. J. W. PO Box 18, Sefton, North Canterbury.

Fitzgerald, B. M. Ecology Division, DSIR, Private Bag, Lower Hutt.

Flux, J. E. C. Ecology Division, DSIR, Private Bag, Lower Hutt.

Gibb, J. A. Ecology Division, DSIR, Private Bag, Lower Hutt.

Innes, J. G. Forest Research Institute, Private Bag, Rotorua.

King, C. M. Royal Society of New Zealand, PO Box 598, Wellington.

Lavers, R. B. 223 Milford Rd., Te Anau.

McIlroy, J. C. CSIRO Division of Wildlife and Ecology, PO Box 84, Lynham, ACT 2602, Australia.

Moller, H. Ecology Division, DSIR, Private Bag, Nelson.

Moors, P. J. Royal Australasian Ornithologists' Union, 21 Gladstone St, Moonee Ponds, Victoria 3039, Australia.

Murphy, E. C. Natural History Department, Television New Zealand, PO Box 474, Dunedin.

Nugent, G. Forestry Research Centre, PO Box 31-011, Christchurch.

Pickard, C. R. Conservation Sciences Centre, PO Box 10-420, Wellington.

Rudge, M. R. Ecology Division, DSIR, Private Bag, Lower Hutt.

Sadleir, R. M. F. Conservation Sciences Centre, PO Box 10-420, Wellington.

Taylor, R. H. Ecology Division, DSIR, Private Bag, Nelson.

Tustin, K. G. PO Box 134, Wanaka.

Warburton, B. Forestry Research Centre, PO Box 31-011, Christchurch.

Williams, J. M. Canterbury Agriculture and Science Centre, Ministry of Agriculture and Fisheries, PO Box 24, Lincoln.

Acknowledgements

The authors of the separate chapters in this book have drawn heavily on unpublished data and reports from various official and private sources and have co-operated between themselves in exchanging data, observations, information, illustrations, criticism, copyright permissions, and other kinds of help. The authors here formally recognize and thank those named in the following list whose contributions were especially helpful. In addition, various organizations provided field support and assistance for authors or their informants, without which our knowledge of the biology of mammals in New Zealand would be much the poorer; prominent among these are DSIR and the former NZWS, NZFS, and DLS.

(Numbers refer to species accounts.)

1–6 D. Campbell, J. E. C. Flux, D. Moore, K. A. Wodzicki, the Editor of *Tuatara*.

7 J. E. Berney, F. Knight, J. Merchant.

9–12 A. Cox, J. F. Findlay, J. E. C. Flux, J. E. Hill, J. A. Mackintosh, R. L. Peterson, E. D. Pierson, and the Directors of Museums at Auckland, New Plymouth, Wellington, Christchurch and Dunedin.

13 J. Bell, T. M. Broad, J. E. C. Flux, P. Nelson, B. T. Robertson, D. L. Robson, D. W. Ross.

14 J. A. Gibb, J. P. Parkes.

15 J. G. Innes, P. J. Moors, M. Rae, R. H. Taylor, J. A. V. Tilley.

16 J. Craig, M. C. Crawley, T. K. Crosby, A. M. P. Dick, C. M. King, H. Moller, G. A. Taylor, R. H. Taylor, J. A. V. Tilley.

17 I. A. E. Atkinson, S. M. Beadel, W. A. G. Charleston, D. Cunningham, B. M. Fitzgerald, P. D. Gaze, B. J. Karl, C. M. King, J. R. Leathwick, F. Maplesden, H. Moller, P. J. Moors, K. Smith, R. H. Taylor, World Health Organization.

18 K. N. Bailey, B. D. Bell, D. M. Cunningham, M. J. Daniel, M. G. Efford, B. M. Fitzgerald, M. M. Flux, J. G. Innes, C. M. King, H. Moller, P. J. Moors, R. H. Taylor, M. B. Thompson, S. J. Triggs.

26 J. Davies, R. Newall.

27, 28 M. G. Efford, M. M. Flux, D. V. Merton, R. H. Taylor, Blackwell Scientific Publications.

29 B. M. Fitzgerald, P. J. Moors, R. J. Pierce, B. Springett.

30 B. J. Karl, N. P. Langham, R. J. Pierce, C. R. Veitch.

31 R. J. Holmes, E. J. Kirk.

33 R. McGibbon, M. R. Rudge.

35 N. Boyd, Mr & Mrs G. Joll, J. P. Parkes.

36 I. A. E. Atkinson, B. D. Bell, J. E. Berney, J. M. Clark, A.

Fairweather, A. G. Heath, D. Orwin, J. P. Parkes, M. R. Rudge, A. H. Whittaker, G. Yeates.

The editor also thanks all the contributors, especially those who responded to requests for constructive general comments on the project as it developed (I. A. E. Atkinson, C. N. Challies, B. M. Fitzgerald, J. E. C. Flux, J. A. Gibb, J. G. Innes). A. D. Pritchard drew the maps from information supplied by the authors. A. Beauchamp helped with the illustrations. G. B. Corbet advised on the skull key. C. Duval gave indispensable moral and technical support, helped with proofreading and composed the page layout. At Oxford University Press, J. Olson, A. French, N. Jackson, H. Allan, and J. Rawnsley supported the project throughout its lengthy gestation and worked hard to overcome the many problems it encountered. We gratefully acknowledge financial assistance from the Lottery Board, the John Ilott Charitable Trust, the Ministry of Agriculture and Fisheries, and the Department of Scientific and Industrial Research.

The Mammal Society, New Zealand Branch

The Mammal Society is an international association of mammologists based in London but interested in mammal science all over the world. The New Zealand Branch was established in 1989 in recognition of the completion of *The Handbook of New Zealand Mammals* and the healthy development of mammal science in New Zealand which it represents. For further information, write to The Secretary, Mammal Society, New Zealand Branch, C/o The Royal Society of New Zealand, P.O. Box 598, Wellington.

Part I
The Mammals of New Zealand
General Information

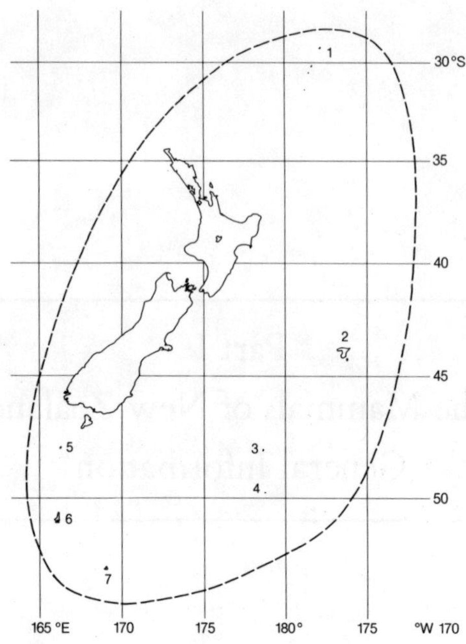

Map 1A New Zealand Region, extending from the Kermadec Is (30°S) to Campbell I. (52°S). Macquarie I., Lord Howe I. and Norfolk I., administered from Australia, are excluded. (1) Kermadec Is.; (2) Chatham Is.; (3) Bounty Is.; (4) Antipodes Is.; (5) The Snares; (6) Auckland Is.; (7) Campbell I.

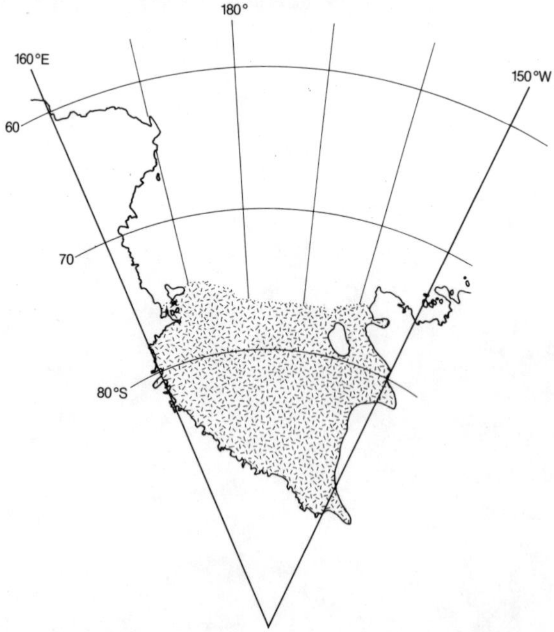

Map 1B Ross Dependency, New Zealand's section of Antarctica. Stippled area: Ross Ice Shelf.

Introduction

The contemporary fauna of land-breeding mammals in New Zealand is completely different, in composition and origin, from that of any other country in the world. Of course, the faunas of all countries are distinctive to some extent; but the mammal fauna of New Zealand is a special case, for three reasons. First, it is an artificial mixture of native and introduced species. That is not in itself unusual, since man has introduced a great variety of mammals to almost every country in the world, and few mammal faunas do not include some introductions (Lever, 1985); but in New Zealand the established introduced species vastly outnumber the native species. Second, this unusual mixture includes species brought from a great variety of natural habitats and from both temperate and tropical climates. Those that have established themselves have adapted in interesting ways to living with each other and in a novel environment. Third, the islands of New Zealand were free of all mammals but bats and seals until a mere 1000 years ago. The arrival of large numbers of introduced mammals (and a huge range of other species) had traumatic effects on the native mammals and on the other fauna, especially the birds and insects, and the environment in which they evolved (Druett, 1983; King, 1984a). Furthermore, both the native and the introduced mammals of New Zealand have had a short but turbulent history of exploitation, in successive, separate waves and using different techniques, by two distinct races of human settlers.

The circumstances of these events are unique in detail, though not in principle, and are paralleled only on other remote islands such as Mauritius, Canary Is., or Hawaii. The ecology, behaviour, interactions, and population dynamics of mammals in New Zealand, both native and introduced, are of particular interest to mammal biologists everywhere; there is no recent, comprehensive account of them; hence the need for this volume.

Geographical Background

The New Zealand region comprises a large archipelago of temperate islands and their surrounding continental shelf, plus the Ross Dependency in Antarctica (Map 1).

The North and South Islands, together known as "the mainland", measure 114 453 km^2 and 150 718 km^2, respectively. The largest of the other islands, Stewart (included by some authors in the phrase 'the three main islands') is 1746 km^2. The Chathams group totals 963 km^2, and all the other islands, from the subtropical Kermadecs to the subantarctic Auckland and Campbell groups, total 824 km^2. The grand total is 268 704 km^2, spread over an enormous

Table 1: The broad vegetation cover of New Zealand in pre-Polynesian, early European and modern times.

	Pre-Polynesian		1840		Present	
	Area[1]	%[2]	Area	%	Area	%
Native forest and light woodland	c.21 102	78	c.14 000	53	6 246	23
Alpine zone	3 725	14	3 725	14	3 725	14
Open country[3]						
'Arid zone' (grassland & scrub)	c.1 225	5 }	7 744[5]	29 }	13 940	52
Induced fern and tussock	—	— }		}		
'Improved' and 'Other' grazing land	—	—	—	— }		
Lakes[4]	324	1	324	1	340	1
Riverbeds	150	0.6	150	0.6	160	0.6
Swamps	c.230	0.9	455	2 }	882[5]	3
Dunes	c.60	0.2	130	0.5 }		
Exotic forests	—	—	—	—	740	3
Cropping land	—	—	—	—	429	2
Towns and cities	—	—	—	—	370	1
Total	**26 832[4]**		**26 528[2]**		**26 832**	

1. In thousands of hectares.
2. Percentage of the land area of New Zealand, including all the offshore islands but not the subantarctic islands. The figures for 1840 are calculated for the area of the North and South Islands only, but the error in the percentages thus introduced is probably immaterial in view of the probable errors in compiling the figures for that time, and also for pre-Polynesian times. Because accurate measures are now impossible, these data are acknowledged to be 'best estimates' only (G. C. Kelly, unpubl.). See also McGlone (1983).
3. Different definitions in each time period unavoidable. Later categories include earlier ones.
4. Natural lakes only; 16 000 ha has been subtracted for artificial (hydro-electric) lakes, so the total for the first column does not add up to 26 832 (cf. Kelly, 1980).
5. By subtraction (includes other habitats classified under 'miscellaneous').

Data from King (1984a) and sources listed therein.

area of the south-western Pacific Ocean from 33°S to 53°S latitude and from 162°E to 173°W longitude.

The Ross Dependency is defined as the segment of the Antarctic south of 60°S and from 160°E to 150°W, a triangular area of about 414 400 km² mostly of sea but also including almost the whole of the Ross Ice Shelf.

The geology, landforms, soils, natural vegetation, and climate of New Zealand are described in Williams (1973) and Wards (1976), and the geological history behind them by Stevens (1980, 1985). The main islands sit astride an important tectonic boundary, and have been shaped by the active geological processes that can still be observed today — including volcanism, large earthquakes accompanied by extensive horizontal and vertical earth movements, glaciation, and rapid erosion. The islands are somewhat elongated, and few inland locations are more than about 100 km from the sea. There are at least 223 named mountain peaks exceeding 2300 m altitude; the highest in the North Island is Mt. Ruapehu (2797 m), and in the South Island Mt. Cook (3764 m). There are also several very large freshwater lakes, including seven of over 100 km^2 area; Lake Taupo is 606 km^2. The climate is temperate with prevailing westerly winds, but the complex topography produces great local variation. Rainfall ranges from less than 500 mm/year in parts of Central Otago to more than 10 000 mm/year in some high-altitude basins in the Southern Alps. Mean annual sea level temperatures range from 15°C in the far north to 12°C around Cook Strait and 9°C in the south of the South Island. The distribution of natural vegetation cover, and the historical changes in it, are summarized in Table 1. The various types of native forest (mostly mixed podocarps on the lowlands, southern beech on the mountains; Fig. 1) are all evergreen. In lowland and northern districts the growing season is long and grass may grow all the year round.

Historical Background

In the late Cretaceous period, about 90–65 million years ago, the fragment of continental crust that later became the New Zealand archipelago split off from the side of Gondwanaland, the ancient southern landmass, and began to drift away eastwards. The animals and plants living on the detached segment were presumably representative of those that occupied the rest of Gondwanaland at the time, but as soon as they became cut off from genetic interchange with their ancestors and relatives they began to evolve independently. Hence, the great majority of known native species of plants and lower animals of New Zealand are endemic, and evolved in the absence of herbivorous and predatory mammals. The evolution of the native fauna is described in detail by Fleming (1979) and Kuschel (1975), among others.

The islands and their surrounding continental shelf were already isolated before the start of the evolutionary radiation of mammals that followed the disappearance of the dinosaurs. Modern mammals such as we would recognize began to evolve from the Oligocene period onwards (about 37–24 million years ago; Kurtén, 1971). So there were no mammals, not even the earliest mammals now extinct, on the land or in the surrounding seas when the islands were first isolated; none existed. Elsewhere in the world, large herbivorous mammals and their predators eventually became the dominant, or at least the most conspicuous, members of most natural ecosystems. In New Zealand, these roles were taken over by large flightless birds: the moas, twelve species, some of which were among the largest birds that ever lived; and at least three species of eagles and hawks, now all extinct. Likewise, in New Zealand the many large flightless insects came to occupy the role filled elsewhere by small herbivorous mammals, especially

Fig. 1 New Zealand was originally almost entirely covered with forest, from coast to treeline; mixed podocarp/hardwood species dominated the lowlands (foreground), and southern beeches (*Nothofagus* spp.) the hills. Paparoa Range, north Westland (NZFS).

Fig. 2 The modern landscape of New Zealand is largely artificial, converted over the century 1850–1950 from forest to pasture for the benefit of the two dominant species of introduced mammals, domesticated sheep and cattle. A scene in Northland, about 1910 (Alexander Turnbull Library).

rodents and lagomorphs. Some of the flightless Orthoptera became the ecological equivalents of rodents (Stevens, 1980).

All the mammals that reached New Zealand are therefore colonizers in the strict sense, since they could not land here without travelling over water at some time since they first evolved in some other environment. They came in three distinct groups.

First came those which, tens of millions of years after the opening of the Tasman Sea, developed from terrestrial ancestors into marine or flying forms (the seals, sea lions, and bats). We know of 11 species; by universal consent, these have earned the name "native", since they made their own way here, and have adapted to local conditions since. *Mystacina* has evolved into two distinct species, one of which has three subspecies, and these can truly be called native. There may have been other species we do not know about, which arrived but did not establish, or established but did not survive into modern times.

Much later, in about AD 850–950, came a second group of mammals: Polynesian settlers arrived, probably from the Marquesas, Cook or Society Islands, accompanied by their dogs and rats (Davidson, 1984). All adapted well to New Zealand conditions, though not to the extent of evolving new subspecies, or even distinguishable races. Some say they have been here long enough to qualify as "native", by length of residence if not by indigenous origin, but this approach leads to inconsistencies and conceals the massive impact that the Polynesian immigrants of all three species had on the original fauna and flora of the islands. For example, the kiore has long been regarded as a harmless vegetarian, in contradistinction to the recently arrived, and certainly very damaging, European rats. This is biological nonsense. All rats are introduced, and the damage done by kiore appears slight because it was long ago and undocumented. We have rather little evidence of what New Zealand was like in pre-human times, so we tend to underestimate the scale of the changes induced during the thousand or so years that the Polynesian colonists and their descendants, the Maori, had the islands to themselves. However, there is no doubt that the pre-European extinctions among the native fauna were considerably more drastic than post-European ones, and they included all the largest, meatiest ground-dwelling birds — all the moas and about 18 other species of waterfowl, rails, and gamebirds. The primary agent of these extinctions was a single species of predatory mammal, *Homo sapiens* (Anderson, 1984; Cassels, 1984; King, 1984a; Trotter & McCulloch, 1984). The kiore had equally extensive effects on the smaller native birds and the lizards and insects (p. 190).

A third phase began in 1769, with the arrival of the Europeans. They came slowly at first, but after the annexation of New Zealand by Britain in 1840, European settlers and the other mammals they brought with them began to invade the country in overwhelming numbers and variety. The speed and thoroughness of the European takeover of what was, after all, not an empty country, is remarkable. It is the last and clearest example of "ecological imperialism", and the reasons for it are elegantly explained by Crosby (1986). The European colonists were outstandingly successful because (1) they arrived as a complete and integrated community — the people plus the animals, crops, parasites, and diseases with which they had evolved; (2) they took over a land from which the dominant large native herbivores had already been removed; and (3) they met a community that had, for geological reasons, evolved separately and had never been exposed to such a comprehensive onslaught before.

The three groups of mammals (including man) are not different in principle, since all have colonized New Zealand from outside; the main difference between them is in how long ago they came. The Europeans and their companions arrived recently enough to be, unlike the Polynesians and theirs (and still more unlike the native species), very obviously alien; yet some of them, such as Norway rats and house mice, have already been here for hundreds of generations. It is time that the native and introduced mammals were treated in practice as resident species of equal status in the scientific sense, and therefore listed together in this book. However, they are not equal in the conservation sense, and in certain protected natural areas whose primary management objective is the preservation of endemic biota, the control or even eradication of introduced mammals may be feasible and desirable. That does not prevent both native and exotic species being accepted in this book as contributing to the present mammalian fauna of New Zealand. The combination is not natural, but it exists as a working, evolving community of mammal species. We may attempt to manage some species according to our present philosophy and needs; but ultimately the community will continue to evolve according to natural processes largely beyond our control.

Species Composition of the Mammal Fauna of New Zealand

Taxonomic and Geographical Distribution

The total list of land-breeding mammals known to have reached New Zealand at some time or other includes 65 species (Tables 2 and 3).

By comparison with the mammal faunas of comparable islands in the Northern Hemisphere, the taxonomic and geographical distributions of the species present in New Zealand are remarkable for two reasons. First, the total number of living, established species is small for the size of the islands, and only 14 of them are common and widespread; yet the biomass of mammals in one well-studied New Zealand forest is high by international standards (25 kg/ha; Brockie & Moeed, 1986). Unfortunately, this high mammalian biomass is gained at the expense of native birds and invertebrates, many of which have been lost to the community; and 99% of it is accounted for by a dense population of possums, whose effects on the forest are detrimental in the long term (p. 95). Second, the 14 most common and widespread mammal species are all introduced. None of the 11 native species is widespread; nine are counted as localized, rare or vagrant (including four

Table 2: Systematic species list of the mammals of New Zealand.

A. Native species	Status[1]	Page
ORDER CHIROPTERA		
Family Vespertilionidae		
1. *Chalinolobus tuberculatus* NZ long-tailed bat	L	117
Family Mystacinidae		
2. *Mystacina tuberculata* Lesser NZ short-tailed bat	L	123
3. *M. robusta* Greater NZ short-tailed bat	E	131
Family Pteropodidae		
4. *Pteropus scapulatus* Little red flying fox	V	136
ORDER CARNIVORA		
Family Otariidae		
5. *Arctocephalus forsteri* NZ fur seal	L	246
6. *Phocarctos hookeri* NZ sea lion	L	256
Family Phocidae		
7. *Mirounga leonina* Southern elephant seal	L	262
8. *Leptonychotes weddelli* Weddell seal	A	268
9. *Hydrurga leptonyx* Leopard seal	A	271
10. *Lobodon carcinophagus* Crabeater seal	A	274
11. *Ommatophoca rossi* Ross seal	A	277

B. Introduced species	Origin	Purpose	Status	Page
ORDER MARSUPIALIA				
Family Macropodidae				
12. *Macropus eugenii* Dama wallaby	Australia	Sport	L	35
13. *M. r. rufogriseus* Bennett's wallaby	Australia	Sport	L	44
14. *M. parma* Parma wallaby	Australia	Sport	L	51
15. *M. dorsalis* Black-striped wallaby	Australia	Sport	E	57
16. *M. robustus* Roan wallaby	Australia	?Sport	NE	35
17. *Petrogale p. penicillata* Brushtailed rock wallaby	Australia	Sport	L	58
18. *Wallabia bicolor* Swamp wallaby	Australia	Sport	L	64
19. Unidentified wallaby	Australia	?Sport	NE	35
20. Unidentified kangaroo	Australia	?Sport	NE	35
Family Phalangeridae				
21. *Trichosurus vulpecula* Brushtail possum	Australia	Fur	W	68
Family Potoroidae				
22. *Potorous tridactylus* Long-nosed potoroo	Australia	?	NE	34
Family Dasyuridae				
23. *Dasyurus* sp. Marsupial cat	Australia	?	NE	34
Family Paramelidae				
24. *Isoodon obesulus* Southern brown bandicoot	Australia	?	NE	34
Family Pseudocheiridae				
25. *Pseudocheirus peregrinus* Ringtail possum	Australia	?	NE	34

(For key to footnotes, see p. 13)

B. Introduced species	Origin	Purpose	Status	Page
ORDER INSECTIVORA				
Family Erinaceidae				
26. *Erinaceus europaeus occidentalis* West European hedgehog	Britain	Pest control	W	99
ORDER LAGOMORPHA				
Family Leporidae				
27. *Oryctolagus c. cuniculus* European rabbit	Britain	Sport, meat	W	138
28. *Lepus europaeus occidentalis* Brown hare	Britain	Sport, meat	W	161
ORDER RODENTIA				
Family Muridae				
29. *Rattus exulans* Kiore, Polynesian rat	Polynesia	?Meat, stowaway	L	175
30. *R. norvegicus* Norway rat	Europe	Stowaway	L	192
31. *R. rattus* Ship rat	Europe	Stowaway	W	206
32. *Mus musculus* House mouse	Europe	Stowaway	W	225
Family Sciuridae				
33. *Tamias striatus* Gray chipmunk	North America	?	NE	173
34. "Brown California squirrel"	North America	?	NE	173
Family Caviidae				
35. *Cavia porcellus* Guinea pig	?South America	?	NE	173
ORDER CARNIVORA				
Family Canidae				
36. *Canis familiaris*[2] Kuri, Polynesian dog	Polynesia	Pet, ? meat	E	281
European dog	Europe	Pet, utility	—	281
Family Mustelidae				
37. *Mustela erminea* Stoat	Britain	Pest control	W	288
38. *M. nivalis vulgaris* Weasel	Britain	Pest control	L	313
39. *M. furo* Ferret	Britain	Pest control	W	320
Family Felidae				
40. *Felis catus* House cat	Europe	Pet, pest control	W	330
Family Viverridae				
41. *Herpestes edwardsi* Indian grey mongoose	?India	?Pet	NE	280
Family Procyonidae				
42. *Procyon lotor* Racoon	North America	?Pet	NE	280
ORDER PERISSODACTYLA				
Family Equidae				
43. *Equus caballus* Feral & domestic horse	Europe	Utility	L	349
44. *E. zebra* Zebra	S. Africa	?	NE	349

B. Introduced species	Origin	Purpose	Status	Page
ORDER ARTIODACTYLA				
Family Suidae				
45. *Sus scrofa* Feral & domestic pig	Europe	Utility	W	358
Family Bovidae				
46. *Bos taurus* Feral & domestic cattle	Europe	Utility	L	373
47. *Rupicapra r. rupicapra* Chamois	Europe	Sport	W	380
48. *Hemitragus jemlahicus* Himalayan tahr	Asia	Sport	L	392
49. *Capra hircus* Feral & domestic goat	Europe	Utility	W	406
50. *Ovis aries* Feral & domestic sheep	Europe/ Australia	Utility	L	424
51. *Connochaetes gnou* Gnu	S. Africa	?	NE	371
52. *Pseudois nayaur* Bharal, blue sheep	Asia	Sport	NE	392
Family Camelidae				
53. *Lama glama* Llama	South America	?	NE	358
54. *L. pacos* Alpaca	South America	?	NE	358
Family Cervidae				
55. *Cervus elaphus scoticus* Red deer	Britain	Sport	W	436
56. *C. elaphus nelsoni* Wapiti	North America	Sport	L	458
57. *C. nippon* Sika deer	East Asia/ Britain	Sport	L	468
58. *C. u. unicolor* Sambar deer	Sri Lanka	Sport	L	477
59. *C. timorensis* Rusa deer	Indonesia/ New Caledonia	Sport	L	483
60. *Axis axis* Axis deer	India/ Australia	Sport	E	488
61. *Dama d. dama* Fallow deer	Britain/ Tasmania	Sport	L	490
62. *Odocoileus virginianus borealis* White-tailed deer	North America	Sport	L	507
63. *O. hemionus* Mule deer	North America	Sport	NE	506
64. *Alces alces andersoni* Moose	North America	Sport	R	514
65. Unidentified South American deer	?	?	NE	506

Section B is updated and corrected, where necessary, from Wodzicki (1950) and Wodzicki and Wright (1984). All known species imported before 1900 are listed, including some which were probably never released in the wild; but the total number is not certain since few accurate records were kept last century, when most liberations were made. Exotic mammals imported but confined to captivity (e.g., in zoos) this century are omitted. Many species were slow to establish and were introduced repeatedly, even some of those which were later extremely successful. Most introductions were organized between 1850 and 1910. No new species were added from 1910 to the 1980s; since then a renewed interest in acclimatization of exotic species for diversification of farming stock (p. 18) has stimulated some new introductions, both of species already present and of new species. None of the latter have been released into the wild, but escapes may be expected in due course.

C. Cetaceans[3]

ORDER MYSTICETI	
Family Balaenopteridae	
Balaenoptera musculus	Blue whale
B. physalus	Fin whale
B. borealis	Sei whale
B. edeni	Bryde's whale
B. acutorostrata	Minke whale
Megaptera novaeangliae	Humpback whale
Family Balaenidae	
Balaena glacialis	Southern right whale
Capera marginata	Pygmy right whale
ORDER ODONTOCETI	
Family Physeteridae	
Physeter macrocephalus	Sperm whale
Kogia breviceps	Pygmy sperm whale
K. simus	Dwarf sperm whale
Family Ziphiidae	
Berardius arnouxi	Arnoux's beaked whale
Hyperoodon planifrons	Southern bottlenose whale
Ziphius cavirostris	Goose-beaked whale
Tasmacetus shepherdi	Shepherd's beaked whale
Mesoplodon layardi	Strap-toothed whale
M. bowdoini	Andrews' beaked whale
M. grayi	Scamperdown whale
M. ginkgodens	Ginkgo-toothed whale
M. hectori	Hector's beaked whale
Family Delphinidae	
Globicephala melaena	Long-finned pilot whale
G. macrorynchus	Short-finned pilot whale
Orcinus orca	Killer whale
Pseudorca crassidens	False killer whale
Peponocephala electra	Melon-headed whale
Grampus griseus	Risso's dolphin
Tursiops truncatus	Bottlenose dolphin
Stenella caeruleoalba	Striped dolphin
S. attenuata	Spotted dolphin
Delphinus delphis	Common dolphin
Lagenorhynchus obscurus	Dusky dolphin
L. cruciger	Hourglass dolphin
Cephalorynchus hectori	Hector's dolphin
Lissodelphis peronii	Southern right whale dolphin
Family Phocoenidae	
Australophocoena dioptrica	Spectacled porpoise

1. A, Antarctic only; E, extinct; L/R/V, localized, rare or vagrant; NE, never established; W, widespread. For summary of taxonomic and status distributions, see Table 3.
2. Includes feral crossbreds; domestic and working dogs excluded.
3. The cetaceans (whales and dolphins) do not breed on land and are not covered in this book. For full descriptions, see Baker (1983) and Daniel and Baker (1986).

Table 3: Taxonomic and geographical distribution of land-breeding mammals in New Zealand.

	Total	NE	W	L/R/V/A	E
Native					
Bats	4	1	0	2	1
Pinnipeds	7	0	0	7	0
Introduced					
Marsupials	14	7	1	5	1
Insectivores	1	0	1	0	0
Lagomorphs	2	0	2	0	0
Rodents	7	3	3	1	0
Fissipeds	7	2	3	1	1
Ungulates	23	7	4	11	1
Totals	65	20	14	27	4
				41	

NE, never established; W, widespread; L/R/V/A, localized, rare or vagrant on the mainland, or confined to the Antarctic; E, extinct. For list of species and classifications of status, see Table 2.

seals confined to the Ross Dependency and common there), one never established (p. 136) and one is extinct (p. 131). All three established bats and both fur seals and sea lions were once much more common and widely distributed than they are now.

These figures reflect the familiar contrast between the effects of human settlement on native and introduced species. Seen with the benefit of hindsight, this effect does not seem surprising; however, we miss much of the impact of the story if we forget that many of the startling consequences of the introductions were not expected by the people responsible for them. First, the early New Zealanders of both races were wholly ignorant of ecology and acted only in what they judged to be their own best interests at the time — as we still do today. To the Europeans, acclimatization seemed the obvious way to populate New Zealand with useful animals, and at least the settlers did try to choose only the best and "innoxious" species (Gibb & Flux, 1973). Second, the behaviour of the introduced species was often not predictable from the attributes of the same or similar species in their native lands: see, for example, the contrast in the biology of brushtail possums in Australia and in New Zealand (p. 93).

Size Distribution of the Mammal Fauna of New Zealand and the Purposes of the Introductions

In the natural or mostly natural mammal faunas of comparable Northern Hemisphere islands there are always many more species of small herbivores (rodents and lagomorphs) than of large ones (ungulates). Britain, for example, has or has had 20 species of rodents and lagomorphs, and 12 of wild or feral ungulates, out of a total of 76 species of living or recently extinct native and introduced land-breeding mammals (Corbet & Southern, 1977). New Zealand has or has had six and 16, respectively, out of 46 (excluding species that never established: Table 3). The reason is, of course, that the people who selected the mammals to be brought to New Zealand were interested only in those they thought would be useful, as farm stock or as game, and these were usually large. Even two of the "small" mammals, the rabbit and the brown hare, are here only because of their sporting and culinary values; and a third, the kiore, was highly prized as food by the Polynesians and may have been loaded on to the voyaging canoes on purpose, rather than arriving as stowaways. The only other rodents here, the Norway and ship rats and the house mouse, are universal commensals of European man, able to maintain thriving "wild" populations in ships and stored food, forage and bedding, and to colonize new countries without deliberate help. By contrast, no less than 23 species (nine kangaroos and wallabies, and 14 wild ungulates) were brought in solely to supply sportsmen with wild game. The disproportionate number of game species introduced illustrates the influence and the motives of the acclimatization societies, the bodies responsible for most of the introductions.

Not all the species that ever reached either country still survive. But even in the contemporary faunas, the size distribution of mammals in New Zealand is strongly skewed towards the larger species, by comparison with that of the mammal fauna of Britain (Table 4). The shortage of small mammals means, in turn, a greatly restricted choice of prey for small carnivores in New Zealand; large carnivores would be better placed, but, not surprisingly, there are none.

Origins of Introduced Species

Of the 54 known introduced species (Table 2), a large number (20) came directly or indirectly from Europe, particularly from Britain, and all of these established successfully. They include all but one of the most widespread and successful introduced species, such as the hedgehog, rabbit, brown hare, ship rat, house mouse, feral cat, stoat, goat, and red deer. Next in frequency are the Australian species (14), of which half established at least locally (several confined to Kawau I.); one of these, the brushtail possum, is now the only common and widespread introduced species that does not come from Europe. Of the 10 species brought from North or South America, seven never established and three (wapiti, white-tailed deer and moose) did so only locally. Of the six Asian species, the tahr and the sambar and rusa deer established locally and so, for a while, did the axis deer, but the other two did not. Both species from Polynesia (rat, dog) were successful at first, though later displaced by equivalent European kinds. The two from Africa (zebra, gnu) were confined to Sir George Grey's menagerie on Kawau I. and were probably never meant to be acclimatized on the main islands.

The present exotic fauna is the combined result of both natural and artificial

Table 4: Size distribution of the contemporary faunas of land-breeding mammals in New Zealand compared with Britain.[1]

Mean body size	<100 g	100 g – 1 kg	1-10 kg	>10 kg	n
New Zealand	4	5	8	19	36
%	11	14	22	53	
Britain	29	12	7	15	63
%	46	19	11	24	

1. Resident species only, native and introduced, including bats and pinnipeds. British data from Corbet and Southern (1977).

selection. People selected the species to be introduced, and the New Zealand environment (helped or hindered by people) selected the species that would succeed. The two processes, in unknown proportions, have made New Zealand one of the few places in the world where such a curious mixture of temperate and tropical, wild, feral and commensal mammals survive together as the dominant fauna in the wild. Some of these are exposed to the possibility of hybridization with close relatives for the first time; for example, the European and North American forms of one species, *Cervus elaphus,* as well as the separate temperate and tropical species of *Cervus*.

Mammals and the Economy of New Zealand

Mammals were not very important to the subsistence economies of the Polynesian colonists or their descendants, the Maori. The only native mammals worth hunting, the seals and sea lions, were welcome but out of reach for much of the year. The Polynesian rats and dogs, which they brought with them, were small and not abundant enough to provide meat in quantity. While the moa and other large ground birds lasted, that did not matter; but from the 15th century onwards, the pre-European Maori were always short of meat. Some European authors have wondered whether the widespread cannibalism characteristic of later Maori culture was an attempt to make up the deficiency, even to the extent of speculating that the prohibition of it after 1840 left "a serious gap in diet" (Owens, 1981). But in fact cannibalism was practised largely because it had strongly spiritual overtones — it represented the ultimate victory over an enemy, to reduce his body to mere food (Metge, 1976) — and although the Maori evidently enjoyed human meat and made the most of it when opportunity offered, its contribution to the normal diet was small (Davidson, 1984).

Since the first arrival of the European explorers, and the introduction of money and a commercial economy, mammals have taken a central role in New Zealand history. They have been responsible for much of the most profitable trade, but also for some of the most profound environmental change, ever seen in this country. The economic wealth of New Zealand has depended on the products of mammals since the very beginning of European times — in fact, even before many Europeans were resident here. Other species of mammals proved much less beneficial.

ECONOMIC SPECIES

Three distinct phases can be distinguished in the exploitation of mammals in New Zealand.

First, from 1792 to 1840 there was a frantic period of uncontrolled, boom-and-bust exploitation of native marine mammals. Commercial expeditions, equipped with far more sophisticated technology than the Maori ever had, competed to collect the largest possible harvest of whales and seals in the shortest possible time. The trade was obviously not sustainable, and it was clear that the stocks would soon be exhausted; but there was no legal authority able to enforce moderation on everyone, and it was in no one's individual interest to hold back.

Second, from 1840 onwards, traditional stock farming was transplanted from Europe *en bloc.* It then proceeded to adapt to local conditions and to develop its own extraordinary productivity and efficiency, and its products were sold, reliably and profitably, almost entirely in Britain. The dominant land use in modern New Zealand is still the introduced ryegrass/clover pasture, established in place of the native forest (Fig. 2), and maintained explicitly for the production of two key species of domesticated mammals, sheep (about 65 million) and cattle (about 9 million). A few other species could also be found, as minority stock (goats, pigs) or essential working partners (dogs, horses). For over 100 years these have been the most conspicuous mammals in New Zealand, although they do not figure much in this book except as sources of feral populations of domesticated species.

The third phase began in the early 1960s. The development of aerial hunting set off a new wave of boom-and-bust exploitation, especially of red deer in the inaccessible ranges of the South Island (p. 455). There is a striking parallel between the two phases of commercial hunting, of marine mammals in the early 19th century and of red deer in the mid-late 20th century. Both followed many years of hunting by inefficient traditional methods; both were set off by technical advances that changed the economics of the chase, and brought an abundant but previously unavailable population within reach. In both cases the animals were removed faster than they could be replaced, the rate of harvest was controlled only by profit, and the boom times were short-lived. The differences were that (1) the marine mammals were native, whereas the red deer were introduced and classified as "noxious animals"; and (2) there was no legislative authority available to control the harvest of mammals before 1840, whereas in the 1960s and 1970s parliamentary authority existed but it did nothing because under the Noxious Animals Act 1956 (repealed in 1977) it was improper to restrict access to noxious animals for the purposes of their destruction.

The same period also saw some profound changes in agricultural practice. Changes in the economics of trade in the staple products of ordinary sheep/

beef/dairy farms (caused by, for example, Britain's entry into the European Economic Community) made continued total dependence on traditional ways unsafe and stimulated research into diversification. With the passing of enabling legislation and the development of new techniques for handling live deer in commercial numbers, deer farming became a growth industry. Helicopter crews changed their rifles for nets and dart guns and began to live-capture deer to supply breeding stock to deer farms at enormous prices. Red deer were suddenly transformed from pest to asset, not because they were any different, but because of a fundamental change in the way people viewed them (Caughley, 1983). Other mammal species made the same transition, and high-quality strains of goats, ferrets, and rabbits are all farmed now. There is also renewed interest in the importation of any new exotic species that could be profitable, e.g., llamas and chinchillas.

Successive New Zealand governments have been aware of the problems caused by introduced mammals in the past, and now have regulations about importing new ones (not stringent enough for some). There is a blacklist of prohibited species which are not allowed in at all; the mammals included on it are the mink *(Mustela vison)*, coypu *(Myocastor coypus)*, grey squirrel *(Sciurus carolinensis)*, and muskrat *(Ondatra zibethicus)*. All these have bad records of causing damage in other countries where they have been acclimatized (Lever, 1985). Also prohibited are any fox or mongoose, or "any other animal likely to cause damage". Permits may be issued by MAF for importation of special quality ferrets to improve the stock on fur farms (p. 329), and a few experimental llamas and alpacas (p. 358), chinchillas (p. 173), and Père David's deer (p. 436). Some of these could eventually become economic species of the future.

PESTS

The other side of the balance sheet is that the introduced mammals have also caused some severe environmental problems. New Zealand's fledgling pastoral industry was nearly wrecked by rabbits; the native vegetation has been profoundly modified by possums, goats, and deer; and human hunters, rats, and cats have caused scores of extinctions among the native birds, lizards, and insects. These are all long-standing problems, and early attitudes to them are well summarized by Wodzicki (1950). Since then, however, we have found solutions (or sometimes merely new attitudes) to some old problems, and many new ones have arisen. The list of challenges posed by the introduced mammals to contemporary management authorities includes some opportunities and many problems unknown in Wodzicki's time.

1. Over most of the country, rabbits are no longer seen as a serious threat to agriculture, and legislation prohibiting the keeping of rabbits was repealed in 1980 in order to allow commercial production of domestic rabbits for meat and fur. In central Otago, where habitat and climate most favour them, rabbits are still a considerable problem, and the cause of furious debate among high-country farmers opposed to the official ban on introducing myxomatosis (p. 160).

2. The most serious mammal pest at present is probably the brushtail possum, not only because it damages the native vegetation but also because it is a natural reservoir of bovine tuberculosis. Neither possums nor tuberculosis can be eradicated from New Zealand with present technology (p. 98).

3. Excessive populations of ungulates (sheep and game) have long been held responsible for the high rate of soil erosion in the high country. The main justification for efforts to control deer, goats, chamois, and tahr, and for the retirement of high altitude grazing land, has been the protection of soil and water resources. But soil erosion in a young, geologically active country is naturally high; and variations in it are now believed to be controlled, at least in part, by subtle changes in climate (Grant, 1985, 1989).

4. Browsing mammals certainly do change the floristic composition of the forests, but this is no longer seen as inevitably leading to destruction of the forests as natural communities. A recent and still controversial suggestion is that the dieback is often natural (Veblen & Stewart, 1982). The goal of eradication has been abandoned in practice. There are now 10 gazetted recreational hunting areas in which the aim of management is to enhance hunting rather than to protect vegetation, soil, or water, although they can be established only where there is no likely conflict between soil and water values and hunting. These changes in attitude could hardly have been predicted a mere 25 years ago (Miers, 1985). On the other hand, the more subtle effects of browsing are now recognized as serious for the survival prospects of some threatened native birds, e.g., the takahe (Mills and Mark, 1977) and North Island kokako (Leathwick, Hay and Fitzgerald, 1983).

5. The national parks are governed by legislation which calls for the eradication of introduced species within their borders, but powerful sporting interests have opposed the application of this principle to the wapiti (p. 467).

6. Introduced predators have certainly caused many extinctions among the native fauna in the past, but our perceptions have changed as to which predators were most responsible. The part played by kiore during the pre-European era has been greatly underestimated (p. 190), and that played by stoats in European times equally overestimated (p. 311). It has always been assumed (e.g., by Thomson, 1922) that ship rats arrived with James Cook; but it now seems more likely that the first European rats to discover the vulnerable ground-nesting birds of New Zealand were probably not ship rats, but Norway rats (p. 197).

7. Ideas are also changing concerning what should be done about predator control in contemporary times. There is little justification for generalized predator control on the main islands, except as part of integrated programmes to manage takahe, kokako and kakapo (King, 1984a). However, on important island reserves the local extermination of pest mammals is now possible and (though very expensive) attainable. Recent notable successes include the eradication of cats from Little Barrier I. and of possums from Codfish

I. and Kapiti I. The gravest threats now are probably the risk of accidental or malicious transport of predators (especially rats) to these and other offshore islands of high wildlife value.

8. Whether or not a species has a history of causing damage in New Zealand, once it is established there seems to be a precedent for keeping it. The probable disappearances of the axis deer and moose are regarded with regret; when the population of sambar declined too far they were protected by a hunting ban (p. 483); the theoretically possible extermination of the tahr would be strongly resisted by sportsmen; the kiore is popularly regarded as a harmless, quasi-native species worthy of protection (Gibb & Flux, 1973); people who witness an encounter between a rabbit and a stoat tend to sympathize with the rabbit (King, 1982b), even though rabbits have done incalculable damage to New Zealand. On the other hand, this conservative attitude probably contributed to the survival of the parma wallaby, "rediscovered" on Kawau I. after it was believed extinct in Australia (p. 52).

9. Feral farm animals have always been regarded as a nuisance, and many herds have been exterminated. Recently there has been renewed interest in the conservation of ancient genetic varieties lost elsewhere through intensive selective breeding, and some herds of feral domesticated species are now protected (Whitaker & Rudge, 1976).

More than ever before, mammals in modern New Zealand are seen as a mixture of asset and nuisance, sometimes simultaneously. Acceptance of them as *de facto* members of the New Zealand fauna does not reduce the difficulty of finding appropriate responses when they begin to trespass on the values that we have chosen to defend.

Adaptation in Action

There are many examples of the impact of human activities on the distribution, population biology, and evolution of mammals around the world, and of the responses of the affected mammals to this disturbance, but New Zealand provides by far the best of them. Most of the introduced species arrived at a known date and place, from a known stock, and dispersed into an environment hitherto innocent of those kinds of animals and, conversely, entirely foreign to them. The opportunity to compare the biology of a familiar species in its own and in an alien environment is irresistible to the evolutionary biologist. New Zealand therefore provides a vast natural laboratory for observing the processes of adaptation, working at both the ecological and the evolutionary levels, both in the native mammals and their environment and in the introduced mammals and their descendants. For example, here we have hedgehogs and brown hares free of some of their most debilitating diseases and parasites, and deer free of the large carnivores which influence their population dynamics elsewhere; we have famous game animals carefully protected by sportsmen overseas but controlled here as pests; we have substantial populations of Bennett's wallabies, brown hares, and some deer derived from very few colonists, with only minor evidence of founder effects, if any; we have house mice, ferrets, and feral domesticated stock living completely independently of human control, in niches occupied by other mammals elsewhere, and brown hares common in alpine regions from which in Europe they would be displaced by mountain hares *(Lepus timidus)*; we have examples of relatively rapid adaptation in body size, some in accordance with Bergmann's Rule (hedgehogs, hares, possums, kiore) and others not (stoats), whilst still other species have remained the same size as their relatives overseas (ship rats, cats). In addition to these, we have classic examples of population irruptions, both of colonizing herbivores on the once-only scale (e.g., rabbits and red deer), and of carnivores and rodents on a short-term, indefinitely repeatable 3–4-year cycle (stoats and house mice in beech forests). Some species find conditions more favourable in New Zealand (e.g., the climate milder, food more abundant, competitors fewer) than in their homelands. Red deer, feral goats, and brown hares breed at a relatively earlier age, and rabbits breed for longer each year, compared with their relatives elsewhere; possums, hedgehogs, and house mice are more abundant here than in their homelands. On the other hand, the vast freshwater resources of New Zealand are not exploited by any mammals, and other niches support fewer species than they do elsewhere and perhaps might here. The problem of working out the environmental and evolutionary bases of these differences provides endless food for thought. Mammalogy in New Zealand is a young science, but its horizons are wide.

C. M. K.

Identification

From the Whole Animal

The mammals described in this book include members of eight orders and 17 families. The general characters of these major groups are listed below. Within these groups there are several sets of similar species which can be distinguished only by reference to particular details of colour, shape or anatomy. The distinguishing marks of these species are listed in separate tables entered in the text under the appropriate group introduction.

Order Marsupialia
Family Macropodidae: kangaroos and wallabies, p. 35.
The shape and gait of wallabies are familiar but not always diagnostic — they hop only when moving at high speed. Of the five living species in New Zealand, the smallest are about the size of a large hare, but with longer tail and shorter ears; the largest are three to four times bigger. Distinguishing marks of wallabies in New Zealand: Table 6.
Family Phalangeridae: possums, p. 67.
The brushtail possum, the only species present, has large eyes with red eyeshine, prominent ears, long prehensile tail, soft fur, and excellent climbing ability; the most similar species are the smallest wallabies, which have much longer hind feet and tails, and the cat, which has green eyeshine and a non-prehensile tail. In the hand, the distinguishing characters of marsupials (p. 34) are unmistakable.

Order Insectivora
Family Erinaceidae: hedgehogs, p. 99.
The European hedgehog, the sole representative of the family, has an unmistakable spiny back.

Order Chiroptera
Family Vespertilionidae: evening bats, p. 115.
Family Mystacinidae: NZ short-tailed bats, p. 122.
Family Pteropodidae: fruit bats, p. 136.
Bats are distinguishable from birds in flight by their more fluttering action and erratic path. Distinguishing marks of the four species recorded in New Zealand: Table 24.

Order Lagomorpha
Family Leporidae: rabbits and hares, p. 138.
Universally familiar animals with long ears, very short tail and hopping or leaping run. Distinguishing marks of the two species present in New Zealand: Table 27.

Order Rodentia

Family Muridae: rats and mice, p. 173.

Rats and mice have pointed faces, short bodies and long, naked tails. Weasels and stoats are the only other small mammals, but they have longer bodies and furred tails. Distinguishing marks of the four species of rodents present in New Zealand: Table 34.

Order Carnivora

Family Otariidae: fur seals and sea lions, p. 243.

Family Phocidae: true seals, p. 262.

The two species of sea lions and five of true seals are the only marine mammals that breed on land in the New Zealand region (including Antarctica). Distinguishing marks: Table 43.

Family Canidae: dogs, p. 280.

Polynesian and European domestic dogs.

Family Mustelidae: stoats, weasels, and ferrets, p. 287.

Small, slender, short-legged hunters, with short, rounded ears flat to the head, pointed faces, and long whiskers. Distinguishing marks of the three species present in New Zealand: Table 46.

Family Felidae: cats, p. 330.

Domestic and feral house cats.

Order Perissodactyla

Family Equidae: horses, p. 349.

Domestic and feral horses.

Order Artiodactyla

Family Suidae: pigs, p. 358.

European (and Asian?) domestic and feral pigs.

Family Bovidae: cattle and allies, p. 371.

Domestic and feral cattle, sheep, and goats, plus two goat-like wild game animals, all with permanent horns in both sexes. The chamois has a short coat and erect, slim, hooked horns; the tahr has a long shaggy coat and stout horns curving smoothly backwards. Distinguishing marks of the five species of bovids in New Zealand: Table 63.

Family Cervidae: deer, p. 432.

Large browsing or grazing ungulates with deciduous antlers in males. Distinguishing marks of the seven living species (one with two distinct subspecies) present in New Zealand: Table 78.

Key to Skulls

Includes all living and extinct native and successfully introduced land-breeding mammals whose skulls could be found on the main islands of New Zealand, including Otariidae but not Phocidae.

1. Teeth either absent or all single-cusped and borne on a long narrow rostrum; nostrils open at top of skull .... **Cetacea**, p. 13
   Teeth present, not all single-cusped; nostrils open at front of skull 2
2. Angle of lower jaw has distinct interior shelf, carrying the pterygoid muscles; tympanum covered by an extension of the alisphenoid bone; lacrymal bone extends beyond orbit to cover part of snout; palate fenestrated .... [Marsupials] 11
   Lower jaw not as above; tympanum covered by tympanic bone; palate entire 3
3. Upper incisors absent; lower jaw with 3 pairs of incisors plus 1 pair of incisiform canines; 6 pairs of cheekteeth (premolars and molars); orbit closed .... [Artiodactyls] 17
   Upper incisors present; orbit open 4
4. Lower jaw with only 1 pair of chisel-shaped incisors, separated from molars by a diastema as long as or longer than molar row 5
   Lower jaw with >1 pair of incisors, not chisel-shaped; diastema either absent or shorter than molar row 6
5. One pair of incisors in upper jaw .... [Rodents] 22
   Two pairs of incisors in upper jaw .... [Lagomorphs] 25
6. Crowns of molars flat, with enamel ridges .... **Horse**
   Crowns of molars with cusps 7
7. Lower canines tusk-like, angular in cross-section; I^3 widely separated from I^2 .... **Pig**
   Lower canines rounded or ovate in cross-section; I^3, if present, close to I^2 8
8. Length of skull >25 mm 9
   Length of skull <25 mm .... [Bats] 26
9. Canines prominent above and below; tympanic bulla inflated; mandibular condyles transversely elongated .... [Carnivores] 10
   Canines not prominent; tympanic bulla incomplete; median spine projecting behind posterior edge of palate .... **Hedgehog**
10. Length of skull <250 mm; post-canine teeth 3-7, very variable in shape and size .... [Land carnivores] 28
    Length of skull >180 mm; post-canine teeth 6, rather uniform in shape and size (large central cusp, small anterior and posterior cusps) .... [Otariids] 31
11. Masseteric fossa shallow, masseteric canal absent; molars with separate cusps .... **Brushtail possum**
    Masseteric fossa deep, masseteric canal present; molars transversely ridged 12

12. I^1 wider than I^3 .... **Brushtailed rock wallaby**
    I^1 narrower than I^3 13
13. M^1 larger than M^2 .... **Swamp wallaby**
    M^1 smaller than M^2 14
14. Groove on I^3 on rear two-thirds of tooth .... **Dama wallaby**
    Groove on I^3 about central 15
15. Rear edge of I^3 curved .... **Bennett's wallaby**
    Rear edge of I^3 angular 16
16. Groove on I^3 posterior to midline .... **Parma wallaby**
    Groove on I^3 anterior to midline .... **Black-striped wallaby**
17. Males with solid, bony, deciduous antlers growing from permanent bony pedicle; upper canines sometimes present .... [Cervids] 32
    Males, and often females, with permanent horns (keratin sheath on a bony core); upper canines absent .... [Bovids] 18
18. Palatine bone occupies about half of space between molar rows; pedicles directed sideways; parietal and occipital bones at acute or right angles to frontal bone .... **Cattle**
    Palatine bone occupies about one-third of space between molar rows; pedicles directed upwards or backwards; parietal and occipital bones at obtuse angle to frontal bone .... [Caprinids] 19
19. Horn cores perpendicular to cranium .... **Chamois**
    Horn cores sweep backwards from cranium 20
20. Lambdoidal suture straight, coronal suture angled; shallow lachrymal fossae present .... **Sheep**
    Lambdoidal suture angled, coronal suture straight; no lachrymal fossae 21
21. A line through the centres of the orbits passes well clear of posterior edge of nasal bone; ear canal directed backwards .... **Tahr**
    A line through the centres of the orbits approaches or touches posterior edge of nasal bone; ear canal directed sideways .... **Goat**
22. Maxillary tooth row >5 mm long .... [*Rattus*] 23
    Maxillary tooth row <5 mm; upper incisors have notch in wearing surface; M^1 with 3 roots; M^3 much smaller than M^2 .... **House mouse**
23. Cranium with rounded margins (temporal ridges widely curved); length of parietal bone along its outer edge exceeds width of cranium 24
    Cranium with straight margins (temporal ridges parallel); length of parietal bone along its outer edge same as width of cranium .... **Norway rat**
24. Mandibular toothrow <6 mm .... **Kiore**
    Mandibular toothrow >6 mm .... **Ship rat**
25. Width of internal nares more than narrowest part of palatal bridge and exceeds length of molar row; no suture between supraoccipital and interparietal bones; supraorbital processes massive, triangular in shape .... **Brown hare**

(continued on p. 30)

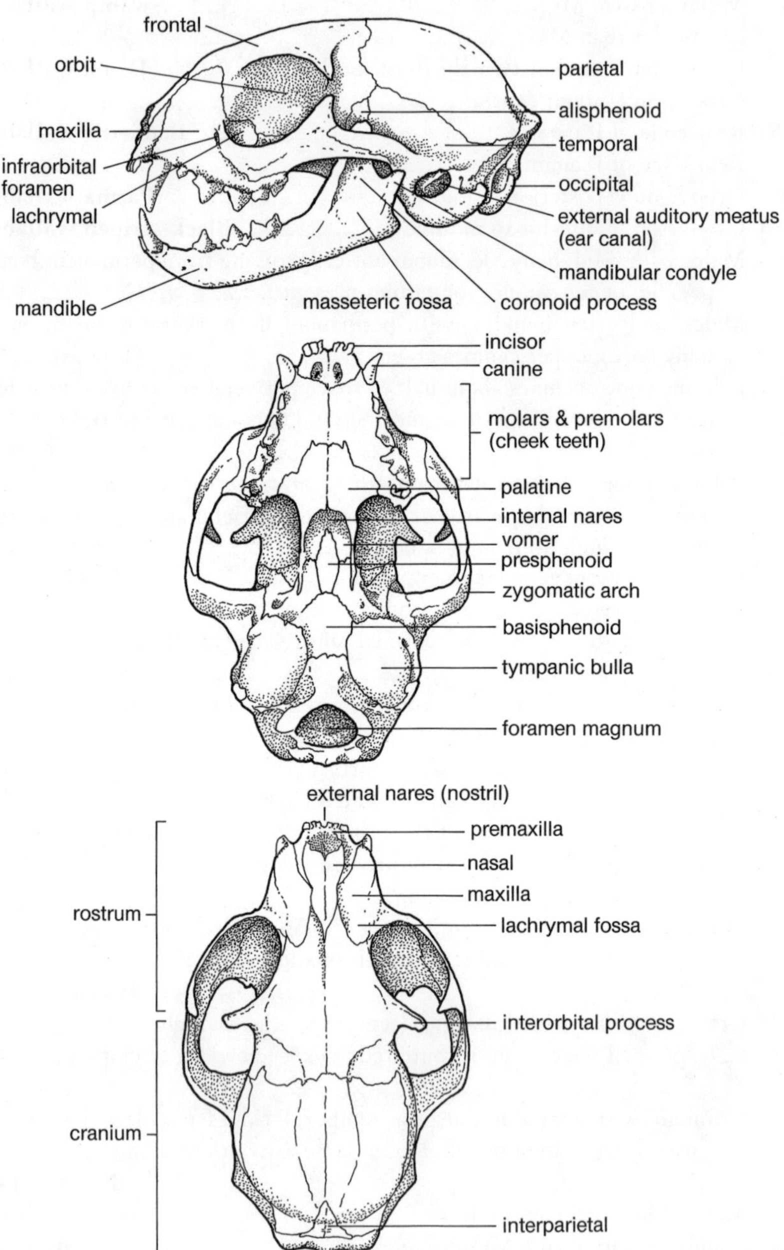

Fig. 3 Skull of cat, showing location and nomenclature of bones (J. Lavas).

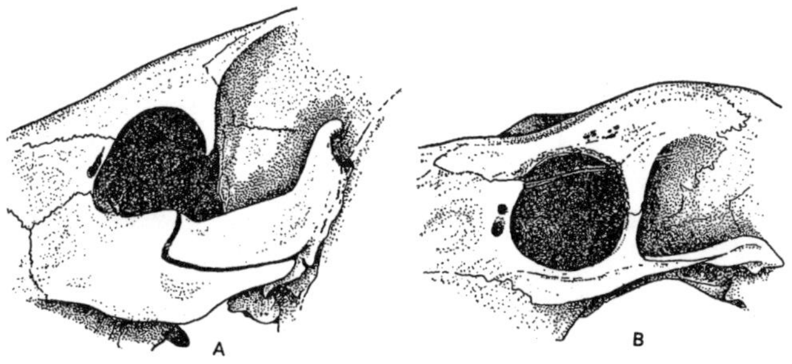

Fig. 4 Structure of orbit in (A) pig and (B) red deer; orbit respectively open and closed behind.

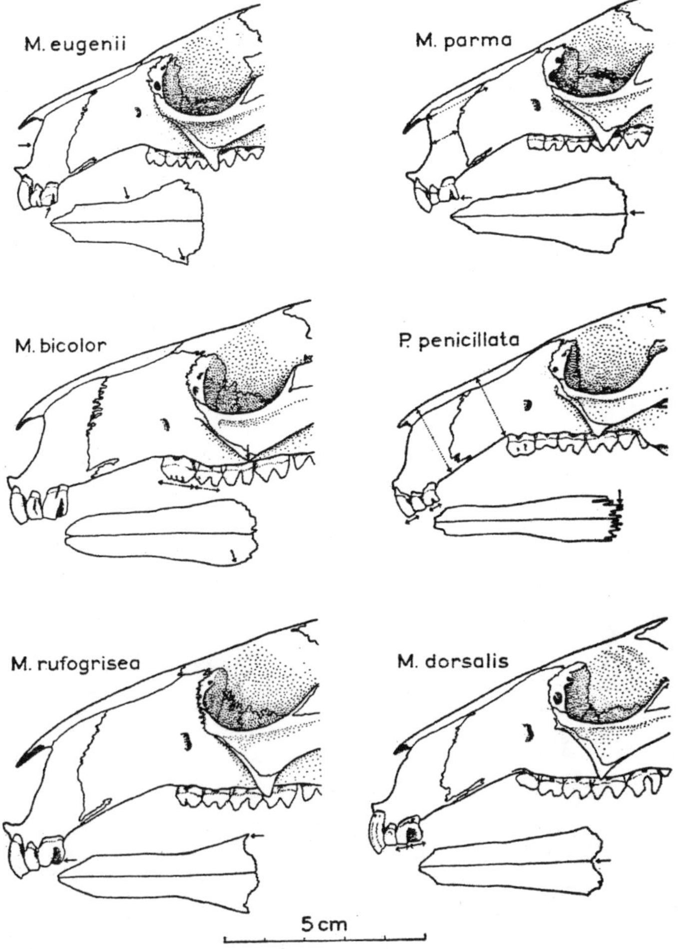

Fig. 5 Lateral views of anterior part of skull, and dorsal views of nasal bones, of the six species of wallaby now or formerly present in New Zealand (from Wodzicki & Flux, 1967a).

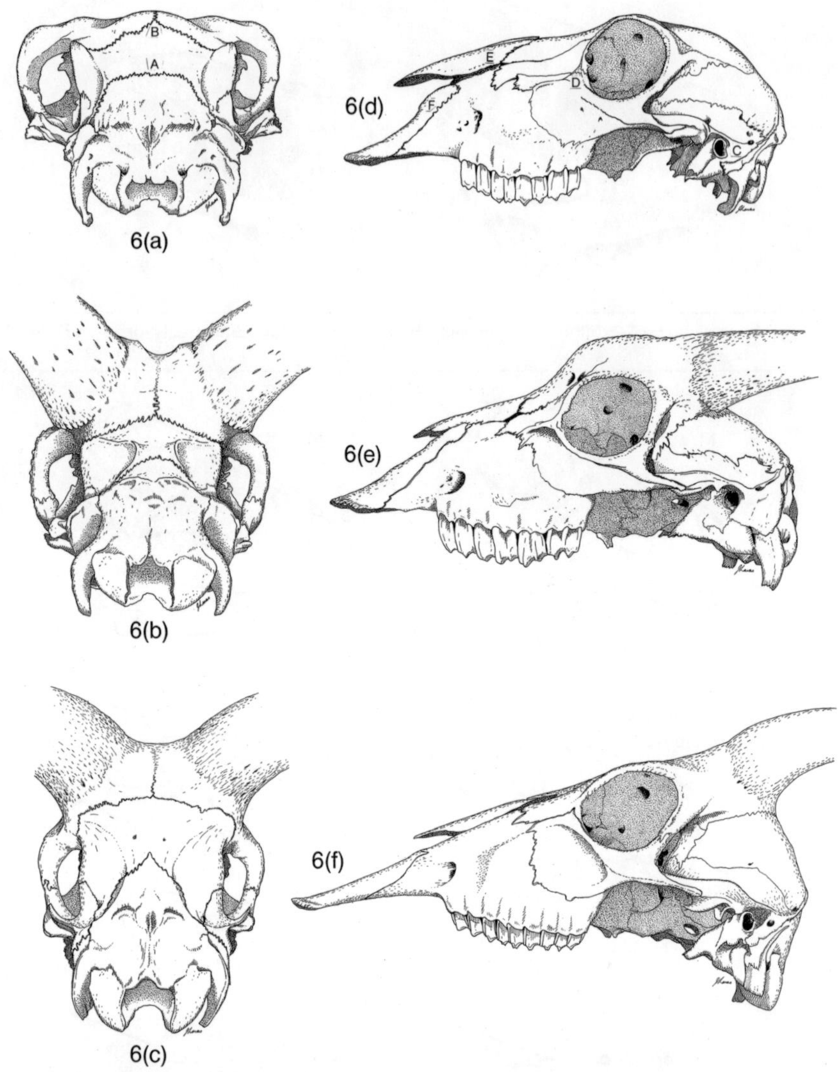

Fig. 6 Characters distinguishing skulls of (6(a), 6(d)) sheep, (6(b), 6(e)) goat and (6(c), 6(f)) tahr, in posterior and lateral view (J. Lavas, from information supplied by M.R. Rudge). Key: A, lambdoidal suture; B, coronal suture; C, ear canal; D, lachrymal fossa; E, nasal bone; F, premaxilla.

Fig. 7 Upper incisor of mouse, lateral view, showing distinct notch in cutting edge.

Fig. 8 Skulls of (8(a)) Norway rat and (8(b)) ship rat. Key: a, maximum length of parietal bone; b, maximum width of temporal ridges.

Fig. 9 Characters distinguishing skulls of (9(a), 9(b)) rabbit and (9(c), 9(d)) brown hare. 9(a), 9(d), dorsal view of occipital area; 9(b), 9(c), ventral view of palate (redrawn by J. Lavas after Corbet, 1964).

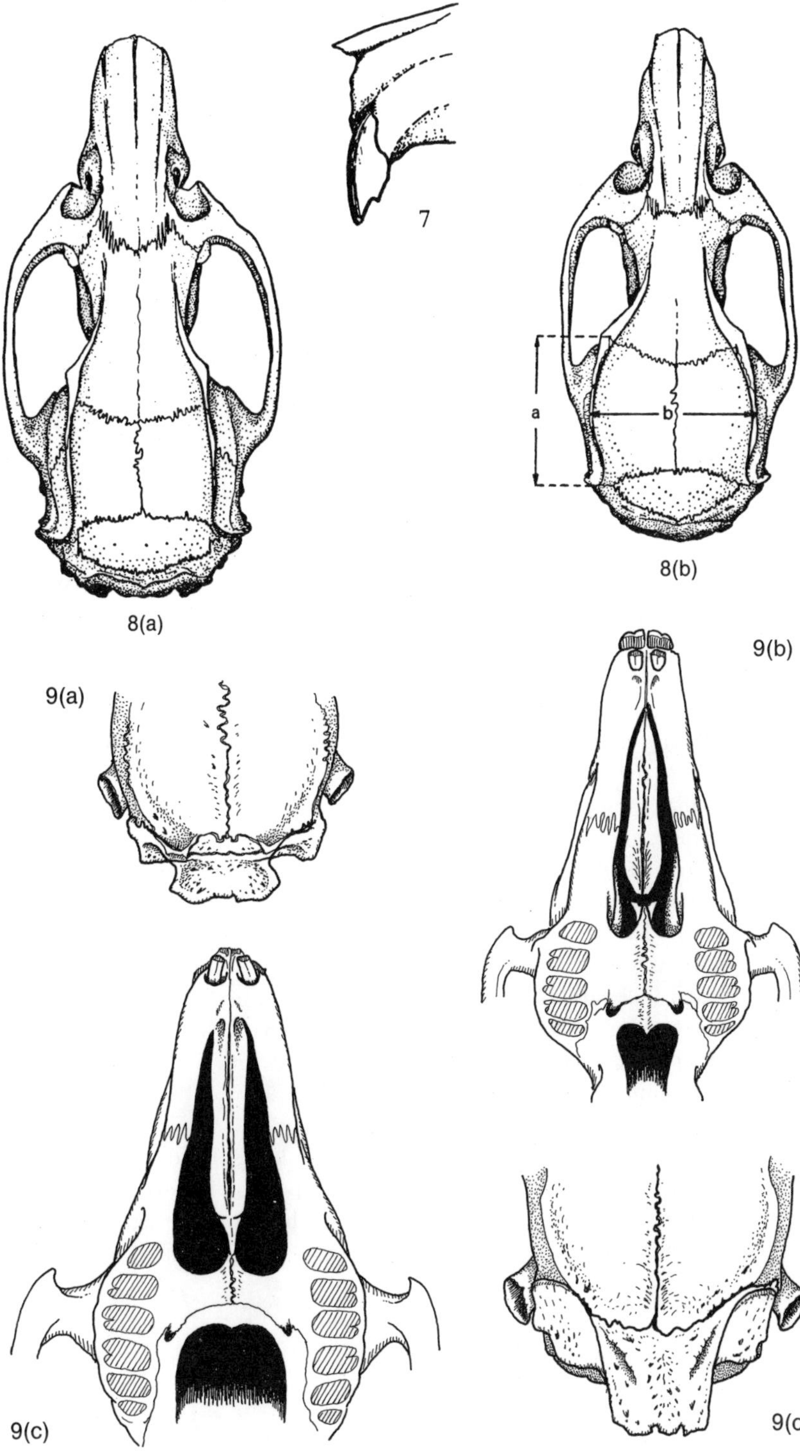

7

8(a)

8(b)

9(a)

9(b)

9(c)

9(d)

25. Width of internal nares less than narrowest part of palatal bridge and same as length of molar row; distinct suture between supraoccipital and interparietal bones; supraorbital processes narrow and slender .... **Rabbit**
26. Skull broad in shape, length <14.3 mm .... **Long-tailed bat**
    Skull narrow in shape, length >17.3 mm 27
27. Length of skull 17.3-19.1 mm .... **Lesser short-tailed bat**
    Length of skull 21.0-22.5 mm .... **Greater short-tailed bat**
28. Premolars 4/4, molars 2/3 .... **Dog**
    Premolars 2-3/2, molars 1/1 .... **Cat**
    Premolars 3/3, molars 1/2 .... [*Mustela*] 29
29. Length of skull >54 mm, mandible >32 mm; rostrum parallel-sided .... **Ferret**
    Length of skull <54 mm, mandible <32 mm; rostrum not parallel-sided 30
30. Length of skull >42 mm, mandible >22 mm; infraorbital foramina distinctly larger than diameter of canines .... **Stoat**
    Length of skull <42 mm but not <30 mm, mandible <22 mm; infraorbital foramina about equal to diameter of canines .... **Weasel**
31. I^3 and all post-canine teeth large, over half the length of the canines .... **NZ sea lion**
    I^3 and all post-canine teeth small, less than half the length of the canines .... **NZ fur seal**
32. Skull very large (>500 mm) with elongate nose (distance from anterior tip of nasal bone to anterior tip of intermaxilla almost the same as that from posterior edge of nasal bone to posterior edge of occipital bone) .... **Moose**
    Skull smaller (<400 mm) with nose shorter than cranium (comparative distances not as above) 33
33. Lachrymal fossae in front of each orbit rather shallow; vomer divides nares into two separate chambers posteriorly; width of auditory bullae equal to length of ear canal .... **White-tailed deer**
    Lachrymal fossae deep; vomer and bullae not as above 34
34. Upper canines usually absent .... **Fallow deer**
    Upper canines usually present .... *Cervus*, *Axis*

Note: There is insufficient information available at the time of writing to complete this key down to the species of *Cervus*. The males, at least for part of the year, can be distinguished from their antlers (Fig. 73).

Preamble to Species Accounts

Taxonomy

World Listings

Each account of a species now or formerly resident in New Zealand is preceded by a brief introduction listing the general characters of the order, family and genus to which it belongs, and the number and distribution of related species. Different authorities disagree as to the number and arrangement of animal groups, so for consistency this information is all taken from one source, Macdonald (1984), unless otherwise stated.

Domesticated Forms

Domesticated species are those which have been tamed and bred under human control for many centuries. Under artificial selection, most have become different from their wild ancestors, although they are still interfertile with them. In remote areas, individual animals of domesticated species may escape human management and establish feral populations in the wild; these are often particularly interesting to observe, since they are thereby exposed to natural selection again, perhaps for the first time in very many generations. In New Zealand there are feral populations of sheep, cattle, goats, horses, pigs, house cats, and ferrets. Also, a domesticated form of the rabbit persists on two islands of the Auckland group, although elsewhere domesticated rabbits have interbred with the wild form; and packs of feral dogs were once common in the South Island.

Nomenclature of these feral populations presents problems. Taxonomists disagree as to whether the names of domesticated forms, which, except in the case of the pig (p. 358), usually have priority, should be applied to their wild ancestors (Corbet & Hill, 1980; Honacki, Kinman & Koeppl, 1982). Nor can the domesticated forms be considered as subspecies, since they have no definable geographic range. Corbet's (1978) suggestion, the use of a vernacular description after the name of the genus, e.g., *Capra* (domestic), has not become standard practice. This book therefore continues to use the names that have been traditionally applied to these animals and that are in common use; the separate names of the wild ancestors are listed in Table 5.

Maps

The species distributions are shown at the largest scale consistent with the information available, so they are not standardized in size or level of detail. Some show accurate range boundaries (e.g., in economically important species, which have been well surveyed); some show only the general area within which a species might be found in suitable habitat.

Table 5: Nomenclature of domestic animals that are feral in New Zealand and their probable wild ancestors.

Domestic form		Probable wild ancestor
Ferret	*Mustela furo* Linnaeus, 1758	*M. putorius* Linnaeus, 1758
Cat	*Felis catus* Linnaeus, 1758	*F. silvestris* Schreber, 1777
Horse	*Equus caballus* Linnaeus, 1758	*E. ferus* Boddaert, 1785
Pig	*Sus scrofa* Linnaeus, 1758	*S. scrofa* Linnaeus, 1758
Cow	*Bos taurus* Linnaeus, 1758	*B. primigenius* Bojanus, 1827
Sheep	*Ovis aries* Linnaeus, 1758	*O. ammon* Linnaeus, 1758
Goat	*Capra hircus* Linnaeus, 1758	*C. aegagrus* Erxleben, 1777
Dog	*Canis familiaris* Linnaeus, 1758	*C. lupus* Linnaeus, 1758

Nomenclature from Corbet and Southern (1977); Corbet (1978); Corbet and Hill (1980); re pig, see p. 358.

Layout of Species Accounts

All species accounts are set out in a standardized form for easy reference. The main headings and subheadings are always entered in the same order, but some may be omitted in shorter accounts or where information is lacking.

Common and Latin name. Brief synonymy.

Description. Distinguishing marks, appearance, pelage, moult, senses, locomotion, non-social behaviour, skull, teeth, skeleton.

Field sign.

Measurements.

Variation, in size, colour, subspecies etc.

History of colonization.

Distribution, in the world and in New Zealand.

Habitat.

Food. Diet, feeding behaviour.

Social organization and behaviour. Activity, dispersion, home range, nests, voice and communication, reproductive behaviour.

Reproduction and development. Male and female organs, age at puberty, seasonal cycle, gestation, birth, growth of young.

Population dynamics. Density, sex ratio, age determination, age structure, mortality, hybridization, competition.

Predators, parasites and diseases.

Adaptation to New Zealand conditions. The consequences of the arrival of a native or introduced species for itself, as reflected in, for example, changes in size, habitat, behaviour, interspecific interactions, diet, numbers, etc., compared with the parent stock.

Significance to the New Zealand environment. The consequences of the arrival of an introduced species for the native communities it entered; also, historical changes in the distribution, density, and population dynamics of native species.

Part II
Species Accounts

Order Marsupialia

Marsupials are native to Australia, New Guinea, and North and South America, but are most plentiful and diverse in Australia (Kirsch & Calaby, 1977).

The features characteristic of marsupials, and which distinguish them from the placental mammals, are as follows.

1. The young are born, after a short gestation, at a very early stage of development, and complete the rest of it in the mother's pouch (but some polyprotodont marsupials do not have pouches, and one monotreme does; Barbour, 1977).
2. The marsupial ovum is larger and has a series of membranes, like that of reptiles and birds and quite unlike that of placentals (Tyndale-Biscoe, 1973).
3. There is no extensive chorio-allantoic placenta (except in koala and wombats). Foetal marsupials absorb nutrients through a yolk sac placenta, which normally does not implant into the womb (Dawson, 1983).
4. The female reproductive system has two uteri and two lateral vaginae (Kean, 1966).
5. The penis is bifid, except in kangaroos, and is posterior to the testes (Barbour, 1977).
6. The pelvis has a pair of epipubic bones, extending forward in association with the pouch and abdominal musculature (Kean, 1966).
7. The angle of the lower jaw has a distinct interior shelf, carrying the pterygoid muscles; the tympanum is covered by an extension of the alisphenoid bone rather than by the tympanic bone; the lacrymal bone extends beyond the orbit to cover part of the snout (Tyndale-Biscoe, 1973).
8. Body temperature and metabolic rates are lower than in placentals (Dawson & Hulbert, 1970).

The earliest fossil marsupials found in Australia date from the late Oligocene (30 million years BP) (Clemens, 1977). During Australia's long isolation, the marsupials evolved forms equivalent to the placental insectivores, rodents, carnivores and herbivores of the rest of the world. There are about 18 living families, with a total of 266 species in 76 genera.

Two families are represented in New Zealand, the Macropodidae and the Phalangeridae. One species each from four other families has been liberated but have not established in the wild. The long-nosed potoroo or rat-kangaroo (*Potorous tridactylus,* Family Potoroidae) was liberated at an unknown location by the Auckland Acclimatisation Society in 1867; the marsupial cat (*Dasyurus* sp., Family Dasyuridae) in 1868; the southern brown bandicoot (*Isoodon obesulus,* Family Paramelidae) in 1873; and the ring-tailed possum (*Pseudocheirus perigrinus,* Family Pseudocheiridae) in 1867 (Wodzicki, 1950).

Family Macropodidae

The macropods ("big feet") are the most familiar animals of Australia; about 50 species in 11 genera are living today, and many more are known as fossils. The distinguishing marks of the five members of the family now living in New Zealand are summarized in Table 6.

Genus *Macropus*

This genus contains most of the larger kangaroos and wallabies, totalling 14 species. At least seven of them have been introduced to New Zealand. Three still live here (*M. eugenii, M.r. rufogriseus* and *M. parma*). Four have disappeared: *M. dorsalis* (p. 57); the roan wallaroo, *M. robustus,* Kawau I., 1860 to 1870; an unidentified kangaroo, *Macropus* sp., Dunrobin Station and Bluff Hills, 1868; an unidentified wallaby, Mt. Nimrod, Timaru, 1903 (Wodzicki, 1950; Wodzicki & Flux, 1967a). All members of the genus are typically grazers, although they occupy habitats ranging from arid grasslands to temperate forests. The wallabies found in New Zealand are the smaller members of the genus, predominantly residents of forest/scrub habitats and tending to be solitary, in contrast to the large kangaroos in Australia, which are gregarious inhabitants of open woodlands and grasslands.

1. Dama wallaby

Macropus eugenii (Desmarest, 1817)

Synonyms *Thylogale eugenii* Desmarest, 1817; *T. flindersi* Wood Jones, 1824; *Protemnodon eugenii* Desmarest, 1817.

Also called tammar, kangaroo island, or silver-grey wallaby.

DESCRIPTION (Fig. 10)

Distinguishing marks, Table 6 and skull key, p. 24. See left-hand front endpaper for colour plates.

The dama is one of the smallest of the wallabies, but has the general appearance and all the characteristics typical of the family Macropodidae. Its pelage is generally grey-brown above, paler grey below, with rufous shoulders; the long tapering tail is uniformly grey; the ears are long and pointed. When hopping the forelimbs are usually held apart and away from the body: damas also lope using all four limbs and tail (pentapedal gait). They frequently lie down on their side, unlike parmas, and also rest sitting with their tails between their legs. When alerted they stand with their forelimbs held out from the chest ready to take flight. They groom, using their tongues, teeth, forepaws and the syndactyl toes of the hind feet (M. Vujcich, 1979).

Dental formula $I^{3}/_{1}\,C^{0}/_{0}\,Pm^{1}/_{1}\,M^{4}/_{4} = 28$. The molar tooth row moves forward throughout life, continually providing new surfaces for chewing coarse herbage.

FIELD SIGN

Grazing or browsing sign due to wallabies cannot easily be distinguished from that of other herbivores, but their almost square and flattened faecal

Table 6: Distinguishing marks of wallabies in New Zealand.

	Dama *Macropus eugenii*	Bennett's *M. r. rufogriseus*	Parma *M. parma*	Black-striped *M. dorsalis*	Brushtailed rock *Petrogale p. penicillata*	Swamp *Wallabia bicolor*
Tail	Tapering, uniform colour, well furred	Tapering, dark towards tip	Tapering, often white tipped, relatively hairless	Tapering, uniform colour, relatively short hair	Bushy, rufous colour, black tip	Tapering, almost black
Fur	Grey-brown above, pale grey below, rufous shoulders	Grey above, pale grey below, rufous-brown on neck and shoulders	Brown-grey above, pale grey below, white cheek stripe	Brown above with distinct dark dorsal stripe, belly fur white, horizontal white stripe on thigh	Bluish-grey above, rufous on rump and belly, dark face with light white-grey cheeks	Dark black-grey above, yellow-rufous below, light yellow cheek stripe, distinct orange band around base of ears
Rank order of size	4	1=	5	1=	3	2
Distribution	Rotorua district, Kawau I.	South Canterbury	Kawau I.	Possibly liberated on Kawau I., now believed extinct	Kawau I., Motutapu I., Rangitoto I.	Kawau I.

pellets are distinctly different from those of other mammals. The long narrow hind feet and dragging tail leave characteristic tracks on soft substrates.

MEASUREMENTS
See Table 7.

VARIATION
Dama wallabies differ significantly in size between habitats, probably because they can reach their full potential size only on good grazing. In the Rotorua region they are generally larger than on Kawau I., and those inhabiting forest/pasture margins in the Rotorua district are larger than those from within forest (Table 7).

HISTORY OF COLONIZATION
The first dama wallabies to reach New Zealand were released on Kawau I. about 1870 by Sir George Grey (Wodzicki & Flux, 1967a). More were liberated in the Rotorua district in 1912, although it is uncertain whether this stock came from Kawau I. or as a separate introduction from Australia. From the liberation point, probably at the southern end of Lake Okareka, damas extended their range mainly north and east. By the late 1940s and early 1950s reports of their spread and high numbers began to appear. Estimates of the area occupied increased from 0.6 km^2 in 1946, 2.3 km^2 in 1956, and 5.7 km^2 in 1966, to 16.2 km^2 in 1984 (R. M. F. Sadleir, unpubl.). Overall their rate of spread in the Rotorua district has been slow (approx. 0.22 km^2/yr), inhibited to some unknown extent by physical barriers (e.g., lakes) and by poison operations in the 1960s.

DISTRIBUTION
WORLD. The dama wallaby was formerly widespread in southern mainland Australia, but is now restricted to two areas (south-western Western Australia and southern South Australia) and several islands off the southern and western coast (Houtmen Abrolhos, Garden I., Recherche Archipelago, and Kangaroo I.; also formerly on Flinders I.) (Calaby, 1971; Smith, 1983).

NEW ZEALAND (Maps 2, 4). On Kawau I. dama are widespread but vary in density between habitat types (V. Vujcich, 1979). The mainland population is still expanding slowly; in recent years there have been increasing numbers of confirmed sightings to the north of the present range, and several in the Urewera ranges to the east (probably the result of illegal releases) and in Taranaki, but no confirmed populations have established themselves outside the area mapped.

HABITAT
The dama prefers "edge" habitats, because it requires relatively dense vegetation for shelter and cover during the day, with access to grasses and pasture species for feeding at night. Kawau I. once had extensive pastures, but farming was abandoned in 1973, and most stock was removed. The pastures are now being invaded by kanuka (*Kunzia ericoides*) and manuka scrub (*Leptospermum scoparium*); there are also some exotic conifers. The coastal cliffs have remnant pohutukawa (*Metrosideros excelsa*) with scattered puriri

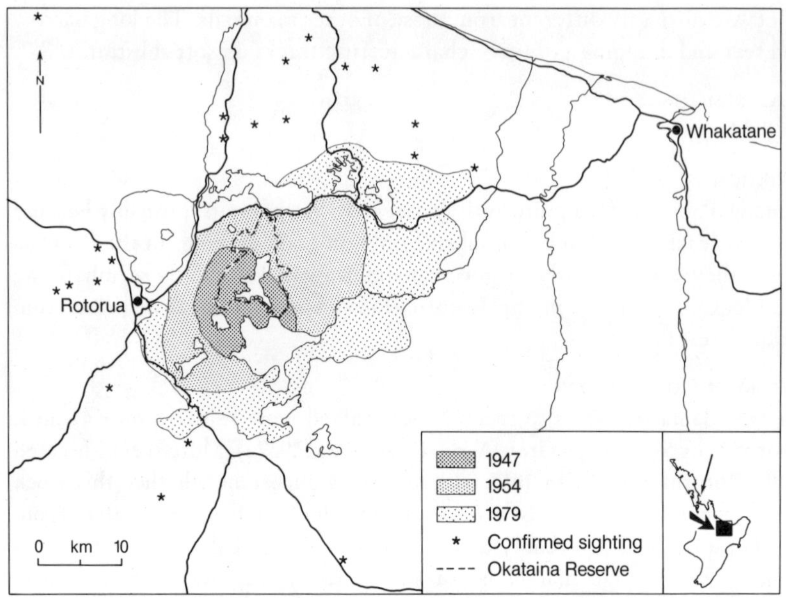

Map 2 Distribution of dama wallabies, Rotorua district, at three stages of colonisation. There is another population on Kawau I. (arrowed on inset, and see Map 4).

(*Vitex lucens*) and taraire (*Beilschmiedia taraire*). Only remnant patches of the once diverse indigenous forest remain, in which there are almost no palatable understorey species (Warburton, 1986).

In the Rotorua district the damas inhabit predominantly podocarp/tawa/mixed hardwood forest with adjoining areas of manuka scrub, bracken (*Pteridium esculentum*) and pasture. A large area of exotic forest also provides suitable cover and access to grass and weeds along roadsides and in young plantations. Many damas are found solely within the forest, but these do less well than those with ready access to grassed clearings or pasture.

FOOD

The dama is primarily a grazer. The diet of damas with access to grassland includes a high percentage of pasture species (V. Vujcich, 1979; Williamson, 1986). On Kawau I. grasses were the main food, although damas there also ate many kanuka leaves (28% of the diet) (see Table 11).

The diet of 55 individuals from the interior of kamahi/mixed hardwood forest included fewer (3%) grasses but more (50%) kamahi (*Weinmannia racemosa*) and mahoe (*Melicytus ramiflorus*); the next eight items in order of frequency were hangehange (*Geniostoma rupestre crassa*), *Coprosma* sp., pigeonwood (*Hedycarya arborea*), manuka/kanuka, ferns, supplejack (*Ripogonum scandens*), rangiora (*Brachyglottis repanda*) and broadleaf (*Griselinia* sp.) (Williamson, 1986). Within exotic forest damas ignore *Pinus radiata*, concentrating on grass and weeds (Williamson, 1986).

Damas feed mainly at night, and sporadically throughout the day (V. Vujcich, 1979). While feeding they stand still, feeding from side to side, then moving forward with their pentapedal gait to resume feeding in an arc pattern. Damas have been noted to regurgitate food and re-eat it (V. Vujcich, 1979). They occasionally drink water but are also capable of withstanding long dry periods (Smith, 1983).

SOCIAL ORGANIZATION AND BEHAVIOUR

ACTIVITY AND DISPERSION. The dama wallaby is largely nocturnal, though it is sometimes active in the late afternoon (Kinloch, 1973). On Kawau I. nest sites for day-time resting were most frequently found under thick short kanuka scrub, usually within 100 m of a grassed area (M. Vujcich, 1979). These sites were often shared by two or more individuals: active damas form groups of up to five individuals of mixed size and age (M. Vujcich, 1979). Their home range often varies depending on the distance from nest sites to feeding area: in the Rotorua district some individuals travelled up to 500 m to pasture (Williamson, 1986). In Australia, the mean home-range size in an island population was estimated to be 42.4 ha during summer (Inns, 1980).

COMMUNICATION. Males threaten each other with displays such as grass pulling with the forepaws or vigorously scratching the chest. Rarely, these displays lead to fights in which the opponents kick at each other's stomach and chest with the hind feet, or spar with the forefeet, grabbing each other's chest, neck or face. Submissive displays involve lowering the head and crouching (M. Vujcich, 1979). Individuals greet each other by nose touching, especially when a new individual joins a group. Damas are constantly alert when feeding, with frequent periods of looking and listening. A disturbed individual will face the intrusion, alerting others nearby by posture or hind foot thumping, and if necessary fleeing, accompanied by all others in the area (M. Vujcich, 1979).

REPRODUCTIVE BEHAVIOUR. Male damas sniff the cloaca and pouch areas of the female, while she either remains passive or pushes him away. Later he moves in front of her, sniffing her chest, head and pouch, signalling his intentions with an erect penis and lashing tail. Dominant dama males often displace sub-dominants attending females (M. Vujcich, 1979).

REPRODUCTION AND DEVELOPMENT

Female damas are sexually mature at 12 months of age (M. Vujcich, 1979), although the breeding success of yearlings depends on habitat (higher in pasture than in forest; Williamson, 1986). There is an immediate postpartum oestrus, and most females mate again 24 hours after the previous young is born. Embryonic diapause, a characteristic of the family Macropodidae, is maintained in the early stages by a lactation stimulus and later by some seasonal cue; in all, the blastocyst can remain quiescent for eleven months. Blastocysts are reactivated by photoperiodic stimuli in December and January (Sadleir & Tyndale-Biscoe, 1977). Gestation is 28 days, and pouch life averages 250 days (Russell, 1974).

On Kawau I. births were restricted to the late summer months of January, February and March (M. Vujcich, 1979). In the Rotorua district in 1983, backdating of pouch young (Murphy & Smith, 1970) showed that births started in mid January. Of 301 estimated birth dates, 50% ranged from 30 January to 8 February and the rest tailed off rapidly until the last birth on 10 July. The mean date of birth was 5 February. There was very little pouch mortality, and most young left the pouch in November and December (R. M. F. Sadleir, unpubl.).

The mammary gland increases slowly in weight from an average of 1 g in February to 6 g in August. From September to December it rapidly increases to a mean weight of 22–25 g, as the large pouch young suckles, then regresses. A few females continue to suckle the previous year's young outside the pouch, so that between January and March one mammary may be 20 g while the newborn young is suckling on the other, 1 g mammary.

Twin joeys are unknown, but in the Rotorua sample a twin embryo (two differentiated embryos in a single vesicle) was found in a female shot in January 1983. This has never been recorded in any macropod before.

It appears that most damas reach full size at about 3 years of age, but the maximum attainable size depends on the suitability of the habitat (Williamson, 1986).

POPULATION DYNAMICS

From March to November, 86% (n = 163) of the Rotorua females had viable quiescent blastocysts. All females born in 1982 bred in the 1983 season.

Age in young damas can be estimated from molar progression. Knowlton (1986) used a device for making accurate estimates of the position of the molars in the upper jaw, considerably reducing the error in age determination by the molar index (Kirkpatrick, 1964). This method was applied to Rotorua dama by Williamson (1986). He distinguished separate annual age classes from 1 to 4 years, and a single class for those 5 years and older. J. E. Knowlton, R. M. F. Sadleir and J. White (unpubl.) analysed age cohort data from Rotorua dama sampled in 1982–84; the distributions of molar index classes by month were very similar each year, indicating a relatively uniform survival rate after the young leave the pouch. The majority of individuals were 1- and 2-year-olds, and very few were in the >5-year class.

There have been no estimates of absolute population density. A faecal pellet survey by Knowlton and Panapa (1982) in part of the Rotorua district found pellet frequencies of >40%, which indicates a moderate to high density of damas. On Kawau I. the density is patchy, but locally high around grassed clearings.

PREDATORS, PARASITES AND DISEASES

There are no published reports of predation on dama wallabies in New Zealand. Casual observers have noted attacks by dogs (D. Moore, unpubl.). In Australia, predation by feral cats is believed to have contributed to the disappearance of damas on Flinders I. (Smith, 1983).

Very little research has been carried out on parasites and diseases of wallabies in New Zealand. In eight damas from the Rotorua district, Bowie (1980) found no trace of parasitic infection. Damas on Kawau I. have at least one unidentified ectoparasite (M. Vujcich, 1979).

Diseases in damas have not been recorded. Their potential as vectors of bovine tuberculosis is unknown.

ADAPTATION TO NEW ZEALAND CONDITIONS

The wallabies are native to the Southern Hemisphere, so when introduced to New Zealand they did not have to adapt to a 6-month change in seasons as did those species that came from the Northern Hemisphere. Like the other wallabies, damas have settled well in New Zealand, retaining their usual patterns of behaviour, seasonal breeding, maturation and growth. Numbers of dama in Australia have been greatly reduced by habitat clearance and predation: but in the Rotorua district their population is still expanding.

Although damas appear to survive within indigenous forest, they have done much better (in terms of age at sexual maturity and body weight) in habitats comprising both forest for cover and pasture for food.

SIGNIFICANCE TO THE NEW ZEALAND ENVIRONMENT

DAMAGE. For many years the Rotorua damas have been considered as potential pastoral pests, but so far there have been no reports of major damage and only occasional browsing in new plantings of *Pinus radiata.* Of far greater importance is the potential effect of damas on protected indigenous vegetation. A vegetation survey in the Okataina scenic reserve showed that regeneration of most palatable species such as hangehange *(Geniostoma rupestre crassa), Fuchsia excorticata,* raurekau *(Coprosma grandifolia),* karamu *(Coprosma robusta),* pate *(Schefflera digitata)* and five-finger *(Pseudopanax arboreus)* was inhibited in sites accessible to wallabies. Deer and pigs were also present, but in low numbers, so most damage was probably attributable to wallabies (Knowlton & Panapa, 1982). Unpalatable species such as mangeao *(Litsea calicaris),* rewarewa *(Knightia excelsa)* and rangiora *(Brachyglottis repanda)* were common in the understorey. It is probable, therefore, that in the few forests they inhabit at high density, damas are as capable of changing the pattern of forest succession, or at least altering the local abundance of different species, as are deer and possums elsewhere (pp. 95, 453).

On Kawau I. damas keep all grassy areas down to a short sward. Palatable species are absent from the understorey of remaining indigenous forest patches, and establishment of their seedlings under the manuka/kanuka canopy is inhibited.

CONTROL. Serious attempts to control damas in the Rotorua district started in 1962–63 with eight aerial control operations using 1080 baits, covering all land tenures. The effectiveness of the poison was not assessed (Warburton, 1986). Since then, control has been restricted to rateable land and carried out by the local pest destruction board, usually by spotlight shooting and with occasional use of 1080 poison. Many are also shot for sport.

Table 7: Body measurements of adult dama wallabies in New Zealand and Australia (mean ± SD).

	Sex	Sample size	Body wt (kg)	Total body length (mm)	Head and body length (mm)	Tail length (mm)	Hind foot length (mm)	Reference
Rotorua	M	170	5.6 ± 1.2	1 000 ± 72	—	414 ± 35	159 ± 7	Williamson (1986)
(pasture)	F	119	4.3 ± 0.8	911 ± 59	—	376 ± 39	151 ± 7	Williamson (1986)
Rotorua	M	42	4.3 ± 0.7	939 ± 66	—	400 ± 34	152 ± 6	Williamson (1986)
(forest)	F	14	3.3 ± 0.5	872 ± 59	—	366 ± 24	144 ± 7	Williamson (1986)
Kawau I.	M	10	3.6(2.8–5.4)	—	460	390	140	Wodzicki & Flux (1967a)
(all ages)	F	11	3.7(2.7–4.4)	—	460	390	130	Wodzicki & Flux (1967a)
Rotorua	M	10	5.5(2.9–7.6)	—	530	420	150	Wodzicki & Flux (1967a)
(all ages)	F	10	4.6(4.0–5.6)	—	490	380	140	Wodzicki & Flux (1967a)
Australia	M	—	7.5(6–10)	—	643(590–680)	411(380–450)	—	Smith (1983)
	F	—	5.5(4–6)	—	586(520–630)	379(330–440)	—	Smith (1983)

Table 8: Body measurements of Bennett's wallabies in New Zealand and Australia (mean ± SD).

	Sex	Sample size	Body wt (kg)	Total body length (mm)	Head and body length (mm)	Tail length (mm)	Hind foot length (mm)	Reference
Canterbury	M	198[1]	12–21	1 435 ± 23	–	–	200–230	Catt (1975)
Canterbury	M	10[2]	13.2(6.8–22.7)	–	650	620	220	Wodzicki & Flux (1967a)
Australia	M	–	19.7(15–26.8)	–	780(710–920)	770(690–860)	–	Calaby (1983)
Canterbury	F	147[1]	10–12	1 275 ± 17	–	–	200–210	Catt (1975)
Canterbury	F	12[2]	10.5(6.8–13.6)	–	610	620	200	Wodzicki & Flux (1967a)
Australia	F	14	14(11–15.5)	–	710(660–740)	690(620–780)	–	Calaby (1983)

1. Over 3 years old. 2. Includes all ages.

Fig. 10 Dama wallaby, Rotorua district, 1969 (J.H. Johns/NZFS).

Fig. 11 Bennett's wallaby (NZFS).

There has been no extensive control of damas in indigenous forests since 1962–63, because wallabies were believed to do minimal damage in them. With Knowlton and Panapa's (1982) report on the effect of wallabies on regeneration, and the belief that wallabies at high density use indigenous forest as refuges and forage on adjoining rateable lands, and are also under pressure to expand their range, poison operations using 1080 have again been undertaken.

On Kawau I. control work was stopped in 1973 when farming was abandoned. From 1967 to 1975 permits were issued for live capture, and about 1824 dama were collected (Warburton, 1986). See also pp. 52, 57.

B. W. & R. M. F. S.

2. Bennett's wallaby

Macropus rufogriseus rufogriseus (Desmarest, 1817)

Synonyms *Wallabia rufogrisea* Desmarest, 1817; *Macropus rufogriseus fruticus* Ogilby, 1838.

Also called red-necked, brush, or scrub wallaby.

DESCRIPTION (Fig. 11)

Distinguishing marks, Table 6 and skull key, p. 24.

Bennett's is the largest wallaby in New Zealand, standing some 800 mm tall and frequently weighing over 15 kg. It has relatively long fur, rufous coloured over the shoulders and greyish brown on the rest of the body, except the pale grey on the chest and belly. It is a very alert animal and when disturbed does not often stop until it has travelled out of sight. It uses all available cover for refuge and constantly scans for danger with its acute hearing and sight.

Dental formula $I^3/_1\ C^0/_0\ Pm^1/_1\ M^4/_4 = 28$.

FIELD SIGN

The faecal pellets are generally a flattened square shape, but can often be more elongated and round. Tracks can be seen leading into tunnels under dense thickets of scrub and flax *(Phormium tenax),* showing the distinctive long-toed footprints on soft substrate. Their browsing cannot be differentiated from that of sheep, which share their habitat.

MEASUREMENTS

See Table 8.

HISTORY OF COLONIZATION

In 1870, a Captain Thomson brought several Bennett's wallabies from Tasmania to Christchurch (Thomson, 1922). In 1874 two females and a male, probably from this stock, were liberated on the eastern Hunters Hills, near Waimate, Canterbury (Studholme, 1954). Numbers increased dramatically, and the subsequent natural spread (augmented by deliberate relocations in the following 30–40 years) extended their range to include the Kirkliston, Grampian, Albury and Two Thumb ranges, reaching altitudes up to 2000 m.

Bennett's wallabies have also established themselves in exotic plantations in Kakahu State Forest and Pioneer Park, and there is a small population near Peel Forest. At Quartz Creek, between Lakes Hawea and Wanaka, a small colony has resulted from an escape from a zoological garden in about 1945 (Warburton, 1986).

DISTRIBUTION

WORLD. The native range is the forests of south-east Australia, Tasmania and the islands of Bass Strait. The subspecies present in New Zealand, *M. r. rufogriseus,* comes from the Bass Strait islands and Tasmania (Johnston & Sharman, 1979).

There are also two wild populations of *M. rufogriseus* in England, derived from escapes in about 1940. One is in the Peak district, just south-east of Manchester, and probably comprises only about 12 animals (Yalden & Hosey, 1971); the second is in Ashdown forest, about 50 km south of the centre of London (Taylor-Page, 1970).

NEW ZEALAND (Map 3). Bennett's wallaby is restricted to the South Island. There is one thriving population centred on the Hunters Hills near Waimate, and four smaller populations in Kakahu Forest, Pioneer Park, and Peel Forest, and at Quartz Creek.

HABITAT

Bennett's wallaby is an "edge" species, requiring cover for shelter and access to grasslands on which to feed (Catt, 1975). The ranges it occupies reach 1000 m or more above sea level, covered with tall tussock (dominated by *Chionochloa rigida*) and dissected by steep gullies. The valleys on the eastern side of the Hunters Hills support remnants of a once widespread podocarp forest, now dominated by broadleaf, mahoe *(Melicytus ramiflorus),* marbleleaf *(Carpodetus serratus),* and fuchsia *(Fuchsia excorticata),* with occasional totara *(Podocarpus cunninghamii)* and matai *(Prumnopitys taxifolia).* The margins of these forests are frequently bordered by areas of flax, *Coprosma* spp. and matagouri *(Discaria toumatou),* with intertwining bush lawyer (*Rubus* spp.) and *Muehlenbeckia* spp. Further west, many gullies support matagouri and *Coprosma* species.

At Quartz Creek the wallabies live in forests of silver beech *(Nothofagus menziesii)* and, especially, manuka (Warburton, 1986). This mosaic of forest, scrub and tall tussock is ideal for Bennett's wallaby.

FOOD

Bennett's wallaby is a grazer, but is often seen taking browse off palatable tree species. In the Hunters Hills area, about 25 species of grasses and herbs made the largest contributions to the diet (McLeod, 1986). The relative proportions of each species in the diet changed slightly from season to season, but not significantly: hawksweed (*Hieracium* spp.) was the favourite in all seasons, and clover was particularly relished in spring (Table 9).

Grasses such as yorkshire fog, browntop, cocksfoot and bluegrass were found more often in the diet than in the pasture, suggesting that the wallabies were actively selecting these species; but McLeod (1986) still considered

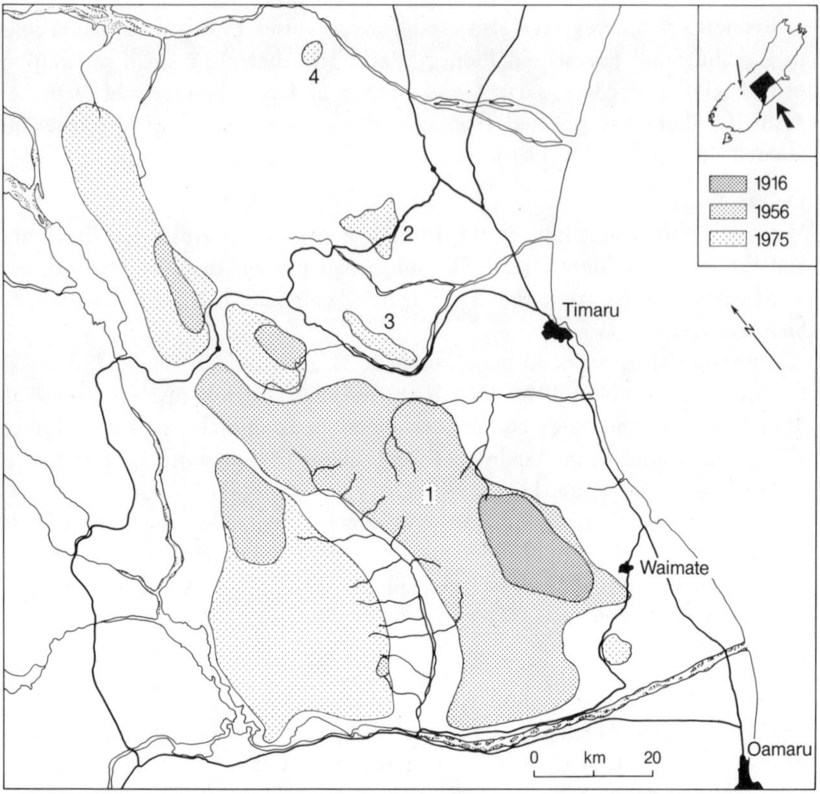

Map 3 Distribution of Bennett's wallabies at three stages of colonisation (compiled by G. Guilford, M. Barnett, L. T. Pracey and staff of APDC). Populations centred at (1) Hunters Hills; (2) Kakahu Forest; (3) Pioneer Park; (4) Peel Forest; and with (5) an outlying population at Quartz Creek (arrowed on inset).

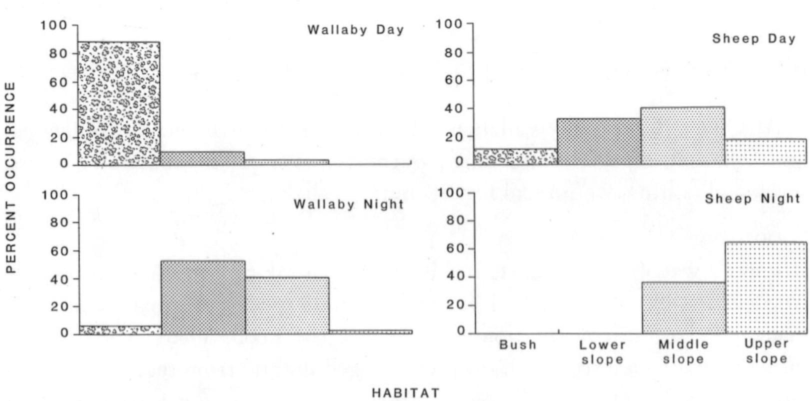

Fig. 12 Distribution of observations of Bennett's wallaby and sheep in relation to habitat type (from McLeod, 1986).

Bennett's wallaby to be an indiscriminate grazer.

Bennett's wallaby in the Hunters Hills eat the same plants at night as sheep do during the day, and in the same areas (McLeod, 1986), with the exception of *Celmisia spectabilis*. Wallabies browse on this species, extracting the soft leaf bases and discarding the coarser leaf blades (Catt, 1975); sheep avoid it.

Wallabies move to feeding areas along well-defined tracks leading from day-time cover to open grass areas. When feeding on grass, they crouch and move their heads in an arc, as in the dama. When browsing on shrubs, the forepaws are used to manipulate food, and females often hand leaves to their pouch young at foot (McLeod, 1986). They may scratch and dig with the forepaws to reach roots; regurgitate (preceded by violent choking) and then reingest their food; and drink occasionally, especially in drier months (McLeod, 1986).

SOCIAL ORGANIZATION AND BEHAVIOUR

ACTIVITY AND DISPERSION. Bennett's wallabies are nocturnal, solitary animals. They keep under cover during the day and feed in the open at night, except in wet conditions, when they remain under or close to cover (McLeod, 1986). There is no information on their home ranges in New Zealand. Marked individuals observed by McLeod (1986) moved 1–3 km in a night, all within their own home range, though seldom further than 150–180 m from the bush edge; they also made occasional sorties of up to 2–3 km to feed on seasonally abundant swede crops. When hunted by dogs they double back on to their own ground after a certain distance (Catt, 1975). Their home ranges overlap those of domestic sheep, which use the same areas but at different times. During the day the wallabies retreat to the lower slopes and bush to find cover, while the sheep feed on the middle slopes; in late afternoon and evening, the sheep move up to higher altitudes and the wallabies emerge to feed on the middle slopes (Fig. 12).

Bennett's wallabies use both rest and den sites, which are neither shared nor defended. Rest sites found by McLeod (1986) were on grass clearings in the bush or among tussock or flax at higher altitudes and were used during the day, more by males than by females. Dens were in more secluded places, often under thick undergrowth or logs, with a depression often covered with leaves and moss, and they were used more often by females.

COMMUNICATION. Members of a local population have very few intraspecific encounters and a low incidence of agonistic behaviours (McLeod, 1986). Individuals identify one another by smell, but contacts are usually restricted to the breeding season (McLeod, 1986). Although solitary, they still react to an alarm with the typical macropod thumping of the hind feet.

Threat behaviour is simple; the aggressor stands erect with ears pricked, chest expanded and forepaws crossed. In a fight, each animal holds the other in a head lock and attacks by biting, punching, kicking and fur pulling (McLeod, 1986). Submission is signalled when one partner turns its head and upper torso away from its opponent, if necessary augmenting the signal

Table 9: Foods of Bennett's wallabies in Canterbury.

Winter 1984		Spring 1984		Summer 1984–85		Autumn 1985	
Species	%	Species	%	Species	%	Species	%
[2]*Hieracium* spp.	13.33	*Hieracium* spp.	16.66	*Hieracium* spp.	9.60	*Hieracium* spp.	11.00
[1]*Holcus lanatus*	7.53	*Trifolium* spp.	9.06	*D. glomerata*	8.40	*A. scabrum*	9.26
[1]*Agropyron scabrum*	6.83	*A. scabrum*	7.30	*A. scabrum*	7.93	*A. odoratum*	8.43
[1]*Dactylis glomerata*	6.03	*H. lanatus*	6.80	*A. odoratum*	7.46	*D. glomerata*	8.26
[1]*Anthoxanthum odoratum*	6.00	*A. odoratum*	6.60	*Carex* spp.	6.96	*H. lanatus*	6.66
[4]*Blechnum* spp.	5.56	*F. novae-zealandiae*	5.90	*F. novae-zealandiae*	6.56	*C. rugosa*	5.92
[3]*Cyathodes colensoi*	4.76	*Carex* spp.	5.10	*Trifolium* spp.	5.66	*F. novae-zealandiae*	5.83
[1]*Carex* spp.	4.60	*V. australis*	5.00	*A. tenuis*	5.20	*Carex* spp.	5.33
[1]*Festuca novae-zealandiae*	4.33	*C. colensoi*	4.30	*Coprosma* spp.	4.80	Moss species	4.23
[2]*Geranium sessiliflorum*	4.26	*D. glomerata*	4.20	*C. holosteoides*	4.10	*Coprosma* spp.	4.06
Moss species	4.10	*C. holosteoides*	3.76	Moss species	4.06	*A. tenuis*	3.80
[2]*Vittadinia australis*	3.63	*Coprosma* spp.	2.90	*Celmisia* spp.	3.23	*C. colensoi*	2.93
[2]*Cerastium holosteoides*	3.36	Moss species	2.83	*H. lanatus*	2.90	*Rumex* spp.	2.93
[2]*Trifolium* spp.	3.20	*Celmisia* spp.	2.40	*Rumex* spp.	2.80	*F. rubra*	2.86
[2]*Rumex* spp.	3.16	[1]*Agrostis tenuis*	2.33	[3]*Coprosma rugosa*	2.36	*C. holosteoides*	2.53
[2]*Brachycome sinclairii*	2.76	*Blechnum* spp.	2.23	[1]*Festuca rubra*	2.26	*V. australis*	2.13
[3]*Coprosma* spp.	2.46	*G. sessiliflorum*	2.10	[2]*Crepis capillaris*	1.90	*Trifolium* spp.	2.10
[2]*Celmisia* spp.	2.23	*Rumex* spp.	1.86	*Blechnum* spp.	1.90	*Celmisia* spp.	2.10
[2]*Hypochaeris radicata*	1.90	*G. antipoda*	1.76	*G. sessiliflorum*	1.73	*G. sessiliflorum*	1.23
[3]*Gaultheria antipoda*	1.70	*B. sinclairii*	1.30	*C. colensoi*	1.63	[1]*Poa laevis*	1.13

Data expressed as percentage frequency of occurrence of the twenty most commonly eaten species by season (from McLeod, 1986).
1. Grasses. 2. Herbs. 3. Trees and shrubs. 4. Ferns.

with ear quivering and tail swinging (McLeod, 1986). The dominant partner usually chases the other off for 20–30 m after a fight (McLeod, 1986).

During the breeding season several males will follow each oestrous female and fight between themselves for possession of her (15% of interactions observed in breeding season, and 1.5% of interactions out of breeding season, were of fighting over females). The successful male will approach the female and sniff her pouch and cloaca. The male may hold and cuff the female about the head. If she is receptive she crouches and raises her tail, and the male will mount her. After mating, both sexes groom or graze together, and then will attempt to mate again after 5 to 15 minutes. If the female is not receptive she will move away as the male approaches, and often coughs and hisses at him (McLeod, 1986).

REPRODUCTION AND DEVELOPMENT

Males in the Hunters Hills were sexually mature by 21–22 months, although some were mature as early as 16 months. All females were found to be mature by 23–24 months; some showed eversion of nipples from as early as 14 months, but the young of these early-maturing females often died (Catt, 1977).

Seasonal variations in the weight of the prostate gland are a useful indicator of male reproductive capacity. The peak of the season is in February-March, declining from April to July; from then until the following February, the prostate glands are at minimum size (Catt, 1977). Merchant and Calaby (1981) reported that Bennett's wallabies are fertile throughout the year; so the seasonal breeding pattern must be determined by the females.

Fertilization at the postpartum oestrus results in a quiescent blastocyst delayed by the suckling pouch young: when it leaves the pouch the quiescent blastocyst remains delayed by environmental factors until the following February-March breeding season. Gestation is about 30 days. The complete gestation of a delayed blastocyst within the breeding season takes about 26 days, but after June, seasonal quiescence inhibits further development (Merchant & Calaby, 1981). Most young are born in February and March, and progressively fewer each month from April to July. In the population studied by Catt (1975), age specific birth rates were estimated at 0.947 births per year in females >2 years old, and 0.581 in those <2 years; the younger females also tended to have their joeys later than the older ones. Pouch young stayed in the pouch for about 274 days, and most emerged in November-December.

At 50–75 days old, joeys are able to release themselves from the teat. Their eyes open at 135–150 days and fur covers the body at 165–175 days. Catt (1981) reported that females were larger than males at 1 year of age, but later the males overtook them, and in adults the males are on average 10–15% larger and 40% heavier.

POPULATION DYNAMICS

Local density varies considerably, from >2–3 per hectare down to almost zero, especially after successful control operations.

Sex ratio in pouch young is 1:1, but males predominate in shot samples of adults (Catt, 1977). Molar progression and cementum annuli are unreliable methods of age determination; annual zonation in the periosteal zone of the mandible is more successful (Catt, 1979). Catt found a typical reversed J-shaped age structure, in which the oldest animal sampled was 9 years. Juvenile mortality was low, with an increase in mortality at 1 year and greater. The mortality of male wallabies was independent of age (Catt, 1975). The exponential birth rate he found (.347) was probably near the maximum possible.

PREDATORS, PARASITES AND DISEASES

There are no published records of predation on Bennett's wallaby, although dogs can readily catch young or weak individuals.

McLeod (1986) reported that 75% of wallabies sampled from the western side of the Hunters Hills carried high numbers of nematodes, which could comprise up to 20–30% of the stomach contents. Six species have been identified: *Labiostrongylus communis, Globocephaloides trifidospicularis, Rugopharynx longbursarius, R. omega, R. australis* and *Pararugopharynx protemnodontis* (Mason, 1975; Bowie, 1980).

No ectoparasites have been recorded. A spongy outgrowth of the lower jaw (lumpy jaw) may result from bruising or dental infection (Catt, 1975).

ADAPTATION TO NEW ZEALAND CONDITIONS

Bennett's wallaby has adapted well to the environment of the Hunters Hills, expanding both in range and in population numbers.

In Australia it inhabits eucalypt forests where there is a moderate shrub component and access to open areas for grazing (Calaby, 1983). Bennett's wallaby in New Zealand has ready access to plentiful grazing, and cover is available in the form of indigenous bush, scrub areas, flax and matagouri, and tall *Chionochloa* tussock associations.

Seasonal breeding, typical of the Tasmanian subspecies, has been retained; by contrast, the Australian mainland *M.r. banksianus* breeds continuously (Johnston & Sharman, 1979). The high mortality of 1-year-olds, and age-independent mortality of older wallabies, contrasts with the usual mortality pattern in many other mammal populations, which have a low mortality in 1–2-year-olds, and then a steady increase with age (Caughley, 1966). The difference is probably due to a high artificial mortality from shooting operations, acting on independent wallabies of all ages. The high exponential birth rate is also a response to culling.

SIGNIFICANCE TO THE NEW ZEALAND ENVIRONMENT

DAMAGE. In the 1940s, Bennett's wallaby was recognized as a pasture pest. Farmers reported them driving sheep off the pasture, fouling sheep feed, damaging fences, and destroying agricultural crops and exotic pine plantings (Warburton, 1986). No quantitative measure of damage is available except from within Waimate State Forest. In 1978, apical buds and up to 20% of needles were removed from many *Pinus radiata* seedlings, and some seedlings were totally destroyed. However, the damage, although locally severe, was restricted to within 50–100 m of the bush edge. In many remnant patches of indigenous forest (e.g., Matata, Nimrod, and Tasman Smith scenic reserves, and Gunn and Hook State Forests) the understorey is severely depleted, and there is no regeneration of palatable species. However, sheep also occasionally feed in these areas, contributing to this damage to some unknown extent.

CONTROL. The DIA was responsible for controlling Bennett's wallaby from 1947 to 1956; during this period about 70 000 wallabies were killed by government cullers, plus an estimated 30 000 by private shooters (Warburton, 1986). NZFS took over control in 1956; in 1959 a Rabbit Board's Wallaby Destruction Committee was formed, which in 1971 became the South Canterbury Wallaby Board, and has continued control to the present.

In 1960, 1080 poison was first used against wallabies, and an extensive and significant reduction in numbers was achieved over the following few years. Unfortunately, use of 1080 poison on farmland creates practical difficulties; but still, where wallabies have increased too much, 1080 carrots, artificial baits or 1080 gel are used (Warburton, 1986). Recently, control has relied heavily on shooting with the aid of dogs to flush wallabies. From 1969 to 1984 the South Canterbury Wallaby Board has shot an average of 2500–3000 each year, although the number of hunters and the areas covered change annually. Catt (1975) estimated that this level of control removes about 20% of the population, and could continue indefinitely as a sustained harvest.

B. W. & R. M. F. S.

3. Parma wallaby

Macropus parma Waterhouse, 1845

Synonyms *Thylogale parma* Waterhouse, 1845; *Protemnodon parma* Waterhouse, 1845.

Also called white-throated or small brown wallaby.

DESCRIPTION (Fig. 13)

Distinguishing marks, Table 6 and skull key, p. 24.

The parma is the smallest member of the genus *Macropus*. It is uniformly light brown, with a light grey throat and chest and a white stripe on the cheek. The ears are short and rounded. When hopping the body is held very low and the forepaws are tucked close to the chest. Parmas do not often lie down, but rest by sitting with the tail between the legs (M. Vujcich, 1979).

When alerted they stand with their forelimbs held close to the chest. They groom with their tongues, teeth, forepaws and the syndactyl toe of the hind feet. Females often clean the inside of the pouch whether or not pouch young are present (M. Vujcich, 1979).

Dental formula $I^3/_1\ C^0/_0\ Pm^1/_1\ M^4/_4 = 28$.

FIELD SIGN

The faecal pellets are distinctively flattened, and square to rectangular (Maynes, 1974), but cannot easily be distinguished from dama pellets.

MEASUREMENTS

See Table 10. Body size appears to depend on the quality of the habitat; females, especially, are significantly smaller in habitats where food is in short supply.

HISTORY OF COLONIZATION

Macropus parma was released on Kawau I. in about 1870 by Sir George Grey, along with five other species of wallabies (Table 6). The distinction between dama and parma was forgotten, although local residents recognized two colour "phases" of small wallabies on the island, calling one the small brown and one the silver-grey. It was not until skins and skulls were examined in 1965 that parma were "rediscovered" (Wodzicki & Flux, 1967b). At that time *M. parma* was regarded as very rare or even extinct in Australia, so in 1969 the IUCN requested that *M. parma* be given protection on Kawau I. This was done by Ministerial gazette, and all killing of *M. parma* was prohibited. From 1967 to 1975, 736 parmas were captured alive to supply the zoos and to establish breeding colonies throughout the world. However, in 1972 Maynes (1974) confirmed that *M. parma* still lives in Australia. He mapped its distribution (Maynes, 1977a) and showed that it is probably in no danger of extinction there, so he recommended that no further Kawau I. individuals should be released into the wild. The parma was then removed from IUCN's red data book, and in January 1984 protection of *M. parma* on Kawau I. was revoked (Warburton, 1986).

DISTRIBUTION

WORLD. *Macropus parma* is found in the Great Dividing Range of coastal New South Wales, where it can be locally common but is generally considered to be rare (Maynes, 1977a). After its rediscovery on Kawau I. many captive colonies of parma were established in overseas zoos (Wodzicki & Flux, 1971).

NEW ZEALAND (Map 4). In New Zealand *M. parma* is found only on Kawau I.

HABITAT

On Kawau I. parmas inhabit the tall kanuka and remnant taraire forests, preferably with a moist tree fern (*Cyathea dealbata*) understorey, and feed on the scattered grass clearings (M. Vujcich, 1979). They can also be found in the dense undergrowth of unpalatable shrubs under the exotic forests.

FOOD

Both parma and dama feed mainly on grass (60% of diet), but on Kawau I. parma also took herbs including those avoided by the dama (Table 11). Only *Hypericum japonicum* and *Oxalis corniculata* were eaten less often than would be expected from their relative abundance in the pasture (V. Vujcich, 1979). Parmas often drink from pools and streams.

Parmas feed in the same manner as dama. Individuals feed independently of other wallabies and form no cohesive groups (M. Vujcich, 1979).

Fig. 13 Parma wallaby (J.H. Johns).

SOCIAL ORGANIZATION AND BEHAVIOUR

The parma is predominantly nocturnal and keeps under dense cover during the day. It is a solitary animal, and groups of two are seen infrequently. It is less aggressive than the dama, seldom making threats or fighting; submissive behaviour is signalled by crouching, lowering the head, nose sniffing, and ear quivering (M. Vujcich, 1979). Alarm behaviour ranges from the standing posture, with attentive listening and looking in the direction of the disturbance, to thumping the hind limbs to alert other individuals.

Table 10: Body measurements of adult parma wallabies in New Zealand and Australia (mean ± SD).

	Sex	Sample size	Body wt (kg)	Head and body length (mm)	Tail length (mm)	Hind foot length (mm)	Reference
Kawau I.	M	24	4.6 ± 0.2	489 ± 6	483 ± 9	140 ± 1	Maynes (1977a)
Kawau I.	M	—	3.3 ± 0.1	—	435 ± 4	137 ± 1	V. Vujcich (1979)
Australia	M	5	4.9 ± 0.4	498 ± 26	502 ± 26	142 ± 3	Maynes (1977a)
Kawau I.	F	31	3.5 ± 0.1	457 ± 5	441 ± 5	129 ± 1	Maynes (1977a)
Kawau I.	F	—	3.0 ± 0.1	—	408 ± 6	133 ± 1	V. Vujcich (1979)
Australia	F	7	4.0 ± 0.4	486 ± 19	475 ± 26	134 ± 2	Maynes (1977a)

Table 11: Foods of parma and dama wallabies on Kawau I.

		Parma	Dama
Herbs	*Hypericum japonicum*	✓	✓
	Oxalis corniculata	✓	—
	Cirsium vulgare	✓	✓
	Veronica plebeja	✓	—
	Soliva anthemifolia	✓	—
	Hydrocotyle americana	✓	✓
	Anagallis arvensis	✓	—
	Trifolium repens	✓	—
	Galium parsiense	✓	—
	Galium propinquum	✓	—
	Centaurium erythraea	✓	—
	Gnaphalium gymnocephalum	✓	✓
	Centipeda orbicularis	✓	—
	Dichondra repens	✓	—
Herbs *(Cont'd)*	*Lotus pedunculatus*	✓	—
	Lotus angustissimus	✓	—
Grasses and sedges	*Lachnagrostis filiformis*	✓	✓
	Nothodanthonia racemosa	✓	✓
	Paspalum digitatum	✓	✓
	Carex inversa	✓	—
Shrubs	*Pomaderris phyllicifolia*	✓	—
Trees	*Kunzia ericoides* leaves	✓	✓
	bark	✓	—
	Pinus radiata leaves	✓	—

Data from V. Vujcich (1979).

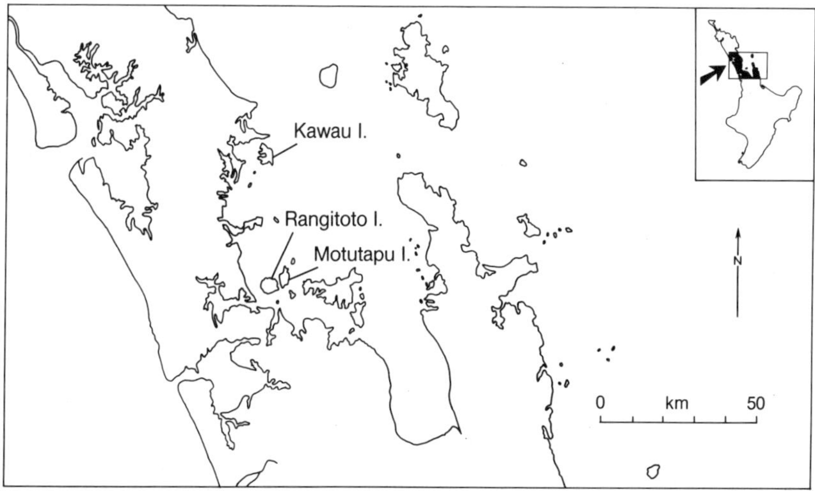

Map 4 Hauraki Gulf, showing location of Kawau I. (dama wallaby, parma wallaby, black-striped wallaby, brushtailed rock wallaby and swamp wallaby), and Rangitoto and Motutapu Is (brushtailed rock wallaby).

Thumping associated with fleeing results in a general exodus. Vocalization is limited to hissing, in adults when threatening, and in pouch young when distressed. Nest sites are usually under tall kanuka with an understorey of tree fern, and often up to 200 m from the grassy feeding areas. They are never shared (cf. dama).

During interspecific encounters with damas, parmas were most often displaced, even by damas of smaller size (M. Vujcich, 1979). Reproductive behaviour is similar to that described for damas (p. 39).

Reproduction and Development

Most females on Kawau I. first breed at 2–3 years old, but males at 1 year (Maynes, 1977b). The breeding season is long but variable; young may be born in any month of the year (Maynes, 1973). In some years on Kawau I., the parma appears to breed almost continuously (e.g., 1970 and 1971); in other years (1972, 1973) only from March to July (Maynes, 1977b). M. Vujcich (1979) also observed a long season in 1976 and 1977 (from February to August, except in May and June) and suggested that the later breeders were probably young females which had just reached puberty. The length of the oestrous cycle averages 41.8 days, and the gestation period, 34.5 days; pouch life is about 7 months (Maynes, 1973).

Population Dynamics

Sex ratio is 1:1 in pouch young and newly independent young (V. Vujcich, 1979) and in shot samples of adults (Maynes, 1977b).

Most animals in Maynes's (1977b) sample were 1–2 years old; one was estimated to be 9.5 and one 7.75 years. From a live-trapping study, Kinloch

(1973) estimated a mean mortality rate of 0.31 per year, and a mean life expectancy (after 6 months) of 2.71 years.

There have been no estimates of density, but the parma is probably the most common wallaby on Kawau I. Compared with other wild grazers in New Zealand, it lives in very high numbers there (B. Warburton, unpubl.).

PREDATORS, PARASITES AND DISEASES

Parmas on Kawau I. have no predators except for infrequent attacks from domestic dogs.

Both internal and external parasites have been collected from parmas in Australia (Johnston & Mawson, 1940; Maynes, 1977a), but the only parasite recorded on Kawau I. was an unidentified "red/brown louse" (M. Vujcich, 1979).

ADAPTATION TO NEW ZEALAND CONDITIONS

The optimal habitat of the parma in Australia appears to be wet sclerophyll forest with thick shrubby understorey and grassy patches (Maynes, 1983). In New Zealand, it has adapted well to pasture with adjoining cover; on Kawau I. its numbers have increased dramatically. The high numbers have, however, resulted in nutritional stress. Parma females on Kawau I. have become significantly smaller than females from less dense populations in New South Wales (Table 10). They also first breed later, at 2–3 years old; female parma in New South Wales breed at 1 year (Maynes, 1977a, b). Maynes suggests that the smaller size of Kawau parmas has become genetically fixed, as stock from Kawau I. raised elsewhere on surplus food still remain smaller than their Australian relatives. Kawau males sampled by V. Vujcich (1979) were significantly smaller than those sampled by Maynes (Table 10); Vujcich suggests that the increase in population since hunting was stopped has led to food shortages and a decline in body condition.

SIGNIFICANCE TO THE NEW ZEALAND ENVIRONMENT

DAMAGE. In the 1800s Kawau I. was covered with a diverse and luxuriant forest (Buchanan, 1876); but there has been extensive clearing of forest, by Sir George Grey and by farmers up until 1973, so little of the island remains unaltered by man. In addition, the parma, plus at least three other wallaby species and the brushtail possum, have significantly curtailed the regeneration of the remaining indigenous forest and eliminated hundreds of species still present on nearby wallaby-free Challenger I. (Esler, 1971). The parma probably contributes more to this damage than the other wallabies, but this is important only if the island has a high priority as a floral reserve. Since the vegetation cover is already significantly modifed by human activities, the effect of the parmas can probably be tolerated in view of their zoological interest. The present high numbers of parmas are acceptable because the island is not farmed, and until recently the parmas were protected. If farming were resumed, or if there were attempts to live-capture and relocate this or any other wallaby species on to the mainland, control of wallabies on Kawau would be necessary again.

CONTROL. Control of wallabies, including *M. parma*, on Kawau I. was carried out as early as 1923, and then periodically (especially in 1964–66) up until 1969. Most of the wallabies were shot, although poison (possibly cyanide) was also used (Warburton, 1986). The parma wallaby was protected by Ministerial gazette in 1969; shooting of the other three wallaby species on the island continued for some years but ceased in 1973 when farming was abandoned. In 1984, protection for the parma was removed, but control of this and other wallabies on the island has not so far been reinstated.

B. W. & R. M. F. S.

4. Black-striped wallaby

Macropus dorsalis (Gray, 1837)

Synonym *Wallabia dorsalis* Gray, 1837.

Also called scrub wallaby.

DESCRIPTION (Fig. 14)

The black-striped wallaby is one of the largest wallabies, attaining a head and body length of 1590 mm and weighing up to 20 kg (Kirkpatrick, 1983). Its light brown-grey fur and very distinctive dark dorsal stripe make it readily identifiable. It is said to have a distinctive gait, characterized by a short hop with the body hunched and head held low (Kirkpatrick, 1983).

MEASUREMENTS

The only known New Zealand specimen, a female with joey, weighed 12.7 kg. The head and body length was 720 mm, tail length 680 mm, hind foot 220 mm, ear 85 mm (Wodzicki & Flux, 1967a).

HISTORY OF COLONIZATION

The black-striped wallaby was presumably among the several species of wallabies liberated on Kawau I. by Sir George Grey about 1870. Nothing is known of its history there, and the few specimens recorded as *dorsalis* (e.g. by Le Souef, 1930) were misidentified (Wodzicki & Flux, 1967a). The only convincing record is of a female with joey, poisoned in 1954, identified by R. Kean, NZFS. The skin and skull have been lost, but the measurements (see above) lie well outside the range of *M. parma*, and there is no other species on Kawau I. with which *dorsalis* could be confused (Wodzicki & Flux, 1967a). Black-striped wallabies therefore survived on Kawau I. for over 80 years, but are now extremely rare or, more likely, extinct.

DISTRIBUTION

WORLD. Eastern Queensland and north-eastern New South Wales.

NEW ZEALAND (Map 4). Kawau I. only; probably extinct.

Nothing is known of the biology of this species in New Zealand. J. E. C. Flux (unpubl.) suggests that the reason it disappeared after so long is that it was a large wallaby inhabiting open scrubland and was therefore more vulnerable to, or selected by, the hunters working on wallaby control on the island from the 1920s.

B. W. & R. M. F. S.

Genus *Petrogale*

This genus contains the rock wallabies, which, as their name implies, frequently inhabit rocky terrain. There are about ten species, but with much variation (in, for example, chromosome number) and many island and mainland subspecies (Kirsch & Calaby, 1977; Hayman, 1977).

They are usually more brightly coloured than other wallabies; their tails are bushy and less tapered, and carried over the back when moving; the fourth toe of the hind foot is shorter than in other macropods; and the soles of the hind feet have a fringe of stiff hairs (Sharman & Maynes, 1983). Suckling young which have left the pouch are often left in a safe location while the mother feeds, presumably an adaptation to decrease the extent to which a young animal has to traverse difficult terrain.

5. Brushtailed rock wallaby

Petrogale penicillata penicillata Griffith, 1827

Synonyms *Petrogale inornata* Gould, 1842; *P. herberti* Thomas, 1926; *P. purpureicollis* Le Souef, 1924; *P. lateralis* Gould, 1842.

Also called blacktailed rock wallaby or rock wallaby.

DESCRIPTION (Fig. 15)

Distinguishing marks, Table 6 and skull key, p. 24.

Petrogale p. penicillata is the only rock wallaby in New Zealand, and its bushy tail and rufous coloured rump make it quite distinct from the *Macropus* species. It is dull to dark brown above, rufous on the belly, and has a whitish patch on the forehead.

It usually moves with short jumps, in an apparently erratic course; the tail serves as a rudder and counterbalance, but is not used to support the body, as in other Macropodidae. Trees are frequently climbed up to 10 m above ground, as long as the trunks are less than 30° from the horizontal (Batchelor, 1980).

Chromosome number 2n = 18?

Dental formula $I^3/_1 C^0/_0 Pm^1/_1 M^4/_4 = 28$.

FIELD SIGN

Faecal pellets (round to pear shaped, sometimes cylindrical) are diagnostic on Rangitoto I. and Motutapu I., as no other wallabies are found there. The wallabies make well-defined pathways to den sites on cliff faces.

MEASUREMENTS

See Table 12.

HISTORY OF COLONIZATION

The brushtailed rock wallaby was liberated on Motutapu I. by J. Reid in 1873 (Thomson, 1922). From here the animals were able to move freely to Rangitoto I. at low tide, when the two islands are joined: by 1912 they had reached high numbers on Rangitoto I. (Warburton, 1986). The few rock wallabies found on Kawau I. presumably result from liberations by Sir George

Fig. 14 Black-striped wallaby (C. Andrew Henley/NPIAW).

Fig. 15 Brushtailed rock wallaby (F. Kristio/NPIAW).

Grey, but whether he obtained new stock from Australia or translocated specimens from Motutapu I. is unknown.

DISTRIBUTION

WORLD. The nominate subspecies, *Petrogale p. penicillata,* is relatively common in suitable habitat throughout its range in Victoria, New South Wales and Queensland (Maynes & Sharman, 1983). Rock wallabies are also found in parts of Oahu I., Hawaii, where they escaped from captivity in 1916 (Lazell, Sutterfield & Giezentanner, 1984).

NEW ZEALAND (Map 4). *Petrogale p. penicillata* lives only on three islands in the Hauraki Gulf. On Kawau I. it is restricted, perhaps by competition or a shortage of den sites, to cliff faces and steeper areas; on Motutapu I. likewise, since the rest of the island has no suitable cover; on Rangitoto I. it is found throughout.

HABITAT

The cliffs of Kawau I. are vegetated with pohutukawa, puriri and taraire. Rock wallabies can also be seen in parts of the pine forests near Bon Accord Harbour.

On Motutapu I. the cliffs are broken by shelves and benches (well tracked by rock wallabies) and vegetated with pohutukawa, tawapou (*Planchonella* sp.), kowhai (*Sophora* sp.), puriri, rewarewa and karaka (*Corynocarpus laevigatus*) and an understorey of *Coprosma robusta,* mapou (*Myrsine australis*), kawakawa (*Macropiper excelsum*) and Mediterranean plant (*Rhamnus alerternes*). The rock wallabies venture on to the adjoining pastures to feed.

On Rangitoto I. the rock wallabies inhabit patches of vegetation dominated by pohutukawa, surrounded by bare scoria yet to be colonized by vegetation. The patches also support other plant species such as northern rata (*Metrosideros robusta*), broadleaf (*Griselinia lucida*), Kirk's daisy (*Senecio kirkii*), and *Astelia banksii.* On the central cone, rewarewa, manuka, kanuka, mingimingi (*Cyathodes* spp.) and *Coprosma lucida* are common. The island has very little fresh water and is typically a hot, dry, rocky habitat.

FOOD

There are no studies of the food of rock wallabies in New Zealand, but Batchelor's (1980) observations show that they are primarily grazers, occasionally also taking fallen leaves from, for example, karaka trees. They were never observed browsing pohutukawa, although they often climbed these trees. In contrast, on parts of Rangitoto I. they have no access to grasses and so they browse extensively, for example, on pohutukawa growing close to the ground and also on *Senecio kirkii* and *Astelia banksii* (Warburton, 1986).

SOCIAL ORGANIZATION AND BEHAVIOUR

ACTIVITY. Mainly crepuscular and nocturnal. On Motutapu I. most rock wallabies feed during the night on pasture and then retreat soon after sunrise to the protection of the cliffs. Here they climb trees to browse, indulge in exploratory behaviour, groom, sun themselves or sleep. They then return to graze on the pasture around sunset (Batchelor, 1980). If the weather is wet they stay in their dens.

DISPERSION AND HOME RANGES. On Motutapu I., rock wallabies are gregarious, with a defined social hierarchy (Batchelor, 1980). Day ranges are basically restricted to the cliffs, and night ranges extend out on to the pasture. Ranges measured on Motutapu I. in 1978–79 were about 0.58 ha and 0.38 ha in male and female adults, respectively; sorties seldom exceeded 150 m (Batchelor, 1980). Well-formed paths were used for traversing to and from the feeding areas.

DENS. The social hierarchy is well enforced around resting and den sites, which are probably a limiting resource. Rest sites are used for sunning, grooming and sleeping, and these are usually at the base of a rock or on the branch of a tree (Batchelor, 1980). Dens are more enclosed than rest sites, have more than one entrance, and provide more shelter. They are often used by more than one individual at different times, and occasionally at the same time. Many dens are excavated under the roots of pohutukawa growing on the cliffs.

COMMUNICATION. In intraspecific encounters, each animal sniffs the snout of the other, and presumably can thereby identify the other individuals of the group. Rock wallabies may huddle in dens, and groom each other, especially the females. Vocalizations are restricted to hissing, e.g., in juveniles separated from their mothers, in chased individuals, and in females when rebuffing males during attempted copulation (Batchelor, 1980).

In agonistic encounters the opponents face each other, pull grass, strike out with the forepaws, and eventually may chase or occasionally fight. In a fight each attempts to pull the fur and bite the neck and head of the other, and to rake the chest and stomach with the hind feet. Fights may be avoided by submissive behaviour, i.e., averting the face or body away from the aggressor.

When alarmed, rock wallabies thump the ground with the hind feet and flee, alerting other wallabies in the area to the danger (Batchelor, 1980).

REPRODUCTIVE BEHAVIOUR. Batchelor (1980) recognized five sexual behaviours, not all observed at every encounter. (1) The male sniffs the female's cloaca, while she usually remains motionless; (2) he follows her about, while keeping subordinate males away; (3) he forces her head on to his sternal gland or genitalia; (4) in the "mating bow", he sniffs her cloaca, then stands upright with head and forepaws pointing upwards, and often with penis everted; (5) copulation, each attempt rarely exceeding 5–10 seconds. Dominant males tend to gain the highest number of successful matings.

REPRODUCTION AND DEVELOPMENT

On Motutapu I., rock wallabies have a continous breeding cycle and young are born in most months (Batchelor, 1980). The receptive period of oestrous females is only 3–12 hours; hence the males constantly check their condition. Age at first breeding varies, but of 15 females classed as 2+ years, 40% produced young. From 11 adult females with 22 pouch young, the mean

Table 12: Body measurements of brushtailed rock wallabies in New Zealand and Australia.

	Sex	Sample size	Body wt (kg)	Head and body length (mm)	Tail length (mm)	Hind foot length (mm)	Reference
Hauraki Gulf[1] (all ages)	M	7	4.8(3.0–5.8)	530	470	140	Wodzicki & Flux (1967a)
	F	6	4.3(2.8–5.6)	520	470	140	Wodzicki & Flux (1967a)
Motutapu I.[2] (adults)	M	13	6.7 ± 0.9	—	—	—	Batchelor (1980)
	F	17	5.7 ± 0.8	—	—	—	Batchelor (1980)
Australia (adults)	—	—	7.5	540(520–580)	610(560–670)	—	Maynes & Sharman (1983)

1. Composite sample from Kawau (n=8) and Rangitoto (n=5).
2. ± SD.

Table 13: Body measurements of swamp wallabies in New Zealand and Australia.

	Sex	Sample size	Body wt (kg)	Head and body length (mm)	Tail length (mm)	Hind foot length (mm)	Reference
Kawau I.	M	2	12.8(11.8–13.8)	780	690	230	Wodzicki & Flux (1967a)
(all ages)	F	9	10.3(5.5–14.5)	670	470	200	Wodzicki & Flux (1967a)
Australia	M	—	17(12.3–20.5)	756(720–850)	761(690–860)	—	Merchant (1983)
(adults)	F	—	13(10.3–15.4)	697(660–750)	692(640–730)	—	Merchant (1983)

annual number of offspring raised to independence was 1.35 per adult female.

POPULATION DYNAMICS

On Motutapu I., Batchelor (1980) estimated that density was about 12/ha in the areas inhabited by rock wallabies. On Rangitoto I. absolute density has not been measured, but relative density is higher near to the coast than inland (Morgan & Copland, 1985).

Sex ratio is 1:1 in both adults and pouch young (Batchelor, 1980).

The age structure of rock wallabies on Motutapu I. was analysed by Batchelor (1980). Pouch mortality was low; all pouch young observed were raised to independence. Juveniles, and young in the 1st and 2nd age classes, were occasionally found dead, trapped in branches of trees. There were many 3-year-olds, but very few in the 4+ year class. Some juveniles showed deformities, which Batchelor (1980) suggested could be due to inbreeding.

PREDATORS, PARASITES AND DISEASES

None known.

ADAPTATION TO NEW ZEALAND CONDITIONS

In Australia the preferred habitat of *P. penicillata* is typically steep rock with many ledges and caves and well exposed to the sun (Short, 1982), usually in sclerophyll forest within reach of grassy clearings (Maynes & Sharman, 1983). Motutapu I. proved suitable for rock wallabies, providing cliffs, ledges and den sites. Den sites are considered to be a limiting factor (Batchelor, 1980). These cliffs adjoin a large area of pasture, providing suitable food. The numbers of rock wallabies increased on Motutapu to the extent that control became necessary.

On Rangitoto I. the terrain is relatively flat, but the extensive areas of scoria provide a rocky habitat and a multitude of den sites. On Kawau I. the numbers and distribution of rock wallabies are relatively restricted, perhaps because of competition from the other three wallaby species, or because den sites are scarce.

The brushtailed rock wallaby appears to be somewhat smaller in New Zealand than in Australia (Table 12).

SIGNIFICANCE TO THE NEW ZEALAND ENVIRONMENT

DAMAGE. On Motutapu I. the rock wallaby is considered to be a pest, because it competes with stock for pasture and accelerates erosion of the cliffs by excavating den sites. In the past it also damaged wind-break plantings, but control and habitat clearance over most of the island have restricted its impact to the cliffs. On Kawau I. the effect of rock wallabies is similar, but to a lesser degree. On Rangitoto, the presence of brushtailed rock wallabies is more serious. The succession of vegetation on Rangitoto since the last volcanic eruption about 250 years BP is recognized internationally, and much concern is expressed about the influence that wallabies are having in altering it and inhibiting the regeneration of established species. Some, such as Kirk's senecio, are heavily browsed, and individual plants of *Astelia banksii* can be seen reduced to stumps (Warburton, 1986). Research is now under way to ascertain how important the wallaby is in influencing the flora on the island.

CONTROL. On Kawau I. rock wallabies were shot and poisoned, along with the other species, from 1923 until 1973. From 1967 to 1975, 210 rock wallabies were live-captured for transfer to zoos (Warburton, 1986). Since 1973 no official control has been undertaken on Kawau I. (see p. 57). On Motutapu I., wallabies have been killed, probably since the late 1940s; however, the first official operation on Motutapu was in 1965, when NZFS undertook control at the request of DLS, who were attempting to establish shelter belts. After another operation in 1967, using shooting and poison, wallaby numbers were considered to be low (Warburton, 1986). Private trappers also live-captured rock wallabies, and numbers were still considered low in 1969. There has been no official control since.

Some control may have been undertaken on Rangitoto I. in the early 1950s but its effects are unknown.

The wallabies on these islands are restricted in their distribution by natural barriers, so they do not threaten to extend to new areas. However, on islands which are, like Rangitoto, open to the public, they pose a special problem: they make nice pets and can easily be captured and released on other islands or on the mainland. Seven individuals were illegally released on Great Barrier I. in 1981, but fortunately all were recaptured or killed. The potential for further illegal liberations still exists.

B. W. & R. F. M. S.

Genus *Wallabia*

The single species in this genus is superficially like the *Macropus* wallabies but has been classified into a separate genus because its small chromosome number and browsing feeding habits set it apart. Its dentition also differs from the larger *Macropus* species, in that the broad fourth (third, Tyndale-Biscoe, 1973) premolar is never shed (Merchant, 1983); and the gestation period is longer than the oestrous cycle, a characteristic of very few if any other marsupials (Merchant, 1983).

6. Swamp wallaby

Wallabia bicolor (Desmarest, 1804)

Synonyms *Protemnodon bicolor* Desmarest, 1804; *Macropus bicolor* Desmarest, 1804.

Also called swamp, black, wallaroo, or black-tailed wallaby.

DESCRIPTION (Fig. 16)

Distinguishing marks, Table 6 and skull key, p. 24.

This is the largest of the wallabies on Kawau I. It has a dark grey back, yellow-buff belly, a light yellow cheek stripe and orange markings around the base of the ears. The fur is generally regarded as coarse and unattractive (Kirkpatrick, 1970).

Fig. 16 Swamp wallaby (E. Beaton/NPIAW).

Chromosome number 2n = 11 in the male, 10 in the female (Hayman, 1977). The male has two y chromosomes with attached autosomes (Tyndale-Biscoe, 1973).

Dental formula $I^{3}/_{1}\ C^{0}/_{0}\ Pm^{1}/_{1}\ M^{4}/_{4} = 28$.

FIELD SIGN

The footprints are considerably larger than those of the other wallabies on Kawau I.; the faecal pellets are large and almost resemble a flattened cube.

MEASUREMENTS

See Table 13.

HISTORY OF COLONIZATION

Swamp wallabies were liberated on Kawau I. in about 1870 by Sir George

Grey (Wodzicki & Flux, 1967a). Little is known about their establishment and increase, but the species has survived there for over a century.

DISTRIBUTION

WORLD. *Wallabia bicolor* is restricted to the eastern side of mainland Australia from Cape York to Victoria. Five subspecies are recognized, one restricted to some islands off coastal Queensland (Merchant, 1983); the one introduced to New Zealand was probably *W. b. apicalis* (Wodzicki & Flux, 1967a).

NEW ZEALAND (Map 4). *Wallabia bicolor* lives only on Kawau I., especially the northern end (Kinloch, 1973).

HABITAT

In Australia the swamp wallaby is found in forest with thick undergrowth, e.g., in Queensland brigalow scrub. Its local distribution is apparently linked with that of dense undergrowth. On Kawau I. its preferred habitat is the thick kanuka scrub on the northern part of the island, which often has an understorey of mingimingi (*Cyathodes fasciculata*), grass in more open areas, and occasional tree ferns (*Cyathea dealbata*) in wetter gully sites.

FOOD

Swamp wallabies are browsers rather than grazers, and grass makes up only a very small proportion of their diet. In Australia they apparently prefer the coarse browse supplied by trees and bushes rather than grass (Edwards & Ealey, 1975). These authors also report that the swamp wallaby readily consumes toxic species such as bracken (*Pteridium esculentum*) and hemlock (*Conium maculatum*), and others high in oil content such manna gum (*Eucalyptus viminalis*). Hollis, Robertshaw & Harden (1986) found that swamp wallabies consumed a high proportion of fungi, a readily accessible source of protein.

Pellets collected on Kawau I. contained a high proportion of coarse unidentifiable material from dicotyledons, and very little grass. Swamp wallabies probably chew bark from kanuka, browse on mingimingi and eat lilies (*Arum* sp.) (Kinloch, 1973). They appear to feed at all times of the day and night (Edwards & Ealey, 1975).

SOCIAL ORGANIZATION AND BEHAVIOUR

Swamp wallabies are solitary. The only intraspecific association is between mother and young, although groups of socially unrelated individuals may gather to share a common feeding area (Kirkpatrick, 1970). They are largely nocturnal, but remain active throughout the day between periods of sitting with their tails between their legs (Edwards & Ealey, 1975).

REPRODUCTION AND DEVELOPMENT

Nothing is known of reproductive behaviour or breeding in the swamp wallaby in New Zealand. In Australia they breed in most months of the year (Edwards & Ealey, 1975). Both sexes are mature by 15–18 months old. The gestation period is 33–38 days, longer than the oestrous cycle; females mate again about 8 days before the first young is born, and the resulting blastocyst becomes quiescent. Pouch life spans 8–9 months, but young at foot, up to 15 months old, can continue to suckle (Merchant, 1983).

POPULATION DYNAMICS

There is no information on sex ratio of pouch young or adults, age structure or mortality of swamp wallabies on Kawau I. or in Australia. Their density in the northern part of Kawau I. is relatively high; on some survey transects in that area, counts of their faecal pellets were higher than those of the dama and parma (Kinloch, 1973), but elsewhere on the island they are less abundant.

PREDATORS, PARASITES AND DISEASES

Nothing is known about predators or parasites; no diseases are recorded apart from dental caries (Kinloch, 1973).

ADAPTATION TO NEW ZEALAND CONDITIONS

Swamp wallabies are primarily browsers and prefer a habitat with a dense understorey. Kawau I. was probably a less than ideal place to release them, since most of the island had already been cleared for farming and/or forestry. What forest was left was reduced to few species, with no dense undergrowth or great diversity of browse, and swamp wallabies do not appear to be able to adapt to feeding on pasture. Perhaps this explains why the Kawau I. swamp wallabies appear to be smaller than their relatives in Australia (Table 13).

SIGNIFICANCE TO THE NEW ZEALAND ENVIRONMENT

The swamp wallaby is confined to Kawau I., where it is a minority species unlikely to contribute much additional damage to indigenous or, as in Australia (Merchant, 1983), exotic trees or seedlings. As long as no animals are relocated on to other islands or the mainland, their presence on the island can be regarded as inconsequential. They are still common in Queensland (Kirkpatrick, 1970).

B. W. & R. F. M. S.

Family Phalangeridae

The phalangerids comprise three genera of cuscuses (*Phalanger*, 10 species), the scaly-tailed possum (*Wyulda*, one species), and the brushtail possums (*Trichosurus*). All are medium sized (up to about 5 kg) and mainly arboreal, found in every kind of forest in Australia, Papua New Guinea and adjacent islands between Sulawesi and the Solomons (Smith & Hume, 1984).

Genus *Trichosurus*

The brushtail possums (three species) all have large ears, pointed faces, close woolly fur and bushy tails. The only member of the family introduced to New Zealand is the common brushtail possum, *Trichosurus vulpecula*. It is the most widespread of the three in Australia, found in wooded and forested areas throughout the eastern, central and south-western parts of the country.

Trichosurus arnhemensis, the northern brushtail possum, is found in the Kimberley and northern parts of the Northern Territory and is generally smaller than the brushtail possum, with a sparsely furred tail; *T. caninus,* the mountain brushtail possum or bobuck, lives in rainforest and wet sclerophyll forests in southern Queensland, eastern Victoria and New South Wales and has much shorter, more rounded ears, which lack white patches of fur at their bases (Strahan, 1983).

7. Brushtail possum

Trichosurus vulpecula (Kerr, 1792)

Synonym *T. fuliginosus* Ogilby, 1831 (A complete list of synonyms is given by Thomas, 1888.)

The Latin name refers to its foxlike ears and pointed snout (*Trichosurus* — hairy tail; *vulpecula* — little fox). In New Zealand, its common name has traditionally been "opossum" (Kean, 1964, 1965; Caughley, 1965a; Troughton, 1965), especially in the fur trade. In Australia, the more frequent usage is "possum", and the standard common name for *T. vulpecula* adopted by the Australian Mammal Society is common brushtail possum (Strahan, 1980) or, as here, simply possum.

DESCRIPTION (Fig. 17)

Distinguishing marks, p. 22 and skull key, p. 24. See left-hand back endpaper for colour plates.

The possum has a thick bushy tail, a pointed snout and a darkly stained sternal gland on the chest. The fur is thick and woolly. The ears are long and narrow, tapering towards their tips, nearly naked inside and on the outer tips. Size, weight and predominant colour vary greatly around New Zealand (see below).

There are two general colour forms, grey and black, each with much variation. Greys are generally a clear grizzled grey on the body, with the face pale grey, darker around the eyes and on the side of the snout, and white at the base of the ears. The back and sides of the body and the outer sides of the limbs are a uniform grizzled grey, often darker medially and posteriorly. The chin is dark, but the throat, chest, belly and inner sides of the limbs are white or dirty yellow. The sternal gland stains the fur a dark rusty red, more prominently in males than females; juveniles have only a faint grey streak. The fur around the pouch of females and at the base of the tail is often also stained brownish. The hands and feet are usually off-white or grey. The tail is thick and cylindrical, with bushy grey fur changing abruptly to black at about midpoint, though the extreme tip may be white in about 5% of animals. In blacks the general colour is a deep umber-brown tinged with rufous, paler on the forequarters, sides and below, darker along the posterior back. The ears have little or no white at the base, and the tail is nearly wholly black (though occasionally also with a white tip). The sternal gland and other sebaceous staining are less obvious.

Fig. 17 Brushtail possum carrying a radio transmitter, Orongorongo Valley (DSIR).

Table 14: Linear measurements of the possum in New Zealand (mean ± SD).

Location	Latitude	Sex	n	Total length (mm)	Tail length (mm)	Head length (mm)
Silverdale,	36°37′	M	40	786	321	89
Auckland		F	54	782	319	88
Bridge Pa,	39°43′	M	25	787 ± 59	322 ± 28	99 ± 7
Hawke's Bay		F	28	786 ± 49	319 ± 19	101 ± 8
Whareama,	40°54′	M	20	780 ± 47	297 ± 18	97 ± 6
Wairarapa		F	20	774 ± 36	300 ± 24	101 ± 8
Orongorongo	41°22′	M	60	789 ± 30	328 ± 15	92 ± 3
Valley		F	60	788 ± 27	326 ± 15	90 ± 3
Mt. Misery,	41°55′	M	62	831 ± 47	356 ± 21	96 ± 5
Nelson		F	45	829 ± 34	351 ± 19	95 ± 4
Taramakau River,	42°45′	M	31	840 ± 36	—	—
Westland		F	50	820 ± 41	—	—

Data from Voller (1969); M. N. Clout, S. Triggs, R. E. Brockie and P. E. Cowan (unpubl.)

Possums moult continuously but with one major peak of shedding (A. J. Nixon, unpubl.), which in the Orongorongo Valley is in August-November (P. E. Cowan, unpubl.).

The tail is naked on its ventral surface for the last 80-150 mm. The iris of the eye is brown, the vibrissae long and well developed. The hand has five digits, each with a strong curved claw and pronounced palmar pads and striations. The foot is hairy under the heel, but otherwise similar to the hand, except that the second and third digits are syndactylous (i.e., joined for most of their length) and bear enlongated claws used in grooming. The first toe is well developed, opposed to the rest of the toes, and bears no claw. There is a fully formed pouch with two mammaries and forward-directed opening (Type 5 of Russell, 1982). The testes descend permanently into the scrotum, which is pendulous and prominently situated anterior to the penis.

Possums have three basic gaits (Jolly, 1976; Winter, 1976): walking, along horizontal and up gently inclined surfaces; half-bounds, in jumping from branch to branch or up steeply inclined surfaces or in long grass; and bounds, up steeply inclined and vertical surfaces, such as tree trunks (Goldfinch & Molnar, 1978; Winter, 1976). The prehensile tail maintains its grip until all feet have shifted from one branch to the next. Possums can swim, but are generally disinclined to enter water.

The skull is stout and heavily built. The nasals are smoothly convex above, with a shallow nasal notch; the forehead is flattened; the interorbital region is narrow, concave along its centre, its edges sharply ridged; the palate is imperfect; the anterior palatine foramina run back level with the canines; the posterior palatal vacuities extend from the back of the first molar nearly to the hind edge of the palate, bounded only by a narrow transverse strip of bone; the bullae are low, flattened and scarcely inflated.

Dental formula $I^3/_2\ C^1/_0\ Pm^2/_1\ M^4/_4 = 34$.

Only 49% of 350 skulls from Banks Peninsula had the "standard" dental formula (Gilmore, 1966). Tooth eruption is described by Kingsmill (1962); molars erupt in sequence and M_4 is visible in most possums by 12–14 months of age.

Chromosome number 2n = 20 (Hayman, 1977).

A recent bibliography by Morgan and Sinclair (1983) listing 808 references on all aspects of possum biology is updated by Brockie *et al.* (1984).

FIELD SIGN

Tracks used by possums ("pads" or "runs") are often most evident where possums emerge from forest out on to pasture to feed, though they are also visible in forest or scrub when possum numbers are high (Pracy & Kean, 1969). Frequently used trees bear extensive surface scratches; occasionally, soft-barked overmature or dead trees are mauled for no obvious reason (play trees of Kean, 1967). Bark biting, done with the large upper incisors and leaving a series of horizontal scars, can be seen on a wide variety of native and introduced trees and shrubs. Often the same trunk, marked repeatedly, becomes heavily scarred from the ground to about 1 m up. Bark biting may have social importance as a visual sign (Kean, 1967). Possums also strip and eat bark from trees, particularly young *Pinus radiata* (Clout, 1977). Pellets may be found singly or in groups; they are usually about 15-30 mm long, 5-14 mm wide, crescent shaped, slightly pointed at the ends (one sample of 100 pellets had mean length and breadth of 24 × 8 mm; P. E. Cowan, unpubl.). Their colour and texture vary greatly depending on diet. Leaves browsed by possums have torn rather than cut edges, with the midrib and lower part of the leaf often partly remaining, unlike insect browse (Meads, 1976a). Possums are destructive feeders, leaving the ground littered with broken branches or discarded leaves. In late winter possums feeding on five-finger *(Pseudopanax arboreus)* eat only the base of the petiole and discard the rest; in summer and autumn they throw down many partly eaten fruits of native trees. Urine trails are sometimes seen, particularly if possums have been feeding on kamahi *(Weinmannia racemosa)* or five-finger, which stain the urine dark brown (Bolliger & Whitten, 1940).

MEASUREMENTS

See Tables 14, 15. Typical adult measurements are as follows: total length 650–930 mm, head length 80–115 mm, tail length 250–405 mm. Adult weights range from 1400 to 6400 g (Fraser, 1979; Triggs, 1982). There is little difference in size or weight between the sexes in field samples, even when corrected for age (Bamford, 1970; Green & Coleman, 1984). In nine separate studies, mean adult female weight was 98.6% that of males. But the asymptotic weight (and presumably size) of males calculated from growth curves consistently exceeds that of females (Clout, 1977).

Table 15: Geographical variation in body weights of adult possums in New Zealand.

Location	Latitude	Habitat[1]	n	Body weight (kg) mean	Body weight (kg) maximum	Reference
North Island						
Waipoua SF	35°40′	N	178	2.51	>3.0	M. D. Thomas & J. D. Coleman (unpubl.)
Silverdale	36°37′	F	94	2.39	>3.3	Triggs (1982)
Urewera Nat. Park	38°13′	N	103	2.56	—	A. Pearson (unpubl.)
Tokoroa	38°15′	E	>150	2.75	4.90	Clout (1977)
Taranaki, SF 91	38°56′	E	105	2.58	>3.60	Clark (1977)
Patea/Waitotara	39°45′	F	100	2.25	3.10	A. J. White (unpubl)
Havelock North	39°54′	F	56	2.40	3.85	G. D. Ward (unpubl.)
Wanganui	40°00′	F	82	2.32	3.95	P. E. Cowan (unpubl.)
Akitio	40°36′	F/N	41	2.40	3.60	L. T. Pracy (unpubl.)
Kapiti I.	40°51′	N	1 043	2.81	4.50	P. E. Cowan (unpubl.)
Whareama	40°54′	F	40	2.40	3.60	P. E. Cowan (unpubl.)
Wainuiomata Valley	41°20′	N	76	2.66	3.55	R. E. Brockie (unpubl.)
Orongorongo Valley	41°21′	N	196	2.40	3.70	Crawley (1973)
Pararaki	41°32′	N	127	2.18	—	J. D. Coleman & M. D. Thomas (unpubl.)
South Island						
Tennyson Inlet	41°06′	F/N	100	2.51	3.73	R. Bray & G. Struik (unpubl.)
Waimangaroa	41°43′	F/N	25	3.17	3.89	Gilmore (1966)
St Arnaud	41°50′	N	97	2.95	4.40	Taylor & Magnussen (1965)
Mt. Misery	41°55′	N	139	3.16	4.45	M. N. Clout (unpubl.)
Kaikoura	42°20′	F/N	19	3.50	4.26	Gilmore (1966)
Claverly	42°36′	F/N	176	3.12	—	B. Warburton, M. D. Thomas, J. D. Coleman (unpubl.)
Mt. Bryan O'Lynn	42°35′	N	224	2.85	4.20	J. D. Coleman & W. Q. Green (unpubl.)
Taramakau River	42°45′	N	96	3.10	4.10	Voller (1969)
Kokatahi	42°55′	N	37	3.10	—	Gilmore (1966)

Hokitika River	42°58′	N	2 177	3.12	>4.5	Boersma (1974)
Mt. Thomas	43°09′	E	20	2.78	>3.30	Yom-Tov, Green & Coleman (1986)
Ashley SF	43°13′	E	264	2.80	—	B. Warburton (unpubl.)
Karangarua River	43°31′	N	35	3.30	4.45	P. D. Gaze (unpubl.)
Copland Valley	43°37′	N	134	3.47	6.30	Fraser (1979)
Banks Peninsula	43°40′	F/N	590	3.53	5.14	Gilmore (1966)
Dunedin	45°52′	F	17	2.72	4.05	Winter (1963)
Waitahuna	45°55′	E	95	2.56	3.40	J. Bowie (unpubl.)
Codfish I.	46°45′	N	34	3.28	—	W. Q. Green (unpubl.)
Stewart I.	47°00′	N	186	3.08	—	J. D. Coleman & C. J. Pekelharing (unpubl.)

1. N, native forest; F, farmland with scrub; E, exotic forest.

Table 16: Geographical variation in proportion of black possums in New Zealand populations *(continued overleaf)*.

Location	Latitude	Habitat[1]	n	% Blacks	Reference
Auckland			4 412	4	Wodzicki (1950)
Kaitaia	35°06′	F/E	85	6	Julian (1984)
Waipoua SF	35°40′	N	50	0	W. Q. Green (unpubl.)
Silverdale	36°37′	F	132	0	Triggs (1982)
Tokoroa	38°15′	E	180	25	Clout (1977)
Taupo	38°40′	E	51	61	S. Triggs (unpubl.)
Gisborne/Hawke's Bay			418	46	Wodzicki (1950)
Motu	38°10′	N	36	66	W. Q. Green (unpubl.)
Bridge Pa	39°43′	F	159	14	G. D. Ward (unpubl.)
Taranaki					
Waverley/Waitotara	39°45′	F	303	7	A. J. White (unpubl.)
Wanganui	40°00′	F	93	4	P. E. Cowan (unpubl.)

Table 16: Geographical variation in proportion of black possums in New Zealand populations *(continued)*.

Location	Latitude	Habitat[1]	n	% Blacks	Reference
Wellington			13 584	38	Wodzicki (1950)
Kapiti I.	40°51′	N	1 296	48	P. E. Cowan (unpubl.)
Whareama	40°54′	F	163	20	P. E. Cowan (unpubl.)
Wainuiomata Valley	41°20′	N	76	29	A. J. White (unpubl.)
Orongorongo Valley	41°21′	N	724	36	P. E. Cowan (unpubl.)
Pounui	41°21′	F	20	32	S. Triggs (unpubl.)
Pararaki	41°32′	N	50	39	W. Q. Green (unpubl.)
Nelson/Marlborough					
Tennyson Inlet	41°06′	F/N	103	40	R. Bray & G. Struik (unpubl.)
Mt. Misery	41°55′	N	141	46	M. N. Clout & P. D. Gaze (unpubl.)
Westland			5 325	90	Wodzicki (1950)
Hohonu	42°40′	N	2 267	97	Cook (1975)
Mt. Bryan O'Lynn	42°35′	N	44	100	W. Q. Green (unpubl.)
Taramakau River	42°45′	N	98	89	Voller (1969)
Karangarua River	43°31′	N	35	100	P. D. Gaze (unpubl.)
Canterbury			1 503	67	Wodzicki (1950)
Mt. Thomas	43°09′	N	38	53	W. Q. Green (unpubl.)
Banks Peninsula	43°40′	F/N	>300	53	Gilmore (1966)
Otago			573	25	Wodzicki (1950)
Dunedin	45°52′	F	31	13	Winter (1963)
Southland			699	30	Wodzicki (1950)
Codfish I.	46°45′	N	35	100	W. Q. Green (unpubl.)
Stewart I.	46°45′	N	47	73	W. Q. Green (unpubl.)
Chatham I.	43°48′	N/F	74	100	P. E. Cowan (unpubl.)

1. N, native forest; E, exotic forest; F, farmland with scrub.

VARIATION

SUBSPECIES. Three subspecies are recognized at present (Strahan, 1983): *T. v. vulpecula,* from south-eastern, central and south-western Australia; *T. v. johnstoni,* from northern Queensland; and *T. v. fuliginosus*, from Tasmania. Only *T. v. vulpecula* and *T. v. fuliginosus* were introduced to New Zealand; they differ only in the larger size and more robust features of the Tasmanian subspecies.

BODY WEIGHT (Table 15). In the North Island, populations on farmland (2.36 ± 0.03 kg) are significantly lighter than those from native or exotic forests (2.54 ± 0.06 kg). North Island populations in general (n = 13; 2.45 ± 0.04 kg) are significantly lighter than South Island ones (n = 17; 3.04 ± 0.08 kg). The north-south cline in body and skull size and body weight is closely correlated with mean annual temperature; larger possums are found in more southerly, cooler regions, as predicted by Bergmann's rule (Yom-Tov, Green & Coleman, 1986).

COLOUR (Table 16). Blacks predominate on the west coast of the South Island and are more common in native forest, whereas greys predominate on farmland and in drier, open country, though there are exceptions (Wodzicki, 1950; Kean, 1971; R. E. Brockie, unpubl.). In Tasmania, the distribution patterns of greys and blacks are closely associated with annual rainfall and vegetation type; blacks predominate in the wetter areas, and in tall, closed forest (Guiler & Banks, 1958): the same relationship is observed in New Zealand (R. E. Brockie, unpubl.), perhaps because the larger black possums are physiologically more suited to the wetter forest areas than the smaller greys (Williams & Turnbull, 1983). The present north-south variation in size and colour may, however, also reflect the original pattern of introductions, but how far is not clear (see below). Obvious founder effects do still remain; for example, only black possums were introduced to and are recorded from main Chatham and Codfish Islands (Table 16). There have been no major changes in the broad pattern of coat colour variation since 1950. In Orongorongo Valley, the proportion of blacks has remained at 35–40% since at least 1953 (Kean, 1971).

The New Zealand fur trade recognizes eight different colours (S. Dyet, unpubl.; see left-hand back endpaper for colour plate): slate, dark brown and red-brown are varieties of the black form, while rusty and pale are varieties of grey. Red-necks may be black, but are usually grey, with a pronounced browning of the shoulder fur. Probably only the grey and black forms are true-breeding colour types; for example, pales are found with brown offspring, and browns with pale offspring.

DENTITION. Highly variable; the most common variants are missing canines or first premolars in the upper jaw (Table 17 and Gilmore, 1966). The lower I_2 is often minute; other minute teeth (probably Pm_1 and Pm_3) may also be present in front of the lower premolar; and missing or extra incisors and molars are often found (Gilmore, 1966; Archer, 1975).

Table 17: Frequency of occurrence of canines and/or first premolars missing from the upper jaws of brushtail possums in New Zealand and Australia.

Location	Latitude	n	Occurrence (%)
Australia (incl. Tasmania)		301	8.3
New Zealand			
Taupo	38°40′	34	20.6*
Wanganui	40°00′	151	8.6
Kapiti I.	40°51′	93	44.1*
Whareama, Wairarapa	40°54′	37	24.3*
Orongorongo Valley	41°21′	203	16.3*
Karangarua River	43°31′	34	11.8
Banks Peninsula	43°40′	350	45.7*

Data from P. E. Cowan (unpubl.).
* Significantly different from Australia.

HISTORY OF COLONIZATION

Possums were introduced to establish a fur trade similar to that which had flourished in Australia since the early 1800s. The first successful liberation was made in the forest behind Riverton in 1858 by Mr C. Basstian, at the place where possums had been unsuccessfully released some 20 years earlier by Captain J. Howell (Pracy, 1974). Most importations were made by the regional acclimatization societies, particularly between 1890 and 1900 (complete list of introductions in Pracy, 1974). The total imports numbered only about 200–300 Australian and Tasmanian individuals. More than half (58%) of recorded shipments came from Tasmania, and 74% of all possums imported were of the black form; Tasmanian blacks, particularly favoured for their large size and superior fur quality, were all released in the South Island, except for a few taken to Paraparaumu and Lake Waikaremoana. Greys, all imported from mainland Australia (Gippsland and New South Wales), were released primarily in the Catlins district and near Dunedin and Auckland (Pracy, 1974). The sources and patterns of these introductions have partly determined the present pattern of possum variation in New Zealand. For example, S. Triggs (unpubl.) found a high correlation between average body weight and proportion of blacks in New Zealand possum populations.

The consequent spread of possums was accelerated by additional liberations of the New Zealand-bred progeny of the original introductions; Australian stock account for only about 8% of recorded liberations. About half the additional liberations were "legal", made with official government approval by the acclimatization societies and private individuals equally; the remainder were "illegal", the unrecorded efforts of private individuals acting without official sanction. Assisted dispersal of possums was most vigorous between 1890 and 1940. Illegal liberations were particularly common from 1920 to 1940 (Pracy, 1974), and continue even today; in one area north of Kaitaia possums were apparently released a few years ago by a local trapper (Julian, 1984).

The patterns of liberation and dispersion of possums in New Zealand were strongly influenced by changes in public attitudes and government policy. Initially possums were considered to be "creating a valuable industry" (Wellington Acclimatisation Society, 1892); in the southern Longwoods and the Catlins, there was extensive uncontrolled trapping for skins by 1890. To restrain the destruction of this valuable resource, the government in 1889 (urged by the Southland Acclimatisation Society) brought the possum under the Protection of Animals Act 1880. Complaints by settlers near Riverton of damage by possums to fruit, vegetables, grain and grass (apparently in support of a case for continued trapping) then prompted the government to enquire in 1891 from the governments of New South Wales, Victoria and Tasmania for "any information as to the destructions of the opossum in orchards, gardens, etc.". All replies were similar to that from Tasmania, that "the damage done by opossums . . . is very small, and amply compensated by the commercial value of their skins" (Wellington Acclimatisation Society, 1893).

The conflict of interests between the acclimatization societies and farmers, orchardists and conservationists resulted in several contradictory changes of legislation. Increasing trapping of possums prompted the government in 1911 (again pressured by acclimatization societies) to declare possums as imported game under the Animal Protection Act 1908. In 1912, protection was withdrawn after counter-pressure from possum trappers. In 1913, further outcry from the societies resulted in an Order in Council absolutely protecting possums in practically all the bush-covered districts of New Zealand; but wholesale poaching remained rife, and thousands of skins were exported, nominally as rabbit skins (Thomson, 1922).

In 1919, after continuing debate, the government requested Professor H. B. Kirk to investigate possum damage to forests. Kirk (1920) concluded, "The damage to New Zealand forests is negligible and is far outweighed by the advantage that already accrues to the community", and "Opossums may, in my opinion, with advantage be liberated in all forest districts except where the forest is fringed by orchards or has plantations of imported trees in the neighbourhood". Kirk recommended an open season from May to July with a licence fee plus royalty. In 1921, legislation laid down the procedure for the taking of possums, and prohibited the harbouring and liberation of possums without permission from the DIA. Skins could be sold only to licensed dealers, only stamped skins could be sold, and a levy was charged for stamping. The administration of these provisions was partly controlled by the acclimatization societies, for which they received part of the royalties.

After 1922, there was increasing evidence of the detrimental effects of possums on native forests. The DIA took a firm stand and, despite Kirk's recommendations, refused to sanction any of the repeated applications for further liberations made from 1920 to 1940 (Pracy, 1974). Uncontrollable illegal liberations therefore increased greatly during this period. The general tide of opinion eventually swung against protection of possums, and in 1946–47 all restrictions on the taking of possums were cancelled, penalties for harbouring and liberation were increased, and limited poisoning was legalized. For post-war control operations, see p. 97.

DISTRIBUTION

WORLD. The brushtail possum is endemic to mainland Australia, Tasmania and some offshore islands, including Kangaroo I. and the islands of Bass Strait (Strahan, 1983). It has the widest distribution of any marsupial in Australia.

NEW ZEALAND (Map 5). Possums are found throughout the North Island except the upper slopes of Mt. Egmont (= Taranaki) and Mt. Ruapehu and in northern Auckland province (Pracy, 1980). They are scarce throughout Northland (Map 6A) and probably absent from the Aupouri peninsula and parts of coastal and forest areas near Warawara State Forest (Julian, 1984); they are spreading westwards and northwards at about 5 km per year, but are restricted by the large drainage outfalls of farmland to the north-west, and the Awapuni River.

Possums are widespread in the South Island, except in parts of South Westland and western Fiordland, and in the upper catchments of a few rivers in South Canterbury and north-west Otago (Pracy, 1980). Dispersal continues; for example, possums invaded the upper Copland and Karangarua Valleys only within the last 10 years (Pekelharing & Reynolds, 1983). Possums are scarce south of Haast and absent south of the Arawata River, except around Milford Sound (M. J. Meads, unpubl.). Possums are now moving down the west side of Lake Te Anau, but have not yet reached the Doon River. They have recently moved from Rowallen State Forest across the Wairaurahiri River to Waitutu State Forest (Map 6B; J. R. von Tunzelman, unpubl.).

Possums have been introduced on to 17 islands and are still present on at least 13 of these (Table 18). They are present throughout Stewart I., though scarce south of Port Pegasus-Doughboy Bay and inland, except around bush margins (Purey-Cusp & McClymont, 1979). They are common in the coastal areas of rata-kamahi forest, from the Tikotatahi-Port Adventure area north around the coast to Paterson Inlet, and from there round to Masons Bay. On main Chatham I., possums were liberated in 1911 at Lake Rotorua bush at the northern end of Hanson Bay; they spread only slowly to the rest of the island. In 1932, they were still largely confined to the Hapupu-Kaingaroa area, though evidently in some numbers, as 1100 were trapped then. Southward spread was impeded by the Hikurangi channel, which periodically connects Te Whanga Lagoon to the sea; nevertheless, some reached the Horns by 1940, although they remained uncommon in the southern half of the island till the 1950s. Possums are now found over the whole island, principally in fragmented forest patches but also in the extensive areas of sand dunes and in the peat communities some distance from the nearest forest. They have apparently become more numerous in

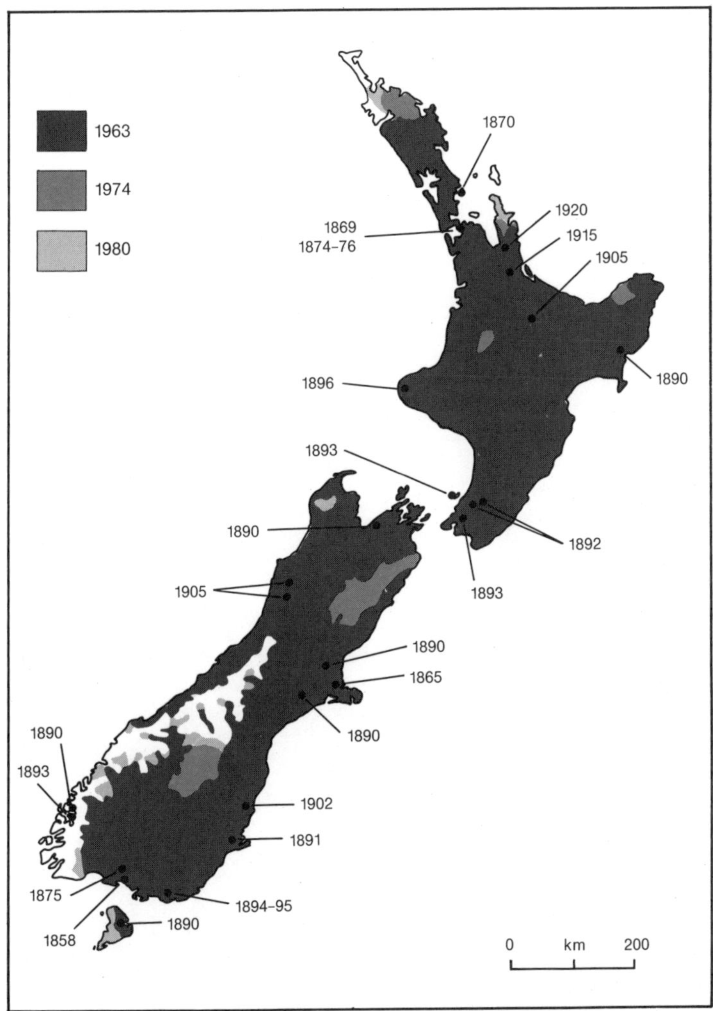

Map 5 Distribution of possums on the main islands of New Zealand, at three stages of colonisation (from Pracy, 1974, 1980; Wodzicki, 1950), with dates and sites of original introductions. For distribution on offshore and outlying islands, see Table 18.

recent years (I. A. E. Atkinson, unpubl.).

The Auckland I. introduction was apparently unsuccessful; on D'Urville I. possums were eradicated by residents soon after their introduction (Buckingham & Elliott, 1979); eradication programmes on Kapiti I. and Codfish I. (Cowan, Atkinson & Bell, 1985) were completed by the end of 1987.

HABITAT

Cover and a suitable and varied food supply are apparently the possums' only requirements. Hence they are found in a diverse range of habitats,

Table 18: Distribution and history of possums on the offshore and outlying islands of New Zealand.

Island	Latitude	Area (ha)	Date of liberation	Number released	Colour	Current status
Northern North Island						
Kawau	36°26′	*c.*2 075	1869	?	grey	++ (1985)
Rangitoto	36°47′	2 333	pre-1900	?	grey	++ (1985)
Motutapu	36°46′	1 509	1868	?	grey	++ (1985)
Whanganui	36°46′	*c.*300	1920	40	black	+ (1979)
Rangipukea	36°50′	37	1920	12	grey	+ (1979)
Central						
Kapiti	40°51′	1 970	1893	10	black }	* (1987)
			?1932	?	grey }	
Northern South Island						
D'Urville	40°50′	16 782	?	?	?	- (1979)
Tarakaipa	41°04′	*c.*30	?	?	grey	+ (1981)
Southern South Island						
Dome (Lake Te Anau)	45°22′		?1926	?	black	?
Ruapuke	46°46′	*c.*1 530	?1915	?	black	++ (1985)
Codfish	46°47′	1 396	*c.*1890	?	black	* (1987)
Stewart	47°00′	174 600	1890	>15	black, grey	++ (1985)
Native } Paterson Inlet	46°58′	*c.*66	?	?	black	+ (1985)
Tommy } Paterson Inlet	46°58′	*c.*15	?	?	black	- (1985)
Bravo } Paterson Inlet	46°58′	*c.*20	?	?	black	+ (1985)
Outlying islands						
Chatham	43°48′	90 650	1911	?	black	++ (1985)
Auckland	50°45′	45 975	1890	9	?	- (1984)

?, not known; -, not now present; +, present at (date), status unknown; ++, established, self-sustaining population; *, eradication completed. Island areas from I. A. E. Atkinson (unpubl.); other data from Pracy (1974 and unpubl.), P. R. Notman (unpubl.).

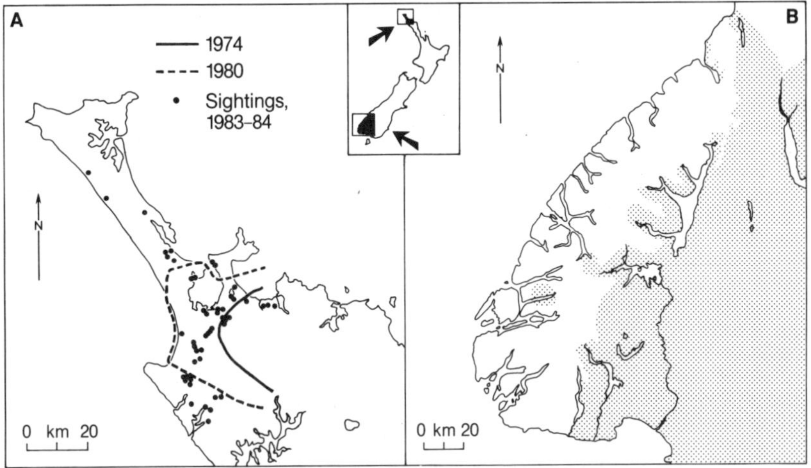

Map 6 (A) Contemporary spread of possums in Northland (Julian, 1984). (B) Distribution of possums in Fiordland, 1984 (J. R. von Tunzelman, unpubl.).

excepting only the high rainfall, mountainous terrain of south-west Fiordland. They live in all types of indigenous forest from sea level to the treeline; in montane scrublands and tussock grasslands; throughout the introduced and indigenous grasslands; in exotic forests, shelter belts, orchards and cropping areas; in thermal regions, swamp and pakihi country; in sand dunes and urban and city areas. Forests are the major habitat, especially mixed hardwood forests, where possum densities are higher than in beech or exotic pine forests; forest/pasture margins often support very dense populations. Possums tolerate rainfall from 350 mm to >8000 mm per year, and altitudes up to 1800 m and 2400 m, respectively, in the South and North Island mountain ranges, regularly ranging above the snow line in beech forest (Pracy, 1975; Green, 1984; M. N. Clout, unpubl.).

FOOD

Possums are best described as opportunistic herbivores, feeding mainly on leaves. They also take buds, flowers, fruits, ferns, bark, fungi and invertebrates, and at times these comprise most of the diet. Kean and Pracy (1953) listed more than 70 native tree species, 20 ferns and a few vines and epiphytes, as well as grasses, herbs and sedges eaten by possums, and this list has subsequently been extended (Green, 1984). Cultivated grain and vegetable crops, horticultural produce, introduced ornamental shrubs and flowers are also eaten (Pracy, 1980), as well as animals, such as small birds and mice, if caught; in captivity, meat is readily accepted.

There is much regional variation, but the local diet concentrates on only a few species. In six separate studies, the six most commonly eaten plants (determined largely by local availability) comprised 65–90% of total diet (Green, 1984; Coleman, Green & Polson, 1985). In the podocarp/broadleaf

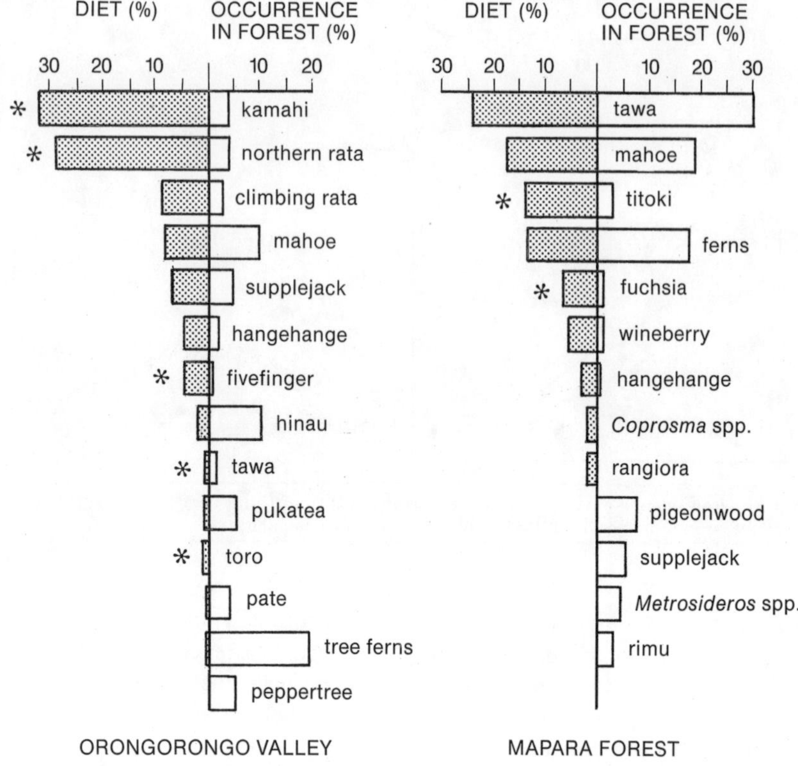

Fig. 18 Preferences of possums for leaves of various forest trees in relation to their availability (from Green, 1984). Asterisks indicate species at risk, or already reduced by browsing (for Latin names, see text).

forests of Orongorongo Valley, leaves of kamahi, northern rata *(Metrosideros robusta),* climbing rata *(M. fulgens)* and mahoe *(Melicytus ramiflorus)* comprised 82% of the leaves eaten; in a *Pinus radiata* plantation, broom, pine pollen cones and blackberry 66%; on farmland, pasture species up to 35%. But possums show pronounced preferences for some plants relative to their availability (Fig. 18; Fitzgerald, 1976; Coleman, Green & Polson, 1985). Fuchsia *(Fuchsia excorticata),* rata, kamahi, five-finger, mahoe, kohekohe *(Dysoxylum spectabile),* wineberry *(Aristotelia serrata),* titoki *(Alectryon excelsus)* and toro *(Myrsine salicina)* are heavily browsed wherever they are found (Fitzgerald, 1981). In one area of Orongorongo Valley, sustained selective browsing killed 18 of 50 mature canopy northern ratas between 1969 and 1984; five other heavily browsed trees have recovered only since possum-excluding metal shields were fitted to their trunks (M. J. Meads, 1976a and unpubl.).

Preferences among browsed plants are inconsistent between areas, for unknown reasons (Fitzgerald, 1981). Selection is influenced by plant species composition and relative abundance in the habitat; plant growth form; physical and nutritional properties of leaves, e.g., digestibility, toxins; and the availability of other foods, e.g., fruits. Diet also varies with sex, season and altitude (Fitzgerald, 1976; Coleman, Green & Polson, 1985).

Fruits of at least 65 species of native plants are eaten, generally in proportion to availability, though with some selection (Kean & Pracy, 1953; Coleman, Green & Polson, 1985; P. E. Cowan, unpubl.). Entire crops of some species, e.g., kaikomako *(Pennantia corymbosa)*, may be destroyed in a few days. Records of individual stomach contents include 209 fruits of kaikomako, 248 of ngaio *(Myoporum laetum)*, 1307 of mahoe, or 37 of kawakawa *(Macropiper excelsum)*; a single night's droppings from individual possums have contained more than 900 seeds of kaikomako, 12 000 of tree nettle *(Urtica ferox)*, or 5000 of kawakawa. About 22% of kaikomako and 44% of kawakawa seeds in the droppings were broken, but at least some intact seeds in the droppings germinate (P. E. Cowan, unpubl.).

Buds and flowers are frequently eaten and seasonally may comprise up to 30% of the diet (Fitzgerald, 1976). Invertebrates were found in 48% of pellets examined by Cowan and Moeed (1987), mainly in those from summer and autumn, but they contributed only a small (<5%) part of total biomass eaten. Fourteen orders of invertebrates were recorded, mostly (75%) stick insects, cicadas, wetas, beetles and dipteran (bibionid) fly larvae (these found only in winter, sometimes in large numbers; Clout, 1977; Cowan & Moeed, 1987).

In captivity, possums thrive and breed on a wide variety of foodstuffs — lucerne, clover, docks, swedes, silver beet, grain, maize, apples, dried peas, cabbage, carrots, bread and rabbit pellets. Their nutritional requirements, other than for energy and protein (Fitzgerald *et al.*, 1981; Harris, Dellow & Broadhurst, 1985), are scarcely known, but variety of diet is important in maintaining good health.

SOCIAL ORGANIZATION AND BEHAVIOUR

ACTIVITY. Possums are nocturnal, though in winter starving or sick animals may emerge to feed in the afternoon. They are generally active in their dens about 1–2 hours before sunset, finally emerging about 30 minutes after sunset (Winter, 1976; Ward, 1978; MacLennan, 1984). They return just before dawn in summer, but often several hours earlier in winter; males return later than females (Winter, 1976). The first 1–2 hours after emergence are usually spent in the den tree, grooming, sitting or moving about (Ward, 1978; MacLennan, 1984). Heavy rain may delay emergence for up to 5 hours (Ward, 1978).

Of total time out of the den, 45% is spent sitting or otherwise inactive, 10–15% feeding, 10–20% grooming and about 20–30% in other activities, e.g., travelling between trees. Little time is spent on active social interaction, except during the breeding season (Winter, 1976; Ward, 1978; MacLennan, 1984).

Possums are largely arboreal, but spend about 10–15% of their time on the ground, feeding and moving about, more on moonlit nights and less in heavy or persistent rain. Feeding occupies about 1–2 hours each night, starting about 2 hours after sunset; possums feed at two to four different sites during the night, in two to three sessions separated by 2–3 hours of inactivity (Winter, 1976; Ward, 1978; MacLennan, 1984).

DENS. Daytime dens are under cover, preferably above ground, in hollow tree limbs or trunks, clumps of vines or epiphytes, or ceiling cavities of buildings; otherwise in clumps of flax, toe-toe (*Cortaderia* sp.), blackberry, gorse, at the base of hollow trees, under logs, underground among tree roots, in burrows of other animals such as kiwis, in haystacks or old wood piles. In Orongorongo Valley, hollow trees and epiphytes are common den sites, and more than 95% of dens located were above ground (P. E. Cowan, unpubl.); on Mt. Bryan O'Lynn, Westland, few such sites were available, and more than 70% of dens were under logs or roots of trees (Green & Coleman, 1987). In forest, possums generally use five to ten different dens at any one time, none exclusive to only one possum, but dens are not usually shared, except with young, unless density is high and den sites few (P. E. Cowan, unpubl.). Males and females use similar types of den sites (Green & Coleman, 1987). Occupied dens are actively defended against other possums. Den sites are usually on the periphery of a possum's nightly range; possums change to another den on average two nights in three, but sometimes one individual may use the same den for weeks, and some den sites are used much more frequently than others (Ward, 1978; P. E. Cowan, unpubl.). On farmland, sharing of dens is more frequent – up to five possums may be found in one hollow willow tree (Fairweather, Brockie & Ward, 1987).

HOME RANGE. Males generally have larger home ranges (mean 1.9 ha, range 0.7–3.4 ha, length 295 m) than females (mean 1.3 ha, range 0.6–2.7 ha, length 245 m) (Green, 1984). Large home ranges (males 24.6, females 18.3 ha) and range lengths (males 880, females 820 m) were observed when possums made downhill movements of up to 1.5 km from dens in native forest to pasture to feed (Green, 1984; Green & Coleman, 1986). Home range sizes calculated by radio-tracking greatly exceed those from trapping, often by more than 50% (Clout, 1977; Ward, 1984), unless inclusive boundary strip methods are used for trap-revealed ranges. Estimates of home range size are also affected by period of observation (range size measured over two years may be double that measured in one: Ward, 1978), and by 0.5–1.5-km forays in search of pasture (Green, 1984), seasonal foods, such as flowers or fruit (Jolly, 1976; Ward, 1978), or during the breeding season (Ward, 1978).

Extensive overlap of home ranges is common, both between and within sexes (Crawley, 1973; Green, 1984). The social organization of possums in New Zealand, as in Australia (Winter, 1976), appears to be a system of mutual avoidance between co-dominants (older males and females); only the area around dens or den trees is actively defended. The exclusive areas of high ranking males and females are extensively overlapped by the ranges of lower ranking individuals (Clout, 1977; Triggs, 1982). Recent re-analysis of older data (Green, 1984) shows that males do not hold territories, overlapped by the ranges of non-territorial females, as originally proposed by Dunnet

(1964), though Clout (1977) found significant separation of male, but not female, ranges in pine forest. No evidence of territorial behaviour has been observed in podocarp-broadleaf forest (Crawley, 1973; Ward, 1978; Green & Coleman, 1981), but in encounters between free-living possums, older, heavier individuals usually win, and females are usually dominant over males (Jolly, 1976; Winter, 1976; Ward, 1978). Social postures and agonistic interactions were analysed by Jolly (1976) and Winter (1976). In captivity, social organization is based on dominance hierarchies, with dominants able to displace subordinates from feeding and den sites (Biggins & Overstreet, 1978).

SCENT-MARKING. Individual recognition is based primarily on smell, and secondarily on vision (Biggins, 1979, 1984). Important scent glands include the labial and chin glands, sternal gland, paracloacal oil and scent glands, and the glands of the pouch region (Green, 1963; Biggins, 1984; Russell, 1985). The sternal and paracloacal glands are larger and more active in males than females (Bolliger & Whitten, 1948; Biggins, 1979). Scent-marking and associated behaviour has been analysed by Biggins (1984) and Russell (1985). Rubbing of branches or the bases of tree trunks with the chin or sternal gland is the most commonly observed type of scent marking (Winter, 1976); the marks may provide information on social status, location, sex, time of marking and age (Russell, 1985). The secretions of the paracloacal glands may be deposited alone (cloacal dragging) or mixed with urine (urine dribbling) or faeces, both actively and passively; copious white oil gland secretion is often produced by animals under attack or disturbed by handling (Thomson & Pears, 1962; Russell, 1985).

VOICE. Possums have a wide repertoire of vocal signals, which includes at least 22 sounds (Winter, 1976). There are screeches, grunts, growls, hisses and chatters (mostly used in aggressive interactions); zook-zooks and squeaks (dependent juveniles and pouch young); shook-shooks and clicks (males during courtship). The shook-shook sound of the male, resembling the call of juveniles, may serve to reduce female aggression (Winter, 1976; Wemmer & Collins, 1978). A unique dilated thyroid cartilage forms a rigid-walled spherical chamber opening into the larynx, and may act as a resonator in sound production (Winter, 1976).

REPRODUCTIVE BEHAVIOUR. During the breeding season, older females may be accompanied by one, sometimes two, consort males for 30–40 days before oestrus; the male may share the female's den (Jolly, 1976; Winter, 1976). Casual matings (without a consort period) are often aggressive and attract other males, several of which may attempt to mate with the female (Winter, 1976; Ward, 1978).

REPRODUCTION AND DEVELOPMENT

Most males mature at 1–2 years old, slightly older than females; at puberty the testes grow rapidly to adult size (Gilmore, 1966). The testis weights of adults do not vary seasonally, but the epididymes are about 25% heavier in the breeding season (Tyndale-Biscoe, 1984). Prostate gland weight increases

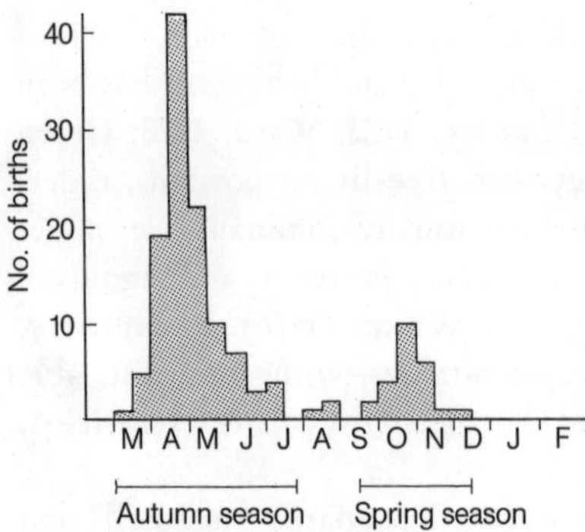

Fig. 19 Seasonal distribution of possum births per month, Tokoroa district (from Clout, 1977).

Table 19: Contribution of spring births to total breeding success of possums living in Australia and New Zealand.

	Latitude	Total births all year	% of births in Sep-Nov	Reference
Australia				
Jabiluku, NT	12°38′	39	19.5	Kerle (1984)
Brisbane, QLD	27°33′	43	18.6	
Bonalbo, NSW	28°44′	27	0	
Clouds Ck, NSW	30°05′	45	6.6	
Tutanning, WA	32°36′	27	33.3	
Sydney, NSW	33°53′	28	25.0	
Yass, NSW	34°51′	29	10.3	
Adelaide, SA	34°56′	121	14.1	
Canberra, ACT	35°17′	113	25.6	
Urana, NSW	35°20′	54	9.4	
Kameruka, NSW	36°45′	37	10.8	
Beaufort, VIC	37°26′	44	29.6	
Healesville, VIC	37°39′	20	30.0	
Geeveston, TAS	43°06′	199	0	
New Zealand				
Silverdale	36°37′	58	15.5	Triggs (1982)
Tokoroa	38°15′	140	16.4	Clout (1977)
Bridge Pa	39°54′	112	16.0	R. E. Brockie (unpubl)
Kapiti I.	40°51′	110	1.0	P. E. Cowan (unpubl)
Whareama	40°54′	272	24.3	P. E. Cowan (unpubl)
Orongorongo Valley	41°21′	394	0.8	B. D. Bell (1981)
Mt. Misery	41°55′	46	0	Clout & Gaze (1984)
Banks Peninsula	42°32′	136	13.2	Gilmore (1966)
Menzies Bay	42°32′	42	2.4	Tyndale-Biscoe (1955)
Ashley SF	43°13′	115	2.6	Warburton (1977)

four-fold between mid February and late March, just before breeding, in response to changes in testosterone levels, probably stimulated by pheromones from oestrous females (Temple-Smith, 1984). Changes in prostate and epididymis size are associated with seasonal peaks of conception (Gilmore, 1969).

Adult females are polygamous and polyoestrous; an oestrous cycle lasts 26 days, about 8 days longer than a pregnancy. There is no postpartum oestrus, but if a young dies during the breeding season, oestrus follows after about 9 days (Pilton & Sharman, 1962). Twins are rare (Kean, 1975, recorded only one set in about 8000 females), but different sized young are occasionally found in one pouch, because of (1) accidental cross-fostering of an older young when several females den together, or (2) lactation failing to inhibit oestrus, leading to a second birth about 30 days after the first (Tyndale-Biscoe, 1955). Age at first breeding in females varies. In established populations in native forest, few females breed as one-year-olds; in colonizing populations or those in exotic forests or on farmland, usually more than half breed as one-year-olds; these proportions may be even higher in Australia (Green, 1984). Fertility declines in older (>10-year) females (B. D. Bell, 1981).

Births are recorded in every month of the year, though in most populations they are highly seasonal (Kerle, 1984). The main season of births is in autumn with a second smaller and more variable season in spring (Fig. 19). In the autumn season, 80–100% of adult females (3–10 years old) produce young each year. The median date of autumn births varies from mid April to early June in different populations, depending on nutrition (reflected in female body weights); when female weights are high, breeding is earlier (Clout, 1977; B. D. Bell, 1981). Assessment and interpretation of body condition in relation to ecological performance suggests that reproduction in possums is regulated by food supplies (Humphreys *et al.*, 1984).

The spring birth season is associated with early breeding in the previous autumn, and both are related to food supply. Females in colonizing populations, and in exotic forest or pasture/scrub, breed in spring more often (mean 13%; range 2–33%) than those in established populations in native forest (mean 1%; range 0–3%) (Green, 1984). Spring breeding is not related to latitude (Table 19), but to habitat and density, both of which affect female condition. Spring births are also recorded in females which missed the previous autumn breeding, or which have matured since then. Females may breed in both autumn and spring (double-breeders), but only in populations with early autumn breeding, and only if in above average condition. The first young of a double-breeder is weaned and independent at 6 months old, as females do not normally suckle two such differently aged young at once. Both spring breeding and double-breeding are more often observed in Australia than in New Zealand, but there is still much natural variation (Green, 1984; Kerle, 1984).

The single young is born weighing 0.2 g, after 17–18 days gestation (Tyndale-Biscoe, 1984). It has well-developed lungs, upper digestive tract, mouth, forelimbs and olfactory epithelium, but rudimentary hindlimbs (Hughes & Hall, 1984). The newborn possum climbs unaided from the urogenital opening to the pouch, perhaps guided by olfaction; it attaches to a teat, the end of which swells up inside its mouth; it remains permanently attached for about 70 days, then releases the teat for increasing intervals.

Growth of various body parts and changes in weight with age are described by Lyne and Verhagen (1957). The relationship between head length and age is linear for about 150 days and has been used to age young and calculate dates of birth (e.g., by Crawley, 1973). However, growth rates differ both between populations and within the same population in different years (Efford, 1981; B. D. Bell, 1981), which affects the accuracy of the age estimates.

Weight increases slowly to about 90–110 days of age, then more rapidly, associated with parallel increases in size, weight, and milk production of the suckled mammary gland (Smith, Brown & Frith, 1969; P. E. Cowan, unpubl.). Fur is evident by 90–100 days; eyes open about 10 days later; homeothermy develops at about 110–120 days (Gilmore, 1966). The young emerge from the pouch for gradually increasing intervals from 120 to 140 days old, first to ride on their mother's back, later as young at foot. Males play no part in the rearing of their young. Most young have left the pouch permanently by about 170 days old, though some continue to suckle for up to 240 days. The young gradually become more independent, and from 240 to 270 days may be found alone but within their mother's home range, denning with their mother or alone (Winter, 1976).

More males are born than females (117 males per 100 females; Hope, 1972). The survival of pouch young is variable but equal between sexes, and lower in spring-born than autumn-born cohorts (B. D. Bell, 1981; Triggs, 1982). Females breeding for the first time, those over 10 years old, or those of less than average body weight are less successful at raising young (B. D. Bell, 1981). Generally, survival of pouch young is >90%, but varies between localities, habitats and populations with differing maternal age and condition (Clout, 1977; B. D. Bell, 1981; Hocking, 1981). For example, in Orongorongo Valley, average survival to independence of young born in 1966-75 was 58%; near Hobart, Tasmania, 78% (B. D. Bell, 1981; Hocking, 1981).

Population Dynamics

Colonization of new areas follows a predictable pattern, described in the Taramakau Valley by Bamford (1972) and in Westland National Park by Pekelharing and Reynolds (1983). There is an initial slow but steady increase in numbers, a relatively short-lived irruption to a peak population, then a sharp, often drastic decline to lower and more stable levels (Pracy, 1975). Peak densities (assessed from pellet counts) may be more than twice those of post-peak populations. Severe damage to vegetation is evident at peak population levels, but some recovery is possible after the subsequent decline, particularly if ungulates are absent (Pracy, 1975; Pekelharing & Reynolds, 1983). Pre-peak, peak and post-peak populations differ in many characteristics; populations in decline show increased average age, an excess of females, later sexual maturity, cessation of spring breeding, reduction of adult body weight and growth rate, and lower fat reserves (Pracy, 1975; Fraser, 1979). These effects are partially reversed by density reduction after trapping or poisoning (P. E. Cowan, unpubl.; Green & Coleman, 1984).

DENSITY. In podocarp-broadleaf forest density averages about 10–12 per hectare (range 7–24 per hectare) (Batcheler, Darwin & Pracy, 1967; Crawley, 1973; Coleman, Gillman & Green, 1980; Brockie, 1982); about 1–3 per hectare in pine plantations (Clout, 1977; Warburton, 1977); about 0.5 per hectare in beech *(Nothofagus)* forest (Clout & Gaze, 1984); and about 1 per hectare in scrubby farmland (Jolly, 1976; Triggs, 1982), though reaching about 5 per hectare in streamside willows and 10 per hectare in a scrub-filled swamp (R. E. Brockie, unpubl.). Density varies with forest association and altitude. In Westland, forest-pasture margins supported about 25 possums per hectare, lower slopes 10 per hectare, upper slopes 5–6 per hectare, and mountain tops (above 800 m) only 2 per hectare (Coleman, Gillman & Green, 1980). On Mt. Misery, Nelson Lakes National Park, the lower areas of mixed podocarp-beech forest had about 0.9 possums per hectare compared with only 0.3 per hectare in pure beech forest above 1400 m, and none at all above the treeline (Clout & Gaze, 1984). Numbers are usually highest in February-May (with the seasonal influx of newly independent young) and lowest in September-October (due to winter mortality) (Brockie, Bell & White, 1981).

DISPERSAL. Movements up to 10 km, and occasionally 30 km, are recorded (Clout & Efford, 1984; R. E. Brockie, unpubl.), and Ward (1985) radio-tracked a juvenile male over >5 km in two nights. Dispersing possums are usually less than two years old, and male (Clout & Efford, 1984), though young females occasionally move long distances too (Ward, 1985). In Orongorongo Valley, 72% of newly independent females settled on their natal area, whereas only 28% of males did so, and even those males tended to shift their ranges away from their natal area in time (Clout & Efford, 1984). The consequences of male-biased dispersal are (1) an excess of males in colonizing or reinvading populations, and (2) the establishment of local groups of closely related females, with moderately high levels of inbreeding (Clout & Efford, 1984; Green & Coleman, 1984; Triggs, 1987).

SEX RATIO in possum populations varies from unity (Tyndale-Biscoe, 1955; Crawley, 1973; Coleman & Green, 1984) to an excess of males (Boersma, 1974; Fraser, 1979; Coleman & Green, 1984) or of females (Clout, 1977; Fraser, 1979). Females increasingly outnumber males at ages greater than 7 years, so that, with the exception of an extinction-trapped sample from Westland (Coleman & Green, 1984), sex ratios decline with age, favouring males early in life and females later (Brockie, Bell & White, 1981; Coleman & Green, 1984). Colonizing or reinvading populations are initially male-biased (Clout, 1977; Fraser, 1979; Green & Coleman, 1984), but gradually revert to parity or an excess of females (Clout, 1977). Small samples probably overestimate the proportion of males, which are more active and range further, and are hence more often caught (Coleman & Green, 1984).

AGE DETERMINATION by molar tooth wear has been used in adults (Kean, 1975; Winter, 1980) and validated on known-age skulls (P. E. Cowan & A. J. White, unpubl.). Accurate age is determined by counting annual rings

Table 20: Numbers and value of possum skins exported from Australia and New Zealand.

Location	Date	No. skins exported	Value NZ$
Australia			
New South Wales	1920–40	12 605 418	
Victoria	1876–1902	10 744 984	
	1924–40	8 263 378	
Queensland	1920–40	8 592 140	
Western Australia	1920–40	745 494	
South Australia	1920–40	266 310	
Tasmania[1]	1973–82	1 852 243	
New Zealand	1921–30	1 109 655	
	1931–40	1 120 057	
	1941–50	3 713 976	
	1951–60	4 128 456	
	1961–70	7 361 778	
	1971	330 568	396 532
	1972	186 475	198 830
	1973	1 240 112	2 020 913
	1974	1 579 782	3 869 470
	1975	1 788 299	4 637 670
	1976	1 589 929	4 384 058
	1977	1 659 102	6 632 451
	1978	2 724 138	12 557 662
	1979	2 617 149	13 532 991
	1980	3 201 981	23 431 550
	1981	2 741 228	19 840 131
	1982	2 032 000	13 907 000
	1983	1 426 626	7 897 915
	1984	1 409 677	7 406 023
	1985	1 371 051	9 231 257
	1986	1 910 830	12 633 461

1. Only Tasmania still allows possum harvesting.
Data from Wodzicki (1950); Gilmore (1966); Johnson (1977); P. E. Cowan (unpubl.).

in lower third molar cementum, despite occasional problems with accessory lines (Pekelharing, 1970; Clout, 1982). Other ageing techniques are discussed by Kingsmill (1962) and Tyndale-Biscoe (1955).

AGE STRUCTURE in samples from 15 areas is reported (Brockie, Bell & White, 1981; Coleman & Green, 1984). Possums under two years old comprised 43% (range 22–66%) of the samples, and in 75% of them, young males outnumbered young females (Brockie, Bell & White, 1981). Variation in the proportion of young animals in the population was related to changes in the survival rate of young or the fecundity of adults (Clout & Barlow, 1982). Maximum longevity of free-living possums is 13 years for males and 14 years for females (Brockie, Bell & White, 1981).

MORTALITY of pouch young is less than 10% in pine forest, mixed scrub and bush patches (Tyndale-Biscoe, 1955; Gilmore, 1966; Jolly, 1976; Clout, 1977; Warburton, 1977; Triggs, 1982), but in Orongorongo Valley podocarp/broadleaf forest, it averages 42% (range 30–68%) (B. D. Bell, 1981): most pouch young die at about 3–4 months of age. Few independent young survive from first pouch emergence at 4–5 months, to 2 years of age (Dunnet, 1964; Crawley, 1973). Ward (1985) followed the fates of 116 pouch young born in Orongorongo Valley in 1979 and 1980: 39% died in the pouch, 38% disappeared (and almost certainly died) between pouch emergence and 9 months of age; of the remaining 23%, 37% died on or near their natal area and 22% dispersed. Thus, of the original 116 young, only 11 survived to 2 years of age on their natal area. Among adults, 26% (34% of males, 17% of females) disappeared over a 13-month period (Crawley, 1973). Other studies also suggest annual adult mortalities of 15–30% on average (Bamford, 1972; Spurr, 1981; Clout & Barlow, 1982), varying with age (generally lowest at 2–4 years, then increasing: Bamford, 1972; Boersma, 1974; Brockie, Bell & White, 1981). Brockie, Bell & White (1981) estimated the further life expectancy of 3–4-year-old possums at about 5 years. Mortality may occasionally be severe; about 40% of adults and all but one pouch young died in Orongorongo Valley during a particularly wet winter in 1977 (A. J. White, unpubl.).

Populations differ in reproductive parameters, mortality rates and productivity (Spurr, 1981). In forest a female, on average, weans 1.5 young by 3 years of age, in pasture/scrub or colonizing populations 2.0 young, or more if double-breeding is taken into account (Green, 1984). The Orongorongo Valley population produces only 0.4 independent young per female per year compared with 1.2 per female per year in Canberra, Australia (Brockie *et al.*, 1979). The dynamics of possum populations have been modelled in relation to general management (Green, 1984), effects of control operations (Spurr, 1981) and controlled harvesting of furs for maximum sustainable yield (Clout & Barlow, 1982; Barlow & Clout, 1983).

PREDATORS, PARASITES AND DISEASES

Feral cats are the most frequent predators of living possums. In Orongorongo Valley, possum hair or bones were found in up to 38% of cat scats in winter and spring, and newly independent young are often eaten in spring (Fitzgerald & Karl, 1979). Stoats often scavenge on dead possums, especially in fur-trapping areas (King & Moody, 1982), and young may be taken occasionally by stoats, moreporks or Australasian harriers. The significance of natural predation in the population dynamics of possums is unknown, but is probably much less than that of hunting. More than 2 million possum skins were exported each year in 1978–82 (Table 20), leading in many accessible areas to a scarcity of possums and a marked reduction in the size of skins offered for sale (Clout & Barlow, 1982).

Infestation with fur mites is common, resulting in irritation and fur loss from the lower back and rump; mites may cause lumbro-sacral dermatitis

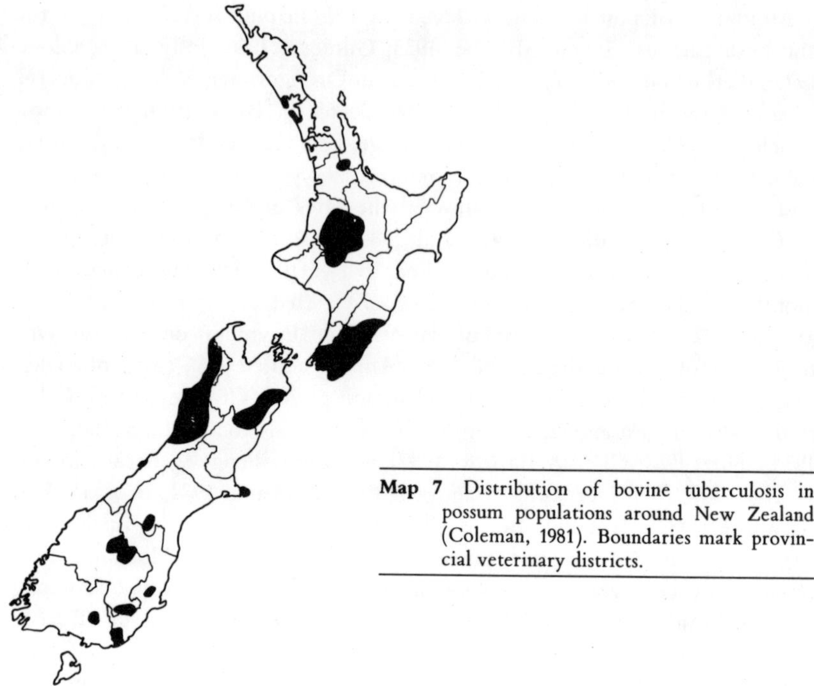

Map 7 Distribution of bovine tuberculosis in possum populations around New Zealand (Coleman, 1981). Boundaries mark provincial veterinary districts.

(rumpiness) (Presidente, 1978). *Atellana papilio* was found on all of 59 possums from Banks Peninsula (Sweatman, 1962), and on 46% of 125 possums from Central Otago, 42% of which also harboured *Trichosurolaelaps crassipes* and 50% *Petrogalochirus dycei* (Bowie & Bennett, 1983). Such mites occasionally cause allergic reactions in possum trappers. Heavy mite burdens are generally found on possums in poor condition or in high density populations where sharing of nest sites is common (Pracy, 1975).

The tapeworm *Bertiella trichosuri* infests 19–31% of possums in the North Island and 0–41% in the South Island, with little difference between forest and pasture feeding populations, contrary to Pracy and Kean's (1968) original suggestion (Khalil, 1970; Presidente, 1984). It is found in possums throughout the western North Island from Wellington and the Wairarapa to Lake Taupo, but not in those in the forests of the Urewera, Aorangi or northern Rimutaka ranges (Pracy, 1975), except in about 30% of possums in Orongorongo Valley (P. E. Cowan & W. A. G. Charleston, unpubl.). In the South Island, it is found in the Golden Bay area, near Westport and in north Canterbury, but not in central Westland or on Stewart Island (Warburton, 1983). Infested females had lower mesenteric fat reserves and reduced fecundity compared with non-infested females (Warburton, 1983).

Liver flukes, *Fasciola hepatica,* are found in stock in both North and South Islands, but they infest possums only in Taranaki, where up to 70% may carry flukes. Fascioliasis is probably the most pathogenic of parasitic infections encountered by possums, causing death within 5–6 months (Presidente, 1984).

Nematodes, *Parastrongyloides trichosuri* and *Paraustrostrongylus trichosuri*, are common, more so in possums from forested areas (100% and 73%, respectively) than in those from farmland (44% and 7%). *Trichostrongylus colubriformis* is a sheep parasite often found (8-70%) in possums from farmland (Presidente, 1984). Other nematodes recorded include *Trichostrongylus axei, T. retortaeformis* and *T. vitrinus*.

Periodontal disease, similar to that of sheep, affects possums in both forest and farmland. The disease (cause unknown) usually attacks the lower jaw; in the worst cases, the bone may be eroded until the teeth sit only on pillars of bone separated by large cavities, or work loose and fall out. Generally, fewer than 10% of skulls are affected (P. E. Cowan & A. J. White, unpubl.)

Possums may carry the protozoan *Eimeria* sp., Whataroa virus, and mycotic infections caused by *Emmonsia* sp., *Trichophyton mentagrophytes, T. ajelloi* and *Microsporum cookei*.

Possums carry leptospirosis and bovine tuberculosis, both infective to man and livestock. Infection with *Leptospira interrogans* serovar *balcanica* is common and widespread in the North Island (up to 80% of adults), but absent in Westland, perhaps because most of the original releases there came from *balcanica*-free Tasmania. Infection of possums with other serovars (*ballum, copenhagenii, pomona, tarassovi*) is rare, and *balcanica* is apparently not readily transmitted to humans, even high-risk groups such as possum trappers (D. K. Blackmore, unpubl.). The effects of *balcanica* infection on possums are not well known, but appear slight (Presidente, 1984).

Bovine tuberculosis (*Mycobacterium bovis*) was first detected in possums in New Zealand only in 1967, but has since been found in 23 general localities (Map 7). It has not been found in Australian possums despite recent extensive surveys (Presidente, 1984), perhaps because they live at much lower density and have only infrequent contact with cattle, or because the *Mycobacterium* strains infecting New Zealand possums are more virulent than those in Australia (Corner & Presidente, 1981). The distribution of TB-infected possums is linked with a high and persistent reactor rate in cattle sharing the same environment, although TB in possums is probably self-maintaining in the absence of infected cattle (Ministry of Agriculture & Fisheries, 1986). The general infection rate is 0.1-12%, higher in Westland and the King Country, and locally to 25%. The natural course of the disease in possums is not known; after experimental infection, possums develop progressive symptoms and die after 1-2 months (Julian, 1981). *Mycobacterium vaccae, M. avium* and *M. fortuitum* have also been recorded in New Zealand possums.

ADAPTATION TO NEW ZEALAND CONDITIONS

Possums in New Zealand have adapted to new diets and habitats, and have fewer parasites, predators and competitors than those in Australia; the consequences of these differences affect density, body size, home range size and possibly social organization and regulation of numbers (Green, 1984; Kerle, 1984; Smith & Hume, 1984), as follows. (1) In Australia, their diet

Table 21: Number of species of parasites and presence or absence of diseases in possums in New Zealand, in the Australian states from which possums were introduced (New South Wales, Victoria, Tasmania), and in the whole of Australia.

Parasite	Number of species in		
	New Zealand	NSW, VIC, TAS	Australia
Ectoparasites			
Mites	3	11	16
Ticks	0	7	11
Fleas	0	4	6
Endoparasites			
Protozoa	1	3	4
Cestoda	1	2	2
Trematoda	1	1	1
Nematoda	6	17	23
Diseases			
Leptospirosis	+	+*	+
Bovine tuberculosis	+	-	-
Other *Mycobacterium*	+	+	+

+, present; -, absent; *, not recorded in Tasmania.
Data from Presidente (1984).

is largely *Eucalyptus* leaves, when little else is available, though they make use of a wide variety of trees, shrubs, grasses, herbs, flowers and fruits if they can (Fitzgerald, 1984a; Kerle, 1984; Proctor-Grey, 1984). The nutritive value of *Eucalyptus* leaves is generally low, and toxic secondary compounds in them may limit possum populations if alternative food is scarce (Freeland & Winter, 1975). (2) The preferred habitat of *T. vulpecula* in Australia is open forest and woodland; in contrast to New Zealand, they are rarely found in rainforest or wet sclerophyll forest, except in the absence of *T. caninus* (Kerle, 1984). However, *T. vulpecula* in New Zealand is common in rainforest and has even reproduced the same relationship between colour phase and habitat (rainfall, vegetation) observed in Tasmania (p. 75). (3) Possums in New Zealand carry fewer parasites than those in Australia (Table 21). (4) There are also fewer predators; in Australia, possums are killed by dingos, feral dogs and cats, foxes, wedge-tailed eagles, lace monitors, and carpet pythons, though they rarely comprise more than 5% of any predator's diet (Croft & Hone, 1978; Brooker & Ridpath, 1980; Jones & Coman, 1981; Newsome *et al.,* 1983). (5) Competition for food and den sites is much more frequent in Australia than in New Zealand; about half of the Australian species of possums and gliders use tree hollows, and most leaf-eating species show some dietary overlap with brushtail possums (Menkhorst, 1984; Smith & Hume, 1984).

The consequences of these differences are a two to twentyfold increase in density in New Zealand and changes in reproduction, body size, home range size and probably in the means of regulation of numbers, as follows. (1) On mainland Australia, density is usually less than one possum per hectare, though it is higher in Tasmania (Kerle, 1984); in New Zealand, such low numbers are found only in scrub or beech forest. (2) Breeding in New Zealand is generally later than in Australia, and in established populations, maturation is delayed, mortality of pouch young is increased and spring breeding is reduced (Brockie *et al.*, 1979; Green, 1984). However, the variation in reproductive characteristics found in New Zealand populations can be matched from within Australian, particularly Tasmanian, samples (Kerle, 1984); and the changes associated with transition from a colonizing to a stable population are closely paralleled in Australian populations inhabiting forest at various ages after recovery from fire (Hocking, 1981). (3) Possums on mainland Australia are generally lighter and smaller than in New Zealand. The average body weight of eight Australian samples was only 2.2 ± 0.2 kg (Kerle, 1983), although in Tasmania the average weight is about 3.4 kg, similar to that in the South Island (Hocking, 1981). (4) Home range sizes in Australia are at least twice as large, averaging 5.4 ha for males and 2.4 ha for females (Green, 1984). (5) Density in Australia is controlled largely by social behaviour (e.g., residents commanding key resources, such as dens, and expelling transients and immatures: Winter, 1976), whereas New Zealand populations are limited more by food than by behaviour (Green, 1984; Humphreys *et al.,* 1984).

Genetic (founder) effects, due to the small number of possums originally imported, are perceptible in dentition (Flux, 1980). Possums in New Zealand more often have missing teeth than those in Australia (Table 17). Within New Zealand, the effect is most pronounced in isolated populations, e.g., on Kapiti I. and Banks Peninsula.

Possum populations around Australia show marked differences in allele frequency for blood serum proteins (Hope, 1970; Kerle, 1983); tissue protein analysis revealed three out of thirteen loci to be polymorphic (Kerle, 1983). Genetic variation among New Zealand populations is currently under study. Nine polymorphic and 36 monomorphic allozyme loci were identified in tissue and blood samples. Predominantly black New Zealand populations were genetically similar to the Tasmanian populations, whereas predominantly grey populations were genetically closer to possums from Victoria and New South Wales. The New Zealand possum populations thus still reflect to some extent the variation in their ancestral stock (Triggs, 1987).

SIGNIFICANCE TO THE NEW ZEALAND ENVIRONMENT

DAMAGE. Selective browsing of preferred plant species (listed above) intensifies the impact of possums on New Zealand forests. The effects are unquestionable (Kean & Pracy, 1953; Howard, 1964; Bathgate, 1973a; Holloway, 1973), although the consequences for forest dynamics and soil erosion are debatable (Mosley, 1978; Veblen & Stewart, 1982; Stewart & Veblen, 1982a,b, 1983; Allen & Rose, 1983; Batcheler, 1983; Jane & Green, 1983; Grant, 1985, 1989). Extensive canopy defoliation and mortality attributable to possums has been described in many areas, e.g., in the podocarp-broadleaf forests of the southern North Island (Zotov, 1949;

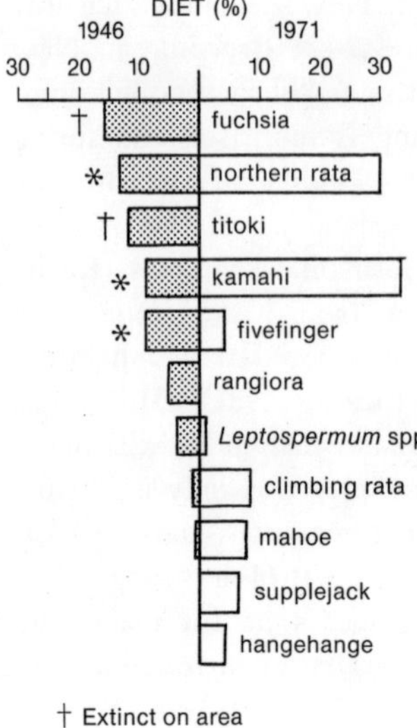

Fig. 20 Changes in diet of possums, Orongorongo Valley, 1947–1976, resulting from browse-induced extinction of highly palatable plants (from Green, 1984).

Holloway *et al.*, 1963; Wallis & James, 1972; Cunningham, 1979; Batcheler, 1983) and in the forests of Westland (Holloway, 1959; Wardle, 1974; Pekelharing, 1979; Pekelharing & Reynolds, 1983). Beech forests are also damaged but less severely (Grant, 1956; James & Wallis, 1969; James, 1974). Selective browsing on particular species and individual trees eliminates some species and favours others less palatable to possums, resulting in a gradual change in forest composition (Fig. 20; Meads, 1976a; Fitzgerald, 1976, 1978; Coleman, Green & Polson, 1985; Cowan *et al.*, 1985). In Orongorongo Valley, fuchsia, titoki, tutu (*Coriaria arborea*), toro and five-finger have disappeared, replaced by tree ferns, pigeonwood (*Hedycarya arborea*) and mahoe; northern rata, kamahi and tawa (*Beilschmiedia tawa*) are under attack, and have already disappeared in some areas (Campbell, 1984).

Many of Westland's forests have shown widespread and progressive canopy mortality over the last 40 years. The consequent accelerated erosion has been attributed principally to possums, especially when combined with depletion of the understorey by ungulates (Wardle, 1974; Coleman, Gillman & Green, 1980; Batcheler, 1983; Pekelharing & Reynolds, 1983). However, canopy trees die for other reasons; from studies of tree population dynamics, Veblen and Stewart (1982; Stewart & Veblen, 1982a,b, 1983) suggest that large-scale canopy die-back is a natural consequence of the stand dynamics of trees such as southern rata, where synchronous mortality of senescent even-aged stands derives from synchronous colonization of catastrophe-created sites (e.g., landslips). Moreover, New Zealand's mountains are generally unstable and subject to frequent earthquakes, rainstorms and tropical storms (Jane & Green, 1983; Shaw, 1983); and the effects of subtle changes in climate are often underestimated (Grant, 1985, 1989). The effects of possums therefore may be secondary to inevitable die-back and erosion. Allen and

Rose (1983) disagree, however, partly because dead canopy trees are not all the same size. Batcheler (1983) reviewed rata-kamahi die-back and concluded that possums were largely responsible, especially in areas where the diet of possums locally includes much southern rata and kamahi (Fitzgerald & Wardle, 1979; Coleman, Green & Polson, 1985). Recent trials in which ratas were artificially defoliated emphasize the role of possums; although younger trees were more able to withstand defoliation, they, too, eventually succumbed to sustained browsing (Payton, 1985). The problems of reaching a conclusion on this matter are made more difficult because nowhere in New Zealand have possums yet attained an equilibrium with native vegetation, so the extent and acceptability of changes to native flora and fauna at equilibrium levels are still unknown.

Secondary effects of possum browsing may be less obvious. Canopies weakened by browsing may be more susceptible to windthrow, salt damage, pathogens, insects or climatic extremes (Green, 1984). Possums may compete with native birds for seasonal resources, e.g., fruit, with consequent risk to the more frugivorous species, particularly the wattlebirds (Williams, 1976). For example, the extensive overlap in diet between possums and the North Island kokako (*Callaeas cinerea wilsoni*) (Fitzgerald, 1984b) may partly explain the decline of this bird (Leathwick, Hay & Fitzgerald, 1983). Possum browsing of leaves may reduce the production of flowers and fruit, with consequent effects on native animals; on Kapiti I., sustained possum browsing of kohekohe trees had prevented any flowering or fruiting for several years before possum control was reintroduced (A. D. Robinson, unpubl.). Possums may compete for nest sites with hole nesting birds, such as kiwis, parakeets and saddlebacks (Jolly, 1983; Lovegrove, 1985) and kill native birds and disturb their incubation to an unknown extent (Lovegrove, 1985).

Possums are responsible for damage and economic losses in exotic forest plantations (Clout, 1977; Warburton, 1977), in catchment plantings (Jolly & Spurr, 1981; Thomas, Warburton & Coleman, 1984), in improved pastures, and in various farm and horticultural crops (Pracy, 1980; Spurr & Jolly, 1981). In catchment plantings alone, annual losses probably exceed $800 000 (Thomas, Warburton & Coleman, 1984).

The reason that the possum is a pest in New Zealand and not in its native Australia is largely because of its much higher density in New Zealand. In Australia possums do only minor damage in introduced and native forest plantations and in some farming areas (Barnett, How & Humphreys, 1976; Clout, 1977; Fitzgerald, 1981; Statham, 1984).

CONTROL. The DIA, though not equipped to deal with large-scale possum control, was left in charge of the problem when restrictions on trapping and poisoning were lifted in 1946–47. It therefore started a national distribution survey and field research programme. In 1951, a bounty of 2s 6d for each possum killed but not skinned was introduced, largely as an interim measure to restrict the spread of possums, but, conversely, it probably promoted the last renewal of illegal liberations (Pracy, 1974). In 1956, responsibility for possum control passed to NZFS under the Noxious

Animals Act 1956. Bounties continued till 1962, but did little to control increasing and expanding possum populations, even though over the 11 years about 8.2 million bounties were paid, costing more than $2 million (Pracy, 1980). After withdrawal of bounties, large-scale possum control was conducted in farming areas by Rabbit Boards, later by APDB's, under the Agricultural Pests Destruction Act 1967, and in Crown-owned forested areas and national parks by NZFS under the Wild Animal Act 1977 (Coleman, 1981). Since April 1987, DOC has administered the Wild Animal Act.

Control of possums is now directed at reducing damage to acceptable levels consistent with the well-being of native forests, plantations and farmland, and eliminating sources of reinfection of bovine tuberculosis.

Control is by 1080 (sodium monofluoroacetate) poisoned carrots or pellet bait dropped by helicopter or fixed-wing aircraft, plus ground control by trapping, shooting or poisoning with cyanide paste. Commercial trappers kill 1–3 million possums each year (Table 20) and are considered a useful adjunct to official control measures (Miers, 1973), but their activity depends on skin prices (Wodzicki, 1950); in periods of sustained high prices, their effects on possum populations are significant (Clout & Barlow, 1982).

Research and development of control techniques has largely been done by NZFS and APDC. Information is available on poisons (Peters, 1972; McIlroy, 1983), on bait development (Morgan, 1981a, 1982; Batcheler, 1982; Morgan, Batcheler & Peters, 1986), on bait delivery (Batcheler, 1982) and on assessment of the effects of control (Bamford, 1970; Bamford & Martin, 1971; Batcheler, 1975; D. J. Bell, 1981; Jane, 1981; Morgan, 1981b).

The identification of possums as a reservoir of bovine tuberculosis stimulated a dramatic increase in expenditure on control by MAF, acting through NZFS and APDC, from about $1.4 million jointly up to 1976, to $3.0 million and $1.9 million, respectively, in 1976 to 1981 (Coleman, 1981; APDC, 1982). In the last three years, more than $4.5 million has been spent on bovine TB/possum control (T. J. Ryan, unpubl.). In few cases was the eradication of local foci of infection successful (Coleman, 1981). The number of infected herds of cattle increased 15% between 1981 and 1982, and in 1983, MAF stated, "Practically the whole of the increase is in areas where there are infected possum populations. It is now recognised that the continued presence of infected possums in some afforested areas makes total eradication of tuberculosis an unrealistic goal". The aim of total eradication, involving the large-scale aerial poisoning operations typical of the 1970s, has been replaced by the aim of cost-effective control in TB-endemic areas, using smaller, more intensive operations, more often including workers on foot. Severe outbreaks are still recorded, e.g., in the central King Country from 1982 to the present (1987). The continuing cost of possum/TB control is high (aerial poisoning cost >$18 per hectare in 1985), and the large ranges of possums at forest-pasture margins make it necessary to extend control work into forest by at least 1 km from the margin to reduce possum-cattle interactions (Green & Coleman, 1986).

P. E. C.

Order Insectivora

The insectivores comprise six families, 60 genera and 345 species of mainly small, ground-dwelling mammals that feed on invertebrates. They were originally found throughout the world except in Australasia and South America, where the smaller marsupials are their ecological replacements (p. 34). They have very different external forms but share a number of primitive characters, including plantigrade feet, five clawed toes on each foot, small brain, large number of simple teeth, long muzzles and abdominal testes. Only one member of this order is represented in New Zealand, the European hedgehog, and this common and familiar animal, with its unique coat of spines, is never likely to be mistaken for any other creature.

Family Erinaceidae

The three genera of spiny hedgehogs are found in the deciduous woodland, steppe and desert zones of Eurasia and Africa. There is also another group of five genera of hairy hedgehogs which live in the evergreen forests of Asia. All 17 species in the family feed on the ground, mainly at night, and many of them have a long fossil record.

Genus *Erinaceus*

There are about four Eurasian species of *Erinaceus*, and another three in Africa. They are by far the most familiar of all the insectivores and the only ones to develop a close and rather friendly relationship with man, figuring in folktales and children's literature for generations. Their distinctive characteristic is their ability to roll up into a protective ball of spines when alarmed.

8. European hedgehog
Erinaceus europaeus occidentalis Barrett-Hamilton, 1900

Description (Fig. 21)
Distinguishing marks, p. 22 and skull key, p. 24.

A grey-brown nocturnal insectivore, with the back and sides entirely covered with spines, and with a pointed face, powerful forefeet and rather long and narrow hind feet. A powerful dorsal muscle, the musculus orbicularis, enables a hedgehog to roll itself into a prickly ball when disturbed, or for several weeks during hibernation. Each spine is a single modified hair about 200 mm long, anchored in the skin by a hemispherical bulb, bent at its narrowest point where it emerges from the skin, and tapering to a fine point. The spines are creamy white, darkening to grey-brown at the base and with a broad dark band just below the sharp tip. The spines are hollow, but have reinforcing ridges running along their length. The underside is covered with grey to pale brown coarse hair, variously patterned and mottled. There are dark markings on the cheeks, around the eyes and on the upper surface of the feet. The tail is about 20 mm long. The skull is robust and box-like with well-developed zygomatic arches.

Dental formula $I^3/_2\ C^1/_1\ Pm^3/_2\ M^3/_3 = 36$.

The canine teeth are short and undeveloped. The sharp cusps on the molars and premolars wear flat after about three years.

FIELD SIGN

The five-toed footprints resemble those of a large rat, but the forefeet are broader and shorter than the hind feet, so a hedgehog leaves two distinctively different prints. The droppings are usually cylindrical in shape, 30-50 mm long, about the thickness of a pencil, and pointed towards each end. Bile stains fresh droppings a dark greenish black. Insect fragments in them, such as head capsules, legs, or millipede segments are often visible to the naked eye. Hedgehogs often advertise their presence by snuffling loudly and walking noisily through undergrowth.

MEASUREMENTS

See Table 22.

VARIATION

The body weight of hedgehogs varies greatly with season, as they lay up fat for the winter and lose it while hibernating. Average body weights for New Zealand animals usually range between 620 and 700 g for most of the year, falling to between 540 and 650 g in winter. Hedgehogs from Wellington lose an average 10% of their weight while hibernating. Individuals starved as juveniles and then overtaken by winter may remain stunted (about the size of a clenched fist) for the rest of their lives. Some small (400 g) animals, marked and recaptured at intervals near Wellington, survived the cold and continued to put on weight through the winter (Brockie, 1974a).

The colour of the spines may vary from dark brown to pale straw colour, and about one in 10 000 is albino. Young animals have darker noses, spines and footpads; older ones pale with age.

HISTORY OF COLONIZATION

The first recorded hedgehogs in New Zealand were a pair received by the Canterbury Acclimatisation Society in 1870 from the purser of the *Hydaspes*. Another was landed at Lyttelton in 1871, and three were liberated in a Dunedin garden in 1885. In 1890 hedgehogs were found at Sawyers Bay (Port Chalmers), and in 1894, 12 recently imported hedgehogs escaped at Merivale, Christchurch (Thomson, 1922). Nothing is known about the exact origins of the New Zealand stock except that one crate of hedgehogs was dispatched from Perthshire, Scotland; however, they almost certainly all came from Britain. At first, hedgehogs were brought to New Zealand to remind settlers of their homeland, but in the 1890s they came to be

Fig. 21 Hedgehog, Stokes Valley (J.H. Johns).

regarded as natural predators of garden pests and so were brought in to control slugs and snails. Reliberations of New Zealand-born hedgehogs were made at Gore, Waimate and Westland by 1909. Between 1907 and 1912, South Island hedgehogs were brought to Wairoa, Napier, Stratford, Hawera and Carterton, to control garden pests. Further liberations, and their natural spread, took them to Blenheim and Hamilton (1914), Lower Hutt (before 1915), Opotiki (1916), Greatford (1924), Raetihi (1927), Auckland city (1927–29), Murchison (1928), Thames (1929), Stewart I. (1930), Waiouru (troops released hedgehogs in the early 1940s), Queenstown (1946), Taupo (1948), Motueka (1950), Takaka (1961) and St. Arnaud (1969) (Wodzicki, 1950; Brockie, 1975). In 1982 the TVNZ Natural History Unit programme *Wildtrack* solicited reports of hedgehog sightings; some 600 viewers responded, but none reported any extension of range beyond the 1972 limits.

DISTRIBUTION

WORLD. Hedgehogs are found throughout Western Europe, from central Sweden and Finland in the north to the Mediterranean, including Sicily, and from Ireland to the Oder river on the Polish-German frontier. The British subspecies, ancestor of the New Zealand stock, is found throughout Britain except on some Hebridean islands.

NEW ZEALAND (Map 8). Hedgehogs are abundant throughout lowland districts, especially near the coasts, but less numerous in the hills and rare in mountainous areas. They are scarce or absent in areas with more than 250 frosty days a year, such as the upland Southern Alps and parts of the central North Island plateau, and in areas with more than 2500 mm of rain a year, such as Fiordland. They are present on Chatham, Quail, Stewart,

D'Urville, Waiheke and Rabbit (Nelson) islands. Occasional hedgehogs have been reported at the Hermitage, Mt. Cook; in the garden of the hotel at Milford Sound; washed up on the shores of Lakes Wanaka and Te Anau; at a height of 1274 m in the Beebe Range, Marlborough; at the foot of the Hooker and Tasman glaciers (Wodzicki, 1950; Brockie, 1975); and at 1500 m in the Garvie Range (J. Black, unpubl.).

HABITAT

Hedgehogs are abundant on lowland and coastal farmland and in sand dune country, e.g., in Manawatu and North Auckland, where frosts are few and mild, and snails (*Helix aspersa*) abundant. They are numerous in dairy country, where slugs are abundant in long pasture. Lowland stream and river sides are favoured habitats. Cities and suburbs also support dense populations of hedgehogs, because invertebrates and dry hibernacula are available, as well as extra food purposely provided by householders. Hedgehogs are less common on dry, open, inland and upland areas, where frosts are harder and there is less invertebrate food available. Because dry nest sites are hard to find, few hedgehogs penetrate New Zealand rainforests. For example, in South Westland hedgehogs live on the coastal sand dunes, but not in the forested townships of Franz Josef and Fox Glacier, 20 km inland. On the Whangamoa Saddle road (between Nelson and the Rai Valley) and the Rimutaka summit road (between Wellington and the Wairarapa) hedgehogs are frequent road casualties at the foot of the pass on either side, but very rare towards the forested summits. Nevertheless, a few hedgehogs do survive in very wet broadleaf-podocarp forest of the Ruahine, Tararua and Rimutaka ranges, and in beech forest of the South Island.

FOOD

Hedgehogs are mainly insectivorous, but will eat any animal substance and even some plant material. In and around Wellington, slugs, beetles, millipedes, garden snails, insect larvae, earwigs and spiders contribute most to their diets, with lesser amounts of frogs, moths, flies, woodlice, ants, bees, wetas, cicadas and fly maggots (Brockie, 1959). A single hedgehog was found to have eaten at least 25 slugs, probably the introduced *Agriolimax reticulatus*, and another was seen eating the native slug *Athoracophorus bitentaculatus* in native forest near Wellington. Among the many species of beetle eaten, huhu (*Prionoplus reticularis*) feature in the diet of hedgehogs from pine forest near Auckland and the Dunedin district. Millipedes were found in 37% of the Wellington sample, and several droppings contained nothing else. Snails are a favourite item; 36% of the droppings of wild hedgehogs examined contained radulae and shredded skin of *Helix aspersa*. Captive hedgehogs were observed to crush small snails before eating the animal inside, but large snails are too strong and can resist the determined grappling of hedgehogs (Brockie, 1959). Hedgehogs also eat the native snail *Wainuia urnula*. Campbell (1973) found that slugs were frequently eaten on dairy pasture near Lincoln College, but not snails: other important items were moth larvae, earwigs,

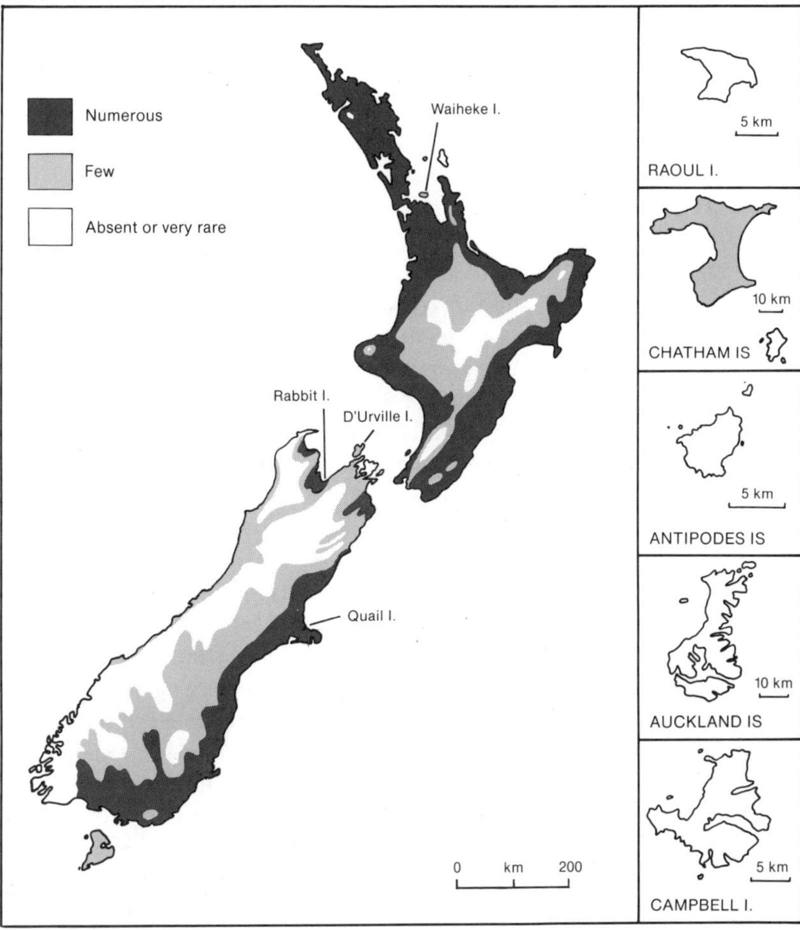

Map 8 Distribution and relative abundance of hedgehogs in New Zealand in 1972 (Brockie, 1975).

beetles, spiders, harvestmen, earthworms and adult grass-grubs (*Costelytra zealandica*).

East (1972) calculated that hedgehogs on pasture land near Christchurch eat up to 850 adult grass-grubs a day: Campbell (1973) calculated that hedgehogs were capable of destroying up to 40% of the available grass-grubs during the two-month flight season. Hedgehogs do not eat indiscriminately. Sand hoppers and woodlice are numerous near Wellington, but few were found in hedgehog droppings. However, hedgehogs frequently take in grass roots and leaves and other leaf fragments, probably incidentally while eating invertebrates. They will eat a little ripe sweet fruit and water melons, and apparently take the young leaves and succulent tips of clover.

It has been claimed that hedgehogs have helped to reduce the numbers of skylarks and ground nesting game birds in New Zealand, but there is no evidence to support this view. Some reports of hedgehogs attacking birds' nests have been published (Wodzicki, 1950; Bull, 1953a) but more intensive investigations reveal that these isolated incidents have a negligible effect on the birds' populations. Westerskov (1955) found only 0.3% of pheasant nests destroyed by predators (harrier hawks and rats). Balham (1949) found that grey and mallard duck nests near Himatangi were not touched by the abundant hedgehogs. Nor has any evidence been found for hedgehogs attacking chukor or quail nests (G. R. Williams, unpubl.).

SOCIAL ORGANIZATION AND BEHAVIOUR

ACTIVITY. Hedgehogs are almost exclusively nocturnal, but are occasionally seen abroad by day. They emerge from their nests at dusk and are most active early in the evening. Laboratory experiments by Herter (1938), Kristoffersson (1964) and Campbell (1973) established that hedgehogs do not feed continuously through the night but concentrate their activities between two and three hours after sunset, and again between eight and nine hours after sunset, with less activity between these times. This nightly rhythm of activity has been confirmed among wild hedgehogs at Lower Hutt (Brockie, 1974a). Daylight sightings are usually of animals in search of water during droughts, or of sickly or underweight animals during winter.

DISPERSION. Hedgehogs are normally solitary, except during brief mating encounters and when nestlings depend on their mothers; but they may crowd together to exploit a rich source of food, such as emerging insect larvae. On a Lower Hutt golf course 67% of tagged hedgehogs were recaptured within 400 m and 95% within 800 m of their point of release. Some parts of the golf course were never visited by hedgehogs, while other paths were well trodden by many animals to form hedgehog roads (Brockie, 1974a). Parkes (1975) reported that the average size of male hedgehogs' home ranges on farmland in the Manawatu was 2.5 ha, and for females 3.6 ha. In some habitats, hedgehogs have separate or adjoining summer and winter home ranges. At Lower Hutt many hedgehogs spent the colder months in suburban gardens, but the summer on the golf course (Brockie, 1974a). While moving about, hedgehogs mark their progress with a thin trail of urine. Hedgehogs do not attempt to defend their home ranges, which usually overlap with many others. The movements of radio-tagged hedgehogs remain unpredictable even when food is regularly laid out for them at the same spot for years. Morris (1985) found that animals did not learn to shorten the distance travelled between their nests and the regularly proffered food, but stumbled across it accidentally. Neither did they attempt to defend the food.

NESTS. Of 35 winter nests reported in New Zealand, eight were in holes among tree roots, four in compost heaps or under heaps of leaf mould, three under haystacks, three under logs, three under houses or sheds, three under tussocks, two in rabbit burrows, two under "red-hot pokers" (*Kniphofia* sp.), and two under sheet iron; other nests were found under rocks, boxthorn, toe-toe, gorse, or simply buried in the soil of a herbaceous border (Brockie, 1974b). Twenty-six out of 30 nests in coastal sand dunes adjacent to farmland in the Manawatu were in marram tussocks. Others were built in *Carex* rush among fallen pine trees, in pine needles and in the base of a large clump of flax. Most of these animals nested on well-drained slopes and avoided flood-prone hollows and slacks. The Manawatu nests housed one hedgehog at a time, and within a month each animal used up to four nests averaging 190 m apart (Moors, 1979).

HIBERNATION. New Zealand hedgehogs begin to hibernate when mean earth temperatures reach 10-11°C. In North Auckland, coastal Manawatu, and probably in other warm and relatively frost-free areas, few hedgehogs enter into hibernation, and then only for short periods (Brockie, 1974a; Moors, 1979). In Palmerston North, southern Hawke's Bay and Wellington, hibernation extends from June or July to about mid September (Brockie, 1974a; Parkes, 1975); males emerge from hibernation several weeks earlier than females (Parkes & Brockie, 1977). The proportion of the population hibernating depends on the severity of the winter, and the period of hibernation is presumably longer in colder inland and upland parts of New Zealand. New Zealand hedgehogs must attain a weight of at least 300 g before they can survive hibernation. While hibernating, the heart slows to about 20 beats/minute, and breathing develops into an erratic pattern; the animal takes 40 or 50 rapid breaths and then ceases to breathe altogether for periods of over an hour. While hibernating it expends only 2% of its normal metabolic energy, but if the environmental temperature approaches 0°C, a hedgehog can increase its metabolic rate to maintain its body temperature above 1°C (Kristoffersson & Soivio, 1964). While preparing for hibernation, winter fat is laid down first on the belly, then the back. When these sites are well covered, fat is then laid down around the kidneys and the greater omentum. During hibernation the fat is used up in the reverse order; the subcutaneous fat, which provides insulation, is the last to be used (Brockie, 1974a). Morris (1969) found that hibernation is very hazardous for hedgehogs, and calculated that 65% of juveniles near London died during their first winter, mostly the smaller ones. The necessity to have a well-made, well-sited winter nest is a significant factor in determining the distribution and habitat of hedgehogs (Morris, 1973).

"SELF ANOINTING" AND REPRODUCTIVE BEHAVIOUR. Like their European relatives, New Zealand hedgehogs occasionally froth at the mouth and plaster the spittle on their spines. This behaviour is normally elicited by hedgehogs of the opposite sex, but can be induced among young animals by removing them from their nests or by placing adult animals close to strong-smelling substances such as tobacco, glue, or soap. Of 929 wild hedgehogs examined in New Zealand, 19 bore signs of recent "self-anointing". The spittle thrown on to the back has a pungent smell and is thought to be a signalling device advertising lost nestlings to their mothers or, among adults, to play some part in mating behaviour (Brockie, 1976). The more usual courting behaviour is "cartwheeling", when a male circles a female for long periods and attempts to bite her feet. Sometimes the pair attract the attention of another male, which may join in.

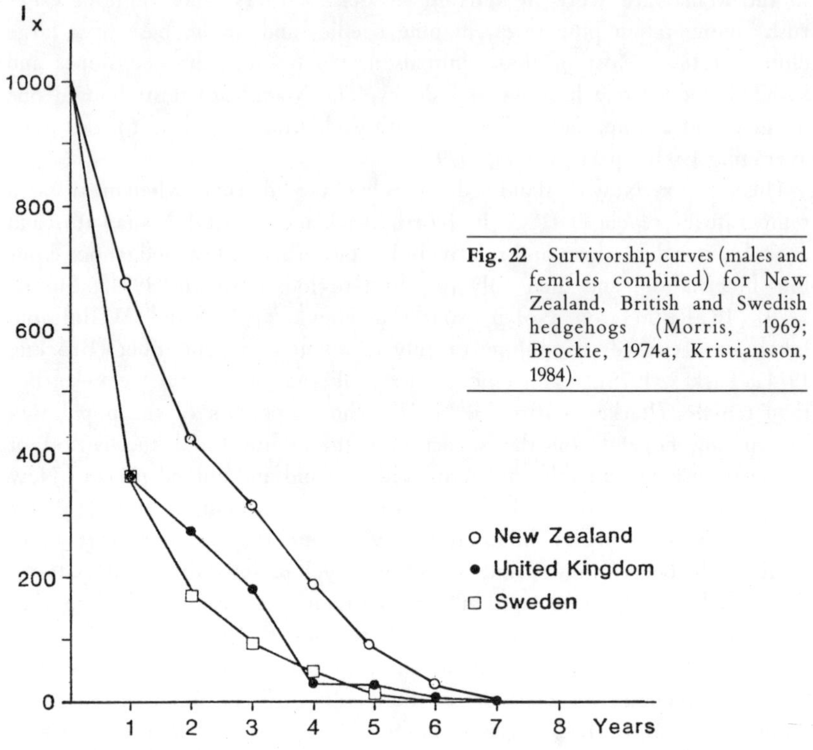

Fig. 22 Survivorship curves (males and females combined) for New Zealand, British and Swedish hedgehogs (Morris, 1969; Brockie, 1974a; Kristiansson, 1984).

REPRODUCTION AND DEVELOPMENT

The reproductive organs of both sexes regress during hibernation, but enlarge again in early spring. Hedgehogs collected in September have large accessory glands in males, and sperm in the vaginas of some females, which suggests that breeding starts as soon as the animals emerge from hibernation. However, none of 15 females dissected in October showed any sign of pregnancy. Hedgehogs near Wellington mate from September onwards, but few pregnant females are found before December (Brockie, 1958). The gestation period is 31-35 days (Herter, 1938). There is no postpartum oestrus (Deanesly, 1934).

The breeding season is prolonged. First litters in the southern North Island are born in late November and December, and second litters through to February. Young may be born as late as May near Wellington, and females in advanced pregnancy have been found at Kaitaia during August. Data collected by Wodzicki (1950) from 85 stock inspectors throughout New Zealand show that the breeding season is longer in northern parts of the country than in the south. There is an almost complete lack of reliable data on the reproduction of hedgehogs in the South Island.

North Island hedgehogs carry 4–7 embryos (mean 4.4), but juvenile mortality is high, as nests contain an average of only 2.7 young. The young are blind at birth and weigh 15-21 g. They are born without spines, but a first coat of white spines grows within a day, through which a second

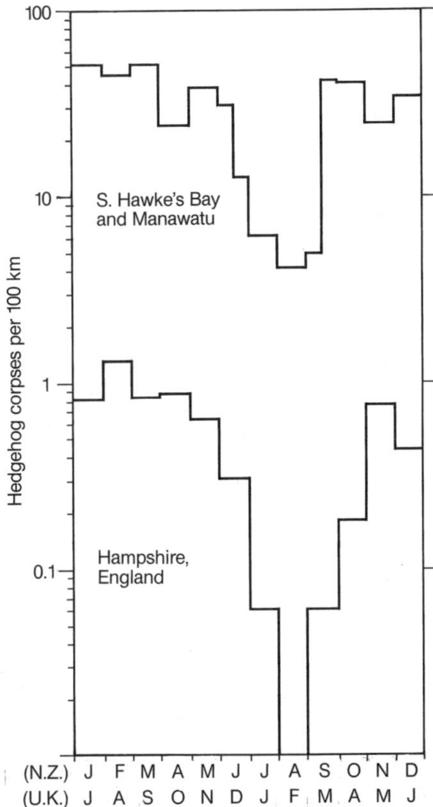

Fig. 23 Comparison of numbers of hedgehogs killed monthly on roads in southern North Island, New Zealand, 1950-1958, and Hampshire, England (after Davies, 1957; Brockie, 1960).

set of darker spines grows within 36 hours. The youngsters can roll up after 11 days; their eyes open at 14 days, and they remain with their mothers for 6-7 weeks. Their deciduous teeth are replaced at about 1-2 months.

Population Dynamics

Density. Very large numbers of hedgehogs may visit small areas over a period of time. For example, Parkes (1975) marked 150 individuals on 16 ha of dairy farm in the Manawatu over 17 months; Brockie (1974a) marked 207 on a 56-ha golf course at Lower Hutt over 2 years. At any one time, however, the population density of the resident animals on the farm never rose above 1.1-2.5/ha, and on the golf course, 1.75/ha. In southern Sweden, Kristiansson (1984) captured up to 80 hedgehogs a year in the 50-ha village of Revingeby.

Survival. The age of a hedgehog can be determined by counting the microscopic growth layers in the teeth and jaws, and, with less accuracy, by examining the wear on the teeth. The oldest hedgehog so far found in New Zealand was 7 years old. The average lifespan of hedgehogs on Manawatu farmland was 1.97 years (Parkes, 1975) and in suburban Hutt Valley, 2.67 years (Brockie, 1974a). Survivorship curves for New Zealand, British and Swedish hedgehogs are shown in Fig. 22.

Table 22: Body measurements of hedgehogs in New Zealand.

	Body weight (g)	Head and body length (mm)		Hind foot length (mm)		Skull (mm) Condylobasal length		Skull (mm) Zygomatic width	
	Both sexes	M	F	M	F	M	F	M	F
Range	300–1300	150–266	190–255	35–44	33–43	51.7–58.0	51.8–58.4	31.0–37.0	30.1–36.1
Mean	684.6	226.0	227.8	39.8	38.7	55.4	54.4	34.0	33.6
Standard deviation	156.7	31.7	20.3	2.6	2.1	1.6	2.1	1.3	1.1
Sample size	177	39	32	44	49	36	25	36	23

Most body measurements are from Wellington Province, and most of the skull measurements are from Christchurch; only animals with fully erupted teeth are included. Data from Brockie (1974a).

Table 23: Parasites and diseases of hedgehogs in New Zealand *(continued overleaf)*.

Name	Prevalence in New Zealand	Notes	Reference
Siphonaptera			
Nosopsyllus fasciatus (rat flea)	3% of Wellington animals	Normal host is *Rattus rattus*.	Brockie (1958)
Ctenocephalides felis (cat flea)	1% of Wellington animals	Normal host is the domestic cat.	Brockie (1958)
Leptopsylla segnis (mouse flea)	Rare: one record, in Orongorongo Valley.	Normal host is *Mus musculus*.	B. M. Fitzerald (unpubl.)
Acarina			
Caparinia tripilis (hedgehog scab mite)	Very common	Scabs may blind, disable, or kill many hedgehogs, particularly males. Spreads hedgehog ringworm.	Brockie (1974b); Smith & Marples (1963)

Notoedres muris	Found on Auckland hedgehogs	Infected animals may lose their spines.	Heath, Rush-Munro & Rutherford (1971)
Hirstionyssus talpae	Isolated from hedgehog nest in Wellington	Blood sucking parasites.	Sweatman (1962)
Haemolaelaps megaventralis	Isolated from hedgehog nest in Wellington	Blood sucking parasites.	Sweatman (1962)
Androlaelaps casalis casalis	Isolated from hedgehog nest in Wellington	Blood sucking parasites.	Sweatman (1962)
Haemophysalis longicornis (cattle tick)	Isolated from hedgehog nest in Wellington	Blood sucking parasites.	Tenquist & Charlston (1981)
Nematoda			
Crenosoma striatum (tracheal worm)	13% of North Island hedgehogs	In trachea and bronchi. Contracted by eating slugs and snails.	Brockie (1958)
Aonchotheca erinacis (stomach worm)	63% of Wellington animals	Found at Rotorua, Kaitaia, Palmerston North. May cause stomach lesions.	Brockie (1958) Marjeed & Cooper (1984)
Fungi and Yeasts			
Trichophyton mentagrophytes var. *erinacei* (hedgehog ringworm)	Very common, e.g. 47% of Dunedin animals	Produces dry scaly skin, occasional baldness in hedgehogs. Accounts for <5% of human ringworm infections. Produces penicillin.	Marples & Smith (1960); Smith, Rush-Munro & McCarthy (1969); Morris & English (1973); Smith (1964)
Trichophyton terrestre	14% of Dunedin animals	A resident on hedgehog skin, non-pathogenic.	Marples & Smith (1962)
Candida albicans	87% of Otago hedgehogs	Isolated from intestinal tract and droppings. An ideal host for this human pathogen.	Smith (1964)
C. krusei *C. guilliermondii*	Occasionally in Otago hedgehogs	—	Smith (1964)

Table 23: *(Continued).*

Name	Prevalence in New Zealand	Notes	Reference
Bacteria			
Salmonella typhimurium	39% of Hamilton hedgehogs	Reservoirs of salmonellosis for farm or domestic stock.	Smith & Robinson (1964)
Staphylococcus aureus	85% of Otago hedgehogs	Grown from various sites including skin, paws, anus, and nostrils. Also isolated from hedgehog scab mites. The hedgehog in New Zealand is a potential reservoir of penicillin-resistant *Staphylococcus aureus* ("H-bug").	Smith (1964)
Proteus mirabilis	53% of Otago hedgehogs	Isolated from gut.	Smith (1964)
P. vulgaris	20% of Otago hedgehogs		
Morganella morganii	45% of Otago hedgehogs		
Providencia rettgeri	3% of Otago hedgehogs		
Escherichia coli	Large proportion of Dunedin hedgehogs	Hedgehogs are healthy carriers of some strains pathogenic to man.	Smith (1964)
Leptospira ballum	Isolated from 6% of North Island animals	Shares this infection with house mice, Norway rats and, rarely, man.	Brockie & Till (1977) Hathaway, Blackmore & Marshall (1981)
Mycobacterium sp.	One of 155 North Island animals examined	Healed scars in lung tissue.	Brockie (1958)

Data updated from Smith (1968).

Hedgehog populations include both resident animals, which may live for several years in the same area, and transients. The transients are mainly young adults, especially young males, seen and marked once in an area but never seen again. They account for the continually changing composition of the population. On the Lower Hutt golf course, transients accounted for 22% of the population in spring, falling to 8% in winter (Brockie, 1974a).

ROAD DEATHS. Hedgehogs are frequent road casualties in New Zealand. Of 570 mammals and birds found dead in the 1950s on the roads between Waikanae and Levin and between Woodville and Takapau, 482 were hedgehogs (compared with 36 possums, 13 rabbits, 12 hares and 4 rats). In October 1951, 61 hedgehogs/100 km were counted between Taihape and Wellington and 41/100 km along the Woodville-Takapau road in February 1956 (Brockie, 1960).

The casualty rates have been lower in recent years in the North Island, and perhaps have always been low in the South Island. Between 1983 and 1987, 65 drives between Woodville and Takapau in all months of the year revealed an average of 11 hedgehogs/100 km of highway — a significant fall from the mean of 25/100 km in the 1950s. A drive from Picton via Reefton to Fox Glacier in February 1985 revealed only four corpses, or 0.003/100 km (R. M. F. Sadleir & J. E. C. Flux, unpubl.). The number of hedgehogs killed on the road also varies markedly with locality and season. They are most frequently run over near bridges across drains, streams and rivers, or in the suburbs of cities and towns. On a drive from Wellington to Kaitaia and back in February 1984 (1406 km) corpses were seen along the length of the main highways except on the Volcanic Plateau (central North Island) where one or two were seen only near the townships of Waiouru, Turangi and Taupo (total 10.8 hedgehogs/100 km). In lowland areas of the southern North Island, numbers fall from May onward; fewest corpses are counted in July, August and the first half of September (0.1-6.0 corpses/100 km), when most of the animals are hibernating (Fig. 23).

PREDATORS, PARASITES AND DISEASES

Adult hedgehogs are well protected against most predators except wild pigs, but wekas have been seen taking nestlings (B. M. Fitzgerald, unpubl.). Stoats and harrier hawks scavenge dead hedgehogs (Carrol, 1968; King & Moody, 1982). Parasites and infections carried by hedgehogs in Europe and New Zealand are listed in Table 23.

ADAPTATION TO NEW ZEALAND CONDITIONS

New Zealand's relatively short mild winters are kind to hedgehogs, especially to young ones during their first winter (Brockie, 1974a). Relatively few New Zealand hedgehogs die while hibernating, their breeding season is longer, and there are fewer nocturnal insectivores to compete for food. It is not surprising, therefore, that even though litter size and nestling mortality are about the same as in Britain (4.6 embryos per litter, 20% mortality before weaning: Morris, 1961, 1977), lowland parts of New Zealand support denser populations of hedgehogs than comparable places in Europe.

Although many factors contribute to the number of hedgehog corpses on the road, including the volume of traffic and heavy rainfall, consistent huge differences in road casualties between areas presumably reflect differences in the densities of their hedgehog populations. In Hampshire, England, during the 1950s, Davies (1957) counted up to 1.3 hedgehog corpses/100 km of road; Holisova and Obrtel (1986) counted up to 0.4 hedgehog corpses/100 km along roads in Moravia, Czechoslovakia; and counts in Scandinavia, France and Switzerland reveal similar numbers. In lowland areas of the North Island, hedgehog road casualties during the 1950s were consistently at least 20 times more numerous than this (Fig. 23), in some areas up to 60 times more numerous, and are still usually at least 10 times as high.

On average, New Zealand hedgehog skulls are between 2% and 3% shorter than those of British animals, and they are also lighter in body weight. The mean weight of 277 New Zealand animals was 685 g, and the heaviest individual weighed 1300 g. The mean weight of British hedgehogs is 733 g, and weights of >1500 g are not uncommon (Morris, 1984). Before hibernation, Swedish hedgehogs reach a mean weight of 1568 g (Kristiansson, 1984). Herter (1938) found a German hedgehog weighing 1800 g. On the other hand, in the milder climate of New Zealand, hedgehogs lose less of their body weight while hibernating (10%), compared with those in Britain (25%: Morris, 1964) or Sweden (40%: Kristoffersson & Suomalainen, 1964).

Earthworms (probably mainly *Allolobophora caliginosa*) contribute little to the diet of New Zealand hedgehogs, but are frequently eaten by British animals. Wroot (1985) argues that British hedgehogs do not feed casually on anything they find, but select food to maximize the energy they take in while minimizing the expenditure of energy. He showed that hedgehogs passed up readily available beetles in order to feed on more nutritious earthworms. He also showed that earthworms were eaten in direct proportion to their availability, which reflected the amount of rainfall. British hedgehogs can be serious predators of colonial nesting seabirds (Axel, 1956; Kruuk 1964), but that problem has not been recorded in New Zealand.

Of 77 New Zealand hedgehogs examined, 51% showed some dental abnormality, usually the absence or faulty eruption of lower incisors or premolars. It seems that a heritable dental condition, present in about 16% of British hedgehogs but absent in continental Europe, was transmitted to New Zealand in the colonizing stock (Brockie, 1964).

New Zealand hedgehogs suffer much more seriously from the effects of mange mites than their European relatives. Although *Caparinia tripilis* has been recorded as killing a few animals in Germany, deaths from these parasites do not reach nearly the same proportions in Europe as they do in New Zealand. On the other hand, the New Zealand animals are entirely free of fleas. European hedgehogs are notorious flea-bags; nearly every specimen carries very large numbers of the specific hedgehog flea, *Archaeopsylla erinacei*, but this flea was apparently lost on the long voyage to New Zealand last century, as it has never been seen here.

SIGNIFICANCE TO THE NEW ZEALAND ENVIRONMENT

The main economic significance of hedgehogs in New Zealand is their capacity for carrying and spreading human and stock infections. Hedgehogs are a potential reservoir of penicillin-resistant *Staphylococcus aureus* ("H-bug") and of potentially pathogenic strains of *Escherichia coli*, the yeast *Candida albicans*, various ringworm fungi (dermatophytes), and *Leptospira ballum*; and they would be potential carriers of foot and mouth disease if it ever reached New Zealand. The transmission of hedgehog ringworm to human contacts has been verified, but so far there have been no proven cases of infection from the other diseases. Hedgehogs also carry *Salmonella typhimurium*, which may infect farm stock and pets, and bovine tuberculosis. However, the prevalence of TB in wild hedgehogs is well below 1%; possums are a much more serious barrier to eradication of TB in New Zealand (p. 98).

R. E. B.

Order Chiroptera

Bats are the only mammals capable of sustained flight. They are easily distinguished from all other mammals by the extension of the finger bones to support the wing membranes (the name Chiroptera means "hand-wing"). Although basically similar in structure to other mammals, bats have evolved many other structural and physiological modifications, in addition to wings, to adapt them to their unique way of life. Some of their most unusual features are the specialized ears and elaborate facial skin growths of the many bats which have evolved echolocation. The size range of bats is enormous; from the very small bumble-bee bat (*Craseonycteris thonglongyai*), which weighs only 1.7 g, to the giant flying fox (*Pteropus giganteus*), which weighs over 1.2 kg and has a wingspan of 1.7 m.

Little is known about the origin and evolution of bats. The earliest known fossils, from the early Eocene (55 million years ago), are in some characters as advanced as many living species of the suborder Microchiroptera. Therefore the earliest bats must have arisen long before this, possibly in the mid to late Cretaceous (70–100 million years ago). Their immediate ancestors might have been small arboreal insectivores possessing gliding membranes and a primitive form of echolocation (Hill & Smith, 1984).

There are 951 living species of bats in the world, divided into two suborders. The Megachiroptera contains only one family, the Pteropodidae, found in the Old World tropics from Africa to the Pacific. These include the flying foxes, and all members of this family feed on fruit, pollen and/or nectar. They have large eyes and, except for one genus (*Rousettus*), navigate by vision rather than echolocation. The Microchiroptera contains 18 modern families, found almost worldwide. Whereas the majority feed on insects, others feed on a range of foods from fruit, nectar and pollen, to blood, birds, small mammals and even fish. Most species of Microchiroptera have small eyes and all are capable of navigating by echolocation.

The New Zealand bat fauna is unlike that of any other country, because only three species are present, all endemic; because they, with the seals, are the only native land-breeding mammals; and because the Mystacinidae, the longest-resident bats, have evolved in isolation for so long, and possess characters not found in bats elsewhere in the world.

The origin and phylogenetic affinities of the endemic Mystacinidae (short-tailed bats) have been a puzzle since last century. They are the most terrestrial of all bats and have a unique assemblage of morphological characters which has led to taxonomic confusion since 1843. At one time or another, *Mystacina* has been placed in 7 of the 18 extant families of microchiropteran bats and in 3 of the 4 microchiropteran superfamilies. Recent immunological evidence has now unequivocally placed *Mystacina* as an early offshoot of the superfamily Phyllostomoidea, with close relationships only to the South and Central American bat families Noctilionidae, Mormoopidae and Phyllostomidae (Pierson *et al.,* 1982, 1986). The separation of the *Mystacina* and *Noctilio* lineages is estimated to date from about 35 million years ago, in the early Oligocene (Pierson *et al.,* 1986).

The ancestors of the long-tailed bats are believed to have reached New Zealand from Australia by chance dispersal across the Tasman during the Pleistocene (Daniel, 1979), as have those of many of our birds. It is surprising that not more Australian species (there are over 60 species of bats there) have been blown to New Zealand in the past. At least three species have reached Lord Howe Island, 500 km from Australia, and two have reached Norfolk Island, about 1300 km from Australia (Daniel & Williams, 1984); but, aside from *Chalinolobus tuberculatus,* there is only one isolated record of a single vagrant flying fox reaching New Zealand from Australia (p. 136).

The distinguishing characters of the three species of native New Zealand bats are summarized in Table 24. All three are, like other members of the endemic vertebrate and invertebrate fauna of New Zealand, protected by the Wildlife Act 1953. Live bats must not be handled or disturbed in any way. When hibernating bats are disturbed, death from starvation usually follows. If a dead bat is found, it is against the law to keep it as a curio. It must instead be sent for scientific study, either to one of the major museums or to the nearest office of the Department of Conservation, together with adequate information on date and locality where found.

Two exotic species of bats have arrived dead in New Zealand as stowaways in ships' cargo. A Japanese pipistrelle (*Pipistrellus javanicus abramus*) arrived in a cargo of car parts from Japan (Daniel & Yoshiyuki, 1982), and an Australian lesser long-eared bat (*Nyctophilus geoffroyi*) arrived in a cargo of Australian timber (Daniel & Williams, 1984).

Suborder Microchiroptera
Family Vespertilionidae

The Vespertilionidae (evening bats) is a family of about 319 species in 42 genera, the largest of the 18 extant families of microchiropteran bats. Vespertilionid bats are among the most widely dispersed of mammals, found almost worldwide (except for the polar and near polar regions), including remote oceanic islands.

Genus *Chalinolobus*

There are six species of Australasian wattled or lobe-lipped bats: five in Australia (including Tasmania and Norfolk Island) plus New Caledonia and New Guinea, and one species, *Chalinolobus tuberculatus*, restricted to New Zealand (Dwyer, 1960a; Hall & Richards, 1979).

The Australasian wattled bats are small to medium sized bats whose lower ear margin terminates in a fleshy lobe. There is often an additional lobe in the corner of the mouth. The ears are short and broad and the tragus

Table 24: Distinguishing marks of bats in New Zealand.

<table>
<tr><th></th><th>Long-tailed bat
Chalinolobus tuberculatus</th><th>Lesser short-tailed bat
Mystacina tuberculata</th><th>Greater short-tailed bat
Mystacina robusta</th></tr>
<tr><td>Flying activity starts</td><td>Before or just after sunset</td><td>After dark</td><td>After dark</td></tr>
<tr><td>Roosts</td><td>Native and exotic trees, caves</td><td>Native trees and caves</td><td>Native trees, caves, and seabird burrows</td></tr>
<tr><td>Tail length
Tail position</td><td>Almost as long as head and body. Included in large ‘v’-shaped interfemoral membrane along most of length</td><td colspan="2">Very short, free, penetrates interfemoral membrane on dorsal surface</td></tr>
<tr><td>Fur</td><td>Glossy black to dark brown, long and silky, no guard hairs</td><td colspan="2">Dark brown flecked with white, short and velvety, guard hairs over underfur</td></tr>
<tr><td>Jaws</td><td>Fleshy lip-lobule at corner of mouth</td><td colspan="2">No lip-lobule</td></tr>
<tr><td>Hind feet</td><td>Small, delicate</td><td colspan="2">Large, robust</td></tr>
<tr><td>Claws</td><td>Without spurs on toes and thumbs</td><td colspan="2">With spurs</td></tr>
<tr><td>Ears</td><td>Small, broad, and rounded</td><td>Large, pointed, extend to or beyond muzzle when laid forward</td><td>Large, pointed, do not reach muzzle when laid forward</td></tr>
<tr><td>Nostrils</td><td>Small</td><td>Prominent and narrow</td><td>Short and broad</td></tr>
<tr><td>Forearm length</td><td>37.4–44.4 mm</td><td>40.5–45.4 mm</td><td>45.3–48.0 mm</td></tr>
</table>

Fig. 24 Long-tailed bat, Rotorua (NZWS).

is pointed inwards. The bats have a high forehead, and the muzzle has well-developed elevations or a glandular appearance (Hall & Richards, 1979).

9. New Zealand long-tailed bat

Chalinolobus tuberculatus (Gray, 1843)

Synonyms *Vespertilio tuberculatus* Gray, 1843; *Vespertilio tuberculatus* Forster, 1844; *Scotophilus tuberculatus* Tomes, 1857; *Chalinolobus tuberculatus* Peters, 1866; *Chalinolobus morio* Thomas, 1889.

Also called New Zealand short-eared bat (English); pekapeka (Maori).

Description (Fig. 24)

Distinguishing marks, Table 24 and skull key, p. 24.

A small, delicately proportioned bat, in contrast to the robust and "chunky" *Mystacina.* The upper parts are glossy black to dark brown and the under parts grey-black, except for the yellowish-brown pubic region. The limbs and membranes are almost naked. The fine and silky dorsal hair is up to 7 mm long, with no differentiation into overhair and underhair (Dwyer, 1960a). The hair of preserved museum specimens is often bleached to a light reddish brown. The wing and tail (interfemoral) membranes are grey-black.

The head is short, broad and moderately hairy, with the nostrils set on low prominences. The eyes are very small and almost covered by hair and by the well-formed fleshy lids. The small, broad ears and short, broad tragus are rounded distally, and the outer margin of the ear continues along the face, beneath the eye, as an antitragus which terminates just behind the lip-lobule (Fig. 25). The more pronounced tragus extends from within the ear above the antitragus. It is narrow at the base, but widens and is rounded distally (Dwyer, 1960a).

The bones of the forelimb are, with the exception of the thumb, very long and slender. The small thumb projects free from the wrist and carries a long curved claw. In contrast to *Mystacina,* there is no bony phalanx

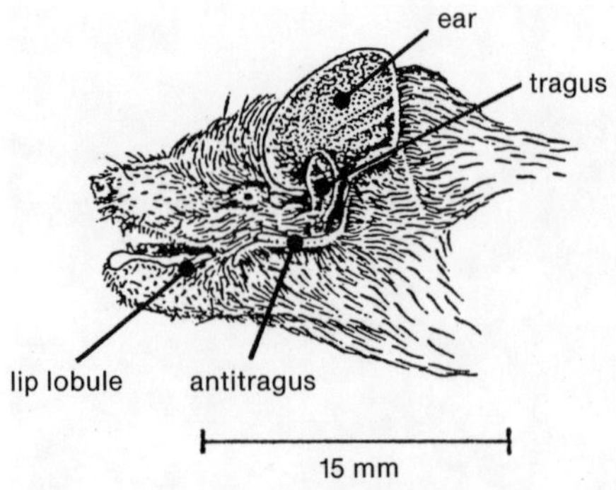

Fig. 25 Head of long-tailed bat, showing shape of ear, tragus and lip lobule (after Dwyer, 1960a).

Table 25: North-south variation in forearm length in the New Zealand long-tailed bat.

	Latitude	Mean (mm)	SE	Range	n
North of Auckland	35-37°S	39.00	± 0.347	37.4–40.6	9
Auckland to Oamaru	37-45°S	39.98	± 0.192	38.0–41.5	29
Oamaru to Bluff	45-47°S	42.73	± 0.86	41.5–44.4	3

Data from M. J. Daniel (unpubl.).

associated with the second metacarpal, and only two are present in each of the remaining digits. The short, pendant penis of the male makes for easy distinction of the sexes. The leg bones are long and slender, and the very small, delicate hind foot is turned outwards. The calcar extends from the heel as a strong process and supports almost half of the posterior border of the large interfemoral membrane. A small, rounded post-calcareal lobe is present near the base of the foot.

Dental formula $I^2/_3$ $C^1/_1$ $Pm^2/_2$ $M^3/_3$ = 34.

FIELD SIGN

Bats leave little sign of their presence or activity, and, because they fly only at dusk and after dark, are rarely encountered. However, in areas where bats have been seen flying regularly, a diligent search can sometimes reveal a hollow tree with accumulations of bat guano, either inside the tree or spilling out on to the forest floor. Individual droppings of both long-tailed and short-tailed bats are small and mouse-like, ranging in size from about 4 × 1 to 5 × 2 mm. Rain or urine-soaked guano smells strongly of ammonia and often dissolves into wet, brown peat-like material full of insect fragments.

Roosts of long-tailed bats have been found inside or under the bark of a surprisingly wide range of native and exotic tree species, as well as in caves and buildings (Dwyer, 1962; Daniel, 1981; Daniel & Williams, 1981, 1983, 1984). In bat caves, often the only positive clue in daytime is one or more piles of guano under a crack or fissure occupied by sleeping bats.

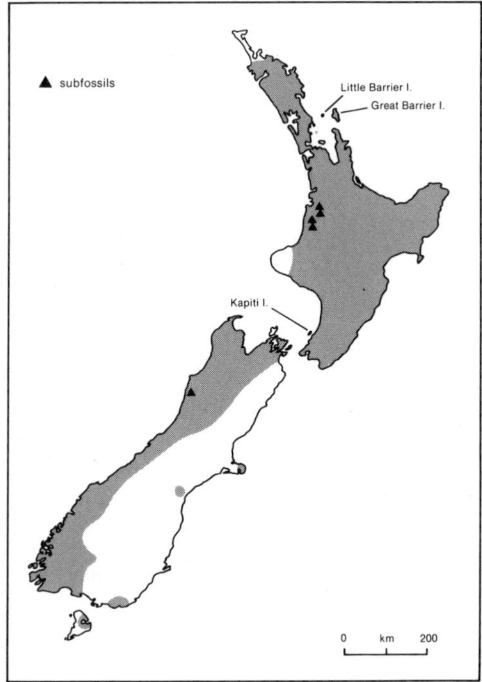

Map 9 Distribution since 1961 of long-tailed bats and unidentified bats which were probably long-tails (Daniel & Williams, 1984); solid triangles, locations of known subfossils.

MEASUREMENTS

Head and body length 45–50 mm; forearm 37.4–44.4 mm; wingspan 250–280 mm; tail 40–43 mm; ear 8–14 mm; condylobasal length 13.6–14.3 mm; weight of adults 7–10 g; weight of juveniles 3–6 g (Dwyer, 1960a,b, 1962; Daniel, 1981 and unpubl.).

VARIATION

Insufficient numbers of adults of both sexes are available to determine whether there is distinct sexual or local variation in body size, but from what can be seen, there could be a north-south cline of increasing body size as measured by forearm length (Table 25). Measurements of additional specimens from both ends of the cline, particularly from Northland, Fiordland and Stewart I., are needed. Dwyer (1960b) noted considerable variation in colour of preserved museum specimens, but this may be just the effect of museum preservatives.

HISTORY OF COLONIZATION

Chalinolobus tuberculatus is probably derived from *C. picatus,* or possibly *C. gouldii,* or a common ancestor with these, windblown across the Tasman Sea some time during the Pleistocene (Daniel, 1979). It has probably evolved in isolation in New Zealand for at least a million years.

DISTRIBUTION (Map 9)

The long-tailed bat is found only in New Zealand, where it is the only common species of bat. It is widely distributed from the north of the North Island (35°S), through the South Island, to Halfmoon Bay on Stewart I.

(47°S). It is also present on Great Barrier I., Little Barrier I. and Kapiti I. (Dwyer, 1960a, 1962; Daniel, 1970; Daniel & Williams, 1981, 1984).

Surprisingly few subfossil or recent skeletons of the long-tailed bat have been reported, and all are from localities within the present range (Cody, 1981; Daniel & Williams, 1983; M. J. Daniel, unpubl.). The only South Island specimen was found completely enclosed in calcite inside a stalagmite in Metro Nile River cave. A total of eight North Island specimens is known from (with numbers in parentheses) Grand Canyon cave (3), Waipuna cave (1), Gardner's Gut cave (3) and Porthole cave (1).

HABITAT

The long-tailed bat is, or was, found in most habitats from sea level to the limit of tree growth in the mountain ranges (*c.* 1000 m asl), including farmland, near water, in limestone caves and outcrops, and in indigenous and exotic pine forests and farm shelter belts of eucalypts and other exotic species (Dwyer, 1962; Daniel, 1970, 1981; Daniel & Williams, 1983, 1984). It invariably roosts near the forest edge and feeds along forest margins, over farmland, streams and lakes; it is also occasionally reported roosting in houses, farm buildings and mountain huts.

FOOD

The long-tailed bat is apparently solely an aerial insectivore, feeding mainly on small moths, midges and mosquitoes (Dwyer, 1962; Daniel & Williams, 1983). Colenso (1890) noted that small flies were readily accepted by a captive bat, and Cheeseman (1894) observed a long-tailed bat catching insects on the wing. Roach & Turbott (1953) recorded that a captive specimen ate mealworms and liver fragments and even a praying mantis (*Orthodera ministralis*), which suggests that medium-sized or even large prey items might also be eaten in the wild. Hand-thrown grass-grub beetles (*Costelytra zealandica*) have been taken by free-flying *Chalinolobus* in the wild (Miss P. Lewis, in Daniel & Williams, 1983). The foods of four Australian species of *Chalinolobus* (*gouldii, picatus, nigrogriseus* and *morio*) consisted mainly of Lepidoptera, but several species of Hymenoptera, Diptera, Coleoptera, Hemiptera and Orthoptera were also eaten (Vestjens & Hall, 1977). Arthropods belonging to some or all these orders may also be consumed by *C. tuberculatus* in New Zealand.

The feeding area used by a colony, and the distance flown by the bats from the roost to feed each night, are practically unknown. Dwyer (1962) records bats feeding at distances up to 1.2 km from the nearest forest edge, and other records of up to 2–3 km have been reported (M. J. Daniel, unpubl.). New techniques, such as the use of micro-radio transmitters (weighing 0.5 g), will provide valuable information on the feeding ranges and seasonal movements (if any) of this species.

SOCIAL ORGANIZATION AND BEHAVIOUR

ACTIVITY. The long-tailed bat emerges from its roost about 10–30 minutes after sunset, while it is still light, and may be observed flying intermittently through the night until just after dawn. Flight is relatively fast and twisty, and often low to the ground or to water, but it also flies above tree canopy height, often along a regular beat (Dwyer, 1962; Daniel & Williams, 1984). It may be mistaken for other species such as fantails (*Rhipidura fuliginosa*), swallows (*Hirundo tahitica*), and even puriri moths (*Aenetus virescens*) in fading light (Daniel & Williams, 1984). The frequency of ultrasound with maximum energy used when "cruising" is 40–45 kHz, in contrast to 60–65 kHz in *Mystacina* (Daniel & Williams, 1984). It is normally first observed on fine evenings in spring and after dusk through summer to autumn.

HIBERNATION. The long-tailed bat hibernates for up to 4–5 months over late autumn and winter in colder parts of New Zealand, but has been observed flying in winter in warmer regions (Dwyer, 1962; Daniel & Williams, 1984). Sites used for hibernation include hollow trees, under tree bark, caves, and less commonly houses and farm buildings (Daniel & Williams, 1981, 1983, 1984). Roosting posture is invariably head down, hanging from the claws of one or both hind feet.

DISPERSION. The long-tailed bat is, like many bats, colonial; often large numbers roost and breed within a small area. Last century several bat-tree-roosts cut down contained "hundreds" or "thousands" of bats (Buller, 1893; Cheeseman, 1894). Few large roosts have been reported within the last 30 years; the largest, in Grand Canyon cave, had an estimated 200–300 bats (Daniel & Williams, 1983). The majority of roosts found have contained from 1 to about 50 bats, on average only about 10 individuals. Few colonies have been studied in detail. One summer roost, in a *Pinus radiata* tree, contained only pregnant females and was obviously a nursery colony. Adult males probably roost in separate sites at this time (Daniel, 1981).

REPRODUCTION AND DEVELOPMENT

Little is known about the breeding biology of the long-tailed bat. The time of mating is not known, but is probably in autumn before hibernation, after which sperm is probably stored in the oviducts and uterine glands of females until ovulation and fertilization in spring, as in *C. morio* and *C. gouldii* (Kitchener, 1975; Kitchener & Coster, 1981). Daniel and Williams (1983) reported that *C. tuberculatus* is monoestrous and females give birth to only one young per year. However, four Australian species of *Chalinolobus (gouldii, dwyeri, morio* and *picatus)* regularly produce twins (McKean & Hamilton-Smith, 1967; Kitchener, 1975; Young, 1979); a larger sample of *C. tuberculatus* may show that twins are sometimes produced here as well. Parturition, at least in parts of the North Island, is in summer from December to February, but may be later in the south, as in *C. gouldii* in Australia (Kitchener, 1975; Daniel & Williams, 1983).

Dwyer (1960b) described the external anatomy of a 12-mm foetus. Juvenile development has not been studied, but in many other temperate vespertilionid bats early growth is remarkably rapid; juveniles fly at 4–6 weeks and almost reach adult size in about 6–8 weeks (Stebbings, 1977).

POPULATION DYNAMICS

Nothing is known about the sex ratio at birth, juvenile and adult mortality or adult longevity of long-tailed bats.

PREDATORS, PARASITES AND DISEASES

Morepork owls *(Ninox novaeseelandiae)* and feral and domestic cats are known to kill New Zealand bats (Dwyer, 1962; Daniel & Williams, 1984). Of 36 freshly dead *Chalinolobus* studied, 10 had been killed and partly eaten by feral or domestic cats (Daniel & Williams, 1984). One particular domestic cat near Geraldine killed three *Chalinolobus* (two adult females and one juvenile) under an outside house light, where the bats were apparently feeding on moths attracted to the light (Daniel & Williams, 1981). Dormant bats in accessible roosts are probably killed by ship rats, Norway rats and stoats, but the consequences for the bat population are unknown.

The long-tailed bat flea *(Porribius pacificus)* described by Jordan (1947) is common on *C. tuberculatus* (Dwyer, 1962) and appears to be specific to this host.

Several species of mites have been observed on the long-tailed bat. Manson (1972) described one new species as *Ornithonyssus spinosa;* Dwyer (1962) reported a large species from a juvenile bat as probably belonging to the family Trombiculidae; and a new species of spinturnicid mite (*Spinturnix* sp.) has been found on a long-tailed bat from Fiordland (A. C. G. Heath, unpubl.). A small sample of *C. tuberculatus* eyelids were all negative for *Demodex* (C. Desch, unpubl.).

A new species of cestode, *Hymenolepis chalinolobi,* was described from a long-tailed bat by Andrews and Daniel (1974).

SIGNIFICANCE TO THE NEW ZEALAND ENVIRONMENT

The long-tailed bat is the only one of the three endemic species of bats which is still common. It has special significance to observers of New Zealand mammals, because it is the only species of bat likely to be seen by the general public.

M. J. D.

Family Mystacinidae

The small endemic family of New Zealand short-tailed bats contains only one genus with two species; until recently only one species was described (Hill & Daniel, 1985).

Genus *Mystacina*

The two species of New Zealand short-tailed bats are small to medium sized bats with a range of unusual characters, some unique in the order Chiroptera. The fur is short, velvet-like, dark brown speckled with white, and separated into overhair and underhair. The elongate head tapers forward to prominent nostrils; the simple, pointed ears have a long, narrow tragus; the second digit of the wing is reduced to a metacarpal and a single minute phalanx. The wing membranes near the body are remarkably thick and leathery. When the wing is tightly folded for terrestrial activity, the delicate proximal portions of the wings are protected within a pouch formed between the thickened wing membrane and a cutaneous flap extending from the side of the body. The short tail penetrates the uropatagium on its dorsal surface; the legs are short and the feet stout, short and broad; the claws on the toes and thumbs have unique spurs (illustrated in Daniel & Baker, 1986). *Mystacina* has sturdy, chisel-like incisors and a partially extensible, papillated tongue. The total number of teeth is reduced, by loss of incisors and premolars, to 28.

10. Lesser short-tailed bat

Mystacina tuberculata Gray, 1843

Synonyms *Mystacina tuberculata* Gray, 1843; *Mystacina velutina* Hutton, 1872; *Mystacina tuberculata* Dobson, 1875; *Mystacops tuberculatus* Lydekker, 1891; *Mystacops velutinus* Thomas, 1905; *Mystacops tuberculatus* Miller, 1907; *Mystacina tuberculata* Simpson, 1945.

Also called northern short-tailed bat, New Zealand long-eared bat (English); pekapeka (Maori).

DESCRIPTION (Fig. 26)

Distinguishing marks, Table 24 and skull key, p. 24.

A small, robust bat, with characteristic stocky body shape and prominent pointed ears which extend to or beyond the muzzle when laid forward. It has longer, narrower nostrils and proportionately longer forearm and wing elements than *M. robusta.* The rostrum is lighter and less massive than in *M. robusta,* and the braincase rises more abruptly from the rostrum (Hill & Daniel, 1985).

Chromosomes 2n = 36 (Bickham, Daniel & Haiduk, 1980).

Dental formula $I^1/_1\ C^1/_1\ Pm^2/_2\ M^3/_3 = 28$.

FIELD SIGN

Mystacina tuberculata leaves little evidence of its presence in a forest, apart from the typical musty smell of bat guano in holes in hollow trees. Sometimes colonies of lesser short-tailed bats can be located from up to 100 m away, both in daylight and at night, by the high-pitched audible noises made by the bats inside their roost (M. J. Daniel, unpubl.). On Codfish I., and probably elsewhere, a hollow tree containing a colony of *M. tuberculata* can sometimes be identified as an active bat roost by a swarm of small black fanniid flies hovering round the entrance. On closer inspection, fanniid larvae can be seen in the guano below the roosting bats (M. J. Daniel, unpubl.). This fly is specific to roosts of the short-tailed bat, and is referred to as *Fannia* sp. 1 by Holloway (1984).

MEASUREMENTS

The range of measurements for the species are: total length 60–68 mm; forearm 40–45 mm; wingspan 280–290 mm; length of tail penetrating through tail membrane 6–8 mm (total tail length 10–12 mm); ear 17.5–19.1 mm;

Table 26: Subspecific variation in body measurements of the New Zealand lesser short-tailed bat.

	Range and mean (mm)		
	M. t. aupourica	*M. t. rhyacobia*	*M. t. tuberculata*
Latitude	35°- ?	37°30′?-39°30′?	40°30′-46°40′
Range	Northern North Island	Central North Island	South North Island, north South Island, and Codfish I.
Ear length	18.5-19.1 (18.7)	18.2-18.7 (18.4)	17.5-18.9 (18.1)
Forearm length	40.5-41.2 (40.7)	42.6-45.4 (44.0)	40.8-42.5 (41.7)
Greatest skull length	18.7-19.5 (19.0)	19.2-19.8 (19.5)	18.9-20.0 (19.4)
Condylobasal length	17.3-18.0 (17.9)	18.3-19.1 (18.7)	17.9-18.6 (18.3)
Basal length	15.3-16.2 (15.7)	16.1-16.9 (16.5)	16.1-17.0 (16.3)
Palatal length	8.4-8.9 (8.6)	8.6-9.3 (8.9)	8.5-9.4 (8.8)

Data from Hill & Daniel (1985).

condylobasal length 17.3–19.1 mm; weight 11–15 g (pregnant females to 18.5 g) (Dwyer, 1962; Daniel, 1979; Hill & Daniel, 1985).

Variation

There is considerable variation both in the overall body size of the lesser short-tailed bat and in the size, length and proportions of certain morphological characters. Three subspecies have been described (Hill & Daniel, 1985), showing a general increase in body size southwards, but this is not a simple cline (Table 26). The smallest race with the largest ears, *M. t. aupourica,* lives in the north of the North Island; the largest race, with the smallest ears, *M. t. rhyacobia,* lives not in the extreme south, as might be expected, but on the high volcanic plateau of the central North Island. This race is the largest, both externally and cranially, and has the largest forearm and wing elements. The most southerly subspecies, *M. t. tuberculata,* is intermediate in size between the other two.

History of Colonization

The ancestors of *Mystacina* probably reached New Zealand in the middle or late Tertiary (Daniel, 1976). The likely route was from South America through Antarctica, with over-sea dispersal to New Zealand (Koopman, in Daniel, 1979). This colonizing stock then evolved within New Zealand into a small, northern (less cold-tolerant?) species *(tuberculata),* and a larger, southern (more cold-adapted?) species *(robusta),* perhaps in the late Pliocene or early Pleistocene (Daniel, 1979; Hill & Daniel, 1985). The necessary isolating conditions were probably provided during the glacial periods, when bats surviving in the kauri, podocarp and *Nothofagus* forests of northern New Zealand were separated from those surviving in forest refugia in the South Island and southern North Island. During the warmer interglacials, both species would have dispersed with the re-establishment of lowland

Fig. 26 Lesser short-tailed bat, Little Barrier I. (NZWS).

forest. From the end of the last ice age to European settlement, the lesser short-tailed bat probably occupied all suitable lowland forest habitat on the three main islands, and on many of the offshore islands.

DISTRIBUTION (Map 10)
The lesser short-tailed bat presumably once occupied the whole mainland, but it has disappeared from large areas of its former range. The distribution of New Zealand bats to 1960 was mapped by Dwyer (1960a, 1962), and from 1961 to 1984 by Daniel & Williams (1984).

The kauri forest short-tailed bat, *M. t. aupourica,* is now found only in Omahuta kauri forest, Northland, but probably also lives in the adjacent Puketi kauri forest and in Waipoua kauri forest. The short-tailed bats on Little Barrier I. are probably of this subspecies. No *Mystacina* have been positively reported from the kauri forests of the Coromandel Peninsula since 1895.

The volcanic plateau short-tailed bat, *M. t. rhyacobia,* is found in several podocarp-hardwood and *Nothofagus* forests on and adjacent to the volcanic plateau. It ranges from Waitaanga State Forest in the west to Pureora Forest, Tongariro National Park, Kaimanawa Forest Park, Mamaku Plateau, Urewera National Park, Whirinaki State Forest, and East Cape. A possible report from the Raukumara Range has not been confirmed.

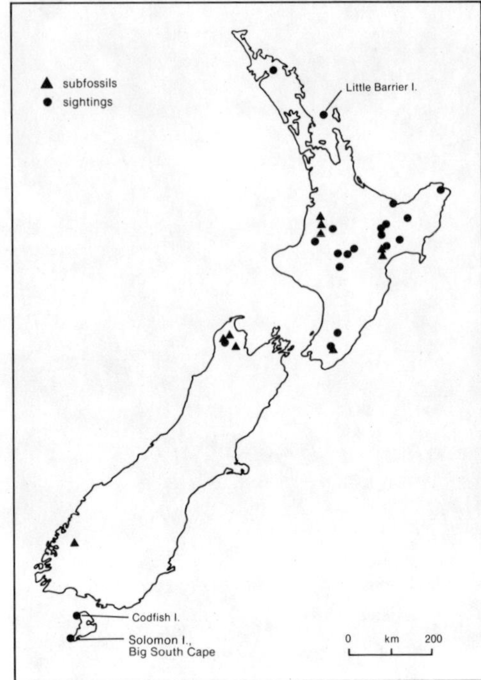

Map 10 Confirmed sightings since 1961 (Daniel & Williams, 1984) of lesser short-tailed bats in New Zealand (solid circles), and locations of known subfossils (solid triangles).

The southern short-tailed bat, *M. t. tuberculata,* is now found only in Tararua Forest Park in the North Island, in North-West Nelson Forest Park in the South Island, and on Codfish I. (Daniel & Williams, 1984; Hill & Daniel, 1985). *Mystacina tuberculata* was sympatric with *M. robusta* on Big South Cape and Solomon Is. until 1965. Neither species has been positively reported there since then. *Mystacina tuberculata* also lived on Jacky Lee I., north-east of Halfmoon Bay, Stewart I., in the 1930s and 1940s (Stead, 1936; Daniel & Williams, 1984), but has not been reported since then. There have been no confirmed records of this species from Stewart I. since European settlement (Daniel & Williams, 1984). Unconfirmed reports of bats seen at Puai and on Poutama I. in 1985 may have been lesser short-tailed bats (M. J. Daniel, unpubl.).

Subfossil skeletons of *M. tuberculata* have been found in caves, on rock ledges and in sinkholes (tomo) in both the South and North Islands as far north as Waitomo (*c.* lat. 38°S) (Dwyer, 1970; Cody, 1981; Millener, 1981; Daniel & Williams, 1984; Worthy, 1984; Daniel, unpubl.). At least four individuals have been found in the South Island and 25 in the North Island (Map 10), as follows: Aurora cave (Te Anau) (1), Gouland Downs rock ledge (1), Heaphy River rock ledge (1), Nettlebed cave (1), Gardner's Gut cave (4), Self Respect cave (1), Hukanui no. 7 rock ledge (1), Bushface no. 1 rock ledge (1), Aussie cave (2), Karamu cave (6), Virginia cave (1), Pukeroa cave (2), Papamaru cave (1), Coincidence cavern (1), F1 cave

(Waitomo) (1), and Ruakokapatuna caves (4). The ages of these subfossils are not known.

HABITAT

The lesser short-tailed bat is restricted to the interior of indigenous forest and has never been reported from exotic pine forests. It has been seen flying near fishing boats anchored off Codfish I., and the highest record, still below the bush line, is from Whakapapa village, Mt. Ruapehu, at about 1100 m (Daniel & Williams, 1984). Roosts have been found in kauri *(Agathis australis),* rimu *(Dacrydium cupressinum),* totara *(Podocarpus totara* and *P. hallii),* southern rata *(Metrosideros umbellata),* kamahi *(Weinmannia racemosa)* and beech (*Nothofagus* sp.) trees. Several roosts have been found in granite sea-caves and pumice caves, and single bats in houses and mountain huts (Daniel & Williams, 1984). On Codfish I., one of the main feeding areas for the southern subspecies is on the precipitous western and southern cliffs, where there are many thousands of petrel and muttonbird burrows in the peat under a low canopy of muttonbird scrub (*Olearia* sp.). The roost sites for these bats are, however, in podocarp-rata-broadleaf forest (M. J. Daniel, unpubl.).

FOOD

The lesser short-tailed bat eats a remarkably wide range of foods: arthropods collected from the air, on the ground and on tree trunks, plus fruit, nectar and pollen (Daniel, 1976, 1979). It is also partly carnivorous, and was formerly persecuted (with *M. robusta*) by muttonbirders on Big South Cape and Solomon Is., for chewing fat and meat off plucked muttonbirds hung up overnight to dry (Dwyer, 1962). Nestling and adult bush birds such as bellbirds and yellow-crowned parakeets are also eaten (Stead, 1936; M. J. Daniel, unpubl.). In captivity *Mystacina* eats a wide variety of foods, including oranges and mealworms *(Tenebrio molitor)*; the average nightly consumption per bat is about 20 mealworms (Daniel, 1979 and unpubl.).

The unexpected discovery that *Mystacina* eats plant foods as well as insects (Daniel, 1976) was the first clue that the Mystacinidae might be closely related to the South American Phyllostomidae, the only other microchiropteran family which feeds on plants (Daniel, 1979; Pierson *et al.,* 1982, 1986). This relationship has recently been confirmed by molecular techniques (Pierson *et al.,* 1986).

SOCIAL ORGANIZATION AND BEHAVIOUR

ACTIVITY. The lesser short-tailed bat emerges from its roost 1–2½ hours after sunset, when it is totally dark in the forest. It may be observed flying intermittently through the night, returning to the roost for the last time about an hour before sunrise (M. J. Daniel, unpubl.). Preliminary observations of radio-tagged and microlight-tagged male bats on Codfish I. showed that in September the number of nightly feeding flights was 4–10 in the first 3–4 hours of activity (M. J. Daniel, unpubl.). Flight is not as fast nor as twisty as that of *Chalinolobus*; in forest, flight is invariably low, usually within 2–3 m of the ground, but bats disturbed during the day will unhesitatingly fly in bright sunshine at canopy height, some 20–50 m above ground. The frequency of ultrasound with maximum energy used when "cruising" is 60–65 kHz (Daniel & Williams, 1984).

Lesser short-tailed bats observed at the same tree roost on Codfish I. (in a Hall's totara, *Podocarpus hallii*), have moulted from November to December for at least the last 7 years. This moulting roost is apparently occupied only at this time of year; the accumulated guano is full of moulted hair and resembles "candyfloss" (M. J. Daniel, unpubl.).

BURROWING. This is the most remarkable behaviour of the species, which was shared by *M. robusta. Mystacina* scuttles up tree trunks and along branches with rodent-like agility, and burrows in and under leaf-litter and humus on the forest floor in search of food; it also burrows into the rotten wood inside hollow trees, and excavates its own tunnels and roost sites (Daniel, 1976, 1979; Daniel & Williams, 1984). The "floor" of one fallen hollow kauri was occupied by a nursery colony. Although *M. tuberculata* has not yet been observed using muttonbird or petrel burrows, as *M. robusta* did, it probably does regularly enter seabird burrows on Codfish I., either in search of arthropods in the guano-enriched peat soil, or for temporary roost sites. The fur of bats on Codfish I. mist-netted on their return to their roost by M. J. Daniel (unpubl.) was frequently dirty and stained with wet, peaty soil, probably from seabird burrows.

HIBERNATION. Neither species of *Mystacina* hibernates, in contrast to *Chalinolobus,* but they continue to fly during winter when weather is suitable, in air temperatures as low as –2°C on Codfish I., although in the milder climate of Northland, no bats have been observed flying at below 9°C (Dwyer, 1962; Daniel, 1979; Daniel & Williams, 1984; M. J. Daniel, unpubl.).

REPRODUCTIVE BEHAVIOUR. In February to April, the breeding season of lesser short-tailed bats on Codfish I., adult female bats establish a nursery colony in a single, large hollow tree, while the males roost separately in one or more other trees. After dark, individual male bats fly to separate small hollow trees in the forest and begin to call, a repetitive, high-intensity "song" which is audible to the unaided ear for about 50 m and lasts all night, possibly throughout the breeding season (M. J. Daniel, unpubl.). Preliminary observations of radio-tagged female bats in March 1986 showed that they fly at least 8–10 km, probably several times a night, visiting groups of calling males in different parts of the island, before flying to the cliffs to feed (M. J. Daniel & E. D. Pierson, unpubl.). The possibility that this represents a lek mating system for *Mystacina* is being further investigated. Lek mating behaviour has been recorded in only two species of Megachiroptera; it has not been reported in any of the approximately 778 species of Microchiroptera, to which *Mystacina* belongs (Bradbury, 1977, 1981). The timing of the breeding season in Northland is not known, but may be in autumn and winter.

REPRODUCTION AND DEVELOPMENT

The lesser short-tailed bat is monoestrous and has only one breeding season per year. Mating in captive bats has been observed in autumn in holding bags in Omahuta kauri forest, and in spring in Wellington Zoo, but in the wild it may continue through winter to spring. Parturition at Omahuta (*c.* lat. 35°S) is in summer (December to January), but is believed to be in autumn (April to May) in southern populations on Codfish I. (*c.* lat. 47°S). On Big South Cape and Solomon I., parturition in the lesser and greater short-tailed bats was apparently synchronous. The season of births in other populations, on the volcanic plateau, in the Tararua Range and in North-West Nelson Forest Park, is not known. Only one young is produced per pregnant female. It is not known whether the reproductive cycle of these bats involves normal direct development, or a delay in development, fertilization or implantation (Daniel, 1979 and unpubl.).

Post-natal development is rapid; juveniles fly at about 4–6 weeks and reach almost adult size (but not weight) at 8–12 weeks (M. J. Daniel, unpubl.).

POPULATION DYNAMICS

Nothing is known about the sex ratio at birth, juvenile or adult mortality, or adult longevity.

PREDATORS

The only avian predator at present is the morepork (*Ninox novaeseelandiae*), although the crevice-dwelling laughing owl (*Sceloglaux albifacies*) (now extinct) may well have been a predator in the past (Dwyer, 1962; Williams & Harrison, 1972; Daniel & Williams, 1984).

Domestic cats are known to kill bats, possibly when they are attracted to moths flying round outside house lights. Of 23 dead lesser short-tailed bats handed in for study, six had been killed by domestic cats. The effect of feral cats and stoats on the forest bat population is not known (Daniel & Williams, 1984). Stoats are excellent tree-climbers (p. 290) and have been reported within 20 m of a large nursery colony of *M. tuberculata*, but there is no evidence that they prey on bats (Daniel & Williams, 1984).

Predation on lesser short-tailed bats by the three species of introduced rats (kiore, ship rat and Norway rat) must certainly have been serious throughout the bats' former range and may still be, but there is little direct evidence to quantify it, apart from the tragedy of Big South Cape and Solomon Is. Both species of *Mystacina* disappeared from these islands by 1965, following an irruption of ship rats, but whether the immediate cause was direct predation, competition for food, or disturbance to nursery roosts by ship rats, is not known (Daniel & Williams, 1984). Ship rats are also believed to have exterminated two species of bats on Lord Howe I., the endemic *Nyctophilus howensis* and *Eptesicus pumilus*, after a shipwreck in 1918 (McKean, 1975).

Ship rats have frequently been seen inside two hollow fallen kauri trees which were occupied at the time by several hundred lesser short-tailed bats. Of some 40 bat skeletons collected from these roosts, 29 (72%) had a hole chewed in the back of the skull and the brains removed, presumably by ship rats. However, this could be either evidence of predation, or the result of scavenging by the rats on bats which had died of cold or starvation (Daniel, 1979; Daniel & Williams, 1984).

The effect of kiore on the lesser short-tailed bat must have been considerably less than that of ship rats, because high populations of this bat still coexist with kiore on Little Barrier I. and Codfish I. However, if either ship or Norway rats ever reached these islands, their value as a sanctuary for many vulnerable and endangered species, including bats, would be destroyed (Daniel & Williams, 1984).

Two other potential dangers for this fruit-eating and terrestrial bat arise from control operations against possums (p. 98). One bat is known to have been killed by a fruit-lured cyanide possum bait; and air-dropped carrot baits (dyed green to deter birds) containing 1080 poison (sodium monofluoroacetate) could be a hazard to short-tailed bats on the forest floor (Daniel & Williams, 1984; M. J. Daniel, unpubl.).

PARASITES AND ASSOCIATED INSECTS

The lesser short-tailed bat has no host-specific fleas, nor does it share any fleas with other native or introduced species (Daniel, 1979). A single record of the long-tailed bat flea (*Porribius pacificus*), recorded on a museum specimen of *M. tuberculata* by Dwyer (1962), is believed to have resulted from accidental association at the museum (Daniel, 1979). However, it is host to an undescribed tick of the genus *Argas* (*Carios*). Four other species of *Argas* (*Carios*) ticks are found on several species of Australian bats (G. W. Ramsay, in Daniel, 1979).

About four to six undescribed, probably host-specific, species of mites have been found in the fur of *M. tuberculata* (Dwyer, 1962; G. W. Ramsay, in Daniel, 1979). A new species and genus of mesostigmatid mite (*Chirolaelaps mystacinae*) was implicated in the death of six *M. tuberculata* in Wellington Zoo (Heath *et al.*, 1987a, b). A second species of *Chirolaelaps* remains to be described (A. C. G. Heath, unpubl.). A new species (*Chirophagoides mystacops*), new genus and new subfamily of sarcoptic mite was described from the wing of a single *M. tuberculata* in the British Museum (Natural History) collected on Solomon I. (Fain, 1963, 1968). Nutting *et al.* (1975) did not find any demodicid hair follicle mites in two *M. tuberculata*, but two undescribed demodicid mites have recently been found on this host (C. Desch, unpubl.).

Holloway (1976) described a new family (Mystacinobiidae), genus and species of batfly (*Mystacinobia zelandica*) from a roost of lesser short-tailed bats. This remarkable species, unlike all other batflies of the families Streblidae and Nycteribiidae which feed on blood, is not parasitic on *Mystacina* itself, but feeds on the bat guano and is completely dependent on the bat for its development, survival, and transportation to other roosts. Specimens of an undescribed new species of *Mystacinobia* have been found on *M. robusta* from Solomon I., and fragments in roosts of *M. tuberculata* on Codfish I. (Daniel, 1979; B. A. Holloway, unpubl.).

Andrews and Daniel (1974) examined two preserved *M. tuberculata* and five *M. robusta* for cestodes and nematodes; none were found. No parasites were found in blood smears from 35 *M. tuberculata* at the Omahuta colony in 1977 (S. S. Desser, in Daniel, 1979).

SIGNIFICANCE TO THE NEW ZEALAND ENVIRONMENT

Mystacina tuberculata is listed in the New Zealand Red Data Book (Williams & Given, 1981) as "vulnerable". Because predation by rodents, feral cats and perhaps stoats still continues in the few indigenous forests inhabited by this bat, as well as other potential dangers from cyanide possum baits and 1080 poison, Daniel and Williams (1984) believe that the North Island populations of the lesser short-tailed bat will move into the "endangered" category within a decade, and that the only South Island population, in North-West Nelson Forest Park, is already in that category. If *M. robusta* is extinct, then *M. tuberculata* is the only surviving species of the endemic family Mystacinidae.

M. J. D.

11. Greater short-tailed bat

Mystacina robusta Dwyer, 1962

Synonym *Mystacina tuberculata robusta* Dwyer, 1962.

Also called southern short-tailed bat, Stewart Island short-tailed bat (English); pekapeka (Maori).

DESCRIPTION (Fig. 27)

Distinguishing marks, Table 24 and skull key, p. 24.

A medium-sized and extremely robust bat, about one-third larger than the lesser short-tailed bat (*M. tuberculata*). By comparison, the greater short-tailed bat has wider, shorter nostrils and proportionately shorter forearm and wing elements; the rostrum is slightly heavier and more massive; and the braincase generally rises a little less abruptly from the rostrum (Hill & Daniel, 1985). The fur colour and texture are as in the smaller species.

Dental formula $I^1/_1$ $C^1/_1$ $Pm^2/_2$ $M^3/_3$ = 28.

MEASUREMENTS

Total length to 90 mm; forearm 45-48 mm; wingspan 290-310 mm; tail 15 mm; ear 18.0-18.6 mm; condylobasal length 21.0-22.5 mm; estimated weight 25-35 g (no live specimens have been weighed) (Dwyer, 1962; Hill & Daniel, 1985).

VARIATION

Some of the subfossil material, particularly the skull of NZMA 412 from Flowers Cave, Puketiti (Dwyer, 1970; M. J. Daniel, unpubl.), is considerably larger than the holotype *robusta* and the other specimens measured by Hill & Daniel (1985). These larger specimens are believed to represent an undescribed new subspecies of greater short-tailed bat (M. J. Daniel & J. E. Hill, unpubl.).

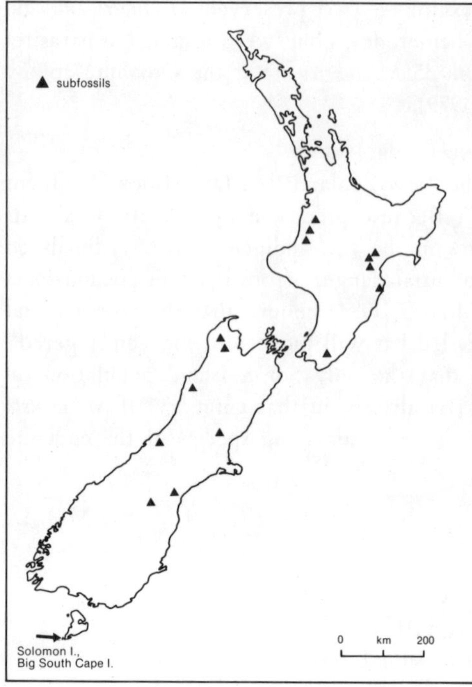

Map 11 Distribution of known subfossils of greater short-tailed bats in New Zealand. Arrowed: Big South Cape I. and Solomon I., the only known locations of living greater short-tailed bats until their probable extinction in 1965.

HISTORY OF COLONIZATION

See p. 124.

DISTRIBUTION (Map 11)

The greater short-tailed bat is, or was, confined to New Zealand. No live or freshly dead specimens have ever been collected from the three main islands (North, South or Stewart) since the beginning of organized European settlement in 1840 (Dwyer, 1962; Daniel, 1979; Daniel & Williams, 1984; Hill & Daniel, 1985).

Subfossil skeletons of *M. robusta* have been found in caves, on rock ledges and in swamps in both the South Island and in the North Island as far north as Waitomo (*c.* lat. 38°S) (Dwyer, 1970; Cody, 1981; Daniel & Williams, 1981, 1984). At least 18 individuals have been found in the South Island: Nettlebed cave (8), Whitcombe Mountains cave (1), Gouland Downs (2), Hermit cave (1), Waitaki Gorge rock shelter (1), Pyramid Valley swamp (4) and Gould's Opihi rock shelter (1). The skeletons from Pyramid Valley swamp were found associated with moa skeletons dated at 2000–3000 years BP and were assumed to be of similar age (R. J. Scarlett, in Daniel, 1979). The ages of the other South Island skeletons are unknown. At least 18 individuals have been found in the North Island: Gardner's Gut cave (4), Puketiti Flowers cave (2), Moa Nest cave (1), Hukanui cave and rock shelters (5), Te Waka rock ledge (1), Ocean Beach Maori midden (1), Dartmoor rock shelter (1) and Ruakokapatuna caves (3).

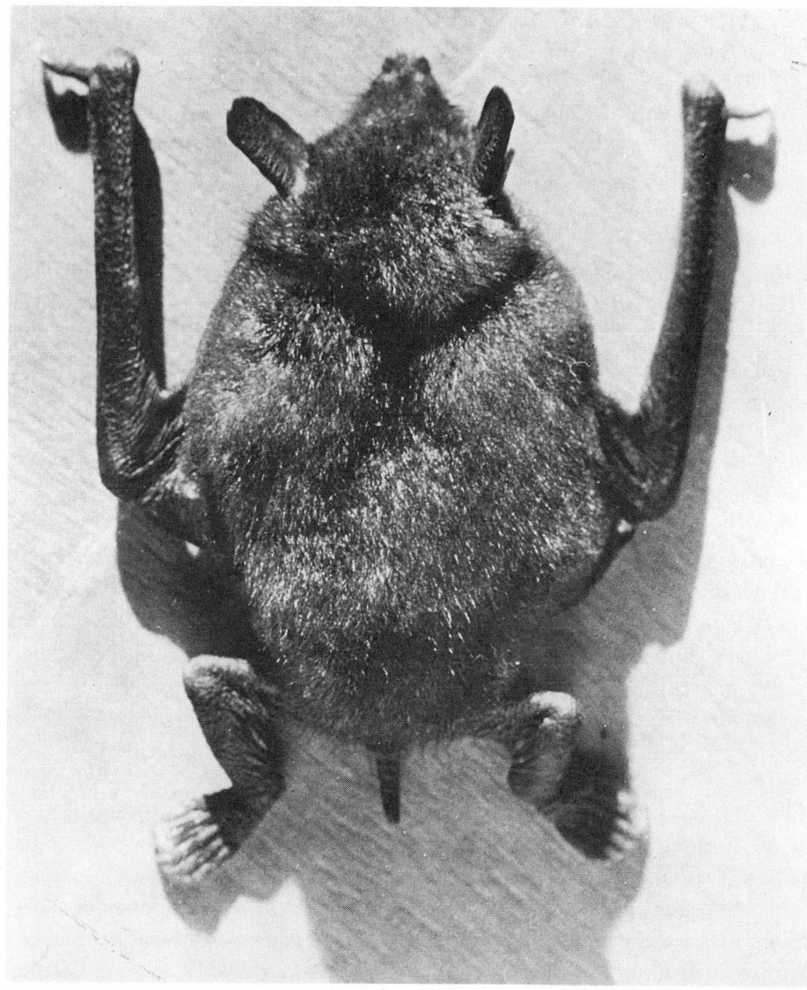

Fig. 27 Greater short-tailed bat (extinct) (J.A. Mackintosh).

The only estimated ages available for the North Island material are for the Hukanui and Dartmoor specimens, which were located on rock ledges in inland Hawke's Bay above the Taupo ash layer of 186 AD (Wilson *et al.*, 1980); and for the single skeleton found in an Ocean Beach Maori or moa hunter midden in Hawke's Bay, which is estimated to be from 150 to 500 years old (R. J. Scarlett, unpubl.).

From 1840 until the mid 1960s, living greater short-tailed bats were known only on two rat-free muttonbird islands off Stewart I. where they were sympatric with *M. tuberculata. Mystacina robusta* has not been seen on either island since 1965 and is believed to be extinct (Williams & Given, 1981; Daniel & Williams, 1984; Bell, 1986).

HABITAT

Big South Cape I. (= Taukihepa, 930 ha) is the largest, and adjacent Solomon I. (= Rerewhakaupoko, 32 ha) one of the smaller, of a group of muttonbird islands lying some 2-10 km off the rugged south-west coast of Stewart I. Both are granite islands with a deep mantle of peat, and are visited each year from March to May by Maori hunters for traditional muttonbirding. Forest vegetation on Big South Cape I. consists mainly of muttonbird scrub (*Olearia lyallii* and *O. angustifolia*) and southern rata (*Metrosideros umbellata*), with a belt of podocarp-broadleaf forest in the middle of the island. The island rises to 240 m, and its top consists of pakihi and associated scrub. The vegetation of Solomon I., highest point 75 m, is also mainly muttonbird scrub and southern rata (Fineran, 1973). The granite shorelines of both islands are rugged and contain numerous caves; the largest, at Puwai on the south coast of Big South Cape, was occupied by both species of *Mystacina* until 1965. Two sea caves at Bat's Cave Landing on the south-east coast of Solomon I. were occupied by *Mystacina* (species not known) until at least the late 1920s (Stead, 1936; Daniel & Williams, 1984). These caves were probably the main roosts for both species of short-tailed bats before human disturbance forced the bats to leave some time in the late 1920s (Stead, 1936). The caves were thoroughly searched by M. J. Daniel in March 1982 and March 1984. He found no bats, bat guano or skeletons, and concluded that bats had not used these caves for many years. Ship rat droppings were seen throughout the three caves.

FOOD

The greater short-tailed bat probably ate the same wide range of foods as *M. tuberculata*: aerial, ground and tree-trunk arthropods, fruit, nectar and pollen (Daniel, 1979). Pollen analysis of stomach contents of two *M. robusta* found both rata pollen and fern spores (Daniel, 1976). Both species of short-tailed bats were partly carnivorous and used to eat fat and meat off plucked muttonbirds hung up to dry overnight; *M. robusta* possibly also ate nestling birds (Stead, 1936; Dwyer, 1962; M. J. Daniel, unpubl.).

SOCIAL ORGANIZATION AND BEHAVIOUR

Little was recorded about this species before it disappeared in 1965; what there is (J. A. Mackintosh, unpubl.) comes from Solomon I. in the late 1950s to 1966.

ACTIVITY. Greater short-tailed bats emerged from their roost well after dark, about 1-2 hours after sunset. Flight was relatively slow, less twisty than that of *Chalinolobus*, and rarely more than 2-3 m above the ground. Stead (1936) suggested that the bats did not fly higher than this because of the thousands of petrels and shearwaters flying to their burrows during the breeding season. However, *M. tuberculata* also flies close to the forest floor in kauri forest in Northland and on Codfish I.; so low flight appears to be characteristic of both *Mystacina* species (Daniel, 1979). On burrowing behaviour, see p. 128.

HIBERNATION. There was no prolonged period of hibernation, even at high latitudes; *M. robusta* was seen flying in March, April, May, July and August (J. A. Mackintosh, unpubl.).

REPRODUCTION AND DEVELOPMENT

Little is known. There are only three pregnant greater short-tailed bats in museum collections, dated December 1931, May 1951, and August 1964, but these could be the dates of accession, not of collection. Field observations are scanty. J. A. Mackintosh collected a juvenile specimen (1-2 weeks old) on Solomon I. (47°S) in late May and also observed several nursery colonies with young, in hollow southern rata trees and in the burrow of a sooty shearwater, from late April to mid May of 1963–65. This suggests that *M. robusta*, like *tuberculata* on Codfish I. some 50 km to the north, was monoestrous and had only one young a year, in autumn (April-May). This is about 4-5 months later than the season of births in *M. tuberculata* in a Northland kauri forest at latitude 35°S (Daniel, 1979). The time of mating and the details of the reproductive cycle are not known.

POPULATION DYNAMICS

Nothing known.

PREDATORS

Stead (1936) reported an unidentified short-tailed bat killed by a morepork on Solomon I. In former times, some bats may have been killed by the laughing owl in the North and South Islands, before both bats and laughing owls became extinct there (Williams & Harrison, 1972).

The kiore (*Rattus exulans*) is believed to have caused or assisted the extinction of the greater short-tailed bat in the North and South Islands and Stewart I. (Daniel & Williams, 1984). These bats used muttonbird burrows as roost sites and may, because of their greater size and weight, have been even more terrestrial than *M. tuberculata*; if so, this could explain why the larger species apparently disappeared from the North and South Islands first.

Ship rats are believed to have caused the final extinction of the greater short-tailed bat on its last two island refuges. In 1962 or 1963, ship rats were accidentally introduced on to Big South Cape and Solomon Is. The rats reached plague numbers from 1964 to 1967, with tragic consequences both for the endemic avifauna and for *M. robusta* and *M. tuberculata* (Atkinson & Bell, 1973; Bell, 1978; Daniel & Williams, 1984). In 1961 several hundred *Mystacina* were reported in Puai Cave on Big South Cape, and regular sightings of flying bats were made on both islands. The last confirmed *M. robusta* from Big South Cape was collected in August 1964, and by 1965 very few flying bats were seen (one was mist-netted at Murderer's Bay and filmed by NZWS); the last confirmed *M. robusta* on Solomon I. was mist-netted by J. A. Mackintosh in April 1965. In 1966 no bats were reported from either island. In November 1967 no flying bats were seen, although four unidentified bats were observed roosting in Puai Cave. No positive sightings of bats have been made on either island from 1967 to the present, apart from one possible sighting of an unidentified bat on Solomon I. in 1971 and a possible sighting at Puai and on Poutama I. in 1985 (Daniel & Williams, 1984; M. J. Daniel, unpubl.). It is not known if the effects of ship rats on the bats was the result of direct predation, continual disturbance in roost sites, or competition with rats for arthropod food.

PARASITES AND ASSOCIATED INSECTS
Several unidentified mites, an unidentified argassid tick and an unidentified species of batfly (*Mystacinobia* sp.) have been found on museum and mist-netted specimens of *M. robusta* (Daniel, 1979).

SIGNIFICANCE TO THE NEW ZEALAND ENVIRONMENT
Mystacina robusta is listed in the New Zealand Red Data Book (Williams & Given, 1981) in the "indeterminate" category, because little was known about it at that time: it is now believed to be extinct (Daniel & Williams, 1984; Bell, 1986). If it is still extant, it is New Zealand's rarest endemic mammal and must be placed in the highly endangered conservation category.

M. J. D.

Suborder Megachiroptera
Family Pteropodidae

This family, the only one in the Megachiroptera, contains about 173 species in 44 genera. These fruit, nectar, and blossom feeding bats are found in the tropics of the Old World, in Africa, Asia, and the Pacific including Australia.

Genus *Pteropus*

There are some 65 species of Old World giant fruit bats or flying foxes; four are found in Australia, and the rest in Africa, Madagascar, India, SE Asia and the Pacific. These are large to very large bats with heads and faces resembling those of foxes, prominent eyes, large simple ears with no tragus, and no noseleaf. All have a claw on the second finger as well as a claw on the thumb.

12. Little red flying fox
Pteropus scapulatus Peters, 1862
Also called little red fruit bat, collared flying fox.

DESCRIPTION (Fig. 28)
The smallest of the four species of Australian flying foxes, with reddish brown fur, a light brown-yellow mantle and pale fur round the eyes. When in flight, the wings appear partly transparent.

Dental formula $I^2/_2\ C^1/_1\ Pm^3/_3\ M^2/_3 = 34$.

MEASUREMENTS
Head and body length about 220 mm, tail very short, and forearm length 118–132 mm (Hall & Richards, 1979; Hyett & Shaw, 1980).

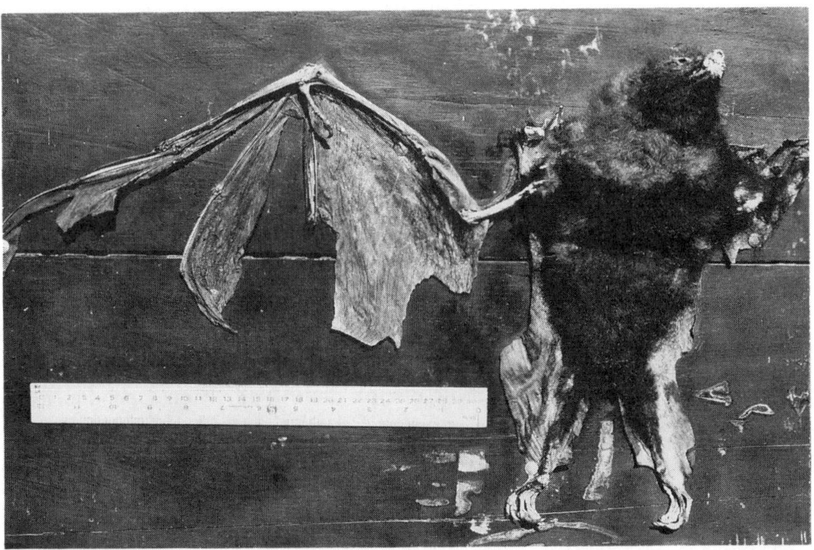

Fig. 28 Little red flying fox – the only specimen known to have reached New Zealand (J.E.C. Flux).

DISTRIBUTION

Widespread in eastern and northern Australia, from dry inland areas to the coast. Absent in Tasmania. This nomadic species can migrate long distances within Australia, following the irregular flowering of eucalyptus blossom.

There is only one vagrant record from New Zealand; a specimen was found electrocuted under powerlines after a storm in Hamilton East about 1927–29 (Daniel, 1975). Unusually large migrations of this species into New South Wales and Victoria were observed a few years after a severe drought in Queensland in 1926–27 (Ratcliffe, 1931); this specimen may have been the only survivor of a group blown over the Tasman Sea at this time.

This is the only documented record of an Australian bat reaching New Zealand alive since European settlement last century, although undoubtedly other Australian bats have been blown to New Zealand in the past but died out unrecorded. For example, Stock (1875) reported large unidentified bats near Wellington, Wanganui and the Clarence River.

M. J. D.

Order Lagomorpha

Lagomorphs are small to medium sized herbivores, which, like rodents, have a pair of large, chisel-shaped incisors separated by a long diastema from the grinding cheek teeth. In other respects they are distinct from rodents: cheek teeth as well as incisors grow continuously throughout life; there is a second pair of small, peg-like incisors directly behind the first pair; the nostrils are covered by a retractable flap of skin; the gut has a large caecum with a spiral septum inside; and the tail is very short. They have a unique way of getting the most out of their food. Faeces produced by day are soft and green and are eaten and passed through the gut a second time before reappearing as the familiar hard fibrous pellets produced in the evening and at night. This process is called reingestion, refection or coprophagy.

There are two families: the Ochotonidae (14 species in one genus), the short-eared pikas of montane Asia and western North America, and the Leporidae (44 species, 10 genera), which includes all rabbits and hares.

Family Leporidae

The world-wide family Leporidae contains one large genus, *Lepus*, the open country hares, and nine smaller genera of mainly burrowing rabbits, including *Oryctolagus*. All have long ears, hind legs and feet. Two species of leporids are present in New Zealand, distinguished as in Table 27.

Genus *Oryctolagus*

The genus *Oryctolagus* has only one species, the European rabbit. It is clearly distinct from the hares, but superficially resembles some other genera of rabbits such as the American cottontails (*Sylvilagus*).

13. European rabbit

Oryctolagus cuniculus cuniculus (Linnaeus, 1758)
Synonym *Lepus cuniculus* Linnaeus, 1758

The subspecies *O. c. cuniculus*, native to Iberia, lives wild in New Zealand; it is also the ancestor of all the domestic breeds which have been spread by man to most parts of the world. The only other, smaller, subspecies, *O. c. huxleyi* (Haeckel, 1874) (adult body weight <1 kg), was probably introduced from southern Spain to the Canary Is. and Madeira.

The common name "rabbit" was originally used in Britain only for the young; adults were called "conies". Males are "bucks", females "does"; young in the nest are usually called "kittens", and those recently out of the nest "runners".

Description (Fig. 29)
Distinguishing marks, Table 27 and skull key, p. 24.

The rabbit is a small to medium sized herbivore, usually grey-brown (agouti) in colour. Other colour varieties are infrequent: on the mainland, fewer than 1% are black, and fewer still russet or ginger; some silver-greys, descendants of domestic rabbits released on Swyncombe Station in 1859, persist between the Seaward and Inland Kaikoura ranges. The French domestic breed Argente de Champagne remains on Rose and Enderby Is. (see Table 29) and probably Macquarie I. Domestic breeds of various colours were released on the mainland and some islands in the mid to late 19th century, but by 1916, 95% of skins exported through Dunedin were agouti.

The testes of adult bucks are scrotal for most of the year, except in late summer and autumn. Does usually have four pairs of nipples, easily visible when swollen during lactation, but otherwise inconspicuous. The surrounding fur on the belly is heavily plucked to line the nest.

All rabbits have scent glands under the chin and near the anus; their size is related to age, sex, social status and reproductive condition (Mykytowycz & Dudzinski, 1966; Fraser, 1985). The chin gland is conspicuous on dominant bucks as a pea-sized lump, often rubbed nearly bare on objects used to mark territories.

Normally silent, rabbits squeal shrilly when seized by a predator; some grunt softly when handled. Occasionally when alarmed they thump their hind legs loudly on the ground. Their eyes shine pink in a spotlight at night. Rabbits can swim wide and turbulent rivers when pressed, and climb sloping trees if forced by flooding.

The juvenile coat is uniformly grey, lacking the adults' ochraceous buff underfur on the nape; it is shed at 2–3 months. The annual moult of adults starts in spring and ends in autumn. Dark patches inside a fresh skin mark its progress.

Dental formula $I^{2}/_{1}$ $C^{0}/_{0}$ $Pm^{3}/_{2}$ $M^{3}/_{3}$ = 28.

FIELD SIGN

Faecal pellets vary in size, colour and consistency according to content; they are smaller, usually darker, less friable and less spherical than those of hares. Unlike hares, rabbits (especially bucks, less often does and young) typically use special dungheaps ("buckheaps") or latrines, which may contain >1000 pellets. They serve as olfactory information centres and may smell strongly of urine. Most dungheaps are on open ground where rabbits congregate at dusk before dispersing after dark; fewer, smaller heaps are made where they feed at night. Dungheaps are often placed on low mounds or beside small rocks or logs; they may be ringed by lush vegetation (fertilized by leachates from the dung), which is kept well grazed. Some are used for years.

Shallow scrapes in loose soil are made by rabbits digging for rootlets, but more often by bucks displaying in territorial disputes; the latter are often topped off with a few droppings and/or a squirt of urine.

Rabbits living at high density on well-drained, light soils may dig large communal warrens, with fresh soil scattered around. Large warrens, a labyrinth of underground tunnels extended piecemeal over the years, may

Table 27: Distinguishing marks of lagomorphs in New Zealand.

	Rabbit *Oryctolagus cuniculus*	Brown hare *Lepus europaeus*
Weight (adult male)	1.3-2.1 kg	2.4-4.8 kg
Length of ears	60-70 mm	90-105 mm
Length of hind feet	75-95 mm	130-155 mm
Body colour	Grey-brown	Tawny
Tips of ears	Narrow black rim	Black patch at tip
Colour of eyes	Brown	Yellow
Gait	Bobbing, tail up, showing white underside when unhurried; scuttling rush, tail down, when alarmed	Loping, tail down showing black upper surface

occupy about 0.5 ha, with >100 entrances and as many occupants. They are now rare in New Zealand. Even where still numerous, as in Central Otago, rabbits tend to live in small scattered warrens. Where above-ground cover is sufficient (e.g., scrub, rocks, logs), there are no permanent warrens, only breeding "stops". Rabbits sometimes lie out through the day in forms or "squats", mainly in fine weather.

Footprints left in damp sand, mud, or snow are smaller than those of hares. Rabbits form well-defined tracks in long grass, as between daytime cover and night feeding grounds; some consist of well-worn pads, like stepping stones, about 300 mm apart.

Close-cropped herbage or hedged shrubs can betray the presence of rabbits to a practised eye. Rabbits graze young crops around paddock margins close to cover. Traces of fur may be found on the ground or on snags.

MEASUREMENTS

See Tables 27 and 28.

VARIATION

The body weights of rabbits vary regionally, seasonally (Table 28) or annually, often inversely with population density (Gibb, Ward & Ward, 1978; Fraser, 1985). However, over most of New Zealand, where populations are sparse and food is abundant, there is little variation in carcass weight with sex or season; weights from the western North Island (Table 28) are similar to those from many areas during the last 35 years. The heavier rabbits in North Canterbury (formerly including Motunau I.; Cox, Taylor & Mason, 1967) are probably descended from original liberations of large domestic stock, though all are agouti. In Central Otago, where dry summers and food shortages are common, bucks become slightly heavier than does in autumn and winter, but both sexes are heavier than North Island rabbits except in winter. Seasonal changes in body weight are partly due to changes in the amount of body fat, but variable recruitment of young to the adult population also influences average weights. Adult rabbits do not require

Fig. 29 Wild rabbit, northern Canterbury (Andris Apse).

green vegetation to maintain or increase body weight if they can obtain sufficient moisture from dead vegetation, but young rabbits require green feed for normal growth. The average length of adult rabbits varies little between sexes, despite regional differences (Table 28).

HISTORY OF COLONIZATION

Ocean-going sailing ships often carried domestic rabbits for food. Some were put ashore as food for castaways on remote islands, but most of these died out. The first rabbits (two pairs) known to have been released in New Zealand were placed by James Cook on Motuara I., in Queen Charlotte Sound, in 1777 (Beaglehole, 1967) (Table 29). When permanent shore stations were established, as on Mana I., rabbits were a regular item of trade. Exactly when or where domestic rabbits first became established ferally is unknown

Table 28: Body measurements of adult rabbits (>6 months old) in three districts of New Zealand.

	Western North Island		North Canterbury		Central Otago	
	Male	Female	Male	Female	Male	Female
Carcass Weight[1] (g)						
Spring Mean	1 344	1 371	1 560	1 583	1 453	1 441
n	3 792	3 820	3 384	2 916	296	294
SE	2.1	2.1	2.2	2.4	10.2	10.7
Summer Mean	1 370	1 360	1 521	1 536	1 551	1 546
n	1 138	972	1 795	1 468	235	166
SE	3.9	4.2	3.1	3.4	10.9	12.6
Autumn Mean	1 353	1 325	1 506	1 477	1 415	1 358
n	2 335	2 067	2 908	2 342	267	254
SE	2.7	2.9	2.4	2.7	11.7	10.3
Winter Mean	1 366	1 388	1 528	1 530	1 312	1 239
n	2 570	2 344	1 520	1 162	444	390
SE	2.6	2.7	3.3	3.8	9.7	10.0
Length[2] (mm)						
All year Mean	484	490	492	494		
n	9 565	7 937	966	822		
SE	1.9	2.2	0.7	0.9		

1. North Island and Canterbury carcass weights include kidneys and associated fat (J. M. Williams, unpub.); Otago weights exclude them (Fraser, 1985). 2. Tip of nose to tail.

(Wodzicki, 1950; Lever, 1985; B. T. Robertson, unpubl.). Acclimatization societies devoted great care and expense to importing stock from Britain; this proved embarrassing later. For example, in 1866 the Canterbury society reported breeding and distributing rabbits, but by 1889 it was denying responsibility for introducing "that great scourge". Most releases of domestic breeds were slow to establish; others thrived where the modified habitat suited them, and became pests well before wild rabbits were imported. The spread of rabbits was limited at first by the scarcity of suitable habitat.

Rabbits were taken to offshore islands (see Table 29) and kept in semi-confinement on the mainland from the 1820s onwards (Williams & Williams, 1822; Wakefield, 1842; Godley, 1951). Known early records include Mana I., 1834 (McNab, 1913), Bay of Islands, 1838 (Thouars, 1841), Riverton near Invercargill, 1838, 1839, Queenstown, 1839 (Wilson, 1976), Nelson, 1842 (Anon., 1842; Saxton, 1849), Rabbit I. near Nelson, 1843 (Adams, 1851); Waitarangi Station, Wairarapa, 1847, Carterton, 1857. Further liberations followed in both main islands during the 1860s, although domestic breeds living ferally were a problem before the arrival of wild rabbits. Wild rabbits rapidly supplanted domestic breeds, except on islands they never reached (e.g., Enderby); they were commonly sold in Southland by July 1867, and by 1868 Riverton farmers complained of damage to crops; in 1869, 10 000 skins were taken from Swyncombe Station, near Kaikoura, from six months' trapping.

Once established, rabbits became enormously abundant and very destructive to pastoral farmland. From 1878 extensive grazing lands in Otago were abandoned as owners gave up struggling against rabbits (Thomson, 1922). Rabbits reached peak numbers in Otago in about 1890 and extended on to mountain lands never occupied since, as between Manapouri and Te Anau, where annual rainfall exceeds 2500 mm (Riney *et al.*, 1959) and above the bushline at 1350 m on the Kaimanawa and Tararua ranges. The subsequent decline may have been accelerated by the release of hundreds of mustelids (p. 293), by the gassing of burrows and trapping for skins, as well as by overgrazing.

The early distribution of rabbits followed the development of farming. They spread rapidly from Otago and Kaikoura, especially with the advance of sheep grazing on South Island tussock lands. Most suitable ground in the South Island was occupied by 1900. In the North Island rabbits spread more slowly, and continued to do so through the first half of the 20th century as forest and scrub were cleared. Most suitable habitat was occupied by 1948 (Wodzicki, 1950), although parts of North Auckland were not colonized until the 1950s.

DISTRIBUTION

WORLD. About 2300 years ago the rabbit was confined to the Iberian Peninsula (Lever, 1985). Phoenician traders distributed rabbits around the Mediterranean; domestic forms were bred in French monasteries between AD 600 and 1000. Domestic rabbits probably reached Britain in the 12th

Table 29: Distribution and history of rabbits on the offshore and outlying islands of New Zealand.

Island	Area (ha)	Lat.° S	History	Present status[1]
Allports	12	41°14′	Formerly common	X
Auckland	45 975	50°35′	Introduced by 1847	✓
Bells (Fellows)		41°18′	Silver-greys introduced 1867	X
Bests		41°18′	Silver-greys introduced 1867	X
Browns (Motukorea)	60	36°50′	Introduced 1972-73, increased to 5000-7000 by 1978; still common despite control	✓
Campbell	11 331	52°30′	Introduced *c.*1883?	X
Chatham	90 650	44°00′	Possibly introduced	X
Clarence		42°10′	Island in river. Blacks introduced before 1865, then spread	?
D'Urville	16 782	40°50′	A few reported	?
Enderby	710	50°30′	French breed(s) probably introduced 1840, 1865; over-run 1880	✓
Friday		50°30′	Rabbits transferred to Rose I. *c.*1850	X
Great Barrier	28 510	36°11′	Common	✓
Happy Jack (Motukahaua)	c.24	36°39′	Common	✓
Haulashore	<10	41°16′		✓
Inner Chetwode	242	40°54′	White rabbits introduced 1914-18, present to *c.*1935	X
Junction Group (largest of three)	c.7	36°14′	Present *c.*1936	X
Kawhitihu (Stanley)	120	36°38′	Introduced 19th century	✓
King Billy		43°33′		✓
Korapuki (Rabbit)	11	36°40′	Introduced 19th century, not common	✓
Leper	2	41°15′	Introduced *c.*1946, exterminated 1949-50	X
Mahurangi (Goat)	32	36°49′		?
Mana	217	41°05′	From 1834, but probably only in captivity	X
Mangere	113	44°17′	Introduced *c.*1890, exterminated by cats *c.*1895	X
Maud	309	41°02′	Common *c.*1912	X
Motuara	57	41°05′	James Cook liberated 2 pairs 1777	X
Motuihe	179	36°49′	Common despite control	✓
Motukawao Group (see Happy Jack, Moturoa and Motuwi)				

Motukiore		35°48′	Connected to mainland at low tide	✓
Motumaire	5	35°16′	Present 1974, 1975	✓
Motunau	3.5	43°04′	Exterminated 1962	X
Moturoa	146	35°12′	Present 1975	✓
Moturua (Rabbit)	26	36°42′	Common 1969	✓
Motutapu	1 509	36°46′	Common before 1883, present 1980	✓
Motuwi (Double)	26	36°40′	Present 1979	✓
Native (Rabbit)	60	46°55′	Recorded 1872, probably exterminated by 1950	X
Ngawhiti	5	40°49′	Introduced *c.*1910, exterminated *c.*1912	X
Ohinau	45	36°44′	Introduced 19th century	✓
Okokewa (Green)		36°09′		✓
Otata	22	38°42′	Present in 1930s	X?
Penguin	8	37°04′	Present 1974	✓
Puangiangi	69	40°46′	Present before 1957	?
Puketutu		36°58′	Present 1940s, 1950s	?
Rabbit		41°16′	Silver-greys introduced 1843, still present	✓
Rabbit		44°14′	Probably never present, despite name	X
Rakino	150	36°43′		?
Rangitata		44°03′	Island in river; rabbits present before 1864	✓?
Rose	75	50°31′	Introduced *c.*1850 from Friday I.	✓
Rough		41°17′	Silver-greys introduced 1867, formerly common	X
Shoe	40	36°59′		✓
Slipper	210	37°03′	Doubtful record	X
Stewart	174 600	47°	First recorded 1860s, exterminated 1946	X
Sugarloaf	<1	39°03′	Formerly common	X
Tahoramaurea (Browns)	60	40°53′	Present *c.*1920, exterminated by *c.*1925	X
Taieri		46°03′	Connected to mainland at low tide; common	✓
Waiheke	9 459	36°47′	Common	✓
Waitaki		44°50′	Island in river; common	✓
Whale (Motuhora)	140	37°52′	Introduced *c.*1967; still common despite control	✓

1. ✓ Present, X absent, ? unknown. After J. E. C. Flux (unpubl.).

century (the young were considered a delicacy) and were later spread throughout the British Isles including Ireland, Orkney and Shetland, and to other islands in the north-east Atlantic. Much later, rabbits were put ashore from ocean-going sailing ships in South Africa (from Holland) in 1654, Chile in the mid 18th century, Falkland Is. in 1764, New Zealand in 1777, and Australia in 1788. Rabbits have been recorded on more than 550 islands worldwide (Flux & Fullagar, 1983).

NEW ZEALAND (Map 12). Though much less abundant now than formerly, rabbits still live in most suitable habitat in the North and South Islands, from the coast up to 1000 m altitude in the South Island and higher (e.g., above bushline on Mt. Egmont in 1960) in the North Island. They have died out or been exterminated on several offshore and outlying islands, but still survive on about 24 (listed in Table 29).

Surveys of control operations in 1984 showed that the rabbit's distribution had changed little on the two main islands since 1946 (P. C. Nelson, unpubl., cf. Wodzicki, 1950). Over most of the country rabbits are now almost scarce: only in limited areas of Central Otago, the Mackenzie Basin, North Canterbury and Marlborough do they still compete seriously with stock, threaten plant cover, or accelerate soil erosion. Large numbers persist only where the climate resembles that of the western Mediterranean.

HABITAT

Desirable features of ideal rabbit habitat include an annual rainfall of <1000 mm, a sunny aspect, light soil, and adequate cover close to feeding grounds kept closely grazed with or without other grazing animals. Although rabbits can live with higher rainfall (as on Haast River flats and in bush clearings in the King Country), they do so only on light soils and where other animals help to maintain a close-cropped sward. Rabbits can maintain their own short sward even on productive soils (Gibb, Ward & Ward, 1978), but only under ideal conditions. Rabbits especially favour dunelands and dry, stony riverbeds, limestone hills with outcropping rocks, and sunny coastal slopes; they usually avoid cold and wet conditions. Small groups persist on dunes behind remote beaches in Westland and on rough ground inside city boundaries. Only alpine lands, the interior of indigenous forest, unbroken scrub and heavily built-up areas are without rabbits.

The slow establishment and spread of rabbits after introduction suggests that the natural habitats then available were not ideal. Farming, with the burning of tussock grassland and bush, sowing with introduced grasses, and heavy grazing by sheep, created suitable habitat. Their spread was slower in the North Island, presumably because much of the lowland had been forested and took longer to convert to pasture than the tussock grasslands of the south; it was also wetter in the north.

FOOD

There has been no detailed study of the diet of wild rabbits in New Zealand, so we rely on indirect information from studies of rabbit-induced changes in vegetation (p. 158). Rabbits eat a wide range of plants including indigenous

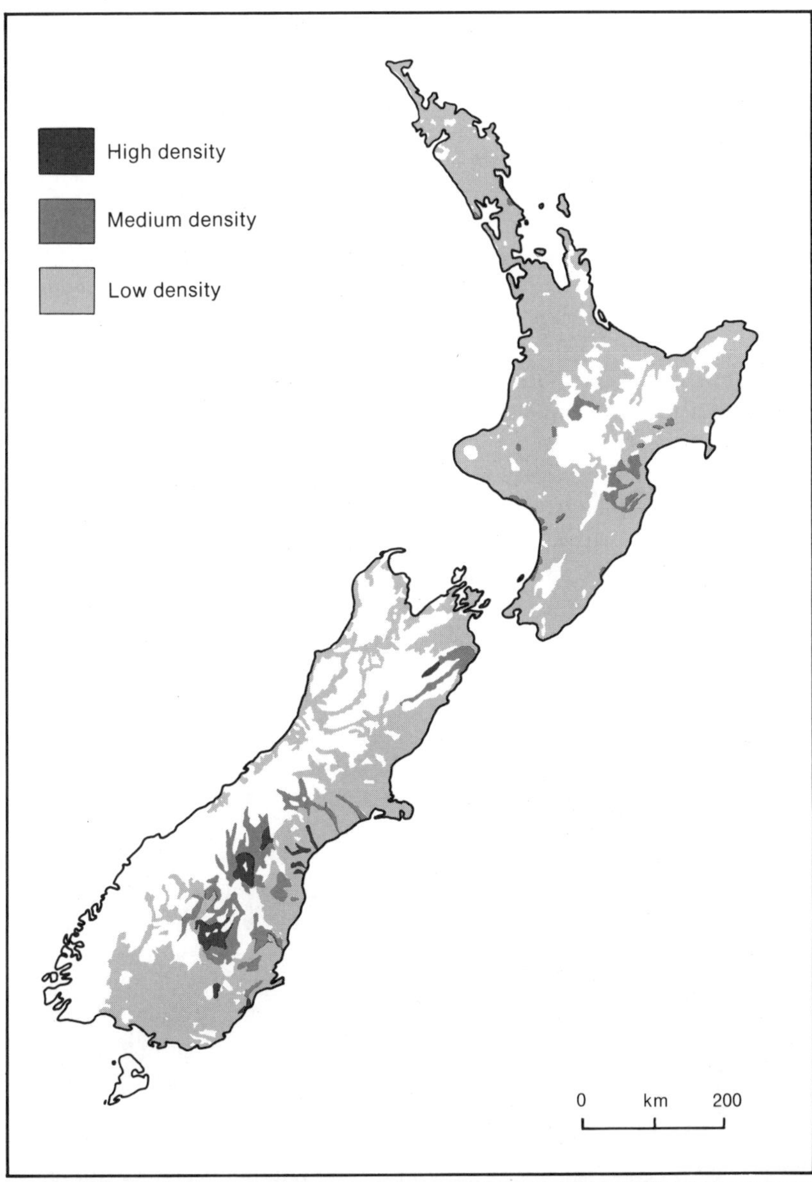

Map 12 Distribution and relative abundance of rabbits in the North and South Islands. High density: populations that regularly increase to carrying capacity if left uncontrolled. Medium density: populations that occasionally reach carrying capacity and require some control. Low density: populations that seldom increase to carrying capacity and normally require no control. Blank areas (land over 1000 m asl, forest or lakes) have few or no rabbits. Map drafted by E. J. Johnson from a national survey of rabbit control requirements by APDC (P. C. Nelson, unpubl.) plus an evaluation of the risk of rabbit infestation on all land types using Land Resources Inventory Maps (J. M. Williams, unpubl.). For distribution on offshore and outlying islands, see Table 29.

and introduced grasses, clovers, and the foliage, flower buds and seeds of many annuals. Ultimately the food available depends on the total grazing pressure by rabbits and other animals, and the selective removal of palatable plants unable to withstand this pressure.

Clovers are probably selected for their high nutritive value, but they may be less tolerant of grazing than many grasses. A diet of ryegrass (*Lolium perenne*) and white clover (*Trifolium repens*) alone is too low in fibre for domestic rabbits (Joyce, Rattray & Parker, 1971). A minimum intake of about 18% crude protein is also required, hence minor additional high-fibre components are essential. The average ryegrass-clover pasture is probably sufficiently diverse to provide a balanced diet, judging from the fast growth of young rabbits fed on it (Gibb, Ward & Ward, 1978).

Rabbits need food containing more than 55% water to maintain themselves. This requirement influences their diet and survival in semi-arid regions of Australia, where they concentrate their urine (Cooke, 1982), but rarely in New Zealand. Rabbits and hares maximize digestion of cellulose by coprophagy or refection. Special soft pellets enclosed in a mucous membrane are taken direct from the anus and swallowed whole, mainly between 0800 and 1200 hours (Watson, 1954). Young rabbits begin reingesting as soon as they are weaned and eating grass. The soft pellets contain high concentrations of proteinaceous bacteria, phosphorus, sodium and potassium, which slowly dissolve when returned to the stomach and are absorbed (Griffiths & Davies, 1963).

The total food intake of wild rabbits, compared with sheep, their main grazing competitor, has received scant attention in New Zealand. Early estimates were largely subjective, or based on extrapolations from the requirements of domestic rabbits. Munro (1917) reckoned that five rabbits equal one ewe; Wodzicki (1950) estimated that 10 wild rabbits ate as much as one sheep. Modern work shows that rabbits grazing improved pasture consume about 70 g of dry matter per day, so 13 rabbits will eat as much as one 50 kg ewe (J. M. Williams, unpubl.). A subjective Australian estimate by Myers and Poole (1963) equated 7–10 rabbits with one sheep. However, in arid parts of Australia the ratio of absolute food intake of rabbits and sheep is 80:1275 g per day, so 16 rabbits equal one sheep (Short, 1985).

SOCIAL ORGANIZATION AND BEHAVIOUR

ACTIVITY. As a rule, rabbits spend most of the morning and early afternoon resting in cover and emerge in late afternoon or evening. In mild districts they are more active during the afternoon in summer than in winter; but frozen ground at night can compel them to feed through the middle of the day in winter (R. J. Pierce, unpubl.). Usually they do not move far before dusk. Most rabbits are active all night, except in strong winds and heavy rain. They continue feeding in daylight when short of food, but normally fewer are out at dawn than at dusk. Does spend longer feeding than bucks, especially when breeding. Members of a pair commonly spend the day apart. The doe is the more sedentary and often appears at the same place each evening, where the buck joins her. The pair spend much of the night together (Gibb, Ward & Ward, 1978; J. A. Gibb, unpubl.).

HOME RANGE. Size of home range depends on topography, the distance to feeding grounds, and the dispersion of other rabbits. In an 8-ha enclosure of good pasture in the Wairarapa, 80% of sightings of tagged rabbits fell within 1 ha (Gibb, Ward & Ward, 1978). Both sexes had larger ranges in summer and autumn than in winter and spring, and larger at night than by day. Bucks had larger ranges than does by day, but not at night (cf. Fraser, 1985). Both sexes occupied smaller ranges at high than at low population density, and first-year rabbits of both sexes had bigger ranges than older animals. Home ranges may be larger outside an enclosure and are very much larger where rabbits commute some distance (up to 500 m) between daytime cover and night feeding grounds (J. A. Gibb, unpubl.). The small ranges typical of high density are expanded rapidly among survivors when a population is reduced, as by poisoning (Fraser, 1985).

Dominant bucks frequently patrol their territories, challenging trespassers. Neighbours meeting at a territorial boundary may display, parading back and forth a few metres apart with tails (scuts) held high, digging scrapes, rolling on loose soil, "chinning" prominent objects, and leaping up skittishly, twisting in mid air and enurinating at each other. Rabbits rarely actually fight, but can inflict serious wounds by lashing out with their back legs or (less often) by biting.

DISPERSION. Rabbits usually live in small groups. At low density, single bucks may even live alone for a year or more until they can infiltrate a group or are joined by others. The composition of large groups is complicated: each usually includes a recognizably dominant pair, with a strict social hierarchy among the others (Fraser, 1985). Pairs may remain together for a year or more if both survive (Gibb, Ward & Ward, 1978); intact pairs do not wander far from their accustomed range. Social groups, hierarchies and ranges continually change by recruitment and mortality; pairs may break up at the end of the season as bucks court newly mature does, and bucks may adjust their ranges to include new partners.

REPRODUCTIVE BEHAVIOUR. In spring and early summer bucks spend much time socializing and keeping their does from other bucks. They spend less time feeding than do does, and they lose weight. Bucks attempting to mount unreceptive does are rebuffed. Attentive bucks stay close to their does, waiting for their opportunity; they circle the doe to prevent her straying, and watch her building the nest, but do not help.

The doe digs a special short burrow or "stop", 100 m or more from her daytime resting place and often hours, days, weeks, or even months before it is needed; many are never used. Most are about 1 m long, with the roof of the chamber approximately 150 mm below the surface. Rabbits may also (but rarely in New Zealand) breed in offshoots of permanent burrows. The doe makes a nest inside the stop, taking in mouthfuls of dead grass every few minutes, sometimes for spells of 30 minutes, and eventually lines the nest with fur plucked from her belly. The entrance to the stop

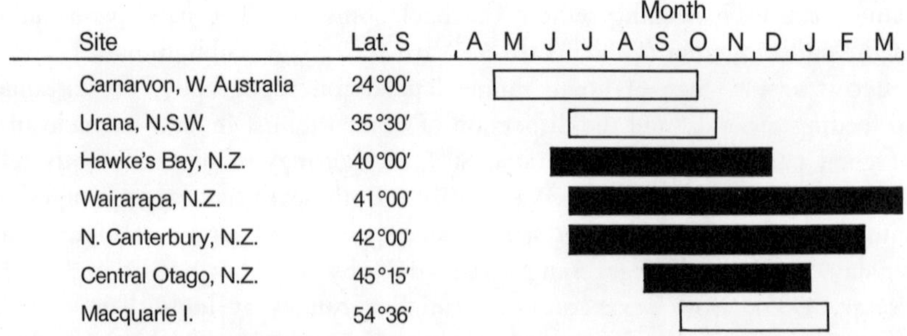

Fig. 30 Main breeding seasons of rabbit (>50% of females pregnant) at localities in Australia and New Zealand, according to latitude. Population densities were low in the Wairarapa, medium-low in North Canterbury, medium-high in Hawke's Bay and high in Central Otago (after Watson, 1975; Parer, 1977; Bell, 1977; Skira, 1978; King, Wheeler & Schmidt, 1983; Gibb, White & Ward, 1985; Fraser, 1985; J. M. Williams, unpubl.). White bars = Australian sites; black bars = New Zealand sites.

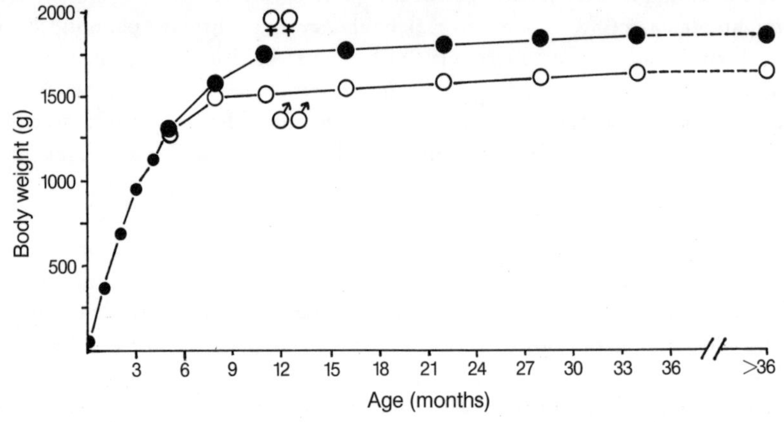

Fig. 31 Total body weights of rabbits relative to age. Young up to 5 months old (living at medium-high density) were weighed in Hawke's Bay (Tyndale-Biscoe & Williams, 1955); older rabbits (living at low density) in the Wairarapa (Gibb, White & Ward, 1985). Adult does tend to be heavier than adult bucks because they are likely to be pregnant; carcass weights of the two sexes are alike.

is kept blocked with soil except when the doe visits the young. R. J. Pierce (unpubl.) found nests above ground in rank vegetation between small tussocks growing on hummocky shingle of a South Island riverbed, which may have been difficult to dig into.

Mating is brief, and follows parturition within hours or minutes. Does in oestrus are sometimes chased by a mob of frenzied bucks and mounted by several in quick succession before gaining cover.

Reproduction and Development

Studies of reproduction in Hawke's Bay (Watson, 1957), Wairarapa (Gibb, White & Ward, 1985) and Wanganui (Williams & Robson, 1985) in the North

Island, and in North Canterbury (Bell, 1977; J. M. Williams, unpubl.) in the South Island, agree closely. More than 50% of adult bucks have scrotal testes in all months except January to April, and more than 50% of adult does are pregnant in all months except March, April and May. Mean litter size at birth ranges from 4.5 in June to 6.5 in October. On average through the year, adult does are pregnant for about 70% of the time, with a mean litter size of 5.2. Annual productivity is 45-50 young per adult doe, the highest ever recorded. Peak fertility is reached at 9-12 months in does, >36 months in bucks. By contrast, in Central Otago conditions are harsher (Fraser, 1985), and the breeding season is short and sharply defined from September to January-February. Litter size is as in other districts, but annual productivity is only about 22 young per adult doe.

The breeding season begins later in the year at higher latitudes (Fig. 30). Early-born does may themselves breed later in the same season, though few of their young survive. Where rainfall is erratic (as in parts of Australia) breeding may be opportunistic, following rain and pasture growth at any time of year (Wood, 1980). This was observed once in autumn, in does of all ages, living at low density in the Wairarapa, and resulted in substantial recruitment (Gibb, Ward & Ward, 1978). Opportunistic breeding has also been observed in Central Otago (J. Bell, unpubl.).

Wild rabbits often resorb some embryos (singly or as whole litters), mainly during early gestation. This may allow them to match litter-size at birth to a failing food supply and so reduce losses of whole litters after birth. The incidence of prenatal mortality is naturally variable: in the Wairarapa it was detected in up to 28% of pregnant does, reducing their litter-size at birth by about 30%. It was more frequent in old does than in young ones (Gibb, White & Ward, 1985).

Non-pregnant does have a 7-day oestrus cycle (Myers & Poole, 1962), and pregnant does an immediately postpartum oestrus. Gestation lasts about 30 days. Young are born sparsely furred and blind; their eyes open at approximately 7 days. The doe visits the nest to suckle them for about 5 minutes, apparently only once per night. They weigh 30-35 g at birth and about 250 g when ready to leave the nest at 21 days.

Young rabbits grow rapidly, aided by extremely efficient digestion (Fig. 31). They gain 10-12 g per day from weaning (at 3 weeks) until 7 weeks of age, and 6-8 g per day from then until 16 weeks (Tyndale-Biscoe & Williams, 1955; Whittle, 1955; Fraser, 1985; J. M. Williams, unpubl.). Young of both sexes are sexually mature at 3-4 months and reach adult weight at about 6 months. They continue to gain weight slowly until >3 years old. Length also increases rapidly with age to a maximum at about 6 months. When food is short, young grow more slowly or die; retarded young may never attain full adult weight (Gibb, Ward & Ward, 1978).

POPULATION DYNAMICS

DENSITY. Rabbits were very abundant from the 1870s to the late 1940s. They were drastically reduced during the 1950s by vigorous control (the "killer policy", see p. 159); but then relatively stable, low-density populations developed during the 1960s. There has been little change since then apart from a resurgence in Central Otago since 1980, the probable result of favourable weather and poor acceptance of baits.

There are few estimates of total rabbit density anywhere in New Zealand. The numbers poisoned or trapped early this century give some clue and were sometimes astonishing. For example, in 1912-13, 49 rabbits per hectare were trapped on a 2800-ha property in the Wellington district; one 445-ha block yielded 76 rabbits per hectare. These tallies excluded poisoned rabbits (Deem & Jenkinson, 1914).

Estimates of relative numbers, made by spotlighting at night, were obtained from numerous sites in 1975-85 (Williams, 1983; Williams & Robson, 1985; Williams *et al.*, 1986; Kerr *et al.*, 1986). Average numbers counted on 17 transects in Central Otago ranged from 4 to 70 per km, maximum 560 in a single km. In North Canterbury numbers ranged from 2 to 6 per km, and in the western North Island 1 to 8 per km (see also Map 12). During 1982, at several riverbed and hill country sites in Canterbury, the number of rabbits spotlighted ranged from 3.2 to 11.4 per ha (Williams *et al.*, 1986). A mark-recapture study on a 4-ha site in Central Otago recorded densities of 10-35 per ha (Fraser, 1985).

DISPERSAL. Young may continue to shelter in the breeding stop for a few days after first emerging, or they may move away at once to feed and never return. Some of a litter may stay together for weeks, others scatter singly. The parental bond is weak after weaning and many young become separated or are driven away at an early age. Small young keep more closely to cover than do adults, so are less often seen or shot. Few young rabbits have been marked and released in New Zealand. Of seven small young marked in Hawke's Bay, two were eventually shot 3 km away (Tyndale-Biscoe & Williams, 1955); movements of up to 1.5 km were recorded in Central Otago by Fraser (1985). In the Wairarapa enclosure young rabbits shifted an average of only 50 m from the spring of their birth to the following winter, and only 20% moved >100 m (Gibb, Ward & Ward, 1978). Adult rabbits rarely shift their ground.

SEX RATIO at birth is 50:50. Thereafter, in the Wairarapa, the percentage of bucks in shot samples increased to 65% among rabbits >3 years old (Gibb, White & Ward, 1985). Bucks also predominated in areas regularly shot in the Wanganui district (Williams & Robson, 1985), but they comprised only 52% of populations in Central Otago usually controlled by poisoning (Fraser, 1985). Overseas (Myers, 1971; Rogers, 1979), an excess of does suggests that there are differences in mortality factors, and hence sex ratios, between habitats.

AGE DETERMINATION. The first method of ageing rabbits used in New Zealand was based on epiphyseal fusion of long bones and vertebrae, which estimates age, ± 1 month, up to 3 years and then simply as >36 months (Watson & Tyndale-Biscoe, 1953; Gibb, White & Ward, 1985). Supplementary information can be obtained from the tibia length of the very young (Gibb,

White & Ward, 1985), and from the carcass weight of juveniles <36 weeks old (Wodzicki & Darwin, 1962). The method usually employed now is based on the dry weight of the eye lens, which increases geometrically with age, is reasonably accurate up to 24 months, and is rather insensitive to season, food supply and other environmental factors (Myers & Gilbert, 1968). This method is widely used to measure the effect of control on rabbit survival.

AGE AND SURVIVAL. The age structure of a population provides a measure of age-specific survival. There is no such thing as *the* age structure of rabbit populations (Gilbert *et al.*, 1987). Most developed farmlands in New Zealand support a kaleidoscope of local populations fluctuating more or less independently of each other (Gibb, White & Ward, 1985; Williams & Robson, 1985), and the proportion of young rabbits in each varies with the length of the breeding season and with juvenile survival. In any stable population a high reproductive rate must be balanced by equivalently heavy mortality, usually of the young. For example, in the Wairarapa, only about 10% of young born from May to November (Fig. 32), and 1% of those born in other months, survived to 6 months of age (Gibb, White & Ward, 1985).

Beyond the age of 6 months, on average about 40% of rabbits survive each successive year. In a declining population in the Wairarapa enclosure, first-year rabbits survived less well than 2-year-olds, and 2-year-olds less well than older rabbits (Gibb, Ward & Ward, 1978). Most rabbits die long before they become senescent. The maximum physiological lifespan in captivity is 10-12 years; the oldest (tagged) rabbit known alive in the wild (in Orongorongo Valley, Wellington) was 7 years old when last seen (J. A. Gibb, unpubl.).

POPULATION REGULATION. Population regulation is the net result of variations in the mortality rate and (less critical) of annual productivity. The mechanisms maintaining the balance are not well understood, despite much research. The variable age structure of local populations reflects the relative importance of different mortality factors. Two extremes are illustrated in Fig. 33. The stable population from Wanganui, with fewer young and more older rabbits, is typical of rabbits living in small pockets on improved farmland in the North Island and probably much of the South Island. The unstable population from inland Kaikoura, with many young and few older rabbits, is more typical of drier rangeland in the South Island, where conditions favour juvenile survival (though in the Wairarapa, there were fewer young (<1 year old) in samples of rabbits from low rainfall districts than in those from higher rainfall districts, probably because of a shorter breeding season in the drier districts: Gibb, White & Ward, 1985). Unstable populations, sometimes approaching carrying capacity, are also characteristic of populations on islands and in experimental enclosures, where predation is reduced and dispersal is prevented.

The potential importance of predation in dictating adult as well as juvenile survival was demonstrated in the Wairarapa enclosure (8 ha), where predators (mainly feral cats) reduced the number of rabbits from >1000 to only 13 in 3.5 years; they also prevented any young surviving after leaving the nest

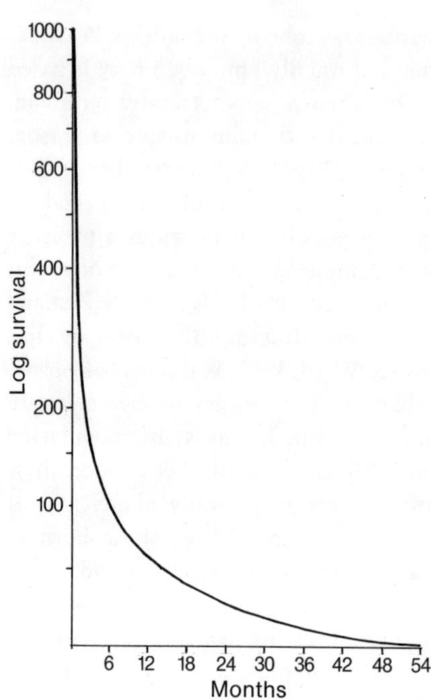

Fig. 32 Survival of young rabbits, born May–November, in the Wairarapa. Young born at other times of year, out of the main breeding season, survive less well (about 1% reach 6 months) (after Gibb, White & Ward, 1985).

during 2.5 breeding seasons. When most cats and ferrets were removed, young rabbits again survived above ground and the population exceeded carrying capacity 2 years later. The increase in the numbers of cats and ferrets, until they were removed, lagged behind the rabbits' increase, thereby generating a predator-prey cycle. The predators could reduce the rabbits so low because some predators survived, subsisting mainly on other foods but still occasionally catching a rabbit (Gibb, Ward & Ward, 1978). This relationship may also hold outside enclosures, but an alternative scenario is possible.

Where contemporary populations are stable, rabbits have a very long breeding season (see above) and provide a reliable supply of young for predators. The prominence of rabbit in the predators' diet, even where rabbits are scarce (Fitzgerald, 1988), suggests that the rabbits' prolific response to a perennial food supply on developed farmland may be important in sustaining the numbers of predators. However, carnivores seem unable to reduce dense, unconfined populations of rabbits, as in Central Otago, where dry conditions favour juvenile survival. The relatively short breeding season of rabbits in these drought-prone districts may also hinder the development of stable predator-prey relationships.

At high density in the Wairarapa enclosure the predators' task was made easier because the rabbits became hungry, though by no means starving. As the rabbits' food supply dwindled, they spent longer hours feeding and ventured further from the warrens to get enough; they also became less watchful of predators.

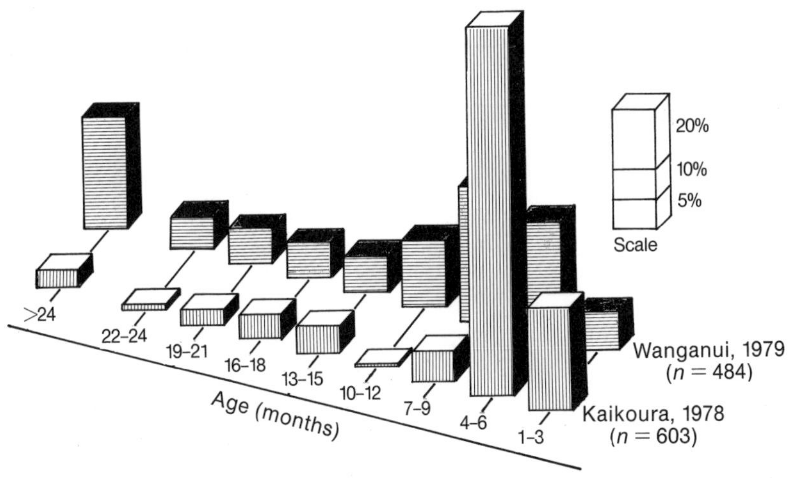

Fig. 33 Percentage age composition of rabbits collected in February, (a) from an unstable population with many young and few old rabbits near Kaikoura, and (b) from a stable population with fewer young and more old rabbits near Wanganui (after Williams & Robson, 1985).

In Australia, up to 80% of young rabbits may be killed by predators within 2 weeks of leaving the nest (Richardson & Wood, 1982), and reduction of predators has resulted in increased numbers of rabbits (Newsome, Catling & Parer, 1983). Equally heavy predation is likely in New Zealand. For example, in Orongorongo Valley, Wellington, most adult as well as juvenile rabbits probably die, eventually, through predation. But identifying the cause of death does not by itself explain the severity of the resulting mortality.

It is much more difficult to see what limits the numbers of rabbits in the contemporary sparse and stable populations, whose numbers fluctuate within narrow limits. In the Wairarapa and in Orongorongo Valley, peak density in autumn is scarcely twice the late winter minimum, and annual fluctuations in the size of the breeding population are minimal (Gibb, White & Ward, 1985; J. A. Gibb, unpubl.). Density-dependent mortality must be finely tuned to regulate numbers so unobtrusively.

Field experience suggests that rabbit numbers respond to the presence or absence of a mix of favourable conditions. These include (1) adequate cover above or under ground; (2) a quantitatively and qualitatively sufficient year-round food supply; (3) modest predation; (4) a benign climate, not too wet; and (5) a warm aspect, well drained and on an easily dug soil. Dominant rabbits may regulate their intolerance of neighbours so as to reduce density-dependent mortality and/or enhance reproductive success.

PREDATORS, PARASITES AND DISEASES

When rabbits were first introduced, the harrier *Circus approximans* was the only predator capable of killing even small young up to 500 g. There were

Table 30: Parasites of wild rabbits in New Zealand.

Species	Status
Coccidia	
Eimeria stiedae	Common in liver of young; sometimes fatal
E. perforans	Widespread, common; may kill
E. pyriformis	Widespread and common
E. irresidus	Uncommon
E. flavescens	Uncertain identification; probably common
E. media	Uncertain identification; probably rare
Trematoda	
Fasciola hepatica	Local in North Island; usual host sheep
Cestoda	
Cysticercus pisiformis	Widespread, in small numbers
Coenurus serialis	Two records, Wairarapa
Nematoda	
Passalurus ambiguus	Mainly South Island
Trichostrongylus retortaeformis	Widespread and common
T. axei	Single record; usual host sheep
Graphidium strigosum	Widespread in South Island and southern North Island; perhaps absent in north of North Island
Nematodirus spathiger	Single record Hawke's Bay; usual host sheep
Pentastomida	
Linguatula serrata	Rare, confirmed only in South Island; usual host dogs
Ixodoidea	
Haemaphysalis longicornis	Single old record; usual host cattle
Acarina	
Listrophorus gibbus	Widespread and common
Cheyletiella parasitivorax	Confirmed in North Island only
Anoplura	
Haemodipsus ventricosus	Uncommon; confirmed in North Island only
Siphonaptera	
Nosopsyllus fasciatus	Single record; usual host rats
Ctenocephalides ? felis	Old record; usual host cats

After Bull (1953b); Tenquist and Charleston (1981).

no native carnivores. Cats were present before rabbits, but were not at first common in the wild. When rabbits became a pest, live mustelids (p. 293) were imported and bred for release, and farmers brought cats from towns. Ferrets, stoats and cats prospered, with no appreciable effect on the numbers of rabbits.

Individual rabbits are susceptible to predation above and below ground. Feral cats kill both adult and young above ground, by ambush or pursuit (when flushed from cover rabbits can sprint a short distance but soon tire); mustelids can follow them into their burrows. Of these predators, feral cats are the more important. Domestic cats and stoats take mainly young (King & Moody, 1982; Fitzgerald, 1988); weasels rarely eat rabbits. Man is also an important predator, affecting the numbers of rabbits both directly and indirectly (see p. 159) and releasing carnivores to prey on them. The combined effects of human and animal predation can be substantial (p. 159, 329, 348).

Of the 21 species of parasites listed in Table 30, three (? four) coccidia, one tapeworm, three roundworms and one mite are reasonably common in the North and/or South Island. Several other species sometimes turn up on rabbits, but have other regular hosts. The specific rabbit flea (*Spilopsyllus cuniculi*), recently introduced to Macquarie I. and the mainland of Australia as a vector for myxoma, is absent from New Zealand.

Severe coccidiosis of the liver by *Eimeria stiedae* can kill young rabbits when they are 6–11 weeks old and living at high densities, but does not affect adults and is inconsequential at low density (Bull, 1953b). *E. perforans* may sometimes kill young rabbits.

Local parasite faunas may still be changing. In the 1950s the nematode *Graphidium strigosum* was spreading north around Lake Taupo at an average rate of 8 km per year (Bull, 1964). In general, parasite loads are probably lighter now than when rabbits were more abundant.

Myxomatosis is absent.

ADAPTATION TO NEW ZEALAND CONDITIONS

Pure-bred descendants of the original domestic stock survive on Rose and Enderby Is. in the Auckland group (Table 29) and "coloured" rabbits (presumably part-domestic) persist inland of Kaikoura; rabbits on the north Canterbury mainland (and formerly on Motunau I.) are unusually heavy and presumably also descended from domestic stock (Table 28). Elsewhere no trace of the early domestic breeds remains, and existing populations resemble those in Britain, whence they came. However, the New Zealand rabbit differs from its relatives in Australia (also introduced from Britain), where the harsh environment has produced distinctive local populations (Myers, 1971).

The initial population "explosion" of rabbits in New Zealand drove them to places where none now survive, such as dense evergreen forest and mountain tops. How rabbits existed in such habitats, and what triggered their demise (before carnivores were widespread), is not known. Nor is it clear today, where the habitat has changed very little, why some populations that had run riot until the 1950s are now sparse and stable.

In Spain, the rabbit's native country, summers are hot and dry. Rabbits evolved opportunistic breeding, which permits them to exploit the flush of green feed that follows fickle autumn rain. The native fauna, and the rabbits, can do the same in semi-arid Australia, but it is less advantageous in the more predictable conditions of Britain and New Zealand. Opportunism in the rabbit contrasts with deterministic breeding in the hare (p. 168).

Worldwide, as in Australasia (Fig. 30), rabbits start breeding later and finish earlier in the season with increasing latitude. New Zealand's temperate, oceanic climate and extensive managed pasture permit a long plant growing season, and hence food to sustain a long breeding season for rabbits over most of the country. Litters are only slightly larger here than elsewhere, but annual productivity is exceptionally high. With so long a breeding season, does born early may themselves breed later in the same season. This enhances the rabbit's potential rate of increase, though in fact very few young born late in the season survive to maturity. The exception is Central Otago, which has a low rainfall and a continental climate. The breeding season is shorter and more sharply defined, and productivity is reduced, yet population density is highest there (p. 146).

The rabbit's high reproductive rate is balanced in most places by equivalently heavy mortality, mostly of young. The causes of this high juvenile mortality are being investigated. Despite the few species of predators, most New Zealand rabbits probably die from predation; but the numerical relationship between predator and prey is poorly understood.

New Zealand rabbits have a generally similar parasite fauna to those in Britain, with two important exceptions. The rabbit flea and the intestinal tapeworm *Cittotaenia* sp. are absent, probably because of the long quarantine enforced by transit in sailing ships. The absence of the flea and the paucity of native vectors probably explain why attempts to introduce myxomatosis in the early 1950s failed (Filmer, 1953). New Zealand is now one of the few countries with rabbits but no myxomatosis.

Rabbits have adapted well to life in New Zealand and are firmly established. They no longer threaten agricultural production on a national scale, because in most districts numbers are naturally controlled at low levels. However, appropriate management is still needed to alleviate damage in some districts, notably Central Otago.

SIGNIFICANCE TO THE NEW ZEALAND ENVIRONMENT

DAMAGE. The impact of rabbits on the grazing lands of 19th and early 20th century New Zealand is legendary. Native grasslands had evolved without grazing mammals; new farms, with soils suitable for burrowing, provided extensive tussock grasslands, burned or heavily grazed by sheep, which suited rabbits well. Rabbits had no natural enemies. The resulting population explosion by rabbits, combined with overstocking of sheep (not recognized at the time), changed the composition of the vegetation. This was perceived by Cockayne (1919) and is well illustrated from contemporary examples of recovery of vegetation after removal of rabbits.

Rabbits were exterminated from Motunau I. in 1962. Silver tussock plants (*Poa caespitosa*) increased in size, and much previously bare ground became covered with ice plant (*Disphyma australis*) and barley grass (*Hordeum murinum*); perennial grasses (*Dactylis, Festuca, Lolium*) and clovers (*Trifolium* spp.), and six species of woody plants, cabbage trees (*Cordyline australis*) and winged thistles (*Carduus tenuiflorus*) reappeared; sorrel (*Rumex acetosella*) and nettles (*Urtica ferox*) decreased. Rabbits had evidently prevented some plants (e.g., boxthorn, *Lycium ferocissimum*) from establishing, while encouraging sorrel and nettles (Cox, Taylor & Mason, 1967).

On Molesworth Station, Marlborough, rabbits were decimated from about

1952. Sweetbriar (*Rosa rubiginosa*), previously checked by rabbits, spread unhindered; *Poa colensoi*, *Holcus lanatus* and sorrel, formerly close-grazed, flourished as never before (Moore, 1976). While the reduction in rabbits was an important element in these changes, the simultaneous change from sheep to cattle grazing probably also contributed.

Similar differences in plant composition were induced by rabbits in the Wairarapa enclosure (Gibb, Ward & Ward, 1978): clover, yarrow (*Achillea millefolium* and hawkbit (*Leontodon taraxacoides*) became less common, and winged thistles more so. However, selective grazing is not the only determinant of species composition: exposure of bare ground encourages annuals, and accumulations of droppings permit the growth of plants demanding high nitrogen levels (e.g., nettles).

Damage to pastures by rabbits crippled farming in many areas. Some farmers, impoverished by declining sheep numbers and wool yields, turned instead to "farming" rabbits: 33 000 skins were exported in 1873, one million in 1877, 9 million in 1882, and nearly 18 million in 1884. Numbers remained high for a long period. A total of 149.5 million skins were exported in the years 1920–29 (Munro & Wright, 1933). Export of rabbit products (skins and canned and frozen meat) continued until the rabbit was "decommercialized" in the mid 1950s, but never offset the damage done by rabbits and did nothing to halt degradation of the land.

CONTROL. The first Rabbit Nuisance Act of 1876 heralded a long battle against the rabbit. Several rabbit fences were built in an unsuccessful effort to halt their northward spread from Southland. Parts of the Hurunui fence are still intact today; it began on the North Canterbury coast and extended 80 km inland. In the 1880s to 1890s much effort was devoted to distributing rabbit predators. From the 1880s to the 1940s rabbits were hunted with guns and packs of dogs, trapped and poisoned. Unfortunately, some of these methods worked against each other: poison laid to kill rabbits also killed carnivores scavenging poisoned meat, while gin trapping and gassing of burrows killed cats and mustelids too. Shooting, though inefficient in controlling rabbits, at least spares the carnivores, which may then make their own contribution to control (p. 329, 348). In the mid 1940s, when World War II interrupted control and hunting, rabbits became more numerous than ever before.

A post-war Government Review resulted in legislation setting up a national Rabbit Destruction Council empowered to make important changes. Trained men were to be employed to control rabbits, the government would provide at least 50% of the finance needed, the rabbit was progressively "decommercialized" and a "killer" policy was adopted. By 1960, 210 rabbit boards had been established, covering 15.1 million hectares.

Spectacular initial success was largely due to the treatment of dense populations with air-dropped carrots and oats, loaded at first with arsenic and then with compound 1080. This is possible in New Zealand, where few vulnerable native species of wildlife inhabit agricultural land. The present distribution and abundance of rabbits is the product of this initial success followed by 30 years of land development (scrub clearance and the maintenance of a rich sward) unfavourable to rabbits. Only where conditions specially favour them, in drier districts with light soils, do rabbit numbers still approach carrying capacity and threaten pastoral production. Such conditions are very local (Map 12).

In May 1985 an Order in Council permitted the import and farming of 10 specified breeds of domestic rabbits, provided that they were free of disease, securely housed, and prevented from grazing pasture. Thus, after a lapse of 30 years, rabbits again became commercial, as well as popular pets. Fears that they might escape and breed with wild rabbits were soon forgotten.

Myxomatosis is absent from New Zealand. Attempts to establish the myxoma virus at several localities in the North and South Islands in the early 1950s failed, presumably because of a lack of vectors (Filmer, 1953). Its absence may permit higher densities of rabbits in a few districts in some years, but its introduction now would probably reduce numbers significantly only in restricted areas and for less than 20 years. It might also destabilize currently stable low-density populations. Myxomatosis could not be contained within present problem areas, and it could jeopardize future use of the wild rabbit as a valuable resource. There is strong public opposition to its introduction on humanitarian grounds (Gibb & Flux, 1983).

In 1985 a proposal to introduce the rabbit flea as a vector and then the myxoma virus from Australia, to control rabbits in Otago, was withdrawn after an adverse Environmental Impact Report (Bamford & Hill, 1985); but it was amended and resubmitted in 1987. The Parliamentary Commissioner for the Environment advised against these introductions. Instead, an integrated $28 million Rabbit and Land Management Programme began in 1989, principally to protect lands severely affected by rabbits.

National extermination of rabbits, once considered possible and adopted as official policy until 1971, is recognized as impracticable and unnecessary (Howard, 1959; Gibb, 1967). However, direct control will always be needed in a few especially rabbit-prone districts; elsewhere rabbits are more susceptible to habitat manipulation. Rabbits are no threat where good soils, easy topography and adequate rainfall have allowed land development to capitalize on initial successful control. Problems persist only where physical conditions restrict land development while enhancing rabbit survival.

J. A. G. & J. M. W.

Genus *Lepus*

Lepus is the largest genus of lagomorphs, containing about 21 species of open-country hares distributed throughout the world. Two species live in Britain, with almost mutually exclusive distribution: *L. timidus*, the mountain or arctic hare, in Ireland and highland Scotland, and *L. europaeus*, the brown hare, in lowland Scotland, Wales and England. Only *europaeus*, highly regarded as a game animal, was brought to New Zealand.

Fig. 34 Brown hare, half alert in form (J.E.C. Flux).

14. Brown hare

Lepus europaeus occidentalis de Winton, 1898

Synonyms *Lepus timidus* (*nec* Linnaeus) Erxleben, 1777; *Lepus europaeus* Pallas, 1778; *Lepus medius* Nilsson, 1820; *Lepus capensis* (*nec* Linnaeus) Petter, 1961.

Recent workers have used *L. capensis*, but from DNA relationships (Schneider & Leipoldt, 1983) and lack of natural hybridization where *europaeus* and *capensis* meet in Russia (Angermann, 1983), they are again considered distinct species (Corbet & Hill, 1986). Hares in New Zealand are the British subspecies (*L. e. occidentalis*), which is darker and smaller than European subspecies living at the same latitude.

Also called common hare, European hare, field hare.

Males are "jacks" or "bucks"; females are "jills" or "does"; young are "leverets".

DESCRIPTION (Fig. 34)
Distinguishing marks, Table 27 and skull key, p. 24.

The hare is about the size of a domestic cat, but with much longer ears and hind legs. It sits upright or crouches like a cat, but never lies curled up on its side.

PELAGE AND MOULTS. The summer coat is tawny on the cheeks, chest, shoulder, sides and haunch, and mottled black-and-fawn on top of the head and back, where the hairs are longer and marked with variable bands of black and brown. The belly is pure white; below the chin, up to a line from the nose to the base of the ear through the eye, near white. The tail is white below and black above with a white margin, and the ears have pale edges and black tips. By late summer the coat appears buffer and less tawny, from bleaching and from abrasion of the black tips on each hair. In winter the coat is thicker but the same colour as the summer coat, except that it has more white hairs around the base of the ears, on the stripe through the eye, and over the haunch, which may become predominantly grey.

Leverets (Fig. 35) are born with a thick coat of pile hair the same colour as an adult's, but as they grow, more dark underfur becomes visible, giving a characteristic dusky look.

Young animals first moult when they reach a body weight of 1.25–1.75 kg, apparently irrespective of time of year. The second moult is to winter pelage in April, when adult males and females are also moulting (Flux, 1967b). The spring moult in adults is more variable and is earlier in males (peak in August) than females (November). Some moulting hares can be found in all months. In Scotland the spring moult starts along the back and proceeds downwards, but finishes on the head. The autumn moult is in the reverse order (Hewson, 1963).

SCENT GLANDS. The anal gland in hares is much smaller than in rabbits and is in two parts, the anterior one non-secretory. The posterior part increases in weight and activity during the breeding season, in both sexes (Mykytowycz, 1966a). The inguinal glands open into bare skin pouches either side of the genitalia. They are larger, and differ less between the sexes and seasons, than in rabbits (Mykytowycz, 1966b). In contrast, the lachrymal and Harder's glands, situated in the eye socket, are larger than in rabbits and vary with the breeding season (Mykytowycz, 1966c). Harder's gland is larger in male than in female hares but its function is unknown; it might produce a secretion for application to twigs or vegetation via the nasal pad (Braestrup, Thorsonn & Wesenberg-Lund, 1949; Flux, 1970), which is also larger and rougher in males (Westlin, Jeffsson & Meurling, 1982). The submandibular gland is undeveloped in hares compared with rabbits (Mykytowycz, 1965). In general, the glands in hares seem to function for individual or sexual identification, not for territorial marking as in rabbits.

THE SENSES of smell, hearing and sight are acute, although hares cannot easily discriminate stationary objects. Long whiskers on the muzzle, cheeks,

Fig. 35 Leveret (young) of brown hare (J.E.C. Flux).

and above the eyes provide tactile sense, as may the nasal pads (Westlin, Jeffsson & Meurling, 1982).

VOICE. Hares are normally silent, but may grunt and grind their teeth, and drum with the feet; when caught they scream loudly.

SKULL. The arched shape of the skull is characteristic, and hares can also be distinguished from rabbits by the larger size, wider internal nares (Fig. 9) and absence of an interparietal bone (see Barrett-Hamilton, 1912). Many skull bones are fenestrated for lightness, an adaptation for fast running; 15 skulls of adult hares averaged 23.8 g, range 19.7–27.9 g (additional calcium is deposited with age, most obviously on the supraorbital bones). By comparison, a possum of the same average body weight (3.6 kg) had a skull weight of 33.3 g.

Dental formula $I^2/_1\ C^0/_0\ Pm^3/_2\ M^3/_3 = 28$.

FIELD SIGN

Hare footprints are distinctive because the large hind feet overreach the forepaws, and at slow speed they are normally placed asymmetrically, unlike rabbits, which have the elongated hind footprints side by side. At top speed the gait becomes completely symmetrical. On soft surfaces, such as snow, the five hind toes are spread, leaving prints like a dog's. The four toes on the small forepaws are kept close together to give a pear-shaped print, and forefoot prints follow each other in line, about 100–200 mm apart.

Table 31: Body measurements of adult brown hares in New Zealand (mean ± SE).

	Waikato (38°S)	n	Wairarapa (41°S)	n	Canterbury (43°S)	n	Southland (46°S)	n	All areas	n
Males										
Ear (mm)	95.4 ± 0.3	100	96.5 ± 0.3	100	97.6 ± 0.3	100	95.4 ± 0.3	100	96.2 ± 0.1	400
Hind foot (mm)	145.2 ± 0.4	100	143.3 ± 0.5	100	143.8 ± 0.5	100	144.0 ± 0.4	100	144.1 ± 0.2	400
Weight (kg)	3.25 ± 0.02	103	3.29 ± 0.02	129	3.29 ± 0.03	107	3.38 ± 0.03	106	3.30 ± 0.01 (2.4–4.4)	445
Skull length (mm)			98.6 ± 0.4 (96.9–99.9)	10	96.1 ± 0.9	10			97.3 ± 0.5	20
Skull width (mm)			46.7 ± 0.4 (45.1–48.4)	10	45.4 ± 0.7	10			46.1 ± 0.4	20
Females										
Ear (mm)	94.5 ± 0.3	100	95.5 ± 0.3	100	97.4 ± 0.4	100	95.4 ± 0.3	100	95.7 ± 0.2	400
Hind foot (mm)	145.2 ± 0.5	100	142.1 ± 0.4	100	144.1 ± 0.4	100	143.7 ± 0.4	100	143.8 ± 0.2	400
Weight (kg)	3.73 ± 0.03	104	3.66 ± 0.02	192	3.77 ± 0.04	123	3.88 ± 0.03	133	3.75 ± 0.02 (2.4–4.8)	552
Skull length (mm)			96.7 ± 0.5 (94.4–98.7)	10	96.6 ± 0.5	10			96.6 ± 0.3	20
Skull width (mm)			46.8 ± 0.3 (44.5–48.0)	10	45.5 ± 0.3	10			46.1 ± 0.3	20

Data from Flux (1967b).

The same paths or "runs" are used habitually and in deep grass may become conspicuous depressions 100–200 mm wide, running along ridges and up and down slopes. Groups of pellets accumulate at favoured stopping places. "Forms" are oval, 200 × 400 mm depressions in vegetation, or sometimes scratched into a bank, in which the hare sits all day. They are often in the shelter of rushes or shrubs; in unsheltered forms hares face downhill or with their backs to the wind. Forms may be temporary, or well-used traditional sites. In winter they may become closed-over snow tunnels.

The pellets are typically flattened spheres, 15 mm × 10 mm, with a slight "tail" on one side. They tend to be larger, paler and more fibrous than rabbit pellets, but this reflects the diet: hares fed on clover produce small, dark, pear-shaped pellets very like a rabbit's. Captive hares and rabbits on the same diet produce identical pellets, scaled to the size of the animal. Pellets normally disintegrate in 1–3 months, but in alpine areas may have a half-life of 3 years (Flux, 1967a), giving the impression that the animals are very abundant. On valley flats in Canterbury, pellet densities of 100/m^2 are normal (J. P. Parkes, unpubl.).

Hares clip vegetation with a characteristic 45° cut, and the tips of the plant eaten are often left on the ground with a few pellets nearby. Scraping to reach plant roots is less common than in rabbits.

MEASUREMENTS (Tables 31, 32)

Among 800 hares collected from four representative locations on the mainland, body weight showed a significant ($P < 0.001$) increase of 30 g per degree of latitude, as expected from Bergmann's Rule (Flux, 1967b). At high density in a subalpine valley in Canterbury, 410 adult males averaged 3.15 kg, and 353 adult females averaged 3.3 kg (Douglas, 1970a).

VARIATION

Grey phase individuals (which appear white at a distance but have a sprinkling of dark hairs, dark eyes and black ear tips) are relatively common (1:5000). True or partial albinos are less common (1:50 000), and melanics are extremely rare (1:500 000). No other colour forms have been recorded in New Zealand. Four of 1603 hares examined had a white band round the right wrist (Flux, 1966), an abnormality otherwise reported only from the Netherlands (Broekhuizen & Kalsbeek, 1969). Unusual animals, sometimes thought to be hare × rabbit hybrids (which are impossible), usually turn out to be aberrant domestic rabbits.

HISTORY OF COLONIZATION

The first hares came on the *Eagle* in 1851, jumped from a porthole and swam ashore at Lyttelton (Donne, 1924), where they apparently died out. Most of the hares now established in New Zealand came later from England via Phillip Island, Victoria, where six hares increased in two years to 200 by 1865 (Rolls, 1969). Recorded introductions prior to 1876, by which time hares were causing "widespread damage" (Lamb, 1964), include: Otago,

Table 32: Comparison of body weights of adult hares in New Zealand and Britain (kg ± SE).

	New Zealand (general)	n	Scotland (Kincardineshire)	n	Significance of difference
Males	3.30 ± 0.01	445	3.43 ± 0.07	35	ns
Females	3.75 ± 0.02	552	3.34 ± 0.06	51	$P < 0.01$
Source	Table 31		R. C. Flux, unpubl.		

Samples collected all year round, but most Scottish ones were from the shooting season, when female hares are lighter.

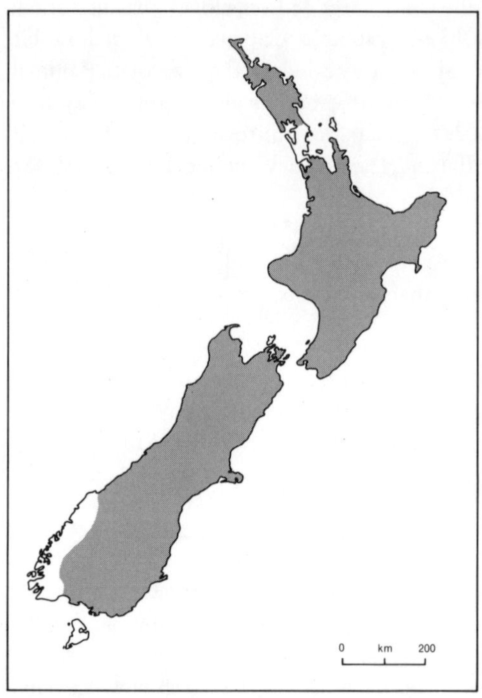

Map 13 Distribution of brown hares in New Zealand.

3, 1867; 1, 1869; 3, 1875; Southland, some, 1869; 2, 1871; 2, 1874; Canterbury, 1, 1867; 2, 1868; 1, 1869; some, early 1870s; 3, 1871; 9, 1873; Nelson, 2, 1868; 1, 1869; 19 "liberated to date", 1870; some, 1872; 9, 1873; Wellington, 2, 1874; 14, 1875; Auckland, 2, 1868; some, 1870; 5, 9, 1871 (Thomson, 1922; Lamb, 1964; Ashby, 1967; Sowman, 1981). The Protection of Certain Animals Act of 1861 gave hares protection until 1870, but an amendment of 1866 allowed landowners or tenants to control them. The subsequent spread was rapid but largely unrecorded. Guthrie-Smith (1969) described hares swarming at Peel Forest in the early 1880s, hemmed in by mountains and forest and the Rangitata River, and Burnett (1927) says they were first noticed in alpine valleys in Canterbury in 1896, reached maximum numbers in 1903, and declined for no apparent reason thereafter. In the North Island, hares reached Tutira in 1893 (Guthrie-Smith, 1969) and were causing "much trouble" among pines at Kaingaroa in 1899 (McKinnon & Coughlan, 1960). Because of the complex topography and repeated liberations in New Zealand, the rate of spread cannot be measured; in Australia it was about 60 km a year (Jarman & Johnson, 1977).

DISTRIBUTION

WORLD. The brown hare probably originated as a large mountain form of the desert hare, *L. capensis*, in Israel or Iran. During the postglacial period it spread rapidly over open country throughout lowland Europe, following human clearance of forests for agriculture. For at least 2000 years the distribution of brown hares has been increased by liberations for sport (Barrett-Hamilton, 1912). The more recent of these were in the Falkland Is. about 1740 (Harris, 1982), Barbados 1842 (Feilden, 1890), Ireland 1852 (O'Rourke, 1970), Tasmania 1854 (Guiler, 1968), Australia 1859 (Rolls, 1969), Sweden 1886 (Frylestam, 1976), South America 1888 (Grigera & Rapoport, 1983), USA 1888 (Osborn, 1933), Canada 1912 (Dymond, 1922), Siberia 1935 (Berger, 1944), and Irkutsk 1938 (Il'in, 1962).

NEW ZEALAND (Map 13). Brown hares are now spread throughout both main islands on all suitable habitat from sea level to 2000 m, except for parts of South Westland, most of Fiordland (Parkes, Tustin & Stanley, 1978), and an area from Auckland to about 80 km north. They are absent from all the other islands. New areas, e.g., South Westland (P. D. Gaze, unpubl.) and Urewera river flats (W. Shaw, unpubl.) have been colonized within the last 20 years as road construction provided access.

HABITAT

Hares occupy practically all kinds of grassland or open country, including coastal sand dunes, cropland, pasture, clearings in scrub or forest, marshes and moorland. Alpine tussock grassland from the timberline to the upper limit of vegetation is occupied even in winter, when hares tend to seek cover for the day at the timberline and feed at night on vegetation exposed by the wind. Even where forest cover is readily available within the home range, only 2% of the time is spent there (Parkes, 1984a).

FOOD

On alpine grassland in Cupola Basin, Nelson Lakes National Park, hares ate short grasses in summer, and shrubs and tussocks in winter; over the whole year the percentage occurrence in the diet averaged 31% *Poa colensoi*, 26% *Chionochloa* spp., and 11% *Celmisia*. The preference order of the 20 commonest plants eaten (estimated from the number of bites at each) was: *Poa colensoi, Chionochloa rubra, C. pallens, Schoenus pauciflorus, C. flavescens, Aristotelia fruticosa, Celmisia allanii, Hymenanthera alpina, Celmisia coriacea, Coprosma pseudocuneata, Oreomyrrhis colensoi, Dracophyllum uniflorum, Pittosporum divaricatum, Anisotome filifolia, Wahlenbergia albomarginata, Trifolium repens, Nothofagus solandri* var. *cliffortioides, Celmisia spectabilis, Coprosma brunnea,* and *Gaultheria depressa* (Flux, 1967a).

At 1100–1600 m on Mt. Ruapehu, hares showed little seasonal variation in diet, and differences with altitude generally reflected availability. The average annual diet over 10 locations was 44% *Chionochloa rubra*, 24% *Celmisia spectabilis*, 15% *Senecio bidwillii*, 7% moss, 4% *Poa colensoi*, 2% seeds, 2% grass, and traces of *Notodanthonia setifolia, Schoenus pauciflorus, Aciphylla squarrosa*, bark, *Hebe odora, Calluna vulgaris*, and *Nothofagus solandri* (Horne, 1979). Hares at Mt. Cook browse *Carmichaelia* and *Celmisia lyalli* (Wilson, 1986).

There is no information on the diet of hares in other habitats in New Zealand, but in Europe hares on agricultural land eat 90.1% soft green feed, 5.5% woody plants, 2.2% root crops, 1.7% grain crops, and 0.5% forest plants (Zörner, 1977).

SOCIAL ORGANIZATION AND BEHAVIOUR

Hares are largely solitary and nocturnal. By day they crouch in a form, reingesting (p. 138) about 10 times per day (Watson & Taylor, 1955; Flux, 1981a). True sleep occupies only 2–4 minutes per day, usually in the afternoon. At dusk, or earlier in spring, hares start feeding and may move some distance to find grazing (e.g., 1.7 km in Scotland; Hewson & Taylor, 1968). Above the timberline, in snow, they may travel 15 km while feeding in one night (Flux, 1967a). Many may congregate on favoured crops such as lucerne, and they often feed in pairs, of the same or opposite sexes, 1 m apart. In a subalpine valley west of the Craigieburn Range, Canterbury, hares moved on average 280–640 m (Douglas, 1970a) and five adult home ranges averaged 53 ha (Parkes, 1984a), compared with 29 ha in the Netherlands (Broekhuizen & Maaskamp, 1982), 25–100 ha in England (Barnes, Tapper & Williams, 1983) and 330 ha in Poland (Pielowski, 1972). Feeding takes about a third of a hare's total time; towards dawn it returns to its daytime range, grooms for 5–15 mintues, and settles down in one of its forms at almost the same time day after day. Before entering the form, it "back tracks", approaching the form by a different route each time (Flux, 1981a). Other tricks used to deflect predators are circling, leaping sideways, chasing a fresh hare from its form, and using a well-practised "race track". Rabbits dominated hares in 45 of 55 encounters on a communal feeding area (Flux, 1981a).

REPRODUCTIVE BEHAVIOUR is most intense in September (mad "March" hares). Several males congregate round a female in oestrus; the dominant male attempts to chase off rivals, and this leads to vigorous boxing matches if his authority is disputed. The male then approaches the female cautiously; she threatens by crouching with ears lowered, and strikes out with her forefeet if the male comes too close before she is receptive (Schneider, 1976, 1978). A female may thus rebuff males for many hours. Copulation may be at any time of day, and is frequent: one pair mated 23 times in 1 hr 24 min (Flux, 1981a).

REPRODUCTION AND DEVELOPMENT

The breeding season starts soon after the shortest day in all nine countries studied except Argentina (Flux, 1965a; Amaya, Alsina & Brandani, 1979), and in New Zealand over 90% of females are pregnant from August to

February. The season starts a week or two earlier in mild than in cold winters, and young females breeding for the first time start three weeks later than adults. Mating is promiscuous, and ovulation is induced, but because sperm may remain viable for as long as the 42-day gestation period, the paternity of subsequent litters is never certain (Martinet & Raynaud 1972). Superfoetation (fertilization of a female already pregnant, leading to the simultaneous development of embryos of different ages) is common in captivity, where the mean interval between births was 38 days (Martinet, 1977), but only 3 of 24 females with full-term embryos showed superfoetation in the wild (Flux, 1967b).

In an extensive study of breeding in New Zealand hares, Flux (1967b) found that up to 70% of young females heavier than 2.7 kg (to about 5 months old) bred in the same season in which they were born. Only in April were no males fecund, and only in May were no pregnancies recorded. Embryos tended to occupy alternate uterine horns in successive pregnancies. The mean number of corpora lutea per litter in adult females (2.8) rose from 1.0 in June to 3.8 in November before declining slightly. The average litter size (with embryos over 30 g, or after about 32 days of gestation) was 2.14 and the average number of successful litters per year was 4.59, giving an annual production of 9.8 young per female, allowing for pre- and post-implantation loss. Young females had small litters (1.4) and contributed little to the population in their year of birth (Flux, 1967b). In inland Canterbury, the non-breeding season for females was longer (21 March–3 July), none of 36 juvenile females examined bred in their year of birth, and the first juvenile bred on 22 July (J. P. Parkes, unpubl.).

The mean litter size of *L. europaeus* in 11 different countries is inversely correlated with average annual temperature and ranges from 2.8 at 7°C (Sweden) to 2.2 at 12°C in North Island, New Zealand (Broekhuizen & Maaskamp, 1981). North and South Island hares fit this trend roughly, although there was no significant difference in either ovulation rate or survival of embryos in a large sample of hares from an inland valley in Canterbury (Douglas, 1970a) compared with low-ground hares.

At birth leverets weigh 123 (100–165) g and then grow at 18.8 (16–27) g per day until 3 months old. In the Netherlands they are nursed once a day for a few minutes about an hour after sunset (Broekhuizen & Maaskamp, 1980), but in New Zealand the only female observed, for two nights, fed her young twice, at dusk and 0200 hours (Flux, 1981a). The leverets begin to eat some vegetation when they reach 200 g (about a week old). Most are weaned at 1 kg, but some continue to suckle until they reach 1.4 kg. Adult size is reached in 5 months (Flux, 1967b).

POPULATION DYNAMICS

Only in inland Canterbury do hare populations reach the high densities found in Europe. Here Douglas (1970a) snared 977 different hares on 1300 ha in two years, so the population density must have been at least 1/ha (Parkes, 1984a, suggests 2–3/ha) or ten times the rough estimate for average New Zealand populations (Flux, 1981b).

Table 33: Counts of rabbits and hares in Britain and New Zealand, made from a car by the same observer on tour.

		Britain 1977	New Zealand	
			North Island 1965-66	South Island 1964
Rabbits	Live	396	5	3
	Dead	435	20	19
Hares	Live	27	0	18
	Dead	12	19	24
Miles travelled		2 300	2 200	2 400

Data from J. E. C. Flux (unpubl.).

The sex ratio at birth (117 males : 96 females, or 55% males) slowly declines with age to 45% males among shot samples of adult hares, and 33% male in the oldest age group (Flux, 1967b). This indicates that female hares live longer than males, as Pielowski (1971) found in Poland, where the maximum physiological age was assessed at 12-13 years. Among 778 snared adult hares from Canterbury, Douglas (1970a) recorded 54.1% males, perhaps reflecting the fact that males move further than females. Reviews of several European studies (Abildgård, Andersen & Barndorff-Nielsen, 1972; Broekhuizen, 1979; Zörner, 1981; Kovács, 1983) gave average adult mortality rates of 45% to 66%, so less than 1% of the population would reach 8 years old, and extremely few would die of old age.

Leverets are subject to high mortality, and the highest proportion of young in the population is only 56% in March (Flux, 1967b). Over March-May, figures comparable with those for the hunting season in Europe show 46% of the New Zealand population as juvenile, compared with 14-50% in Germany (Rieck, 1956) and 33-60% in Poland (Pielowski, 1976).

PREDATORS, PARASITES AND DISEASES

Adult hares are remarkably free from predation in New Zealand. When first introduced they were attacked by harriers (*Circus approximans*) and "not infrequently could be noticed hirpling about the run half-plucked with two or more birds in hot pursuit" (Guthrie-Smith, 1969); but they soon learned to take avoiding action (e.g., Sagar, 1979). Many leverets are taken by harriers, whose diet in high-country Canterbury included 78% hare (Douglas, 1970b), although for New Zealand as a whole it is only 5-6% (Carroll, 1968; Redhead, 1968), and 0.5% in Manawatu (Baker-Gabb, 1981). Other recorded predators of (usually young) hares include stoats (King & Moody, 1982), ferrets and weasels (Douglas, 1970a), and cats (Gibb, Ward & Ward, 1969; Collins & Charleston, 1979a). Marples (1942) records a little owl feeding on a dead hare, probably as carrion.

Tenquist and Charleston (1981) list six species of ectoparasites from New Zealand hares: *Demodex* sp., *Haemophysalis longicornis, Listrophorus gibbus, Linguatula serrata, Haemodipsus lyriocephalus,* and *Ctenocephalides canis canis*. Flux

(1967b) also recorded one *Ctenocephalides felis* in 1513 hares examined. There has been no specific study of endoparasites in hares, but Bull (1964) records that 98% of 41 hares examined had *Trichostrongylus retortaeformis*, and 35% of 20 hares had *Graphidium strigosum*. Undoubtedly hares must carry other species of parasites, for example *Fasciola hepatica* and *Eimeria* spp. Brucellosis in hares is normally caused by *Brucella suis* (not recorded in New Zealand); five hares examined for leptospirosis were negative (Hathaway, Blackmore & Marshall, 1981); and two of 52 hares examined carried *Yersinia pseudotuberculosis* (Mackintosh & Henderson, 1984).

ADAPTATION TO NEW ZEALAND CONDITIONS

The body weights of hares in New Zealand and Britain are compared in Table 32.

Reduced genetic variability would be expected in New Zealand hares, because most of the individuals brought here came from breeding stock established from only six founders (p. 165). This may explain the high incidence of ovarian tumours (5.6% in 588 adult female hares, Flux, 1965b; none found in far larger samples examined in Europe), and of missing rear upper molars (7.9% of 140 skulls, Flux, 1980; 9.8% of 1059 skulls in inland Canterbury, J. P. Parkes, unpubl.; cf. 1% of 235 skulls from Europe in the British Museum (Natural History)). Otherwise the hares seem very similar to the original English stock, with none of the drift in characters that might be expected — indeed, was expected by early naturalists (Thomson, 1922).

Hares in New Zealand are relatively free of parasites, as are many other mammal species here. The long quarantine on sailing ships bringing hares to New Zealand resulted in relatively clean stock; by comparison, Austrian hares carry 45 species of parasites (Kutzer & Frey, 1976). Many of the diseases which affect or can be transmitted by hares in Europe, for example plague, rabies, tularemia, brucellosis and myxomatosis, are not present in New Zealand.

Two distinct habitats are occupied in New Zealand: agricultural pasture largely composed of the same plant species as in Britain, to which the hares would be pre-adapted; and alpine tussock grassland, with a completely new range of plants, and a climate and topography far more suited to the mountain hare, *L. timidus*. Although relatively few New Zealand brown hares live at high altitudes, they are resident and appear to maintain their numbers, so they can obviously survive in such habitat in the absence of competition.

Over the whole country the density of hares is comparable with that overseas, but they tend to be more nocturnal than in Britain, so are less often seen (Table 33). In the Netherlands, peak populations on an airfield reached 2.35/ha, but 0.25/ha is the average (Broekhuizen, 1976); in Poland the best 20% of hunting areas have 0.5/ha, the others 0.1-0.3/ha (Pielowski, 1976); and in Roumania the density range is 0.1-0.3/ha (Almasan & Cazacu, 1976). The highest accurately recorded density, on a 100-ha island in Denmark, was 3.39/ha, but this population normally fluctuated around 1–2/ha (Abildgård, Andersen & Barndorff-Nielsen, 1972). The best hare habitat, mixed farming with small fields, plenty of cover, and no agricultural pesticides or mechanization, is not available in New Zealand, but the estimated potential harvest of hares (1.9/km^2) is similar to that in Europe as a whole (1.8/km^2, ranging from 9.3/km^2 in Denmark to 0.1/km^2 in Spain; Flux, 1981b). Despite the absence of any significant predation on adults, hares did not stage an enormous population irruption when first introduced, as did rabbits, and contemporary hare densities here are no higher than in Europe. The reason is that hare populations are self-limited by some behavioural mechanism. This observation supports the suggestion that removal of predators in Sweden would probably increase the number of hares available for hunters without necessarily affecting their breeding density (Erlinge *et al.*, 1984).

SIGNIFICANCE TO THE NEW ZEALAND ENVIRONMENT

DAMAGE. Hares always live at relatively low density, hedge palatable plants without killing them, graze a few leaves from many plants over a wide area, and do not dig burrows. Damage to fruit trees, pine plantations and horticulture by hares is significant but less conspicuous than rabbit damage and can be controlled by shooting, fencing and repellents. Conversely, hares may benefit mountain land by carrying nutrient from their feeding grounds on the valley floor to higher slopes where they spend the day. In enclosures in Poland, hares ate 144 g dry weight of fresh vegetation daily, but their damage to grain crops was shown to be more than offset by their fertilizing and weeding the crop (Kaluzinski & Bresinski, 1976). Although the Rabbit Nuisance Amendment Act (1947) gave rabbit boards power to control hares in New Zealand, they have never been officially listed as "noxious animals". There may be a case for reducing their numbers to protect native alpine vegetation; elsewhere they fit into the New Zealand environment better than most other introduced mammals.

MANAGEMENT. Hares used to have value as a sporting animal, but coursing declined in popularity and was eventually outlawed by the Animals Protection Act of 1960. Beagling is apparently still legal but rare (Pitt-Turner, 1974). Hunts on horseback, held by 25 clubs, attract up to 100 riders in fox-hunting uniform during the season from April to July; and "when the district is known to lack hares . . . the hunt then spends the day chasing aniseed" (Steel, 1975). Organized hare-drives are held in Canterbury and Southland, generally to collect funds for charity by selling the hares, but most hare shooting is casual. Hares can occasionally be bought in game meat shops at \$5/kg, half the price of domestic rabbits (1987 prices); the numbers exported vary widely, from 10 000 to 20 000 in most years with a peak of 130 000 in 1966. The game packing regulations introduced in 1966 required hares to be obtained 5 miles (8 km) from the nearest rabbit-poisoning operation, and this curtailed exports from 1967 onwards. Live hares can be captured (by netting from helicopters); the technique was developed to supply live hares for export to France and Italy, but so far MAF regulations have prevented this trade.

J. E. C. F.

Order Rodentia

The Rodentia is the largest order of mammals, comprising 30 families, 389 genera and about 1702 species world-wide (out of a total of about 4070 species of living mammals). The rodents as a group have been extremely successful, adapting to practically every terrestrial and freshwater habitat, and some species are extremely abundant. Their main distinguishing character is their gnawing incisors, a single pair in each jaw with permanently open roots (i.e., they grow continuously) and chisel-shaped cutting edges. The long sockets of these teeth, and the large muscles developed to operate them, determine the shape of the skull; the other pair of incisors, the canines and some of the cheek teeth are absent. Most rodents eat seeds or vegetation, but some are very adaptable omnivores, and a few have developed a close commensal relationship with man.

The four species of rodents now living in the wild in New Zealand are all members of the same family, the Muridae. In addition, two individuals each of two members of the Family Sciuridae (the gray chipmunk, *Tamias striatus*, and the "brown Californian squirrel") were liberated in about 1906 in Dunedin. Guinea-pigs (*Cavia porcellus*, Family Caviidae) are popular pets which must often escape, and were deliberately released in 1869 (Wodzicki, 1950) but have not established in the wild. In 1985, MAF permitted the import of a single shipment of 30-40 chinchillas (*Chinchilla laniger*, Family Chinchillidae, originally from South America), a small rabbit-like animal which produces a fine, valuable fur. Few New Zealand fur farms stock chinchillas yet, as they are expensive and their potential is still being assessed; if they become widespread, escapes are likely in future.

Family Muridae

The murids form a very large family of rats and mice (241 genera, 1082 species) mainly in tropical Africa, Asia and Australasia, but members of two genera (*Mus* and *Rattus*) have spread world-wide in association with man. The four species present in New Zealand were all introduced, accidentally or deliberately; one by the Polynesians about 1000-1200 years ago, and the others by European explorers and settlers in the late 18th and early 19th centuries. The distinguishing characters of the four species are shown in Table 34.

Genus *Rattus*

Most of the numerous kinds (over 60 species and 450 named races) of *Rattus* live in Asia and Australasia. None is native to New Zealand, but *R. exulans* has been here so much longer than the other two, it has almost acquired the status of an "honorary native" species, especially as it used to seem

Table 34: Distinguishing marks of rodents in New Zealand.

	Kiore *Rattus exulans*	Ship rat *Rattus rattus*	Norway rat *Rattus norvegicus*	House mouse *Mus musculus*
Adult weight (g)	Normally 60-80, up to 180	Normally 120-160, up to 225	Normally 200-300, up to 450	Normally 15-20, up to 30
Max. head-body length (HBL) (mm)	180	225	250	115
Tail	Slightly shorter or longer than HBL	Much longer than HBL	Clearly shorter than HBL	Slightly shorter or longer than HBL
	Thin and uniformly dark all over	Uniformly dark all over	Thick with pale underside	Uniformly grey-brown
Ears	15.5–20.5 mm; cover eyes when pulled forward	19.0–26.0 mm; otherwise as for kiore	14.0-22.0 mm; do not cover eyes when pulled forward	12.0–15.0 mm
	Fine hairs do not extend beyond edge of ear		Obvious hairs extend beyond edge of ear	
Adult hind foot length	24.5–31.0 mm	28.0–38.0 mm	30.0–41.5 mm	15.0–21.0 mm
Colour of upper side of hind foot	Outer edge dark near ankle, rest of foot and toes pale	Uniform colouring over whole foot, usually dark	Always completely pale	Uniformly grey
Fur on back	Brown	Grey-brown or black	Brown	Dull grey-brown
Fur on belly	White-tipped grey giving irregular colour	Generally uniform grey, white or creamy-white; rarely irregular colour	Similar to kiore	Uniformly grey or white
Length of droppings	6.4–9.0 mm	6.8–13.8 mm	13.4–19.1 mm	3.9–7.6 mm
Number of nipples	8	10-12, usually 10	12	10-12
Habits	Agile climber, digs small holes; nests mainly on ground; feeds on ground and trees; infrequent swimmer	Very agile and frequent climber; rarely burrows; nests mainly in trees and shrubs; infrequent swimmer	Burrows extensively; climbs much less frequently than other rats; strong swimmer; nests underground; very wary	Mainly ground-dwelling though capable climber; nests in small holes

NB: Identifications should be made from the aggregate of characteristics.

Fig. 36 Kiore (NZFS).

to fit better into the landscape and to cause less damage than the two European species. This opinion is understandable but wrong: we are only now beginning to appreciate the disruption to the native fauna caused over hundreds of years by *R. exulans*. All three species are aliens that have, in their different ways, transformed the ecosystems they colonized.

15. Kiore

Rattus exulans (Peale, 1848)

Synonyms *Mus exulans* Peale, 1848; *Mus maorium* Hutton, 1877, 1879.

Also called Maori rat, Pacific rat, Polynesian rat.

DESCRIPTION (Fig. 36)

Distinguishing marks, Table 34 and skull key, p. 24.

The kiore is the smallest of the three species of rats in New Zealand, but similar in appearance to the ship rat. Its fur is brown above, darker along the spine, and intermingled grey and white on the belly. The hind foot is dark at the outer edge near the ankle but otherwise pale. For external sexual characters, see Fig. 39. No melanic forms have been reported from New Zealand, although some are present in Samoa (Marples, 1955).

The tail is evenly dark all over, slim and about the same length as that of the head and body together. There are eight nipples, arranged as two pectoral and two inguinal pairs. Quay and Tomich (1963) have described a specialized midventral sebaceous gland of narrow linear shape, primarily characteristic of adult males but of uncertain function.

Chromosome number 2n = 42 (Yosida, 1973).

Dental formula $I^1/_1\ C^0/_0\ Pm^0/_0\ M^3/_3 = 16$.

Table 35: Body measurements (mean and range) of adult kiore in the New Zealand region.

Sex	Head and body length (mm)	n	Tail length (mm)	n	Condylobasal length of skull (mm)	n	Weight (g)	n	Reference
Macauley I., Kermadec group (30°S)									
Males	133 (122–144)	16	158 (148–164)	15	32.5 (31.1–33.9)	14	84.0 (74–104)	16	Moors (1981 and unpubl.)
Females	127 (118–135)	12	149 (140–156)	12	31.1 (29.7-32.0)	9	72.1 (57–85)	12	
Marotiri (Lady Alice) I., Chickens group (35°53′S)									
Males	158 (123–175)	93	165 (141–180)	89	36.6 (35.7–37.8)	4	134.5 (80–187)	76	I. McFadden (unpubl.)
Females	146 (112–165)	80	159 (115–181)	80	35.4 (34.1–36.0)	5	104.4 (57.5–150)	68	
Tiritiri I., inner Hauraki Gulf (36 °36′S)									
Males	156 (133–184)	82	157 (136–180)	82	33.9 (30.1–35.3)	50	108.5 (76–145)	82	H. Moller (unpubl.)
Females	147 (118–165)	67	156 (122–173)	67	33.5 (32.3–34.5)	50	95.1 (47–124)	67	
Kapiti I. (40°52′S)									
Males	149 (128–177)	37	154 (129–171)	35	—		91.1 (55–130)	39	A. M. P. Dick (unpubl.)
Females	139 (87–162)	37	145 (105–165)	36	—		84.7 (30–150)	36	
Stewart I. (47°S)									
Males	126 (109–144)	31	140 (125–163)	28	31.6 (30.0–33.8)	10	76.9 (55–97)	31	P. J. Moors & D. M.
Females	121 (99–141)	41	137 (124–166)	34	31.1 (22.6–33.8)	13	67.2 (51–87)	39	Cunningham (unpubl.)

These five populations were chosen to represent the range of latitudes occupied by kiore in the region.

Kiore can swim, but not willingly or well. They can stay afloat for up to 160 minutes in tropical waters (Jackson & Strecker, 1962b) or up to 17 minutes (average 10 minutes), swimming up to 130 m (average 65 m), in New Zealand waters (17°C) (n = 9, A. H. Whitaker, unpubl.).

FIELD SIGN

Food remains can be found in small, dry, enclosed spaces in fissures, rocks, tree roots and trunks where kiore retreat to eat (Campbell *et al.*, 1984). The accumulated bits of chewed hard parts of seeds, insects and snail shells bear characteristic rodent gnawing marks. Kiore sometimes strip bark and uproot seedlings, and sometimes dig small scrapes up to 10 cm deep in search of food. They shelter underground among rocks and in tree stumps and roots, but apparently do not excavate tunnels or underground nests. Nests have been found in tree holes behind loosely attached bark and at the bases of low bushes. A nest illustrated by Daniel (1969) appears to be formed mainly from grass leaves and stalks. Dry pellets of kiore tend to be smaller than those of the other rats (Daniel, 1969), but vary enormously in dimensions when fresh.

MEASUREMENTS

See Table 35. The measurements given should be regarded as indicative rather than definitive, because the age and size of kiore at reproductive maturity ("adulthood") vary greatly between populations, as well as from year to year (see below). Moreover, the methods of measurements used by different observers are not always identical. Cunningham and Moors (1983) describe a standard technique, recommended for future work.

VARIATION

The kiore from Marotiri I. are heavier than any recorded elsewhere, in New Zealand or overseas (Table 35), but it is not known whether this is a founder effect inherited from the rats that originally colonized the island, an effect of subsequent selection, or a local environmental effect related perhaps to food supply. There is some variation in age at maturity, apparently related to habitat (p. 187).

HISTORY OF COLONIZATION AND DECLINE

Kiore were brought to New Zealand by the Polynesians, although whether deliberately or accidentally is not known. Whatever the reason, kiore bones are present in archaeological sites of all ages (Davidson, 1984), so kiore must have become widespread in suitable habitats fairly soon after the Polynesians arrived in New Zealand, between AD 800 and 1000.

The Maori trapped kiore extensively for food. Two main methods were used: pit traps (1.2–1.5 m in depth) baited with berries, and spring snares skilfully constructed with supplejack (*Ripogonum scandens*) and other pliant plant materials. Details of construction of these and other kinds of trap are given by Wilson (1878), Downes (1926) and Best (1942). Towards the end of the trapping season (August), when kiore tended to move from the forest to scrub and fernland, the shorter vegetation was sometimes burnt and the animals dug from holes in the ground. Kiore were then plucked of their fur, cooked in an *umu* (steam oven) and, if not eaten immediately, packed in their own fat in kelp-bags or gourd-vessels. They could then be carried from place to place and were sometimes eaten during ceremonial feasts (Best, 1942).

The care taken in trapping and preparing kiore suggests that they were much esteemed as food, even though they did not contribute a major part of the diet. On the other hand, kiore also probably caused some losses to stored food; Best (1942:420) mentions *kumara* (sweet potatoes) being damaged by kiore. One reason for the widespread pre-European use of storehouses mounted on poles (*pataka*) may have been to protect food or belongings from damage, although later this proved a useful defence against European rats as well.

The use of kiore as food has led to the belief (e.g., by Stead, 1937) that they were deliberately carried to offshore islands to establish stocks. However, the absence of kiore from some important islands settled by the Maori, including the Three Kings and Poor Knights, considered with the generally irregular distribution of this rat among offshore islands, suggests that island introductions around New Zealand may have been accidental rather than purposeful (Atkinson, 1986).

Mainland populations of kiore varied greatly in numbers from place to place and from year to year. Both Cook and Banks noted that kiore were scarce at places they visited in New Zealand during 1769 and 1770. The Maori knew that kiore periodically irrupted to high numbers, and that these events coincided with years in which beech trees (*Nothofagus* spp.) produced plentiful seed (Best, 1942). This kiore/beech seed relationship appears to have paralleled the one that can be observed today in mice in New Zealand beech forests (King, 1983a). The present-day absence of a similar response to beech seed by ship and Norway rats supports the suggestion that mice and kiore have niches similar to each other and distinct from those of the two larger rats. Although small populations of kiore still remain in parts of Fiordland where beech forest is present (e.g., Hollyford Valley), they no longer irrupt.

Plagues of rats encountered by the first settlers at New Plymouth in 1841 and 1842 may have been kiore (Atkinson, 1973a). If so, then very high numbers of rats sometimes also developed in non-beech mainland forest, as they still do on many offshore islands without beech forest. However, the best-known kiore irruptions on the mainland in historic times (1884, 1888) were associated with heavy seedfalls of beech in the South Island (Meeson, 1885; Rutland, 1890). According to these observers, most rats leaving the forest during the irruptions were males. Earlier irruptions were noticed in the north of the South Island in 1872, 1878 and 1880 (Thomson, 1922:80), but this is certainly an incomplete record.

Kiore disappeared from most of the North Island by about 1850–60, and from the northern part of the South Island in the 1890s (Atkinson, 1973a).

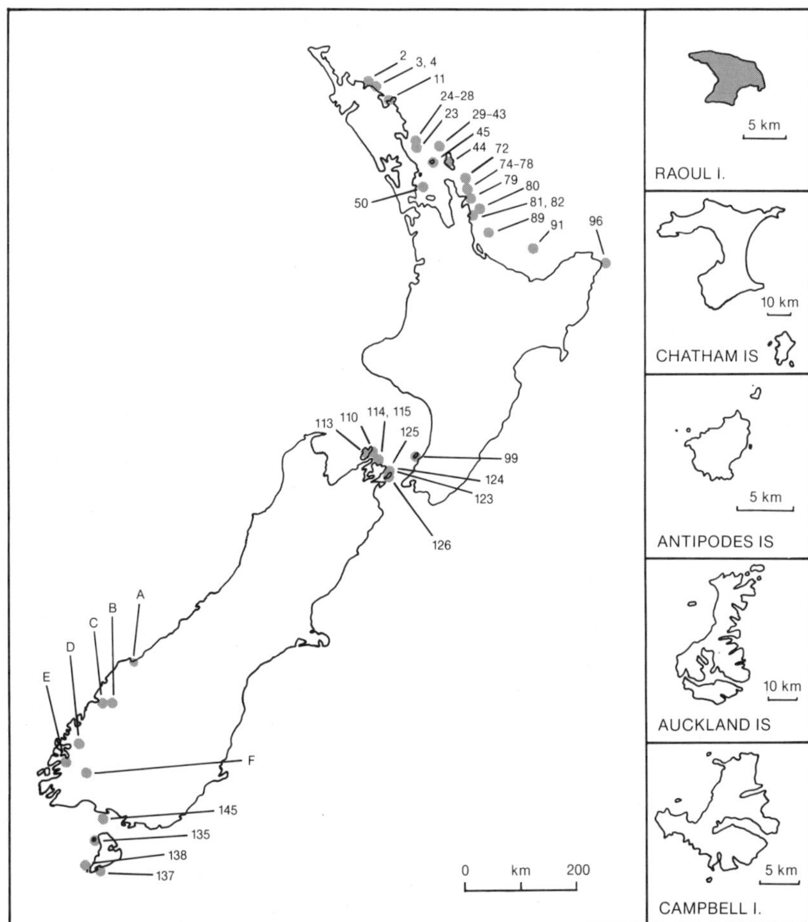

Map 14 Present distribution of kiore on the main and inshore islands of New Zealand. Inshore islands numbered as in Table 36; mainland locations A–F listed in text (p. 180). During the Polynesian period (c. AD 1000–1800), kiore were probably present throughout the main island plus Chatham (now extinct there), Raoul and Macauley Is., where they are still present.

The decline was attributed to the spread of Norway and ship rats by Dieffenbach (1843), Wodzicki (1950) and Watson (1956). Although the spread of these rats may have contributed to the decline of kiore, this explanation is insufficient alone; kiore coexist with either Norway or ship rats on a number of offshore islands (Atkinson, 1978) and in mainland Fiordland, and with both these rat species on Stewart and Pearl Is. (R. H. Taylor, unpubl.).

A second explanation is that kiore were susceptible to trypanosomes (*Trypanosoma lewisi*) introduced by rats from Europe (Doré, 1918). However, *T. lewisi* is found in an abundant population of kiore on Little Barrier I., which is not known to have been reached by other rat species (Ford-Robertson & Bull, 1966).

A third explanation is that kiore cannot survive competition with mice, since the decline of kiore is correlated with the spread of mice in both the North and South Islands (Taylor, 1975a; p. 227). The four introduced rodents have not been found to coexist anywhere in the New Zealand region (although they do on many Pacific islands), suggesting strong competitive exclusion between the species in New Zealand conditions (Taylor, 1975a, 1978a, 1984). Ship rats were the last of the four rodents to colonize New Zealand, and their spread coincides with the final disappearance (but not the earlier decline) of kiore from most of the mainland (Atkinson, 1973a).

DISTRIBUTION

WORLD. *Rattus exulans* ranges through the tropical zone from continental and insular south-east Asia (eastern Bangladesh, Andaman Is.) eastwards across numerous islands in the western and central Pacific as far as Easter I.; north to Burma and to Kure Atoll of the Hawaiian group; and south to Stewart I., New Zealand (Wodzicki & Taylor, 1984). The species is unknown on mainland Australia, but has been collected from two offshore islands north of that continent (Taylor & Horner, 1973).

NEW ZEALAND (Map 14). During the Polynesian period the kiore became widespread throughout the three main islands of New Zealand, and reached many offshore islands and a few outlying islands. Its bones have been found in coastal middens, rock shelters and limestone caves in many localities where it no longer lives; it was probably once present throughout all suitable habitats.

At present kiore in the New Zealand region are largely confined to offshore and outlying islands including Stewart I. (Table 36). Mainland localities are restricted to parts of South Westland and Fiordland: (A) MacFarlane Mound, 7 km south-east of the mouth of the Arawata River near Jackson Bay (Robertson & Meads, 1979) and possibly at Jackson Bay (Choate, 1965); (B) Hollyford Valley (Taylor, 1975a); (C) coastal and inland valleys of Milford Sound including the Sinbad, Esperance and Tutoko Valleys (Taylor, 1975a); (D) Takahe Valley in the Murchison Mountains; (E) Doubtful Sound and Lake Manapouri (Watson, 1956); (F) Borland Valley, north of Lake Monowai (King, 1983a).

Kiore populations can persist on islands at least as small as 6 ha without recolonization, and possibly on even smaller ones. Considering only offshore islands >5 ha in area, kiore are present on at least 28 (22%) of the islands north of latitude 38°; eight (26%) of the islands lying between latitudes 38° and 45°S; but are on only five (6%) islands south of latitude 45°S (I. A. E. Atkinson, unpubl.).

Among the outlying islands (>50 km from the mainland), kiore are now found only on Raoul and Macaulay Is. of the Kermadec group, but skins and skeletons held by the British Museum (Natural History) and Canterbury Museum (NZ) show that kiore were formerly present on Chatham I. Their original distribution included three (7%) of the New Zealand outlying islands >5 ha.

Table 36: Distribution and history of rodents on the offshore and outlying islands (>0.8 ha) of New Zealand. Numbers in column 1 refer to Map 14.

No.	Island	Area (ha)	Kiore	Norway rat	Ship rat	Mouse
Northland (35° S)						
1	Motuopao	47		UR[1]		
2	Stephenson	123	+			
	Cavalli Is.					
3	Motukawanui	355	+			
4	Haraweka		+			
	Bay of Islands					
5	Motuarohia	58		+		
6	Urupukapuka	208		+	+	
7	Unnamed (N11/584619)				+	
8	Unnamed (N11/591619)			+		
9	Motuapo	1		UR		
10	Moturoa	146		+	+	+
11	Moturua	135	+	+		
12	Motukiekie	29		+		
13	Okahu	21		+		
14	Motumaire			UR		
15	Motuoi	1		UR		
16	Motungarara	1		UR		
17	Waewaetorea	47		UR		
18	Poroporo	4		+		
19	Kohangaatara	1				+
20	Moturahurahu	2			+	
21	Motuwheteke	1			+	
22	Rimariki	7			+	+
Hauraki Gulf (36°S)						
23	Hen	476	+			
	Chickens					
24	Coppermine	72	+			
25	Whatupuke	90	+			
26	Mauitaha	22	+			
27	Motumuka (Lady Alice)	138	+			
28	Araara	<1	+			
	Mokohinau Is.					
29	Motukino (Fanal)	75	+			
30	Burgess	52	+			
31	Atihau (Trig)	16	+			
32	Horomea (Maori Bay)	10	+			
33	Motupapa	2	+			
34	Motuharakeke (Flax)	1	+			
35	Stack B	<1	+			
36	Stack C	1	+			
37	Stack D	1	+			
38	Stack E	<1	+			
39	Stack F	<1	+			
40	Stack G	<1	+			
41	Stack J	<1	+			
42	Lizard	<1	+			
43	Arch Rock	1	+			

continued overleaf

Table 36: *(Continued).*

No.	Island	Area (ha)	Kiore	Norway rat	Ship rat	Mouse
44	Great Barrier	28 510	+		+	+
45	Little Barrier	2 817	+			
46	Kawau	2 257			+	
47	Rangitoto	2 333		+	+	+
48	Motutapu	1 509		+	+	+
49	Ponui	1 851		UR		
50	Tiritiri Matangi	220	+			
	Noises group					
51	Motuhoropapa	10		(+) EX.[2]		
52	Otata	22		(+) EX. 1980-85		
53	David Rocks	1		(+) EX. 1964		
54	Maria	2		(+) EX. 1964		
55	Waiheke	9 459		+		+
56	Motuihe	179		UR		
57	Browns	60				+
58	Goat	9			+	
59	Arid	350			+	
60	Rakino	150		+		
Coromandel–Bay of Plenty (36-37°S)						
	Motukawao group					
61	Motukahaua	24			+	
62	Motumakareta	4			+	
63	Motuwhakakewa	3			+	
64	Motuwi (Double)	26		+		
65	Motukaramarama	18		+		
66	Motuwinukenuke	3		+		
67	Moturua (Rabbit)	26		+		
68	Ngamotukaraka	3		+		
	Three Kings					
69	A	3		+		
70	B	2		+		
71	C	2		+		
72	Cuvier	181	+			
73	Mahurangi	32		+		
	Mercury group					
74	Great	1 718	+		+	
75	Red	203	+			
76	Stanley (Kawhitihu)	120	+			
77	Double	43	+			
78	Korapuki	18	(+) EX.1986			
79	Ohinau	45	+			
	Aldermen group					
80	Middle Chain	22	+			
	Slipper group					
81	Slipper	210		+		
82	Rabbit	11	+			
83	Penguin	8	+			
84	Shoe	40		+		
85	Hauturu	7		+		

Table 36: *(Continued).*

No.	Island	Area (ha)	Kiore	Norway rat	Ship rat	Mouse
86	Whenuakura	2		(+) EX.1985		+
87	Moturua (Rabbit)	26		+		
88	Motumorirau (Paul)	3		+		
89	Mayor	1 277	+	+		
90	Rurima	6	(+) EX. 1985			
91	White	238	+			
92	Motuhora (Whale)	140		(+) EX. in progress		
93	Motiti			UR		
94	Rangipukea	37		UR		
Eastern and Central (38-41°S)						
95	Mokoia (L. Rotorua)	135		+		+
96	East	8	+			
97	Portland	150				+
98	Bare	15		+		
99	Kapiti	1 970	+	+		
100	Tauhoramaurea	5		+		
101	Motungarara	7		+		
102	Somes	23			+	
103	Leper (Mokopuna)	2			+	
104	Mana	217			(+) EX. 1989?	
105	Taputeranga	5		+		
Nelson-Marlborough (40-41°S)						
106	Motu	5			+	
107	Ngawhiti	5			+	
108	Adele	88				+
109	Haulashore	<10		+		
110	Whakatere-Papanui	74	+	+		
111	Puangiangi	69		+		
112	Tinui	100		+		
113	D'Urville	16 782	+			+
114	Inner Chetwode	242	+			
115	Te Kiore[3]	7	+			
116	Titi	32		(+) EX. 1975		
117	Forsyth	775			+	+
118	Tawhitinui	22			(+) EX. 1982	
119	Awaiti	7			(+) EX. 1982	
120	Tarakaipa	35				+
121	Allports	12				+
122	Blumine	377				+
123	Pickersgill	103	+			+
124	Long	142	+			
125	Motuara	57	+			
126	Arapawa	7 785	+		+	+
Fiordland (45-46°S)						
127	Resolution	20 860		+		
128	Breaksea	182		+ EX. 1987		
129	Hawea	8		+		

continued overleaf

Table 36: Distribution and history of rodents on the offshore and outlying islands (>0.8 ha) of New Zealand. Numbers in column 1 refer to Map 14.

No.	Island	Area (ha)	Kiore	Norway rat	Ship rat	Mouse
130	Long	1 960				+
131	Spit	5			UR	
132	Coal	1 622				+
133	Weka	108			UR	
Stewart I. group (46-47°S)						
134	Stewart	174 600	+	+	+	
135	Codfish	1 396	+			
136	Native	60			+	
137	Pearl	500	+	+	+	
138	Putauhinu	141	+			
139	Ulva	259		+		
140	Bench	121		+		
141	Big South Cape	940			+	
142	Solomon	26			+	
143	Pukeweka	4			+	
144	Rosa (Port Pegasus)	2		+	+	
145	Centre	110	+			
146	Ruapuke	1 525				+
Outlying islands						
147	Raoul (29°16′S)	2 938	+	+		
148	Macauley (30°13′S)	306	+			
149	Chatham (44°S)	90 650	Extinct	+	+	+
150	Pitt (44°16′S)	6 203				+
151	Antipodes (49°43′S)	2 025				+
152	Auckland (50°45′S)	45 975				+
153	Enderby (50°30′S)	710				+
154	Masked	5				+
155	Campbell (52°33′S)	11 331		+		
156	Folly	<10		+		
157	"Disappointment" (50°37′S)	2		+		
158	"Low-lying"	2		+		

Data from Atkinson, (1978, 1986 and unpubl.) and R. H. Taylor (unpubl.). Areas given are approximate only and taken from several sources, some of which disagree.
1. UR = Unidentified rats.
2. EX = Exterminated at date shown. Note that some may have since recolonized
3. Joined to Inner Chetwode (= Nukuwaiata) at low tide.

HABITAT

Kiore can live in a wide range of habitats, including grasslands of danthonia (*Rytidosperma* spp.), cocksfoot (*Dactylis glomerata*) and prairie grass (*Bromus willdenowii*); stands of bracken (*Pteridium esculentum*); shrublands and scrub of *Leptospermum scoparium* and kanuka (*Kunzea ericoides*), *Dracophyllum longifolium* and dwarf podocarps; broadleaved coastal scrub; secondary forests of kanuka, pohutukawa (*Metrosideros excelsa*) and southern rata (*M. umbellata*); and mature hardwood forests of varied composition both with and without podocarps. South Island habitats of kiore include mountain and silver beech forests and, formerly, a range of climates from very wet (Westland) to very dry (Otago).

Within New Zealand, kiore reach higher densities in grassland than in forest (Taylor, 1975a, 1978a; Moller, 1977; Moller & Tilley, 1986). Density and reproductive performance in bracken is particularly low (Bunn, 1979). Other rodent species may restrict kiore to grassland or thick ground cover (Taylor, 1978a, 1984). For example, ship rats dominate kiore in enclosures (McCartney & Marks, 1973), and Norway rats dominate ship rats (Barnett & Spencer, 1951); so Norway rats would almost certainly dominate kiore, though this has not been tested. Kiore may be able to escape social interference in grassland, or in dense ground cover, and so persist there whether or not these habitats offer them better food or living conditions. Moller (1977) suggests selective removal of one species across the bush-grass boundary to test this hypothesis.

Beech forests were an important mainland habitat for kiore during the Polynesian period and, at least during population irruptions, kiore moved from them into adjacent tussock grasslands (*Chionochloa rubra*) and bracken fernlands as well (Best, 1942; Elder, 1962). Evidence given to the Maori Land Court (Opotiki Minute Book No. 1, 1878) reveals that the Maori formerly hunted kiore on the southern part of the Kaingaroa plateau. At that time these pumice plains were dominated by silver tussock (*Poa cita*) and monoao (*Dracophyllum subulatum*).

The altitudinal limit for kiore was at least 1300 m, in both the North (Maungapohatu, Best, 1942) and South Islands (Nelson district, Meeson, 1885; Bell & Bell, 1971). By contrast, they can also feed in the rocky intertidal zone along the shore (Moller, 1977).

One of the more extreme habitats occupied by kiore is on White I., where the sparse woody and herbaceous vegetation is exposed to toxic fumes from the volcano (Wodzicki & Robertson, 1959).

Kiore are commensal in Asia (Harrison, 1957; Searle & Dhaliwal, 1957), and they commonly inhabit buildings on New Zealand offshore islands.

FOOD

Kiore eat a wide range of animal foods, including lepidopteran larvae, wetas, centipedes, spiders, earthworms, ants, beetles, weevils, cicadas, snails, and, less often, lizards and birds (Bettesworth, 1972b; Bunn, 1979; Campbell *et al.*, 1984; Meads, Walker & Elliott, 1984); and also the flowers, part or all of the fruit, stem, leaf and roots of many forest plants (Campbell, 1978; Campbell *et al.*, 1984; Dick, 1985).

On Tiritiri I., insects predominated in the kiore diet in winter, reaching 90% (by volume) of stomach contents in one July sample (Bunn, 1979). Dick (1985) found that more insects were eaten in spring than in summer or autumn in lowland grassland on Kapiti I., but he did not sample that area in winter. Insect consumption was also higher there in spring than winter in one area of kanuka and coastal forest, but not in another similar habitat.

A single winter sample of stomach contents from kiore living in coastal forest and shoreline on Red Mercury I. contained 60% (by volume) animal food (Bettesworth, 1972b). Seed formed the main food for kiore in grassland on Tiritiri in summer and early autumn, while grass stem and leaf material predominated from late winter to early summer (Bunn, 1979).

SOCIAL ORGANIZATION AND BEHAVIOUR

Kiore are predominantly nocturnal, but become active just before dark during times of high density.

Home ranges on Tiritiri I. in 1979 and 1980 averaged at least 37–60 m across, but were very variable (Nicholas, 1982). The distances moved by these kiore between successive recaptures were low in winter but increased dramatically during the spring breeding season, particularly in males (Moller, 1977; Bunn, 1979; Nicholas, 1982). Mature males tended to move further than mature females, and immature kiore of both sexes were most sedentary (Moller, 1977; Nicholas, 1982). Similar patterns have been observed in Pacific and Asian studies (Jackson & Strecker, 1962a; Tomich, 1970; Tamarin & Malecha, 1971; Wirtz, 1972; Lindsey *et al.,* 1973; Williams, 1974a; Dwyer, 1978). On Tiritiri I., rats moved from forest towards grassland when the grasses began seeding (Nicholas, 1982).

Long-range dispersal is difficult to detect, but Moller (1977) snap-trapped four male and three female kiore which had moved an average of 268 m since their last live-capture.

Kiore can nest and feed in trees (Campbell *et al.,* 1984; H. Moller, unpubl.). All of the kiore trapped in bush on Kapiti I. by Daniel (1969) were caught in snap traps set up in trees (M. J. Daniel, unpubl.), although on this island interference from the larger and behaviourally dominant *R. norvegicus* may cause *R. exulans* to feed more often than usual in trees. Arboreality of kiore was studied in detail by McCartney (1970) in the Marshall Is.

In the breeding season adult females strongly avoid one another and adult males associate with them only for mating (Nicholas, 1982). After breeding adult males strongly avoid females, which are dominant over males, at least in captive colonies (Egoscue, 1970; L. S. Davis, 1979). Family ties are loose and brief; immature rats at first avoid and are subordinate to adults, but as they reach sexual maturity they become dominant to the older (already sexually mature) rats of their parents' generation. The older rats disappear faster than younger rats at this time (Moller, 1977). Vigorous and noisy chasing, sometimes of four rats at once, was often seen on Tiritiri I. during peak population periods.

REPRODUCTION AND DEVELOPMENT

On Tiritiri I. (Moller & Craig, 1986) and Cuvier I. (Craig, 1986), where there are extensive areas of grassland, pregnant females are found in only 3–4 months over spring and early summer. On predominantly forested islands, the breeding season is longer, e.g., Little Barrier (6–8 months; Craig, 1986), Hen (5–10 months; Craig, 1986), Kapiti (at least 6 months; Dick, 1985) and Stewart I. (7 months; P. J. Moors, unpubl.).

The highest average litter sizes are recorded on Tiritiri I. (6.7, n = 15, range 4–9; Moller & Craig, 1986) and Motukawanui I. (6.7, n = 7, range 4–8; Hitchmough, 1980). Smaller average litters are produced on Kapiti I. (4.7, n = 6; Dick, 1985) and on Little Barrier I. (4.7, n = 12, range 2–6; Watson 1956). Gestation takes 19–21 days, and there is a postpartum oestrus (Wirtz, 1973), so females may be carrying the next litter before weaning the previous one at 4 weeks. Moller and Craig (1986) calculated that mature females average about three litters per year and so produce 19–21 young annually on Tiritiri I.

The young can see and hear at 2 weeks old and are weaned by the end of the fourth week (Wirtz, 1973). The majority of young kiore on Tiritiri I. do not mature to breed in the same season in which they are born. Most reach maturity at very large sizes (95 g for males, 93 g for females), some 8–12 months after birth (Moller, 1977). In contrast, many young on Little Barrier I. mature earlier and at smaller sizes (48 g for females) and are able to breed in the season of their birth (Watson, 1956; Moller, 1977). At least some kiore on Kapiti (Dick, 1985), Arapawa (Moller, 1978) and Moturua Is. (Moller & Tilley, 1986) also reach sexual maturity at very small sizes.

Population Dynamics

Seasonal variations in density of kiore in New Zealand result from the regular annual pulse of births over a limited period. Density reaches its lowest levels in spring and early summer, and peaks in autumn.

Strong seasonal patterns have been monitored for six successive years on Tiritiri I., where breeding is particularly concentrated and short (Moller, 1977; Bunn, 1979; Nicholas, 1982). Annual maximum densities in the years 1975 to 1980 were 100-170 kiore/ha in grassland, and 70-80/ha in forest; minimum densities were 6-50/ha in grassland and 10-15/ha in forest. The pattern of fluctuations between these peaks and troughs varies between years and habitats. In some years, when breeding finishes early, the rats lose weight in early autumn and survive less well; density drops rapidly and remains relatively stable and low throughout winter. In other years the rats maintain weight through the autumn, survive better, and the population decline is more gradual. Dietary studies and measurements of grass seed abundance suggested that the seasonal availability of grass seed causes the strong pattern in breeding and resulting population fluctuations on Tiritiri I. To test this hypothesis, Bunn (1979) added supplementary food to an experimental trapping grid while monitoring a control grid in grassland. On the experimental grid, the rats bred longer, maintained weight, matured earlier and survived longer.

Kiore density fluctuates less on forested islands, such as Little Barrier and Hen Is., than on Tiritiri (Craig, 1986). Repeated snap trapping on Kapiti and Stewart Is. and the South Island (all on forested sites) also suggests low and relatively stable kiore numbers there. Breeding is more prolonged on the forested islands, and the seasonal pulse in numbers, though still evident,

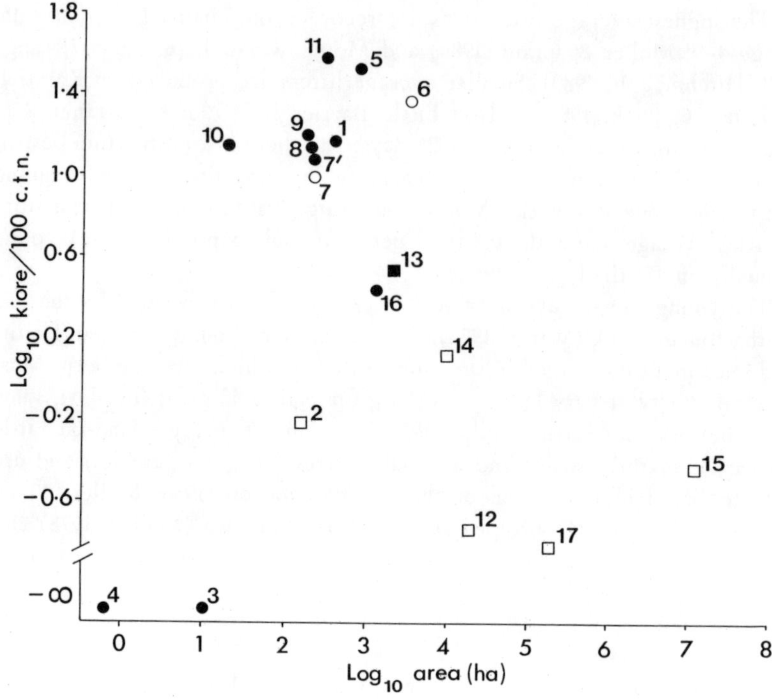

Fig. 37 Log of snap trap indices (see Fig. 42) for kiore plotted against log of island area for seventeen offshore islands around New Zealand. The data on which this figure is based are given in Appendix 2 of Moller & Craig (1986). A single average is given for islands which have been repeatedly sampled. Habitat, seasonal coverage, trap type and bait vary considerably. Key: squares indicate that rodents other than kiore are also present on the island; circles indicate that kiore alone is present; solid symbols indicate that cats and mustelids are absent, open symbols that cats or stoats are present. The number beside each symbol indicates the island's name in rank order of latitude, as follows: 1, Motukawanui; 2, Moturua; 3, Hokoromea; 4, "Stack G", Mokohinau group; 5, Hen; 6, Little Barrier; 7, Cuvier, with cats; 7′, Cuvier, after cats removed; 8, Tiritiri; 9, Red Mercury; 10, Korapuki; 11, White; 12, D'Urville; 13, Kapiti; 14, Arapawa; 15, South; 16, Codfish; 17, Stewart.

is less pronounced than on Tiritiri. Density tends to be higher on islands not inhabited by cats, stoats or, especially, other rodents (Fig. 37). Taylor (1975a, 1978a) suggests that the kiore is particularly sensitive to competition from other rodents, and especially from mice. Since the size of the island and the presence of other rodents and their predators are all correlated, the relative importance of each in determining kiore numbers cannot yet be determined.

PREDATORS, PARASITES AND DISEASES

Avian predators of kiore in New Zealand include moreporks (*Ninox novaeseelandiae,* Atkinson & Campbell, 1966), harriers (*Circus approximans,* Moller, 1977) and almost certainly wekas (*Gallirallus australis*). Kingfishers (*Halcyon sancta*) are known to take mice and may eat young kiore. Harriers fly to Tiritiri I. each autumn when kiore are abundant, and leave again by late winter to breed elsewhere (Moller, 1977).

Kiore predominated in the diet of cats on Little Barrier I. (Marshall, 1961). Mustelids, particularly stoats, are the only other potential mammalian predators, but they are absent from nearly all the islands inhabited by kiore except the South Island. Killing and eating of mice in large enclosures by ship rats was recorded by Lidicker (1976), so direct predation of kiore by the larger rats, particularly at the nest, is possible.

Parasites carried by kiore in New Zealand include the following. Mites: *Demodex* sp., *Gymnolaelaps annectans, Hypoaspis nidicorva, H. claviger, Mesolaelaps australiensis, Notoedres muris, Radfordia ensifera*. Lice: *Hoplopleura pacifica, Polyplax serrata*. Fleas: *Leptopsylla segnis, Nosopsyllus fasciatus, Parapsyllus longicornis, Pygiopsylla hoplia, P. phiola, Xenopsylla vexabilis*. Trypanosome: *Trypanosoma lewisi*. Cestode: *Hymenolepis diminuta*. Nematode: *Mastophorus muris*, and/or *Physaloptera* sp. (Ford-Robertson & Bull, 1966; Smit, 1979; Tenquist & Charleston, 1981; Gibson, 1986; Gibson & Pilgrim, 1986; Moller & Craig, 1986). W. A. G. Charleston (unpubl.) has re-examined Gibson's material and considers that the *Mastophorus* nematode was probably *Physaloptera*. The material has now decayed, so confirmation is possible only from new specimens.

Adaptation to New Zealand Conditions

Detailed studies of diet in tropical Pacific croplands or simple atoll ecosystems all showed that kiore ate mainly vegetable matter, with animal remains never constituting more than 8% of the volume of stomach contents (Strecker & Jackson, 1962; Kami, 1966; Fall, Medina & Jackson, 1971; Mosby, Wodzicki & Thompson, 1973; Temme, 1982). The highest-latitude island studied, other than New Zealand, is Kure Atoll, where Wirtz (1972) found insects averaged 30% of the diet and were eaten more frequently when kiore were scarce and plant food was least available. New Zealand populations inhabit more complex ecosystems, with more seasonal variation in food supplies than those of the Pacific islands, so invertebrates become an important food resource in some months (Bettesworth, 1972b; Bunn, 1979; Dick, 1985).

The maximum weights of kiore recorded in New Zealand (187 g in males and 150 g in females) are more than twice the maximum weights recorded in populations closest to the equator (e.g., in New Guinea, where the heaviest of 152 males was 62 g and the heaviest of 171 females was 60 g; P. D. Dwyer, unpubl.). There is a significant increase in body size with increasing latitude (Moller, 1977), and Efford (1976) found that kiore in New Zealand have bigger teeth than those in tropical populations. These findings are predicted by Bergmann's Rule.

The strongly seasonal patterns of reproduction and population density in kiore in New Zealand contrast with those observed in the tropics, e.g., on Java, where females breed for most or all of the year and populations are more constant in spatial organization and in numbers (Koeppl, Slade & Turner, 1979). Litter size also increases significantly with latitude, but production and survival do not (Moller & Craig, 1986; Moller, 1977). Young *R. exulans* mature at smaller sizes and younger ages in the tropics than in New Zealand. The pattern in New Zealand is most similar to that on Kure Atoll, where there was an 8-month breeding season and some young did not mature in the season of their birth (Wirtz, 1972).

SIGNIFICANCE TO THE NEW ZEALAND ENVIRONMENT

Apart from the use of kiore for food (p. 177), almost nothing is known about their past significance to man in New Zealand, either as despoilers of food or vectors of disease. Much more is known about possible effects of kiore on the indigenous fauna and flora, although the magnitude of these effects is uncertain.

DAMAGE TO NATIVE FAUNA. Earlier writers considered the kiore to be a "harmless vegetarian", but more recent studies have shown that kiore eat a wide range of smaller animals (p. 185). They could have had a major effect on the New Zealand biota, and caused local or total extinctions of some species of flightless beetles and weevils, giant wetas, land snails, frogs, lizards, tuatara, small seabirds and landbirds, and bats (Atkinson, 1964a, 1978, 1985, 1986; Atkinson & Bell, 1973; Whitaker, 1973, 1978; Crook, 1973; Ramsay, 1978; Worthy, 1984, 1987; Daniel & Williams, 1984).

Circumstantial evidence that kiore have caused local extinctions or permanent reductions of particular animal species comes largely from study of contemporary disjunct distributions, especially on islands. In these studies, the missing or rare native species is assumed to have been present or more abundant before kiore arrived, because suitable habitats appear to be available. Whitaker (1973) found fewer lizard species, and markedly fewer individuals, on islands inhabited by kiore compared with those without. Crook (1973) found reduced densities of tuatara on kiore-inhabited islands compared with islands where kiore were absent. Atkinson (1978) showed that white-faced storm petrels, northern diving petrels and fairy prions were abundant as breeding birds only on islands without kiore, and that snipe and bush wrens in southern New Zealand had become entirely restricted to kiore-free islands well before their final extinction on Big South Cape (p. 224). Watt (1986) demonstrated that three species of large flightless weevil restricted to a very few of the rat-free Poor Knights Is., and previously thought to be endemic to that group, were originally distributed more widely, including on the mainland. As suggested by Craig (1986), experimental reintroductions of some native species, thought to have been present originally on an island now inhabited by kiore, might be a useful method of testing such hypotheses if all other conditions are still unchanged.

A second approach, taken by Crook (1973), is to observe the age structure of several populations of a vulnerable native species resident on islands with and without kiore. Crook measured snout-vent lengths of tuatara larger than 130 mm and found few animals under 200 mm on kiore-inhabited islands. He suggested that kiore were causing a failure of recruitment in tuatara populations. An experimental test of this conclusion would be to eliminate kiore from a tuatara-inhabited island.

More rarely, there is direct evidence of local extinctions caused by kiore. An example is that of Lizard I., Mokohinau group, where five lizard species had been recorded before kiore reached the island in 1977 (Atkinson, 1986). A 90-minute survey in 1979 failed to find any lizards (McCallum, 1980).

On the mainland, evidence for kiore-induced extinctions of small native animals is wholly circumstantial, but nonetheless often compelling. Two kinds of evidence can be used: first, stratigraphic studies demonstrating that a particular extinct species disappears from the sequence at the same time that kiore remains enter it. For example, Worthy (1984) suggested that the extinction of a large flightless fungus beetle and two species of large (*c.* 20 mm long) weevils can probably be attributed to kiore predation. Although one such case of temporal correlation may not be convincing, repeated findings of a similar kind would make a substantive case for a kiore effect, especially when linked with a second kind of evidence, reasonable inferences as to how vulnerable the extinct species might have been to predation or competition from kiore. Such inferences were used by Atkinson (1985) in suggesting that kiore may have been responsible for the disappearance of several small New Zealand passerine birds, especially those of the family Ancanthisittidae (rifleman, bush wrens, etc.), whose extinction on the main islands during Polynesian times cannot be explained by hunting or habitat destruction. For example, the Stephens Island wren (*Xenicus lyalli*) was until recently known only from a single island in Cook Strait, but fossil evidence now shows that it lived on both main islands before the arrival of the Polynesians (Millener, 1981 and unpubl.). At least two other species of wrens, discovered only recently, became totally extinct in the same period (Millener, 1988).

The impact of kiore predation on native fauna was probably much more subtle than that of merely reducing prey density. Moller's (1985a) study of tree weta behaviour, in populations with and without rats and other ground predators, suggests marked changes in the sex ratio and social organization of tree wetas coexisting with these predators. Moller found that tree wetas fed, moved freely, and nested on the ground only on Maud I., the only population he studied which had few or no ground predators.

It is therefore probable that kiore have been responsible for extinctions or reductions on the mainland of a quite significant number of animal species. These include an unknown number of larger invertebrates (for example, an undescribed species of giant weevil, *Anagotis* sp., the giant weta *Deinacrida heteracantha*, a few species of cave weta and land snail), three species of native frogs (Worthy, 1987), several skinks (particularly large nocturnal species of *Cyclodina*) and larger geckos, the tuatara, the greater short-tailed bat (p. 132) plus some species of smaller seabirds such as Cook's petrel and small landbirds such as the New Zealand snipe. All these examples should be treated as working hypotheses to be tested further with additional information. For some species, competition for the same food supply may have been a worse problem than direct predation.

Kiore probably had little effect on the larger birds in New Zealand, such as the moa (but see Fleming, 1962). On Kure Atoll, kiore were found to eat into the thoracic cavities of live albatrosses and great frigatebirds on their nests (Kepler, 1967; Woodward, 1972), but food resources are much more limited there than in New Zealand.

DAMAGE TO NATIVE FLORA. Evidence for negative effects by kiore on the regeneration of plant species is so far restricted to a very few coastal species, e.g., milk tree *Paratrophis banksii* (Atkinson, 1986), but further studies, including the use of exclosures as demonstrated by Campbell (1978), are needed.

CONTROL. Eradication of kiore from small islands, such as Rurima (6 ha) and Lizard I. (Mokohinau group, <1 ha) and Korapuki (Mercury group, 18 ha) has apparently been successful. By August 1987 no sign of rats had been seen on Rurima for 3 years, and on Korapuki for 9 months (I. McFadden, unpubl.). On large islands only temporary, localized reductions are possible, e.g., at the ranger station on Little Barrier I., using break-back traps and "Talon WB" anticoagulant rodenticide. Bait trials carried out by McFadden (1984) showed kibbled grains and flavoured petrogel paste to be effective baits for poisoning operations. Acceptable lures include aniseed, banana, coconut, clove, eucalyptus and vanilla. The NZWS used bromadialone anticoagulant on Rurima I., because sodium monofluoroacetate (1080) poison at 0.8% concentration is avoided by kiore (I. McFadden, unpubl.).

Kiore are agricultural pests in the Pacific because they damage coconuts, cocoa, sugarcane and a variety of other crops (Wodzicki, 1969; Williams, 1974a, b, c, 1975; Temme, 1982). They are also a public health risk, because they carry leptospirosis (Tomich, 1979), plague (Tomich, 1970; Turner, Padmowirjono & Martoprawiro, 1975), lungworm and a variety of other pathogens and parasites. Wodzicki (1978–79) reviews the relationships between rats and man in the central Pacific.

I. A. E. A. & H. M.

16. Norway rat

Rattus norvegicus (Berkenhout, 1769)

Synonyms *Mus norvegicus* Berkenhout, 1769; *Mus decumanus* Pallas, 1778; *Mus hibernicus* Thompson, 1837; *Epimys norvegicus* Miller, 1912; *Rattus norvegicus* Hinton, 1913.

Also called brown rat, common rat, water rat, sewer rat (English); pouhawaiki (Maori).

DESCRIPTION (Fig. 38)

Distinguishing marks, Table 34 and skull key, p. 24.

This is the largest rat in New Zealand. It has a stout body, heavy tail and relatively small ears. The adult coat of both sexes is coarse and rather shaggy, greyish brown on the flanks, grading to a darker umber brown (sometimes with a russet tinge) along the back. The belly and throat are pale grey. Juveniles have a sleeker look with finer, greyer fur and disproportionately large feet. Mature males have a prominent hairy scrotum

Fig. 38 Norway rat, Broad Bay, Otago Peninsula (NZWS).

at the base of the tail. Females of all ages have a small patch of hairless skin behind the urethral papilla, which distinguishes them from immature males (Fig. 39). Nipples are visible only on mature females; there are typically three pelvic and three pectoral pairs.

Norway rats have acutely-developed senses of smell, touch and hearing. Scent signals play a crucial role in a rat's life, enabling it to distinguish between familiar and strange individuals, detect sexually active mates, and find food and water. The facial vibrissae and long guard hairs of the pelt are very sensitive to touch and are important aids to orientation and movement in the dark (Brooks & Rowe, 1979). Rats can hear and emit sounds at frequencies up to about 90 kHz (Anderson, 1954). Social vocalizations include a range of shrill squeals and whistles and a screaming alarm call. Nestlings use ultrasonic calls at 40–65 kHz to communicate with their mother (Noirot, 1968). The eyes are specialized for nocturnal vision and are very sensitive to light, but they are not visually acute and lack colour vision. They give a red reflection in a beam of light.

Norway rats swim readily and well, hence "water rat" is a common alternative name. This skill enables them to reach new islands unaided. Depending on water temperature and sea conditions, channels up to 600 m wide may be crossed in northern New Zealand (Atkinson, 1986). Dagg and Windsor (1972) report a swimming speed of about 1.4 km/h over short distances in laboratory trials. Norway rats can also climb with agility when necessary (e.g., Hill, Robertson & Sutherland, 1983), but do so more rarely than ship rats. Where they coexist, Norway rats usually remain on the ground or in the basements and ground floors of buildings, whilst ship rats occupy the trees or upper storeys.

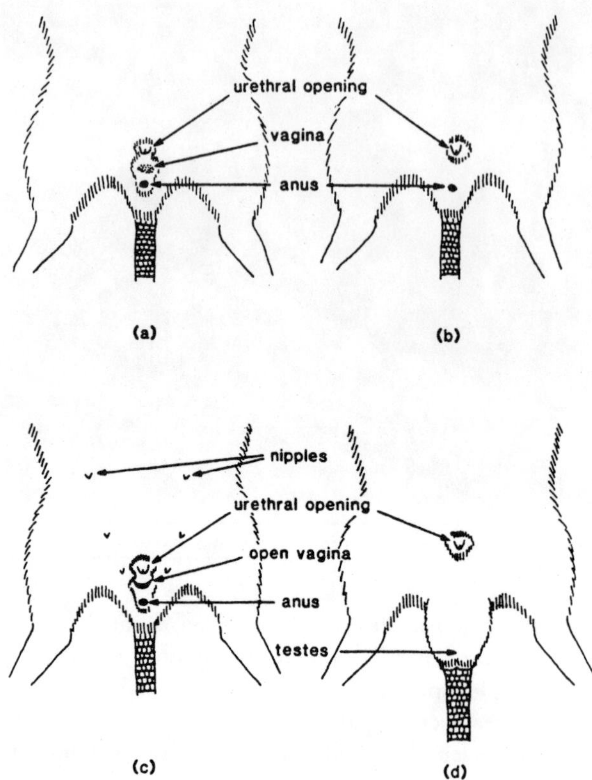

Fig. 39 External sexual features of immature and mature rats.
(a) Immature female: urethral opening close to anus; vagina covered with a hairless patch of translucent skin.
(b) Immature male: urethral opening further from anus than in immature females; no hairless patch. In both sexes, urethral opening on a small protrusion.
(c) Mature female: vagina open or secondarily closed. Nipples present, but may be hidden in fur if animal not breeding.
(d) Mature male: testes in scrotum, which hides anus and is bald at rear.
Reproduced with permission from Cunningham & Moors (1983).

The adult skull is angular and robust. The parietal ridges are roughly parallel and give the adult cranium a rectangular appearance from above, compared with the oval shape of the skull of the ship rat (Fig. 8) and kiore. Also, the parietal length exceeds occipital width, but the reverse generally holds in the other two species. There are one pair of incisor teeth and three pairs of molars in each jaw. The incisors grow continuously, are yellowish orange on their front surfaces, and are separated from the molars by a diastema. The anterior pair of molars in each jaw is the largest and first to erupt.

Chromosome number 2n = 42.

Dental formula $I^1/_1$ $C^0/_0$ $Pm^0/_0$ $M^3/_3$ = 16.

FIELD SIGN

Norway rats characteristically excavate burrows 60–90 mm in diameter, often in sloping ground beside watercourses or beneath cover such as rocks, tree roots and tussocks. They excavate large volumes of soil and are able to remove stones 150–200 g in weight. In Fiordland, Reischek described Norway rats "digging holes round the hut" and "They dug up the potatoes in the garden and dragged them away . . . I found a great birch [southern beech, *Nothofagus* sp.] tree quite undermined with rat holes. The bark had been gnawed away up to 50 inches above ground level . . ." (Reischek, 1930: 251-2). Around towns and farms they burrow beside the foundations of buildings and under heaps of rubbish. Pathways 50-100 mm wide link burrows with feeding sites, sometimes several hundred metres away. In damp places, or if rats are abundant, the paths become well-worn trails of bare compacted earth. Trails in buildings may appear as dark smears on the floor. The footprint is broad in relation to its length, though not specifically identifiable; the print shows the tips of the four toes on the forefoot and the five on the hind, sometimes accompanied by tail marks. The running stride is 90-100 mm long, and bounding strides up to 600 mm long are taken when fleeing or carrying food (Lawrence & Brown, 1973).

Remains from hard foods, such as empty mollusc shells, crab carapaces and bones, are sometimes scattered around burrow entrances. Food may be hoarded in burrows and sheltered places. Other feeding sign is generally similar to that of ship rats and kiore. Droppings are deposited as separate pellets in small groups along runways, at feeding sites and in other haunts. They are cylindrical and stubby, sometimes tapering to a point; on average they are about 16 mm long, roughly twice the length of other rat pellets (Daniel, 1969). Pellets from Norway and ship rats can usually be distinguished by their length (L) and width (W) measured to the nearest 0.1 mm, using the expression $y = 100\ L/W^3$: if the average of y from a sample of at least five pellets is less than 20.0, then in 95% of cases the pellets are from Norway rats (Huson & Davis, 1980).

MEASUREMENTS

See Table 37. Males are usually larger than females of the same age, but precise data on the size of Norway rats in New Zealand are limited, and virtually confined to island populations. Comparisons between the measurements given should be made cautiously. Criteria of maturity vary between the samples, and in any case size at maturity is greatly influenced by season, food supply and population density (see Davis, 1953). Few wild adults weigh more than 400 g, and most are in the range 150-300 g. The largest rats are usually the oldest; however, reports of especially large Norway rats are likely to refer to individuals from commensal populations, where weights exceeding 500 g, and exceptionally 600 g, have been reported in Britain (e.g., Barrett-Hamilton & Hinton, 1910-21; Perry, 1945). The largest of three commensal females caught near Christchurch by Gibson (1973a) weighed 364 g. Cunningham and Moors (1983) describe a standard technique for taking measurements, recommended for use in future work.

Table 37: Body measurements (mean and range) of wild adult Norway rats in the New Zealand region.

Location	Sex	n	Head and body length (mm)	Tail length (mm)	Weight (g)	Reference
Waewaetorea I. (35° 12′S)	M	31	213 (188–246)	175 (160–190)	240 (173–343)	H. Moller & J. Tilley (unpubl.)
	F	44	204 (174–244)	167 (144–201)	210 (134–332)	
Mokoia I. (38° 05′S)	M	66	187 (166–234)	156 (121–194)	233 (124–380)	Beveridge & Daniel (1965)
	F	75	192 (160–233)	162 (125–192)	245 (141–422)	
Pureora Forest (38° 31′S)	M	22	203 (160–238)	163 (130–180)	224 (104–328)	C. M. King, M. Flux & J. G. Innes (unpubl.)
	F	8	187 (172–209)	153 (121–175)	183 (150–230)	
Stewart I. (47° 08′S)	M	18	179 (157–192)	162 (128–188)	214 (115–288)	P. J. Moors & D. M. Cunningham (unpubl.)
	F	22	171 (138–201)	156 (131–181)	182 (103–280)	
Campbell I. (52° 33′S)	M	50	189 (167–213)	172 (102–197)	226 (142–360)	G. A. Taylor (unpubl.)
	F	50	183 (163–205)	166 (147–182)	194 (130–307)	

NB: Commensal animals can be larger.

VARIATION

The degree of variation in body size between populations in the north and south of New Zealand, at high and low altitudes, or on islands compared with the mainland, is unknown. There is little variation in coat colour or pattern over the entire world range of the species, and albino and melanistic individuals are rare. Some genetic variation is known, for example polymorphism in 22 out of 49 skeletal characteristics in a sample from Singapore (Berry & Searle, 1963), and in the structure of the X chromosome and chromosome pairs 3 and 12 in some commensal populations (Yosida, 1980; Diaz de la Guardia, Camacho & Ladron de Guevara, 1981). However, these forms of variation have not been investigated in New Zealand populations.

The common laboratory rat is a domesticated form of this species. Rigorous artificial selection has produced a multitude of colour variants, together with differences from the wild type in many aspects of behaviour, growth and relative organ weight (Barnett, 1975). Natural genetic variation has been greatly reduced in inbred strains.

HISTORY OF COLONIZATION

Norway rats were the first of the European rodents to become established in New Zealand. Almost certainly they got ashore in the late 18th century from visiting European or North American sailing ships, especially the many sealers and whalers which began calling from 1792 onwards. Norway rats might even have got here by 1772, because in that year the French explorer Crozet (1891) saw rats in the Bay of Islands which he thought looked "the same species as those we have in our fields and forests". The common species in France then was the Norway rat.

There are very few unequivocal records from the first half of the 19th century, at least partly because naturalists were confused about which species of rats were present. However, comments by Williams (in Rogers, 1961), Cruise (1823) and Darwin (1845) indicate that Norway rats had spread throughout the northern districts of the North Island by the 1830s, and by mid-century they were apparently common in both the North and South Islands. Wakefield (1848) referred to their extensive migrations all over the islands, and they were sufficiently numerous for White (1890) to hunt them for sport on the Canterbury Plains in 1855. For the next few decades, until ship rats began to spread, Norway rats were the species encountered most often in town and bush over much of mainland New Zealand.

By the end of the century, however, these huge populations were fast disappearing. Douglas (1894), for example, remarked that in South Westland "The Norwegian rat . . . which swarmed in the country at one time, is now becoming extinct from some cause or other". This drastic change in status, which is still reflected in today's discontinuous distribution, has often been attributed to competition from invading ship rats. However, Taylor (1978a) has argued that it was caused by predation from mustelids, principally stoats, which were released in large numbers in both the North and South

Islands beginning in 1884. More recent analyses by Taylor (1984) suggest that predation and competition could both have played a part in the decline.

Many of New Zealand's islands were colonized at an early stage. Kapiti, Chatham, Stewart and Campbell were all bases for large sealing and whaling fleets, and almost certainly became infested during these operations. Norway rats were definitely present on Chatham I. by 1840 (Dieffenbach, 1841) and on Campbell by 1867 (Atkinson, 1985). At least eight infested islands are concentrated in the Bay of Islands, possibly a legacy from the busy port established there early last century. The spread to new islands continues: Norway rats reached Raoul I. in 1921 (Watson, 1961) from the wreck of the schooner *Columbia River;* the Noises Is. were colonized about 1956 (Moors, 1985a); and Whenuakura I. between late 1981 and early 1984 (Newman, 1986). Norway rats still get aboard ships now and then, and this remains a potential source of new island invasions. For example, a pregnant female was trapped in 1984 on a fishing boat in Port Pegasus, Stewart I.

DISTRIBUTION

WORLD. The extensive world distribution of the Norway rat is almost entirely the result of accidental dispersal by man. The species is thought to have originated in north-eastern China (Johnson, 1962), from where it spread westwards, first reaching western European cities in the early decades of the 18th century (Barrett-Hamilton & Hinton, 1910-21). From then until about 1850 it was the common rat aboard all European and North American sailing ships (Atkinson, 1985) and so was transported to countries around the world. It is now found across Europe, Asia Minor and southern Siberia to the Pacific coast, in China, Korea, Japan, temperate North and coastal South America, and locally in Africa and southern Australia. In tropical countries it has colonized many ports but generally not spread inland. It has also become established on many oceanic islands, from the tropics to polar regions (see Atkinson, 1985). The common and scientific names derive from the mistaken belief that the rats first reached Britain from Norway.

NEW ZEALAND (Map 15). In mainland New Zealand Norway rats have a wide but patchy distribution. The most extensive populations are commensal with man in virtually all towns and cities, around farms and on cropland. Away from habitation they are much less common, normally living only as isolated populations in wetlands and along watercourses. There is no comprehensive, recent information on their distribution. Extensive trapping has produced a few positive records from Pureora Forest Park and the Marlborough Sounds and Nelson, but none from western Fiordland and South Westland as far north as Franz Josef Glacier (C. M. King, R. H. Taylor, unpubl.). Last century Norway rats were very widespread, and were even reported in snow at an altitude of 1200 m in the Southern Alps (Reischek, 1888). They are now probably absent at heights above 1200 m in the North Island and 900 m in the South Island (R. H. Taylor, unpubl.). Many offshore islands have also been colonized (Table 36), and this list will probably lengthen as rats known to be on other islands are positively identified, and as additional

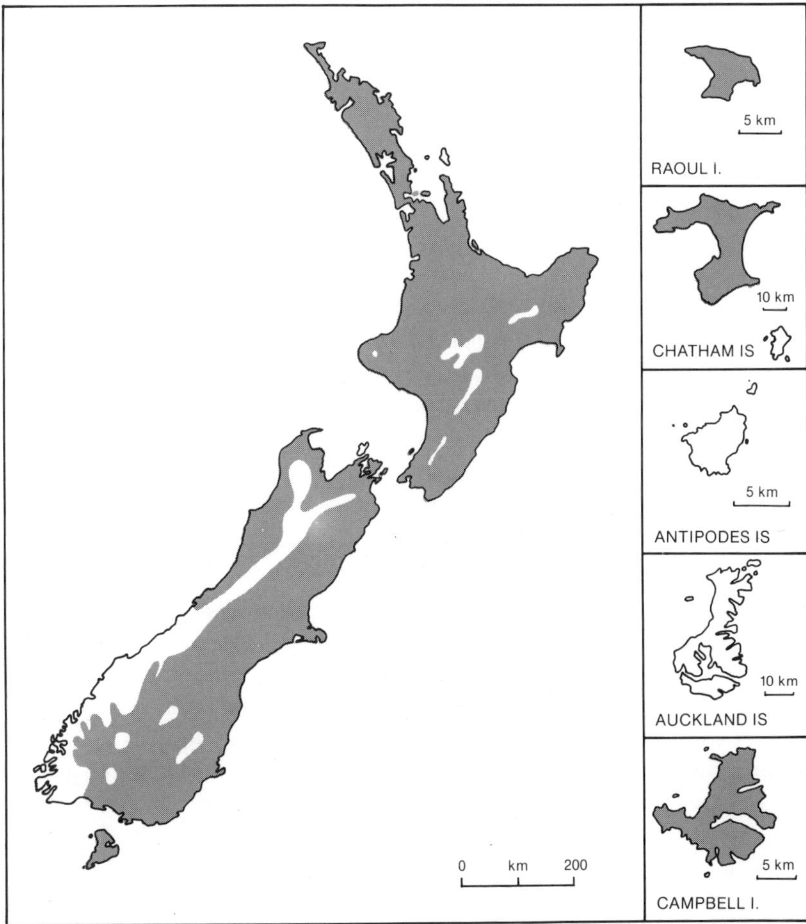

Map 15 Distribution of Norway rats on the main islands of New Zealand. Local populations may be found in pockets of suitable habitat throughout the marked area, but are more widespread on Stewart I. and in most towns. For distribution on offshore islands, see Table 36.

islands are surveyed. For example, a single adult male was caught on Codfish I., Foveaux Strait, in 1984, although intensive trapping has not so far produced any other evidence of a permanent population.

HABITAT

Wild populations in the North and South Islands are now mostly associated with wetland habitats such as estuaries, lagoons, lakes, marshes, rivers and streams. On braided riverbeds they favour stable side-streams and swamps and tend to avoid expanses of bare shingle.

Commensal populations will occupy virtually any type of building to which they can gain entry, and where there are suitable burrowing conditions and adequate food. Typical urban haunts include rubbish tips, sewers, wharves, sea walls, land reclamations and industrial sites where food is processed, stored or disposed of. On farms they are commonly found in barns, granaries, pigsties, poultry yards, stores of stock feed, cereal and root crops, and along irrigation and drainage ditches. Some field-living rats move into farm buildings in winter (Huson & Rennison, 1981).

During their period of ascendancy on the mainland last century, Norway rats lived in all types of habitat from coastline to mountain top. Such ubiquity is now found only on islands, particularly those on which Norway rats are the only rodents present. Nonetheless, not all habitats are equally preferred. On islands such as Waewaetorea (Moller & Tilley, 1986), Mokoia (Beveridge & Daniel, 1965), Whale (Bettesworth, 1972a) and Kapiti (Daniel, 1969; Dick, 1985), populations are generally most dense in grassland, scrub and damp forest at low altitudes, or near the shoreline. In the scrub and forests of southern Stewart I. they are found only near watercourses. Almost all the wild Norway rats collected there have been trapped within 10 m of running water, regardless of the type and structure of the surrounding vegetation.

FOOD

DIET. An important reason for the Norway rat's successful spread around the world is its omnivorous and opportunistic attitude to potential food. In urban areas these rats eat virtually everything humans do, and more besides. Cereals, meat, fruit and vegetables are eaten fresh, cooked or as garbage. Offal and other animal by-products, edible flotsam along the waterfront and sewage residues are not scorned. Farm rats eat stored grains, silage, processed stock feed, cereal and root crops, and weed and grass seeds.

Dietary studies of wild rats in New Zealand have identified mainly seeds, fruits, leaves, fern rhizomes, insects, molluscs, crustaceans and annelids (Beveridge & Daniel, 1965; Moors, 1985a), with some seasonal and regional variations (Dick, 1985). Lepidopteran larvae, wetas, beetles and other invertebrates are important high-protein components of the diet, even though they are usually present only in small amounts. Populations on small islands rely to a considerable extent on shoreline foods collected at low tide or by diving. Vertebrate items such as eggs, birds and lizards are sometimes taken; on Whale I. the eggs and newly hatched chicks of grey-faced petrels *Pterodroma macroptera* are an important winter food (Bettesworth, 1972a). Carcasses of dead animals are readily scavenged, including other rats caught in traps. Sheep and seabird carrion on Campbell I. is rapidly cleaned up.

FEEDING BEHAVIOUR. Each day adult rats eat the equivalent of about 10% of their body weight in dry grain (Leslie & Ranson, 1954), and on a dry diet they need to drink about 25 ml of water (Chitty, 1954). When feeding they typically rest on their hindquarters and use one or both forepaws to manipulate the food. Seeds and other transportable items are often carried away to be eaten under cover or hoarded. Dominant individuals are able to control access to food, and caches left by these rats may provide secondary sources for subordinates excluded from primary supplies (Calhoun, 1963).

Laboratory trials have shown that wild Norway rats are able to select nutritionally appropriate foods and avoid harmful ones (Barnett, 1975), and that the feeding behaviour and preferences of adults are crucial in determining those of their young (Galef & Clark, 1971). Transmission of learned eating habits within colonies accounts for the development of characteristic feeding patterns, including "bait shyness" during poisoning operations.

Norway rats readily chew through any material softer than the enamel on their incisors, including building timber, aluminium, soft mortar, plastic and asphalt. They also have a habit of gnawing the angular edges of wooden objects, leaving distinctive twin grooves in the timber. This feeding sign has been utilized to detect the continued presence of rats during and after eradication campaigns (Moors, 1985b).

SOCIAL ORGANIZATION AND BEHAVIOUR

ACTIVITY. Norway rats are mainly nocturnal and are most active several hours after sunset and again before sunrise (Calhoun, 1963). Disturbance, low social rank and the availability of food can greatly modify this pattern, however. There is a strong behavioural drive to explore unfamiliar surroundings, and in this way a rat learns the detailed topography of its home range and the locations of burrows, trails and food supplies. Opposing this curiosity is an avoidance of strange objects encountered in otherwise familiar surroundings (the "new object reaction"). This reaction lasts at most a few days, but it can, for example, substantially reduce initial trapping success or acceptance of poison baits (Shorten, 1954). The combination of exploration, avoidance and learned habits is elegantly adapted to provide a rat with maximum information about its environment at minimum risk (Barnett, 1975).

DISPERSION. Colonies are usually founded by a pair or a pregnant female. Colony members can recognize each other by odour, and strangers are aggressively driven off. Social behaviour is regulated by a complex system of stereotyped postures, body odours and vocalizations which Barnett (1975) considers to be fixed rather than learned. Calhoun (1963) closely observed the behaviour and social organization of a colony in a 0.1-ha enclosure. He found social rank was determined by success in combat, body size, place of birth, migration within the enclosure, and reproductive success. In part of the enclosure, high-ranking males vigorously defended territories, each of which contained several adult females occupying well-maintained burrows. These females mated only with their resident male, and were able to breed regularly and successfully. As numbers rose, subordinate juveniles (mainly males) were evicted into the remainder of the enclosure, where the rats lived in large, loosely organized groups. Territories were not set up, and oestrous females were incessantly harried by numerous males. These stressful conditions resulted in poorly maintained burrows, low breeding success and increased mortality. Dense populations tolerate intruders more readily than sparse ones, possibly because individual recognition becomes less practicable as numbers grow (Telle, 1966).

HOME RANGE. Routine movements are confined to a network of trails linking burrows and points of regular activity. Whenever possible the trails run alongside vertical surfaces like rocks or fallen timber and are often concealed in vegetation (Calhoun, 1963). Males generally travel more extensively than females, although the extent of daily movements and home ranges varies with environmental conditions. On Motuhoropapa I. the mean distance between consecutive captures was 49 m for females and 113 m for males, with a maximum of 330 m; three male home ranges varied from 0.8 to 1.8 ha (Moors, 1985a). On English farmland rats regularly travel hundreds of metres to feed, and one was radio-tracked making a round trip of 3.3 km (Taylor & Quy, 1978). Distance travelled and home range size both increase as food becomes more dispersed in these rural habitats (Hardy & Taylor, 1979). On the other hand, rats in buildings seldom travel more than 100 m and their home ranges are small (Davis, Emlen & Stokes, 1948; Hardy & Taylor, 1979). Territorial defence has been observed in captive populations (Calhoun, 1963), but at present it is unclear whether and under what conditions free-living Norway rats display this behaviour.

NESTS. Norway rats spend most of their time on or below the ground. They are industrious burrowers and may occupy and enlarge the same warren over many years. A burrow dug up near Christchurch extended for 10.3 m and had seven entrances and two nesting chambers (Gibson, 1973a). Nests are constructed from whatever materials are readily at hand, such as grass, leaves, shredded paper or fabric.

REPRODUCTION AND DEVELOPMENT

Norway rats in New Zealand generally breed from spring (August) until autumn (April). Winter pregnancies are infrequent, except in particularly favourable habitats. For example, in well-fed commensal populations in Britain the pregnancy rate fluctuates around 30% throughout the year (Leslie, Venables & Venables, 1952). The length of the breeding season depends on the timing of ovarian activity and regression in females, since both wild and commensal males remain fertile year round (Perry, 1945; Moors, 1985a).

Females are polyoestrous, ovulate spontaneously, and in captivity have an oestrous cycle lasting 7-14 days (McClintock & Adler, 1978). Counts of corpora lutea in the ovaries showed that females on islands in Hauraki Gulf shed an average of 10.8 ova at each ovulation (Moors, 1985a). Oestrous females may be mounted by many different males, although most copulations end without ejaculation (McClintock & Adler, 1978). Gestation lasts 21–24 days, followed within a day or so by postpartum oestrus. The placental implantation sites remain as permanent dark scars on the uterus, which are useful indicators of a female's past breeding history. Litter size in New Zealand averages 6–8, but published information is sparse. Elsewhere pregnancies become more frequent and litter sizes larger among heavier, older females (Leslie, Venables & Venables, 1952).

Kits are naked and blind at birth and weigh 4–6 g. By the end of the first week their external ears have opened and short hair covers the body;

a week later their eyes are beginning to open and co-ordinated movements can be made. They are generally weaned at about 28 days of age.

The size difference between males and females is already evident in captive rats by 5 weeks of age (Hirata & Nass, 1974). Newly independent juveniles have been trapped at weights as low as 21 g near Christchurch (Gibson, 1973a) and 23 g on Campbell I. (P. J. Moors, unpubl.), but usually they weigh 30–40 g. Growth is rapid at first, but slows down with age, although it never ceases entirely.

Early-born juveniles mature in the season of their birth, whereas late-born ones remain immature until the next breeding season. Sexual maturity is usually assessed from external signs: a perforate vagina in females and scrotal testes in males (see Çunningham & Moors, 1983). These stages may be reached by captive females and males at weights of 84–114 g and 83–143 g respectively (Hirata & Nass, 1974), but there is considerably wider variation among commensal and wild rats (Davis, 1953; Moller & Tilley, 1986). In addition, external indicators of maturity misclassify a small proportion of rats; the only certain signs are internal, i.e., the presence of corpora lutea or uterine scars in females, and mature sperm in males.

POPULATION DYNAMICS

DENSITY. Population density fluctuates widely with season and environmental conditions. Numbers are generally least in spring and greatest in autumn and early winter, when juveniles considerably outnumber adults. Extremely high densities can be reached if food and cover are abundant (e.g., at rubbish tips) and if predators are absent (e.g., on some islands). Trapping success provides an indirect estimate of abundance, and on islands is typically in the range 5–20 C/100TN, up to 123.5 C/100TN on Campbell I. (G. A. Taylor, unpubl; see also Fig. 42). On Motuhoropapa I., density was estimated to be 2.6–4.2 rats/ha (Moors, 1985a).

The average annual production of young rats on Kapiti I. was 33.5 young/adult female (Dick, 1985). These figures do not take account of nestling mortality, so the number of young actually reaching independence must have been much lower. As well, not all females necessarily breed to their full capacity: on islands in Hauraki Gulf, 13 of 27 females examined had ovulated repeatedly but failed to conceive, even though fertile males were present (Moors, 1985a). Average annual productivity was 22.5 young/adult female on St. Clements I., Maryland (Lattanzio & Chapman, 1980), and 38.0 young/adult female in 15 commensal populations outside New Zealand (Brooks & Rowe, 1979).

The longevity of wild Norway rats has not been studied, but few urban or rural individuals survive for more than 12 months (Davis, 1953; Bishop & Hartley, 1976). Females have a longer life expectancy than males, and hence the sex ratio among heavier, older rats is biased towards females (Schein, 1950).

Nowhere in New Zealand are all four introduced rodents known to be sympatric. Taylor's (1984) analysis suggests that their distributions, on the smaller offshore islands at least, may be affected by competitive exclusion. Ship rats or house mice are statistically more likely to inhabit an island which lacks Norway rats; similarly, Norway rats are more likely to be present if ship rats or house mice are absent. Norway rats are sometimes able to prevent pioneer house mice from colonizing islands (Taylor, 1978a) and can exclude kiore from favoured habitats on Kapiti I. (Dick, 1985). Active predation may be part of these competitive interactions, because elsewhere Norway rats are known to kill house mice (R. A. Davis, 1979). Similarly, observations of Norway rats killing kiore last century (e.g., Meeson, 1885) led naturalists to blame Norway rats for the decline in kiore populations in the North and South Islands (but see p. 179).

Predators, Parasites and Diseases

The main predators of Norway rats in New Zealand are cats, stoats, ferrets and harrier hawks (*Circus approximans*). Adult rats may be able to defend themselves against the smaller predators, but juveniles would be vulnerable to them all, and possibly to weasels as well. On Campbell I. skuas (*Catharacta lonnbergi*) commonly kill and eat rats of all ages. In Britain, farm cats can stop reinfestation of buildings, although they apparently cannot eliminate a rat population once it is established (Elton, 1953). In New Zealand, free-living rats may be more susceptible to predators; Taylor (1984) has shown that if stoats are present on an island, Norway rats are significantly less likely to be there too.

Ectoparasites reported from Norway rats in New Zealand include the mites *Haemaphysalis longicornis* and *Notoedres muris;* the louse *Polyplax spinulosa;* and the fleas *Leptopsylla segnis, Nosopsyllus fasciatus, Pulex irritans, Pygiopsylla hoplia, P. phiola, Xenopsylla cheopis* and *X. vexabilis* (Tenquist & Charleston, 1981). Endoparasites include the nematodes *Mastophorus muris, Nippostrongylus brasiliensis* and spiruroid species; the tapeworms *Cysticercus fasciolaris* and *Hymenolepis diminuta;* and the trypanosome *Trypanosoma lewisi* (Carter & Cordes, 1980; Moors, 1985a).

In various parts of the world Norway rats are involved in the transmission of serious human diseases like leptospirosis (Weil's disease), salmonellosis, trichinosis, toxoplasmosis, murine typhus, lymphocytic choriomeningitis, and sometimes plague. In New Zealand they are a reservoir for the spirochaete *Leptospira icterohemorrhagiae,* which lives in the kidneys and is excreted in urine. In the Waikato region 45% of 91 urban rats and 39% of 41 rural rats showed signs of infection with two leptospiral serotypes (although not the serotypes causing most human infections in that area) (Carter & Cordes, 1980). *Salmonella typhimurum* and *S. bovismorbificans* were isolated from 36% of 42 rats from farms near Rotorua where salmonellosis was endemic in sheep (Robinson & Daniel, 1968).

Adaptation to New Zealand Conditions

We do not know whether the decline of the abundant Norway rats on mainland New Zealand last century was related to the coincidental increase in ship rats, but it contrasts with events in Europe and much of North America: there it was ship rats which declined as invading Norway rats multiplied. Possible causes include the different conditions for predation and competitive interactions in the different countries (Taylor, 1978a, 1984), and the different habitat requirements and food preferences of the two species.

Differences in diet between Norway rats in New Zealand and elsewhere reflect the particular foods available rather than intrinsic changes in feeding habits. Other potential differences in morphology, breeding, behaviour and ecology have not been studied in sufficient detail to allow generalizations.

SIGNIFICANCE TO THE NEW ZEALAND ENVIRONMENT

DAMAGE. Locally, Norway rats can be serious urban and agricultural pests and harbourers of disease, but nationally these dangers are overshadowed by the rats' past and present role as a predatory pest of native wildlife.

The terrestrial habits of Norway rats make native animals which live, roost or nest on or near the ground particularly vulnerable. At the beginning of European settlement there were still many such species which had survived the depredations of kiore during the Polynesian period, but which then failed to come to terms with the newly invading Norway rats. In contemporary times most of the native birds still in this position are seabirds, and 27 of the 53 bird species known world-wide to be prey of Norway rats are seabirds (Atkinson, 1985). The rats take eggs and nestlings, and are also large enough to kill adults, not often but more frequently than do ship rats or kiore. The effects of this predation depend on the behaviour and ecology of both the birds and the rats (see Moors & Atkinson, 1984) and range from negligible to causing complete failure of nesting (e.g., Imber, 1985), or even local extinction. For example, at Campbell I. the pipit *Anthus novaezeelandiae* and all the small petrels have disappeared from the main island and are now restricted to rat-free offshore islets.

Many invertebrates, usually the larger ground-dwelling species, have suffered from the introduction of these rats. For example, predation has decimated many populations on Breaksea I., where beetles, wetas and harvestmen were most affected (Bremner, Butcher & Patterson, 1984); and the large endemic insects of Campbell I. are now confined to islets lacking rats (T. K. Crosby, unpubl.). The impact of Norway rats on native reptiles and plants has been given little attention. However, there is evidence that kiore have affected these organisms (p. 190); so Norway rats may have done so too.

Problems with Norway rats as commensal pests arose within a few years of their arrival. By 1832 food stores in Maori settlements in the Bay of Islands were being seriously damaged (Williams, in Rogers, 1961). Attempts to grow grain near Riccarton in 1840 failed, largely because of the scale of losses due to Norway rats (Thomson, 1922). Initially the Maori prized the large Norway rats as food, but eventually came to revile them because of their unclean habits (Best, 1942).

CONTROL. Modern protection measures rely mainly on rodenticides (poisons), usually anticoagulants which interfere with blood clotting. Warfarin was the first and best known of these, but genetic resistance to it has developed in some rat populations overseas, and may do so here. The new anticoagulants brodifacoum and bromadialone remain toxic to warfarin-resistant rats, and have the added advantage of being lethal after a single dose (a lethal dose of warfarin must be accumulated over several days). Many campaigns to control rat numbers are ineffective because rats rapidly reinvade from surrounding unpoisoned populations. However, total eradication can be achieved if the target population is isolated, and in New Zealand Norway rats have been successfully eliminated from the Noises Is. (Moors, 1985b), Titi I. (R. H. Taylor, unpubl.), and Breaksea (R. H. Taylor & B. W. Thomas, unpubl.). However, the largest island on which eradication is currently feasible is only 200–300 hectares, which means that, at the moment, many islands of high conservation value (e.g., Kapiti, Campbell) cannot be cleared of Norway rats.

P. J. M.

17. Ship rat

Rattus rattus (Linnaeus, 1758)

Synonyms *Mus rattus* Linnaeus, 1758; *Mus alexandrinus* Geoffroy, 1803; *Musculus frugivorus* Rafinesque, 1814; *Mus novaezelandiae* Buller, 1870, plus numerous others.

Also called roof rat, black rat, European house rat.

The subspecific status once given to the three colour morphs of *Rattus rattus* is now known to be unjustified (Tomich & Kami, 1966). Some authorities refer to commensal ship rats of European origin (2n = 38) with the trinomial *Rattus rattus rattus*, regardless of coat colour or locality, in order to distinguish them from Asian (2n = 42) subspecies (e.g., Marshall, 1977; Musser & Califia, 1982).

DESCRIPTION (Fig. 40)

Distinguishing marks, Table 34 and skull key, p. 24. See right-hand back endpaper for colour plate.

The body is sleek and slender, with a scaly, sparsely haired tail and large, thin, almost hairless ears. The tail is dark grey all round, and longer than the head plus body in adults. All four feet are dorsally hairy and uniformly coloured, but naked beneath. The forefeet have four clawed toes, the hind feet have five. The eyes are small, dark and beady, very sensitive to light but not visually acute. Ship rats are normally clean, well-groomed animals, with smooth fur and a characteristic, not-unpleasant smell.

There are three colour forms. (1) The "rattus" morph has a black back merging to a slate grey belly; young rats may show a bluish tinge. (2) The "alexandrinus" morph has a grey-brown back and a slate-grey belly. The dorsal pelage consists of brown-tipped grey hairs overlain by long black

Fig. 40 Ship rat (FRI).

guard hairs. (3) The "frugivorus" morph also has a grey-brown back, but the ventral surface, from the chin to the anus, is pure or lemony-white.

The three morphs interbreed freely (Feldman, 1926). Most North Island rats are "frugivorus", and most South Island rats are "alexandrinus" (Table 38). Intermediates are rare, but possible. Individuals of the "rattus" (black) morph are more docile to handle (Worth, 1950; Daniel, 1972) and more abundant in buildings.

The normal number of nipples is 10, two pectoral pairs and three inguinal pairs. However, one or both second pectoral nipples may be doubled, giving a total of 11 or 12. For external sexual characters, see Fig. 39.

Ship rats are skilful climbers in buildings or forests and can scale rough vertical surfaces, traverse fine wires (Ewer, 1971), and run through lattices of fine branches. One was dislodged from an epiphyte over 18 m up a rimu tree in the Orongorongo Valley (Daniel, 1972). Live trapping and tracking indicate that such individuals do not live entirely up trees, but use the forest floor as well (Daniel, 1972; Innes & Skipworth, 1983). In contrast to Norway rats, commensal ship rats often live in ceilings of buildings (hence the alternative name of "roof rat"); also they rarely burrow, and are unwilling swimmers. However, studies on islands west of Great Barrier I. suggest that some have swum up to 750 m to reach new islands (J. McCallum, unpubl.). In cooler, rougher South Island waters, maximum swimming distance would be about 300 m (R. H. Taylor, unpubl.).

The skull and teeth are specialized for gnawing and grinding. The large incisors grow throughout life and are self-sharpening. The grinding molars also grow continuously and are subject to continuous wear, which is used as an index of age (Fig. 41). The cranium of a ship rat skull (Fig. 8) has round margins (cf. straight margins in the Norway rat).

Table 38: Geographical variation in proportions of colour morphs in ship rat populations in New Zealand.

	Latitude (°S)	% "rattus"	% "frugivorus"	% "alexandrinus"	n	Reference
North Island						
Waipoua State Forest, Northland[1]	35°38′	27	39	24	201	K. Smith (1986)
Great Barrier I.	36°11′[2]	17	83	0	53	Daniel (1972)
Pureora State Forest Park	38°28′	13	87	0	372	C. M. King, J. G. Innes & M. Flux (unpubl.)
Egmont National Park	39°20′	21	79	0	77	J. G. Innes (unpubl.)
Tiritea Valley, Manawatu	40°26′	11	89	0	307	Innes (1977)
Orongorongo Valley, Wellington	41°21′	13	87	0	280	Daniel (1972)
South Island						
Waimangaroa	41°43′	35	6	59	114	Best (1968)
St. Arnaud, Nelson Lakes National Park	41°50′	21	0	79	44	J. G. Innes (unpubl.)
Westland National Park	43°20′	58	0	42	45	J. G. Innes (unpubl.)
Christchurch grainstores[3]	43°32′	57	9	34	176	Best (1968)
Banks Peninsula	43°47′[2]	21	6	73	162	Best (1968)
Eglinton & Hollyford Valleys, Fiordland	44°50′	32	10	58	106	C. M. King & H. Moller (unpubl.)
Stewart Island						
Paterson Inlet & Halfmoon Bay	46°55′	12	12	76	25	R. H. Taylor (unpubl.)
Chew Tobacco Bay	47°00′	0	39	61	100	Gales (1982)
Robertson River	47°08′	0	62	38	79	P. J. Moors & D. Cunningham (unpubl.)

1. 10% classed as "intermediate". 2. Approximate only, since exact source of specimens unknown.
3. Non-forest habitat.

All New Zealand ship rats examined so far have had 2n = 38 chromosomes (but see below).

Dental formula $I^1/_1C^0/_0Pm^0/_0M^3/_3 = 16$.

FIELD SIGN

Ship rats are among the most widespread mammals on the New Zealand mainland, yet are seldom seen and little known, largely because they are nocturnal, often arboreal, and shy. It is difficult to confirm their presence without actually setting traps, but certain field signs may give them away.

Ship rats collect the hard seeds of miro (*Prumnopitys taxifolia*) and hinau (*Elaeocarpus dentatus*) and eat them under cover, such as in hollow logs. Several hundred seeds, each with a small hole neatly chewed in one end or side and the contents removed, may be found in one cache (Beveridge, 1964). In trees, rats may feed on any available flat surface, and leave pellets, seeds and insect remains (e.g., weta legs and heads) on the tops of large limbs, or in branch crotches or epiphyte clumps. Rat nests and feeding platforms are described on page 000. Pellets averaged 8.6 mm length (range 6.8-13.8) and 3.6 mm width (range 2.7-5.0) in one sample (Daniel, 1969).

Rat tracks may be found in fine mud or sand but cannot be identified to species. Burrowing is characteristic of Norway rats, not ship rats.

In farm sheds, warehouses, factories or houses, ship rats might be detected from their scuffling at night or from gnaw-marks or pellets at feeding sites. Inside buildings their movements along beams or pipes can be traced by the smears made by generations of rats ducking under transverse beams (McIntosh & Adams, 1955).

The scats of predators are likely to contain rat fur, bones and teeth if rats are present. Through most New Zealand forests the ship rat will be the only rat species present in scats.

MEASUREMENTS

See Table 39. The figures given are indicative only, since not all the samples were taken at the same time of year, or measured in exactly the same way. Cunningham and Moors (1983) describe a standard method, recommended for use in future work. Furthermore, the age and size of ship rats at sexual maturity may vary from year to year, or between areas.

VARIATION

Commensal ship rats taken from Christchurch grainstores were significantly heavier (average 185 g for males and 163 g for females) than wild ones from Banks Peninsula forest (160 g; 137 g) although body lengths were very similar (Best, 1968). The relative proportions of the three colour morphs vary dramatically between islands; the "alexandrinus" morph is rarely recorded from the North Island, except in Waipoua State Forest, Northland, and likewise the "frugivorus" morph is virtually absent from the South Island, and the "rattus" morph from Stewart I. (Table 38). The reasons are more likely to be connected with physiological characters related to the coat colour genes than with the coat colours themselves (Tomich, 1968).

HISTORY OF COLONIZATION

The first voyages by European ships to New Zealand in the mid 18th century were made at a time when, in Britain and other parts of Europe, the Norway rat had temporarily displaced *R. rattus* as the common commensal rat in stores and warehouses. For up to 150 years after AD 1700, the Norway rat was the species most commonly carried, as an unwelcome and accidental passenger, aboard ships plying world trade and exploration routes from Europe. This is not to say that some ships did not carry *R. rattus* during that time, but it is only since about 1850 that they have been the commonest rats aboard ships (Atkinson, 1985).

It is possible that ship rats got ashore at some ports or towns in New Zealand in the early 19th century, but could not get further because Norway rats were already established. Whatever the reason, historical records indicate that ship rats did not spread, in either the North or South Islands, until after 1860 and 1890, respectively (Atkinson, 1973a). At that time, Norway rats were common in all habitats, including forests, and kiore were still present but declining on both main islands. Kiore later disappeared from the North Island and most of the South Island (p. 178); but whether this was the effect of competition from the spreading ship rats (Watson, 1956; Atkinson, 1973a) or mice (Taylor, 1975a) is still unknown. The relative distributions of the four species of rodent in New Zealand, especially on islands, suggest some competitive exclusion (Taylor, 1975a). Two or three of them may coexist on some larger islands, but nowhere in the New Zealand

Fig. 41 Age determination of rats according to the index developed by Karnoukova (1971), drawn by A. Beauchamp from New Zealand material classified and lent by B.M. Fitzgerald and B.J. Karl. Their definitions of the age classes are as follows:

Age class I All molars not fully emerged. M^1 and M^2 present, M^3 absent or only partly emerged.

Age class II Tubercles fused in incomplete transverse loops. Tubercle 7 of M^1 united with Tb 4 in a complete enamel loop without fusion of dentine. (Early II Tb 7 M^1 enamel essentially not connected to Tb 4; Late II Tb 7 M^1 enamel distinctly connected to Tb 4 but dentine usually not connected between Tb 7 and Tb 4.) M^2 and M^3 show little wear. Enamel of M^3 curved inwards to centre of tooth. Dentine lake narrow, showing little wear.

Age class III Tubercles fused in complete transverse loops, usually in dentine as well as in enamel on all upper molars. M^2 and M^3 with anterior and posterior cusps worn but not connected, usually not in enamel and almost never in dentine. If they are connected, this is only in M^3 (cusps on M^3 fuse before M^2).

Age class IV Transverse loops narrow to average width (Tb 7 in M^1 fused to Tb 4 in dentine). Cusps on M^2 and M^3 fused, always in enamel and often in dentine as well. (Cusps on M^2 fused on one side on at least one M^2.)

Age class V Enamel on outer edge fused longitudinally between all transverse loops except first loop of M^1 in Early V; all loops fused longitudinally, sometimes in dentine as well as enamel in Late V.

Age class VI Dentine fused longitudinally in all loops except the first in M^1 in Early VI; dentine forming a trident in Late VI.

Age class VII Eroded crowns, with traces of enamel loops in Early VII; dentine lakes in Late VII.

Note that the distinction between age classes III and IV, which we find the most difficult, is made chiefly on the degree of wear on M^2 and M^3 — in class III the anterior and posterior cusps are not usually fused, and in class IV they are fused.

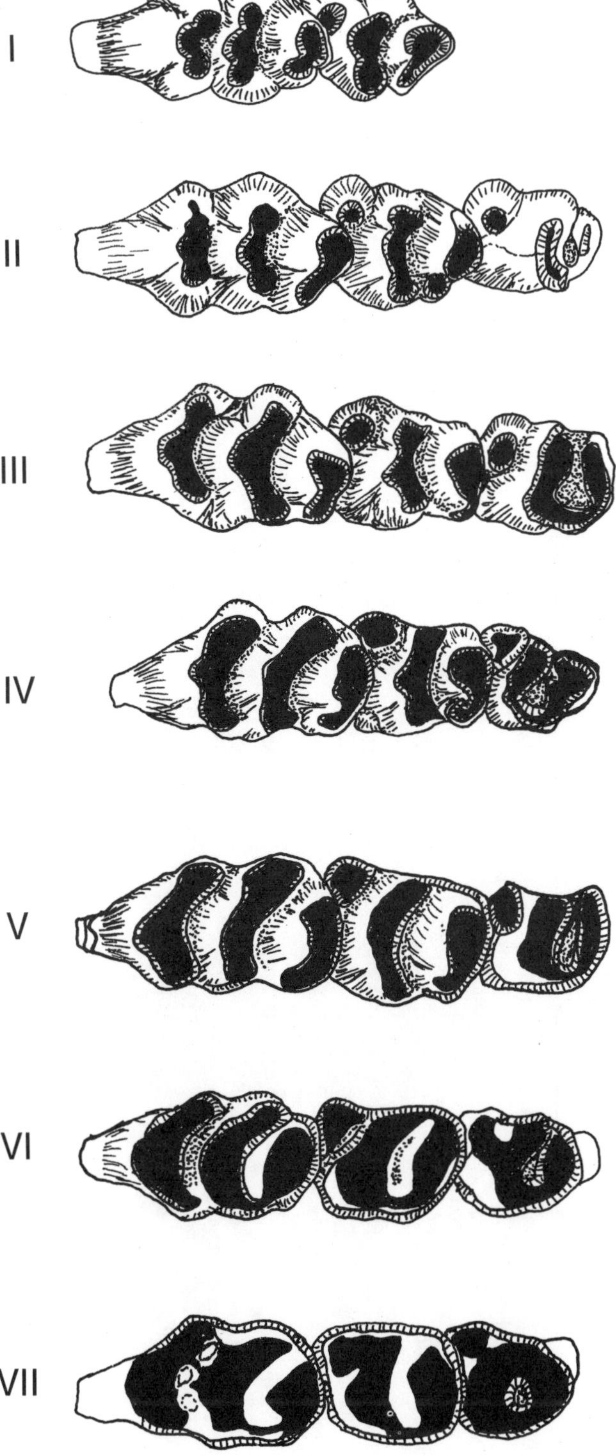
I
II
III
IV
V
VI
VII

Table 39: Body measurements (mean and range) of adult ship rats in New Zealand.

Area	Latitude (°S)	Sex	Head and body length (mm)	n	Tail length (mm)	n	Condylobasal length of skull (mm)	Weight (g)	n	Reference
Waipoua SF Northland	35°38′	M	175 (108-212)	40	220 (115-260)	40	—	166 (52-210)	24	K. Smith (1986)
		F	165 (121-210)	40	207 (130-255)	40	—	132 (53-200)	31	
Pureora SFP	38°28′	M	177(129-204)	200	194 (166-232)	200	—	144 (81-204)	200	C. M. King, J. G. Innes & M. Flux (unpubl.)
		F	170 (131-190)	200	194 (143-219)	200	—	132 (56-184)	200	
Egmont NP	39°20′	M	188 (155-230)	43	198 (179-222)	39	—	157 (113-200)	42	J. G. Innes (unpubl.)
		F	175 (138-198)	32	199 (160-230)	32	—	140 (78-174)	32	
Tiritea Valley, N. Tararua Range	40°26′	M	182 (155-201)	103	199 (173-227)	105	—	154 (89-212)	103	Innes (1977)
		F	166 (125-196)	116	194 (130-228)	116	—	127 (54-173)	114	
Orongorongo Valley, Wellington	41°21′	M	180 (160-203)	49	205 (166-240)	49	39.0 (34.3-42.0) n=31	147 (106-186)	50	B. M. Fitzgerald & B. J. Karl (unpubl.)
		F	174 (153-195)	49	199 (164-245)	49	38.3 (34.1-41.5) n=30	127 (82-479)	49	
St. Arnaud, Nelson Lakes NP	41°50′	M	184 (155-210)	24	187 (172-205)	23	—	144 (196-188)	24	J. G. Innes (unpubl)
		F	165 (123-184)	18	189 (146-207)	18	—	123 (92-184)	18	
Westland NP	43°20′	M	178 (144-210)	22	194 (165-222)	22	—	140 (86-183)	22	J. G. Innes (unpubl.)
		F	173 (149-195)	23	203 (165-230)	23	—	139 (75-168)	23	
Hollyford Valley, Fiordland	44°45′	M	183 (163-203)	36	198 (182-217)	34	—	152 (104-199)	36	C. M. King & H. Moller (unpubl.)
		F	170 (143-191)	26	195 (160-210)	27	—	131 (83-186)	27	
Southern Stewart I.	47°08′	M	152 (121-187)	29	195 (152-223)	32	37.4 (33.5-40.4) n=6	138 (77-205)	31	D. M. Cunningham & P. J. Moors (unpubl.)
		F	148 (122-174)	40	196 (159-234)	40	37.3 (32-40.8) n=12	124 (62-180)	40	

All measurements given are for snap- or Fenn-trapped adults. Criteria for maturity were perforate vaginae for females and scrotal testes with visible epididymal tubules for males.

region do all four live together as they do on warmer Pacific Islands such as Hawaii. This hypothesis needs experimental testing, since there are other factors, such as predators, to consider. For example, ship rats are more likely to live on an island from which stoats are absent (Taylor, 1984).

DISTRIBUTION

WORLD. Ship rats are found throughout the world, from Sweden at 63°N (Brooks & Rowe, 1979) to Macquarie I. at 45°S (Atkinson, 1985). Over its total range, *R. rattus* is divided into five chromosomal races of uncertain taxonomic status (Baverstock *et al.*, 1983). The two most widespread are the Asian form (2n = 42) and the European or Oceanian form (2n = 38) (Yosida *et al.*, 1974), which differ in some external features as well as in diploid number and other chromosomal properties. They may be separate species, but their genetic compatibility in the wild is little known (Musser & Califia, 1982). Only the Oceanian form has spread with human trading activities throughout the world, from its evolutionary homeland in India (Yosida *et al.*, 1974), to Britain by the third century AD (Anon., 1984a) (well before the Norway rat, which has largely displaced it there since the 1700s; Bentley, 1964), and thence to New Zealand and many other oceanic islands. The major period of spread in the Pacific Ocean was from about 1830 to the present day, especially during and since the Second World War (Atkinson, 1985).

NEW ZEALAND (Map 16). Ship rats are by far the most uniformly distributed of the three rat species on the mainland: they are found throughout North, South and Stewart Islands, virtually wherever suitable habitat is found. The offshore islands they are known to be on are listed in Table 36.

All New Zealand ship rats karyotyped to 1985 were the Oceanian form, 2n = 38 (P. J. Moors, unpubl.). This means that the established New Zealand ship rats differ in karyotype, and perhaps in behaviour (Schwabe, 1979), from the Asian form (2n = 42), which has recently reached this country on Asian fishing vessels. Asian form rats have been identified from ships in port at both Auckland and Wellington (P. J. Moors, unpubl.), and one karyotyped in 1985 was 2n = 42.

HABITAT

Ship rats live wild in forests and in a wide range of other habitats, and also as commensals of man in and around buildings.

In the wild they are most abundant (density index up to 22 C/100TN) in mature, diverse, lowland podocarp-broadleaved forests, and absent or very scarce (<0.5 C/100TN) in pure beech (*Nothofagus*) forests, especially silver beech and mountain beech (Daniel, 1978; R. H. Taylor, unpubl.). They are found from the coast to the treeline, but are scarcer at higher altitudes. On a transect up Mt. Misery, in Nelson Lakes National Park, ship rats have not been found above the upper limit of red beech (*c.* 1025 m asl), despite considerable trapping effort in silver and mountain beech forest further up, and in alpine tussock tops (R. H. Taylor, unpubl.). One was seen by

S. M. Beadel (unpubl.) at 1130 m in silver beech-dominated forest on the crest of the Huiarau Range, Urewera National Park.

The only mainland locations so far recorded where extensive trapping of apparently suitable forest habitat has caught no or few ship rats are Jackson Bay (Choate, 1965), Caswell and George Sounds and Lake Monk areas in Fiordland (Daniel, 1978) and between Fox Glacier and Paringa in South Westland (P. D. Gaze, unpubl.). However, they are common in the lower Hollyford Valley (King, 1983a). They are abundant in some *Pinus radiata* plantations (Clout, 1980) and exotic parklands, but may be scarce in very young plantings (C. M. King, unpubl.) and grassland.

Brockie (1977) trapped more *R. rattus* than *R. norvegicus* on dairy farms, but the reverse was true at rubbish tips. Wodzicki (1950) stated "habitats such as cornricks, cornfields, haystacks, barns and sheds, orchard stores and homesteads appear to be shared by both species, although *R. norvegicus* seems to prevail. River and stream banks and shores of ponds and lakes seem to be occupied almost entirely by *R. norvegicus*".

Ship rats are agile climbers, gaining access to buildings via drains or wires and living under roofs or in walls. They live in factories, warehouses and business premises if suitable food, cover and water are available.

Vessels visiting and plying the New Zealand coasts should not be forgotten in a compendium of habitats. *Rattus rattus* is a common seaport species throughout much of the world, including New Zealand, and translocation of individuals is probably continual, despite de-ratting procedures. Consequences of these movements can be serious (p. 224) for island faunas.

FOOD

DIET. Ship rats are omnivorous generalists, yet can be very selective feeders (D. A. Clark, 1981).

Commensal ship rats in New Zealand, as elsewhere, will feed on and damage almost any animal or grain product, fruit or any edible stored product or refuse to which they have access (Wodzicki, 1950).

Ship rats living in native forests eat both plant and animal foods all year round, but the proportions vary seasonally. In two North Island studies (Orongorongo Valley, Wellington and Tiritea Valley, Manawatu) animal foods predominated in spring and summer, and plant foods in autumn and winter (Daniel, 1973; Innes, 1979). Rats collected from Banks Peninsula, and from Waimangaroa, West Coast, South Island, ate mainly plant food in all seasons (Best, 1969). The main animal foods are arthropods, especially wetas, but also beetles, spiders, moths, stick insects and cicadas (Best, 1969; Daniel, 1973; Innes, 1979; Gales, 1982). Important fruits in the diet are those of *Coprosma* spp., karaka (*Corynocarpus laevigatus*), rimu (*Dacrydium cupressinum*), hinau (*Elaeocarpus dentatus*), kiekie (*Freycinetia baueriana* ssp. *banksii*), pigeonwood (*Hedycarya arborea*), kawakawa (*Macropiper excelsum*), miro (*Prumnopitys ferruginea*), matai (*Prumnopitys taxifolia*), supplejack (*Ripogonum scandens*), nikau (*Rhopalostylis sapida*), pate (*Schefflera digitata*), and kohia (*Passiflora tetrandra*)

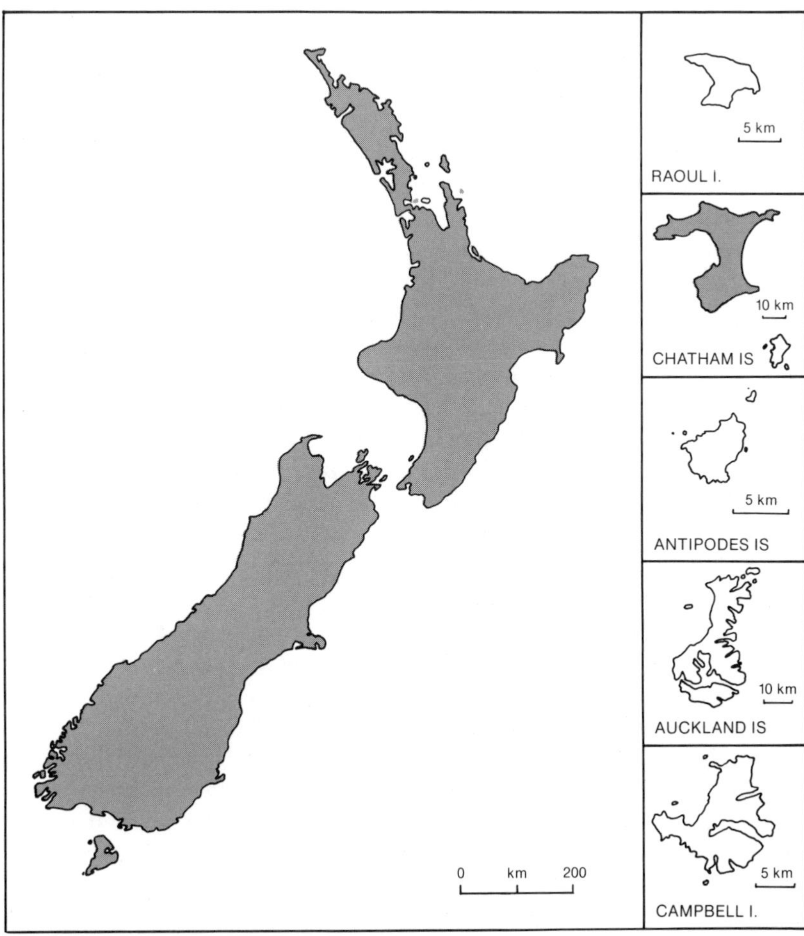

Map 16 Distribution of ship rats on the main islands of New Zealand. For distribution on offshore islands, see Table 36.

(Beveridge, 1964; Best, 1969; Daniel, 1973; Innes, 1979). In an exotic plantation forest where fruits were scarce, rats ate mainly invertebrates in winter (Clout, 1980). Other prey items of conservation significance are native snails (Daniel, 1973; Meads, Walker & Elliott, 1984), slugs (Best, 1969), birds and probably lizards (Whitaker, 1978). Ship rats are known predators of birds' eggs and young, especially of forest passerines; eggs up to at least 61 mm long are taken (Atkinson, 1978, 1985). Rodents (mostly ship rats) were responsible for 16% of 99 known losses of passerine nests at Kowhai Bush, a remnant lowland forest near Kaikoura. Rodents destroyed proportionately more nests with eggs than with chicks (Moors, 1983b). However, birds are now a minor diet item for rats living in established mainland populations, and the significance of their predation on contemporary mainland bird populations remains unclear (Moors, 1983b; King, 1984a; and see p. 311).

On Big South Cape I. (off Stewart I.) in 1964, ship rats at high density (p. 224) browsed punui (*Stilbocarpa lyalli*) to ground level and defoliated five-finger (*Pseudopanax arboreus*); they also quickly eliminated five endemic bird species, one bat species (p. 135) and a large flightless weevil (Bell, 1978; Ramsay, 1978). The diet and ecological impact of rats arriving in a pristine island environment are quite different from those of established populations on the mainland.

There is evidence of some differences in diet between males and females, and between adults and juveniles, from the Galapagos Is., Ecuador (D. A. Clark, 1980), and from Stewart I. (Gales, 1982). Ship rats can be very salt-tolerant, but probably cannot drink seawater (Dunson & Lazell, 1982). They cannot digest cellulose, but may eat green plants as a moisture source (Norman, 1970; Yabe, 1982).

FEEDING BEHAVIOUR. In the laboratory, adult ship rats eat about 15 g (10% of their body weight) of dry food a night (Bhardwaj & Khan, 1974). They prefer to feed under cover and often carry food items to a sheltered place rather than eat them immediately (Ewer, 1971). Studies on the Galapagos Is. (D. A. Clark, 1981, 1982) showed that although rats took a broad range of plant and animal foods, they ate only certain parts. Clark noted that nearly all stomachs she examined contained traces of food from a large number of items, indicating that ship rats sample many potential foods but are very selective about what they accept.

Feeding trials (Moors, 1978) show that ship rats are "messy" predators of eggs and chicks. They lick up spilt yolk or albumen in the nest, disturbing the nest lining; they eat the flesh off the bones of the chicks but leave limbs attached; they open the brain cavity, and eat the pectoral muscle and sometimes the stomach contents. In contrast, mustelids cleanly removed eggs or nestlings from the nest.

SOCIAL ORGANIZATION AND BEHAVIOUR

ACTIVITY. Ship rats are nocturnal and have excellent senses of smell, touch, hearing and taste. They display a strong exploratory drive within their home range (Barnett, 1975) but cautiously approach strange objects such as traps or baits (Cowan & Barnett, 1975; Advani & Idris, 1982).

DISPERSION. Very little is known of the social life of forest-dwelling ship rats, in New Zealand or elsewhere. Trapping studies and nest observations show that they are not colonial, but that individuals or family groups are dispersed rather evenly through available habitat (Daniel, 1972; Innes, 1977).

Commensal rats living in buildings at Accra, Ghana, had a dynamic but cohesive colonial organization in which a few adult females were largely responsible for defence of a group territory, although they were themselves subordinate to a "top male". The territorial boundaries were fixed, but resident rats frequently explored beyond the boundaries and so became familiar with a larger area. Emigration, mainly by subadults, was caused not by shortage of food but by social pressure; it balanced recruitment into the population and continued until surrounding buildings became saturated with rats, and breeding declined (Ewer, 1971).

HOME RANGE. In podocarp-rata-broadleaf forest of the Orongorongo Valley, adult males moved on average 55.4 m (range 0-190 m), and adult females, 38.7 m (range 0-117 m) between successive cage captures (Daniel, 1972). In similar habitat near Palmerston North, comparable figures were 38 m (range 0-67 m) for males and 36 m (range 0-45 m) for females (Innes, 1977). These figures give only an indication of the rats' real movements, since they also reflect the distance between the set cage traps, and they combine data from all seasons. In coastal rimu-rata-broadleaf forest on Stewart I., six ship rats ranged widely in summer (range lengths averaging 142 m ± SE 17 m) (Hickson, Moller & Garrick, 1986).

Adult females tend to occupy exclusive areas in the breeding season (Innes & Skipworth, 1983; Hickson, Moller & Garrick, 1986). Males have larger ranges than females, a result found also in overseas studies (Spencer & Davis, 1950; Watson, 1951; Gomez, 1960; Jackson & Strecker, 1962a; Tomich, 1970). Home ranges are of course three-dimensional, since the rats frequently climb trees. Footprint tracking shows that forest-dwelling rats may traverse most of their home range each night. They know their neighbours well, and if one disappears, invaders take over the empty range within days. This means that removal trapping or poisoning is not a useful long-term control method in areas of continuous habitat, since reinfestation is rapid during and after control programmes (Barbehenn, 1974; Innes & Skipworth, 1983).

NESTS. Ship rat nests are difficult to find in mature forest, since many sites are available in epiphyte clumps or tree hollows. They may build roughly spherical, sparrow-like nests in young trees or hedges if other sites are not available (White, 1897; Innes, 1977). These are substantial, stable structures, 30 to 100 cm across, loosely woven of twigs and leaves cleanly bitten from branches near the nest site. They also construct smaller feeding platforms up trees, often based on old birds' nests and containing pellets and food remains. Nests and feeding platforms in use have new leaves and twigs added to them regularly. Rats observed at a nest in forest in 1976 emerged to feed just on dark; they were very alert to unusual sounds, and constantly tested the air for scents (Innes, 1977).

REPRODUCTION AND DEVELOPMENT

The reproductive cycle of the ship rat is well documented from laboratory populations. The oestrous cycle is 4-6 days long, and the gestation period 20-22 days. Breeding is polyoestrous; there is postpartum oestrus (Asdell, 1964; Rowett, 1974) and perhaps delayed implantation of ova fertilized during lactation (Mantalenakis & Ketchel, 1966). Litter size is 3-10, averaging 5-8 (Bentley & Taylor, 1965; Southwick, 1966; Hirata & Nass, 1974; Cowan, 1981); the interval between litters averages 32 days (range 27-38) (Cowan, 1981). Growth is rapid and the pups are weaned at 21-28 days (Cowan, 1981), when they weigh about 38 g, although there is considerable variation (Bentley & Taylor, 1965). They may reach sexual maturity at 3-4 months old (Watts & Aslin, 1981). The uterus of the female bears a record of the number of young born, as a series of dark spots or placental scars, fading to grey with age.

New Zealand studies published so far were all done in mature podocarp-broadleaf forests on the three main islands, and all describe a strongly seasonal breeding pattern. Pregnant or lactating females were trapped only between mid September and mid April, occasionally to mid winter following heavy fruit fall (Daniel, 1972; Best, 1973; Innes, 1979). Mean litter size was 4.95 (SD 1.31, n = 19) in the Tiritea Valley, north Tararua Range (Innes, 1979); 6.1 (SD 1.79, n = 26) in the Orongorongo Valley (Daniel, 1972); 5.9 (SD 1.6, n = 14) on Banks Peninsula; and 5.9 (SD 1.90, n = 7) at Waimangaroa, West Coast, South Island (Best, 1973).

Differences in development betweeen *R. rattus* and *R. norvegicus* probably reflect the arboreal habits of *R. rattus*. The length of the tail (a climbing aid) at birth is only about one-third of head and body length in both species; however, by day 18 it equals head and body length in the ship rat, but only 60-70% of head and body length in the Norway rat. Young ship rats, whose first forays may be from a tree-borne nest, have a more marked cling response than young Norway rats (Cowan, 1981).

POPULATION DYNAMICS

DENSITY. The seasonal breeding of ship rats causes a corresponding seasonal change in density, from low numbers in spring and early summer to a peak usually in autumn (Daniel, 1978; Moors, 1978). Absolute density is difficult and time-consuming to measure, but relative density indices (Fig. 42) are simpler and more frequently used. The exact timing and extent of the peaks and troughs in density vary between years (Fig. 43) and habitats. Normally the amplitude of these fluctuations is much less dramatic than in mice (p. 234). Daniel (1972) suggested that the variable winter food supply (seeds) and predation by feral cats were the main factors controlling ship rat density in the Orongorongo Valley (29-month average = 1.7/ha). Heavy seedfalls of hinau (*Elaeocarpus dentatus*) and pigeonwood (*Hedycarya arborea*) preceded both extended (winter) breeding by females and high winter and spring rat density (Daniel, 1978). In red beech-dominated forests, numbers also increase significantly (from <0.5 up to 5.0 C/100TN) after good seeding

Fig. 42 Method of assessing relative density of rodent populations used in a standardised form throughout New Zealand since 1971.
(a) A trap station comprising a wood or metal tunnel, wired at the ends to exclude possums, hedgehogs etc., under which are set one rat trap and one mouse trap, facing in opposite directions.
(b) A ship rat cleanly caught, while eating the bait of peanut butter and rolled oats.
(c) Mice may be caught in rat traps (exceptionally, two at a time), but rats larger than juveniles or kiore are not caught in mouse traps; hence the number of trap-nights per session is twice as many for mice as for rats (Pureora Forest Park; C.R. Pickard).
The formula for calculating trapping success (used as a relative density index) is:

$$[\text{Captures}/(\text{traps set} - \text{half no. of traps sprung})] \times \text{nights set} \times 100$$

i.e., the number of animals caught per trap per 24 hours, corrected for unavailable traps (Nelson & Clark, 1973).

(a)

(b)

(c)

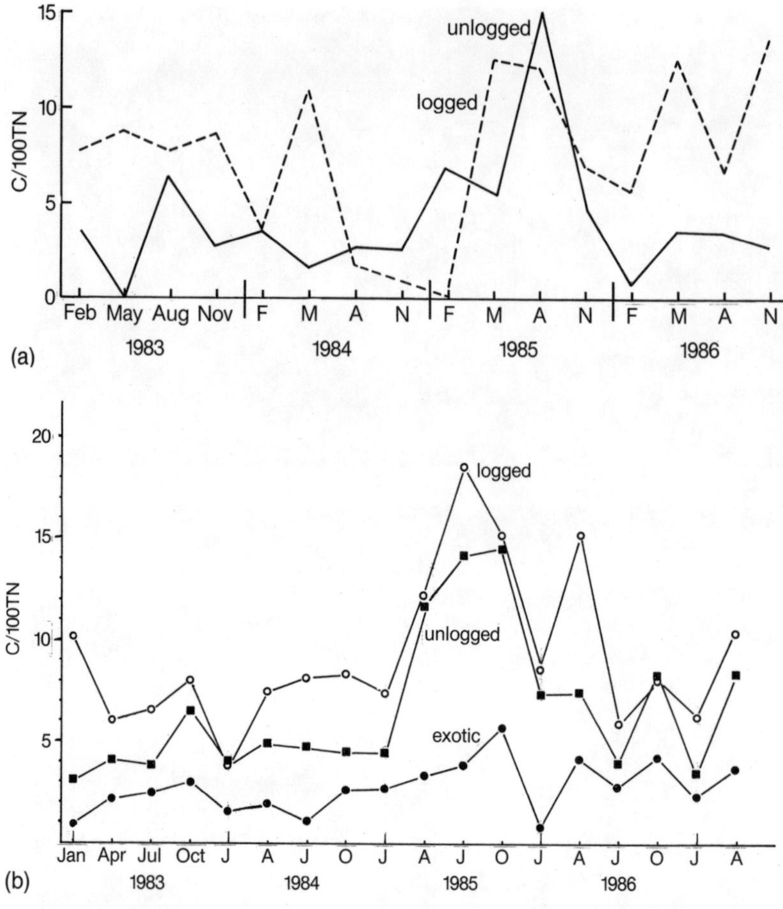

Fig. 43 Density indices (C/100TN) of ship rats recorded at Pureora Forest Park, central North Island (C.M. King, J.G. Innes & M.M. Flux, unpubl.): (a) using rodent traps (Fig. 42) set at 50-m intervals in February, May, August and November in podocarp/broadleaved native forest (solid line unlogged, dashed line logged); (b) using Fenn traps set at 300-m intervals in January, April, July and October in podocarp/broadleaved native forest (squares unlogged, open circles logged) and in exotic forest (closed circles).

of red beech and its associated species (R. H. Taylor, unpubl.). The hypothesis that seed abundance controls rat density could be tested experimentally by adding extra food to a study population. In the Orongorongo Valley between 1971 and 1985, rat numbers increased as cat numbers declined, suggesting an important role for predation in ship rat population dynamics (p. 348) (B. M. Fitzgerald & B. J. Karl, unpubl.).

SEX RATIO. The sex ratio in snap-trapped samples was 49.7% male (n = 163) on Banks Peninsula, and 51.0% male (n = 104) at Waimangaroa (Best, 1973); 57% male (n = 304) in the northern Tararua Range (Innes, 1977); and 53.2% male (n = 543) in the Orongorongo Valley (B. M. Fitzgerald & B. J. Karl, unpubl.).

PRODUCTIVITY. In the Orongorongo Valley, some female rats first breed in the season of their birth, and again in the following year, but they are unlikely to survive for a third breeding season (B. M. Fitzgerald & B. J. Karl, unpubl.). Daniel (1972) calculated that the average production of the Orongorongo population was 10.9 young per female per year. Throughout the world, indoor populations of ship rats breed all year round, although there is a reproductive peak in the warm season in countries with severe winters. Their annual production of young per female is therefore much higher (average 38.5) than that in outdoor populations (average 14.8) (data from Southwick, 1966, and Daniel, 1972).

MORTALITY. Annual disappearance rates exceeded 90% for both males and females in the Orongorongo Valley; i.e., few rats survive more than a year. During a two-year cage trapping study in the Orongorongo Valley, maximum longevity recorded was 11 months for males and 17 months for females (Daniel, 1972), much less than the 3.9 years for males and 3.4 years for females recorded in the laboratory (Bentley & Taylor, 1965). Females live longer, on average, than males (Tamarin & Malecha, 1971; Daniel, 1972).

PREDATORS, PARASITES AND DISEASES

Predation by feral cats is believed to be an important source of mortality limiting the density of ship rat populations in parts of New Zealand (p. 348) (Fitzgerald & Karl, 1979; B. M. Fitzgerald & B. J. Karl, unpubl.). Rats are the staple diet of feral cats in New Zealand forests (Fitzgerald & Karl, 1979; Karl & Best, 1982). In the Orongorongo Valley, rats were found in 40-50% of cat scats in most seasons over a three-year study period (Fitzgerald & Karl, 1979). Rats are also eaten by stoats (King & Moody, 1982) and, rarely, by moreporks (St. Paul, 1977).

Eleven ectoparasite species recorded from New Zealand ship rats are listed by Tenquist and Charleston (1981). They are the mites *Eulaelaps stabularis, Haemaphysalis longicornis* and *Notoedris muris*; the louse *Polyplax spinulosa*; and the fleas *Ctenocephalides felis felis, Leptopsylla segnis, Nosopsyllus fasciatus, N. londiniensis, Pulex irritans, Pygiopsylla phiola* and *Xenopsylla cheopis*. The sarcoptid burrowing mite *Notoedres* produces conspicuous lesions on the nose, tail and scrotum of its host. At Tiritea, north Tararua Range, 19% of 170 males and 7% of 131 females showed these lesions (Innes, 1977). Fleas, mostly cosmopolitan species, are common, especially *Nosopsyllus fasciatus* and *Leptopsylla segnis*, whose primary hosts are rats and house mice, respectively. *Xenopsylla cheopis*, the main flea vector of bubonic plague, is found on ship rats in Auckland (Smit, 1979). Twenty-one cases of plague were reported between 1900 and 1911, mostly from the Auckland district; there were nine deaths (MacLean, 1955).

A variety of helminth parasites have been recorded, including the nematodes *Capillaria hepatica, Mastophorus muris, Nippostrongylus brasiliensis, Syphacia muris* and *Heterakis spumosa*, and the cestodes *Hymenolepis diminuta* and *H. nana* (Gibson, 1972). The spiruroid nematode *Mastophorus muris* was recorded from ship rats examined by Daniel (1973). In a study of 191 rats from native forest near Palmerston North (Charleston & Innes, 1980), the majority were found to harbour spiruroid nematodes in their stomachs; about 64% were infected with *Physaloptera getula* and about 20% with *M. muris* (most of which were also carrying *P. getula*). Most infections comprised fewer than 10 nematodes, but some rats carried larger numbers, occupying most of the stomach lumen. Both these nematodes have obligatorily indirect life cycles involving arthropods, perhaps wetas, as intermediate hosts.

The ship rat is one of the maintenance hosts for serovar *ballum* of the parasite *Leptospira interrogans*, which can cause leptospirosis in man (Brockie, 1977; Hathaway & Blackmore, 1981).

Sarcocystis sp., which is also known to infect *R. norvegicus*, has been found in ship rats collected from Stewart I. (Collins, 1981a).

ADAPTATION TO NEW ZEALAND CONDITIONS

The ship rat is "one of the most successful mammalian weed species" (Patton, Yang & Myers, 1975), since it is established around the entire globe; it has adapted easily to life in New Zealand.

No strong founder effect, or genetic drift, is likely in ship rat populations on the main islands, because of the multiple introductions. On the smaller offshore islands, a detailed study of genetic and phenotypic variability of ship rats, such as that conducted in the Galapagos Is. (Patton, Yang & Myers, 1975), would be interesting. The Galapagos research suggested that phenotypic, rather than genetic, flexibility accounts for the colonizing success of the species.

The widespread distribution of ship rats in New Zealand is typical of that on Pacific islands which never had an indigenous fauna of small mammals. By contrast, in Australia, America, Asia, Africa and Europe, ship rats are mainly commensal or confined to disturbed habitats (Ecke, 1958; Medway, 1969; Jeffrey, 1977; Musser, 1977; Arnold, 1978; Watts & Aslin, 1981). This is presumably because they are outcompeted by native species elsewhere, although in the Galapagos Is., endemic rat species (*Oryzomys* and *Nesoryzomys* spp.) disappeared from islands reached by ship rats (Brosset, 1963). In New Zealand, ship rats arrived later than Norway rats and replaced them as the common rat in the wild, whereas in Britain, ship rats arrived first, were widespread for centuries and were then replaced by the later-arriving Norway rat, now the common countryside rat. The reason for this difference is unknown.

New Zealand ship rats are much the same size as those in non-commensal populations elsewhere in the world (Johnson, 1962; Marshall, 1977; Yosida, 1980; Senzota, 1982). Some authors report very heavy maximum weights, e.g., 350 g (Maser *et al.*, 1981, from Oregon, USA) and 340 g (Watts & Aslin, 1981, from Australia); or average weights, e.g., "adults usually weigh 150-200 g" (Corbet & Southern, 1977, from Britain). These are probably from commensal individuals, which tend to be heavier.

The keys to the flexibility of the ship rat are breeding and diet. The 6-7 months' breeding season in wild populations studied in New Zealand (latitude 40°-43°S) is short by comparison with the 10 months' season in

New Caledonia (20°S; Nicholson & Warner, 1953), 12 months in Caroline I. (8°N; Jackson, 1962) and Venezuela (10°N, and Gomez, 1960), and 10 months in Puerto Rico (18°N; Weinbren *et al.*, 1970) and Hawaii (20°N; Tamarin & Malecha, 1971). Litter size in New Zealand (4.9-6.1 young per litter) is greater than in the tropics (3.8-5.3) (Jackson, 1962; D. B. Clark, 1980; Tomich, 1981), as predicted by Jackson (1965), and observed also in *R. exulans* (p. 189). Population density is generally low (1-3 per hectare) in New Zealand compared with the 12-15 per hectare found in Cyprus macchie scrub (Watson, 1951), up to 64 per hectare in Hawaiian kiawe forest (Tamarin & Malecha, 1971) and 6-19 per hectare in a range of forest and scrub habitats in the Galapagos (D. B. Clark, 1980). Home ranges in coastal forest on Stewart I. are much larger than in tropical forest or shrubland habitats (Spencer & Davis, 1950; Jackson & Strecker, 1962a; Weinbren *et al.*, 1970; Tomich, 1970; Tamarin & Malecha, 1971; Tomich, 1981).

Plant matter consistently comprises about 80% (by overall volume) of the diet of ship rats around the world, regardless of latitude (Kami, 1966; Norman, 1970; D. A. Clark, 1981; Yabe, 1982; Advani, 1984; Copson, 1986). Some insects are always taken when available, but only in New Zealand (and perhaps Big Green I., Tasmania; Norman, 1970) is there a seasonal predominance of arthropods in the diet. This may be because fruit is relatively scarce in New Zealand forests in spring and summer (Leathwick, 1984), or perhaps it reflects the unique availability to New Zealand ship rats of abundant wetas. These large flightless orthopterans were found in 39-76% of ship rat stomachs in New Zealand studies (Best, 1969; Daniel, 1973; Innes, 1979; Gales, 1982; Moller, 1985a).

SIGNIFICANCE TO THE NEW ZEALAND ENVIRONMENT

Unlike the kiore, the ship rat has never been used for food in New Zealand. There is, therefore, nothing good to be said for its presence here!

Improved sanitation conditions around buildings have considerably lessened the threat posed by commensal rodents to public health. In the wild, however, ship rats have been in the past, and in places still are, serious threats to wildlife conservation.

PAST DAMAGE. It is extremely difficult retrospectively to isolate the ecological impact of the 19th century ship rat populations that colonized the mainland, since their spread was contemporary with massive habitat destruction and, in the South Island, also with the spread of mustelids. Also, they were the fourth (and last) rodent species to establish. None the less, their spread in the North Island was "more or less coincidental with declines of the bellbird (*Anthornis melanura*), robin (*Petroica australis*), stitchbird (*Notiomystis cincta*), saddleback (*Philesturnus carunculatus*) and thrush (*Turnagra capensis*). In the South Island, in addition to these species (excepting stitchbirds, which were never recorded in the South Island), declines of the yellowhead (*Mohoua ochrocephala*), South Island kokako (*Callaeas cinerea cinerea*) and red- and yellow-crowned parakeets (*Cyanoramphus novaezealandiae* and *C. auriceps*) occurred during the period when ship rats were spreading" (Atkinson, 1973a).

However, no other rodents or predators were present when ship rats reached Big South Cape I., off Stewart I., in 1962 or 1963. The rat population irrupted and remained high for three years, by which time nine landbird species had declined or disappeared; five, including the Stewart I. forms of the robin, fernbird (*Bowdleria punctata stewartiana*), and snipe (*Coenocorypha aucklandica iredalei*), as well as the wren (*Xenicus longipes variabilis*) and South Island saddleback, became extinct on the island. The only known colony of the greater short-tailed bat, *Mystacina robusta*, was much reduced and later disappeared (p. 135). Several species of large invertebrates were reduced and at least one, a flightless weevil, became extinct. However, some native birds (South Island tomtit, tui, morepork) and introduced birds (blackbird, chaffinch, hedge sparrow) increased after the irruption (Bell, 1978; I. A. E. Atkinson, unpubl.).

Other extinctions probably induced by ship rats have been recorded on Hawaii (Atkinson, 1977), Lord Howe I. (Hindwood, 1940) and Midway I. (Fisher & Baldwin, 1946). Like Big South Cape, these islands had no native rodents before ship rats arrived; by contrast, Christmas I. and the Galapagos Is., which had endemic rat species before ship rats arrived, have lost no native birds so far. This suggests that the birds there were already adapted to coping with predation.

Other evidence is provided by correlations between extinctions or disjunct distributions of birds, invertebrates or reptiles and the arrival or presence of rodents. For example, many large, flightless, ground-dwelling invertebrates such as large weevils and giant wetas have either been eliminated or are confined to rodent-free islands (Ramsay, 1978).

PRESENT DAMAGE. Native plant seeds and animals, especially invertebrates, make up the bulk of the ship rat diet (p. 214), but the impact of ship rat predation on the complex mainland ecosystems of New Zealand remains unclear.

Campbell (1978) reviewed the effects of rats on New Zealand vegetation and concluded that seed damage by ship rats probably replaced that of kiore, which in turn may have replaced that by wetas. No impact of ship rats on regeneration has yet been demonstrated, although it is well known that they destroy many seeds.

Similarly, an observed predation, or the finding of unmistakable remains in a ship rat stomach, is not sufficient evidence that predation is harming the prey population. Moors (1983b) found that 70% of observed nests of eight species of common native birds at Kowhai Bush, Kaikoura, were destroyed by mustelids and rodents, especially ship rats. However, this does not prove that those bird populations are in any danger from predation, or that predator control would assist them. On the other hand, the North Island kokako (*Callaeas cinerea wilsoni*), which suffers heavy predation at least partly from ship rats (Hay, 1981), is steadily declining in numbers. While competition from browsing mammals and habitat destruction both adversely affect the kokako (Leathwick, Hay & Fitzgerald, 1983), ship rats may be a significant additional threat.

Ship rats as prey may indirectly affect important endemic species. For example, rats are the staple diet of cats on Stewart I., and by implication may control the numbers of cats. Cats, in turn, prey on the endangered kakapo (*Strigops habroptilus*) (Karl & Best, 1982).

CONTROL. Clearly, for nature conservation reasons it is important to stop ship rats from establishing on rodent-free islands, although a biologically significant island which already has one rat species still requires protection from colonization by others (Atkinson, 1986).

Strict precautions should also be taken to prevent the establishment of Asian-type ship rats, which are genetically and perhaps ecologically different from the resident Oceanian type (p. 213), and which could present new problems for nature conservation in New Zealand.

Large-scale eradication of ship rats has never been attained anywhere. Small-scale poisoning operations unaccompanied by habitat management have been undertaken; temporary population reduction in small areas to protect particular values such as a snail population (Meads, Walker & Elliott, 1984), a fishing boat mooring point (Hickson, Moller & Garrick, 1986) or bird nests (J. Innes, unpubl.) is achievable, although reinvasion is rapid.

J. G. I.

Genus *Mus*

Mus is a large genus of about 40 species of mice living mainly in Asia and Africa; one species, the house mouse (*Mus musculus*), is among the most widespread of all mammals. Like its relatives the ship and Norway rats, the house mouse has developed a close commensal relationship with man, and thereby gained a ticket to the world.

18. House mouse

Mus musculus Linnaeus, 1758

Also called field mouse, wood mouse (English, popular usage only); kiore-iti (Maori).

The house mouse in New Zealand is thought to be descended from European stock, previously assumed to be referable to one species (Schwarz & Schwarz, 1943; Nowak & Paradiso, 1983). Recently, the taxonomy of the house mouse in Europe has been re-examined, and at least two separate species are now recognized: *Mus domesticus* Rutty, 1772 in Britain and western Europe, and *Mus musculus* L. in Scandinavia and eastern Europe (Marshall & Sage, 1981; Corbet, 1984; Sage, Whitney & Wilson, 1986; Sage *et al.,* 1986). Mice in New Zealand share morphometric characteristics of both species (M. G. Efford, B. J. Karl & H. Moller, 1988), but as yet there has been no genetic analysis of New Zealand populations.

DESCRIPTION (Fig. 44)

Distinguishing marks, Table 34 and skull key, p. 24.

The house mouse has a long tail (about equal to its head and body length), large prominent black eyes, round ears and a pointed muzzle with long whiskers. Wild-living mice are a dull brownish grey colour; the belly is usually grey but can be brown or white (E. C. Murphy, unpubl.). Mice can be distinguished from juvenile rats by their larger and more rounded ears and their smaller feet, which are grey on the upper side. Laboratory and domesticated mice have been bred into a variety of colours including and combining white, brown, grey and black.

Smell and hearing play a large part in predator avoidance, intraspecific recognition and food location and selection (J. C. Smith, 1981). Mice can distinguish between at least 18 other individual mice on the basis of smell alone (Kalkowski, 1967); they have acute hearing (Rowe, 1981) and communicate at high sound frequencies (Sales, 1972). They can see and identify objects at least 15 m distant (Hopkins, 1953) and use visual landmarks in assessing the position of territorial boundaries (Mackintosh, 1973) and for homing (Newsome, Cowan & Ives, 1982). Mice are basically ground dwelling but can swim (Evans *et al.,* 1978) and also climb well (Dewsbury, Lanier & Miglietta, 1980). Mice were frequently caught in hinau (*Elaeocarpus dentatus*) trees 6 m above the ground in the Orongorongo Valley (M. J. Daniel, unpubl.). They intensively examine new objects and surroundings after an initial reconnaissance, and constantly re-examine old ones (Crowcroft, 1959).

The chromosome number is usually 2n = 40, all acrocentric. However, populations may have from one (2n = 38) to nine (2n = 22) pairs of metacentrics (Sage, 1981).

Dental formula $I^1/_1C^0/_0Pm^0/_0M^3/_3 = 16$. M^3 is rudimentary.

FIELD SIGN

The presence of mice in buildings is usually revealed by large numbers of small dark brown or black pellets (5–7 mm long) and a strong "soiled" smell, produced by an acetate in the urine. Mice are messy eaters; they leave tooth marks in soft foods and grains, and footprints on dusty surfaces. Outside, they make networks of tunnels in long grass and dig burrows, often with several branches and chambers and as many as four exits (Berry, 1968). Nests can be found under wood piles, tunneled into banks, or between walls in buildings.

MEASUREMENTS

See Table 40.

VARIATION

Mice continue to grow throughout life, but the rate of growth slows during winter and after the first six months. Therefore, mean body size (weight) in any local population of wild mice varies considerably throughout the year, owing to seasonal changes in age structure (Badan, 1979; Pickard, 1984; M. G. Efford, B. J. Karl & H. Moller, 1988). There are no significant differences in mean body weight or length between male and female mice of the same age from the same population (Badan, 1979; Pickard, 1984).

Fig. 44 House mouse, white-bellied form, Marlborough Sounds (J.E.C. Flux).

Two measuring techniques have been generally used in New Zealand: the "hanging tail" and the "British Museum new method" (Jewell & Fullager, 1966; Cunningham & Moors, 1983). Some of the variation seen in Table 40 may be due to different measuring methods, as the mouse body is flexible and the base of the tail is not an easily detected skeletal point (Jewell & Fullager, 1966).

HISTORY OF COLONIZATION

Mice, like Norway rats and ship rats, were carried to New Zealand as stowaways on Australian and European ships. Mice were first recorded in New Zealand on Ruapuke I. in Foveaux Strait, after a shipwreck in 1824 (McNab, 1907). Dieffenbach (1843) was the first to record the presence of the "common domestic mouse of Europe" in the North Island, but they were known to be at the Bay of Islands Mission Station by about 1830 (Guthrie-Smith, 1969). At this time, merchant ships were reaching New Zealand with cargoes of machinery, paper, clothes, seeds and cereal. These cargoes and their packing materials afforded mice shelter and safety, especially from rats, during the long voyage. Once unloaded at ports around the country, mice were carried inland amongst stores for sheep stations and settlements.

Mice were not found in the South Island until the mid 1850s. Gillies (1877) states that there were no mice in Otago in 1852, but that a "year

Table 40: Body measurements of adult[1] mice in New Zealand (mean ± SD).

Location	Weight (g)[2]	n	Total length (mm)	n	Head – body length (mm)	n	Reference
Woodhill							
(pine forest)	19.0 ± 1.0	114	—		86.9 ± 2.0	115	Badan (1979)
Pureora							
(logged podocarp forest)	16.7 ± 2.3	6	170.0 ± 9.0	7	84.4 ± 7.5	7	C. M. King, J. G. Innes & M. Flux (unpubl.)
(young pine forest)	17.3 ± 2.2	14	169.5 ± 9.3	13	86.1 ± 5.9	15	
Orongorongo Valley							
(mixed & beech forest)	18.3 ± 2.1	242	174.1 ± 8.2	226	—		B. M. Fitzgerald & B. J. Karl (unpubl.)
Marlborough Sounds							
(mixed & beech forest)	20.1 ± 1.9	26	174.2 ± 7.1	26	88.5 ± 3.8	26	E. C. Murphy (unpubl.)
Craigieburn							
(beech forest)	20.0 ± 2.2	14	179.2 ± 3.9	14	94.6 ± 3.5	14	C. M. King[3] (unpubl.)
Hollyford Valley							
(mixed & beech forest)	19.1 ± 2.5	66	175.5 ± 9.5	68	90.1 ± 6.0	68	C. M. King[3] (unpubl.)
Eglinton Valley							
(beech forest)	21.1 ± 2.7	77	178.8 ± 10.8	77	93.0 ± 6.3	76	C. M. King[3] (unpubl.)
Mana I.							
(coastal scrub)	21.1 ± 1.7	102	162.2 ± 3.5	101	79.3 ± 1.8	101	Pickard (1984)
(coastal scrub & pasture)	22.2 ± 3.3	265	178.2 ± 9.8	221	92.8 ± 6.2	221	M. G. Efford, B. J. Karl & H. Moller (unpubl.)
Allports I.							
(mixed forest)	21.6 ± 3.3	58	172.2 ± 11.3	58	86.9 ± 7.6	58	E. C. Murphy (unpubl.)
Mabel I.							
(coastal scrub)	25.7 ± 2.0	11	190.3 ± 3.5	11	99.5 ± 1.7	11	E. C. Murphy (unpubl.)
Antipodes I.							
(tussock)	22.1 ± 3.5	59	172.9 ± 9.7	36	84.8 ± 5.0	47	P. J. Moors (unpubl.)
Enderby I.							
(tussock & rata)	24.9 ± 1.7	16	180.7 ± 6.9	16	93.8 ± 3.0	16	D. M. Cunningham & P. J. Moors (unpubl.)

1. Only mice with tooth wear class 6 and above (see Fig. 46) are included. 2. Pregnant females excluded or weight of embryos subtracted, except for Orongorongo Valley. 3. Means broken down by age and density given by King (1982a).

or two" later they were seen in Dunedin. They then rapidly spread to the Taieri Plains, and across the Taieri and Clutha Rivers. White (1890) recorded three cases of colonization through the mass migration of large numbers of mice, two in Canterbury between 1856 and 1864 and one on the shore of Lake Wakatipu, Central Otago, around 1864. By the turn of the century mice occupied most suitable habitats throughout the North and South Islands.

DISTRIBUTION

WORLD. The present world distribution of the house mouse is probably more extensive than that of any other mammal apart from man. From its original beginnings on the steppe zone of southern Eurasia, the house mouse has accompanied man to tropical, temperate, semi-desert and subantarctic regions throughout the world (Schwarz & Schwarz, 1943; Sage, 1981). Most of the possible places it could colonize are now occupied, but new opportunities still arise occasionally; for example, mice were first recorded on Guadalcanal, Solomon Is., in 1965 (Rowe, 1967). In Europe, Africa and the Americas, further spread, especially in remote areas, is often limited by competition from indigenous rodents of the genera *Apodemus* (Berry & Tricker, 1969), *Peromyscus* (Briese & Smith, 1973) and *Microtus* (Lidicker, 1966). By contrast, in Australia the mouse has displaced some indigenous species, for example *Pseudomys hermannsburgensis,* and *P. desertor* (Newsome & Corbett, 1975).

NEW ZEALAND (Map 17). Mice are found throughout the North and South Islands and on many offshore islands (Table 36). This distribution has not changed significantly over the past 35 years (Taylor, 1978a; Wodzicki & Wright, 1984). Mice are found from the coast to high altitude, e.g., at 1200 m in the Marlborough Sounds (Wodzicki, 1950) and to over 1300 m in Craigieburn Forest Park (King, 1982a). They have colonized offshore islands mainly through accidental transport by man (Taylor, 1984).

On some islands mice have failed to become established or have remained a commensal only. For example, mice were observed around the buildings on Campbell I., but died out after the island was abandoned in 1931 (Taylor, 1978a); on Mokoia I. (Lake Rotorua) mice are confined to the vicinity of huts (Beveridge & Daniel, 1966). Taylor has suggested that mice are prevented from dispersing on these islands, and on Stewart, Kapiti, and Raoul Is., by the presence of high numbers of Norway rats. This suggestion can be tested by examining present day combinations of rodents on islands; islands without Norway rats are significantly more likely to have mice (Taylor, 1984).

HABITAT

The house mouse is usually thought of as typical of urban habitats, but in New Zealand it is also found in native and exotic temperate forests, pasture, croplands, and subalpine tussock (Taylor, 1978a). It also inhabits more traditional places such as houses, stores and factories (especially those dealing with food products), rubbish tips and farm buildings. In Pureora Forest Park, mice are more common in disturbed habitats, such as road edges and exotic forests, than in virgin forests (C. M. King, J. G. Innes & M. Flux, unpubl.).

FOOD

In New Zealand, wild mice eat both invertebrates and plant material. In Hunua, a kauri (*Agathis australis*) forest, adult arthropods and seeds were eaten throughout the year; in Woodhill, a pine (*Pinus radiata*) forest, the main foods were adult arthropods and lepidopteran larvae, plus seeds from spring to autumn (Badan, 1979, 1986). Other items eaten in both areas were leaves, fungal spores, annelids and arthropod eggs. On Mana I. (mainly grassland), insect larvae and seeds were the most important food items. Seeds were eaten mainly in spring and summer, and lepidopteran larvae in winter (Pickard, 1984). Lizard and bird remains were found in mouse stomachs, but they comprised only a very small part of the diet (some 6% by volume).

Studies of feral mice outside New Zealand also show that invertebrates and seeds are important foods. In cold and/or wet habitats, more invertebrates than plant material are eaten (Berry & Tricker, 1969, in the Shetland Is.; Watts & Braithwaite, 1978, in wet heathland, Australia; Copson, 1986, on Macquarie I.), whereas in desert (Watts, 1970) and cultivated fields (Whitaker, 1966; Houtcooper, 1978) seeds are more important. Commensal mice have access to a wide range of different foods throughout the year, but prefer cereals over foods containing higher proportions of fats and proteins (Southern, 1954; Rowe, Bradfield & Redfern, 1974).

Mice have two main feeding periods, at dusk and dawn, but continue to feed less intensively throughout the night and day (Crowcroft, 1966). Foraging mice continually sniff the substrate and occasionally rear up to sample airborne information (Mackintosh, 1981). They are agile climbers, but usually forage on the ground.

Although a mouse consumes only 3–4 g of food daily (Crowcroft, 1966) it wastes much more, especially when feeding on grain, as it fragments the kernels and eats only a portion of each (Southern, 1954).

A large part of the water requirement of mice is met by the moisture content of their food (Fertig & Edmonds, 1969). This, combined with their ability to concentrate their urine, allows mice to adjust to long dry periods (Koford, 1968), although a prolonged diet of dry food reduces fertility (Rowe, 1981).

SOCIAL ORGANIZATION AND BEHAVIOUR

Mice are mostly nocturnal, although daytime feeding is not uncommon, particularly when food is scarce or density is high (Sidorowicz, 1975; Rowe, 1981). In summer, mice have been seen foraging during the day in the Orongorongo Valley and on Mana, Allports and Mabel Is. (Pickard, 1984; M. J. Daniel, unpubl.; E. C. Murphy, unpubl.).

Mice can be either territorial or colonial, regardless of environment (laboratory, commensal or wild) (Bronson, 1979; Nowak & Paradiso, 1983).

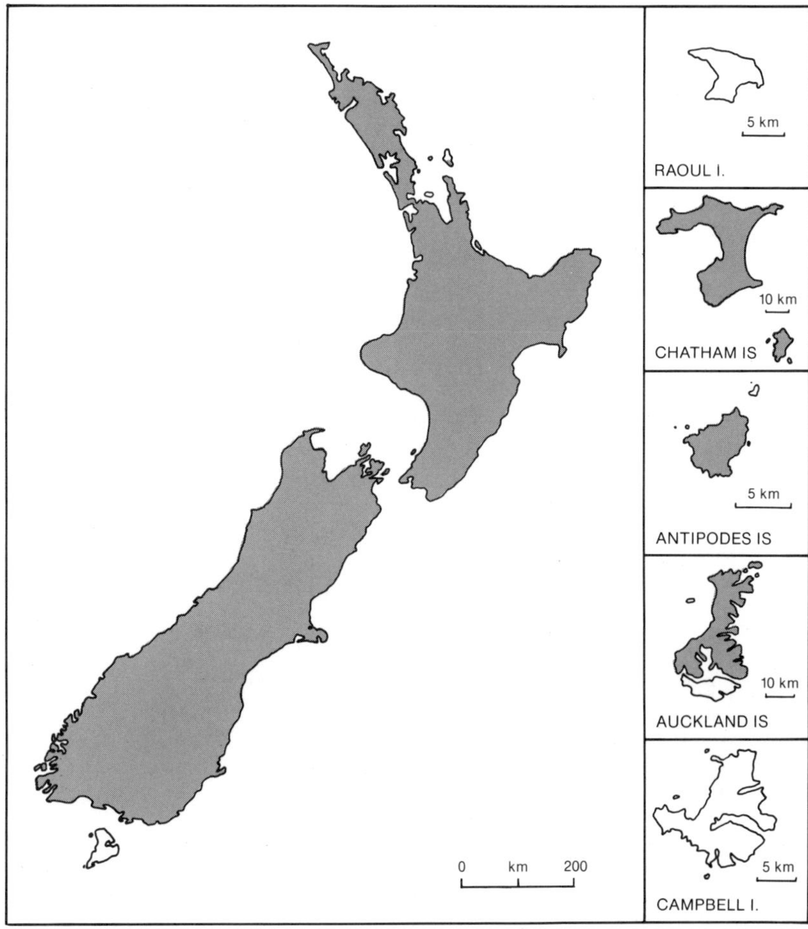

Map 17 Distribution of feral house mice on the main islands of New Zealand. For distribution on offshore islands, see Table 36.

There are three main types of social organization. (1) Dominance hierarchies: one male is dominant over several males and females; the dominant male sires most of the young and patrols and defends the territory (Crowcroft & Rowe, 1963; Singleton & Hay, 1983). Dominance hierarchies have been found only in artificially enclosed populations. (2) Group or shared territories: several males, not necessarily of the same family group, share a territory on an equal basis (Anderson, 1970; Busser, Zweep & van Oortmerssen, 1974). This type of organization is usually found in stable commensal populations. (3) Individual territories: resident mice exclude other mice of the same sex from their home ranges (Berry & Jacobson, 1974; Myers, 1974).

Studies in New Zealand have shown both individual and group territories. In the Orongorongo Valley both males and females maintained individual territories (Fitzgerald, Karl & Moller, 1981). Female home ranges sometimes

Table 41: Range and average litter size in wild populations of mice in New Zealand.

Location	Range	Mean[1]	SD	n[2]	Reference
Woodhill (pine forest)	4-8	5.8	1.1	40	Badan (1979)
Hunua (kauri forest)	2-7	5.4	0.5	9	Badan (1979)
Pureora (podocarp forest)	2-8	5.5	1.9	11	C. M. King, J. G. Innes & M. Flux (unpubl.)
(pine forest)	4-8	5.5	0.9	15	
Orongorongo Valley (mixed & beech forest)	2-8	5.1	0.4	49	B. M. Fitzgerald & B. J. Karl (unpubl.)
Marlborough Sounds (mixed & beech forest)	4-9	6.1	1.5	24	E. C. Murphy (1986 and unpubl.)
Mt. Misery (beech, tussock)	2-7	5.0	1.1	27	R. H. Taylor (unpubl.)
Canterbury (commensal)	—	5.9	—	17	Gibson (1973b)
Craigieburn (beech forest)	4-8	5.8	1.2	12	C. M. King[3] (unpubl.)
Hollyford Valley (mixed & beech forest)	4-8	6.1	1.2	26	C. M. King[3] (unpubl.)
Eglinton Valley (beech forest)	3-8	5.7	1.2	27	C. M. King[3] (unpubl.)
Mana I. (coastal scrub)	5-9	6.4	1.2	24	Pickard (1984)
(coastal scrub & pasture)	2-12	6.9	2.1	49	M. G. Efford, B. J. Karl & H. Moller (unpubl.)
Allports I. (mixed forest)	4-8	5.9	1.1	12	E. C. Murphy (1986 and unpubl.)
Antipodes I. (tussock)	4-7	5.8	0.8	18	P. J. Moors (unpubl.)

1. Obvious resorbing embryos are excluded.
2. n = number of pregnant females.
3. Means broken down by season and year given by King (1982a).

overlapped those of males living in the same area. Individual mice had minimum home ranges averaging 0.6 ha, and they visited many parts of them each night. One animal covered 2.62 ha in three nights.

In high density populations on Mana and Allports Is., mice were not exclusively territorial, as the ranges of same-sex neighbours overlapped. There was no evidence of dominance hierarchies or group territories, as the centres of activities showed no clustering (Pickard, 1984; E. C. Murphy, unpubl.). However, in a low density mainland population in the Marlborough Sounds, adult males and females maintained a group territory for 2.5 years (E. C. Murphy, unpubl.).

On an irrigated cereal farm in Australia, there was evidence of exclusive territories one year but group territories the next (Redhead, 1982). The group territory was associated with higher fertility rates and a more stable population. Although dominance hierarchies and group territories can restrict gene flow (Lidicker, 1976; Singleton & Hay, 1983), population turnover in the wild is so high that this effect is unimportant and genetic diversity is maintained (Berry & Jacobson, 1974; Singleton, 1983).

REPRODUCTION AND DEVELOPMENT

As a laboratory animal, the house mouse has become a model for many reproductive studies; hence, there is a very large literature on reproduction of caged and commensal mice, but much less concerning wild mice.

MATURITY. In grain stores and suburban and pastoral areas near Christchurch, both male and female mice matured at about 8 weeks old and 10.6–12.5 g body weight (Gibson, 1973b). In both Woodhill (pine) and Hunua (kauri) forests, and on Mana I., male and female mice matured at about that age during the summer, but mice born at the end of summer did not mature until the following spring (Badan, 1979; Pickard, 1984).

BREEDING RATE AND SEASON. Breeding normally ceases in winter (Fitzgerald, 1978; Badan, 1979; Pickard, 1984; E. C. Murphy, unpubl.), except in commensal mice (Gibson, 1973b) and in beech forests after moderate to heavy beech seeding (Fitzgerald, 1978; King, 1982a: but see below). As some adult males remain fertile throughout the year (Badan, 1979; Pickard, 1984), the breeding season is determined by some factor acting on the females. Supplementary food (wheat) was added to island and mainland habitats in the Marlborough Sounds to determine whether food supply could be such a factor (E. C. Murphy, unpubl.). The mainland population remained so low that no change could be detected. On the island a few adult females did breed during the winter, but sexual maturation of young females was still delayed. This experiment tested food quantity, not quality; the possibility remains that reproduction in winter is limited by low quality food (Bomford & Redhead, 1987).

It is not clear whether density has any predictable effect on pregnancy rate. Fitzgerald (1978) reported that, at low population densities in several localities throughout New Zealand (<5 C/100TN), 35–40% of the females were pregnant, whereas at higher densities (>7.5 C/100TN), less than 10% of females were pregnant. On the other hand, in South Island beech forests the pregnancy rate was 35–50%, independent of mouse density (King, 1982a). The difference may be due to pooling of data from the beginning and end of the breeding season, since pregnancy rates are not constant throughout. In beech forests, breeding is more intense earlier in the summer (November) and declines earlier in autumn, in proportion to density (King, 1982a). By contrast, at Woodhill (pine forest) the overall pregnancy rate was 33% in the season August 1976 to April 1977, but 18 out of 28 females were pregnant in March (Badan, 1979).

FERTILITY. In female mice the oestrous cycle lasts 4–6 days, and pregnancy 19–21 days. Ovulation is spontaneous, and copulation leaves a vaginal plug which lasts for 18–24 hours. The oestrus cycle stops during lactation, except for one oestrus 20 hours postpartum (Snell, 1941). At Woodhill, 45% of

Table 42: Density indices (C/100TN) of mice recorded in New Zealand.

Location	C/100TN	Trapping sessions	Month	Year	Reference
Woodhill					
(pine forest)	0–47.5	18	*c.* Monthly	1976–77	Badan (1979)
Hunua					
(kauri forest)	1–21.7	8	*c.* Monthly	1977	Badan (1979)
Pureora					
(podocarp forest)	0.6	1	—	1962	Beveridge (1964)
(virgin podocarps)	0–3.7	19	F, M, A, N[1]	1982–87	C. M. King, J. G. Innes & M. Flux (unpubl.)
(logged podocarps)	0.5–3.8	18	F, M, A, N	1983–87	
(young pine forest)	0.5–41.2	19	F, M, A, N	1982–87	
Farewell Spit					
(scrub/sand)	26–57	1	Mar	1976	Taylor (1978b)
Orongorongo Valley					
(mixed & beech forest)	0–12	28	F, M, A, N	1971–78	Fitzgerald (1978); Fitzgerald & Karl (1979)
Marlborough Sounds					
(mixed & beech forest)	0–40	19	*c.* Monthly	1984–87	E. C. Murphy (unpubl.)
Mt. Misery					
(beech forest & tussock)	1–16	10	F, M, A, N	1974–76	Fitzgerald (1978)
Craigieburn					
(beech forest)	0–23.4	11	F, M, A, N	1974–77	King (1983a)
Jacksons Bay					
(podocarp/kamahi forest)	0	1	May	1965	Choate (1965)
Hollyford Valley					
(mixed & beech forest)	0–25.4	17	F, M, A, N	1975–80	King (1983a)
Eglinton Valley					
(beech forest)	0–24.0	22	F, M, A, N	1973–80	King (1983a)

Grebe & Borland valleys (beech forest)	15.8-73.3	2	Aug/Nov	1979	King (1983a)
Lake Monk (beech forest)	0.4	1	Mar/Apr	1957	Riney *et al.* (1959)
Caswell & George Sounds (beech forest)	0.3	1	Mar	1949	Wodzicki & Bull (1951)
Monowai-Grebe Valley (beech forest)	1.5	1	May	1967	Choate (1967)
Mokoia I. (mixed forest)	0.7	—	—	—	Beveridge & Daniel (1966)
Adele I. (mixed forest, coastal scrub)	1-7	9	F, M, A, N	1981-83	Taylor & Tilly (1984)
Arapawa I. (mixed forest, coastal scrub)	10.4	1	Apr	1978	Moller (1978)
Mana I. (coastal scrub)	0-130.2	12	Monthly	1981-82	Pickard (1984)
Allports I. (mixed forest)	0-85	12	*c.* Monthly	1984-86	E. C. Murphy (unpubl.)
Mabel I. (coastal scrub)	100	1	Mar	1986	E. C. Murphy (unpubl.)
Antipodes I. (tussock)	40-77	1	Nov	1978	P. J. Moors (unpubl.)
Auckland I. (rata forest)	0-22	1	Feb	1973	Taylor (1975b)
Enderby I. (rata forest, scrub)	5.0	1	Feb	1973	Taylor (1975b)
(tussock & rata)	24.5-35.8	1	Apr	1980	D. M. Cunningham & P. J. Moors (unpubl.)
Masked I. (rata forest)	11	1	Feb	1973	Taylor (1975b)

1. February, May, August, November.

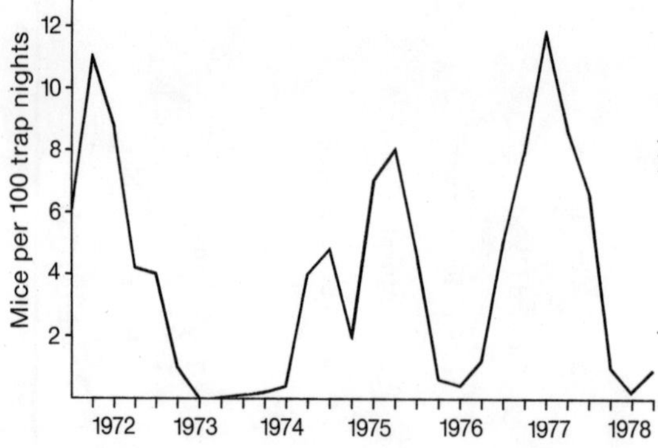

Fig. 45 Density index for mice from August 1971 to May 1978 in Orongorongo Valley (from Fitzgerald, 1978; Fitzgerald, Karl & Moller, 1981). See also Fig. 58.

mice pregnant in the August 1976 to April 1977 season were lactating, indicating postpartum fertilization (Badan, 1979). When this happens there is a brief period of delayed implantation extending the pregnancy by several days (Snell, 1941). Litters can be produced every 20–30 days.

Available data on litter size in New Zealand populations are summarized in Table 41. There is no relationship between mouse age and litter size (Gibson, 1973b; Badan, 1979; King, 1982a). Newborn mice weigh about 1 g, are naked except for short vibrissae, and their eyes and ears are closed. By 14 days they are fully furred, with their eyes and ears open and incisor teeth erupted. The young start to leave the nest after weaning at 20–23 days old, weighing about 6 g (Snell, 1941; Pelikan, 1981).

POPULATION DYNAMICS

DENSITY. The numbers of mice fluctuate seasonally, peaking in autumn and declining through winter, in Woodhill and Hunua forests (Badan, 1979), on Mana I. (Pickard, 1984) and on Allports I. (E. C. Murphy, unpubl.). Density indices (see Fig. 42 for method of calculation) also vary considerably both within and between different areas (Table 42). In beech forests there are irregular population irruptions, following years of abundant beech seed (Fitzgerald, 1978; King, 1983a). The highest recorded density index in a year of good beech seeding is 77 C/100TN in the Grebe Valley in November 1979 (King, 1982a), although in other beech forests the highest recorded figure was much lower (e.g., 24 and 25 C/100TN in the Eglinton and Hollyford Valleys in February 1977 (see Fig. 58). Density can also be high in pine forests; e.g., to 47.5 C/100TN at Woodhill (Badan, 1979) and 41.2 C/100TN at Pureora (C. M. King *et al.*, unpubl.). The snap-trap record for mice in the Orongorongo Valley, from 1971 to 1978 (Fig. 45) never exceeded 12 C/100TN. A live trapping study in the same area gave densities of 3.3 to 0.55 mice per hectare between July 1977 and February 1978 (Fitzgerald, Karl & Moller, 1981).

SEX RATIO. Gibson (1973b) caught significantly more females (83) than males (59) in commensal populations of mice, and this predominance of females was even greater in heavier and older mice. By contrast, trapping of wild populations usually records more males than females (Badan, 1979; King, 1982a; Pickard, 1984; M. G. Efford, B. J. Karl & H. Moller, 1988). On Mana I., males outnumbered females by 1.22 to 1 (n = 704), but this difference was thought to be due to a bias in trappability, rather than to an uneven sex ratio (M. G. Efford, B. J. Karl & H. Moller, 1988). The proportion of males did not change seasonally in a pine forest (Badan, 1979) or with population density in beech forests (King, 1982a).

AGE. Methods of age determination used for mice include body weight (Delong, 1967), head and body length (Newsome, 1969), wear of molar teeth (Lidicker, 1966) and dry weight of the eye lens (Rowe *et al.,* 1985). Only the first two are possible in live mice. Most studies in New Zealand have used toothwear (Fig. 46) to classify mice into age groups as defined by Lidicker (1966).

The age structure of a mouse population changes with time since the last breeding season. After breeding stops in autumn, the whole population ages over winter. At the start of the new breeding season the minimum age is 3–4 months, averaging 6 months. Over the spring and summer months young animals enter the population and the mean age drops to 2–4 months old (Pickard, 1984; Fig. 47). By contrast, in years of good seeding in beech forests, juvenile survival is greatly increased over the three winter months after the seedfall. There is little recruitment after spring, although breeding continues (Fitzgerald 1978; King, 1982a). The population comprises mostly mice born in winter, which become progressively older over the summer and are not replaced (Fig. 47). The ultimate decrease in density is due to failure of recruitment and the disappearance of ageing, non-breeding mice. In the Orongorongo Valley, on one occasion, mice averaged almost 11 months old before the population reached its lowest density (Fitzgerald, 1978). It is not known whether the lack of recruitment is due to emigration or mortality of young mice.

MORTALITY. Mice can live up to three years in the laboratory (Snell, 1941), but rarely survive 18 months in the wild (Pickard, 1984; E. C. Murphy, unpubl.). Predation, diseases, food shortage and bad weather all contribute to mortality. Between March and September on Mana I. there was an 80% decrease in the population (Pickard, 1984). Embryonic mortality can also be high. On Mana I., 18% of pregnant females were resorbing at least one embryo, and 6% of all embryos were being resorbed. Older females had higher resorption rates than young females (M. G. Efford, B. J. Karl & H. Moller, 1988).

COMPETITION. The disappearance of kiore (p. 178) has been linked to the appearance and increase of mice, and it was once a common saying that the mice were driving the kiore out (Gillies, 1877). White (1890) reported that if mice were in the top of a wheat stack, there would be few or no kiore in the foundation. However, Strange (1850) and Tancred (1856) stated that the Norway rat, which by then had overrun all parts of New Zealand, was "destroying" the kiore. All of the introduced European rodents must have interacted with kiore, but mice were the nearest in size, and so competition from them might have been most detrimental (Taylor, 1975a, 1978a).

PREDATORS, PARASITES AND DISEASES

Mice are eaten by a wide range of native and introduced animals, but their main predators are cats and stoats. The number of mice eaten by cats in the Orongorongo Valley varied with the density of mice (Fitzgerald & Karl, 1979). In a year when mice were plentiful, 64% of cat scats examined contained mouse remains, but <20% of scats in the next year, when mouse numbers had dropped. In 1973, although only three mice were caught in 2784 trap-nights, 8-21% of cat scats still contained remains of mice. These figures do not measure the intensity of predation by cats on mice, but they imply that cats may be exerting a heavy pressure on low-density mouse populations, holding numbers down, or perhaps even further reducing them.

In a study of stoats collected throughout the North and South Islands, mice were found in 19% of 1250 guts from all habitats (King & Moody, 1982) and 22% of 866 guts from forests (see Table 49). Fewer mice were found in stoats collected from podocarp or mixed forests than from beech forests, and fewer again in the non-forested areas of Mt. Cook and Kaikoura. Although predation by stoats does not prevent the population irruptions of mice in beech forests after a heavy seedfall, it may affect the size and timing of the peak and decline (King, 1983a, 1985a), because by then the mice have already ceased to add recruits to the peak populations.

Mice were found in 37% of weasel stomachs collected throughout New Zealand (King & Moody, 1982), and in 16.3% of ferret scats from the Manawatu dune country (Roser & Lavers, 1976).

Fig. 46 Index of molar tooth wear devised by Lidicker (1966) and used in most New Zealand studies to classify mice into age groups. The method gives consistent results, but has never been calibrated against known-age material from different habitats in New Zealand. Drawn by A. Beauchamp from New Zealand material classified and lent by B.M. Fitzgerald and B.J. Karl. Lidicker's definitions of the age classes are as follows:

Age class 1 (0–1 month) – body weight <8.8 g; permanent teeth not all fully emerged. This age class does not ordinarily enter the trappable population.

Age class 2 (1–2 months) – body weight 8.8–12.0 g; initial lake development of cusps, but essentially no wear on 3rd anterior molar (M^3).

Age class 3 (2–4 months) – body weight >12.0 g; slight cusp wear, including M^3; oldest mice in this category show beginnings of distinct connections between three lobes of anteriormost cusp of 1st anterior molar (M^1).

Age class 4 (4–6 months) – three lobes of anterior cusp of M^1 distinctly connected, but lake development still narrow; three lobes of M^1 widely spaced.

Age class 5 (6–8 months) – three lakes of M^1 broader and approaching each other, particularly on labial side; oldest mice in this category with posterior two lakes of M^1 sometimes connected on labial side.

Age class 6 (8–10 months) – cusps broadly worn, and three lakes of M^1 usually interconnected labially; enamel edges projecting but rounded; some variability in timing of merging of lakes, but even in the youngest 6s, posterior two lakes connected on at least one side.

Age class 7 (10–14 months) – inter-lake enamel still present, but lingual connections present between lakes as well; outer rim of enamel showing indentations.

Age class 8 (>14 months) – each molar composed almost entirely of one large lake; outer rim approaching a smooth curve.

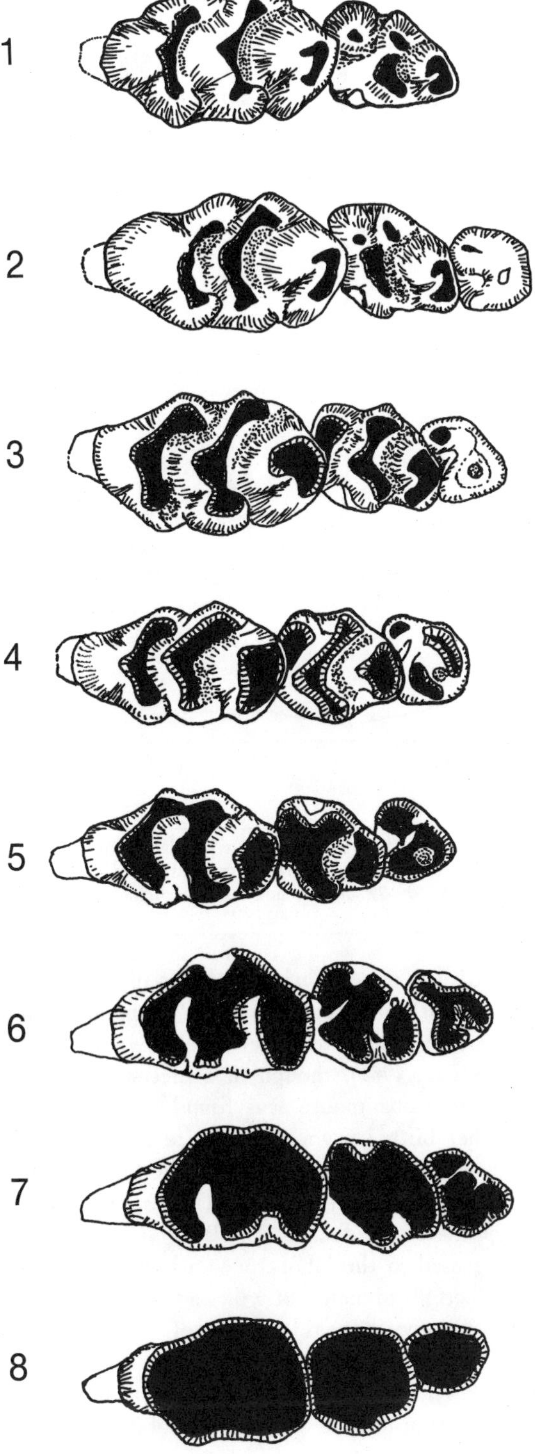
1
2
3
4
5
6
7
8

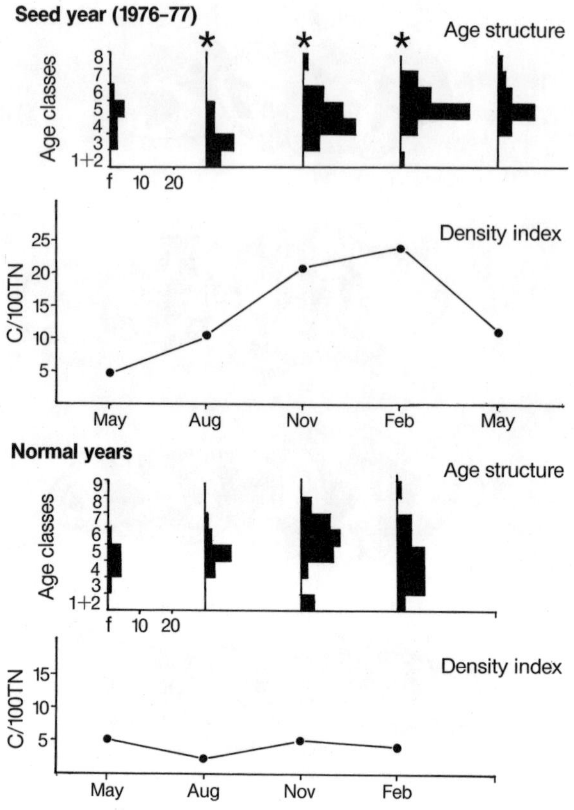

Fig. 47 Age structure and relative density of mouse populations in beech forest in the Eglinton Valley, in normal years (below) and in a seedfall year (1976–77, above). *Significant difference between 1976–77 and the non-seed years pooled, in separate Kolmogorov-Smirnov 2-sample tests (after King, 1982a).

Some native birds eat mice. During a plague of mice in western Otago, weka (*Gallirallus australis*) were reported to be "snapping them up" (Philpott, 1919). Moreporks (*Ninox novaeseelandiae*) are known to feed mice to their young (Hogg & Skegg, 1961), though in a diet study of moreporks from the North Island only one mouse was found in 25 stomachs (Lindsay & Ordish, 1964). Other birds known to eat mice are the kingfisher, *Halycon sancta* (Fitzgerald, Meads & Whitaker, 1986), the falcon, *Falco novaeseelandiae* (Oliver, 1955), the little owl, *Athene noctua* (Marples, 1942) and the Australasian harrier, *Circus approximans* (Redhead, 1969; Baker-Gabb, 1981).

Mice have been used in the laboratory to breed and study a variety of parasites, often infectious to man, but wild and commensal mice are thought to be relatively unimportant as vectors of infectious human diseases (Blackwell, 1981). Salmonella is the most serious disease spread to man by mice (Southern, 1954), but mice are known to be natural reservoir hosts of rickettsial pox (transmitted through the bite of the sucking mite

Allodermanyssus sanguineus), Haverhill rat-bite fever (*Actinobacillus muris*), and the rat tapeworm (*Hymenolepis diminuta*). The mouse also shares, in common with man, murine typhus (*Rickettsia mooseri*), rat-bite fever (*Spirillum minus*) and the dwarf tapeworm (*Vampirolepis nana*). In New Zealand, Brockie (1977) found that mice are renal carriers of *Leptospira ballum*, but the incidence of human leptospirosis caused by this serotype is low.

Three species of flea have been recorded on mice in New Zealand: *Leptopsylla segnis*, the mouse flea; *Nosopsyllus fasciatus*, the rat flea; and *N. londiniensis* (Smit, 1979; Tenquist & Charleston, 1981; Gibson, 1986). Both *L. segnis* and *N. fasciatus* are widely distributed throughout the North and South Islands, but *N. londiniensis* is found mainly in or near ports, and is not as successful a colonizer as *N. fasciatus* (Smit, 1979). On Allports I. in the Marlborough Sounds all 104 fleas identified from mice were *N. fasciatus*; on a nearby mainland area, 78 (94%) were *L. segnis* and only 5 (6%) were *N. fasciatus* (E. C. Murphy, unpubl.); at Craigieburn Forest Park, 43 of 45 fleas found on 97 mice were *L. segnis*, plus one *N. fasciatus* (King & Moody, 1982: 142); only *N. fasciatus* have been recorded from mice on Stewart and Antipodes Is. (Smit, 1979), and from Fiordland (King & Moody 1982: 142).

Fourteen species of mites have been identified on mice: *Myobia musculi, Radfordia affinis, Myocoptes musculinus, Eulaelaps stabularis, Haemogamassus pontiger, Hirstionyssus latiscutatus, Myonyssus decumani, Proctolaelaps hypudaei, Haemaphysalis longicornis, Acarus* sp., *Caloglyphus* sp., and unidentified specimens from the families Acaridae, Cheyletidae and Oribatidae (Gibson, 1972; Tenquist & Charleston, 1981), plus one species of louse, *Polyplax serrata* (Tenquist & Charleston, 1981; Gibson & Pilgrim, 1986). *Polyplax serrata* is rare but widely distributed.

Nematodes identified from mice are *Heligmosomum polygyrum, Syphacia obvelata* (Gibson, 1972), *Physaloptera* spp. (Badan, 1979; Pickard, 1984), *Nematospiroides dubius* (Badan, 1979) and *Mastophorus muris* (Pickard, 1984). Mouse stomachs from Woodhill and Hunua forests had few nematodes, 5% and 1.6% respectively (Badan, 1979), whereas on Mana I. 42% of stomachs were infested (Pickard, 1984).

Three species of tapeworm have been recorded: *Hymenolepis diminuta* (Gibson, 1972; Pickard, 1984); *Vampirolepis straminea* (Murphy, 1988); and *Hydatigera taenaeformis* (strobilicerci) (Gibson, 1972).

ADAPTATION TO NEW ZEALAND CONDITIONS

House mice in New Zealand are much more widespread than in Europe, and they live in a variety of habitats (forests, grasslands) occupied there by species of *Apodemus, Clethrionomys* and *Microtus*. Feral house mice living independently in bush and farmland in New Zealand, and sometimes locally called "field mice" or "wood mice", are quite different from the rodents known by these names in the Northern Hemisphere.

Explosive population increases in mice can be set off when a higher than usual number of young mice survive an unfavourable period and then breed early in the following season; this pattern is observed in New Zealand beech forests. Such irruptions in New Zealand are, however, on a small scale compared with the mouse plagues experienced occasionally in Australia. The reasons for the Australian plagues are not clear, but may include winter refuge availability, prior weather conditions (drought followed by above average rainfall) and food quality. The various hypotheses are discussed by Redhead, Enright and Newsome (1985) and Singleton and Redhead (1989).

SIGNIFICANCE TO THE NEW ZEALAND ENVIRONMENT

DAMAGE. Commensal mice have long been regarded as pests, destroying and fouling foodstuffs of man and stock, but it is much harder to quantify the effect that mice have in the wild. At a 1978 symposium on the ecology and control of rodents in New Zealand nature reserves, concern was expressed about our lack of knowledge on the impact of mice on the local flora and fauna (Dingwall, Atkinson & Hay, 1978), but since then very little new information has been published.

Mice have been known to kill lizards (Whitaker, 1978), some of which were larger than themselves, but they do not often eat them. Lizards were seldom recorded in mouse stomachs on Mana I. (Pickard, 1984) and never at Woodhill or Hunua (Badan, 1986).

Laboratory trials confirm that mice can eat small eggs and nestlings (Moors, 1978). However, in a study at Kowhai Bush, Kaikoura, mouse tracks were only once found in association with predation of a robin's (*Petroica australis*) nest (Moors, 1975). On Allports I., where mice are common, there was no evidence of any mouse predation on eggs or nestlings of robins (Flack & Lloyd, 1978).

Mice may have, or have had in the past, a much more significant impact on the invertebrate fauna. During one study on the large New Zealand weevil, *Lyperobius huttoni*, predation by mice was heavy; 25-33% of the estimated population in one study area was eaten (Bull, 1967). When mice first arrived in New Zealand, such large and vulnerable native insects were probably more common than now. Mice may also contribute indirectly to predation on other species, by sustaining populations of predators when other food is scarce.

At Hunua, kauri was the main seed species eaten by mice between April and June, suggesting that mice may be affecting the regeneration of kauri (Badan, 1979, 1986). Rimu (*Dacrydium cupressinum*) seed baits were eaten (both receptacles and contents) by mice in a study at Pureora forest, though the seeds and fruits of other podocarp species were not eaten (Beveridge, 1964). Mice have not yet been shown to eat beech seed in the wild, but they are known to eat the seeds of hard beech (*Nothofagus truncata*) in the laboratory (E. C. Murphy, unpubl.).

CONTROL. Commensal mice are controlled by the use of commercial poisons, fumigation, trapping and repellents. DOC's Mana I. campaign (p. 183) is the only known attempt to control mice in the wild anywhere in the world.

E. C. M. & C. R. P.

Order Carnivora

The carnivores are nearly all true predators, feeding largely or entirely on other vertebrate animals. They vary greatly in form according to the habitat to which they are adapted; for example, weasels and cats are all fast and graceful on land, while the heavy, lumbering seals are awkward on land but streamlined for speed in water. All have one pair of stabbing canine teeth and one pair of shearing carnassial teeth on each side of the mouth; all have binocular vision, essential for judging the distance to the strike.

New Zealand has seven native species of carnivores, all marine, and five introduced, all terrestrial. Although all twelve are members of the same order and have much in common (warm blood, well-developed brain and senses for hunting, young fed on milk, etc.), the differences between life in water and on land are so great that the marine species are placed in a separate suborder, the Pinnipedia (wing-footed or flippered), from the land-dwellers, the Fissipedia (split-footed or toed).

Cats and dogs are universally familiar, and so problems of identification are likely to arise only in distinguishing members of the suborder Pinnipedia and the family Mustelidae. Tables 43 and 46 list the distinguishing marks of the members of these groups.

Suborder Pinnipedia

Because of their distinctive appearance and largely aquatic mode of life, the seals, sea lions and walruses were until recently treated as a distinct order. The current view is that the three pinniped families are more closely related to terrestrial carnivores than to each other, and so they are now included in the order Carnivora.

There are three families of pinnipeds: the Otariidae (fur seals and sea lions: 14 species in 7 genera); the Phocidae (true seals: 19 species in 10 genera); and the Odobenidae (a single species, the walrus). Both the Otariidae and the Phocidae are represented in the New Zealand region (taken here to include the nearby subantarctic islands and the Ross Dependency, Antarctica; Map 1). Table 43 lists their distinguishing characters.

Family Otariidae

Fur seals and sea lions (otariids) are distinguished from true seals mainly by their prominent external ear pinnae, large fore-flippers and smaller hind-flippers, which can be rotated forwards under the body to assist in locomotion on land.

Of the nine species of fur seals, eight belong to one genus (*Arctocephalus*, the southern fur seals) and the ninth, the only northern fur seal, to the

Table 43: Distinguishing marks of seals of the New Zealand region.

	NZ fur seal *Arctocephalus forsteri*	NZ sea lion *Phocarctos hookeri*	Southern elephant seal *Mirounga leonina*
Body form	Bulky; adult males much larger than adult females	Bulky; adult males much larger than adult females	Largest seals in world; adult males much larger than adult females
Coat colour	Dark grey-brown; pups black	Males dark brown; females creamy-grey	Dark grey; pups black, then silvery-grey (3 weeks)
Fur	Dense underfur; heavy mane in adult males	No underfur; sparse hair; heavy mane in adult males	Short, stiff hair
Head	Snout pointed; luxuriant whiskers	Snout blunt; long whiskers	Inflatable proboscis in adult males
Ears	External ears present	External ears present	No external ears
Teeth	Large canines in males	Large canines in males	Incisors small; post-canines peg-like
Limbs	Hind limbs can be rotated forwards	Hind limbs can be rotated forwards	Hind limbs trail behind body

Measurements	Wt[1] (kg)	Length (m)	Wt (kg)	Length (m)	Wt (kg)	Length (m)
Adult males	185*	2.5*	410*	3.25*	3700*	5.0*
Adult females	70*	1.5	230*	1.9*	400*	2.5*
Pups (birth)						
Males	3.9	0.6	7.9	—	45	0.13
Females	3.3	0.5	7.2	—	45	0.13
Source(s) of data	Crawley & Warneke (1979); Mattlin (1981)		Cawthorn *et al.* (1985); Walker (1964)		Ling & Bryden (1981)	

1. Average or * maximum.

separate genus *Callorhinus*. Sea lions have five species (three southern, two northern), all in separate genera. They can be easily distinguished from fur seals in the field because they have a relatively blunt snout; darker coloration in males and lighter in females; very sparse underfur beneath a coarse outer coat of guard hairs; the first digit of the fore-flipper longer than the second; and the inner digit of the hind-flipper shorter than the rest.

Genus *Arctocephalus*

There are eight species of *Arctocephalus*, the southern fur seals, of which the New Zealand fur seal (*A. forsteri*) is the only pinniped which can be

Weddell seal *Leptonychotes weddelli*	Leopard seal *Hydrurga leptonyx*	Crabeater seal *Lobodon carcinophagus*	Ross seal *Ommatophoca rossi*
Large, fat seal	Long, sinuous body	Lithe	Small, with thick neck, slender body; graceful
Blue-black (dorsal), white spots (belly). Pups grey-brown	Black-blue (dorsal), blue-silver (ventral)	Dark brown (dorsal) to blond (ventral); flippers dark	Dark brown (dorsal), silvery (ventral) Spots, streaks
Body well-furred (3 cm)	Body well-furred (1 cm)	Body well-fured	Very short body hairs
Small head; short muzzle and whiskers	Large head; wide gape; few, short whiskers	Small head; few, short whiskers	Large eyes; short snout; few, short whiskers
No external ears	No external ears	No external ears	No external ears
Incisors, canines forward-projecting, robust	Long canines, complex molars	Complex cheek teeth; small canines	Incisors small, sharp; cheek teeth much reduced
Hind limbs trail behind body	Hind limbs trail behind body. Large fore-flippers	Hind limbs trail behind body	Hind limbs trail behind body

Wt (kg)	Length (m)	Wt (kg)	Length (m)	Wt (kg)	Length (m)	Wt (kg)	Length (m)
c.425	2.79*	500*	3.0*	200–300	2.6*	173	2.0
c.425	2.92*	500*	3.6*	200–300	2.6*	186	2.1
29	—	29.5	0.16	—	0.11	—	—
29	—	29.5	0.16	—	0.11	—	—
Bertram (1940); Kooyman *et al.* (1973)		Brown (1952, 1957); Hamilton (1939); Kooyman (1981c); Ray (1970)		Bertram (1940); J. E. King (1983); Racovitza (1900)		Hofman, Erickson & Siniff (1973); Hofman (1975)	

called a permanent resident of the New Zealand mainland. Two related species, the subantarctic (Amsterdam I.) fur seal (*A. tropicalis*) and the Antarctic (Kerguelen) fur seal (*A. gazella*), are rare visitors to only the southernmost subantarctic islands, mainly Macquarie I. and Antipodes Is.; they are easily distinguishable in the field by their colouration. *Arctocephalus tropicalis* has a distinctive bright yellow or pale cream chest and face (Rand, 1956; King 1959) and a conspicuous tuft of hair on the forehead (Bonner, 1968); and *A. gazella* has a creamy coloured throat and gingery tones on the belly. All fur seals have a fine, dense underfur, much in demand by commercial sealers in the early 19th century (p. 255).

19. New Zealand fur seal

Arctocephalus forsteri (Lesson, 1828)

Synonyms *Otaria forsteri* Lesson, 1828; *Arctocephalus forsteri* Gray, 1871.
Also called Antipodean fur seal, sea bear (early sealers), kekeno (Maori), Australian or Western Australian fur seal (in Australia).

DESCRIPTION (Fig. 48)

Distinguishing marks, Table 43 and skull key, p. 24.

The New Zealand fur seal is dark grey-brown dorsally, shading to paler grey-brown ventrally. The colour observed in the field is very variable, depending upon the stage of moult, the condition of the fur (wet or dry, clean or soiled), and the lie of the guard hairs and underfur. Young pups are black or very dark brown all over, but moult to silvery grey when a few months old.

In adults the thick, soft, chestnut underfur is overlain by coarser, darker guard hairs tinged with white at the tips, which give a silvery sheen to the dry coat. The dorsal fur is longer and more abundant than that of the belly and reaches its fullest development in the prominent mane of adult males.

The fore-flippers are large and powerful and are held at right angles to the body for use in support and locomotion; the hind-flippers, when rotated forwards, support the pelvic region and maintain stability. Fur seals walk on land by moving the fore-flippers alternately and bringing the hind-flippers up together. When galloping at high speed (*c*. 20 km/h), adult males bound forwards by moving the fore-limbs simultaneously, arching the back and swinging the hind-limbs forward together; they are very agile and rarely slip, even on wet, jagged rocks, because the serrated ridges of the soles of the flippers provide a sure grip. When swimming they use their fore-flippers to provide the power, but at times they adopt the phocid style and scull with the rear end of the body, keeping the fore-flippers folded out of the way. When swimming rapidly they often "porpoise", projecting themselves repeatedly out of the water in shallow arcs.

The upper surfaces of the flippers are furred to the metacarpals on the fore-flippers, and to the base of the claws on the hind-flippers. The exposed skin is intensely black. Nails are rudimentary on the fore-flipper, but well-developed on the middle three digits of the hind-flipper, and are used for grooming.

Adult males have a well-developed scrotum, but when moving about on land they can withdraw the testes into the body for protection. There are four abdominal nipples in both sexes.

The head resembles that of a dog or an otter. The snout is sharply pointed, the rhinarium inflated and bulbous. Both sexes have abundant whiskers, especially luxuriant and long (to 500 mm) in adult males. The eyes are large and brown; the fur below the eye is constantly moistened by overflowing tears, since there is no tear duct. The ear pinnae form elongated scrolls,

Fig. 48 N.Z. fur seal, adult male and females (M. C. Crawley).

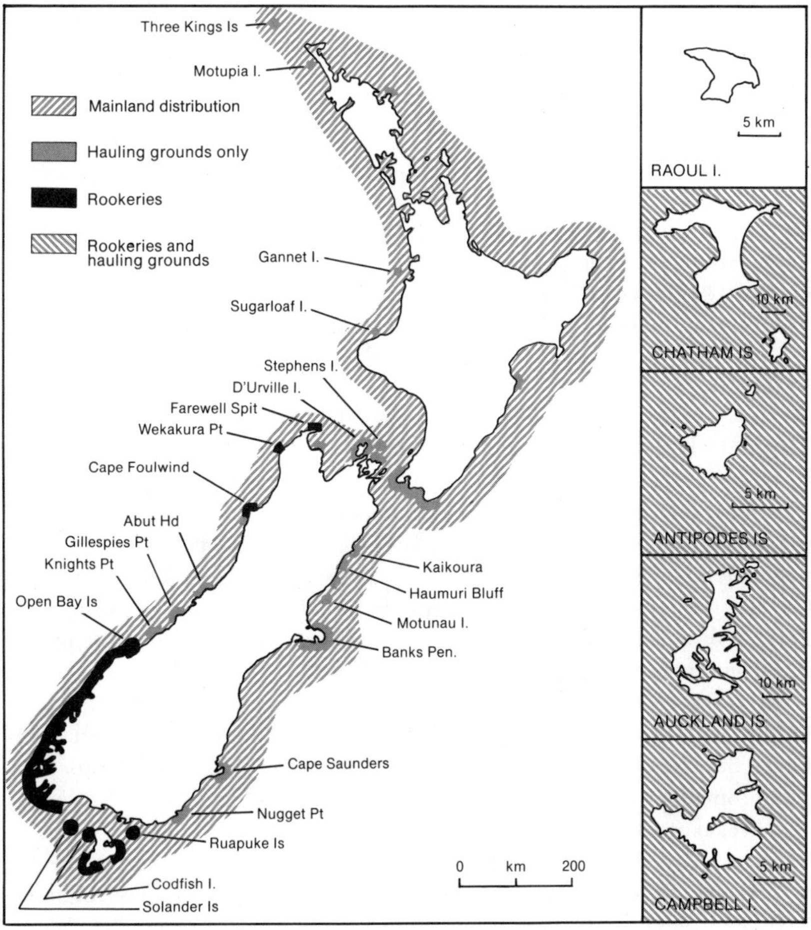

Three Kings Is
Motupia I.
Mainland distribution
Hauling grounds only
Rookeries
Rookeries and hauling grounds
Gannet I.
Sugarloaf I.
Stephens I.
D'Urville I.
Farewell Spit
Wekakura Pt
Cape Foulwind
Abut Hd
Gillespies Pt
Knights Pt
Open Bay Is
Kaikoura
Haumuri Bluff
Motunau I.
Banks Pen.
Cape Saunders
Nugget Pt
Ruapuke Is
Codfish I.
Solander Is
0 km 200
5 km
RAOUL I.
10 km
CHATHAM IS
5 km
ANTIPODES IS
10 km
AUCKLAND IS
5 km
CAMPBELL I.

Map 18 Distribution of New Zealand fur seals in the New Zealand region.

have limited movement and can be closed by muscular action. Their vision, hearing and sense of smell are excellent. Smell is especially important for adult males in detecting the onset of oestrus in females, and for mothers in identifying their own pups.

The skull has a slightly convex forehead, a moderately long rostrum, and long nasals which flare widely; the palate is broad and arched; the tooth rows are parallel; the canine teeth are large in males, but the post-canine teeth are generally small and simple in pattern with high-pointed cusps, weak internal cingula and only a slight tendency to develop anterior accessory cusps.

Dental formula $I^3/_2\ C^1/_1\ Pm^4/_4\ M^2/_1 = 36$.

FIELD SIGN

The presence of fur seals is often indicated by dark, oily stains on the rocks, where seals haul out; white or grey droppings, 40-50 mm in diameter and 100-200 mm long; regurgitations of squid sinews and beaks, fish bones and stomach worms; and small piles of stones (gastroliths) voided from the stomach.

MEASUREMENTS

Sexual dimorphism is very marked; adult males reach up to 185 kg in weight and 2.5 m in length, whereas females seldom exceed 50 kg (maximum 70 kg) and 1.5 m (Crawley & Warneke, 1979). The relatively more massive musculature of males, particularly around the neck and shoulders, is emphasized by a dense, shaggy mane. At birth, male pups average 3.9 kg and 558 mm; females 3.3 kg and 540 mm (Mattlin, 1981).

DISTRIBUTION

WORLD. The New Zealand fur seal is a coastal species with a limited range. There are both Australian and New Zealand populations, but they appear to be geographically isolated and genetically distinct (Shaughnessy, 1970). The Australian population (*c.* 7000) is widely distributed among islands off the southern coast from 117° to 136°E (King, 1969). The New Zealand population (*c.* 50 000) occupies suitable rocky terrain from Three Kings Is. (34°10′S) (Singleton, 1972) to Macquarie I. (54°30′S) (Csordas & Ingham, 1965), a distance of about 2700 km. Within these latitudes, the distribution is discontinuous and seasonally variable.

NEW ZEALAND (Map 18). There are no rookeries in the North Island. In the South Island the main rookeries are in the south and west from Open Bay Is. (43°52′S) round to Green I. (46°46′S), one of the Ruapuke group. Small, isolated rookeries at The Steeples (41°44′S; near Cape Foulwind) and Wekakura Point (40°55′S; near Cape Farewell) have been known of for many years (Wilson, 1974, 1981; Crawley & Wilson, 1976), but as recently as 1984 (M. W. Cawthorn, unpubl.; G. J. Wilson, unpubl.), new rookeries have been found, at sites previously identified only as hauling grounds (Wilson, 1974, 1981), at Tauranga Bay (41°46′S; near Cape Foulwind), Farewell Spit,

Archway Is. (40°30′S; Nelson), Cape Saunders (45°53′S; Otago) and Nugget Point (46°27′S; Otago). The main rookeries are in Fiordland, near Dusky Sound and the Five Fingers Peninsula; on the Ruapuke Is.; on Big Solander I. (the largest rookery, with at least 5000 seals); around Stewart I., particularly on the islands off the east coast; on the Chatham Is.; and on all the subantarctic islands of the New Zealand region, i.e., Snares (Crawley, 1972), Auckland, Campbell, Bounty, Antipodes and Macquarie. Fur seals were exterminated from the Antipodes Is. in the early 1800s, and in 1969 there were still no rookeries there (R. H. Taylor, in Sorensen, 1969), but pups were seen there in 1984 (R. H. Taylor, unpubl.).

Hauling grounds are found throughout the range, either adjacent to rookeries, or nearby but separate; some are well north of the breeding range. Hauling grounds, particularly northern ones, are used most often in winter and spring. In the South Island, large numbers of adult and subadult males gather at Capes Foulwind and Farewell, and at Nugget Point, Otago Peninsula, Haumuri Bluff and the Kaikoura Peninsula. In the North Island, the largest numbers are found on the four main headlands of the Wellington coast, particularly Turakirae Head; and smaller numbers on the Sugar Loaf Is. (near New Plymouth), Gannet I., Motupia I. and the Three Kings Is. Only a few seals have been reported from the east coast of the North Island, mainly at Cape Brett, near Cape Kidnappers, and at Castle Point. Winter aggregations of adult and subadult males, which have moved northwards from rookeries to hauling grounds, begin in May and reach peak densities in July or August. From September onwards numbers decline rapidly; by December none remain on the most northerly sites. At Turakirae Head, 500-600 are present in July, but only 20-30 in December. Because of the demands of reproduction and pup rearing, adult females are restricted to the rookery and the waters nearby between November and August, but in September and October they range widely at sea. Fur seal movement patterns between the subantarctic islands are unknown.

HABITAT

Fur seal rookeries are found mainly on exposed west or north-west coasts. Site requirements include: (1) some form of protection from heavy seas, e.g., off-lying reefs or stacks, rocks, or eroded rock surfaces with sharp ridges, cracks, and guts; (2) retreat areas above the storm splash zone for females and pups (Crawley & Wilson, 1976); (3) rock pools (desirable, but not essential), used by seals in warm weather for cooling off and by pups for practising swimming; (4) some form of protection from interference by land animals, as found, for example, on rocky beaches at the bases of cliffs.

The habitat requirements of non-breeding seals are less stringent; the only critical one for a hauling ground appears to be easy access. Most preferred are shelving rocky ledges and beaches with angular boulders; small or rounded boulders and stones are generally avoided. Hauling grounds are found on suitable points, spurs, headlands, reefs and islets, and, less commonly, in bays. Tussock and scrub adjacent to rookeries are used extensively for resting and shelter; the ground is often highly compacted and the vegetation cleared in areas frequented by seals.

FOOD

New Zealand fur seals feed mainly on cephalopods and fish, though they are known to take penguins, particularly at the subantarctic islands (Bailey & Sorensen, 1962; Csordas & Ingham, 1965; R. A. Falla, in Sorensen, 1969). Stomach contents of 70 seals taken from Kaikoura Peninsula, Banks Peninsula, Cape Saunders, Nugget Point and Bench I. in Foveaux Strait contained 28.8% octopus (*Octopus maorum*), 23.9% squid (primarily *Nototodarus sloanii*), 38.1% barracouta (*Thyrsites atun*) and 9.2% other fish by weight (Street, 1964). Rapson (in Sorensen, 1969) found food in only 25 of 91 stomachs examined in 1946; octopus or squid were found in 22 of the 25 stomachs and fish in only three. Off Otago Peninsula, squid, octopus, hoki (*Macruronus novaezealandiae*) and barracouta are the major components of the seals' diet (Tate, 1981), but there are dramatic seasonal changes; from February to May squid is virtually the only cephalopod eaten, but in July octopus predominates, judging from contents of faeces and regurgitations.

Fur seals eat an average of 4.1-4.5 kg per meal (Street, 1964). Those in captivity at Napier Marineland are fed a diet of mixed fish and squid; large males receive 10 kg per day, large females 7.5 kg, and smaller seals 2.5-3.0 kg per day (K. Newcomb & R. MacDonald, in Cawthorn *et al.*, 1985). Fur seals probably feed in surface waters at night, when they take squid and surface-dwelling fish, and near the sea floor during the day, when they take octopus. Seals have been reported by fishermen to bring large fish, such as ling (*Genypterus blacodes*) and blue cod (*Parapercis colias*), to the surface during the day and break them up for eating.

Access to fresh water is probably not essential for fur seals, but adult and subadult males do drink from brackish pools, particularly on hot days during the breeding season. Territorial males could suffer from thirst during the time (up to 70 days) they spend ashore without feeding in midsummer.

SOCIAL ORGANIZATION AND BEHAVIOUR

ACTIVITY. Seals on the rookery spend most of their time in one or other of four activities — lying down, sitting alert, and interacting with their own or with the opposite sex (Stirling, 1971a,b; Crawley, Stark & Dodgshun, 1977). Territorial males spend about 75% of their time lying down, and divide the rest about equally between sitting alert and all other activities (fighting, mating, herding and grooming); only 5% of their time is spent on intra- and intersexual activities. Lying down is also the predominant activity of adult females (63%), pups (75%) and yearlings (67%); mother-pup interactions (11%) are the next most important. Activity budgets change as the breeding season progresses; males interact with males a great deal during territory establishment and defence early in the season, and with females when they are in postpartum oestrus during December. Swimming becomes an important component of the activity budgets of females, pups and yearlings after the main breeding period is over.

TERRITORIES. During the breeding season, fur seal rookeries have a distinct social organization, imposed by the territorial breeding males. These fight to divide the rookery up between them in late October-early November, before the females arrive, and then defend their territories vigorously until mid January. Adult females come ashore in November but do not distribute themselves evenly; so in due course some males acquire several females within their territories while others have none. This situation leads to more fighting at the height of the breeding season, as disadvantaged males attempt to gain access to oestrous females. Males gaining possession of prime territories early in the season have to face challenges from neighbours or from late arrivals throughout the breeding season and participate in many fights, some of which are fierce and bloody.

Adult males defend their territories for 6-10 weeks without eating, and usually also without drinking, and so their success depends on their stamina, strength, fighting ability, experience and temperament. Challengers advertise their intention by sitting alert and making threat calls before proceeding directly to the territory of their choice and confronting the owner. During these disputes, the bulls meet chest-to-chest, push hard against one another and seek to grip their opponent's face, neck or upper back with their jaws or slash with their massive lower canines (Miller, 1971; Stirling, 1971a). The victorious bull either damages his opponent (the best means of preventing further attacks) or outlasts him in strength. Defeated males break off encounters by ceasing aggressive behaviour, emitting submissive screeches and backing away. Often they are savaged by other territory-holders as they seek to escape to the safety of the sea. Overall, about 30% of male-male encounters involve fighting; the remainder are resolved by use of threat displays.

The territorial system gradually breaks down during January, the end of the main birth and mating period, and by the end of the month most males have gone to sea.

REPRODUCTIVE BEHAVIOUR. Territorial males attempt to retain females within their territories by herding them, generally without success (Miller, 1974); however, once females have given birth they remain close to the birth site for up to 10 days. The territory-holder frequently investigates these sedentary females, checking on their sexual condition, but is resisted until they come into oestrus about 8 days postpartum. Oestrous females are sexually receptive for only about 24 hours and usually mate only once; they move in a strange, jerky manner and become less aggressive towards males. When a male realizes a female is in oestrus he whimpers, smells her and pushes against her with his chest or head and neck. When he mounts she is pressed prone and remains passive while he whimpers, shifts his weight around and thrusts periodically. After 5-45 minutes, possibly with multiple mountings, the female resists by squirming, growling and biting the male's chest, and eventually he dismounts.

Females prefer to settle in seaward territories with suitable resting ledges, shaded areas and good access to the sea, but will move inland to less crowded areas as numbers build up (Crawley & Wilson, 1976). Favoured sites for births (e.g., near boulders, cliffsides or washed-up logs) are used by many females. Shortly before giving birth, females become restless and highly aggressive (McNab & Crawley, 1975). As labour starts, most females lie down, strain forward on their fore-flippers and lift their hind-quarters and flippers clear of the ground. Delivery takes only about two minutes and pups are equally likely to be born head or tail first. Placentas are expelled about an hour after birth. Mothers sniff their pups excitedly immediately after they are born, but do not bite the umbilical cord, lick the pups clean or eat the placenta. They keep the pup very close to them for the first hour.

The pups are very active at first, looking around, sniffing or nuzzling their mothers, and sniffing the rocks. They call within the first 30 minutes and frequently cough and shake their heads to rid the nose, throat, ears and fur of amniotic fluid. Young pups nuzzle and rest as well as suckle, but as they grow older they nuzzle and rest less and suckle more.

Mothers discipline their pups with open-mouth threats, growls and even bites. They pick them up in their mouths to retrieve them or to rescue them from waves or aggressive neighbours. Mother-pup play is initiated by the pup and includes whisker-mouthing, gentle nipping and mouthing. Pups are harassed by other females and ignored by males. When their mothers are away from the rookery they congregate in groups or pods. Mothers returning from feeding trips locate their pups by returning to the place they left them and making pup-attraction calls. Often several pups will respond (pups will attempt to feed from any female) but females feed only their own pups, after identifying them by smell. Young pups avoid water, but can swim in calm pools at 10-20 days old. The first swimming strokes consist of simultaneous movements of the fore-flippers downwards and backwards. The head is held high up out of the water and the hind-flippers are trailed.

Fur seals are highly vocal when ashore, uttering whimpers, barks, growls, screams, moans and whines (Stirling & Warneke, 1971; Miller, 1974); they can probably recognize other individuals from their calls. Subadults are much quieter than adults. Their limited repertoire of postures and calls is both general (used, to differing degrees, by all sex and age classes) and specific to certain functional groups (e.g., the adult females' "pup-calling" posture and "pup-attraction" call).

Territorial males threaten other males with growls and screams, often as a preliminary to fighting; they use the various postures and calls to indicate status, define territorial boundaries and convey their readiness to fight. In increasing intensity of threat the postures used are: "full-neck display", "neck-waving", "oblique stare" and "horizontal neck stretch" (Stirling, 1970). Threat calls and guttural challenges usually accompany these postures. Subordinate animals avoid confrontation by adopting a submissive posture, with lowered head and neck. Adult females use an "open-mouth threat" in an aggressive way between themselves, but as an appeasement posture, with submissive whines, in encounters with territorial males.

REPRODUCTION AND DEVELOPMENT

Examination of ovaries (Mattlin, 1981) and resighting of known-age individuals show that females reach sexual maturity at 4 years of age and bear their first pup at 5 years. Males probably reach sexual maturity at 8 or 9 years, but do not gain a territory until at least 10 years (Mattlin, 1978b).

Adults form polygynous breeding aggregations with a very distinctive social organization, based on strict male territoriality, established in late October or early November. Adult females come ashore in November and give birth to their single pups from late November through to the end of December (mean pupping date 9-10 December; Miller, 1975; Mattlin, 1981). Females come into oestrus 8 days postpartum and generally mate with the nearest territory holder. Matings may be observed between mid November and mid January; the earliest ones probably involve virgin females. Development of the fertilized egg is interrupted by a period of delay during which the blastocyst lies quiescent in the uterus; in April or May, it implants and continues to develop normally.

Females remain near the birth site with their pups for 9-10 days. Then, a day or two after mating, they depart on a feeding trip for 3 to 4 days and return to suckle their pups for a similar period. From then on, females alternate feeding and suckling until the pups are weaned in July or August, after which the females feed at sea until November. Pup growth is greatest within the first 55-60 days after birth; males gain 45-74 g/day and females 46-61 g/day, depending on the year (Crawley, 1975; Mattlin, 1981). From birth to 240 days, pups of both sexes gain an average of 24 g/day (Crawley, 1975).

POPULATION DYNAMICS

NUMBERS. The best estimate of the present (1986) size of the fur seal population within the New Zealand region is 39 000 (range 30 000–50 000), based on the sum of counts made at colonies between 1971 and 1974 by Wilson (1981) plus estimated numbers from published papers (e.g., Crawley & Brown, 1971; Crawley, 1972) describing colonies not visited by Wilson. Taylor (1982) estimated that there were 16 000 seals at the Bounty Is., a figure three times larger than that used by Wilson (1981). The true number of fur seals in the New Zealand region is probably >55 000. Overall, the population is increasing at about 2% per year, but on the Bounty Is. (Taylor, 1982), and perhaps also elsewhere, the rate of increase has been close to 5% since the early 1900s. The increase dates from the beginning of protection (see

Table 44: Parasites of seals in the New Zealand region.

	NZ fur seal	NZ sea lion	Elephant seal	Weddell seal	Leopard seal	Crabeater seal	Ross seal
Trematoda							
Ogmogaster antarcticum				*		*	
Cestoda							
Baylisiella tecta			*				
B. baylisi						*	
Diphyllobothrium tectum			*	*	*		
D. quadratum				*	*		
D. lashleyi				*			
D. archeri				*			
D. perfoliatum				*			
D. rufum				*			
D. wilsoni				*	*		*
D. mobile							*
D. scotti							*
D. ventropapillatum					*		
D. scoticum					*		
Phyllobothrium delphini				*			
Nematoda							
Anisakis sp.	*						
A. patagonica			*				
A. similis			*		*		
Contracaecum osculatum	*?		*	*	*	*	
C. radiatum			*	*	*	*	*
C. rectangulum			*	*	*	*	
C. stenocephalum				*	*		
C. ogmorhini					*		
Filaria sp.			*				
Parafilaroides hydrurgae					*		
Phocascaris hydrurgae					*		
Porrocaecum decipiens	*?	*	*	*	*		*
Uncinaria hamiltoni			*				
Acanthocephala							
Corynosoma antarcticum				*			
C. hamanni				*	*	*	*
C. sipho				*			
C. bullosum			*			*	
C. australe					*		
C. clavatum					*		
Acarina							
Halarachne sp.			*	*			
Anoplura							
Antarctophthirius microchir		*					
A. ogmorhini				*	*		
A. mawsoni							*
A. lobodontis						*	
Lepidophthirius macrorhini			*				

Data from Dailey and Brownell (1972), with additional records from Hurst (1984) and Marlow (1975).

below) and is not at present constrained by habitat requirements; it may be expected to continue until habitat becomes limiting or protection is withdrawn because of political pressure from fishermen.

AGE AND MORTALITY. Little is known about the dynamics of the population. The sex ratio at birth is 1:1, but among adults on rookeries (estimates derived from counts of territorial males and pups), the male:female ratio ranges from 1:5 (Crawley & Wilson, 1976) to 1:7.3 (Mattlin, 1978a). Maximum ages determined so far, calculated from counts of tooth annuli, are 15 years for males and 14+ years for females (Mattlin, 1978a). Pup mortality is about 20% from birth to 50 days and 40% from birth to 300 days (Mattlin, 1978b). Starvation probably causes 70% of the mortality to 50 days; other causes of death include stillbirths, suffocation, drowning, trampling and predation.

PREDATORS, PARASITES AND DISEASES

Fur seals have no natural predators on land, but at sea they may be killed by sharks, killer whales and, offshore from subantarctic island rookeries, sea lions and leopard seals (Mattlin, 1978a). Predation by man was catastrophic last century (see below) but is illegal now, though accidental deaths in trawls, drift-nets and set-nets are still recorded and set-nets are still recorded. It is unknown whether these or other forms of predation or accidental death have any regulatory effect on the population.

The parasites of the New Zealand fur seal are poorly known (Table 44).

SIGNIFICANCE TO THE NEW ZEALAND ENVIRONMENT

Fur seals were killed by the Maori for food and clothing, and the teeth were used to make composite fishhooks (Trotter, 1972). James Cook's crew killed fur seals in Dusky Sound in 1773, using the flesh for food, the skins to repair rigging and the fat for lamp fuel (McNab, 1907).

Commercial sealing began in the New Zealand region on a small scale in 1792-93, when a gang based in Dusky Sound took 4500 skins. Serious, systematic, large-scale exploitation started in the early 1800s, after the exhaustion of the Australian Bass Strait sealing grounds. During the first 20 years of the 19th century, English, Australian and American sealers killed many hundreds of thousands of seals on the rookeries and hauling grounds of the South Island, Stewart I., the islands of Foveaux Strait, the Chatham Is. and the subantarctic islands, and thereby brought the New Zealand fur seal close to extinction. The discovery of each new coastline or island group abounding with seals was followed by a period of intensive, indiscriminate, destructive sealing which continued until few or no seals survived and the enterprise became uneconomic. Because of the competitive nature of the sealing industry, the numbers of seals killed and skins taken were never published, but the magnitude of the slaughter is indicated by records of cargoes of 60 000 skins from the Antipodes Is. in 1804-05 and 53 000 in 1906-07 (McNab, 1907), and a total of more than 180 000 skins from Macquarie I. between 1810 and 1813 (Csordas, 1958). By the 1830s, the seal populations had been eliminated on Antipodes and Macquarie Is., and reduced to mere remnants on the Snares, Bounty, Chatham, Auckland and Campbell Is. (McNab, 1907).

The sealers also introduced dogs, cats and Norway rats all round the New Zealand coast and to the offshore and subantarctic islands. The consequences of these introductions were neither foreseen nor intended, but proved, ultimately, to be catastrophic for the New Zealand native fauna as a whole. The depredations of these (and other) introduced animals are fully documented by Druett (1983) and King (1984a). Many of the sealers and whalers settled in New Zealand, intermarried with the Maori and established a permanent European presence. Seals (and whales) were therefore responsible for stimulating the first wave of European colonists, with momentous consequences for the history of the country.

The New Zealand fur seal was saved from extinction only by the increasingly poor economics of sealing and the imposition of a system of closed seasons and hunting permits from 1875 to 1916. Except on Campbell I. in 1924 and 1926, and in parts of southern New Zealand in 1946, no killing of fur seals for other than scientific purposes (under permit) has been allowed since 1916 (Sorensen, 1969). Protection is now guaranteed under the New Zealand Marine Mammals Protection Act (1978), except where the Act allows for commercial fishermen to kill seals actually caught in the act of robbing nets or damaging fishing gear. Although fishermen complain from time to time about the depredations of fur seals on commercial fish species, fur seals are generally considered worthy of continued protection. There is considerable public interest in them because they are large, easily observed, reasonably tolerant of considerate visitors and can be relied upon to be present at their hauling-out grounds in winter and at their rookeries in summer and autumn. Seal colonies add considerably to the wildlife values of an area, and the colonies at Kaikoura, Gillespies Beach (South Westland), Cape Foulwind (North Westland), Otago Peninsula and Turakirae Head (Wellington) are visited and photographed by large numbers of tourists.

M. C. C.

Genus *Phocarctos*

The three genera of Southern Hemisphere sea lions, each with only one species, occupy separate geographic ranges. The only one ever seen in New Zealand waters is *Phocarctos*; its nearest relative is *Neophoca*, the Australian sea lion (distinguished by a cream-coloured cap on the top of the head). The third southern genus, *Otaria*, is confined to South America.

20. New Zealand sea lion

Phocarctos hookeri (Gray, 1844)

Synonyms *Neophoca hookeri* Gray, 1844; *Arctocephalus hookeri* Gray, 1844. Also called Hooker's sea lion, hair seal.

DESCRIPTION (Fig. 49)

Distinguishing marks, Table 43 and skull key, p. 24.

Fig. 49 N.Z. sea lion, adult male, female and pups (H. A. Best).

Adult males are dark blackish-brown all over, with thicker coarser hair forming a mane over the shoulders. Adult females are white to buff to creamy grey dorsally, darker around the muzzle and flippers, and pale yellow ventrally. Newborn male pups are light chocolate brown, paler on the nose, the top of the head and nape of the neck. Female pups are predominantly light-coloured, darker on the head and nape. At 9-18 months, "small juvenile males" (Marlow, 1975) come to resemble subadult females in colour, with a silvery grey dorsum and yellow belly. Older males become progressively darker, while females remain paler.

In water, the sea lion is fast and manoeuvrable, using its fore-flippers to provide thrust. On land it moves by means of a slow walk or fast gallop, using all four flippers and the rear end of the body.

Thermoregulation in air is assisted by the posture of the flippers: they are tucked beneath the body when it is cold, but extended to dissipate heat when it is warm, or used to flip damp sand over the body.

The skull has a posterior prolongation of the tympanic bulla, which distinguishes adult *Phocarctos* skulls from those of the Australian sea lion, *Neophoca*.

Dental formula $I^3/_2\ C^1/_1\ Pm^4/_4\ M^2/_1 = 36$.

FIELD SIGN

Signs of the presence of sea lions are similar to those described for the fur seal, i.e. dark, oily stains on rocks used for resting, droppings and regurgitations of gastroliths, squid beaks and tentacles, small fish and stomach worms. The gastroliths are usually irregular in shape and less than 30 mm in diameter (Marlow, 1975). These signs are likely to indicate sea lions rather than fur seals if they are found (1) in vegetation more than about 50 m from beaches, or (2) on or adjacent to sandy beaches, or those with rounded pebbles or small boulders.

MEASUREMENTS

New Zealand sea lions are sexually dimorphic from birth, both in size and appearance. At birth male pups average 7.9 kg and females 7.2 kg, and at 20 days males are significantly heavier (13 kg) than females (11 kg). Adult males may reach up to 3.25 m standard length (Cawthorn *et al.*, 1985) and 410 kg (Walker, 1964); females, at up to 1.89 m (Cawthorn *et al.*, 1985) and 230 kg (Walker, 1964), are substantially smaller.

DISTRIBUTION

WORLD. The world population of the New Zealand sea lion is now confined to a small, roughly triangular range centred on the Auckland Is. (50°50′S, 166°E), extending northwards to the southern half of the South Island (to 46°S), south-westwards to Macquarie I. (54°40′S, 158°50′E), and eastwards to Campbell I. (52°32′S, 169°08′E).

NEW ZEALAND (Map 19). The largest rookeries are found in the Auckland Is. group, on Enderby, Dundas and Figure-of-Eight Is. (Best, 1974); smaller ones exist on The Snares (48°S, 166°35′E) (Crawley & Cameron, 1972) and on Campbell I. (Bailey & Sorensen, 1962). In 1826, sea lions bred at Port Pegasus (Stewart I.) and females suckled their young at the top of nearby wooded hills (Shepherd, 1940), but this area is now used only as a hauling ground (Wilson, 1979). Stragglers, including some subadult males, visit Macquarie I. regularly (Gwynn, 1953a, b; Law & Burstall, 1956; Ingham, 1960; Csordas, 1963), and all the other subantarctic islands occasionally. Hauling grounds, used mainly by males, especially in winter, include the coasts and offshore islets of Stewart I. and the South Island as far north as 46°S (Falla, 1965; Gaskin, 1972; Wilson, 1979). Some of the subadult and young adult males seen by M. W. Cawthorn (unpubl.) hauled out at Nugget Point, Otago Peninsula, and south of Banks Peninsula were tagged animals from the Auckland Is.

The prehistoric range was considerably wider. A fossil jaw was found at Hawke's Bay and teeth and skull fragments (<1000 years old) were collected in 1922 from Cape Kidnappers (30°38′S, 177°06′E) (Weston, Repenning & Fleming, 1973); and skeletal remains have been recovered from pre-European Maori middens on the east coast of the North Island from as far north as Houhora, 50 km south of North Cape (R. H. Taylor, unpubl.), but few were present in the North Island after the 18th century (Cawthorn *et al.*, 1985).

HABITAT

The New Zealand sea lion, unlike the fur seal, prefers flat, sandy beaches to rocky shores. The main breeding colonies in the Auckland Is. are on sand, but females and young also occupy areas of grassy sward and *Metrosideros*

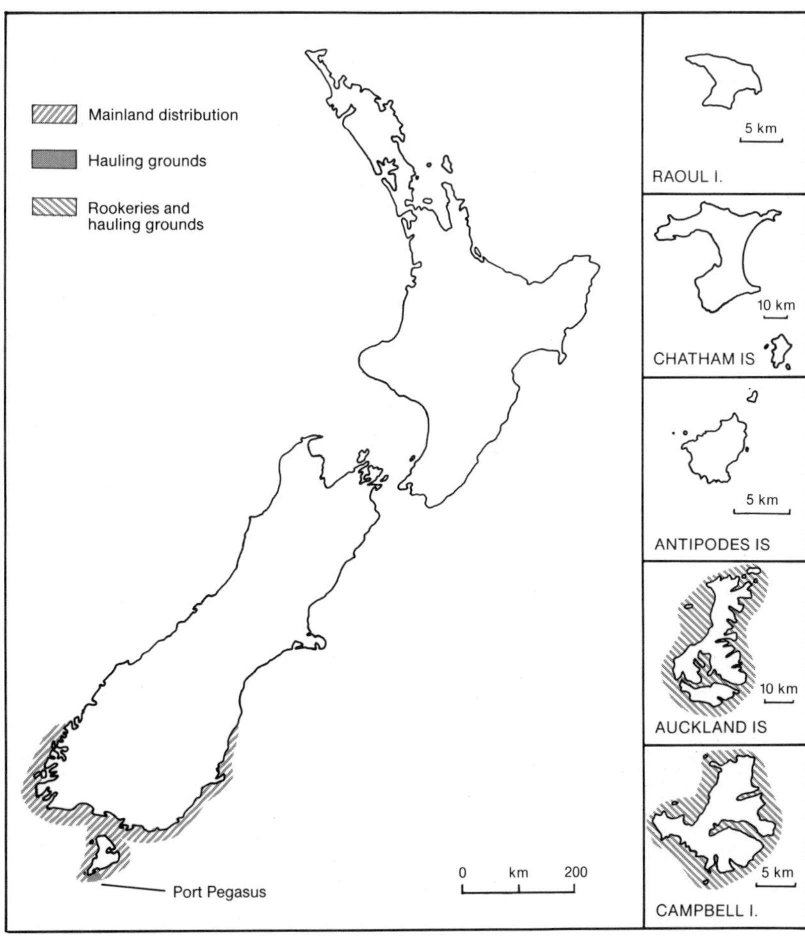

Map 19 Distribution of New Zealand sea lions in the New Zealand region. There are rookeries on Auckland, Dundas, Enderby, Figure-of-Eight, The Snares and Campbell Is.

forest adjacent to the beach (Marlow, 1975). On the Snares Is. (Crawley & Cameron, 1972) and near Port Pegasus, Stewart I. (Wilson, 1979), sea lions of both sexes and all ages move well inland to lie up in dense forest, often at the tops of hills, up to 1 km from the sea. Hauling grounds occupied by small numbers of sea lions, mainly males, may be found on shingle and gently sloping rocky beaches throughout their range.

FOOD

New Zealand sea lions have a distinct seasonal preference for feeding on squid, but will take fish, octopus, krill, crabs, elasmobranch eggs, algae, bivalves and gastropods (Cawthorn *et al.*, 1985). Some have been known to prey on penguins, sea birds, and pups of fur seal and elephant seal (King 1964a; Gaskin 1972; Cawthorn *et al.*, 1985).

SOCIAL ORGANIZATION AND BEHAVIOUR

Marlow (1975) has given a detailed account of the social behaviour of the New Zealand sea lion on Enderby I.

TERRITORIES. Adult males defend territories vigorously; fierce fights often follow if ritualized posturing, similar to that described for the New Zealand fur seal, fails to deter challengers. Adult females have no discernible social hierarchy and are very gregarious. Within harems, females are tolerant of young pups but less tolerant of neighbouring females, especially shortly (2–3 days) after giving birth.

REPRODUCTIVE BEHAVIOUR. Adult males haul out between October and early November and establish territories (Bailey & Sorensen, 1962); at about the same time pregnant females congregate at isolated haul outs some distance from the main rookeries. In early December the pregnant females move to the breeding beaches and become loosely associated with territorial males. Not all territorial males breed; in 1972-73 on Enderby I. only 19% of males mated (Marlow, 1975).

Adult males do not gather harems by herding females, but instead attempt to occupy the best areas within and between dense masses of females on the beach. Bachelor males defend territories without females, on the fringes of the main colony. Females come into oestrus 7-8 days after giving birth and are mated by the nearest territorial male, usually on the sandy beach or along the surf-washed zone, not in the water (Best, 1974; Marlow, 1975). Courtship is initiated by the male, who mounts from the rear while the cow lies prone, but terminated by the female, who bites and thrashes around when she has had enough.

All pups are born on the beach. Females about to give birth isolate themselves from others, become restless and flip sand over themselves to keep cool. Births are rapid (within 10 minutes of rupture of the chorionic sac) and the placenta is passed within an hour. Placentas are often devoured by southern skuas (*Catharacta lonnbergi*). Pups take their first feed within 30 minutes. Females and pups identify one another by sound and smell, and in the hours immediately after birth mutual sniffing and calling are frequent (Best, 1974; Marlow, 1975). The "pup-attraction" call of the female resembles the "mooing" of a cow; the response of the pup is a lamb-like bleat.

Pups congregate into pods, located along harem or territorial boundaries, from 10 days old, and at 20 days are largely independent of their mothers except when suckling. Females make feeding excursions of 2-3 days at first, returning to suckle their pups for 3-4 days between trips. Feeding excursions lengthen as the pups grow older.

At 4-6 weeks, females lead their pups to grassy swards and forest behind the beach, where the pups remain while the females go to sea to feed. Pups learn to swim in tidal rock pools and are later led to sea by their mothers. Marlow (1975) reported that by mid February on Enderby I. the breeding beach was almost deserted. Pups are suckled for up to a year but may stay with their mothers after weaning (M. W. Cawthorn, unpubl.).

REPRODUCTION AND DEVELOPMENT

Pups are born throughout December (mean date between 17 and 20 December), usually within 10 days of a female's arrival at the rookery. Females remain with their single pups continuously for 7-8 days after birth, come into oestrus, and mate, generally with the nearest territorial male, between mid December and mid January (Marlow, 1975).

Nothing is known about the reproductive physiology of the New Zealand sea lion, but there is almost certainly delayed implantation of the blastocyst, as in the New Zealand fur seal and other otariids.

POPULATION DYNAMICS

NUMBERS. Estimates of total population size range from 4000 (Cawthorn, 1975), or >6000 (Falla, Taylor & Black, 1979), to 50 000 (Scheffer, 1958). The present provisional estimate, based on tagging studies by Fisheries Research Division of MAF, is 6500-7000 animals (Cawthorn *et al.*, 1985). Marlow (1975) estimated that there were 1000 sea lions at Sandy Bay, Enderby I. in 1972-73, and Best (1974) reported 2000 on Dundas I. in January 1973. On Campbell I. there were 150 present in 1943 and 100, mainly bachelor males, in 1948 (Bailey & Sorensen, 1962). Crawley and Cameron (1972) counted 37 adults and 7 pups on The Snares in 1970-71.

AGE AND MORTALITY. The maximum ages so far determined from counts of tooth annuli are 23+ years for a male and 18 for a female (M. W. Cawthorn, unpubl.). Females become sexually mature at 3 years of age and produce their first pups at 4 years. They can pup annually thereafter for at least 4 years. Males are sexually mature at 5 years, but not socially mature until 8 years.

Pup mortality is 10-15% in the first 2 months, rising to 20% in cold, wet weather, or in areas where pups may be trapped in rabbit burrows or seal wallows. The first-year mortality has been estimated to be about 35% (M. W. Cawthorn, unpubl.).

PREDATORS, PARASITES AND DISEASES

The only natural predators of sea lions are sharks, killer whales and leopard seals; human predation is now restricted to accidental drownings in trawl nets used by squid fishermen and in pelagic drift-nets. This toll may amount to many hundreds a year. Conservationists are concerned about the level of mortality imposed during commercial fishing, and MAF staff are monitoring the accidental by-catch of sea lions and studying ways of reducing or eliminating it. The parasites are poorly known (see Table 44).

SIGNIFICANCE TO THE NEW ZEALAND ENVIRONMENT

In pre-European times, the New Zealand sea lion ranged up the east coast of the North Island and was exploited by the Maori for food. From the discovery of the Auckland Is. in 1806 to the end of the main sealing period in the 1830s, large numbers of sea lions were killed by sealers and whalers for hides, oil and food (McNab, 1907). Because of their sparse fur, sea lions were not as sought after as fur seals, and at first were taken only opportunistically; but by 1815, fur seals were in very low numbers and sea lion skins became an important cargo. There are no reliable figures on pre-exploitation numbers of sea lions, nor on how many skins were taken, but the kills were undoubtedly greatest on the Auckland Is. and Campbell I. where mere remnants of the original populations survived (McNab, 1907). The indirect consequences of the sealing trade for the land fauna of New Zealand have been described above (see p. 256). Sea lions are now fully protected by law (Marine Mammals Protection Act, 1978).

M. C. C.

Family Phocidae

True seals (or phocids) have an elongated, streamlined body covered in short, coarse hair. They have little or no underfur. The young are born with a woolly natal coat (lanugo), usually of a different colour from the fur of the adult. The hind limbs are used for swimming but are seldom any help on land. The head is dog-like, without external ear pinnae; the post-canine teeth have multiple cusps.

Two subfamilies of phocids are recognized by most authorities: the Phocinae, 10 species, all living in the Northern Hemisphere; and the Monachinae, 9 species, comprising the Antarctic seals, the monk seals and the elephant seals. All five New Zealand phocids belong to the Monachinae. For distinguishing marks, see Table 43.

Genus *Mirounga*

There are only two species in this genus, living at opposite ends of the world: the northern (*M. angustirostris*) and southern (*M. leonina*) elephant seals. As their name suggests, they are distinguished from all other seals by their great size and the large proboscis developed by the mature males.

21. Southern elephant seal

Mirounga leonina (Linnaeus, 1758)

Synonym *Phoca leonina* Linnaeus, 1758.

Also called sea elephant.

DESCRIPTION (Fig. 50)

Distinguishing marks, Table 43.

Adults have short, stiff hair, generally dark grey dorsally and lighter ventrally; pups are born with a black coat, replaced after 3 weeks by a short, silvery grey coat, darker above and lighter below (Ling & Bryden, 1981). The skin (dermis plus epidermis) varies in thickness from a few mm to 40 mm (in the throat region of males). During the moult, large ragged patches of cornified epidermis are sloughed off (Ling, 1968). Subadult elephant seals moult from November to January. They live for about 3 weeks in the tussock above beaches and wriggle around in the dry sand to slough off old skin. Adult females moult over about 16 days in December to late February, usually in deep, muddy wallows, shared with many others. Adult

Fig. 50 Southern elephant seal, adult male wallowing on Campbell I. (G. Wilson).

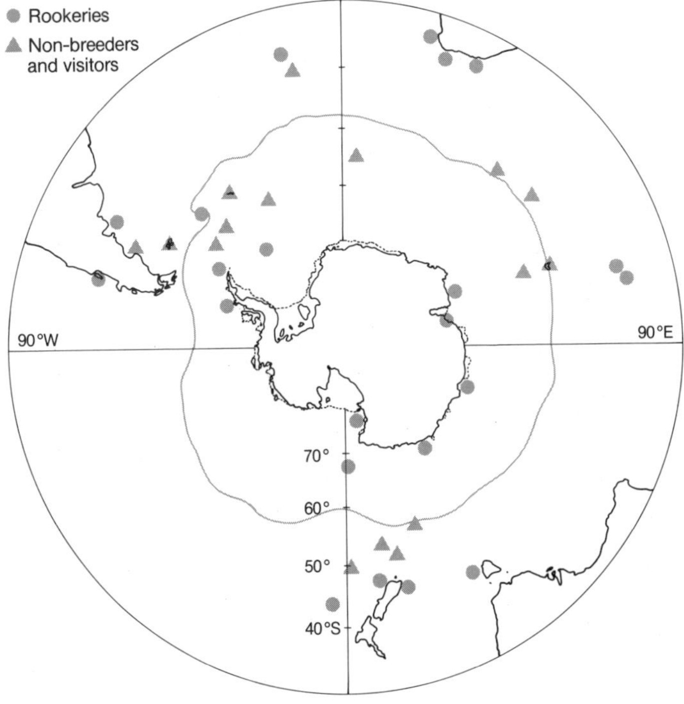

Map 20 Circumpolar distribution of southern elephant seals.

males moult between February and July (Ling & Bryden, 1981). Mud wallows and close company could alleviate itching (Carrick *et al.*, 1962) and/or maintain the high skin temperatures required for moulting (Ling, Button & Ebsary, 1974).

Elephant seals are very cumbersome on land, especially the massive adult males, and progress slowly using a caterpillar-type locomotion accomplished by flexing of the body; the flippers are of no assistance. Compared with most other seals, they are not particularly agile in the water either, although they can swim at speeds up to 20-25 km/h (Ling & Bryden, 1981), using the caudal end of the body and the hind-flippers for propulsion. Their high blood volume and blood oxygen capacity suggest that they are capable of deep diving (Scholander, 1940; Bryden & Lim, 1969); dives of 30+ minutes have been recorded (Matthews, 1952).

The skeleton is similar to that of other phocid seals, except for modifications to the skull of adult males due to the proboscis (King, 1972). In males, the skull grows continuously in length, width and height up to 11 years of age, more extensively in the snout and facial region than in the cranium. In females, the skull grows rapidly in all dimensions up to 30-40 months of age; growth in facial height and width then ceases, but in cranium and facial length it continues very slowly throughout life.

Dental formula $I^2/_1\ C^1/_1\ Pm^4/_4\ M^1/_1 = 30$.

The incisors are much reduced but the canines are well developed, particularly in males. The post-canine teeth are peg-like and probably used only to crush large fish.

Field Sign

Moulting elephant seals leave strips of epidermis and hairs in tussock adjacent to sandy beaches or in deep, muddy wallows. On beaches, trails are formed in the sand as the cumbersome seals haul themselves along.

Measurements

On Macquarie I., adult males reach up to 5 m standard length and 3700 kg; on Signy I. they are somewhat longer (5.8 m), and presumably heavier (Ling & Bryden, 1981). Females are much smaller, reaching 2.5 m and 400 kg (3.0 m at Signy I.). Breeding males (beachmasters) are up to ten times the weight of breeding females. Newborn pups are 1.3 m long and 45 kg at Macquarie I., and a similar length but heavier (50 kg) in the Falkland Island Dependencies.

Distribution

WORLD (Map 20). The southern elephant seal has a circumpolar distribution, mainly in subantarctic waters, ranging from 16°S at St. Helena to the southern limit of open water at 78°S. Most rookeries and hauling-out places are on subantarctic islands between 40° and 62°S in the South Indian and South Atlantic Oceans (Scheffer, 1958; Carrick & Ingham, 1962a); at higher latitudes they haul out on sea ice (Laws, 1953). There is probably little, if any, mixing between seals from the three main breeding stocks centred in the South

Atlantic Ocean (South Georgia, Falkland Is., Scotia Arc), South Indian Ocean (Kerguelen and Heard Is.), and South Pacific Ocean (Macquarie I., New Zealand and its subantarctic islands) (Ling & Bryden, 1981).

NEW ZEALAND. The New Zealand population is concentrated on the Antipodes Is. (Turbott, 1952) and on Campbell I. Some adult females are known to move from Macquarie to Campbell I. during the breeding season. In winter, elephant seals frequently visit the Auckland, Antipodes and Snares Is., less often the Chatham Is., and occasionally various places on mainland New Zealand, from Stewart I. to the Bay of Islands. Pups have been born along the east coast of the South Island at Kaikoura (Bowring & Stonehouse, 1968; Mills *et al.*, 1977); between Oamaru and Nugget Point, pups have been recorded frequently since 1965 (M. W. Cawthorn, unpubl.).

HABITAT

Southern elephant seals haul out on sand or gravel beaches to breed, moult and rest. Easy access is essential because, for the massive males in particular, locomotion on land is difficult. On subantarctic islands they lie on sand, in mud wallows or among the tussock grasses (*Poa* sp.) adjacent to the beaches. In the south, however, they have to lie out on snow or ice (Laws, 1953).

FOOD

The diet of elephant seals at sea is unknown, but fish, cephalopods and crustaceans are thought to be important items (Gaskin, 1972; Laws, 1979). Ashore, they fast for long periods. Adult females lose up to 135 kg over 4 weeks while suckling their pups (Ling & Bryden, 1981). Pups fast for 6 weeks after weaning. Many seals fill their stomachs with sand and small pebbles when ashore; one large male on Macquarie I. had 35 kg of pebbles in its stomach (Ling & Bryden, 1981).

SOCIAL ORGANIZATION AND BEHAVIOUR

Detailed studies of elephant seal behaviour have been made by Angot (1954) and Paulian (1953) on Kerguelen, Amsterdam and St. Paul Is.; by Laws (1953) at South Georgia; and by Carrick *et al.* (1962) and Carrick, Csordas & Ingham (1962) on Macquarie and Heard Is. Elephant seals are gregarious and have well-defined seasonal cycles and a wide range of behaviours appropriate to each age, sex and season.

SOCIAL HIERARCHY. At the beginning of the breeding season, in early August, sexually and socially mature males haul out on the beaches. Agonistic behaviour is intense and involves vocal exchanges, posturing and combat. No territories are established, but when pregnant females arrive ashore, the dominant males herd them together into harems and fight other males to establish dominance over a number of females. Breeding males are classified as beachmasters, challengers and assistant beachmasters, all of which would be 14 years old or more; bachelors are slightly smaller males (3.3–4.2 m), without a fully formed proboscis, and aged 6-13 years.

REPRODUCTIVE BEHAVIOUR. Almost all females which pup on land are impregnated in harems while ashore. Any that escape the attentions of beachmasters are mated at sea by patrolling subordinate males. During mating, the male lies across the neck and shoulders of the female, grips her neck in his mouth and draws her towards his belly using a fore-flipper. The penis is then inserted in the adjacent vulva.

VOICE. Elephant seals make a wide variety of snorts, sneezes, whistles, grunts and yawns which do not appear to have any significance in communication. Socially significant sounds are concerned with territorial protests or threats, or with communication between mother and pup. Pups emit sharp barks and yaps; females moan and yodel after giving birth; and males produce a bubbling roar of great power, possibly amplified by the proboscis (Laws, 1956).

REPRODUCTION AND DEVELOPMENT

Adult males occupy the breeding beaches between August and December and may stay ashore for up to 8 weeks. Females are ashore for about 28 days in September-October. Pregnant females give birth about 5 days after landing and suckle their pups for up to 25 days. They come into oestrus 18 days postpartum and are mated in November, but the implantation of the blastocyst is delayed until late February or early March (Ling & Bryden, 1981).

Pups are born and suckled within the harem. The female about to give birth becomes irritable and moves to the edge of the harem. Mild straining may proceed for several hours, but once the foetal membranes appear birth is rapid (2-15 minutes). The placenta appears after 15-60 minutes and is usually eaten by birds. Twin births are very rare (Carrick, Csordas & Ingham, 1962). Lactation lasts an average of 23 days. Very young pups (1-2 days) can swim, but they do not usually enter the water until they are 4-5 weeks old.

Young seals stay close to their mothers at first, but after weaning move away from harems up the beach; they go to sea at 8-10 weeks old. Mothers feed only their own pups, but pups try to suckle from any female.

Pups grow rapidly; on Macquarie I. pups are born at 45 kg and add about 3 kg per day until weaning at 110 kg (Carrick, Csordas & Ingham, 1962).

POPULATION DYNAMICS

NUMBERS. The breeding stock at Macquarie I. is thought to be stable (Carrick & Ingham, 1962a) at 36 000 females and 2500–4000 males; the total population, including pups and subadults, is 110 000 animals (Warneke, 1982). The South Georgia population is about 300 000 and that of Kerguelen and Heard Is. about 200 000 (Laws, 1960). Laws (1960) estimated the world southern elephant seal population to be 600 000 in mid year (excluding the 50% of pups that do not survive their first winter).

The New Zealand breeding stock is small; the Campbell I. population numbered only 417 in the late 1940s (Sorensen, 1950) and has since declined by 97% (R. H. Taylor, unpubl.).

AGE AND MORTALITY. Males generally live to 18 years and females 10-13 years, but both sexes may survive to 20 years (Laws, 1953; Carrick & Ingham, 1962b, c). In the unexploited population at Macquarie I., females reach puberty at 3-5 years; males are sexually mature at 5 years, but do not reach full reproductive status until 12-14 years. At Macquarie I., the survival of branded, weaned pups to the fourth year of life is over 40%; 20% of females survive to the eighth year, but few exceed 12 years of age.

The sex ratio among weaned and suckling pups averages 53% males on the main rookeries (Ling & Bryden, 1981); there is a slightly greater pre-weaning mortality of females (Carrick & Ingham, 1962c). Causes of pup mortality include abandonment and starvation, drowning and injuries sustained in harems.

PREDATORS, PARASITES AND DISEASES

Killer whales were thought to be the only possible predators on adult elephant seals, and leopard seals on young (Laws, 1979), but M. W. Cawthorn (unpubl.) has seen wounds caused by sharks, leopard seals and killer whales, and observed New Zealand sea lions killing and eating yearling elephant seals at Campbell and Auckland Is. For parasites, see Table 44 and J. E. King (1983). The tapeworm *Baylisiella tecta* is specific to the southern elephant seal. The louse *Lepidophthirius macrorhini* makes burrows in the skin of the hind-flippers (Murray & Nicholls, 1965).

SIGNIFICANCE TO THE NEW ZEALAND ENVIRONMENT

The southern elephant seal was harvested in great numbers on Campbell and Macquarie Is. and elsewhere throughout its range in the 19th century, primarily for its blubber oil but occasionally for hides. This trade finished in 1830. There are now no known threats to stocks in the New Zealand region. Because elephant seals generally live in remote areas, they have little potential as a recreational resource, although there is always keen local interest when elephant seals come ashore near populated areas. One particular old male, locally named "Humphrey", has returned to the Coromandel Peninsula and the Bay of Plenty every summer for the five years to 1989/90, visiting coastal farms and caravan parks and apparently attempting to convert a herd of dairy cows into a harem. Elephant seals are protected fully, both in New Zealand waters (Marine Mammals Protection Act, 1978) and south of 60°S (Convention for the Conservation of Antarctic Seals, 1972).

M. C. C.

Genus *Leptonychotes*

The only species in this genus, *L. weddelli*, is one of the most fully aquatic seals, perfectly adapted to life in the cold Antarctic seas. It inhabits the inshore waters of the Antarctic continent and hauls out on beaches or fast ice.

22. Weddell seal

Leptonychotes weddelli (Lesson, 1826)
Synonym *Otaria weddelli* Lesson, 1826.

DESCRIPTION (Fig. 51)
Distinguishing marks, Table 43.

The Weddell seal is one of the largest of all seals, but the head is exceptionally small relative to the size of the body. The face is benign, with a short muzzle and short vibrissae, and the eyes are large and deep brown. The entire body is covered with fur about 10 mm thick, except for small portions of the underside of the flippers. After moulting the back is blue-black, grading to a silver-white spotting on the belly; the fur colour fades with age to a rust brown on the back just before moulting. At birth, pups have grey-brown hair, much longer than the fur of the adult, shed 9-21 days later (Wilson, 1907).

Weddell seals can dive to 600 m and stay submerged for up to 73 minutes (Kooyman, 1966, 1975, 1981b). They can swim at 4-5 knots and are theoretically capable of travelling between breathing holes up to 9 km apart (Kooyman, 1968). Most dives are made during daylight, and the night hours are spent sleeping on the ice. In summer, when there are 24 hours of daylight, the peak resting period is in the afternoon (Stirling, 1969a; Siniff, Tester & Kuechle, 1971).

The skull is distinctive, because it has exceptionally thin, light bones for such a large mammal. The canines are robust and project forward, as do the incisors, forming an effective ice reamer with which the seals (both sexes) maintain breathing holes in the sea-ice throughout winter. The canines may eventually be worn through to the pulp cavity, which can lead to infection and ultimately death (Stirling, 1969b). In contrast to those of the elephant seal, the post-canines are strong.

Dental formula $I^2/_2\ C^1/_1 Pm^4/_4\ M^1/_1 = 32$.

MEASUREMENTS
The longest Weddell seals measured by Bertram (1940) were 2.92 m (female) and 2.79 m (male) from nose to tail tip.

In early spring both males and females weigh 400-450 kg (Kooyman *et al.*, 1973). The average weight of pups at birth is 29 kg.

DISTRIBUTION (Map 21)
Weddell seals are circumpolar in distribution. They are most abundant near the Antarctic coast, inhabiting both pack and (more often) fast ice (J. E. King, 1983). Small breeding populations may be found on some subantarctic islands, as far north as South Georgia, where 25-30 pups are born per year (Kooyman, 1981b). There are a few records of stragglers venturing north of the Antarctic Convergence, including sightings off New Zealand (Turbott, 1949; Ingham, 1960); the Auckland Is. and Macquarie I. (De Master, 1979), Australia (Turbott, 1949; Kooyman, 1981a), Heard and Kerguelen Is. (De Master, 1979) and (the northernmost at 35°S) Uruguay (Vaz Ferreira, 1956).

Fig. 51 Weddell seal, female and pup (M. C. Crawley).

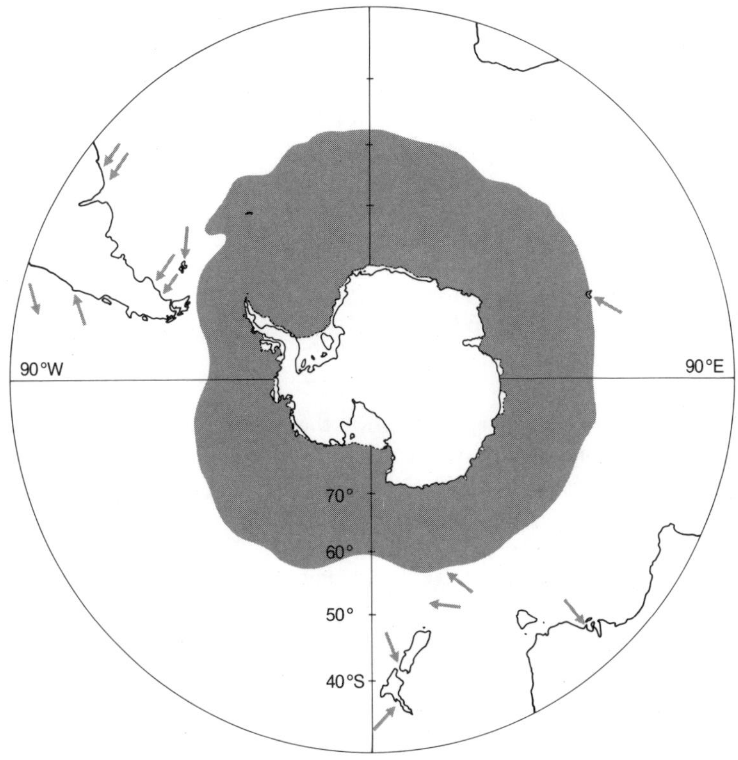

Map 21 Circumpolar distribution of Weddell seals. In shaded area, south of Antarctic Convergence, Weddell seals may be seen at any time; sightings outside this area are arrowed.

Weddell seals probably move locally over distances of 30 km or so, e.g., along the eastern coast of McMurdo Sound from Cape Armitage to Cape Evans. Between November and February, when the northern pack ice breaks up (Kooyman, 1981a), many immature seals join the breeding adults in McMurdo Sound, although few animals aged under 6 years are found in the pupping areas, and young of 1-3 years of age are rare in eastern McMurdo Sound.

HABITAT

Weddell seals inhabit the sea-ice region of the Antarctic, hauling-out on the ice to breed, and holding underwater territories centred on their breathing holes. They go ashore on the beaches of some Antarctic and subantarctic islands and on the Ross ice shelf, e.g., at White I. (Stirling, 1972; Kooyman, 1981a).

FOOD

Weddell seals in McMurdo Sound feed almost entirely on fish, especially the large (35 kg) Antarctic cod *Dissostichus mawsoni* in December, and smaller fish such as *Trematomus* sp. and *Pleuragramma antarcticum* in February (Bertram, 1940; Dearborn, 1965). They also take cephalopods.

SOCIAL ORGANIZATION AND BEHAVIOUR

Adult males establish underwater territories in breeding areas and defend them against other males; many conflicts take place near the vitally important breathing holes (Kooyman, 1968; Siniff *et al.*, 1975).

A wide array of calls is made, ranging from low-pitched buzzes to trills and whistles (1-6 kHz; 25-70 kHz) (Schevill & Watkins, 1965, 1971; Kooyman, 1981a).

REPRODUCTION AND DEVELOPMENT

In McMurdo Sound, pregnant females arrive at pupping areas in early October (Kooyman, 1968); pups (usually singles, twins are rare) are born in October, and females stay with them on the sea ice for the first 12 days; for the next 13 days, females spend 30-40% of their time in the water (Siniff *et al.*, 1975). Lactation lasts 45 days (Bertram, 1940) and is followed by ovulation. Territorial males (present since October) mate with the adult females in the water during December, and implantation follows in January or February.

Pupping begins sooner at lower latitudes; the first pups are born in late August at Signy I., South Orkneys (60°43′S) and at South Georgia (55°S) (Mansfield, 1958).

Males are mature at 3 years, but do not mate until 6-8 years old. Females reach sexual maturity at 2-3 years (Mansfield, 1958), and by age 3 years 97% have ovulated and 80% are pregnant (Stirling, 1971d); however, only 60% of mature females pup each year (Siniff *et al.*, 1975).

Pups weigh about 29 kg at birth, double their weight in the first 10 days, and reach about 110 kg when weaned at 6 weeks old (Stirling, 1969a). At 7 weeks old they can remain submerged for 5 minutes and dive to 92 m (Kooyman, 1968).

POPULATION DYNAMICS

NUMBERS. From aerial and shipboard surveys, the total world population has been estimated at between 750 000 and 1 000 000 (Erickson & Hofman, 1974; De Master, 1979; Kooyman, 1981b), concentrated in the Weddell Sea (100 000) and in McMurdo Sound (2500-3000) (Siniff *et al.*, 1977).

AGE AND MORTALITY. The minimum adult male survival rate of Weddell seals in McMurdo Sound has been calculated at 76% (1966-68) by Stirling (1971d) and about 50% (1971-75) by Siniff *et al.*, (1977). The survival rate of adult females is about 80-85% (Stirling, 1971c; Siniff *et al.*, 1977). The average age of adults in McMurdo Sound is 8-9 years; 10-12 years is common; the maximum recorded is 18 years (Stirling, 1969a). The oldest animal so far found dead was a female aged 22 years (Stirling & Greenwood, 1972).

PREDATORS, PARASITES AND DISEASES

Killer whales and leopard seals prey on Weddell seals, to an unknown extent (Stirling, 1971d; Kooyman, 1981b). For parasites, see Table 44.

SIGNIFICANCE TO THE NEW ZEALAND ENVIRONMENT

In McMurdo Sound, Weddell seals have been killed since the turn of the century for food for men and dogs. The United States stopped the seal kills in 1959, but New Zealand continued until 1986, when public opinion and scientific arguments against the practice (Moller, 1985b) forced a change of policy. The number of seals killed decreased from about 350 in 1956 to about 40 per year in the 1980s (Stirling, 1971e; Crawley, 1978; Kooyman, 1981b). There are no obvious threats to stocks at present, but in future extended harvesting and environmental damage arising from mineral exploitation in Antarctica could present problems unless carefully controlled. Under the Convention for the Conservation of Antarctic Seals (1972), up to 5000 Weddell seals may be killed or captured in any one year.

Genus *Hydrurga*

A genus with a single species, *H. leptonyx*, confined largely to Antarctic and subantarctic waters.

23. Leopard seal

Hydrurga leptonyx (Blainville, 1820)

Synonym *Phoca leptonyx* Blainville, 1820.

Also called sea leopard.

DESCRIPTION (Fig. 52)

Distinguishing marks, Table 43.

The leopard seal is built for speed. It has a long, slim body, comparatively large fore-flippers, large head, wide gape and a serpentine appearance, all of which make it easily recognizable (Kooyman, 1981c). In colour, the leopard seal shades from almost black to almost blue on the flanks, with a distinct boundary between the dark dorsal and the lighter ventral coloration. The lower flanks and belly are nearly silver.

The skull is very long (up to 431 mm) and possesses unusual, complex teeth. The molars have three prominent tubercles with narrow clefts between, and the canines are exceptionally long.

Dental formula $I^2/_2\ C^1/_1\ Pm^4/_4\ M^1/_1 = 32$.

MEASUREMENTS

Yearlings range from 1.6 to 2.3 m in nose-to-tail-tip lengths (Brown, 1957) and adults reach up to 3.0 m (males) and 3.6 m (females) (Hamilton, 1939). Adult weights have been variously estimated at 275 kg (King, 1964a), 450 kg (Ray, 1970) and, for some quite fat animals, 500 kg (Kooyman, 1981c). The only recorded measurements for a newborn pup are 29.5 kg and 1.57 m (Brown, 1952).

DISTRIBUTION (Map 22)

Leopard seals are found throughout the pack ice and southwards to the Antarctic continent. They are year-round residents on the subantarctic islands of South Georgia (Matthews, 1929) and Heard I. (Brown, 1957), gathering in large numbers (100+ animals) on Heard I. in midwinter (July). They are seasonal visitors to other subantarctic islands, especially Macquarie I., where they are common from July to November (Gywnn, 1953a; Rounsevell & Eberhard, 1980). Vagrant individuals have been seen off the Cape of Good Hope, Cape Horn, Tristan da Cunha, New Zealand and Australia (Scheffer, 1958); the northernmost record is from Lord Howe I. at 31°31′S (Hamilton, 1939).

FOOD

Leopard seals eat a variety of foods, including krill (about 30% of the diet), penguins (26%), fish (13%), seals (8%) and cephalopods (8%), and an assortment of marine organisms (15%) (Øritsland, 1977). They probably take whatever is readily available, so the precise diet depends upon location and time of year. They are the only seals that frequently take warm-blooded prey. The most unusual animal reported to be eaten by a leopard seal was a fullgrown platypus (*Ornithorhynchus anatinus*) (Troughton, 1951).

Leopard seals frequent ice-floes and waters adjacent to Adelie penguin rookeries and are adept at catching penguins after underwater pursuits or as they fall back into the water after missing their footing on the ice (Penney & Lowry, 1967). The seal skins a captured penguin by gripping the skin with its incisor teeth and shaking the bird until the skin tears away. One seal was seen to eat six penguins in 70 minutes (Kooyman, 1965).

SOCIAL ORGANIZATION AND BEHAVIOUR

Mating has never been seen in the wild; the season is unknown, probably January to March. Underwater vocalizations are of low to medium frequency (300-3500 Hz) and long duration, producing sonorous drones (Poulter, 1968; Ray, 1970). The lowest frequency call is particularly powerful and resonant and can be heard at the surface and felt through the ice.

Fig. 52 Leopard seal, hauled out on icefloe, Antarctica (M. C. Crawley).

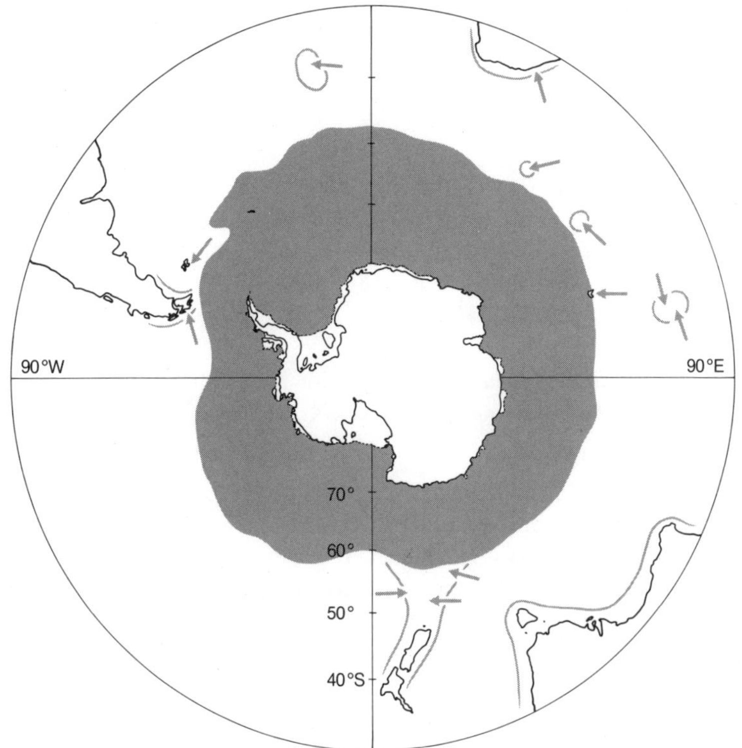

Map 22 Circumpolar distribution of leopard seals, which may be found anywhere within shaded area marking Antarctic Convergence. Leopard seals are common visitors to the coasts and islands arrowed.

REPRODUCTION AND DEVELOPMENT

Males are sexually mature at 4 years. Females ovulate in their second year and commonly become pregnant in their third year (Hamilton, 1939); pups are probably born in November (Gwynn, 1953a; Brown, 1957). A pup was born near Owenga, Chatham Is., in October 1972, but died shortly afterwards (G. J. Wilson, unpubl.).

POPULATION DYNAMICS

The most recent estimate of the total pelagic population of leopard seals in Antarctica is 222 000 (Gilbert & Erickson, 1977). Previous estimates have ranged from 100 000 to 300 000 animals (Scheffer, 1958; Erickson & Hofman, 1974).

PREDATORS, PARASITES AND DISEASES

There are no known predators of leopard seals. For parasites, see Table 44.

SIGNIFICANCE TO THE NEW ZEALAND ENVIRONMENT

Leopard seals have never been systematically exploited and it is unlikely that they will be. They would be vulnerable to uncontrolled human interference in the Antarctic (e.g., krill fishery, mineral exploitation), if it were ever allowed, but currently the Convention for the Conservation of Antarctic Seals (1972) limits any kill to 12 000 in any one year and there are prohibitions on taking seals in closed seasons and in seal reserves.

M. C. C.

Genus *Lobodon*

The sole species of this genus, *L. carcinophagus*, confined to the Antarctic, is probably the most abundant seal in the world.

24. Crabeater seal

Lobodon carcinophagus (Hombron & Jacquinot, 1842)

Synonym *Phoca carcinophaga* Hombron & Jacquinot, 1842.

DESCRIPTION (Fig. 53)

Distinguishing marks, Table 43.

Crabeater seals are predominantly dark brown dorsally, with large brown patches interspersed with lighter brown, grading to blonde ventrally. The flippers are dark. As the fur ages it gradually lightens to a uniform blonde, hence the old name "the white Antarctic seal" (Wilson, 1907). Crabeater seals are lithe in appearance, and fast and agile on snow and ice. Speeds of up to 25 km/h have been reported (O'Gorman, 1963).

The cheek teeth are extremely complex, with several pronounced tubercles separated by deep spaces (Kooyman, 1981d). The main cusps of the upper and lower teeth fit between one another, and the occlusion is perfect. These presumably function as sieves to strain invertebrates, especially euphausiids, out of the water. The canines are small.

Fig. 53 Crabeater seal on icefloe (M. C. Crawley).

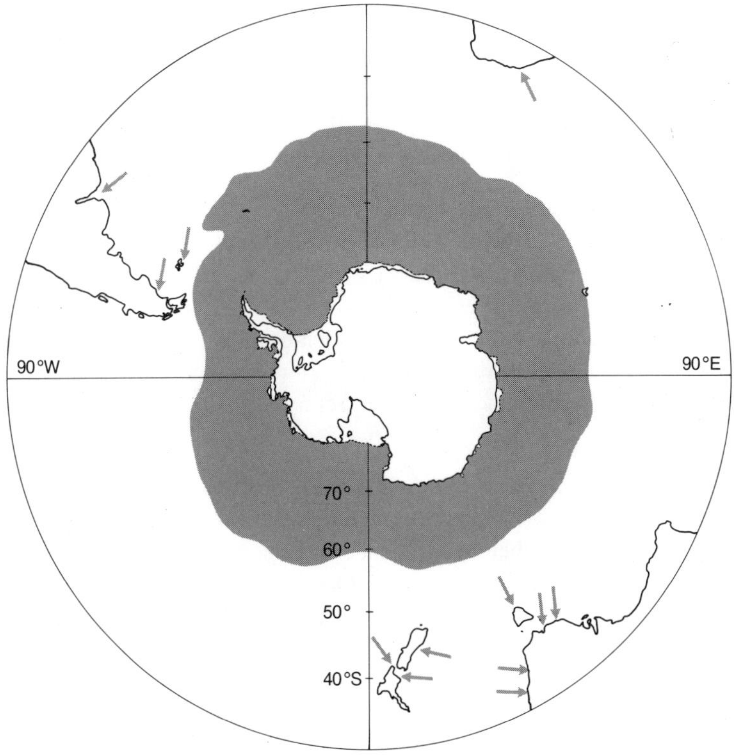

Map 23 Circumpolar distribution of crabeater seals, which may be found anywhere within shaded area marking Antarctic Convergence. Crabeater seals are occasional visitors to the coasts and islands arrowed.

Dental formula $I^2/_2\ C^1/_1 Pm^4/_4\ M^1/_1 = 32$.

MEASUREMENTS

Adults reach up to 2.6 m in nose-to-tail length and range in weight from 200 to 300 kg (Bertram, 1940; J. E. King, 1983). One newborn pup measured by Racovitza (1900) was 1.14 m long.

DISTRIBUTION (Map 23)

Crabeater seals live almost exclusively on the Antarctic pack-ice, but in summer some travel south to the Ross Sea. Lindsey (1937) observed them in the Bay of Whales (79°S) and they are seen every year in McMurdo Sound (77°50′S). Some wander inland and die; carcasses have been found in the Dry Valleys and on the Ferrar Glacier at 1100 m altitude (Wilson, 1907). A live male pup was found in December 1966 on the Crevasse Valley Glacier (76°50′S, 145°30′W), 113 km from open water (Kooyman, 1981d). Crabeater seals are also seen occasionally in New Zealand, Tasmania, southern Australia and South America (Scheffer, 1958). The most northerly sighting recorded is from South Africa (Courtenay-Latimer, 1961).

FOOD

At least 95% of the crabeater's diet is krill (Øritsland, 1977), which is sieved out of the water through the intricately trilobed cheek teeth (King, 1964a). Estimates of annual consumption of krill by crabeater seals range from 63.2 million tonnes (Laws, 1977a,b) to 72 million tonnes (Øritsland, 1977). Other foods include fish, cephalopods and crustaceans.

SOCIAL ORGANIZATION AND BEHAVIOUR

Erickson *et al.* (1971) recorded group sizes of crabeaters in the Weddell Sea in 1968–71; 563 were alone, 408 in pairs and 327 in trios. Large groups are uncommon, although Siniff *et al.* (1979) recorded aggregations of up to 1000. Trios often include a pup (Siniff & Reichle, 1976). Fighting is common (O'Gorman, 1963). When disturbed, crabeaters hiss and blow through their noses. No underwater sounds have been reported.

REPRODUCTION AND DEVELOPMENT

Both sexes are sexually mature in their second or third years, though some may breed at 12–14 months (Lindsey, 1937; Bertram, 1940). Pups are born on ice-floes from late September to November. Lactation lasts about 4–5 weeks (Laws, 1958). The mating season is from October to December, inclusive. The gestation period is 9 months, and about 80% of adult females pup each season (Bertram, 1940). Pups grow rapidly and are almost adult size at 2 years (Laws, 1958).

POPULATION DYNAMICS

The crabeater seal is probably the most abundant pinniped in the world; population estimates range from 8 million to 50 million (Laws, 1973). The currently accepted best estimate of total population size is 14.86 million, based on strip censuses made from ships and helicopters (Gilbert & Erickson, 1977).

The maximum ages so far recorded have been 19 years (Laws, 1958) and 29+ years (Øritsland, 1970b).

PREDATORS, PARASITES AND DISEASES

Leopard seals are thought to prey on first-year crabeater seals (Siniff & Bengston, 1977) and many other seals have parallel scars on their flanks which could be due to attacks by leopard seals. For parasites, see Table 44. The cestode *Baylisia baylisi* and the sucking louse *Antarctophthirius lobodontis* are specific to crabeaters (J. E. King, 1983).

SIGNIFICANCE TO THE NEW ZEALAND ENVIRONMENT

Crabeater seals were taken in small-scale pilot pelagic sealing expeditions in 1892–93 and 1964 (Øritsland, 1970a; Laws, 1973) and a few have been killed for dog food (Laws & Christie, 1976). Because it is so abundant, it is likely to be a target for future harvesting. Other possible threats involve human disturbance, contaminants and competition for their staple food, krill.

M. C. C.

Genus *Ommatophoca*

The sole member of this genus, *O. rossi*, is the most recently discovered (1840) and least well known of the pinnipeds.

25. Ross seal

Ommatophoca rossi Gray, 1844

Also called big-eyed seal, singing seal.

DESCRIPTION (Fig. 54)

Distinguishing marks, Table 43.

The Ross seal is a graceful animal, with a large head, thick neck and slender body. The back is dark, with little spotting, and the ventral surface is silvery. There are streaks on the sides of the head and throat, which may be a chestnut or chocolate colour (Ray, 1970), and spots and streaks on the sides of the flanks. In summer, unmoulted seals are tan to brownish. The moult is in January.

The snout and mouth are very short, and the head is wide. Although not apparent exteriorly, the eye (70 mm in diameter) is proportionately larger than in any other seal (King, 1968). The name *Ommatophoca* refers to the huge orbits, after the Greek *omma* for "eye". The fore-flippers are considerably modified from the usual phocid pattern, with reduced claws and greatly elongated terminal phalanges (King, 1964b, 1968), possibly because they are important in swimming.

The skull is unmistakable, because of the enormous eye sockets. The teeth are "degenerate" (Wilson, 1907), with small, sharp, recurved incisors and canines and much-reduced cheek teeth (King, 1964b).

Dental formula $I^2/_2\ C^1/_1\ Pm^4/_4\ M^1/_1 = 32$.

MEASUREMENTS

The Ross seal is the smallest of the Southern Ocean phocids, averaging 2.0 m standard length for males and 2.1 m for females; estimated weights average 173 kg (males) and 186 kg (females) (Hofman, Erickson & Siniff, 1973; Hofman, 1975).

DISTRIBUTION (Map 24)

Fewer than 200 sightings of Ross seals were reported before 1972 (Hofman, Erickson & Siniff, 1973). Recent evidence suggests that they prefer heavy pack-ice (Hofman, 1975). In the King Haakon VII Sea, Ross seals constituted 18.1% of seals seen (Hall-Martin, 1974; Wilson 1975; Condy, 1977), but elsewhere their representation ranged from none (Siniff, Cline & Erickson, 1970) through 1.9–2.4% (Gilbert & Erickson, 1977) to 5% (Erickson *et al.,* 1974).

FOOD

Scheffer (1958) reported that Ross seals ate ". . . soft-bodied cephalopods and fishes" and, more recently, Øritsland (1977) determined that one sample of the diet comprised 64% of cephalopods, 15% other invertebrates (including some krill) and 22% fish. King (1964a, 1968) suggested that Ross seals can consume squid up to 76 cm long and 6.8 kg in weight.

SOCIAL ORGANIZATION AND BEHAVIOUR

Ross seals seem to be solitary in habit, since only 3% (Hall-Martin, 1974) to 9% (Wilson, 1975) are seen in pairs; but they could be more sociable than they appear, because lone seals on ice are often associated with several others underwater, communicating by sound over long distances.

Ross seals emit characteristic and unmistakable loud cries, both in water and in the air, by inflation of the larynx and enormous soft palate, which appear to act as resonating chambers (Ray, 1970, 1981). Many of the sounds produced have been likened to those produced by bagpipes (King, 1964a).

REPRODUCTION AND DEVELOPMENT

Ross seals pup in November and mate in early December, although implantation is delayed until early March (Tikhomirov, 1975). Females become sexually mature at 2–4 years old, and males at 3–4 years (Øritsland, 1970b). The oldest known male reached 21 years and the oldest female 19 years (Tikhomirov, 1975).

POPULATION DYNAMICS

Although once thought to be rare, recent estimates suggest that the total population of Ross seals could be of the order of 220 000 animals (Gilbert & Erickson, 1977). However, Ray (1981) considers that population estimates are little better than guesses with our present state of knowledge of Ross seal distribution and behaviour. Counts should be made during the January moulting period, when more seals than usual are hauled-out on the ice.

PREDATORS, PARASITES AND DISEASES

Ross seals probably have no predators, since their habitat preferences make it unlikely that they would be vulnerable to killer whales or leopard seals. For parasites, see Table 44. Two species of cestodes, *Diphyllobothrium scotti* and *D. antarcticus* and the sucking louse *Antarctophthirius mawsoni* are specific to the Ross seal (J. E. King, 1983).

Fig. 54 Ross seal (G. W. Johnstone/NPIAW).

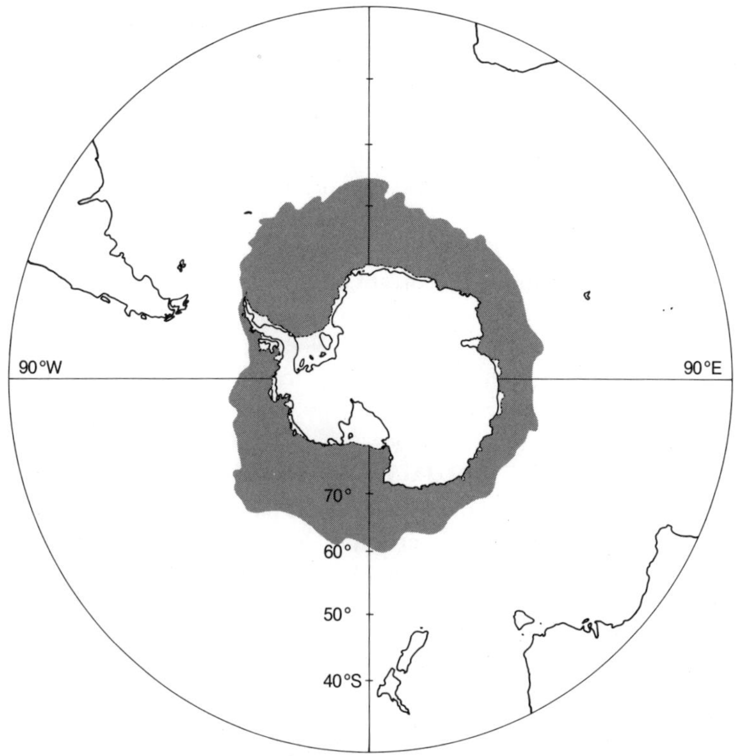

Map 24 Circumpolar distribution of Ross seals. Shaded area defines outer limit of winter pack ice, to which Ross seals are normally confined. Occasional stragglers may reach the subantarctic islands (53°S).

SIGNIFICANCE TO THE NEW ZEALAND ENVIRONMENT

Ross seals occupy areas of the ice pack inaccessible to man except by ice-breaker or aircraft; no future exploitation is likely. For the same reason, their potential as a tourist attraction is low. Currently, the Ross seal is totally protected under the Convention for the Conservation of Antarctic Seals (1972).

M. C. C.

Suborder Fissipedia

The fissipedes, or land carnivores, comprise seven families and about 93 genera and 231 species of mammals ranging in size from weasels to polar bears (a difference of about × 25 000); from bamboo-eating pandas to strictly carnivorous lions; otters in water, martens in snow, cheetahs in savannah, fennec foxes in desert. Two carnivores, the cat and the dog, were among the earliest of the domesticated mammals, and have been carried around the world by man, including to New Zealand. Members of the family Mustelidae had, in 19th century England, great reputations as rabbiters; hence, when rabbits overran New Zealand, three species of mustelids were imported in an attempt "to restore the balance of nature". None of the other 226 species of fissipedes have become established in the wild here, though two others might have done. The Indian grey mongoose (*Herpestes edwardsi*: Family Viverridae) was imported by a retired Indian Army officer who had a farm and menagerie in the Wangapeka-Tadmor district. Escaped or released mongooses were sighted in the wild on three occasions at Glenhope in 1912, and in the Clark River valley between 1914 and 1918 (A. H. McConachie, C. M. Clarke, unpubl.). There were also unconfirmed reports of mongooses in the Sherry River-Hope Range areas until 1920. They presumably died out, but not before some had dispersed 20 km or more. In addition, Wodzicki (1950) records that two raccoons (*Procyon lotor*: Family Procyonidae) escaped from captivity in Rotorua about 1905, but were never seen again.

Family Canidae

The 10 genera (35 species) of living members of the dog family evolved for fast pursuit of prey in open country, probably during the Eocene (54–38 million years ago) in North America. By the Miocene (26-7 million years ago) there were 42 genera. Familiar members of the family include foxes (21 species), wolves (2 species) and jackals (4 species), plus of course the domestic dog. Canids are now found throughout the world, as native species or as introductions aided by man.

Genus *Canis*

The nine species of *Canis* are the wolves (*C. lupus, C. rufus*), the coyote (*C. latrans*), four species of jackals, the dingo (*C. dingo*) and the domestic dog (*C. familiaris*). The domestic dog was probably derived from *C. lupus* at least 10 000

Fig. 55 The kuri in Maori art: a rock shelter drawing from South Canterbury (left) and a small wooden carving from Monck's Cave, Banks Peninsula (right).

years ago. Canids characteristically have a lithe build, long legs, long bushy tails and digitigrade four-toed feet with non-retractile claws.

26. Kuri

Canis familiaris Linnaeus, 1758

Also called guri (southern Maori), kurio (Tuamotuan), uli (Samoan), peto (Marquesan) and hence possibly also pero (Maori); Maori dog, native dog, Polynesian dog, wild dog (English). This account excludes contemporary domestic and farm dogs, mostly of European origin, but includes the wild kuri-European cross-breeds of last century.

DESCRIPTION (Figs 55, 56)
Distinguishing marks, see below and skull key, p. 24.

Kuri were first depicted in rock-shelter drawings and in carvings by the Maori (Fig. 55). These generally show an upright, alert dog with solid forequarters (Anderson, 1981). Early European descriptions were remarkably consistent. In 1769, James Cook described kuri as very small and ugly, and in 1772, Crozet regarded them as ". . . a sort of domesticated fox, quite black or white, very low on the legs, straight ears, thick tail, long back, full jaws, but more pointed than that of the fox . . ." (Allo Bay-Petersen, 1979). In 1773, George Forster thought them a ". . . rough haired sort with pricked ears and much resembled the common shepherd's cur . . ." while his father (Johann) commented on the ". . . prodigious large head . . . [and] . . . remarkable little eyes" of Polynesian dogs in general (Luomala, 1960). The whaler Edwin Palmer (Hocken, 1879) described the dogs of the southern Maori as long in the body, short-legged, bandy and stout in the forequarters. Murison (1877) described "wild dogs" from Central Otago as ". . . low-set with short prick-ears, broad forehead, sharp snout and bushy tail . . .". The similarity of these descriptions over 80 years of European observation suggests that kuri probably survived much longer as a recognizable breed than is often supposed. The

Table 45: Size and shape of pure-bred kuri in comparison with representative European breeds of dogs.

	Kuri	Border collie	Poodle	Grey-hound	Bulldog	Pekinese
Skull size						
Mean occipital length (mm)	168 (14)[1]	185				
Skull shape						
Cephalic index (shape of cranium)	55.9 (9)			50-55	75-90	
Craniofacial index (relative length of muzzle)	10 : 7.9 (8) 10 : 4.8 (1)*			10 : 7		10 : 3
Palatal index (relative width of snout)	57-65	58	65			
Limb bones (mm)						
Humerus	121.0	145.5				
Radius	110.2	147.3				
Ulna	136.7	174.0				
Femur	139.3	162.4				
Tibia	131.2	163.6				
Height at shoulder (mm)	*c.*370	470				

1. Number measured in parentheses.
Data from Allo (1970), except single measurement marked * from Leach (1979).

general picture (Fig. 56) is of a dog about the size of a small collie but much shorter in the leg, and with a longer and sharper muzzle; more heavily built about the shoulders and neck; long-haired, prick-eared and with a bushy tail. The dominant colours were black and white, separately, or in various combinations of patches and spots, but yellow coats also existed before crossbreeding with European breeds began (Anderson, 1981; A. J. Anderson & L. J. Williams, unpubl., cf. Allo Bay-Peterson, 1979).

In comparison with common European breeds, the cranium of the kuri was only slightly broader than in the greyhound, with a prominent sagittal crest and the nasal bone ending level with the posterior borders of the premaxilla; the muzzle was very long and narrowed abruptly at the third premolar to form a sharp, terrier-like snout; the teeth were well-spaced with prominent carnassials; there were, quite often, supernumerary alveoli in the tooth row (Allo, 1970, 1971; Allo Bay-Petersen, 1979); and there was a forward extension of the anterolateral border of the ascending ramus of the mandible (Watt, 1975).

Most of the rest of the body was similar to that of any European dog, except that there was a striking difference in the shortness of the leg bones, especially

Fig. 56 The kuri as sketched by Sydney Parkinson and redrawn from Luomala (1962), left, and a reconstruction based on measurements in Allo (1970), right.

in the lower leg; an average adult kuri had limbs about 20% shorter than in a comparable border collie, a breed it otherwise resembled (Table 45).

After European settlement, a great variety of other breeds of domestic dogs was brought to New Zealand. In time the typical kuri characteristics became swamped by uncontrolled interbreeding, though they were still recognizable in wild crossbreeds late into the 19th century (Thomson, 1922).

Dental formula $I^3/_3$ $C^1/_1$ $Pm^4/_4$ $M^2/_3$ = 42.

FIELD SIGN

Kuri tracks, where preserved (Nichol, 1981), are identical to those of other small to medium-sized dogs. Fossilized faeces (coprolites) range in mean maximum diameter from 16 to 26 mm (Byrne, 1973; Williams, 1980). Little evidence of kuri feeding remains has been reported from archaeological sites, but this may reflect a general unfamiliarity with the characteristic signs; certainly Allo (1970), M. Taylor (1984) and B. Kooyman (unpubl.) have found clear evidence of gnawing by kuri on moa, seal and kuri bones. Scrapes in Maori campsites where dogs were once tethered are probably the rarest signs of kuri (R. Cassels, unpubl.).

MEASUREMENTS

See Table 45.

VARIATION

Allo (1970) found evidence of a few particularly large individuals among South Island kuri; large skulls tend to have better developed sagittal and nuchal crests (for the attachment of heavier muscles). Gollan (in Anderson, 1981) suggested that South Island kuri had been selected for larger size and stronger necks, presumably to hold larger game. Kuri from later sites, particularly in the North Island, tend to be relatively smaller and sharper in the snout; otherwise, Allo found few significant variations in size with time or place.

Kuri skulls show great variation in dental wear and abnormalities. Periodontal disease was confined to South Island kuri, and extreme tooth wear and pre-mortem tooth loss were also much more prevalent in South Island specimens (Allo, 1971) than in those from Palliser Bay (Leach, 1979) and the Auckland and Coromandel areas (I. W. G. Smith, 1981a, b).

The most common dental abnormality was a supernumerary third molar alveolus (or a single-rooted fourth molar; where teeth are missing it is very difficult to tell) in the mandible and, less often, in the maxilla. This was found in few areas, e.g., at Wairau Bar and nearby early Maori sites, and throughout the pre-European period in the district around Otago Peninsula (Allo, 1971; F. J. Teal, unpubl.), but in those it was relatively common. This uneven distribution could be a consequence of deliberate selection for a linked genetic trait, such as particular coat colour (Davies, 1980). Other dental abnormalities are discussed by Allo (1971), Teal (1975), Leach (1979), Davies (1980, unpubl.), and I. W. G. Smith (1981b).

HISTORY OF COLONIZATION

Kuri arrived as domestic animals with the Maori, at about AD 900, and kuri bones are found in Maori sites throughout New Zealand from that time onward.

In Maori mythology kuri are descended from Maui's brother-in-law, Irawaru. Incensed by a minor deception which Irawaru practised while the two were fishing, Maui crushed him beneath the outrigger of their canoe, as it was landing, and then moulded his broken body into the form of the first dog.

DISTRIBUTION

WORLD. In 18th-century Polynesia the dog was found on the Tuamotus, Society Is., Hawaiian Is. and New Zealand. Archaeological evidence shows that it was formerly more widespread and once lived, for instance, in Tonga and the Marquesas (Cassels, 1983). Pure-bred kuri are now extinct as a recognizable breed, and their former relationship to other varieties of domestic dog in the South Pacific is uncertain. The dingo (*Canis dingo* Meyers, 1793), and the free-ranging Highlands dog of Papua New Guinea (*Canis hallstromi* Troughton, 1957) were once regarded as separate species, and kuri were assigned to *Canis otahitensis* (Reichenback, 1836), then *Canis australis* (Youatt, 1846), and later to various distinct subspecies (Luomala, 1960). Today, however, all the Pacific dogs are included within the universal domestic species, *Canis familiaris* (see Stains, 1975), or simply the non-specific class, *Canis* (domestic) (Clutton-Brock, Corbet & Hills 1976); but see p. 31.

NEW ZEALAND (Map 25). Kuri were once present on all three main islands, plus D'Urville and Great Barrier at least, and perhaps other smaller islands too. They were absent from the Chathams and the subantarctic islands, but present on Raoul I. in association with human settlement (Anderson, 1980).

HABITAT

As the only domestic mammal of the pre-European Maori, kuri shared the human habitat. Whether there were ever any independent populations of feral,

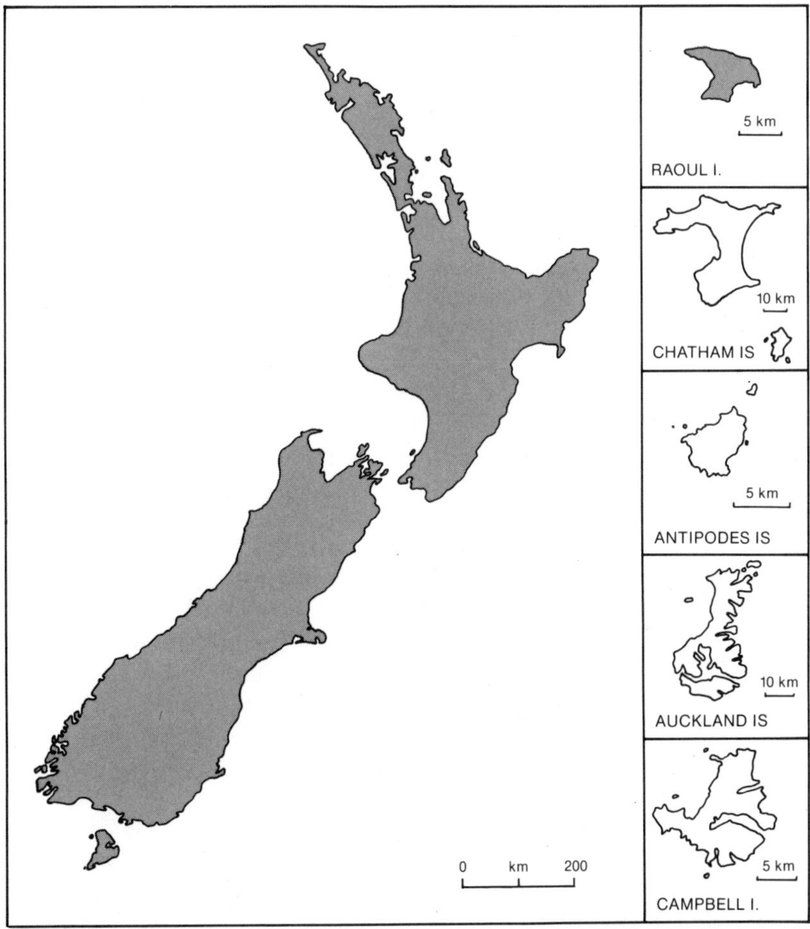

Map 25 Distribution of kuri in pre-European times (c. AD 1000–1800).

pure-bred kuri, either before or after the arrival of Europeans, is unknown. However, in the early years of European settlement, free-ranging packs of kuri-European crossbreeds were a great nuisance on the open grazing lands of Otago and Canterbury.

FOOD

Eighteenth century observers concluded that kuri were fed largely or entirely upon fish (Titcomb, 1969), a view confirmed by contemporary research. Two samples of coprolites have been analysed: in one, from the North Island, remains of small fish, especially snapper of about 300 g liveweight, predominated (Byrne, 1973); the other comprised 51.3% by weight fish bone (Williams, 1980). Large amounts of grit and charcoal in the coprolites indicate scavenging about Maori campsites, and crushed bivalve shell and remains from very small (<30 cm) fish suggest that discarded guts from larger fish were among the pickings. In the South Island, pure-bred kuri were probably fed on bones and scraps from larger game, at any rate in the early Maori period; in the early settler times, packs of wild kuri-European crossbreeds hunted ground birds and pigs, and frequently attacked flocks of sheep (Thomson, 1922).

Social Organization and Behaviour

Very little was recorded about kuri behaviour, which probably means that they generally behaved like most other domestic dogs. The one unusual characteristic which impressed most early observers was that kuri did not bark (nor, generally, do other Pacific dogs), but rather howled somewhat like a European fox (Titcomb, 1969). Johann Forster also claimed that kuri had a poor sense of smell and were exceedingly stupid and lazy (Luomala, 1960). However, domestic kuri, and the packs of feral kuri-European crossbreeds of the 19th century South Island, were skilled hunters of birds, including kiwi, weka, kakapo and ducks. Pure-bred kuri may also once have been taught to co-operate with their owners in hunting moa (Anderson, 1981).

Reproduction and Development

Unknown. Domestic bitches usually have a non-seasonal reproduction cycle of 172–200 days, but feral dogs often revert to a seasonal pattern. Dingo breed only once a year in the wild (Corbett & Newsome, 1975) and also develop more rapidly than European dogs. The details of breeding, growth and development in the kuri may have been rather different from those of modern European dogs, so direct comparison may not be valid. However, kuri were fully interfertile with all European breeds.

Population Dynamics

Kuri were domestic animals, bred and culled (usually for meat) by their human owners. Their growth rate is unknown, but pups probably reached mean adult live weight (12.5 kg) at or soon after 8 months old (Bay-Petersen, 1984). The archaic Maori were strongly carnivorous, and often 40% or more of individual kuri represented in sites of that age had been killed when immature; older dogs are more often represented in later Maori sites (Allo, 1970; Davies, 1980 and unpubl.; I. W. G. Smith, 1981a,b). There is no information on natural mortality or litter size. Pure-bred kuri disappeared during the decline of Maori culture throughout the second half of the 19th century, but in some areas the contemporary domestic dogs may still retain a proportion of kuri ancestry.

Predators and Parasites

Apart from their Maori owners, pure-bred kuri had no predators. The Maori valued kuri not only for companionship and food value, but also made extensive use of their skins for cloak decoration and their bones for the manufacture of spearpoints, awls, and fishhooks. Nothing is recorded of kuri parasites.

Considerable numbers of kuri-European (crossbred) dogs roamed wild in the South Island in the second half of last century and were always shot on sight by shepherds. Some settlers kept packs of hunting dogs especially for running them down (Thomson, 1922). As settlement proceeded, wild dogs were gradually exterminated. Domestic and farm dogs now have to be licensed and are subject to certain compulsory health checks (e.g., for hydatids, an important disease of sheep transmitted through dogs); unlicensed individuals are impounded and/or destroyed.

ADAPTATION TO NEW ZEALAND CONDITIONS

The main physical differences between kuri and other Polynesian dogs were that the latter, especially in Hawaii (Wood-Jones, 1931), had shorter, rounded skulls without strong development of sagittal and nuchal crests; shorter and more rounded muzzles; more markedly bandy legs, and often some evidence of dental caries. They were fed mainly on fruit and vegetables, especially poi (taro). Tuamotuan dogs were said to have had particularly long white hair and to have caught fish in the lagoons (Titcomb, 1969).

SIGNIFICANCE TO THE NEW ZEALAND ENVIRONMENT

Kuri were capable of hunting ground birds, lizards, frogs, and insects, etc.; but whatever impact they may have had on the indigenous fauna can hardly, now, be distinguished within the broader effects of human colonization.

A. J. A.

Family Mustelidae

The mustelids are a large and rather mixed group of small to medium-sized carnivores, adapted to a wide range of habitats and distributed world-wide. The 26 genera and 67 species range from the least weasel (*Mustela nivalis nivalis*) of northern Scandinavia and USSR, the smallest living carnivore, to the large, bear-like wolverine (*Gulo gulo*). The family includes the river-otters (*Lutra*), sea-otters (*Enhydra*), martens and sables (*Martes*), badgers (*Meles, Taxidea*) and skunks (*Spilogale, Mephitis*). Three members of the family, all species of *Mustela*, are established in New Zealand.

As far as is known, no other mustelid has ever arrived here; yet there are reports, in both popular and technical publications, that there is or once was a New Zealand "otter", known to the Maori as "waitoreke" (Watson, 1960; Pollock, 1970, 1974; Daniel & Baker, 1986). Most of these stories can be traced to sightings of unidentified animals in dubious circumstances. No specimen or convincing photograph has ever been presented for proper examination. Pollock's theory, that it is one of the small species of Asian otters, brought to New Zealand by shipwrecked fishermen hundreds of years ago and still present in the South Island, is attractive but wholly unsupported. Daniel and Baker's suggestion, that it is a mythical Maori creature unwittingly elevated into a real animal, seems much more likely.

Genus *Mustela*

The best known of the 16 species of *Mustela* are the European and North American weasels and stoats (*nivalis, erminea* and *frenata*), the domesticated ferret (*furo*) and the North American mink (*vison*). The rest are mostly little-

Table 46: Distinguishing marks of mustelids in New Zealand.

	Stoat *Mustela erminea*	Weasel *M. nivalis*	Ferret *M. furo*
Colour	Summer: brown above, white below Winter: usually same, but may be white or pied	As stoat summer coat, all seasons	Creamy white underfur, usually with black-tipped guard hairs; no distinct white belly
Markings	Brown face Thin brown tail with bushy black tip Legs brown	Brown face Thin brown tail with no black tip Legs brown	Dark face mask Thick tail, usually all black Legs black
Mean length head & body (mm)	M 284 F 256	217 182	417 350
Mean length tail (mm)	M 106 F 91	53 42	165 130
Mean body weight (g)	M 324 F 207	126 57	1 200 600

known Asiatic or South American species, including one first discovered only in 1978 (*felipei*). All have the characteristic long body, short legs and sharp-pointed faces typical of the group. The distinguishing marks of the three species present in New Zealand are shown in Table 46.

27. Stoat

Mustela erminea Linnaeus, 1758

Also called ermine (in Eurasia), short-tailed weasel (in North America).

DESCRIPTION (Fig. 57)

Distinguishing marks, Table 46 and skull key, p. 24.

The stoat has the typical mustelid shape: a long thin body, a smooth, pointed head, short legs, and five toes on each foot, furred between the pads. The claws are sharp and non-retractile. The ears are short, rounded, and set almost flat into the fur. The eyes are round, black and protruding; the whiskers are very long, and the muzzle is black and dog-like. The body fur is short, normally chestnut brown on the head and back, and white or cream (sometimes shading to yellow or even to apricot) on the underside. The tail is much longer than the extended hind legs, and always tipped with a conspicuous tuft of long black hair, which may be bristled out into a "bottlebrush" at moments of great excitement. Males are always much larger than females (Table 47).

Fig. 57 Stoat, male (NZWS).

The scrotum of adult males is furred, regressed in winter (March to August) but still visible, whereas that of the undeveloped subadult males is scarcely visible until their first spring (September). Only adult females have visible nipples, usually 4 or 5 pairs. Body temperature is 38-40°C, respiration rate 86-100 breaths per minute, pulse 400-500 beats per minute (Tumanov & Levin, 1974).

The spring moult (August to November) starts on the head and sweeps backwards over the back and flanks, ending on the belly; the autumn moult (March to June) follows the same route in reverse (pattern illustrated by van Soest & van Bree, 1969). Each cycle is initiated when the seasonal change in day length reaches a critical threshold ratio of light to dark hours, irrespective of weather or altitude. Stoats from the south of the South Island are always the last to start the spring moult and the first to grow their winter coats (King & Moody, 1982: 123). Skins of animals in the process of moulting are marked with black flecks on the inside.

The tail tip is black and bushy at all seasons, but the body fur is denser in winter, and the dorsal fur ranges from entirely brown to entirely white (the "ermine" condition), depending on how cold it is at the time the winter fur is growing. Below a critical minimum temperature, the new hair grows out white; if this condition is met during only part of the period of hair growth, the winter fur will grow out part white and part brown (the "pied" condition). There is great local variation in the proportion of white and pied stoats (see below).

The ears and nose are acutely sensitive; sight probably includes some colour perception, although the eyes are adapted for both day and night vision. The eyes have a vivid green tapetum lucidum (eyeshine) and a brown iris with a horizontal slit pupil.

When keeping under cover, stoats hold the body close to the ground and run with many short steps; when moving fast in the open, they arch the back and

Table 47: Body measurements of adult stoats in New Zealand (mean ± SE).

	Sex	Head & body length (mm)	n	Tail (mm)	n	Body weight (g)	n
Beech forests/grasslands[1]							
Craigieburn FP	M	290.3 ± 1.44	50	107.4 ± 1.89	50	356.2 ± 7.04	47
	F	261.2 ± 1.04	51	92.9 ± 0.82	51	221.5 ± 3.37	49
Tongariro NP	M	288.0 ± 2.60	22	101.5 ± 1.81	22	354.6 ± 12.52	18
	F	254.9 ± 2.31	12	87.2 ± 1.56	12	218.5 ± 7.79	12
Fiordland NP (Eglinton Valley)	M	286.0 ± 1.13	74	108.5 ± 0.80	77	331.4 ± 6.48	62
	F	256.3 ± 1.35	57	93.3 ± 0.95	57	206.0 ± 4.23	52
Mount Cook NP	M	288.0 ± 1.65	39	103.1 ± 1.18	39	331.2 ± 9.60	29
	F	260.9 ± 1.71	53	91.6 ± 0.87	54	226.8 ± 4.12	48
Podocarp/mixed forests							
Egmont NP	M	280.3 ± 1.52	42	103.9 ± 1.18	42	339.1 ± 7.78	38
	F	257.7 ± 1.47	21	89.4 ± 0.95	21	221.0 ± 6.93	20
Fiordland NP (Hollyford Valley)	M	278.1 ± 2.8	36	103.3 ± 1.77	36	302.68 ± 10.11	34
	F	249.6 ± 2.57	46	87.3 ± 0.87	47	200.0 ± 3.32	43
Westland NP	M	274.9 ± 1.69	30	100.8 ± 1.30	29	284.7 ± 7.46	26
	F	253.1 ± 1.18	47	88.4 ± 1.02	47	195.0 ± 3.86	44
New Zealand[2]	M	284.5 ± 0.58	461	106.2 ± 0.38	463	324.4 ± 2.63	395
	F	256.2 ± 0.54	440	90.7 ± 0.32	441	207.2 ± 1.52	398

1. Representative collection areas with best samples. The Eglinton and Hollyford Valleys are on opposite sides of the Main Divide, about 20 km apart; so are Mount Cook and Westland National Parks, about 40 km apart.
2. Total collection

Data from King and Moody (1982 and unpubl.)

bound along (up to 500 mm per jump) with tail held high. They are small enough to enter the burrows of rabbits and rats, and sinuous enough to turn round inside. Their sharp claws and light weight enable them to climb trees with ease, even to great heights (one was observed at canopy level, 15–20 m, from a tree platform in the Orongorongo Valley; this and many other observations are summarized in King, 1975a, 1982b), and run down again head first. They can also climb narrow saplings, and cross between adjacent trees by running along connecting branches (Moors, 1983a); they are sometimes caught in arboreal rat traps. They swim readily and well, in both fresh and salt water, and have reached, unaided, many offshore islands within 1 km of the mainland coast (Table 48); one was seen swimming in Lake Taupo >1 km from shore (Anon., 1949a).

The skull is flattened, scarcely wider across the fragile cheek bones than across the brain case; total (condylobasal) length averages 50.1 mm in males, 45.7 mm in females.

The teeth are, like those of the ferret (p. 323), strongly adapted to an exclusively carnivorous diet.

Dental formula $I^3/_3$ $C^1/_1$ $Pm^3/_3$ $M^1/_2$ = 34.

The skeleton is a generalized mammalian type, remarkable only for the great flexibility of the joints, and the disproportionate length of the backbone. The baculum is 20-30 mm long, slender and gently curved; the proximal end develops a smooth knob in adults (see below).

Chromosome number 2n = 44 throughout the species range (samples from Canada, Sweden and Japan tested: Meylan, 1967; Mandahl & Fredga, 1980; Obara, 1982).

Field Sign

Tracks in soft mud may show the full length of the foot, averaging about 20 × 22 mm in front and 42 × 25 mm behind. Tracks on a harder surface show only the pads, arranged in a shallow semicircle. Scats are long (40-80 mm) and thin, hard and black when dry, and full of hair, feathers and bone fragments. Scats often accumulate in or near a den and are also used as signals, e.g., carefully deposited on a stone in the middle of a path. Dens are usually taken over from prey and are well hidden, used for only short periods at a time, and extremely difficult to find in the field.

Measurements

See Table 47. Sexual dimorphism is very pronounced; on average, male stoats in New Zealand are longer than females by 10% and heavier by 57% (King & Moody, 1982: 74, 90).

Variation

Size. The range of mean condylobasal lengths in adult male stoats within New Zealand (48.9–51.5 mm, difference 2.6 ± 0.45 mm) is almost as great as in continental Europe (46.8–49.6 mm, difference 2.8 ± 0.33 mm). This variation is clearly correlated with habitat; adult males from podocarp and mixed forests average 3% smaller in skull length than those from beech forests and grassland, and 4% smaller in head and body length (see Fig. 59). The same pattern reappears, less clearly, in other body measurements; in females and young as well as males; and even over short geographical distances, e.g., from one side of the Southern Alps to the other (Table 47). Weight is more variable than length (Table 47).

Winter whitening is more common in females than in males, and in stock acclimatized to cold conditions (even after transfer to warmer), which suggests that winter whitening is at least partly controlled by a sex-linked gene. White or pied stoats of either sex are found more often in the south and at high altitude. There is a positive correlation between the proportion of a sample showing some degree of whitening and both the average July temperature and the annual number of days of ground frost (King & Moody,

Table 48: Distribution and history of stoats on the offshore islands of New Zealand.

Island	Area (ha)	Water crossing to nearest mainland (m)
Northland (35°S)		
Motukawanui	355	2 400
Waewaetorea	47	300
Moturoa	146	250
Urupukapuka	208	600
Motuarohia	58	1 000
Rimariki	7	400
Hauraki Gulf (36° S)		
Motuoruhi	70	750
Waiheke	9 459	1 100
Ponui	1 851	1 100
Nelson-Marlborough (40-41°S)		
D'Urville	16 782	500
Forsyth	775	300
Adele	88	800
Fisherman	4	700
Maud[1]	309	900
Tawhitinui	22	400
Pickersgill	103	300
Blumine	400	400
Arapawa	7 785	700
Fiordland (45-46°S)		
Secretary	8 140	650
Resolution	20 860	600
Elisabeth	74	150
Whidbey Point	5	200
Cooper	1 886	200
Anchor	1 525	800
Small Craft Harbour	48	500
Chalky	475	1 100
Weka	108	700
Coal	1 622	400
Bauza	*c.*500	650

Data from Taylor and Tilley (1984), with additions from R. H. Taylor (unpubl.); not all records refer to permanent populations.
1. Exterminated 1983.

1982: 124). In Westland, at 43°S and 120-150 m, fewer white or pied stoats were collected in June-November 1972-76 (11% of 9 females, 6% of 22 males) than near Mt. Cook, a higher altitude valley (700-900 m) at the same latitude (38% of 8 females, 17% of 29 males). On Mt. Egmont, at 39°S, fewer white or pied stoats were collected (5% of 19 females, 0 of 40 males) than in the Eglinton Valley at 44°S (71% of 34 females, 47% of 62 males), even though the southern area is at lower altitude (270-550 m, cf. 460-900 m) (King & Moody, 1982: 124).

HISTORY OF COLONIZATION

By the middle 1870s the damage caused by rabbits to sheep pasture (p. 158) was having a serious effect on the economy of the colony. All attempts to control them failed, and huge areas of grazing land were being ruined. The runholders pressed the government to introduce the "natural enemies" of the rabbit, which they assumed were normally responsible for keeping rabbit numbers down. Despite the protests of ornithologists (Buller, 1877), the Chief Rabbit Inspector recommended in 1883 that stoats and weasels should be liberated, in addition to the ferrets which were already being turned out in huge numbers (p. 324). This policy was officially endorsed by the government, and by some of the acclimatization societies; others objected, because of the risk to introduced game birds, but "as the societies were themselves directly responsible for the rabbits, their protests were ineffective" (Thomson, 1922). The government immediately sent an agent to Britain to procure mustelids; runholders and other private individuals also organized shipments. In 1886 an association of landowners was formed in Masterton for the purpose of funding the supply of natural enemies. Large sums of money were subscribed, and "hundreds of stoats and weasels were introduced into the district" (Thomson, 1922: 72).

The first shipment to arrive was brought in by a Mr Rich of Palmerston in 1884 (number and destination unknown). In 1885, 55 stoats consigned to the government were released at Lake Wakatipu and in the Ashburton district. In 1886 two more government shipments were received, and 82 stoats were released in the Wilkin River, Makarora, Lake Ohau, and Waitaki areas and 32 in Marlborough and west Wairarapa. A prominent Wellington runholder, Mr Riddiford, brought in another 55 in 1886 for his run in the Orongorongo Valley.

According to Wodzicki (1950), all the consignments were arranged through the New Zealand Agent-General in London. It is not certain that all the animals were collected in Britain, but since there were at that time about 17 000 gamekeepers employed on "vermin control" on British game estates (King, 1984a), there would have been no need to search elsewhere. Stoats were easily caught in box traps and housed in small hutches, but losses in transit were high.

Stoats were first released, with high hopes and the full protection of the law, on farmland where rabbits were extremely abundant. But within six years there were reports of stoats spreading into the forests of Fiordland and to other districts far from any known releases, including to places where there were as yet no rabbits (e.g., to Okuru and Waiatoto, on the Westland coast, by the summer of 1889-90; Müller, 1890). Their progress across the two main islands is unrecorded, but stoats were liberated in sufficient numbers, and in so many places, that they could well have spread throughout both main islands by or soon after the turn of the century.

Even after it was already too late to stop the importations, ornithologists continued to protest against them (Martin, 1885; Buller, 1895, 1896), but it was another five years at least before the government changed its policy. Starting

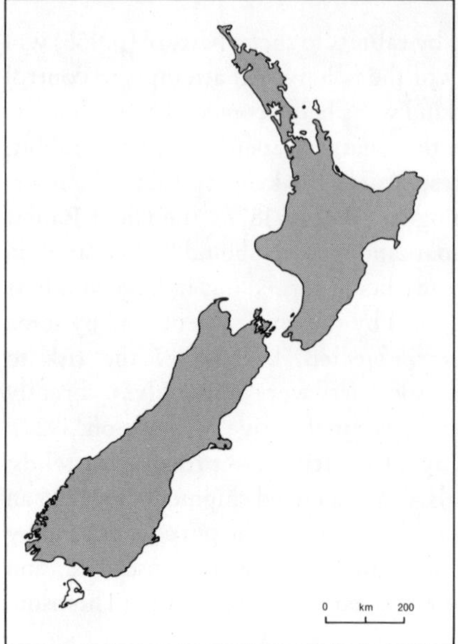

Map 26 Distribution of stoats on main islands of New Zealand. For distribution on offshore islands, see Table 48.

in 1903, successive amendments to or overriding the Rabbit Act (Thomson, 1922) eventually removed all legal protection from mustelids by 1936 and restricted further imports; all three species (especially stoats) have been regarded as serious pests ever since. In recent times, however, this opinion is being reassessed (King, 1984a).

The questionnaire survey conducted in 1948 by Wodzicki (1950) established that stoats were widespread and abundant in both main islands. Over the next 10 years, however, new techniques in aerial distribution of poison baits allowed, for the first time, substantial reductions in the numbers of rabbits throughout the country (p. 159). By the time the next such survey was done in 1961, there were strong indications that stoats had declined in numbers, and possibly also in distribution, by comparison with the 1940s (Marshall, 1963). No comparable survey has been done since then.

DISTRIBUTION

WORLD. The stoat originated in Eurasia in the early to middle Pleistocene and spread to North America across the Bering Bridge during one of the later glacial periods. It is now found throughout the northern Holarctic, from the shores of the Arctic Sea south to about 40°N, including the larger offshore islands of higher latitudes (e.g., Ireland, Newfoundland) but excluding most of the Mediterranean region. Its natural distribution has often been extended by deliberate introductions made in the belief that stoats would be able to control pest populations of small mammals. Known cases include, for example, mainland Shetland Is., north of Scotland (in 17th century or before); New Zealand (1884); Terschelling I., off Holland (1931); and Strynoe Kalv, Denmark (1980) (Venables & Venables, 1955; van Wijngaarden & Bruijns, 1961;

Kildemoes, 1985). On the other hand, continuous persecution from gamekeepers, at least since the 1870s, has never made any permanent impression on the distribution of stoats in Britain (King & Moors, 1979a).

NEW ZEALAND (Map 26). The latest national-scale survey (Marshall, 1963) reported sightings or specimens of stoats from throughout the two main islands, from Northland to Bluff; the present distribution is probably about the same. In the middle 1970s systematic trapping in all the national parks, and in several other areas, always produced at least some specimens (King & Moody, 1982). Stoats were never taken to any of the offshore islands, so they are still absent from Stewart, Kapiti, Great Barrier, Little Barrier and the Chathams; but they have reached many of the nearer inshore islands unassisted, including several of high value for wildlife, e.g., Resolution and Maud (Table 48). Taylor and Tilley (1984) suggest that the maximum swimming range of a stoat in seawater is about 1.1 km. Permanent populations of stoats are unlikely on small islands (Taylor & Tilley's suggested lower limit of <1 km^2 seems too small; in Britain it is around 60 km^2: King & Moors, 1979b), but stoats may make temporary visits to small islands near to the mainland often enough to compromise their value as wildlife refuges.

HABITAT

Stoats live in any habitat in which they can find prey. In New Zealand they can be found anywhere from beaches to remote high country, at any altitude up to and beyond the tree-line; in any kind of forest, native or exotic; in scrub, duneland, tussock grassland and farm pastures (Gibb & Flux, 1973). In open country they keep to cover as much as possible, e.g., scrub-filled gullies, ditches and piles of brush left after land clearing, and are much less common than ferrets (over 5 years' observations at Kourarau, Gibb, Ward & Ward (1978) recorded 7 stoats and 63 ferrets), but they are much more common than ferrets in forests. They tolerate both extremely wet (6000+ mm rain/year in parts of Westland and Fiordland) and moderately dry conditions (<1000 mm in Otago and Canterbury). They do not avoid human settlements and are occasionally seen in rural villages and suburban gardens (King, 1982b).

FOOD

DIET. The most frequently eaten prey of stoats in New Zealand today are birds, feral house mice, lagomorphs (rabbits and hares, not distinguishable from small remains), rats, possums and insects (mostly wetas of the genera *Hemideina, Hemiandrus* and *Gymnoplectron*). Minor items include lizards (mostly *Leiolopisma* and *Hoplodactylus*), crayfish (*Paranephrops*), carrion, and rubbish. This is the general pattern shown by all studies so far done, which include national surveys from field observations by Marshall (1963) and King (1982b), and from gut analysis by Fitzgerald (1964) and King and Moody (1982); and more localized studies based on scat and/or gut analyses by Fitzgerald (1964, Birdlings Flat), Lavers and Mills (1978, Takahe Valley), NZWS staff and associates (D. V. Merton unpubl., Milford area) and King (1983a, Fiordland and

Table 49: Foods of stoats in New Zealand forests.

	Number found	% of 866
Large items		
Carrion	12	1.4
Possum	107	12.8 (of 837)
Lagomorph	113	13.4 (of 845)
Unidentified mammal	94	10.9
Hedgehog	1	0.1
Rat	60	6.9
Medium-sized items		
Bird	364	42.0
Mouse	190	21.9
Fish	2	0.2
Freshwater crayfish	23	2.7
Skink	9	1.0
Gecko	2	0.2
Bird's egg	7	0.8
Small items		
Tree weta	31	3.6
Ground weta	138	15.9
Large cave weta	66	7.6
Small cave weta	49	5.7
Unidentified weta	16	1.8
Lepidoptera	55	6.4
Carabid beetle	36	4.2
Other arthropods	98	11.3
Unidentified	11	1.3

The list gives the number and percentage frequency of occurrence of prey items identified in 866 guts, omitting empty ones. For possums and lagomorphs, the sample sizes are adjusted to exclude all stoats caught in traps baited with these items. "Unidentified" mammals are mostly possums and rats. Data broken down by sex, age, habitat, season, location, etc., available in source publication, King and Moody (1982).

Craigieburn, and 1975a, Orongorongo Valley). The actual proportions of each prey type eaten depend on, among other things, habitat, sex (but not age), season, and even year; so generalizations are difficult. King and Moody (1982) attempted to control for these variables in samples rendered as homogeneous as possible (Table 49 shows the results from forested areas), collected from all the national parks in 1972-76, and found the following common patterns.

In all seasons, female stoats ate more small prey (e.g., mice, wetas) than males, but were still as capable as males of taking lagomorphs, rats and birds; males ate more possums. Large prey provided most of the nutritional needs of both sexes, because small prey, though frequently eaten, contain relatively little net food value.

Differences in diet between habitat types and/or seasons often reflected real changes in the availability of prey. For example, in general, possums were eaten more often in mixed podocarp-hardwood forests, and lagomorphs in beech forest-grassland areas, which reflects the reciprocal distribution of these species. In forests supporting large numbers of possum trappers, male stoats ate possums slightly more often in winter, suggesting that some at least were taken as carrion; by contrast, fewer lagomorphs were eaten in winter, when young rabbits and hares are not available. Similarly, rats were more often eaten in mixed podocarp-hardwood forest, and mice in beech forests, which again reflects the distribution of the two; rats were eaten all the year round, but mice especially often in autumn, the time when (at the end of the breeding season) they are usually most abundant.

Birds were the most frequently taken class of prey in all seasons and forest types, especially in summer, but there was always a sudden drop in autumn corresponding to the seasonal increase in consumption of mice. However, later work (King, 1983a) showed that the number of birds found per 100 stoat guts examined did not decrease when there was plenty of alternative food (e.g., in habitats where larger mammal prey were available, or during a population peak of mice); it seems that mammals of the right size are never sufficiently common in New Zealand to serve as sole prey for stoats. Birds' eggs were eaten in spring, although shell fragments were found less often than expected; but stoats may be able to open whole eggs quite cleanly, since empty shells may be found punctured with needle-like toothmarks about 8–10 mm apart, corresponding to the placing of the canines in a stoat's jaw.

Lizards and insects were eaten all year round, but stoats may have more opportunity to catch them in winter if they are slower moving, or even dormant, in cold weather. June-August records included single guts containing up to 6 skinks or 53 wasps in all stages of development.

Most freshwater crayfish were found in northern sample areas, sometimes with soft shells, as if they had just moulted.

Stoats sometimes search for food in rubbish bins or tips, and will take fresh carrion provided by trappers, hunters or road-kills, although not after it has become fly-blown.

The daily food requirements of New Zealand stoats are not known: they must be at least as much as those of the slightly smaller British animals, estimated by Day (1968) as averaging about 23% of body weight for males and 14% for females. This translates to about 75 g/day for adult New Zealand males (the equivalent of 5 mice), and 30 g/day for females (2 mice) except during lactation, when the food requirements of breeding females increase by 200–300% (Müller, 1970).

FEEDING BEHAVIOUR. Stoats are active hunters and search for prey through all possible cover, down every accessible hole and up every likely tree (Moors, 1983a) in the course of each hunting excursion. Prey are detected by sight (especially if moving) or sound (broods of young birds are taken more often than clutches of eggs: Moors, 1983a). Small mammals and birds are killed in seconds with a swift and accurate bite at the back of the neck: larger prey with well-muscled necks (e.g., adult lagomorphs) take longer, and the actual cause of death is as likely to be fright as physical damage (Hewson & Healing, 1971). This means that, when necessary, female stoats are able to kill larger prey, although the process may demand more energy and incur more risk for them than it does for males. On the other hand, stoats of either sex are extremely bold hunters and have been seen tackling animals very much larger than themselves. (Morrison (1980) observed one hanging from the neck of a Fiordland crested penguin, *Eudyptes pachyrhynchus*.) When a fresh kill has been made, a stoat will drag it to the nearest cover, and start licking the blood exuding from the wound, but it does not suck blood. Small prey are usually consumed whole, although if the stoat is already well fed it may discard the hard parts (teeth, tail, feet, stomach of mammals, and flight feathers of birds); with larger prey, a stoat is better able to avoid the indigestible fur and bones. Killing behaviour is independent of hunger; if opportunity arises, a stoat will kill any suitable prey it can, and cache the surplus for future use. This strategy evolved in Northern Hemisphere ecosystems, in which stoats depend on unreliable resources (fluctuating populations of voles); it explains why a stoat in a hen-run or guinea-pig run (both recorded by King & Moody, 1982) will kill every live animal in sight before stopping to eat.

SOCIAL ORGANIZATION AND BEHAVIOUR

ACTIVITY. Stoats may be seen at any time, day or night (King, 1982b). Radio-tracking studies overseas show that their daily activity is usually divided into short hunting periods of usually less than an hour at a time, interrupted by longer periods resting in one of several dens scattered through the home range (Erlinge, 1979a). They are active for longer, and range further, when prey are scarce.

DISPERSION. Adults live on separate home ranges for most of the year. Males search for females very actively during the breeding season, roughly from September to November, but they do not establish even a temporary pair bond, and take no part in the rearing of the young (Erlinge, 1979b). Females and their young may be seen moving about in family parties (groups of three or more, up to 7+) from late October, when North Island young are first able to leave the nest, until early February, when the last of the South Island litters disperse (King, 1982b; King & McMillan, 1982).

HOME RANGE. The winter ranges of males are large and may include those of several females. Males are dominant over females, except when the females have young, but generally, animals of each sex defend their ranges against those of the same sex (Powell, 1979). The size, length of tenure and pattern of use of home ranges depend largely on the density and distribution of prey. In the Northern Hemisphere, ranges of males are usually about 10–50 ha (Lockie, 1966; Erlinge, 1977a; Debrot & Mermod, 1983), increasing to over 100 ha when food is short (Nasimovich, 1949; Vaisfeld, 1972). The home range system appears to break down in spring, when the males enter into an intensive search for females (Erlinge & Sandell, 1986). Detailed information on the home ranges and social organization of stoats in New Zealand is not available, but in the Orongorongo Valley (near Wellington) in early 1972, one tagged female was able to cross from one side to the other of a circle of live-traps enclosing 14 ha within an hour (King, 1975a); this area was obviously only a part of her total range. In Fiordland in the summer of 1979–80, on two straight lines of live-traps 400 m apart, adult stoats of both sexes were well spaced out (about one per 2–4 km), and each individual tended to visit only the same 2–5 adjacent traps (spanning a total of 0.4–1.6 km) (King & McMillan, 1982). Both these studies were done when food was relatively abundant; in years of scarcity, very few stoats can be caught at all. Home ranges of stoats in New Zealand forests are therefore probably relatively large, at least 60 ha and perhaps up to 100–200 ha in most years (King, 1975a).

DENS. Stoats do not make their own nests, but take over those of other animals such as rats, rabbits or possums. The long thin shape and short fur of stoats make them liable to excessive heat loss while resting (Brown & Lasiewski, 1972); so a warm nest is important even in the mild climate over much of New Zealand; it is an essential condition of survival in colder areas.

VOICE. When nervous, a stoat makes a low hissing sound; if it becomes extremely frightened (e.g., when caught in a trap, or threatened by a dominant or a larger predator) this intensifies into a loud wailing squeal. Friendly encounters between family members or mates are accompanied by an excited, high-pitched trilling. A subordinate individual encountering but not distressed by a dominant gives a submissive trilling sound, which helps to mollify the reaction of the dominant. A sharp explosive chirp or shriek is a defensive threat, very effective against other stoats (Erlinge, 1977b) and even unwary humans.

COMMUNICATION. Stoats mark their home ranges with scent from two types of glands, which have different chemical compositions and have different meanings. The large anal glands under the tail produce a strong-smelling, thick yellow fluid containing several sulphurous compounds, identified from New Zealand material as mixtures of thietanes and dithiolanes (Crump, 1980a; Brinck, Erlinge & Sandell, 1983). The principal components can now be synthesized in the laboratory. The scent produced is individually distinct (Erlinge, Sandell & Brinck, 1982) and is deposited, by a characteristic "anal drag" action, at strategic sites throughout the home range; it signifies ownership of the ground by a particular individual. There are also small scent glands in the skin, whose odour is deposited by "body rubbing", especially in response to an encounter with another stoat or its scent marks. The skin gland secretion contains mainly proteinaceous lipophilic compounds carrying smaller molecules of high volatility (Erlinge, Sandell & Brinck, 1982); it signifies a threat to intruders. Dominant individuals mark in both ways more often and more vigorously, and mark over places previously marked by other stoats; subordinates show signs of fear when scenting strange marks. Hence, it seems likely that scent marks convey both identity and social status, so allowing subordinate individuals to avoid damaging conflicts.

REPRODUCTIVE BEHAVIOUR. European studies show that, in spring, the dominant males change from mainly nocturnal, territorial behaviour to a mainly diurnal wandering over a much wider area (Debrot *et al.,* 1985). This may be because the decisive resource for males during the breeding season (receptive females) is more widely dispersed and less defensible than the decisive resource during the rest of the year (food); dominant males can probably take over any female they can find, so can achieve most matings by searching a large area (Erlinge & Sandell, 1986; Sandell, 1986). Younger males remain on their home ranges, where they have a chance of reaching any resident female(s) first. Females aggressively reject any male until they reach full oestrus, when the vulva becomes enlarged and its orifice reddened, moist and elevated upon a papilla. Courtship is brief, but copulation vigorous, prolonged (2–20 mins) and frequent (2–5 times an hour: Müller, 1970), because ovulation can be induced only by coitus. All attempts to stimulate ovulation by injection of gonadotropins have failed (Gulamhusein & Thawley, 1974). There are two periods of intense sexual activity in males: in September-October, when the adult females have a postpartum oestrus (adult females which fail to produce a litter may be ready to mate earlier, while the others are still pregnant); and October-November, when the precocious juvenile females become receptive. Adult males are apparently able to serve the juvenile females without damaging them, even though they are so small. Ternovsky (1983) observed one 17-day-old female, only 112 mm long, which weighed 18 g (13% of the weight of the mother, and 6% of that of the adult male) and was blind, deaf, helpless and almost immobile; she was mated for one minute, and in the following season produced 13 kits and fed them successfully. It is possible that an adult male may serve his own young conceived the previous year, but the chances of this happening are reduced by the short lifespan, lack of a year-round pair bond and rapid replacement of locally resident males.

REPRODUCTION AND DEVELOPMENT

MALE ORGANS. The testes are simple, about the size of a pea in the mature adult. The penis is held in an internal sheath, with the urethra lying in a groove on the underside of the baculum, a small bone attached to the pelvis at its proximal end, which acts as a rigid support during the vigorous copulations necessary to ensure induction of ovulation. The development of the baculum is controlled by androgens (Wright, 1950); so the distinctive proximal knob characteristic of adults does not develop in young or castrated animals.

FEMALE ORGANS. The uterus is a simple U-shape, of uniform colour (i.e., does not show placental scars): the flattened ovaries have obvious corpora lutea of delay, visible as yellow spots, for 9–10 months of the year. Each represents one ovum released at ovulation, and each persists until spring, regardless of the fate of that ovum, and then degenerates. Successive generations of corpora lutea do not overlap, since those of one set are replaced a few weeks after the young are born by a fresh set representing next year's litter. Since the corpora lutea are found in females of all ages for most of the year, they provide a simple means of estimating potential fecundity, although the actual number of young born may be very different (see below).

AGE AT PUBERTY. Young males do not mature until 10 months old, i.e, in July-August of the season after that of their birth. Young females are extremely precocious and are reproductively mature in October-November, as unweaned nestlings only 3–5 weeks old. Both sexes continue to grow in body size after puberty. The young females are already carrying blastocysts (see below), but these demand little energy to maintain, since further development is delayed until well after the young females have completed their own growth in March-April; young males reach adult size after a sudden, substantial spurt of growth in September and October.

SEASONAL CYCLE. The testes of males of all ages begin to enlarge in July, reach full size in October, and regress slowly from December to a minimum in May (King & Moody, 1982:103). The period of increase begins after the winter solstice, several weeks after the onset of spermatogenesis, and is accompanied by high levels of plasma testosterone (Gulamhusein & Tam, 1974). Fully fertile adult males may be found from August to February.

Females of all ages caught between November and August inclusive are almost always (>99%) in the inactive phase of pregnancy; they carry the small (0.4–0.6 mm) corpora lutea of delay, and support a set of 8–10 diapausing blastocysts in a quiescent (0.05–0.15 g) uterus. In July both corpora lutea and uterus begin to enlarge in preparation for implantation, and between August and October these reach 0.9–1.4 mm and 0.15–0.36 g respectively. Visibly pregnant and post-pregnant (lactating and/or in oestrus) females may be found only in September and October, and by November almost all females are already fertilized and carrying a new generation of corpora lutea of delay. The ecological explanation for delayed implantation is under current debate (King, 1983b, 1984b; Sandell, 1984; Powell, 1985; Stenseth, 1985).

The seasonal cycle in both sexes is controlled by day length. The onset of spermatogenesis and the implantation of the blastocysts both tend to be later in the far south (44–45°) by a short period (about 10 days) corresponding to the lag in the date at which the southern regions reach the critical day length that sets off these processes in July-August. Hence, most North Island litters are born from late September to early October, and most South Island ones from mid to late October. The period of oestrus is brief; among 46 adult females collected from all over New Zealand in the months of September or October, only one was in full oestrus and five in the immediately preceding stage, while not one of 38 newly independent young females collected in December showed any sign of recent oestrus (King & Moody, 1982:117). However, this means only that the males are very efficient at finding receptive females; in captivity, unmated females remain on heat for months, although they do not ovulate and their ovaries contain no corpora lutea (C. M. King, unpubl.).

GESTATION. Active gestation is divided into two phases. From fertilization to the onset of delay (about 2 weeks), the zygotes develop only as far as the blastocyst stage and then float free in the uterus for 9–10 months. The following spring, they implant in the normal way, and the embryos develop to full term in about 4 weeks. The approximate date of birth can be estimated by

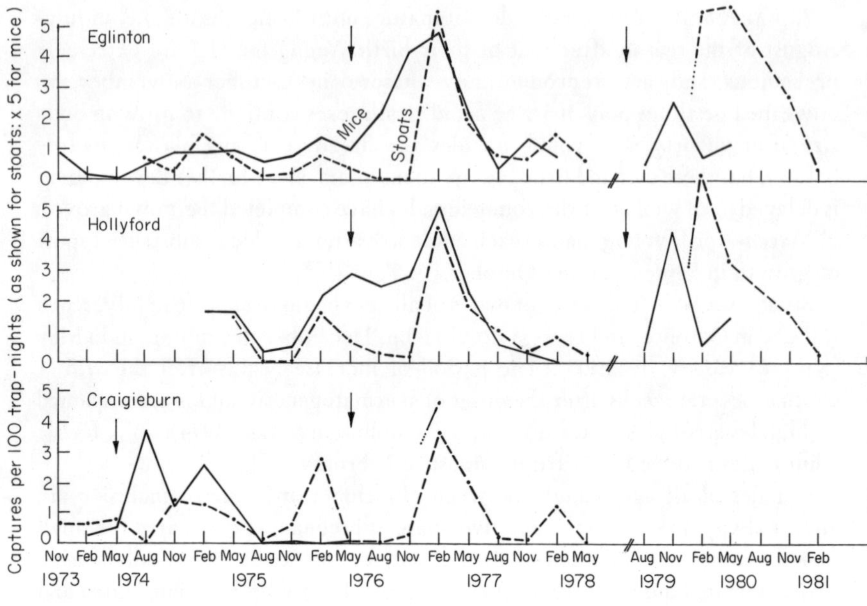

Fig. 58 Population changes in mice and stoats after moderate or heavy seedfalls of beech (*Nothofagus* sp., arrowed) in Eglinton and Hollyford Valleys (Fiordland) and Craigieburn (Canterbury) (from King, 1983a).

calculating t (a negative value, days before birth) from the mean weight of the embryos (W) by a formula given by King and Moody (1982:105):

$$t = (\checkmark W/0.063)+(16-40)$$

The blastocysts distribute themselves evenly between the two horns of the uterus and are often fewer than the number of corpora lutea.

Female stoats are generally less often caught than males at all seasons, but especially in spring; pregnant females, the most difficult to catch of all, comprised only 13 of >1600 stoats collected in 4 years by King and Moody (1982). The mean number of embryos was 8.8 per set (range 6–13), but in at least 5 sets there were some embryos being reabsorbed — in one female, only one of the 8 embryos was viable. Fitzgerald (1964) recorded 3 pregnancies with 7, 12 and 12 embryos. In any given place or year, however, the number of young born is adjusted to prevailing food supplies (see below).

BIRTH AND DEVELOPMENT. The young weigh about 3–4 g at birth, are blind, deaf and toothless, and are covered with a fine whitish down. By 14 days they develop a prominent but temporary brown mane. While the female is away, very young kits (under 5–7 weeks old) are not able to maintain their own body temperature; they huddle together, and if the temperature in the nest drops below about 10–12°C they go into a temporary, reversible cold torpor, and their pulse and breathing slow down (Segal, 1975). They attain full control over body temperature when their fur is fully grown, at the age of about 8 weeks (the black tail tip appears at about 6–7 weeks). Other functions are also developed by then: the milk teeth erupt at 3 weeks, solid food is taken from 4 weeks on, the eyes open at 5–6 weeks (females first), and lactation stops after 7–12 weeks (Hamilton, 1933; East & Lockie, 1965; Müller, 1970). The permanent carnassial teeth, P^4 and M_1, are last in place. The typical prey-killing behaviour is instinctive, but improves with practice from first attempts at 10 weeks to full maturity at about 12 weeks. The female rears the young alone, and hunts for and with them from the time they are active and weaned but still dependent, at about 6–8 weeks, to the time the young disperse, at about 12–14 weeks. Young of the year can be caught in traps from mid to late November, when they first venture out in family groups, but most often in January, after the families break up. Young females usually stay near their natal area, but young males are capable of astonishingly rapid dispersal. In Fiordland, seven young males marked with eartags travelled between 6 and 23 km within a few weeks of independence in December-January 1979–80 (King & McMillan, 1982).

POPULATION DYNAMICS

Stoats have the general characteristics of opportunistic species: small size, short lifespan, and high and variable rates of birth and death, which result in unstable populations whose density and distribution are controlled primarily by food (King & Moors, 1979a; King, 1983a,b).

PRODUCTIVITY. The mean ovulation rate (indicated by the number of corpora lutea) is 10, range 3–20; the average tends to be slightly higher in juvenile females in all years, and in females of all ages in years when mice are very abundant (King & Moody, 1982:113; King, 1981a), but otherwise the mean is relatively constant across a wide range of conditions. There is an inverse correlation between the number of corpora lutea in the two ovaries of one individual.

The number of young born is closely related to food supplies in spring. This correlation is probably quite general, but can best be demonstrated in beech forests after a good mast year (Fig. 58). During the winter following the seedfall, mice increase greatly in density; in the early summer (8–9 months after the fall), huge numbers of young stoats appear (Riney *et al.*, 1959; King, 1981a, 1983a; R. H. Taylor *et al.*, unpubl.). The adjustment of productivity to resources can be made only by variations in the extent of intrauterine and nestling mortality, not (as in many other mammals) by additional productivity, because the maximum number of young born is already fixed at ovulation, in the previous year. In poor years, mortality gradually increases at every stage from implantation to weaning; many females resorb some of their embryos (see above), or even all of them, and others bear their young but fail to rear them. The latter females can be identified, because the normal reproductive season is synchronized by day length. Females collected in September, October or November, which are about to produce or have already produced a litter, can be classified at some stage between early pregnancy and lactation; those that are not, and show no signs of having borne young (enlarged nipples, worn belly fur), have probably abandoned all attempts to breed during that year.

These failures happen most often in beech forests in the "crash" years after a seedfall. At least 11 of 19 females collected in such years (1972 at Nelson Lakes and Arthur's Pass, 1975 at Craigieburn, and 1977 in Fiordland) were classed as failures; in average or good years in the same areas, total failures are very few (King, 1981a). Less disastrous shortages are met with less drastic measures. The average rate of loss is about 13% of blastocysts failing to implant, and about 12% of embryos resorbed. In any one year the extent of this process can be estimated from the ratio of young to adult stoats caught in summer. After a good year, young of the year comprise 80–90% of the catch; in the worst years, 0–10% (King, 1983a).

DENSITY. On Birdlings Flat, Canterbury, Fitzgerald (1964) found at least seven males and two females living on approximately 750 ha. In the Orongorongo Valley, two males and three females (total of 30 captures and recaptures) were live-trapped between January and April 1972, after the good seedfall of 1971; none in November-December 1972; and two males and one female (once each, including one of the males marked in February 1972), between August 1974 and January 1975 (King, 1975a). Actual density was not measured, but Brockie and Moeed (1986) estimate that the average biomass of stoats in the Orongorongo forest is unlikely to exceed 0.002 kg/ha.

Relative density is easier to estimate. From standardized traplines set in beech forest in the Eglinton and Hollyford Valleys (Fiordland National Park) in 1973–78 and 1979–80, and in Craigieburn Forest Park in 1973–79, King (1980a, 1981a, 1983a) showed a regular seasonal variation in average capture rate on an established trapline, from <1 C/100TN per 3 months in winter and spring, increasing to 2 C/100TN per 3 months in summer. Superimposed on this seasonal pattern there was an irregular variation in the height of the summer peak, from 1–2 C/100TN per 3 months in average years to 5–6 C/100TN per 3 months in post-seedfall years (Fig. 58). Higher figures may be obtained for short periods from previously undisturbed populations trapped for the first time; e.g., 9.3 C/100TN per month in the Hollyford in February 1975 (King, 1983a) and 16 C/100TN per month on Adele I. in August 1980 (Taylor & Tilley, 1984). Erlinge (1983) calibrated relative density figures against known density, and showed that C/100TN indices are reliable.

The additional numbers caught in summer are all newly independent young, which quickly disappear over the winter; by the following spring the normal low density is restored (King & McMillan, 1982). There is a significant three-stage correlation between seedfall and mouse density (p. 236) and between the density indices for mice and stoats in Fiordland forests (Fig. 58). The increase in numbers of stoats in post-seedfall summers is obvious enough to be noticed by casual observers, e.g., visitors to national parks (King, 1982b). At Pureora Forest Park, the only podocarp forest sampled in the same systematic way, stoats were at low density (equivalent to, or less than, in a non-seed year in a beech forest) throughout 1983–87 (C. M. King, J. G. Innes & M. Flux, unpubl.). In non-forest habitats, variations in density of stoats may be traceable to fluctuations in rabbits; for example, the abundance of stoats can be reduced by successful control of rabbits, both locally (e.g., at Mt. Cook; King, 1982b) and nationally (Marshall, 1963). The same relationships between food supplies and the density, productivity and population dynamics of stoats have also been demonstrated in the Northern Hemisphere, where the key prey are watervoles (*Arvicola*) in Eurasia (Aspisov & Popov, 1940; Erlinge, 1983; Debrot, 1981, 1983), rabbits in Britain (Sumption & Flowerdew, 1985), and *Microtus* in Canada (Raymond & Bergeron, 1982). Stoat populations in Britain were drastically reduced for 15–20 years after myxomatosis almost eliminated rabbits in 1953–56, but are now recovering (Tapper, 1982; Sumption & Flowerdew, 1985).

SEX RATIO at birth is 1:1 (Müller, 1970; Ternovsky, 1983), but trapping does not usually sample both sexes equally. There is a strong seasonal variation in the proportion of males caught; among 939 stoats collected from December to May inclusive, males comprised between 43% and 53% of the monthly catch, rising through early winter (June, 58% of 95, July, 64% of 95) to between 70% and 77% of the 335 collected from August to November inclusive (King & Moody, 1982:53). This is largely the result of seasonal changes in the activity patterns of the two sexes (see above). In addition, fewer males are caught when the traps are close together (catch comprises <50% males at trap spacings under 800 m) than when they are further apart (>70% at spacings averaging >2 km: King, 1980a). This is mainly because males tend to occupy larger home ranges than females, and so have more opportunity to find traps. Sex ratio is not, however, greatly affected by the huge variations in density of stoats after beech seedfalls; there was no significant difference in the proportion of males collected in Fiordland (40–63%) in seed and non-seed years (King & McMillan, 1982). The real sex ratio of an undisturbed population, when observed by live-trapping or snow-tracking, is likely to be about 1:1, although the individuals most often recaptured or encountered are always males (Vaisfeld, 1972; King & McMillan, 1982; Erlinge, 1983).

AGE DETERMINATION. The ages of living stoats can be estimated only in spring and summer, and even then only two classes (adult and young of the year) can be distinguished. Adult females have visible nipples, large if they have borne young, small if they have not; the nipples of young females are invisible. Over 99% of adult males have enlarged testes in summer, but no young males. These distinctions have been checked later from the teeth, and found reliable (Erlinge, 1983; Grue & King 1984). Body weight is not a reliable criterion after early January, since young females reach 97% of adult weight by February, and young males 80% (King & Moody, 1982:86).

The skulls of young and adult stoats are quite different in appearance. The juvenile form, distinguishable in both sexes until February, has thin chalky bones, a rounded cranium with undeveloped mastoid and occipital condyles, and clearly visible nasal sutures; the sagittal crest is a wide-open V-shape, and the ratio of interorbital to postorbital widths is under 1.15 in males and 1.10 in females. Adults have stronger, smooth and shiny skull bones, a narrower cranium with well-developed condyles and crests, invisible nasal sutures, and postorbital ratios exceeding the above figures.

The baculum of subadult males is thin and delicate and weighs <38 g throughout the winter from March to August inclusive. There is a rapid transition at puberty (starting in July) from the juvenile form to the heavier, more robust adult form with its characteristic proximal knob; and the baculum weight of adults continues to increase slowly for another 4 years (Grue & King, 1984).

If the date of capture is known, and the skull and baculum are available, stoats can be classified into three age classes without histology (King & Moody, 1982:82). (1) Young: young of the year of either sex, collected in November-February inclusive, distinguished on skull characters (note that the females are reproductively mature, whereas the males are not). (2) Subadult: immature males with a baculum weight of <38 g, collected in March-August inclusive. (3) Adult: all stoats other than young in November-February, and males other than subadults in March-August. All stoats collected in September and October are nominal adults.

If histological equipment is available for sectioning canine teeth, the adults can be subdivided into year classes. The layers visible in these sections are laid down annually during winter (Grue & King, 1984).

AGE STRUCTURE. The proportion of each age class represented in a sample can vary enormously, depending on the reproductive success of the years represented. The large cohort of young produced during a post-seedfall year may be identifiable in annual samples for the following 2–3 years (C. M. King, unpubl.). In more stable environments these variations are probably less marked. In New Zealand samples collected by sustained Fenn trapping, most stoats are <2 years old, but a few may be 6–8 years old. One male marked in the Orongorongo Valley in February 1972 was recaptured, still wearing its eartag, in September 1974, when it must have been at least 3 years old (King, 1975a). In Sweden, the proportion of young varied from 31% to 76% and the average life expectancy from independence was estimated as 1.4 years in males and 1.1 years in females (Erlinge, 1983); in Switzerland, the proportion of young was 55–67% and the mean age 14.4 months (Debrot, 1984).

MORTALITY. The mortality rate in beech forest populations is extremely high (>80%) in the first year, then lower (20-30%/year) in adults 1-5 years old (C. M. King, unpubl.). The main controlling factor is probably food. On established traplines, over 5 years (1974–78) the number of adults caught in spring averaged 90% fewer than the number of young caught in the previous summer, regardless of density or food supplies (King, 1981a). In 1979–81, the disappearance rate of marked stoats after the 1979 seedfall was very high; of 107 available at the end of February 1980, only 14 (13%) were recovered over the following year (King & McMillan, 1982). Of these, a much higher proportion of adult than young males was recovered (5 of 11 adults, but 4 of 47 young), whereas about the same proportions of young and adult females were recovered (1 of 7 adults, 4 of 40 young). These figures imply a large difference in the behaviour and/or survival of young compared with established adult males, which is less marked or absent in females. Mortality rates of adults in undisturbed populations in Europe are also high: 0.40–0.78 in males, and 0.54–0.83 in females, according to age and year, in Sweden (Erlinge, 1983) and 0.68 (pooling sexes and years) in Switzerland (Debrot, 1984).

Competition between stoats and other introduced carnivores in New Zealand is possible but so far never proved. Taylor (1984) found an apparent negative correlation between the distributions of stoats and cats on small (<900 ha) islands, and suggested that stoats are able to exclude cats when food is short; King and Moors (1979b) suggest that stoats have a competitive advantage over weasels in New Zealand conditions (see p. 319), but avoid ferrets (p. 310).

PREDATORS, PARASITES AND DISEASES

Stoats are liable to be killed by falcons (*Falco novaeseelandiae*) and wekas (*Gallirallus a. australis*) (Thomson, 1922; Morrison, 1980; King, 1982b), and feral cats (Gibb & Flux, 1973; Fitzgerald & Karl, 1979); but the main predator of stoats in New Zealand, as elsewhere, is man. Many New Zealanders are particularly hostile towards stoats and take every chance to kill any that they see (King, 1982b, 1984a). Many stoats are incidentally killed by traps or poison laid for possums and rabbits. In the 5 years 1944–48, >50 000 stoat skins were sold as a by-product of rabbit trapping, mostly in the South Island (Wodzicki, 1950). The general numbers of stoats are not, however, affected by predation, trapping or accidental deaths.

The only parasite of stoats that has been studied in any detail is *Skrjabingylus nasicola,* a large red nematode which invades the nasal sinuses and causes distortion of the skull bones (King, 1974; King & Moody, 1982:131). Among 1005 stoats over 6 months old, incidence ranged from 0 to 37% (mean 10%); affected individuals carried 1–73 worms each (mean 12.9 in females, 14.2 in males) but were not smaller or lighter than uninfested ones. Subadult males 6–10 months old were already infested as often as adult males, and there was no difference in incidence between the sexes. The geographical distribution of positive cases was not random: incidence was highest in beech forest-grassland habitats with annual rainfall <1600 mm, and apparently nil in Westland and on Mounts Ruapehu and Egmont. The life cycle of the parasite includes two larval stages passed in an obligatory intermediate host, which can be almost any kind of terrestrial mollusc. Stoats do not often eat these directly, but pick up the invasive third stage larvae encysted in a paratenic host. In Europe the paratenic hosts were once thought to be shrews, a theory weakened by the persistence of skrjabingylosis in New Zealand in the absence of shrews (King, 1974); now Weber and Mermod (1983) have demonstrated invasive larvae in wood mice (*Apodemus*) and have produced experimental infestations from them within 24 days. The transmission route of the disease in New Zealand is still unknown.

Among 680 fleas collected from stoats by King and Moody (1982:141), 662 were *Nosopsyllus fasciatus,* the most common flea on rats. The others were 8 *Leptopsylla segnis,* 1 each of *Ceratophyllus gallinae* and *Parapsyllus n. nestoris,* and 8 unidentified. Most fleas were found on stoats which had been caught in leg-hold traps, since fleas soon leave a body after it cools. The louse *Trichodectes*

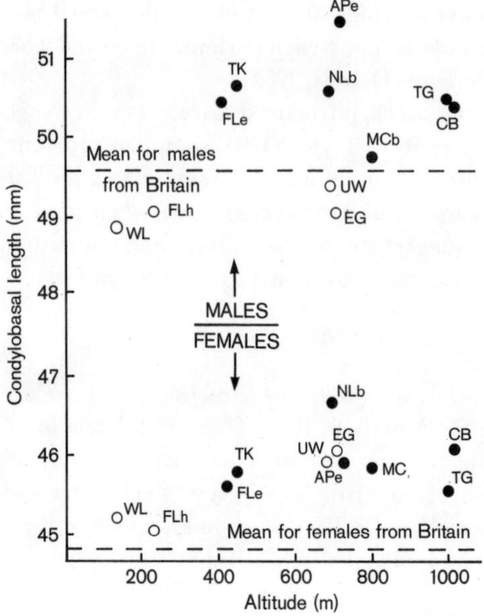

Fig. 59 Variation in size with habitat and altitude of male and female stoats, and a comparison with the mean size of males and females from Britain. Spearman rank correlation coefficients for size and altitude: males r_s = 0.31, NS; females r_s = 0.55, P<0.05. Key: open circles, podocarp and mixed forests; closed circles, beech forests and grasslands. Abbreviations refer to study areas, mostly in national parks. Slightly modified from King & Moody (1982).

ermineae, and the mites *Eulaelaps stabulans* and *Hypoaspis nidicorva* (both normally found on birds) have been recorded on female stoats in the Orongorongo Valley (King, 1975a). Tenquist and Charleston (1981) record the mites *Demodex erminae, Gymnolaelaps annectans, Haemaphysalis longicornis* and *Listrophorus mustelae.* Few other parasites and pathogens are recorded. Stoats are known to be susceptible to distemper and tularaemia (Lavrov, 1944) and bovine tuberculosis (Anon., 1975). Disorders of the gonads, including cysts, were noted by King and Moody (1982:116).

Adaptation to New Zealand Conditions

Since the reproductive cycle of stoats is controlled by day length, regardless of food supplies and population density, the colonizing stock would have switched immediately to Southern Hemisphere seasons. However, the fecundity of female stoats here is the same as in Britain, where the mean ovulation rate is also 10 (range 6–17; Rowlands, 1972), and the mean embryo rate 9 (range 6–13; Deanesly, 1935). This constancy is remarkable considering the great difference in environment and conditions in the two countries. The reason could be that counts of corpora lutea and early embryos represent *potential* productivity, which could be relatively species-specific, rather than *actual* production of young, which is under close environmental control. The full potential is realized only in the areas (Erlinge, 1981) or years (King, 1981a, 1983a) with the highest density of favoured prey.

The frequency distribution of body sizes of mammals in New Zealand is quite different from that in Britain (Table 4); the small mammals of mouse and vole size, which are important prey of stoats throughout the Northern Hemisphere, are relatively scarce. This shortage of small prey might be expected to force changes in the hunting strategy of stoats in New Zealand. For example, stoats might make up for the absence of voles by concentrating more on birds; and the smaller female stoats, which in Europe tend to specialize on small rodents and avoid larger prey (Erlinge, 1979b), might be at a particular disadvantage. Neither prediction has turned out to be correct. Stoats in New Zealand forests do not eat birds more often than on British game estates (King & Moody, 1982:75), although they do eat more insects (found in 7% of guts with prey by Day (1968), cf. 41% in New Zealand). Female stoats here do not avoid larger prey, but eat lagomorphs and rats as often as males. The only other large prey are possums, which are more often eaten by males; but it is not certain whether this is because only the larger males can kill a possum (in areas where possums are made available to stoats as carrion, females can eat them just as well), or because males need more food and are therefore more likely to take advantage of a large carcass. It would be interesting to test the implications of these data for the currently accepted explanation for sexual dimorphism in small mustelids (Erlinge, 1979b; Moors, 1980). Northern Hemisphere studies suggest that dimorphism is a consequence of the different ways in which males and females maximize their individual reproductive success; larger size is an advantage to males in competition for mates, whereas smaller size gives females greater efficiency in hunting small rodents for the young. This explanation may not apply in the totally different conditions in New Zealand, where small rodents are a much less reliable resource for females and young.

Several studies on the diets of mustelids living on fluctuating populations of rodents in the Northern Hemisphere have shown that fewer birds are eaten when rodents are numerous (Dunn, 1977; Tapper, 1979); hence, birds may suffer less than normal from predators during a rodent peak, but more so during the following decline. This hypothesis was applied to New Zealand beech forests by Riney *et al.* (1959), and tested in Fiordland by King (1983a). In the summer after the seedfall, although there were high numbers of mice available, there was no drop in consumption of birds (found in >40% of stoat guts at all densities of mice); hence, the combination of unusually high numbers of stoats and the lack of any buffering effect by mice must have led to additional predation in the *same* summer that the mice were abundant. In the following autumn, when the mice were declining and the analogy with the Northern Hemisphere work predicts that the stoats would switch from mice to birds, the number of stoats had already greatly decreased, and those remaining were definitely not concentrating on birds. The different results in New Zealand beech forests compared with the Northern Hemisphere may perhaps be due to the lack of alternative small prey for female stoats here. Mice are the only representatives of the female stoat's most favoured food; the post-seedfall peaks provoke an intense breeding effort, somewhat like the flowering of a desert, and with results similarly short-lived.

Stoats in New Zealand have become, on average, significantly larger than their British ancestors (Fig. 59). The variation among local populations within New Zealand (Table 47) must have developed from colonizing stock distributed without reference to body size, but now shows consistent and significant differences in relation to sex and habitat (King & Moody, 1982:99). In beech forests, both males and females are significantly larger than in Britain; in podocarp-mixed hardwood forests, females are larger but males are unchanged, or possibly smaller. The exact explanation for this rather dramatic adaptation, established in <100 years, is unknown but probably related to the shortage of small mammal prey (King, 1989). This problem would be expected to be especially acute for females, which have in fact shown the strongest response. However, there is as yet no evidence as to whether the adaptive mechanism involved is genetic or not.

In the Northern Hemisphere, stoats apparently prefer diverse habitats such as farmland, wetlands, scrub and second-growth woodland interspersed with clearings, and avoid unbroken mature forest (Stubbe, 1973; Fitzgerald, 1977). In New Zealand, stoats are practically ubiquitous in all kinds of forest. It would be interesting to know whether this difference is a positive choice, perhaps because New Zealand forest holds more food than the open, hedgeless paddocks that have replaced it, especially for hunters that can climb well; or a negative avoidance of interference competition from cats and/or ferrets, which are generally more common on farmland (pp. 325, 336).

The range of density indices for stoats in New Zealand is comparable with those reported from Europe and North America, commonly from <1 to 7 C/100TN, depending on season (highest in late summer and autumn), year, habitat, prey density, and length of period of calculation (Simms, 1979; Tapper, Green & Rands, 1982; Erlinge, 1983; Debrot & Mermod, 1983). Occasional extraordinary peak numbers of stoats after beech seedfalls may, for very short periods, beat all Northern Hemisphere records: in the Eglinton and Hollyford Valleys during the week 1-4 January 1980, indices of 18.1 and 23.2 new captures per 100 live trap-nights were recorded (King & McMillan, 1982), but the means for the month were 8.2 and 10.7 respectively, and for the 3-month summer season, 5.5 and 6.2 C/100TN (Fig. 58).

Skrjabingylosis in the Northern Hemisphere tends to be more prevalent in high-rainfall areas; the opposite is true in New Zealand (King & Moody, 1982: 134), for unknown reasons.

SIGNIFICANCE TO THE NEW ZEALAND ENVIRONMENT

The introduction of stoats is commonly regarded as one of the worst mistakes ever made by European colonists in New Zealand. The rapid disappearance of native birds since first European contact is well known (8 species and subspecies from 1769 to 1884, counting North and South Island populations separately, and 13 from 1884 to 1984: lists given by King, 1984a), and it is certainly due in part to introduced predators. Many people regard stoats as the most obvious threats to wild birds of all the imported mammals, and blame them both for the historic extinctions and for reducing the density of contemporary populations; others argue that stoats have a compensating value in controlling rabbits and rats. These attitudes have been examined by King (1984a, 1985b).

DAMAGE. Contrary to popular belief, stoats had very little to do with the historic extinctions, since they were never in contact with the great majority of the 153 distinct populations of native birds that are known to have gone from the main and offshore islands since AD 1000. They could have contributed to the loss of only 5 of the 135 that are now extinct and 11 of the 18 that are still threatened. This is because stoats arrived late, after the most vulnerable mainland birds had already been removed by Polynesian and European hunters, rats, cats, dogs, and various drastic habitat changes; and also because stoats have never reached the most important offshore islands on which many endemic birds have survived the disturbances on the mainland. Only in Westland and Fiordland was the ancient fauna still largely intact when stoats arrived, and there they certainly contributed, with ship rats (p. 223), to the final disappearance of the South Island subspecies of the bush wren (*Xenicus l. longipes*), NZ thrush (*Turnagra c. capensis*), laughing owl (*Sceloglaux a. albifacies*), saddleback (*Philesturnus c. carunculatus*), and kokako (*Callaeas c. cinerea*), and perhaps also aided the already advanced decline of the kakapo (*Strigops habroptilus*), takahe (*Notornis mantelli*) and little spotted kiwi (*Apteryx owenii*). The periodic increases in predation by stoats after beechmast years may explain why the South Island kokako disappeared while the North Island kokako (*Callaeas cinerea wilsoni*), living mostly in non-beech forests, has so far survived (Clout & Hay, 1981).

The extent to which predation by stoats now affects the density or distribution of the remaining birds is unknown. Only two out of about 27 populations of threatened species (the takahe and the North Island kokako) live on the mainland. Although stoats may at times take their eggs or chicks, or even attack the adults, there is no evidence as to what effect this has on their chances of survival; the prediction of Lavers and Mills (1978), that takahe are at serious risk after beechmast years, has not been confirmed (J. A. Mills, unpubl.). On the other hand, both takahe and kokako are known to be at substantial risk from reproductive failure, due in part to interference competition with red deer and possums (Mills & Mark, 1977; Leathwick, Hay & Fitzgerald, 1983). All other threatened species live on offshore islands, and those isolated by >1.2 km of open water from the nearest stoats are probably safe; those nearer are at risk of invasion (Taylor & Tilley, 1984). Bell (1983) describes the consequences of the arrival of a stoat on Maud I. in 1982.

The impact of predation by stoats on the non-threatened species is also unknown. At Kowhai Bush, Kaikoura (240 ha) in 1975-77, mustelids (mostly stoats) robbed 51.7% of 149 nests of native and introduced birds. Endemic birds might be expected to be more vulnerable than the introduced species, but the losses were about the same for each (Moors, 1983a,b); however, the native birds included in this study were endemic at a lower level (species or subspecies), and therefore at lesser risk (Spurr, 1979) than most of those now extinct or threatened (many endemic at genus or family level). In Fiordland in the summer after a beech seedfall, there may be a temporary rise in predation on the common bush birds, because the additional numbers of mice permit a huge increase in stoat populations without diverting the stoats' attention from eating birds (King, 1983a). Recent sudden declines in range and numbers of yellowheads (*Mohua ochrocephala*) have been linked to post-seed fall irruptions of stoats (Elliott & O'Donnell, unpubl.). However, reports of attacks on individual birds (Wodzicki, 1950; Morrison, 1980) or of sustained predation on local nesting colonies – e.g., banded dotterels, *Charadrius bicinctus* (Fitzgerald, 1964) and starlings, *Sturnus vulgaris* (Purchas, 1981) are not proof of long-term damage. Predator-free islands commonly support denser populations than the mainland (Brockie & Moeed, 1986), but are different from the mainland in other respects as well.

The impact of stoats on their other prey, mostly introduced mammals, is also unknown. Taylor (1978a) suggested that stoats are able to control populations of Norway rats (p. 205), and King (1983a, 1985a) that they can cut short post-seedfall irruptions of mice, but neither idea has been tested. Kiore disappeared from the South Island at about the time that stoats (and ship rats) arrived there, but this may be coincidence; there are other explanations besides predation (p. 179).

CONTROL. The arguments for and against the control of stoats in protected natural areas are reviewed by King (1984a, 1985b). There is no doubt that stoats eat many birds, but generalized control work is probably unjustified in the great majority of mainland areas for the following reasons: (1) the native bird species most sensitive to predation have already gone, either to extinction or to refuges on stoat-free offshore islands; (2) many other factors influence the population density and dynamics of the remaining birds; (3) many other predators are present in all mainland areas; (4) control is effective only when extremely intensive and sustained, and then only temporarily; (5) effective, temporary control is so expensive that it is likely to soak up all the funds available for any kind of management; (6) other management options are likely to have higher priority.

The only practical control method available at present is the humane Fenn trap. Experiments to test the effectiveness of trapping as a means of reducing populations of stoats, or at least to reduce the potential damage in sensitive areas, show that there is no hope of achieving any permanent, substantial reduction in numbers of stoats over any large area (King & Edgar, 1977; King, 1980a). It is virtually impossible to kill a high enough proportion of the resident population of stoats (>80%) every year, because stoats are naturally short-lived and rapidly replaced, even from 20+ km away (King & Moors, 1979a; King & McMillan, 1982). The most intensive practicable effort (in Fiordland, 1972-78) had no discernible benefit for the local bush birds monitored in 1976–78 (K. Morrison & M. G. Efford, unpubl.). On the other hand, it can sometimes be worth while to include stoat control in an integrated management programme for a small reserve, or as a precautionary measure around an aviary or rearing pen. Fenns set for this purpose should be spaced not more than 200-300 m apart, regularly baited and checked not less than every other day. Leg-hold gin traps cause multiple injuries and should be avoided (King, 1981b).

C. M. K.

Fig. 60 Weasel (H. N. Southern).

28. Weasel

Mustela nivalis vulgaris Erxleben, 1777

The subspecies represented here is *M. n. vulgaris* Erxleben, 1777, native to west and central Europe including Britain. The weasels described by Linnaeus were probably of the northern subspecies (*M. n. nivalis* Linnaeus, 1766), which are extremely small, have the straight colour pattern without gular spots more typical of stoats, and regularly turn white in winter (Stolt, 1979).

DESCRIPTION (Fig. 60)

Distinguishing marks, Table 46 and skull key, p. 24.

The weasel is similar in colour and general appearance to the stoat, but much smaller. The tail is short (little longer than the extended hind legs) and uniformly brown. The white ventral fur adjoins the brown dorsal fur along an irregular boundary, and may include brown spots on the belly and at the corners of the mouth (the latter, called the gular spot, is illustrated in Hartman, 1964). Males are always much larger than females (Table 50).

Males of all ages have enlarged testes throughout spring and early summer; females have 3-4 pairs of nipples, and their ovaries do not contain corpora lutea in winter. The two annual moults follow the same pattern as in the stoat, but winter whitening in weasels is practically unknown in temperate climates. They run, swim and climb as well as stoats, except that their stride is shorter (200-300 mm) and the distances they can cover (on the flat, in water, or up trees) correspondingly less. The skull and teeth are smaller copies of those of the stoat (condylobasal length <42 mm, jaw <22 mm, and with the same dental formula, $I^3/_3\ C^1/_1\ Pm^3/_3\ M^1/_2 = 34$); but the baculum is conspicuously different. It is straighter, shorter (16-20 mm) and stockier, and with a sharply upturned distal hook; and the proximal knob developed by adults is spiny.

Chromosome number 2n = 42 in both European subspecies (Mandahl & Fredga, 1980), but 2n = 38 in Japan (Obara, 1982).

FIELD SIGN

Sign of weasels is very rare and indistinguishable from that of stoats. The five-

Table 50: Body measurements of weasels in New Zealand.

	Males				Females		
	mean	SE	range	n	mean	range	n
Length of head and body (mm)	216.6	1.89	171-239	45	182.8	178-186	4
Length of tail (mm)	56.0	1.19	33-73	43	42.0	38-46	4
Body weight (g)	128.8	3.18	72-185	41	57.5	47-67	4

All ages included. Data from C. M. King (unpubl.).

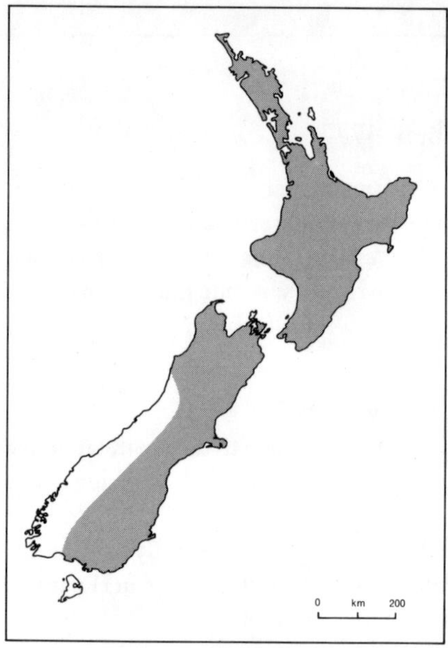

Map 27 Distribution of weasels in New Zealand.

toed footprints (10-13 mm across) and scats (30-60 mm long) are smaller but not reliably so.

MEASUREMENTS

See Table 50.

VARIATION

Weasels are rarely collected in New Zealand, and the samples available are too small to give any information on local variation in size or colour.

HISTORY OF COLONIZATION

Weasels were introduced at the same time as stoats, and for the same reasons (p. 293). The difference was that many more weasels were imported (592 weasels and 224 stoats in 1885 and 1886 alone), probably because in Britain weasels often live at higher density than stoats, and trappers collecting both species indiscriminately would be likely to catch weasels more often. In 1885,

183 weasels arrived; of these 67 were released on a 3200 ha peninsula in Lake Wanaka, "on which they reduced the rabbits, but by no means exterminated them" (Thomson, 1922: 71), plus 28 at Lake Wakatipu and 15 near the Wairau River, Southland; the rest were sold at Wellington, Christchurch and Dunedin. In 1886, one lot of 126 weasels was distributed in the basins of the Wilkin, Makarora, and Waitaki rivers and Lake Ohau; a further 116 were divided between Marlborough and the Wairarapa; plus a private shipment of 167 for Riddiford's station in the Orongorongo Valley (Thomson, 1922). No doubt many other imports went unrecorded.

Early accounts of the spread of weasels are plagued with a semantic problem. The word "weasel" may be used, equally correctly, to refer either specifically to *Mustela nivalis* alone, or to all the small members of the weasel family in general. Guthrie-Smith (1969) observed "the progress of the weasel movement" north to Tutira from the Wairarapa, where "weasels, stoats and ferrets were bred and liberated at Carterton, the first-named thriving on a great scale and quickly overrunning the countryside". He must have known the difference between the three species, so it seems likely that the weasels he described as arriving at Tutira in 1902 (*before* the rabbits!) were *M. nivalis*, rather than small mustelids in general; they went on to reach the Bay of Plenty by 1904. After a very short period of great abundance, the weasels disappeared, and by the time rabbits reached the district, the tide of weasels had passed on.

Müller's report of weasels on the Westland coast in 1889 (Müller, 1890), if it also refers to *nivalis*, is the South Island equivalent of Guthrie-Smith's observation; both suggest a large and rapid but short-lived irruption of weasels soon after the first releases, followed by a massive contraction in both numbers and range during the first half of this century. By the time of Wodzicki's (1950) report, the weasel had become "exceedingly rare". However, the successful control of rabbits in the 1950s seems to have affected weasels less than stoats and ferrets. Marshall's (1963) correspondents reported little change in the distribution of weasels between 1948 and 1961. This is as might be predicted; in Britain, the disappearance of rabbits because of myxomatosis decimated stoats but rather favoured weasels, which have a different feeding strategy (King & Moors, 1979b).

DISTRIBUTION

WORLD. Weasels are sympatric with stoats over most of the Holarctic north of about 40°N, except in Ireland and parts of western USA, which have stoats but not weasels, and central and southern Mediterranean (including Egypt and North Africa), which have weasels but not stoats.

NEW ZEALAND (Map 27). Weasels are patchily distributed over most of the two main islands, except probably in the south and west of the South Island, but they are not known to have reached any of the other islands. Regular Fenn trapping through the middle 1970s produced few weasels (total of 40, compared with >1600 stoats), and these in only 7 of the 14 study areas surveyed by King and Moody (1982). In the most intensively trapped area, Fiordland, none were caught at all. The only two places where they are known to coexist with both ferrets and stoats are the Tasman Valley near Mt. Cook, and in Pureora Forest Park.

HABITAT

Weasels seem to prefer the more disturbed habitats, from suburban gardens to agricultural land, in scrub and cutover or exotic forest, or at the margins between these and open country. The probable reason is that these are more likely to harbour mice than is undisturbed native forest (p. 229).

FOOD

Weasels eat mainly small prey, but very few data are available. In a sample of 26 guts, King and Moody (1982: 70) found that 37% contained mice, 30% birds, 23% geckos, and 13% tree wetas; lesser items included other insects, skinks, and birds' eggs. Weasels can and do kill rabbits, mostly juveniles, but not often (10% of the guts contained lagomorph hair). Some anecdotal accounts are summarized by Wodzicki (1950) and Marshall (1963).

The daily food requirements of British weasels amount to about one-third of their body weight per day (Moors, 1977), i.e., about 40 g/day for New Zealand males (the equivalent of 2-3 mice) and about 20 g/day for females, except during lactation, when a female's food requirements greatly increase. The feeding behaviour of weasels is generally like that of stoats, except that weasels are more closely adapted to exploitation of small rodents (e.g., they are able to follow mice into their burrows and nests, whereas stoats have to wait till they come out) and less capable of tackling larger prey.

SOCIAL ORGANIZATION AND BEHAVIOUR

The social organization of weasels in New Zealand is unknown. In Europe the two sexes live on separate home ranges, much larger in males (5-25 ha) than in females (1-10 ha), and have activity schedules and systems of social hierarchy and communication (mutual avoidance, marking behaviour) much like those of the stoat (Lockie, 1966; King, 1975c; Powell, 1979). The range of sounds made is described by Hartman (1964): a short, sharp chirp (high-level threat), a soft hiss (fear or low-level threat), a wail (distress), and a trill (friendly). Captive weasels are tolerant of each other and of other mustelids while immature, but antagonistic when adult.

Reproductive behaviour was observed in captivity by Hartman (1964). The female, about 6 weeks old when collected in January 1962, was at first kept with two males of about the same age. There was some bickering but no sign of oestrus or mating. She was kept separately after August, and introduced to a male once a week from September. He gave the trilling sounds indicating pleasurable excitement at each meeting, but she refused all contact until 14 October, when they mated for 2-3 hours on each of three consecutive days. On 19 and 29 October, she rejected him; the litter was born on 19 November. They successfully mated again in late December and mid February, each time after "a lively struggle". The next mating, in mid February, was infertile, but by mid September she was again receptive, and the first litter of the spring was born on 19 October. The signs of oestrus are as in stoats.

REPRODUCTION AND DEVELOPMENT

The reproductive organs look like those of stoats, but function quite differently. There is no delay in implantation, so the ovaries contain corpora lutea only when the female is actively pregnant. The fertile season is long, from September to about March (earlier when food is short); when supplied with abundant food (e.g., in captivity), the adult females are physiologically capable of producing up to three litters per season, although they seldom do so in the wild. Litter size averaged 4.5 in four litters recorded by Hartman (1964).

The development of young weasels born in captivity in New Zealand, described by Hartman (1964), is much the same as in Britain (East & Lockie, 1965). The gestation period is 35-37 days, and the young are born blind and totally naked, weighing 1–2 g. By 2 weeks they have acquired a covering of fine white hair, and by 3 weeks, the brown and white colouring of adults. The deciduous teeth erupt at 2-3 weeks, and at 4 weeks they begin to chew on meat, and their eyes open. They can kill efficiently by 8 weeks, and their permanent dentition is complete by 10 weeks. The growth curves of Hartman's captive young rose steeply until about 12 weeks, levelling out at weights far higher than ever recorded in the wild (about 175 g in two males, and 80 g in two females). The family groups break up at about 9-12 weeks, and the young appear in traps from January to May.

The sex ratio at birth is about 1:1. Hartman (1964) recorded 11 males and 7 females in 4 litters born in New Zealand: in larger samples from elsewhere, the average proportion of males at birth is 51% (King, 1975b). The early-born young of both sexes can mature and breed in the season of their birth, at about 3-4 months old (late February to April), but the proportion that do so depends, again, on food supplies. Fitzgerald (1964) collected one first-year female lactating in mid April.

POPULATION DYNAMICS

Weasels share with stoats the general characteristics of opportunistic species (p. 303), but the breeding success of weasels is more strongly influenced by fluctuations in the supply of small rodents, and their average lifespan is shorter (<1 year in Britain: King, 1980b). In Europe the population density of weasels is patchy and very unstable even in the most favourable habitats, and closely correlated with the distribution and numbers of mice and voles (Erlinge, 1974; King, 1980b, 1983b). There is hardly any information on the population biology of wild weasels in New Zealand, except that their numbers are unpredictable. In three areas regularly sampled with Fenn traps (which catch weasels and stoats equally well), weasels appeared only in some years (Table 51). One of these (1976-77 at Craigieburn Forest Park) was a post-seedfall season during which mice and stoats were also very abundant (Fig. 58).

The sex ratio of trapped weasels is always strongly biased towards males; only 7 of 30 collected by Fitzgerald (1964), and 5 of 40 collected by King and Moody (1982:98), were females. This may be because smaller animals tend to occupy smaller home ranges, so widely spaced traps are more likely to catch

Table 51: Variation in numbers of weasels caught.

Year (Nov-Oct)		Mt. Cook NP	Tongariro NP	Craigieburn FP
1972-73	Stoats	13	6	25
	Weasels	0	0	0
1973-74	Stoats	56	20	38
	Weasels	1	0	1
1974-75	Stoats	56	12	61
	Weasels	10	3	1
1975-76	Stoats	62	7	52
	Weasels	2	3	3
1976-77	Stoats	–	–	26
	Weasels	–	–	6
Totals	Stoats	187	45	202
	Weasels	13	6	11

Data from C. M. King (unpubl.).

larger, wider-ranging animals (i.e., males rather than females) living at a given density (King, 1975b).

Age determination in living weasels is practically impossible; by the time the young are independent they are almost the same size as the adults. Skulls can be divided into two classes (young and adult) during summer and autumn in the way described for stoats (p. 305). The shape and weight of the baculum reliably separates young from adult males. Only adults have a well-developed, spiny knob on the proximal end. Adults have distinctly thicker and heavier bacula, but there is no definable minimum weight to distinguish them from juveniles. Eleven bacula collected in December-May, weighing 3-23 mg, were defined by Fitzgerald (1964) as belonging to immature animals; 12 collected in September-May, 32-62 mg, were from adults; but two, of 19 mg and 23 mg, were from pubescent males with enlarged testes. Sections of the canine teeth show growth lines (Jensen, 1978), which are probably annual but have not yet been proven so.

The great majority of weasels caught are in their first year (average 68-83% in samples from five British game estates: King, 1980b), and the mean expectation of life at independence is only a few months even in favourable habitats. The mean annual mortality rate is very high (>0.75 in all ages and both sexes), although weasels in captivity can live for many years. The principal agent of mortality is shortage of food.

PREDATORS, PARASITES AND DISEASES

Weasels are vulnerable to attack by all larger predators, especially cats, stoats, ferrets and harriers (King, 1989), but these do not control the numbers of weasels.

Skrjabingylus nasicola is present in New Zealand weasels (King, 1974), possibly at a higher rate than in stoats. Worms were recovered from one (female) of three weasels from St. Arnaud (Nelson); one (male) of 13 from the Tasman Valley (Mt. Cook); and one (male) of six from near The Chateau (Tongariro) (C. M. King, unpubl.).

Ectoparasites recorded from weasels by Tenquist and Charleston (1981) include *Demodex* sp., *Haemaphysalis longicornis*, *Psorergates mustela*, and *Nosopsyllus fasciatus*.

ADAPTATION TO NEW ZEALAND CONDITIONS

Weasels were brought to New Zealand in great numbers; many more were released than stoats, and certainly more than sufficient to establish a thriving population. At first they seemed to do well (p. 315), but within a few years they were less common than stoats, and now they are among the rarest of the introduced mammals, whereas stoats are among the most successful. An explanation for this difference was suggested by King and Moors (1979b), as follows.

In the Northern Hemisphere, weasels and stoats are both common and widely distributed, even though they both depend on the same prey resources (mostly small mammals and birds). However, there are some crucial differences between them. Weasels are more efficient at exploiting small rodents (especially voles) and can breed rapidly to take advantage of vole peaks, but are vulnerable to local extinction during the following declines; stoats are able to exploit both rodents and larger prey, such as rabbits, but are limited by delayed implantation to producing only one litter a year. Also, stoats dominate weasels when they meet, and may restrict their distribution and inhibit their hunting. In a patchy and variable environment, weasels are always able to find, breed in and disperse from a place where rodents are locally common, and avoid confrontations with stoats, and stoats are able to avoid depending on a single prey resource.

In Europe the two species coexist because their respective advantages over each other are, in the long term, more or less equal; in New Zealand, the absence of voles make the weasel's specialist strategy much less effective than the stoat's more generalist one, so the long-term balance is strongly in favour of stoats. When the two were first liberated, weasels could well have been more common, not only because more weasels were released, but also because they could breed faster to take advantage of whatever small prey (including large native insects and lizards) were then available; but with time the absence of voles, the weasel's key resource, and the stoat's ability to tackle a wider range of prey, proved to be decisive factors favouring stoats. This suggestion has not yet been tested in New Zealand conditions, but the basic idea, as applied in Europe, seems sound; weasels do hunt more effectively in vole tunnels (Pounds, 1981) and do avoid direct meetings with stoats (Erlinge & Sandell, 1988).

The totally different conditions in New Zealand must have induced other changes in weasels, but the samples available at present are too small to detect them. The reproductive cycle is the same as in Britain, but productivity is much less; litters are a little smaller (average 5-6 in British stock; Deanesly,

1944; King, 1980b), and second (summer) litters, the mainspring of weasel population increases during a good vole year, probably less frequent. In Europe, the relationship between weasels and voles (density, nest sites, reproductive rate, etc.) is very close. In poor years, weasels are liable to local extinction, since the females cannot bring up a litter if the density of small rodents is below a certain threshold. In Europe this is about 10-15 voles per hectare (Erlinge, 1974; Tapper, 1979); where there are no voles, mice have to be even more numerous, perhaps because they are more difficult to catch (Delattre, 1984). It is not clear how weasels in New Zealand have survived so long in the complete absence of voles.

SIGNIFICANCE TO THE NEW ZEALAND ENVIRONMENT

Much of what has been said about stoats (p. 310) also applies to weasels, but on a smaller scale. Assessment of the first impact of weasels on native and on other introduced prey is now impossible, because of the constant danger of misidentification in early accounts of "weasels". For example, Thomson (1922:82) stated that "While rats are still very abundant . . . there is no doubt that the spread of weasels throughout the country has vastly diminished their numbers . . .". Thomson may have been using the word "weasel" in its more general sense, although in fact weasels would be good at finding and destroying nestling rats, and thereby could have made some contribution to the historic decline in numbers of rats. But Thomson also quoted, and apparently believed, a report of ". . . a case in which a weasel killed a black swan . . ." (p. 74). At Kaikoura, weasels do contribute to the heavy sum of predation on the smaller bush birds (Moors, 1983a,b); but the "weasel" in the black swan story was probably a stoat.

Control of weasels is unnecessary. They present no known threat to the survival of any native species on the mainland, and are unlikely to reach unaided any offshore island of wildlife value. Their populations are naturally unstable and liable to frequent local extinctions; most of New Zealand offers at best only marginal habitat for them, in which their survival is less certain than is that of the remaining native species with which they coexist.

C. M. K.

29. Ferret

Mustela furo Linnaeus, 1758

Also called fitch, polecat. Contemporary usage is "ferret" for the animal, and "fitch" for its pelt (for "fitch farming", see p. 329).

The ferret is a domesticated form of the polecat, known since at least Roman times (first century AD). Whether it is derived from the European (*Mustela putorius putorius*) or the steppe polecat (*M. eversmanni*) is still unresolved, because traditional taxonomic criteria, e.g., pelage and skull characteristics (Pocock, 1936; Tetley, 1945; Ashton & Thomson, 1955; Rempe, 1970) and measurements (McCann, 1955) may be affected by domestication. Ferrets are fully interfertile with *M. putorius* and have exactly the same karyotype; so some authorities

Fig. 61 Ferret, Lake Tekapo (NZWS).

regard them as a form or subspecies of *M. putorius* (Zima & Král, 1984). For convenience, and to distinguish the domesticated form from its ancestral stock, the trinomial *M. p. furo* is often shortened to *M. furo* (Corbet & Southern, 1977; Corbet & Hill, 1980; Table 5), although that name disguises the important fact that the ferret and European polecat belong to the same biological species.

DESCRIPTION (Fig. 61)

Distinguishing marks, Table 46 and skull key, p. 24.

The ferret has the typical mustelid body plan — a long, narrow body and short legs — but is much larger and relatively more stocky in shape than the stoat. The ears, flattened against the head, and the numerous tactile bristles on the face, are adapted for use in burrows.

The body colour varies from white through light brown to black. The adult coat consists of a short woolly insulating undercoat, creamy white to yellow, covered with dark guard hairs, denser on the shoulders, flanks and legs. The tail is of uniform colour, and its guard hairs give it a bushy appearance, especially when the ferret is excited. The face is creamy or greyish white, with a variable dark "mask", varying in form from a wide dark band across the eyes and nose, through a circle around each eye linked by a V-shaped line over the nose, to merely a dark patch under each eye. These variations can be used to recognize individuals in a population. The belly is paler than the back, but has a dark mid-ventral line. Albinos (creamy yellow or white all over, and with pink eyes) have been recorded in many parts of New Zealand (Fitzgerald, 1964;

Table 52: Body measurements of ferrets in New Zealand.

Location	Sex	n	Head and body length (mm) mean	Head and body length (mm) range	Tail length (mm) mean	Tail length (mm) range	Hind foot length (mm) mean	Hind foot length (mm) range	Weight (g) mean	Weight (g) range	Reference
Pukepuke, Manawatu	M	187	–		–		–		1 114	750–1 500	Lavers (1973b)
	F	118	–		–		–		556	400–800	
Pukepuke, Manawatu	M	31	–		–		–		1 107	730–1 850	Clapperton (1985)
	F	39	–		–		–		627	400–830	
Eketahuna, S. Canterbury, Otago	M	5	418		162		61		–		Wodzicki (1950)
	F	3	358		130		47		–		
Cass River (Sept-Oct only)	M	65	417	361–455	164	142–183	64	57–69	1 312	860–1 800	R. J. Pierce (unpubl.)
	F	38	357	319–380	139	118–163	53	48–59	658	400–1 080	
Otago Pen. (Dec-May only)	M	7	405		–		–		1 103		Robertson (1976)
	F	10	349		–		–		695		
N.Z. general	M	8	408	382–438	162	155–176	62	57–65	–		McCann (1955)
	F	4	355	343–367	129	126–132	48	46–51	–		
N.Z. general	M	16	–		153	134–170	–		1 301	789–1 750	Fitzgerald (1964)
	F	13	–		127	110–145	–		603	403–885	

Robertson, 1976; B. K. Clapperton, unpubl.). Ferrets have only a single annual moult; the old coat is shed and a new one starts to appear in early spring, which is completed by early winter (McCann, 1955).

The skull is long, with a narrow braincase and short upper jaw (less than half the skull length). There are three pairs of small incisors in each jaw. The long upper canines overlap the lower lip, and the lower canines fit into gaps between the incisors and canines of the upper jaw. The premolars and molars are pointed to form shearing blades for cutting meat, especially in the large carnassial teeth (the last premolar in the lower jaw and the first molar in the upper).

Dental formula $I^3/_3$ $C^1/_1$ $Pm^3/_3$ $M^1/_2 = 34$.

There is one pair of anal glands under the tail, which produce a characteristic and strong-smelling musk, stored in sacs that open into the anus. Musk consists of variable combinations of sulphur-containing thietanes and thiolanes (Crump, 1980b; Crump & Moors, 1985) that are sexually and individually distinct (Clapperton, 1985; Clapperton, Minot & Crump, 1988). Ferrets also have odour-producing preputial glands and numerous sebaceous and tubular skin glands (Clapperton, 1985; Clapperton, Fordham & Sparksman, 1987).

Smell and hearing are the main senses; eyesight is poor by day, though better at night (McCann, 1955; Corbet & Southern, 1977). Ferrets walk with the body level to the ground, sniffing continuously for scent, sometimes squatting up on their hind quarters to get elevation. When running, the back is arched and the body sinuous in appearance. Ferrets can swim (Moors & Lavers, 1981), but, unlike stoats, are poor climbers.

Field Sign

Scats are often deposited on conspicuous objects such as rocks. Individual scats can be up to 70 mm long and 10 mm wide, with twisted, tapering ends, usually black when fresh, and obviously containing fur, bone or feathers. Piles of scats form latrines, often at den entrances. Where ferrets are common, they may be seen on roads, illuminated by car headlights, or as road kills; during the day they are occasionally seen near poultry runs and in places with high numbers of rabbits.

Measurements

See Table 52.

Variation

It is possible that the original introductions included genuine wild polecats (*M. p. putorius*) as well as ferrets. There is much variation in coat colour, which Wodzicki (1950) took to mean that there were "at least two wild animals to which the term 'ferret' is frequently applied"; but these variations may be found within one litter (B. K. Clapperton, unpubl.). Most New Zealand ferrets are still docile when trapped, even after generations in the wild, which implies that there is little wild polecat in their ancestry (B. M. Fitzgerald, unpubl.).

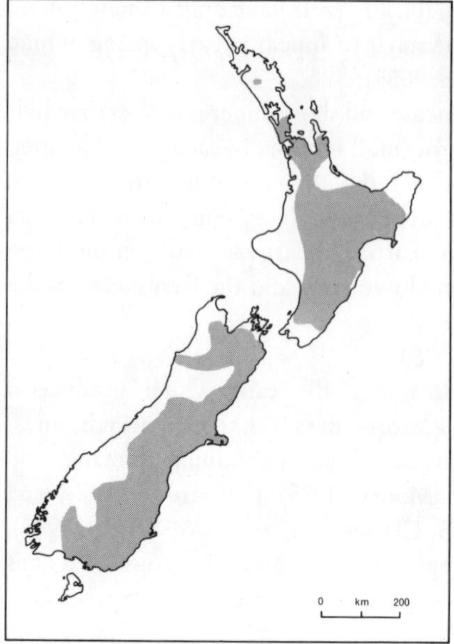

Map 28 Distribution of ferrets in New Zealand.

Male ferrets can weigh twice as much as females (Table 52), but weights vary with season; at Pukepuke Lagoon, resident males weighed on average 1318 g (n = 27) in winter and 886 g (n = 17) in summer, and adult females weighed less than juvenile females after the breeding season (Lavers, 1973a).

HISTORY OF COLONIZATION

By the middle 1870s, rabbits were becoming a serious agricultural pest in New Zealand (p. 158). Farmers demanded that the natural enemies of the rabbit, including ferrets, should be introduced into New Zealand in order to control them. The first releases involved only a few animals (the earliest known, in the valley of the Conway River in 1879, included only five ferrets); but starting in 1882, thousands of ferrets were imported from Australia and Britain, and thousands more were bred locally by the Department of Agriculture until 1897 (Ironside, unpubl. report in Wodzicki, 1950) and by private individuals until 1912. Ferrets were liberated first on the worst-affected pastures in inland Otago (Lake Wanaka, the Makarora and Wilkin valleys, Lake Wakatipu), Southland, Canterbury (Ashburton, Lake Ohau, Waitaki), Marlborough and west Wairarapa. From these areas, they spread throughout the then existing pasturelands, and even into the forests (e.g., west of Manapouri), assisted by enthusiastic government agents and the protection of the law (Melland, 1890; Wodzicki, 1950). By the turn of the century, they were well established in the wild, but had themselves come to be regarded as pests; legal protection was removed in 1903, and the first control campaigns began in the 1930s (see p. 294).

DISTRIBUTION

WORLD. The European polecat, *M. p. putorius,* is found throughout western and central Europe east to the Urals, north to south-eastern Norway, southern

Finland and Sweden, and south to the Black Sea and Mediterranean. In Britain it was widespread up to the mid 19th century, but was then exterminated by gamekeepers over most of the country, surviving only in Wales (Corbet & Southern, 1977). Ferrets may escape into the wild in any country where they are kept, and small independent feral populations are known, for example, on several offshore islands in Britain (Man, Anglesey, Lewis, Arran and Bute; Corbet & Southern, 1977).

NEW ZEALAND (Map 28). New Zealand has the largest known population of fully feral ferrets (Nowak & Paradiso, 1983). In 1948, they were present throughout the two main islands except in Northland (from Kaipara northwards), in eastern Bay of Plenty and Poverty Bay, in large areas of Taranaki, in western parts of Nelson, and the whole of Westland (Wodzicki, 1950). In 1960–62 this distribution was confirmed by Marshall (1963), who pointed out that the areas without ferrets all have high (>1500 mm/year) rainfall and little pasture, i.e., support few rabbits (see p. 147). No ferrets have been reported from Stewart I. or any offshore islands (Fitzgerald *et al.,* 1984), and it is now illegal to introduce ferrets to these places. No distribution surveys have been conducted since 1962.

HABITAT

Ferrets are limited to pastoral habitats, especially pasture, rough grassland and scrubland (Marshall, 1963), and the fringes of nearby forest (King & Moody, 1982). Examples of preferred habitats where ferrets have been studied include the Manawatu dune country (on the west coast of the North Island) where developed farmland lies adjacent to scrubland, sand dunes, swamps and rough grassland (Marshall, 1963; Moors & Lavers, 1981); the dry tussock grasslands of inland Canterbury and Otago (Pierce, 1982); the braided river environments of Canterbury (Robertson *et al.,* 1984); and rough grazing on the Otago Peninsula (Robertson, 1976). On developed grasslands, and in continuous forest, both rabbits and ferrets are generally scarce (Gibb, Ward & Ward, 1978).

FOOD

Ferrets feed mainly on small mammals (lagomorphs, rodents and possums), plus various other items when opportunity offers, including small birds, eggs, lizards, hedgehogs, frogs, eels and various invertebrates (Table 53). Rabbits are important prey; in some areas they are eaten mainly in spring and summer, when young rabbits are abundant (Roser & Lavers, 1976; Gibb, Ward & Ward, 1978), but in the Mackenzie Basin rabbits are important throughout the year (R. J. Pierce, unpubl.).

There is some seasonal variation in diet. Lagomorphs generally predominate in summer, and rodents in autumn and winter (Gibb & Flux, 1973; Roser & Lavers, 1976); birds are eaten throughout the year, though more often in spring and summer; and hedgehogs, eels and frogs are eaten more in autumn (Roser & Lavers, 1976). Females tend to eat more mice, and males more larger prey such as lagomorphs, possums, eels and hedgehogs (Roser & Lavers, 1976).

Rabbits are usually hunted below ground. During 10 years of intensive observation of a rabbit population, Gibb, Ward & Ward (1978) observed

Table 53: Foods of ferrets in New Zealand.

	N.Z.	Kourarau	Pukepuke	Otago Pen.	Cass
Mammals	60.0	77.6	54.7	69.0	90.0
Birds	36.0	7.0	33.5	29.0	10.0
Reptiles, frogs	13.0	0	17.2	9.1	4.0
Fish	10.0	0	13.3	0	2.0
Invertebrates	11.0	16.5	10.3	18.2	10.0
n	55	–	203	55	104
Season	all year	–	all year	Dec–May	all year
Reference	Fitzgerald (1964)	Fitzgerald (1964)	Roser & Lavers (1976)	Robertson (1976)	Pierce (1982)

Data are expressed as % scats or guts containing each item.

rabbits and ferrets only a few metres apart on several occasions, but only twice was a ferret seen chasing a rabbit. Young rabbits are the preferred prey. When the opportunity arises, ferrets gorge themselves and then lie up for several days (Wodzicki, 1950). A radio-equipped ferret remained in one rabbit hole at Pukepuke Lagoon for three successive nights before moving (R. B. Lavers, unpubl.). Ferrets also make and scent-mark caches of surplus food (Wildhaber, 1984; Clapperton, 1985).

SOCIAL ORGANIZATION AND BEHAVIOUR

Ferrets are mainly nocturnal, so are not often directly observed in the wild; studies of captive animals have concentrated on scent communication (Clapperton, 1985) and, overseas, on the behavioural differences between polecats, ferrets and hybrids, and the aggressive behaviour of male ferrets (Poole, 1972, 1973).

The dispersion pattern of ferrets has been observed by live-trapping (Lavers, 1973a,b; Moors & Lavers, 1981). Males and females share the same ground, but exclude other ferrets of the same sex at least from the centres of their home ranges (Moors & Lavers, 1981); this system, found also in stoats and weasels, is described as "intra-sexual territoriality" (Powell, 1979). A ferret may shift its home range during its life, according to the distribution of food and of neighbours. At Pukepuke Lagoon, a swamp/dune/farmland area, home ranges of males averaged 31.3 ha (n = 8, range <3.0–72.2 ha) and that of females 12.4 ha (n = 12, range 1.9–34.4 ha) (Moors & Lavers, 1981); in the Mackenzie Basin, male ranges also reached 70 ha (R. J. Pierce, unpubl.). Fighting among males in the breeding season probably establishes dominance relationships which determine access to females and help to maintain spacing patterns (Moors & Lavers, 1981).

Social relationships are organized by olfactory communication via body odours and scent marks. These provide for individual identification, an olfactory association between a resident ferret and its defended area, and sex recognition and attraction (Clapperton, 1985; Clapperton, Minot & Crump, 1988). Anal gland secretion is laid down by the "anal drag" action that follows defaecation. Other means of scent marking include wiping the preputial glands and the general body fur against prominent objects in the home range (Clapperton, 1985 and in press).

Vocalizations include a range of sounds similar to those of the stoat. When frightened, ferrets make high pitched "barking" noises, which in very submissive animals become a "squeal". There is also a defensive "hiss" and a confident "chatter".

REPRODUCTION AND DEVELOPMENT

The testes of adults are regressed from late February to July, but in August increase rapidly in size and weight (Fitzgerald, 1964). Both sexes are ready for mating in September (males with maximum testis weights, and females with swollen vulvas). Attempts at coitus are not successful until 4–5 weeks after the onset of oestrus, when the vulva is fully swollen. During mating the male grasps the female's neck in his jaws, often drawing blood. The female sometimes squeals and releases anal gland odour if she is unreceptive. Coitus can last for an hour or more. Ovulation is induced, and the female returns to anoestrus about a week after fertilization. Unmated females remain in oestrus, but the follicles degenerate after 2–3 weeks; if they mate after this, they become "pseudopregnant" and show all the signs of a normal pregnancy. There is no delay in implantation, and gestation lasts 41–42 days (McCann, 1955). The first litter of 2–12 (usually 4–8) kittens is born in October or November. Females can return to oestrus, mate again and have a second litter in the same season, but whether and how often they do so in the wild is not known. The breeding season usually ends in late February or March, but wild litters have been found later in the year (R. J. Pierce, unpubl.).

Young ferrets are born with a sparse covering of white hair, replaced by 20 days of age with a denser, darker coat; by 50 days the young have the typical adult appearance. By 30 days, when the eyes open, they are already mobile and are eating meat brought back to the den by the mother. They are weaned by 6–8 weeks, although lactation can last for 10 weeks. They disperse from the natal territory at about 3 months, males first (Moors & Lavers, 1981).

POPULATION DYNAMICS

DENSITY. Moors and Lavers (1981) live-trapped a population of tagged ferrets at Pukepuke Lagoon, Manawatu. The number of ferrets resident in the area (121 ha) was five of each sex in 1970–73, and four males and up to nine females in 1976–77. Males stayed on average for 7.8 and 9.9 months, and females for 14.6 and 20.3 months, respectively. One female, tagged as an adult, stayed for 30 months and was at least 3.5 years old when last caught. There was an annual influx of juveniles in late summer and autumn; more young females than males settled in the area, whereas males continued to wander as transients for at least their first year. Gibb, Ward and Ward (1978) recorded the movements of family groups of ferrets into their enclosed study area in Wairarapa. During the 1961–62 summer, when rabbits in the enclosure, formerly abundant, were rapidly declining, at least six ferrets were seen hunting in the area; after June 1962 none was seen (though two or three were trapped) until February 1963; in October 1963 no fewer than five ferrets were trapped at a time when the rabbit population had been reduced to a total of 15 on 8.5 ha. Ferrets continued to visit the enclosure during this period because there were still more rabbits there than elsewhere, but their scats contained only 23.5% rabbit, compared with 80–95% when rabbits were abundant. Elsewhere, when rabbits are scarce, e.g., after a rabbit-poisoning operation, the local ferrets may be in poor condition and survive only if they can switch to alternative foods (R. J. Pierce, unpubl.), and they are less successful at this than are cats (Gibb, Ward & Ward, 1978). Ferrets are generally much less abundant now than in the era (1880–1950) of massive rabbit numbers throughout the country (Marshall, 1963).

AGE AND MORTALITY. In New Zealand, little is known of age structure or mortality in wild ferret populations, apart from incidental information given by Fitzgerald (1964), Robertson (1976), Gibb, Ward and Ward (1978), and Moors and Lavers (1981). In Britain, the average life expectation for wild male polecats has been estimated at 8.1 months, and there is a high juvenile mortality (Corbet & Southern, 1977). Ferrets in captivity can live for 8–14 years, but probably no more than 4–5 years in the wild.

PREDATORS, PARASITES AND DISEASES

Adult ferrets have few predators in New Zealand except man. Ferrets may be killed on the road, and deliberately or accidentally in traps set for other animals, but these losses are probably relatively unimportant to the population.

Few ectoparasites have been found on feral New Zealand ferrets. Sweatman (1962) identified the mite causing ear canker, *Otodectes cynotis,* and the fur-mite *Listrophorus mustelae.* Tenquist and Charleston (1981) listed, in addition, the hair follicle mite *Demodex* sp., the mite *Sarcoptes scabei,* and the rat flea *Nosopsyllus fasciatus. Demodex* sp. was present in the eyelids of specimens examined by J. R. H. Andrews (unpubl.). The nematode *Skrjabingylus nasicola* was found in skulls of ferrets by King (1974).

ADAPTATION TO NEW ZEALAND CONDITIONS

New Zealand has the largest known population of feral ferrets, so conditions here must suit them well. During the era of greatest abundance of rabbits, when food was virtually unlimited, the breeding and survival rates of wild ferrets were probably much higher than recorded now: since about 1950 the numbers, and presumably also the reproductive and mortality patterns (though not, apparently, the general distribution), of feral ferrets have adjusted to a lower food supply. The earliest releases included many individuals with unnatural coat colours bred up by artificial selection during domestication, but most of these have reverted to the wild-type coloration similar to the polecat.

SIGNIFICANCE TO THE NEW ZEALAND ENVIRONMENT

DAMAGE. The impact of feral ferrets on the New Zealand environment is difficult to assess. They failed to deal with the rabbit problem, since in the country generally the numbers and distribution of rabbits controlled those of the ferrets, not *vice versa* (Marshall, 1963; King, 1984a). However, where conditions favour the predators at the expense of the prey, e.g., in a fenced enclosure, predation by ferrets and cats can be responsible for a huge reduction in numbers of rabbits (Gibb, Ward & Ward, 1978); and in the wild, now that intensive rabbit trapping is a thing of the past, predators are no longer accidentally killed in rabbit traps, and so remain to contribute, perhaps substantially, to keeping the present low number of rabbits in check (Gibb, Ward & Ward, 1969). These ferrets may, at times, survive largely on other prey, but whether they have any effect on native fauna (e.g., birds) is unknown (Fitzgerald *et al.,* 1984).

CONTROL. The first attempt to control ferrets was by a system of bounties, administered by the acclimatization societies and encouraged by the export of dried pelts (Wodzicki, 1950). Feral ferrets would, in theory, be as vulnerable to trapping as the polecat in Britain, which was intensively trapped by gamekeepers, to the point of extinction in lowland England (King & Moors, 1979a); but conditions in New Zealand are too different (the country is far less accessible, and the number of trappers too few) to achieve the same effect on a very large scale. Local problem ferrets attacking the endangered black stilt (*Himantopus novaezelandiae*) in the Mackenzie Basin, and chicks of the royal albatross (*Diomedea epomophora*) at Taiaroa Head, Otago, have been trapped for years, but are always replaced from outside the trapped area (King & Moors, 1979a; Pierce, 1982). The combined effects of predator-exclosure fences and intensive local and seasonal trapping have improved black stilt nesting success (Pierce, 1982), but many nests are still exposed to predation. The use of specific scent lures may increase the effectiveness and efficiency of trapping for ferrets (Clapperton, 1985; Clapperton, Minot & Crump, 1989), but other predators (cats, Norway rats) remain. Elsewhere, the ferret apparently presents little threat to New Zealand wildlife.

FITCH FARMING. The skins of ferrets have, in the past, been valuable in the fur trade. In 1948, at least 4904 superior and first class New Zealand fitch skins were offered for sale at an average price of 16/- each (Wodzicki, 1950). However, skins collected from the wild are less reliable, in quality and supply, than those purpose-bred. Hence, fitch farming is a developing industry, and New Zealand is now one of the main world suppliers of fitch pelts. In 1985, 100 000 fitch skins were exported from New Zealand at an average of NZ$10.00 per skin (top price $45); in 1986, 80 000 at an average of $15.00 (top price $57) (New Zealand Fitch Breeders' Association, unpubl.). In 1986, there were about 127 registered farms (R. Williams, unpubl.); the earliest were stocked with feral animals collected locally, but later ones imported superior stock from Scotland and Finland. Permission to import them depends on the argument that escapees would do no more than temporarily add to the existing population of feral ferrets, and hence that the fitch fur industry has no actually or potentially deleterious environmental effects. This argument is accepted by MAF, despite occasional protests from conservationists. The number of ferrets imported was high at first, declining as the imported strains became self-perpetuating: 2436 in 1982, 1195 in 1983, 1479 in 1984, 0 in 1985 (MAF, unpubl.). However, the importation of other carnivores (e.g., mink) is strictly prohibited: mink look similar to ferrets but have a very bad record of damage to the native biota in countries where they have escaped from fur farms. All ferrets imported into New Zealand are carefully checked to ensure there are no mink among them.

R. B. L. & B. K. C.

Family Felidae

The cat family is a well-defined group of about 36 species, placed by Macdonald (1984) and some others in four genera, although other authorities recognize up to 19 genera (Nowak & Paradiso, 1983). Recent work by Collier and O'Brien (1985), using albumin immunological distances supported by other data, including karyotypes, indicates three major divisions in the family. The ocelot lineage comprises six species of small South American cats, and the domestic cat lineage comprises six species of small cats from Eurasia and/or Africa. The pantherine lineage is the largest and most diverse group, including *Panthera* (lion, tiger, leopard, jaguar, snow leopard), cheetah (*Acinonyx jubatus*), clouded leopard (*Neofelis nebulosa*) and many smaller cats. The family is distributed virtually world-wide except Australasia and most oceanic islands, and is now represented even there by the house cat. Many of the smaller cats are poorly known, and about one-third may be endangered. The felids are the most carnivorous of all land mammals and feed mainly on vertebrate prey, though the smaller cats also feed on invertebrates.

Genus *Felis*

The genus, as recognized by Macdonald (1984), contains about 28 species in 14 subgenera, of which 11 species are found in the New World and 17 in Africa, Europe or Asia. Ewer (1973) restricts the genus to *Felis silvestris* (wild cat), *F. bieti* (Chinese desert cat), *F. chaus* (jungle cat), *F. margarita* (sand cat) and *F. nigripes* (black-footed cat). *F. silvestris* here includes the European wild cat (*F. s. silvestris*) and the African wild cat (*F. s. lybica*), from which the house cat (*F. catus*) was domesticated. The black-footed cat, found in southern Africa, is the smallest felid, weighing 1.5-2.75 kg.

30. House cat

Felis catus Linnaeus, 1758

Also called (1) domestic cat (refers to house cats completely or partially dependent on man), pussy, moggy (English); poti (Maori); (2) feral cat (refers to house cats living independently and breeding in the wild).

DESCRIPTION (Fig. 62)

Distinguishing marks, p. 22 and skull key, p. 24.

Fig. 62 Feral cat, Orongorongo Valley (G. D. Ward).

Table 54: Body measurements of adult[1] feral cats in the New Zealand region.

Locality	Latitude		Weight (kg)		Head and body length[2] (mm)		Tail length (mm)		Reference
			M	F	M	F	M	F	
Mainland									
Orongorongo Valley	41°S	mean	3.67	2.72	514	477	283	247	B. M. Fitzgerald & B. J. Karl (unpubl.)
		range	2.1-5.1	2.2-3.5	501-525	415-505	270-295	180-280	
		n	12	15	3	5	3	5	
Islands									
Raoul	29°S	mean	2.46	1.74	491	447	265	246	B. M. Fitzgerald, B. J. Karl & C. R. Veitch (unpubl.)
		range	1.1-3.9	1.2-2.1	376-565	392-495	165-320	220-270	
		n	59	17	57	17	57	17	
Little Barrier	36°S	mean	2.95	2.23	473	440	283	252	B. M. Fitzgerald, B. J. Karl & C. R. Veitch (unpubl.)
		range	1.6-4.1	1.5-3.8	370-530	380-550	250-320	210-300	
		n	18	35	21	40	21	40	
Herekopare	46°S	mean	3.36	2.75	502	471	247	232	Fitzgerald & Veitch (1985 and unpubl.)
		range	2.8-3.6	2.3-3.3	475-525	445-495	230-260	210-255	
		n	12	8	12	8	12	8	

1. All animals had permanent teeth. Pregnant females excluded.
2. Linear measurements are less reliable than body weight, since observers vary in their method of measuring cats.

The familiar house cat, the only wild felid in New Zealand, is unlikely to be confused with any other animal even at a distance; the nearest shape is that of a possum, which is more hunched, has a longer face, prehensile tail and red "eyeshine".

Many of the colour and pattern variations in the domestic cat are also found in related wild species, but in the house cat they have been retained and spread during domestication (Searle, 1968). In New Zealand, six coat patterns may be distinguished readily: striped tabby (the basic type), blotched tabby, black, grey, ginger and tortoiseshell. White hair is often present on the feet, throat, and belly, and sometimes extends up the legs and sides to meet the shoulders and rump. Entirely white coats, found among domestic cats, are rare in feral cats. Most feral cats are short haired.

Scent glands are present on the chin, at the corners of the mouth, and in the anal region.

Cats have sensitive hearing and respond readily to tones up to 50-60 kHz (the human upper limit is 15-16 kHz). Their excellent night sight is made more sensitive by a well-developed *tapetum lucidum* (origin of their green "eyeshine"), allowing discrimination at levels of illumination only one-sixth of that required by man, but their resolving power (or acuity) is less sensitive. Cats can, with difficulty, be trained to discriminate colours, but they appear not to do this in their daily lives (Ewer, 1973).

Chromosome number 2n = 38. Karyotypes in the family Felidae are morphologically uniform, and the chromosomes of *F. catus* appear to be identical to those of *F. s. silvestris, F. s. lybica* and *F. chaus* (Zima & Král, 1984).

Dental formula $I^3/_3$ $C^1/_1$ $Pm^3/_2$ $M^1/_1$ = 30.

Cats are strongly attracted to the smell of catnip (*Nepeta cataria*), and under its influence behave in a manner similar to that of a cat in oestrus, but both females and males are affected (Palen & Goddard, 1966). However, all cats under 2-3 months old, and about one of three older animals, do not respond (Todd, 1962)

FIELD SIGN

The most characteristic and obvious signs are scats. Domestic cats usually bury their scats, but feral cats often deposit them in conspicuous places on tracks, or on clumps of grass: in the Orongorongo Valley at least 59% were left unburied (Fitzgerald & Karl, 1979). Older unburied scats might be found more easily than buried ones, but the proportion found buried did not change when only fresh scats were considered (B. M. Fitzgerald & B. J. Karl, unpubl.). A scat usually consists of about 3-6 round to elongate segments (illustrated by Veitch, 1985), contains matted fur, feathers and bones, and is dark in colour, not white and chalky like a dog dropping.

Food remains are not specifically diagnostic as field sign of cats. Small birds are eaten entire, except for some body, wing and tail feathers; with larger birds, many body feathers, the tail, wing tips, feet and bill are left (Veitch, 1985). Rodents and young rabbits are usually completely eaten, except sometimes the tail and pieces of skin. Remains of older rabbits often include the

Table 55: Geographical variation in proportions of coat colours and patterns in populations of cats in New Zealand.

Locality	n	Percentage of						Reference
		Tabby		Black	Grey	Ginger	Tortoiseshell	
		Striped	Blotched					
Mainland								
Orongorongo Valley	42	41	26	24	2	2	5	B. M. Fitzgerald & B. J. Karl (unpubl.)
Dunedin area	171	28[1]		45	7	11	2	Marples (1967)
Islands								
Raoul (1972-74)	100	61	–	39	–	–	–	B. M. Fitzgerald, B. J. Karl & C. R. Veitch (unpubl.)
Little Barrier (1978-80)	110	100	–	–	–	–	–	B. M. Fitzgerald, B. J. Karl & C. R. Veitch (unpubl.)
Herekopare (1970)	33	–	–	100	–	–	–	Fitzgerald & Veitch (1985)

1. About one-third of these were striped tabbies.

stomach and the skin, turned inside out over the rabbit's head (Gibb, Ward & Ward, 1978).

MEASUREMENTS

See Table 54.

VARIATION

Body weights, recorded from cats trapped or killed during eradication programmes, are available mainly from island populations. Females average about 70% to 80% of the weight of males, and cats of both sexes from islands to the north of New Zealand tend to weigh less than those from southern localities. This is part of a general geographic trend for cats to be smaller near the equator and larger at high latitudes (B. M. Fitzgerald, B. J. Karl & C. R. Veitch, unpubl.). There is no evidence that feral cats grow much larger than domestic ones; of several hundred feral cats weighed (Table 54), few exceeded 5 kg.

The proportions of different coat colours in a local population vary, but tabbies and blacks usually predominate (Table 55). On the mainland all six colours are represented in various ratios, but on islands often only one or two, presumably those of the founders. Incomplete data are available from other localities besides those listed in Table 55. In farmland in Hawke's Bay 60% are striped tabby, 30% black and 10% other colour patterns (N. P. Langham, unpubl.). In the Wairarapa and in the Mackenzie Basin striped tabbies predominate (Gibb, Ward & Ward, 1978; R. J. Pierce, unpubl.). On Mangere I. the cats were described as tortoiseshell by Thomson (1922); on Auckland I. both black cats and tabby cats are recorded (Taylor, 1975b); on Campbell I. at least some of the cats are black (P. R. Wilson, unpubl.).

HISTORY OF COLONIZATION

Cats were brought to New Zealand in the ships of the early European explorers, from 1769 onwards. All ships of the period were infested with rats and carried cats to control them. The cats must have been kept in large numbers at times, because James Cook left 20 cats at Tahiti in 1774, plus others at Ulietea and Huaheine (Beaglehole, 1961:412). On the *Resolution* a midshipman's cat brought him the rats it caught; he gave the cat the forepart and he ate the back part, cleaned, roasted and peppered (Sparrman, 1953). When the *Resolution* was moored at Astronomers Point, Dusky Sound in autumn 1773, a cat on board "regularly took a walk in the woods every morning and made great havoc among the little birds, that were not aware of such an insidious enemy" (Forster, 1777, 1:128).

Despite this early introduction, cats apparently did not become feral until at least 50 years later. Richard Cruise, who spent 10 months in northern New Zealand in 1820, noted that a cat he was given "must have come from the shipping at the Bay of Islands or from the *Coromandel*. There are no native cats in New Zealand" (Cruise, 1823). Lesson, at the Bay of Islands in 1824, discussed the native and introduced mammals but did not mention cats (Sharp, 1971); nor did Augustus Earle, who was in Northland in 1827-28, though he had already written of the effects of feral cats on birds on Tristan da Cunha (Earle, 1832).

Feral cats had probably become established in the North Island by the 1830s. Polack (1838) suggested that they had been introduced "within the last twenty-five years" but added that they were valued by the Maori as food and for their skin, and any cat belonging to a European was likely to be stolen. By 1840 feral cats were recorded in the Hutt Valley and in the Thames area, and were thought responsible for the "destruction of many indigenous species" (Dieffenbach, 1843). Ensign Best observed that at Taupiri, in the Waikato, in 1842, cats were "made great pets but kept tied up like dogs" (Taylor, 1966). Angas (1847), visiting New Zealand in 1844, noted that domestic animals, including the cat, "are frequently to be met with amongst the natives; especially those of the Waikato and towards the northern districts".

In Canterbury feral cats were present by 1850, when Charles Torlesse recorded killing two wild cats (Maling, 1958) and by 1860 they were "very numerous" (Butler, 1863). Later, when rabbits became a major problem (p. 158), farmers fetched cats from the cities to release on rabbit-infested farmland (Bull, 1969).

Cats were introduced to many offshore and outlying islands by sealers and whalers, settlers, farmers, muttonbirders, and lighthouse keepers, but have since died out on some islands and been eradicated from others (Table 56).

DISTRIBUTION

WORLD. The house cat was domesticated in the eastern Mediterranean at least 3000 years ago, from the North African form of the wild cat (*Felis silvestris lybica*) (Robinson, 1982). It is distinguishable from the contemporary European wild cat (*Felis silvestris silvestris*) by colour pattern, skull characteristics (Schauenberg, 1969) and length of the gut (Schauenberg, 1977). Persian and Siamese cats were probably domesticated independently from Asian forms of *F. silvestris* (Kratochvil & Kratochvil, 1976).

House cats have been taken to most parts of the world, as pets and for rodent control, and are now present in most human settlements from the equator to high latitudes. From these settlements, feral populations have frequently become established (especially in places with few native carnivores), and on islands these have often persisted after the human settlers departed. General information on the distribution of cats on islands is given in Croxall, Evans and Schreiber (1984) and Lever (1985); for the central Pacific see King (1973); for Fiji see Gibbons (1984) and for the subantarctic islands see Johnstone (1985) and Clark and Dingwall (1985).

NEW ZEALAND (Map 29). Feral cats are widely distributed throughout all three main islands (Wodzicki, 1950; Collins & Charleston, 1979a). Some are fully feral; others still visit human settlements for food and shelter. Also, they are, or have been, present on at least 25 islands, ranging from some as small as 50 ha to large, biologically important reserves such as Raoul, Auckland and Campbell Is. (Table 56). Cats coexist with stoats only on islands of 750 ha or more. Taylor (1984) suggests that this is because competition during periods of food shortage may favour stoats, but this idea has not been tested.

HABITAT

Feral cats live in most terrestrial habitats in New Zealand, including sand dunes, pasture, tussock, scrub, exotic plantations and native forest, and from sea level to 3000 m (Wodzicki, 1950; Gibb & Flux, 1973; Collins & Charleston, 1979a). Taylor (1984) considers that cats in the North and South Islands are restricted to areas with rabbits or close to human habitation by competition with stoats: again, this idea remains to be tested.

FOOD

On continental lands with diverse mammal faunas, cats feed mainly on rodents and rabbits; birds are a small component of their diet and reptiles are important prey only at low latitudes. However, cats that have been introduced on to islands are usually limited in their choice of mammalian prey to a few introduced species (mainly rabbits, rats and mice) and they feed to a greater extent on birds (Fitzgerald, 1988). Results of studies of the diet of cats in the New Zealand region are summarized in Table 57.

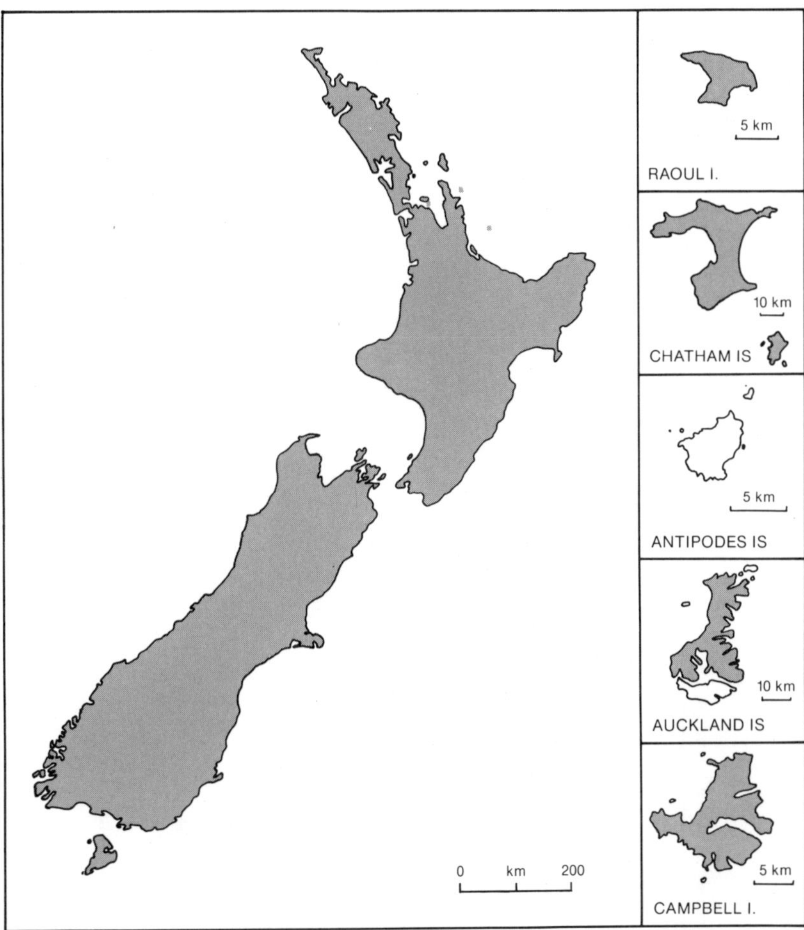

Map 29 Distribution of feral house cats on the main islands of New Zealand. For distribution on offshore islands, see Table 56.

On the mainland of New Zealand cats prey chiefly on small mammals, especially rabbits (mostly young) and rodents, but also hares and possums in some habitats, and occasionally carrion. Adult possums, eaten by cats in the Orongorongo Valley mainly in winter, had probably died from other causes, but young possums of the age when they leave the mother's back were eaten in spring and were probably killed by the cats (Fitzgerald & Karl, 1979).

Birds were present in <15 % of samples in three of the four mainland studies, but 59% in the fourth. Species eaten by cats on farmland were California quail (*Lophortyx californica*) and small passerines including starling (*Sturnus vulgaris*), yellowhammer (*Emberiza citrinella*) and silvereye (*Zosterops lateralis*) (Gibb, Ward & Ward, 1969). In the Orongorongo Valley most of the birds eaten were ground-feeding introduced passerines, but also included silvereyes, fantails

Table 56: Distribution and history of cats on the offshore and outlying islands of New Zealand.

Island	Area (ha)	Introduced by	Dates cats established	Fate of population	Reference
Kermadecs (29°S)					
Raoul	2 938	Settlers	1836–*c.*1870	Still present	Morton (1957)
Northern North Island (36–37°S)					
Little Barrier	2 817	Settlers	1867–1880	Eradicated 1980	Hutton (1868), Reischek (1886, 1887), Veitch (1985)
Great Barrier	28 510	?Settlers	?	Still present	R. H. Taylor (unpubl.)
Tiritiri Matangi	210	Farmers/ lighthouse keepers	Present early 1960s	Eradicated by 1970s	Moller (1977)
Waiheke	9 459	?Settlers	?	Still present	R. H. Taylor (unpubl.)
Kawau	2 257	Settlers	?	Still present	R. H. Taylor (unpubl.)
Rakitu (Arid)	*c.* 345	?	?	Not now present	Bellingham *et al.* (1982)
Rangitoto	2 333	?	?	Still present	R. H. Taylor (unpubl.)
Motutapu	1 509	?	?	Still present	R. H. Taylor (unpubl.)
Ponui (Chamberlains)	1 851	Farmers	Since 1850s	Still present	Bellingham (1979)
Motuihe	179	Farmers	?Last century	Eradicated 1978/79	Veitch (1985)
Cuvier	181	Lighthouse keeper	After 1889	Eradicated 1964	Blackburn (1967), Merton (1972, 1978)
Great Mercury	1 718	Settlers	?	Still present	B. M. Fitzgerald (unpubl.)
Mayor	1 277	Settlers	Before 1926	Still present	Sladden (1926), Veitch (1985)
Whale	140	?	Before 1925	Died out by 1956	Imber (1975)

Central (40°S)					
Kapiti	1 970	Settlers	Before 1905	Eradicated by 1934	Cowan (1907), Veitch (1985)
Nelson-Marlborough (40-41°S)					
Forsyth	775	Settlers	?	Still present	R. H. Taylor (unpubl.)
Arapawa	7 785	?	?	Still present	R. H. Taylor (unpubl.)
Stephens	150	Lighthouse keeper	1892 or soon after	Eradicated 1925	Veitch (1985)
D'Urville	16 782	Settlers	?	Still present	R. H. Taylor (unpubl.)
Southern South Island (45-47°S)					
Anchor	1 525	Sealers	1792	?	Begg & Begg (1966)
Herekopare	28	Muttonbirders	1924-1926	Eradicated 1970	Fitzgerald & Veitch (1985)
Ruapuke	1 525	Settlers	?	Still present	R. H. Taylor (unpubl.)
Putauhina	141	Muttonbirders	?	?Died out or killed by muttonbirders	Veitch (1985)
Outlying islands					
Main Chatham (44°S)	90 650	Settlers	Before 1840	Still present	Dieffenbach (1841)
Pitt (44°S)	6 203	?	Before 1868	Still present	Travers (1868)
Mangere (44°S)	113	Farmers	Shortly before 1893	Died out in 1950s	Forbes (1893), Veitch (1985)
Main Auckland (50°S)	45 975	Sealers	1806-1840	Still present	Ross (1847)
Masked (Auckland group) (50°S)	<5	?	?	Still present	Taylor (1975b)
Main Campbell (52°S)	11 331	Settlers/farmers	After 1916	Still present	Dilks (1979)

After Veitch (1985), with additions from R. H. Taylor (unpubl.).

Table 57: Foods of feral cats in New Zealand.

	Kourarau[1]	Te Wharau[2]	Orongorongo Valley[3]	Mackenzie Basin[4]	Raoul I.[5]	Little Barrier I.[6]	Herekopare I.[7]	Stewart I.[8]	Campbell I.[9]
Mammals									
Possum	0	0	19	0	—	—	—	5	—
Hedgehog	Tr	0	0	0	—	—	—	—	—
Rabbit	77	68	22	81	—	—	—	—	—
Hare	0	10	0	0	—	—	—	—	—
Kiore	—	—	—	—	} 86	39	—	}	—
Norway rat	} Tr	} 3	—	2	}	—	—	} 93	95
Ship rat	}	}	50	0	—	—	—	}	—
Mouse	2	29	43	4	—	—	—	—	—
Stoat	Tr	0	Tr	0	—	—	—	—	—
Sheep	Tr	3	—	Tr	—	—	—	—	0
Goat	0	0	0	0	2	—	—	—	—
Cattle	0	0	—	Tr	—	—	—	—	—
Pig	0	0	0	0	2	—	—	—	—
Birds	4	59	12	14	35	71	93	44	35
Reptiles									
Gecko	Tr	4	} Tr	0	—	0	—	0	—
Skink	0	0	}	3	—	0	0	24	—
Fish	0	0	3	0	0	0	0	0	0
Insects	9	10	57	8	58	17	47	36	60
Spiders	0	0	4	0	0	2	0	0	0
Freshwater crayfish	0	0	3	0	—	—	—	—	—
n =	many	68	677	133	57	94	30	229	20
Gut or scat analysis	scat	scat	scat	gut	gut	scat	gut	scat	scat

In some studies the results were given by season — these have been averaged here to provide a simple, overall figure for the diet. Results are given as percentage occurrence in all studies except Gibb, Ward & Ward (1978), where results are percentage volumes. Tr = trace amount or % occurrence <1%; — = species not present at that locality.

References: 1. Gibb, Ward and Ward (1978); 2. Gibb, Ward and Ward (1969); 3. Fitzgerald and Karl (1979); 4. R. J. Pierce (unpubl.); 5. B. M. Fitzgerald, B. J. Karl and C. R. Veitch (unpubl.); 6. Marshall (1961); 7. Fitzgerald and Veitch (1985); 8. Karl and Best (1982); 9. Dilks (1979).

(*Rhipidura fuliginosa*) and a New Zealand pigeon (*Hemiphaga novaeseelandiae*) (Fitzgerald & Karl, 1979). Cats prey on adults, chicks and eggs of pied stilts (*Himantopus himantopus*) and black stilts (*H. novaezealandiae*) (Pierce, 1986).

Lizards were generally a minor item in the four mainland studies, although in some habitats they may be more important; the gut of one cat from Central Otago contained at least 17 skinks (A. H. Whitaker, unpubl.), and another from the Mackenzie Basin, 23 skinks (R. J. Pierce, unpubl.). Frogs have not been recorded in the diet of cats in New Zealand, and fish only in the Orongorongo Valley, where cats also ate freshwater crayfish (Fitzgerald & Karl, 1979).

Invertebrates (mostly insects but also some spiders) are infrequently eaten by cats in grassland habitats except during outbreaks of pests such as porina (*Oxycanus* sp.) caterpillars and moths, grassgrub (larvae of *Costelytra* sp., Coleoptera) (Gibb, Ward & Ward, 1978) and grasshoppers (R. J. Pierce, unpubl.). In forest of the Orongorongo Valley, many more insects were eaten, including dragonflies, four species of weta, cicadas and carabid and scarab beetles. However, insects, though present in many of the scats, contribute little to the diet (Fitzgerald & Karl, 1979).

On islands in the New Zealand region the choice of mammal prey is even more limited than on the mainland, and rats are the main mammal prey (Table 57). All three species of *Rattus* are eaten, but predation on Norway rats probably falls mainly on young animals if the observations of Childs (1986) in urban Baltimore, USA, apply in New Zealand. Most of the Norway rats that he trapped weighed >200 g, whereas most of those that cats caught weighed <100 g. On Raoul I., where Norway rats and kiore are present, >90% of 236 rats from cat guts and scats in 1972–80 were kiore (B. M. Fitzgerald, B. J. Karl & C. R. Veitch, unpubl.).

Birds usually form a larger part of the diet of cats on islands than at most mainland localities (Table 57) or on the continents (Fitzgerald, 1988). Seabirds are often a large proportion of the birds eaten; species recorded in these studies of diet range in size from diving petrels (*Pelecanoides urinatrix*) up to blue penguins (*Eudyptula minor*) (Marshall, 1961; Karl & Best, 1982; Fitzgerald & Veitch, 1985). Parakeets (*Cyanoramphus* spp.) were important prey on Little Barrier and Stewart Is. (Marshall, 1961; Karl & Best, 1982); yellow-crowned parakeets (*C. auriceps*) disappeared from Herekopare I. soon after cats became established (Fitzgerald & Veitch, 1985); and kakapo (*Strigops habroptilus*) were eaten by cats on Stewart I. (Karl & Best, 1982).

Although most island populations of cats have at least one species of mammal prey available, cats can sometimes live without them. Cats lived on Herekopare I., which had no other mammals, for over 40 years, feeding mainly on seabirds, together with some land birds and invertebrates (Fitzgerald & Veitch, 1985).

Lizards were frequent prey only on Stewart I., where 24% of scats contained remains of skinks (Karl & Best, 1982). Many invertebrates were recorded in cat guts and scats, and although some might have been derived from other prey that was eaten, insects such as wetas, beetles or dragonflies were sometimes present in large numbers without other prey (Dilks, 1979; Fitzgerald & Veitch, 1985; B. M. Fitzgerald, B. J. Karl & C. R. Veitch, unpubl.). Invertebrates were most frequent in the diet of cats on the southern islands; on Raoul I., blowflies, presumably attracted to the trap bait, produced the high frequency of insect remains in cat guts (insects other than blowflies were present in only 19% of the guts).

Changes in the diet of cats with time usually reflect seasonal or longer-term changes in the availability of prey. In the Orongorongo Valley changes in the frequency of mice and rats in cat scats matched changes in the numbers of mice and rats trapped (Fitzgerald & Karl, 1979; Fitzgerald, 1988).

Differences in diet between males and females, or young and adults, are poorly known. Adult females are only 70–80% of the weight of adult males, and some studies show that they take more small prey, and less large prey, than adult males do; other studies show no difference in the diet of males and females (Fitzgerald, 1988). Young cats may eat more insects than adults do. On Herekopare I. juveniles contained significantly more insects (wetas) than did adults, and at Pureora Forest Park, juveniles in a small sample of cats contained predominantly insects.

Vertebrate prey are about 70% water by weight; so healthy cats do not need to drink fresh water, except females suckling young, or when prey is temporarily scarce and cats are living largely on their reserves (Prentiss, Wolf & Eddy, 1959; B. M. Fitzgerald, unpubl.).

SOCIAL ORGANIZATION AND BEHAVIOUR

Cats are often considered to be solitary animals, but their social organization is complex. Domestic cats maintain a small core area of their home range as exclusive property, but tolerate other cats in the rest of it (Leyhausen, 1979). Groups of cats living with a household, or in farm buildings, usually comprise several related adult females, their young of both sexes, and an adult male whose range includes other groups of females. Adult females in a group may help care for each other's young. Young females usually remain in the group and breed there, or leave and establish a new colony elsewhere; they rarely if ever join another group. Young males leave or are driven from the group at 1–3 years old, as they reach sexual maturity (Macdonald & Apps, 1978; Liberg, 1980, 1984).

Less is known about the social organization of feral cats, but it seems to differ little from that of domestic cats. Among feral cats in the Orongorongo Valley, adult females (of unknown relationship) had overlapping home ranges, and adult males probably did too. One young male remained within the home range of an adult female until he weighed >3 kg, and then began to move outside her range. The home ranges observed were rather linear, along the floor of the valley; those of females were about 4 km long, and those of males >6 km. Females with kittens used less than half the length of the ranges they used when they did not have kittens (Fitzgerald & Karl, 1986).

Cats in the Orongorongo Valley did not have permanent dens, except females with young kittens; after the kittens were about 6 weeks old they moved frequently to new places (Fitzgerald & Karl, 1986). However, in habitats where good den sites are scarce, cats may have permanent dens, and the remains of prey accumulate there.

House cats mark their home ranges by spraying urine and leaving scats unburied in conspicuous places, by claw-sharpening on particular trees, and by cheek-rubbing (Verbenne & Leyhausen, 1976; Leyhausen, 1979), but these behaviours are little studied in the wild.

REPRODUCTION AND DEVELOPMENT

The gestation period of domestic cats averages 65 days, and most kittens are born between spring and autumn. In good conditions domestic cats may have two, or even three, litters of kittens per year, with 1–10 kittens per litter (modal litter size 4) (Robinson & Cox, 1970; Jemmett & Evans, 1977).

In the Orongorongo Valley, kittens weighing less than 1 kg were found between October and April (B. M. Fitzgerald & B. J. Karl, unpubl.). On Herekopare I. in June 1970, the smallest juvenile weighed 1.2 kg, and the adult females had not bred recently (Fitzgerald & Veitch, 1985); on Raoul I. eight of 21 adult female cats trapped in spring were pregnant, but four caught in late autumn and winter were not. However, on Little Barrier I., of 36 adult females trapped in May–August of 1978 and 1979, two in June were pregnant, and 11 were lactating in May–August. Two to four embryos per litter were recorded on those islands, and two to four young kittens per litter in the Orongorongo Valley (B. M. Fitzgerald, B. J. Karl & C. R. Veitch, unpubl.). In farmland in Hawke's Bay, litters consist of 2–5 kittens, of which only 1–2 usually survive (N. P. Langham, unpubl.).

In the Orongorongo Valley, kittens remained in the den where they were born until they weighed about 500 g (5–6 weeks old) and were then moved by their mother to a series of temporary dens, staying a few days in each before moving on again (Fitzgerald & Karl, 1986). Their growth rates were similar to those of domestic kittens until they reached 500 g, but thereafter the feral kittens grew more slowly (Dickinson & Scott, 1956; B. M. Fitzgerald & B. J. Karl, unpubl.).

POPULATION DYNAMICS

Population density is primarily determined by food supply, though deliberate trapping, shooting or poisoning may reduce local cat numbers substantially, and island populations have been eradicated in this way (Veitch, 1985). Many cats are caught in gin traps set for possums, and formerly were caught in rabbit traps (Wodzicki, 1950; Reid, 1986). In places intensively trapped for possums, this mortality may substantially reduce the number of cats.

Urban cat populations, obtaining much of their food from man, can reach high densities: 1/ha in Brno, Czechoslovakia (Obrtel & Holisova, 1980), 2/ha in the Portsmouth dockyards, England (Dards, 1983), and 23/ha at a Japanese fishing village (Izawa, Doi & Ono, 1982). Domestic farm cats and feral cats live

Table 58: Parasites of cats in New Zealand.

	Common name	Preferred host	Reference
Siphonaptera			
Ctenocephalides felis felis	Cat flea	Cat	Guzman (1984), Smit (1979)
C. canis	Dog flea	Dog	Guzman (1984), Smit (1979)
Nosopsyllus fasciatus	European rat flea	Rats	Smit (1979), Tenquist & Charleston (1981)
Acarina			
Cheyletiella blakei		Rabbit	Guzman (1982), McKenna (n.d.)
C. parasitivorax	Rabbit mite	Rabbit	Tenquist & Charleston (1981)
Demodex cati	Cat follicle mite	Cat	Tenquist & Charleston (1981)
Haemaphysalis longicornis	Tick	Mammals esp. cattle	Tenquist & Charleston (1981)
Notoedres cati	Cat mange mite	Cat	Sweatman (1962), Tenquist & Charleston (1981)
Otodectes cynotis	Ear mange mite	Cat, dog	Tenquist & Charleston (1981)
Pentastomida			
Linguatula serrata	Tongue worm	Dog	Tenquist & Charleston (1981)
Phthiraptera			
Felicola subrostratus	Biting louse	Cat	Tenquist & Charleston (1981)
Nematoda			
Aelurostrongylus abstrusus			McKenna (n.d.)
Ancylostoma tubaeforme			J. M. Clark (unpubl.)
A. brasiliense			W. C. Clark (1980)
Capillaria aerophila		Cat?	McKenna (n.d.)
C. erinacei (syn. *C. putorii)*		Cat	Collins (1973)
Cylicospirura advena		Cat	W. C. Clark (1981)
Ollulanus tricuspis			Collins (1973), Guy (1984)
Toxascaris leonina			McKenna (n.d.)
Toxocara cati		Cat	Collins (1973)
Trichostrongylus axei & sp.			Guy (1984)
Trichinella spiralis			McKenna (n.d.)
Uncinaria (stenocephala?)			Anon. (1980)
Cestoda			
Dipylidium caninum		Cat	Collins (1973), Anon. (1980)
Taenia taeniaeformis		Cat	Collins (1973)
Protozoa			
Haemobartonella felis			McKenna (n.d.)
Isospora felis		Cat	McKenna & Charleston (1980a)
I. rivolta		Cat	McKenna & Charleston (1980a)
I. lacazei			McKenna & Charleston (1980a)
Toxoplasma gondii		Cat	McKenna & Charleston (1980a)
Sarcocystis spp.		Cat	McKenna & Charleston (1980a)
Eimeria perforans			McKenna & Charleston (1980a)
Klossia sp.			McKenna & Charleston (1980a)
Besnoitia sp.			McKenna & Charleston (1980b)

at much lower densities: about $3/km^2$ in Sweden (Liberg, 1980), $6.3/km^2$ in Illinois (Warner, 1985) and $0.7–2.4/km^2$ in south-eastern Australia (Jones & Coman, 1982b). The highest densities of feral cats are found on islands supporting large populations of breeding seabirds: 1.2/ha on Herekopare I. (Fitzgerald & Veitch, 1985), and $2–7/km^2$ on Macquarie I. (Jones, 1977).

Complaints are often made about people abandoning unwanted cats in the countryside, but most populations of feral cats probably maintain themselves. Kittens born to feral cats in the Orongorongo Valley and in farmland in Hawke's Bay survived to become members of the adult population (Fitzgerald & Karl, 1979; N. P. Langham, unpubl.), and on islands such as Raoul, Little Barrier and Auckland Is. populations of feral cats have survived unaided for >100 years (Table 56).

Feral cats do not live as long as domestic cats, which commonly reach 15 years. Many kittens die of starvation or disease. The oldest known cats in the Orongorongo Valley were two 6-year-old females (B. M. Fitzgerald & B. J. Karl, unpubl.); on Kerguelen and Marion Is. cats lived up to 9 years (Pascal, 1980; Van Aarde, 1983). Free-ranging domestic cats in Illinois farmland rarely survive beyond 3–5 years (Warner, 1985).

PREDATORS, PARASITES AND DISEASES

Cats are occasionally killed by dogs, but their main predator is man. They are shot or trapped in an effort to protect native birds and game birds. Cat "flu" is present in mainland populations of cats, and it kills many kittens.

Parasites of cats in New Zealand (Table 58) have been identified chiefly from domestic cats (e.g., Collins, 1973; McKenna & Charleston, 1980a; Guzman, 1984), but most of those listed are probably also present in feral populations. *Toxocara cati* (arrowheaded worm) has been found in feral cats from Pureora, Orongorongo Valley and Pelorus, and *Toxocara* eggs were found in 65% of 65 cat scats from Hawke's Bay (W. A. G. Charleston, unpubl.). Cats on Raoul I. in 1972 had *T. cati, Ancylostoma tubaeforme* (hookworm) and unidentified worms, probably *Capillaria* spp., but no cestodes (J. M. Clark, unpubl.). *Toxocara cati* was present in large numbers in many of the cats, and cysts containing larval nematodes were found in the stomach wall: *T. cati* has been identified from similar cysts in Australia by Coman (1972). *Ancylostoma brasiliense* was identified from a cat on Raoul I. by W. C. Clark (1980). On Little Barrier I., ascarids (probably *T. cati*), hookworms (probably *Uncinaria stenocephala*) and tapeworms (probably *Dipylidium caninum*) have been recorded (Anon., 1980). *Toxocara cati* has also been recorded from a brown kiwi (*Apteryx australis*) on Little Barrier (Clark & McKenzie, 1982).

Sweatman and Williams (1962) examined 347 feral and farm cats from throughout New Zealand for four species of taeniid tapeworms important in dogs and, as an intermediate stage, in livestock and man. They found none of these tapeworms, although they were able to infect laboratory cats with one of them (*Taenia ovis*).

ADAPTATION TO NEW ZEALAND CONDITIONS

None described. So far as can be seen from scant data, reproduction in feral cats in New Zealand and in south-eastern Australia (Jones & Coman, 1982a) is similar.

SIGNIFICANCE TO THE NEW ZEALAND ENVIRONMENT

Cats have both deleterious and beneficial effects on the native fauna, but on the mainland these are difficult to separate from the effects of other predators (rats, mustelids, and man), diseases, and destruction and degradation of natural habitat (King, 1984a). Their effects are sometimes clearer on islands, where fewer factors are involved, and can be examined in several ways. For example, there are the historical records of species disappearing from islands after cats were introduced; and the species present on cat-inhabited islands can be compared with those on similar, adjacent, cat-free islands. On islands where cats have been eradicated, the subsequent changes in the populations of surviving species, and the successful reintroduction of species that disappeared, are indirect evidence of the effects of cats. The question of why some species, and not others, are affected by cats can be addressed by comparing the habits of species that have been affected with those that have not.

DAMAGE. The Stephens Island wren (*Xenicus lyalli*) was both discovered and exterminated by the lighthouse keeper's cat in 1894 (Buller, 1905), and saddlebacks (*Philesturnus carunculatus*) and other species disappeared from Little Barrier, Cuvier, and Stephens Is. (Turbott, 1961; Veitch, 1985) soon after cats were introduced. On Raoul I., three species of petrel and several other species of birds disappeared after cats, but before Norway rats, were introduced (Merton, 1970; Veitch, 1985). Cats introduced onto Herekopare I., Foveaux Strait, in about 1925 eradicated at least six species of land birds – yellow-crowned parakeet, robin (*Petroica australis*), fernbird (*Bowdleria punctata*), brown creeper (*Finschia novaeseelandiae*), snipe (*Coenocorypha aucklandica*), and banded rail (*Rallus philippensis*) – and large breeding populations of diving petrels and broad-billed prions (*Pachyptila vittata*) (Fitzgerald & Veitch, 1985).

The numbers of some species, especially seabirds, have declined gradually. Cats were introduced onto Little Barrier I. between 1867 and 1880; the grey-faced petrel (*Pterodroma macroptera*) was still surviving heavy predation in the 1940s, but ceased breeding there after 1963 (Turbott, 1961; Veitch, 1980). Cook's petrel (*P. cookii*) and the black petrel (*Procellaria parkinsoni*) were suffering heavy predation during the 1970s, and their populations were still declining, a century after cats were introduced (Imber, 1975, 1987; Veitch, 1980). On Raoul I. the breeding population of sooty terns (*Sterna fuscata*) suffers heavy predation by cats and Norway rats and may be threatened (Taylor, 1979). On Herekopare I., broad-billed prions and diving petrels, still plentiful in the early 1940s, almost 20 years after cats were introduced, had disappeared by 1970 (Fitzgerald & Veitch, 1985). Predation by cats on land birds now rare on the mainland, such as the kakapo (Karl & Best, 1982) and black stilt (Pierce, 1986), has probably contributed to the past and continuing decline in their numbers.

Both lesser short-tailed bats and long-tailed bats are killed by cats, and cats may be a significant cause of mortality, especially at more accessible roosts (Daniel & Williams, 1984; pp. 122, 129).

Cats may also have had a deleterious effect on lizard populations. Lizards were reported to have disappeared rapidly from the mainland after cats were introduced, and some were thought likely to become extinct (Taylor, 1848, 1870; Stack, 1874; Thomson, 1922).

The best contrast between the depleted bird populations on islands occupied by cats (and rats: p. 184) and the dense populations on adjacent cat-free islands is Raoul I. and the Meyer Islets in the Kermadec group (Merton, 1970). Birds that have been eliminated or survive in only remnant populations on Raoul I. but are plentiful on Meyer Islets include the wedge-tailed shearwater (*Puffinus pacificus*), Kermadec petrel (*Pterodroma neglecta*), black-winged petrel (*P. nigripennis*), and Kermadec parakeet (*Cyanoramphus novaezelandiae cyanurus*).

Some birds that persisted at reduced density on islands in the presence of cats have become more common after cats were eradicated; on Little Barrier I. the numbers of stitchbirds (*Notiomystis cincta*) multiplied (perhaps sixfold), and robins and parakeets also increased (Veitch, 1982; Angehr, 1984). Other species that disappeared from islands after cats were introduced have been successfully re-established after the cats were eradicated. These include saddlebacks and red-crowned parakeets (*Cyanoramphus novaezelandiae*) on Cuvier, and saddlebacks on Little Barrier I. (Veitch, 1985).

In mainland forests today birds form a small part of the diet of cats, and these are mostly ground-feeding introduced species such as blackbird (*Turdus merula*), thrush (*T. philomelos*), chaffinch (*Fringilla coelebs*) and hedgesparrow (*Prunella modularis*) that, despite predation, are still among the more common forest birds (Fitzgerald & Karl, 1979). Another indication of the greater vulnerability of ground-feeding birds is that, on Herekopare I., five of the six species that disappeared from the island after cats were introduced are ground-feeders, but of six native species that persisted only two are ground-feeders (Fitzgerald & Veitch, 1985).

Species that have had a long evolutionary history in New Zealand without mammalian predators are more vulnerable than related species that have arrived recently, having evolved with predators. The endemic black stilt is much more susceptible to predation than the widespread pied stilt, because it chooses less protected nest sites, takes longer to fledge, nests solitarily and has ineffective distraction displays (Pierce, 1986).

Cats will also scavenge whenever opportunity arises (Collins & Charleston, 1979a), and thereby may play some part in the transmission of diseases of livestock.

BENEFITS. Cats also have a beneficial role in controlling rabbits and rodents. In a 10-year study of an enclosed population of rabbits at Kourarau (p. 154), cats (and ferrets) reduced the rabbits to very low numbers, and took all young rabbits before, or as soon as, they appeared above ground. The numbers of rabbits began to increase again only when the predators were removed from the enclosure (Gibb, Ward & Ward, 1978). Observations of a sparse population of wild rabbits on 1200 ha of hill pasture and scrub in the Wairarapa showed that, after three years without rabbit control, fewer rabbits were left, they tended to be older animals, and rabbits were the main food of cats. It was suggested that "predation, especially by feral cats, held the rabbits at low density without control by man" (Gibb, Ward & Ward, 1969). Similarly, in north Canterbury, low rabbit populations have remained relatively stable for six years without control by man (Williams, 1983).

Cats may be important in limiting the numbers of rats in some mainland forests. In the Orongorongo Valley the numbers of cats and ship rats from 1971 to 1986 showed a significant negative correlation; as cats became very scarce, probably as a result of intensified possum trapping, the rat population increased substantially, from about 2 C/100TN to more than 8 C/100TN (Fitzgerald, in press; B. M. Fitzgerald & B. J. Karl, unpubl.).

CONTROL. On the New Zealand mainland the interactions between cats and their prey are complex, and reducing the numbers of cats may produce unexpected, deleterious results that minimize or cancel out the expected advantages. Removing cats may allow rat or stoat populations to increase, and their effects may be at least as undesirable as those of the cats (see pp. 223, 311). Control of mainland cats may need to be integrated with control of stoats and rodents to be effective.

However, in the simpler systems on islands the effects of eradicating cats may be clearer. Cats have been eradicated from several islands, including two of >2000 ha (Table 56). The techniques used have generally included trapping, shooting and poisoning; persistence and frequent follow-up checks are required (Veitch, 1985). The most successful bait for traps is fresh fish; cats are attracted by the smell of catnip but it is of limited use as a lure because not all cats respond (p. 333).

In eradication programmes by staff of NZWS, the total population of about 12 cats was removed from Cuvier I. in 1960–64 (Blackburn, 1967; Veitch, 1985), and 33 cats from Herekopare in winter 1970 (Fitzgerald & Veitch, 1985). Eradicating cats from Little Barrier I. was a much larger task, involving 128 people in almost 4000 days of work over 4 years to kill more than 100 cats (Veitch, 1983, 1985). The value of these islands for conservation is now vastly greater; some birds that were present before cats arrived but since disappeared have been reintroduced, and other endangered species can be released there.

B. M. F.

Order Perissodactyla

The odd-toed ungulates or perissodactyls are a small order of mammals today, comprising only three families, six genera and 16 species of horses, asses, zebras, tapirs and rhinoceroses. They appear to run a poor second to the even-toed ungulates, or artiodactyls, in numbers of species, geographical distribution and variety of form. But during the early Tertiary period, 54–25 million years ago, perissodactyls were the dominant large mammals; they were much more abundant, widespread and diverse than now, and they then included an additional three families that are now entirely extinct. The best known members of the order, the true equids, evolved in North America and spread from there into Eurasia during the Pleistocene.

Family Equidae

Genus *Equus*

The single genus in the Family Equidae has seven species. The most familiar of them is the domestic horse, *E. caballus*, spread throughout the world by man. The original native wild horses of North America became extinct about 10 000 years ago, and the last of the true wild horses of Europe and Asia, the Przewalski horse, *E. przewalskii*, sole surviving representative of the ancestors of the domesticated horse, is confined to Mongolia and to zoos (Bokonyi, 1974). The domestic horse is the only member of the family (in fact, of the order) living wild in New Zealand. Also included in the family are the two species of African and Asiatic asses and their domesticated descendants, and the three species of zebras. In 1870 Sir George Grey imported two zebras (*E. zebra*) from South Africa and released them on Kawau I., but they presumably died out.

31. Feral horse

Equus caballus Linnaeus, 1758

Also called wild horse (incorrect: see below), brumby.

Description (Fig. 63)

Distinguishing marks, see below and skull key, p. 24.

Even at a distance and in poor light, horses are not easily mistaken for other large mammals, because their large size, long-muzzled heads and their flowing manes and tails are unique. The sexes are alike, except that stallions tend to be larger and have stronger and more arched necks.

Horses have a smooth, short coat of black, white or brown hair, often with admixtures of white, although horses also whiten as they age (Sandars, 1957) and some old greys may appear completely white. New Zealand feral horses may be black, chestnut, brown, bay, roan, dun or grey, often with small patches of white on the face and legs; large areas of white on the body may give a pied effect (Morrow, 1975). Most feral horses in Aupouri Forest, Northland, are bays or browns, and those of the Kaimanawa ranges are bay, chestnut or brown with black points. White markings on the legs and face are common in both populations (Ricketts, 1979; Hermans *et al.,* 1982; Hermans 1984). The winter coat grows in autumn and, except in old animals, is usually shed and replaced by a shorter summer coat in spring. However, in the Kaimanawas, the smaller immature horses (probably yearlings) keep their winter coats longer, sometimes until February (Hermans *et al.,* 1982). Longer, coarser "horse-hair", often of a different colour from the shorter body hair, grows in tufts at the fetlocks, along the dorsal side of the neck from between the ears (the forelock) to the withers (the mane), and on the whole of the tail.

The legs of the horse are unique. The feet have evolved to carry the whole weight of the body on the tip of the third toe, the only surviving digit on each foot. The bones of the third toe are enlarged, central and symmetrical. There is one main bone, the "cannon bone", between the knee (wrist) or hock (ankle) joint and the toe. The only evidence of second and fourth toes is the two small "splint" bones which lie behind the upper three-fourths of the cannon bone, and two callosities on each leg. One set, called "ergots", are under the fetlocks; the other, called "chestnuts", are on the insides of the legs, above the knees on the forelegs and below the hocks on the hind legs. These horny growths are regarded as the vestiges of the pads of the now-vanished digits (Getty, 1975). Feral horses usually move about by walking, but break into a trot, canter or gallop for short distances if disturbed.

The horse has a large skull, a long diastema, three pairs of upper and three pairs of lower incisor teeth, and (usually) six high-crowned cheek teeth (molars and premolars) on each side of each jaw. Canine teeth are present in males but usually absent in females. The first of the four premolar teeth is usually absent, but may be found in vestigial form on each side of the upper jaw (Getty, 1975).

Dental formula $I^3/_3$ $C^1/_1$ $Pm^{3-4}/_3$ $M^3/_3$ = 42.

FIELD SIGN

The horse is the only large free-ranging mammal in New Zealand with a non-cloven hoof, so its footprints are readily distinguished. Horses without shoes leave almost circular tracks, slightly greater in length than in width (more so in the hind foot), and with a deep indentation at the rear marking the position of the triangular "frog" or soft pad of the foot. Individual tracks vary in size according to age and breeding, but prints are commonly about 80-130 mm across.

Horse droppings are distinctive and familiar. Those of feral horses are often found in large piles; stallions habitually cover the droppings of adult mares and other males with their own (Feist & McCullough, 1976).

Horse hair can be found on certain tree trunks and stumps where feral horses regularly rub their heads, necks and rumps.

Fig. 63 Feral horses, family group near Napier–Taupo road, 1966 (J. H. Johns).

MEASUREMENTS

Mature horses of the Volcanic Plateau were considered to be "the size of large ponies" by Wodzicki (1950), i.e., about 13-14 hands at the withers (1 hand = 102 mm, comprising four points of 25.5 mm: 14.2 hands (1479 mm) is the technical limit to "pony" size). The skeletons of two Kaimanawa horses, examined at Massey University, were assessed as belonging to mature stallions of 12.3 and 13 hands (Ratumaitavuki & Kirk, 1982). Adult feral horses of the Aupouri Forest, Northland, range in size from 14 to 16 hands, averaging about 15 hands (Hermans, 1984).

VARIATION

New Zealand feral horses vary in size and conformation depending on their ancestral breeds. The main physical differences between breeds are in the bulk of bone and muscle, the width of the body, the length of the legs and length and amount of hair. The Kaimanawa horses are primarily of British stock, derived from escaped cavalry horses of the 19th century crossed with locally bred domestic horses containing Exmoor and Welsh pony blood (Anon., 1984b; R. A. L. Batley, unpubl.). About 1945, horses belonging to the Mounted Rifles were liberated near Waiouru. These and farm horses, escaped or released from sheep stations, are thought to have increased the general body size of Kaimanawa feral horses in recent years (Anon., 1982). Some of the Aupouri Forest herd are clearly of thoroughbred descent and have relatively fine bone structure and clean legs. Others are stockier, often with the heavily feathered fetlocks of the draught breeds (Hermans, 1984).

HISTORY OF COLONIZATION

The first horses to reach New Zealand were disembarked at the Bay of Islands by Samuel Marsden in 1814 (Nicholas, 1817). Other early arrivals included 23 landed at Akaroa in 1836 bound for Otago (Holford, 1927), and the first military horses, which landed at Wellington in 1840 (Morrow, 1975). By the

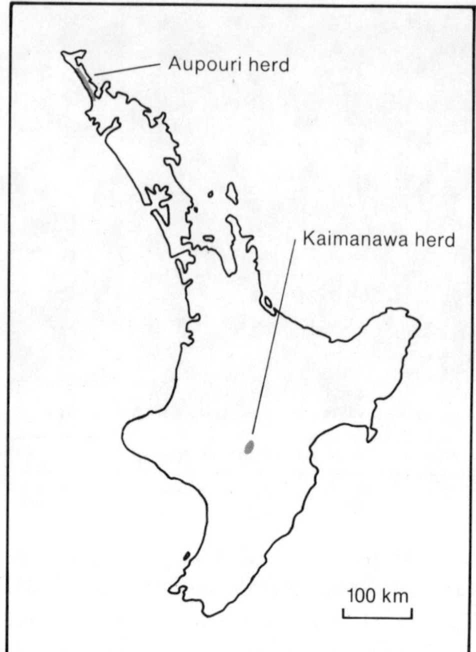

Map 30 Distribution of feral horses in New Zealand in 1985.

1850s horses were the main form of overland transport throughout New Zealand, used by Maori, military forces, travellers, explorers and landowners (Anon., 1984b). By 1861 there were about 500 horses on the Chatham Is., chiefly mares brought from Sydney (Richards, 1952). In all parts of the country escapes and strays gave rise to "wild" herds; some dispersed to remote areas and became truly feral, while others were permitted to range freely and breed in a semiferal state, and were rounded up only when needed (this practice still continues in parts of the North Island).

Feral horses were reported from the Kaimanawa ranges by 1876 (Anon., 1984b) and were plentiful on the Chatham Is. in the 1880s (Richards, 1952); they were probably present in Northland much earlier. They reached their greatest numbers in the late 1800s and early 1900s. Around 1920, feral horses were common in much of the North Island (Thomson, 1922; Morrow, 1975); e.g., on the Volcanic Plateau, in parts of the King Country, Hawke's Bay, Poverty Bay/East Coast, Waikato, Bay of Plenty and the west coast of the Northland peninsula. At that time feral horses in the South Island were largely confined to Marlborough, and there were a few on the Chatham Is. By the early 1950s, they had gone from the Chathams, but were still present in the Marlborough back country (Anon., 1951), and their range in the North Island remained much the same (Wodzicki, 1950; R. H. Taylor, unpubl.). Since then, land development, wild animal control programmes and commercial exploitation have caused their range and numbers to shrink rapidly, and they have now almost disappeared outside the two remaining refuges in the North Island.

DISTRIBUTION

WORLD. The domesticated horse is thought to be derived from the extinct wild tarpan, *E. ferus*, first tamed in central Asia, on the Eurasian steppes and in eastern Europe sometime between 5000 and 3000 BC. Tarpans disappeared from Europe by the early 19th century or before. Some authorities place the Przewalski horse (*E. przewalskii*) as an eastern subspecies of the tarpan and attempts have been made to "reconstruct" the tarpan by selective breeding from primitive domesticated breeds and Przewalski horses (Bokonyi, 1974, 1984; Goodall, 1977; Clutton-Brock, 1981). Over the centuries many national varieties of domesticated horses have developed and been taken to all continents and to a very large number of oceanic islands (Clabby, 1976). Escaped or unwanted horses have established feral populations in many parts of the world including Asia, Europe, North and South America and Australasia (Lever, 1985). The "wild horses" of New Zealand and elsewhere are only feral domesticated horses.

NEW ZEALAND (Map 30). The two remaining North Island populations occupy distinct ranges some 550 km apart (Johns & MacGibbon, 1986). One inhabits an area of 24 000 to 29 000 ha in the Kaimanawa ranges, bounded by the Rangitikei River in the east, the Otamateanui Stream in the north, the Moawhango River in the west and the sheep stations along the Taihape-Napier Road to the south (Anon., 1984b; R. A. L. Batley, unpubl.). The other, in the Aupouri Forest, Northland, occupies about 10 000 ha of coastal dune land adjacent to Ninety Mile Beach, from Sweetwater in the south to the vicinity of The Bluff in the north (Hermans, 1984).

HABITAT

Feral horses prefer open grassland or a mosaic of grassland and open forest or scrub. Native forests and steep mountainous country seem to form natural barriers to their spread (Wodzicki, 1950; R. H. Taylor, unpubl.). In the Kaimanawa ranges, feral horses inhabit plateaux and valleys covered with tall red tussock (*Chionochloa rubra*) and smaller silver tussock (*Poa laevis*) and hard tussock (*Festuca novae-zelandiae*), interspersed with introduced and native grassland and scattered clumps of mountain beech (*Nothofagus solandri*) forest, well watered with spring-fed streams. Some of the horses also graze an adjacent subalpine area up to 1650 m asl, where the main vegetation is *Hebe* spp., mosses, prostrate shrubs and low tussocks. Winters are harsh (snowfalls lie for several days at a time) and the annual rainfall is about 1780 mm (Aitken *et al.*, 1979). The Aupouri Forest herd lives in coastal sand dune country that is being afforested with *Pinus radiata*, first planted in 1963. The horses prefer the open grazing provided by clear flats, too wet for pines, and the oldest pine stands opened up by thinning. Fewer horses are found in dense younger pines or where a thick cover of lupins has been established. Water (swamps, ponds and springs) is freely available to the horses over most of the forest. The climate is subtropical with warm, humid summers and mild winters, and the annual rainfall is about 1200 mm (Hermans, 1984).

FOOD

Horses prefer short tender herbs and grasses. The Kaimanawa herd feeds mainly on tussock and smaller native and introduced grasses; they have not been seen to browse on shrubs or trees (Aitken *et al.*, 1979). In the Aupouri Forest, grasses, clovers, sedges and annual herbs are the preferred foods, but horses have also been recorded grazing on raupo (*Typha orientalis*) and pampas grass (*Cortaderia* spp.). Young Norfolk Island pines (*Araucaria heterophylla*) are also extensively browsed (Hermans, 1984).

Feral horses in the Kaimanawa herd spend approximately half of their time grazing. Periods of feeding (throughout the day and night) are interspersed with resting (standing up or lying down), moving about and other behaviour. A 72-hour continuous watch on a family group of six horses in summer 1982 revealed the following average time budget: feeding 56%, standing resting 19%, lying resting 8%, locomotion 7%, alert standing 6%, mutual grooming 0.6%, self grooming 0.5%, agonistic encounters 0.4%; elimination and sexual behaviour were the least frequent (Hermans *et al.*, 1982). During the middle of the day feral horses often seek shade under trees. In light wind or rain they continue grazing with their backs to the wind, but take shelter from prolonged rain and strong winds (Aitken *et al.*, 1979). Horses can sleep while standing (Clabby, 1976). They require water in large quantities, but need to drink only once or twice a day (Schafer, 1975); circumstantial evidence suggests that feral horses at Aupouri prefer to visit water at night (Hermans, 1984), and very few of the Kaimanawa horses were observed drinking during daylight (Aitken *et al.*, 1979). Feral ponies on Assateague I. (USA) do most of their drinking in the first hour after dark (Keiper & Keenan, 1980).

SOCIAL ORGANIZATION AND BEHAVIOUR

Feral horses have a well-developed social structure (Waring, 1983) and are most commonly found in family groups comprising a stallion, one or more mares and their foals, and immature colts and fillies. There are also less stable "bachelor" groups usually consisting entirely of mature males, and occasional lone individuals, usually stallions. In 1982, the Kaimanawa population included at least 43 discrete groups ranging in size from 1 to 12 individuals. Family groups had an average size of 5.5 horses (Hermans *et al.*, 1982). At both Aupouri and Kaimanawa, the groups can have widely overlapping home ranges, of around 1000-3000 ha at Aupouri Forest (Hermans, 1984), but spacing between groups is usually maintained by ritual encounters between stallions. A stallion drives his mares in the presence of other stallions, but otherwise does not control the daily movement and activities of the group. Although there may be a dominance ranking among the mares, the group does not seem to have any regular leader, either when grazing or when responding to disturbance. When alarmed the stallion invariably places himself between the potential threat and his group, and leaves the scene last. Alarm responses are shown only by mature horses. Foals and subadults always follow their dams (Aitken *et al.*, 1979; Hermans *et al.*, 1982; Hermans, 1984).

Feral horses communicate by visual signals and sounds, and also probably by scent. When alarmed they stand erect, face the disturbance and snort. They squeal when closely challenging other horses. A high pitched "whinny" is used to call a foal or to locate other members of the group. At close range a deep, guttural "nicker" given with ears cocked forward is used as a greeting to others of the group. Dominance and subordination are shown by various postures, including rearing, striking with the foreleg, laying back the ears, swishing the tail, threats to kick or bite and "teeth clapping" (Feist & McCullough, 1976; Waring, 1983).

REPRODUCTION AND DEVELOPMENT

Stallions are fecund throughout the year (Schafer, 1975). Although colts usually reach sexual maturity at about 3 years of age, they do not mate until old and strong enough to compete for mares, probably after they are 5 years old. Oestrus may first appear in fillies in their second summer but few conceive at this age. Most Aupouri Forest mares are thought to breed first at 3 years old (Hermans, 1984). Mares come into season in spring or about 9 days after foaling. The oestrus cycle averages 3 weeks (but is highly variable), and the gestation period about 11 months. At Aupouri foals are born between August and late December, and over 70% of adult mares foal each year (Hermans, 1984).

Normally a single foal is born, most often during the night (Feist & McCullough, 1975). At birth it is fully covered with a fine, short, woolly coat, different in colour from its later coat, which first appears at about 5 months. The foal stands and suckles within an hour or two of birth; can follow its mother within 4-5 hours; and is usually weaned at about 12 months of age (Sandars, 1957; Feist & McCullough, 1976). In the Kaimanawa herd, immature animals up to 3 years old have been seen suckling a mare which had no younger offspring. Foals tend to pattern their daily activity on that of their mothers and copy behaviour such as grazing, resting, seeking shelter and defecating (Aitken *et al.*, 1979). The first milk teeth to be replaced are the first incisors and first premolars, at 2½ years of age; generally, the permanent dentition is completed with the eruption of the third incisor at about 4½ years (Clabby, 1976).

POPULATION DYNAMICS

The adult sex ratio is about 1:1 in both the Kaimanawa and Aupouri herds. The age composition of a sample of 48 Kaimanawa horses observed in 1979 was 31 adults over 4 years old, 12 immatures between 1 and 4 years old, and 5 foals (Aitken *et al.*, 1979). Only 33% of the mares had foals at foot. In 1982, the age composition of 198 Kaimanawa horses was 141 adults, 36 immatures and 21 foals; only 27% of mares had foals at foot (Hermans *et al.*, 1982). Both sets of figures indicate low rates of birth and recruitment in the Kaimanawa herd. By contrast, at Aupouri Forest during the 1983-84 season, 77% of the mares foaled and 46% had a yearling in attendance (Hermans, 1984). The apparent higher reproductive and survival rates of the Aupouri population could be related to the milder climate of Northland and/or human disturbance of the Kaimanawa horses, legally permitted until 1981.

Both the Kaimanawa and Aupouri feral horses are now protected, and have increased significantly in numbers over the last few years. A detailed survey of the Kaimanawa population was carried out, using five helicopters, as part of a defence exercise in November 1984. It revealed 550 feral horses (NZWS, unpubl.), and an average population density of about 2 per 100 ha: this represents at least a doubling of numbers since 1979. Hermans (1984) estimated that the Aupouri population contained 129 horses in 1984, with an average population density of 1.25 per 100 ha: it was then increasing at about 15-20% per year.

Predators, Parasites and Diseases

Feral horses have no predators in New Zealand except man. In the past, feral horses have been shot as vermin and rounded up or trapped for pet food or for breaking in as saddle horses. This toll has ceased in the Kaimanawa ranges since the horses were given legal protection in 1981, but some form of population management may eventually have to be introduced there. In 1986, NZFS was holding the Aupouri population at about 150 horses by periodically arranging for the capture of excess animals (Anon., 1986).

The Kaimanawa horses are known to carry biting lice, and have moderate levels of helminth parasites (Aitken *et al.*, 1979). Cattle ticks abound in the undergrowth of Aupouri Forest and may affect the survival of foals. The Aupouri herd is also infected with moderate to high levels of helminths (Hermans, 1984). There is little evidence of disease in either herd. Excessive or abnormal hoof growth has been noted in some of the more mature Kaimanawa horses, but it is uncertain whether this a genetic trait or simply due to lack of normal wear on the soft terrain (NZWS, unpubl.).

Adaptation to New Zealand Conditions

In New Zealand, feral horses have adapted to a wide range of climatic conditions, and to a diet containing many native plants. Little is known of any physiological, anatomical or behavioural differences that may have evolved.

Significance to the New Zealand Environment

DAMAGE. Feral horses appear to have little effect on forest regeneration, either in New Zealand or overseas. At Aupouri Forest there has been no significant damage to *Pinus radiata* seedlings or trees, although damage to young Norfolk Island pines can be quite severe (Hermans, 1984); in the Kaimanawa ranges, feral horses have not inhibited the regeneration of beech forest or destroyed plant cover or caused erosion in areas of open grassland (Aitken *et al.,* 1979).

CONSERVATION. Intensive land management, improved fencing, and the demand for horsemeat for canned pet food have in the recent past reduced the range and numbers of feral horses much more drastically in New Zealand than in most other parts of the world. The United States and Canada were quicker to develop the protectionist sentiment that has achieved some legal protection for the remnants of their once vast herds of feral horses; conservation in New Zealand was hampered by residual feelings of antipathy towards all introduced browsing mammals, lumped together in the public mind as exotic pests.

However, feral horses are probably now better off here than in all the states of continental Australia, where capture and killing are still the only forms of management practised (McKnight, 1976). Legal protection was given to the Kaimanawa population under a Wildlife Order in 1981; and the Aupouri Forest population has been informally protected by NZFS since the 1960s (Hermans, 1984).

New Zealand feral horses have an aesthetic and historical value (Anon., 1984b). Not only can they be traced back to early European settlement and to the ancient domestic horse breeds of Britain and Europe, but they represent remnants of the large groups of feral horses that were once a very significant part of the New Zealand back country (Morrow, 1975). In recent years, NZWS and NZFS, concerned for the best interests of both the feral horses and their environment, have regularly monitored the two remaining herds for details of population size and range condition, and DOC has so far (1988) continued this policy.

R. H. T.

Order Artiodactyla

The even-toed ungulates or artiodactyls are the largest and most diverse group of large mammals living today; there are about 187 species, in 76 genera and 10 families. Two rather different kinds of animals are included. The pigs and their relatives (peccaries and hippos) are primarily omnivores, with low-crowned molars with simple cusps, large tusk-like canines, short legs and four toes (or two toes and two large dew claws) on each foot; the ruminants (camels, deer, giraffes, cattle, sheep, goats, tahr, and chamois) are specialist herbivores, with ridged, often high-crowned molars, and a progressive tendency to lengthen the limbs and reduce the number of functional toes to two.

Three artiodactyl families are represented in New Zealand, by one of the nine living species of Suidae (pigs), five of the 121 Bovidae (cattle and allies), and seven of the 33 Cervidae (deer) (one with two distinct subspecies). Their distinguishing marks are listed under their respective family headings.

In addition, the llama (*Lama glama*) and the alpaca (*L. pacos*), two South American members of a fourth family, the Camelidae, were unsuccessfully liberated in the last century; the llama in 1865-66 at Otakapu, Rangitikei, and the alpaca in 1878 at Elderslie, Oamaru. Neither established a wild population, but, ironically, both have been reintroduced in recent years as farm stock (p. 18).

Family Suidae

The suids are a small family of nine species classified into five genera: *Potamochoerus* (bush pigs); *Sus* (wild boar, feral and domestic pigs); *Phacochoerus* (wart hogs); *Hylochoerus* (giant forest pigs); and *Babyrousa* (babirusa). All have large, long heads with mobile snouts used for rooting up the ground, short necks and powerful, stocky bodies with coarse, bristly coats.

Genus *Sus*

The only suid present in New Zealand is *Sus scrofa*, which can be distinguished from three of the four other species of *Sus* by its lack of facial warts, and from the fourth, the pygmy hog, *S. salvanius* of Assam, by its larger size and six pairs of mammae instead of three (Cumming, 1984). It is unlikely to be confused with any other large mammal in New Zealand.

32. Feral pig

Sus scrofa Linnaeus, 1758

Also called "Captain Cooker", razorback (English); te poaka (the porker), kuhukuhu, poretere, petapeta, kune-kune (Maori).

Linnaeus applied the name *Sus scrofa* both to the European wild boar and to its distant and very different descendants, the domesticated pigs. Some authorities

Fig. 64 Feral pig (G. S. Roberts).

prefer to distinguish the domesticated form as *Sus domesticus* Erxleben, 1777, but others disagree that feral and domesticated pigs are a separate species, so this name has not achieved common usage. There is no evidence that any true wild boars were ever brought to New Zealand, although the present-day feral pigs look very like them.

The common name "wild pig" should be avoided, since it can apply both to feral pigs of domestic origin and to the true wild boar of Eurasia, as well as to other members of the Family Suidae.

DESCRIPTION (Fig. 64)

Distinguishing marks, see below and skull key, p. 24.

Feral pigs in New Zealand are smaller and more muscular than domestic pigs and have massive forequarters and smaller, shorter hindquarters. They also have longer, larger snouts and tusks, smaller ears and much narrower backs (hence the nickname "razorback"). Old boars usually have well-developed, keratinous plaques or shields up to 90 mm thick on their shoulders, which protect the shoulders and ribs during fighting. Their hair is longer and coarser than that of domestic pigs, and usually black (but see below); the tail is more often straight with a bushy tip than curly as in domestic pigs. Their terminal nostrils set in a strong cartilaginous snout supported by a prenasal bone, combined with their very muscular neck and shoulders, enable them to find and root up food from consolidated ground and partly rotten logs. Their eyes are small and eyesight poor, but their senses of smell and hearing are acute.

Dental formula $I^3/_3\ C^1/_1\ Pm^4/_4\ M^3/_3 = 44$.

Table 59: Body measurements of feral pigs and kune kune in New Zealand.

			Boars	n	Sows	n
Feral pigs	Mainland	Weight[1] (kg)	45-205	56	32-114	31[2]
		Length (mm)	1 100-2 280	21	1 140-1 500	8
		Height (mm)	550-960	22	430-600	11
		Tail length (mm)	200-460	17	200-460	8
	Auckland Is.	Weight[3] (kg)	42	8	37	8
Kune kune		Weight[1] (kg)	<150			
		Height (mm)	<600			

1. Live weight.
2. Sample size possibly greater but not stated in reference.
3. Dressed weight.

Data from Donne (1924); Wodzicki (1950); Roberts (1968); Challies (1975); Tipene (1980); Holden (1982); Willis (1982); J. C. McIlroy (unpubl.); R. H. Taylor (unpubl.).

The permanent teeth are in place by 20-22 months of age. The continuously growing canine teeth (tusks) of adult males are larger than those of domestic pigs and project from the sides of the mouth. The lower tusks are triangular in cross section and curve upwards, outwards and backwards, forming an arc of a circle up to 500 mm in circumference. The total length is 150-300 mm, but up to two-thirds of this is embedded in the lower jaw. A record tusk 318 mm long, described by France (1969), protruded 165 mm from the jaw. The upper canines are shorter (up to 90 mm long) and oblong in cross section. They curve outwards and backwards, functioning as whetstones or "grinders" to the lower tusks. If an upper tusk is broken or deformed, the corresponding lower one can continue to grow in a complete circle, ultimately re-entering the lower jaw.

There is another type of pig in New Zealand (sometimes called the "Maori pig" or kune-kune, meaning "fat and round") which was, at least at one stage, semiferal. It is smaller and shorter-legged (Table 59) and fatter than other breeds, has a turned-up nose, and has either light gold or black and brown hair. Its most distinctive feature is the pair of hairy tubular beads (pire pire, dewlaps or tassels), about 40 mm long, hanging from the lower jaw. The origin and date of arrival of this breed in New Zealand are unknown, except that, contrary to popular belief, they definitely did not arrive with the Maori; other theories are provided by Peirson and Owtram (1945), Tipene (1980) and Willis (1982). Wild kune-kune have been reported from Northland, King Country, Bay of Plenty, Waikato and East Coast in the North Island (Johns & McGibbon, 1986), and about 50 specimens are now being actively conserved in wildlife reserves and on game and private farms (Willis, 1982).

FIELD SIGN

Ground rooted up in pasture adjacent to forest, in large areas of bracken (*Pteridium esculentum*), or in forest, particularly along wet gullies and slopes and in small clearings, on river flats and in swampy areas, is evidence of pigs feeding. Rotten logs are nosed apart, clumps of fern, tussock and sedge are turned over, and large stones and rocks are nosed out of the ground. Cow pats are turned over and turf is peeled back. Remains of lambs killed by feral pigs comprise only the skull (or fragments of it) and the skin of the legs pulled inside out as far as the hocks.

The size and shape of the faeces of feral pigs depend on the size of the animal and its recent diet. They are usually dark, flattened, roughly oval pellets joined in a large cylinder. Pigs feeding on coarse vegetation produce unsegmented cylinders broken into large pieces; on a grass diet, their faeces resemble those of a small pony.

Paths through thick undergrowth, bracken or scrub may be used by any large mammal. Pig tracks usually show only the two rounded cloven hooves, but in soft soil the small, pointed marks of the dew claws may show, behind and to the sides of the main digits. Each hoof is curved on the outside and slightly concave on the inside anteriorly. By comparison, the hoofprint of a sheep tends to be more pointed at the front; that of a goat, kidney-shaped (concave on the inside, convex on the outside); and that of a red deer longer and narrower and straight on the inside, with dew claws straight behind the main hooves rather than to the sides.

Wallows, oval depressions in the ground where feral pigs take mud baths to rid themselves of lice and other ectoparasites, can sometimes be found in wet areas such as swamps, gullies or alongside creeks. Trees and logs handy for rubbing on are often coated with mud, or scarred with tusk marks.

Nests and resting sites range from depressions in the ground lined with grass, bracken or other ferns, often located under a bank, log or tree stump facing the sun, to flattened areas in the middle of a clump of dense bracken.

MEASUREMENTS

See Table 59.

VARIATION

The considerable local variation in the colour of feral pigs in New Zealand is probably a founder effect. Local populations were established from a variety of breeds and have since crossbred to different extents with other populations. For example, the feral pigs of Central Otago are derived from a variety of the Tamworth breed, with a smaller contribution from the Berkshire and Hampshire breeds (Scott, quoted by Thomson, 1922).

Colour patterns vary within and between areas. Black is, and apparently always has been, the most common colour (Angas, 1847; Thomson, 1922). Other colours include rusty red or ginger, sandy brown with black spots, brown and white, white, black with a white stripe, grey, and a smoky blue, particularly in North Canterbury (Thomson, 1922; Roberts, 1968; Guthrie-Smith, 1969). Some piglets are marked with dark, longitudinal stripes, rarely seen in domestic pigs (illustrated in Johns & McGibbon, 1986), which disappear with age. Of the 655 observations made by J. C. McIlroy (unpubl.) in the

Table 60: Distribution and history of feral pigs on the offshore and outlying islands of New Zealand.

Island	Area (ha)	Density	Fate of population
Kermadecs (29°S)			
Raoul	2 938		Eradicated in 1960s
Macauley	306		Died out
Northern North Island (36°S)			
Aorangi (Poor Knights)			Exterminated 1936
Cavalli (Motukawanui)	355		Died out
Great Barrier	28 510	Moderate, esp. northern end	Present
Rakitu (Arid)	350	Low	Eradicated in 1960s
Little Barrier	2 817		Died out long ago
Mayor	1 277	Very low	Continued sporadic control
Motuoruhi (Goat)			Prob. eradicated by 1970
Nelson-Marlborough (40-41°S)			
Arapawa	7 785	High	Present
D'Urville	16 782	High	Present
Blumine	377		Eradicated in late 1960s or early 1970s
Chetwode	242		Eradicated in late 1960s or early 1970s
Stewart (47°S)	174 600	Low, NW area only	Eradicated between 1948 and 1965
Native (Paterson Inlet)	60		Died out in 1940s
Ruapuke	1 525	High	Present
Outlying islands			
Chatham (44°S)	90 650	At least 100, southern end only	Declining due to hunting
Pitt (44°S)	6 203	Low, southern end only	Present
Auckland (50°S)	45 975	Low to moderate, widespread	Present
Enderby (50°S)	710		Died out *c.*1870s
Campbell (52°S)	11 331		Died out

Data from Cuthbertson (1974); Challies (1975); Ogle (1981); Holden (1982); R. H. Taylor (unpubl.); D. Crockett (unpubl.).

Murchison area, 50% were of black pigs, 35% were brown with black spots, 7% were ginger with or without black spots, 3% were totally brown, and the remaining 4% were buff, brown and white, or blue-grey.

History of Colonization

Domestic pigs were common among the Polynesian settlements of the Pacific islands a thousand years ago, but were not carried on, or did not survive, the voyage to New Zealand made by the first colonists. The first pigs to arrive were brought by the French explorer De Surville, who presented two (sexes unknown) to the Maori at Doubtless Bay, Northland, in 1769. Nothing is known of their fate, but it is generally assumed that these were not the progenitors of any later feral population.

In 1773, during James Cook's second voyage to New Zealand, Captain Furneaux released one boar and two sows at Cannibal Cove in Queen Charlotte Sound, but these were caught by the Maori and died without breeding. Later the same year, Cook obtained further pigs from the Society Is. and Tonga and gave two pairs of these to the Maori south of Cape Kidnappers and another sow to the Maori at Queen Charlotte Sound; he also released a boar and three sows at West Bay. Almost a year later, he released another pair near Cannibal Cove, and during his last visit to the area in 1777, he gave the Maori a further pair. In 1793, 10 young sows and two boars of European domestic origin were given to the Maori of the Cavalli Is. by Captain King, Governor of New South Wales; these apparently were released on an island near North Cape, close to Doubtless Bay, where De Surville had given his pigs to the Maori 24 years earlier. In 1805 King returned and gave the same Maori a further 26 sows and four boars.

From the 1790s onwards, New Zealand was visited with increasing regularity by explorers and by European and North American sealing, whaling and trading vessels. Some attempted to establish pigs on the mainland and offshore islands as future food supplies, particularly for castaways. Pigs were liberated on Auckland I. in 1807, on Raoul and Macauley Is. in 1836 and on Enderby and Campbell Is. in 1867 (Thomson, 1922; Table 60).

In later years pigs were both given to the Maori and used as barter by them and by Europeans. For example, European sailors gave pigs to the Maori along the coast of Foveaux Strait in 1824 (Thomson, 1922). American whalers at Russell, in the Bay of Islands, bartered muskets and tobacco for pigs and potatoes from the Maori (Angas, 1847). The Maori greatly valued pigs and helped to spread them both by their custom of giving gifts of food (including pigs) to relatives and friends in other tribes, and by keeping them in semiferal herds from which many escaped into the wild.

Feral pigs were well established around most Maori and European settlements by 1840 (Angas, 1847; Thomson, 1922; Davidson, 1965; Holden, 1982). Their outward spread and enormous increase in numbers and distribution followed the same irruptive pattern observed later in other ungulate populations (Caughley, 1970b; pp. 383, 448). For example, on Auckland I., pigs were released in 1807 at Port Ross and spread over much of the northern part of the island by 1840; they continued to extend their range southwards for another 40 years until most of the remaining suitable habitat was occupied (Challies, 1975). Once established, the population increased rapidly, remained high for several decades and then declined, probably with the depletion of their preferred foods, to the present much lower numbers.

In the North and South Islands the period of maximum numbers appears to have been between 1840 to 1880. Attempts to control them began very early: for example, thousands were killed annually throughout Canterbury last century by private hunters, farmers and station hands (Davidson, 1965). Nevertheless, pigs spread into all suitable parts of the mainland, and escaped or were released on many offshore islands (Table 60) by the end of the 19th century. Pig numbers increased in many areas during both World Wars, particularly on back-country farms reverting to bracken, scrub or forest, because of a shortage of hunters, rifles, ammunition and petrol for transport (Munro, 1927; Mackintosh, 1944), but they have since been reduced by control operations (p. 370), hunting and habitat changes.

DISTRIBUTION

WORLD. Wild boars are native to Eurasia and North Africa, but are now extinct over much of their former range, including southern Scandinavia, the Nile valley and Britain (Tisdell, 1982). They have been introduced into Argentina and the United States of America, where they interbreed with feral pigs. Feral pigs are now found in Australia, New Zealand, the Galapagos Is., Argentina, Venezuela, the United States, (including Hawaii), various Caribbean islands, the Mascarene Is. (Mauritius, Réunion and Rodrigues), the Philippines, parts of Indonesia and Malaysia, New Guinea and various islands in Melanesia and Polynesia.

NEW ZEALAND (Map 31). The latest (1983) survey showed that feral pigs occupy a greater proportion of the North Island than of the South Island (Johns & McGibbon, 1986). Previous surveys, in 1946, 1967, and 1971 (Wodzicki, 1950; Poole & Johns, 1970; Challies, 1976), recorded pigs in many areas from which they have now disappeared. For the present and past distribution of feral pigs on offshore islands, see Table 60.

HABITAT

The key habitat factors for feral pigs are reliable supplies of appropriate food and water and adequate cover. In New Zealand the preferred habitats are native and exotic forests; thick and extensive areas of bracken or gorse (*Ulex europaeus*) adjacent to improved farmland; marginal or reverting farmland on good soils where there is a mixture of regenerating trees, scrub and bracken; and river flats and tussock grasslands containing patches of shelter such as matagouri (*Discaria toumatou*), sweet briar (*Rubus rubiginosa*) or manuka (*Leptospermum scoparium*) (Wodzicki, 1950; Davidson, 1965; Roberts, 1968). Lack of cover and food, and frosts or snow that limit rooting for food, restrict their use of the higher country above 1200 m.

On Auckland I. feral pigs can be found in all habitats, but in summer they favour the open high country above 350 m altitude and the coastal forest and lowland scrub up to 150 m (Challies, 1975). The intermediate belts of short, thick scrub, strips of *Oreobolus* dissecting some low-lying areas of scrub, dense *Chionochloa* tussock and forest might be more fully used during winter.

FOOD

Feral pigs are omnivorous, opportunistic feeders, living mainly on a high-fibre (> 25%), low-protein diet (grasses, legumes, herbs and roots). They also readily eat crops, seeds and other animals when available.

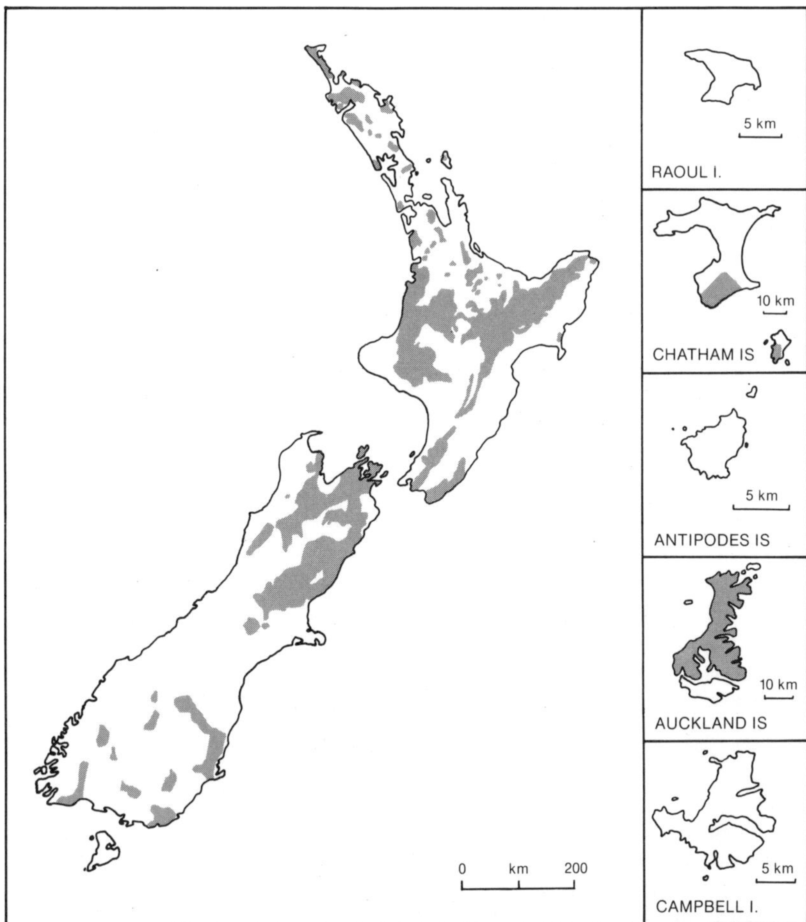

Map 31 Distribution of feral pigs on the main islands of New Zealand, from a survey by NZFS in 1983 (Johns & McGibbon, 1986). For distribution on offshore islands, see Table 60.

On the main islands of New Zealand they eat the berries of hinau (*Elaeocarpus dentatus*) and sweet briar; the roots of *Carex appressa, Aciphylla* spp., nettles (*Urtica* sp.), thistles (*Cirsium* spp.) and bracken (probably the staple item); the base of the fronds of nikau palms (*Rhopalostylis sapida*) and the roots of young palms; and grasses, including introduced species, the tall, tussocky *Microlaena avenacea* and the lower part of the tillers of *Chionochloa* sp. Animal foods include amphipods, centipedes, beetles, earthworms (especially in winter and spring), native land snails (*Powelliphanta* spp.), frogs, lizards, ground-nesting birds and their eggs, young rabbits and carrion (Wodzicki, 1950; Roberts, 1968; Meads, Walker & Elliott, 1984; J. C. McIlroy, unpubl.). The diet of feral pigs in the Ureweras was examined by Thomson and Challies (1988); see Table 61.

Pigs have an alimentary canal similar to that of man, i.e., they have a simple stomach and are unable to fully digest the large amounts of fibre in their diet,

Table 61: Foods of feral pigs in podocarp-tawa forest of the Urewera ranges.

	% of annual diet
Plant material	
Ferns	
Tree fern (*Cyathea* spp.) leaves & stems	8.4
Bracken (*Pteridium esculentum*) roots & rhizomes	5.5
Grasses & sedges	
Miscellaneous grasses & sedges	2.0
Herbs	
Thistle (*Cirsium* spp.)	1.9
Climbers	
Supplejack (*Ripogonum scandens*) roots	11.5
fruits	2.9
Trees & shrubs	
Tawa (*Beilschmiedia tawa*) fruits	21.4
Hinau (*Elaeocarpus dentatus*) fruits	9.5
Tawari (*Ixerba brexioides*) leaves	1.3
Minor plant foods	5.3
Unidentified plant material	2.2
Plant subtotal	**71.9**
Animal material	
Vertebrates	
Possum carrion	10.6
Pig carrion	1.8
Annelids	
Earthworms	10.3
Insects	
Miscellaneous insects	2.7
Minor animal foods	0.2
Unidentified animal material	2.5
Animal subtotal	**28.1**
Total	**100.0**

Most samples were taken from 104 adult feral pigs from the Waimana River catchment, shot between December 1982 and June 1985. Dry weights of stomach content samples were converted to percentages and combined, giving each season (summer n = 26, autumn 26, winter 27, spring 25) equal weight, to give an average annual diet (Thomson & Challies, 1988). Only items having a mean dry weight > 1% of the annual diet have been included.

although some bacterial digestion of cellulose in the caecum helps (Barrett, 1971). Protein is the most critical limiting factor in the nutrition of feral pigs; pigs require protein-rich foods, particularly during late pregnancy, lactation and early growth (Giles, 1980), and sows without a nitrogen intake of at least 15% may resorb foetuses (Pavlov, 1980).

SOCIAL ORGANIZATION AND BEHAVIOUR

ACTIVITY. Pigs are mainly diurnal (Brisbin, Smith & Smith, 1977; Wood &

Brenneman, 1977; Barrett, 1978; K. B. Masters, unpubl.), but may become more nocturnal or restrict their activity to the early morning and late afternoon in hot weather or when subjected to hunting (Pullar, 1950; Hanson & Karstad, 1959) and revert to diurnal activity in autumn and winter if relatively unmolested (Kurz & Marchinton, 1972).

Seven pigs radio-tracked in the Murchison area between October 1983 and April 1984 were active, on average, for about one-third of each 24-hour period, mostly during the evening and night and least in mid afternoon (J. C. McIlroy, unpubl.). Immature pigs were the most active (52-73% of the time) and adult females the least (22-34% of the time).

DISPERSION. Adult boars over 18 months old are invariably solitary, and farrowing sows will temporarily separate themselves from other pigs, but otherwise pigs are normally sociable. Groups comprise (1) two or more (up to six) sows and their piglets; (2) immature pigs of both sexes up to 1 year old, possibly littermates; (3) adult females without piglets; (4) immature males or "bachelor" groups (Anon., 1978a; Giles, 1980; Pavlov, 1980; K. B. Masters, unpubl.). For the number of pigs in groups observed in the Murchison area and on Auckland I., see Table 62.

HOME RANGE. Although Wodzicki (1950) described pigs as nomadic wanderers, they are, in fact, relatively sedentary. In North Canterbury the average linear distance travelled by 10 boars was 3.2 ± 1.2 km (SE), maximum 13 km, significantly further than the average of 0.5 ± 0.07 km, maximum 0.8 km, travelled by 12 sows (Martin, 1975). In the Murchison area, seven feral pigs with ample food and water and subjected to little disturbance travelled an average linear distance per day of 1.4 km (maximum 3.4 km) between October 1983 and May 1984 (J. C. McIlroy, unpubl.). Their home ranges varied from 28 to 209 ha, largest in immatures (mean 130 ha, range 66-209 ha), intermediate in adult boars (mean 54 ha, range 38-69 ha) and smallest in adult sows (mean 33 ha, range 28-39 ha).

REPRODUCTION AND DEVELOPMENT

Feral pigs are polyoestrous: adult females have a 21-day oestrous cycle and a gestation period of about 112-114 days. In New Zealand they probably breed throughout the year, though mainly in spring and summer (Wodzicki, 1950; J. C. McIlroy, unpubl.)

The litter size in New Zealand may be 6-10 piglets, but of these only 3-6 survive (Wodzicki, 1950; Holden, 1982). Litter size in Australia varies from 1 to 10 (Hart, 1979; Giles, 1980; Pavlov, 1980; Boreham, 1981; K. B. Masters, unpubl.) and in the United States from 1 to 11 (Sweeney & Sweeney, 1982).

Farrowing sows lie down to give birth, whereas other hoofed mammals give birth while standing. Each piglet quickly assumes exclusive use of a single teat, and will defend that teat against other piglets until weaned at 2-4 months (Pavlov, 1980). During periods of food shortage the sow's mammae successively dry up, leading to sequential mortality of the piglets.

Young piglets stay within or near the protection of the nest for the first 2-3 weeks after birth and subsequently keep in close contact with the sow by

Table 62: Group sizes of feral pigs observed in the wild.

	Groups	Percent of observations	Group size	Total number of observations	Reference
Murchison area (South Island) (1983-84)	Subadults	33	1-11	765	J. C. McIlroy (unpubl.)
	Piglets	30	1-9		
	Adults only	16	1-4		
	Sow and piglets	13	3-10		
	Mixed sexes and ages	8	2-13		
	Adult sows only	5	1-2		
	Adult boars only	4	1		
Auckland Is. (1972-73)	Solitary	64	1	44	Challies (1975)
	Groups	23	2		
		13	3-5		

frequent nose-to-nose touching, and with each other by frequent squealing. Weaned piglets remain with the sow until the next litter is due. The length of each lactation, the intervals between oestrus cycles, and the number of litters produced each year are highly variable, depending on environmental conditions; the next oestrus may be up to 94 days after parturition, compared with a minimum of 18 days for domestic pigs.

Sexual maturity in feral pigs depends largely on body weight. For example, in Australia feral sows breed only when they have reached 20-30 kg, usually at about 7-12 months of age (Anon., 1978a; Giles, 1980; Pavlov, 1980), cf. 4-9 months in domestic stock. According to Wodzicki (1950), boars in New Zealand reach breeding age at 12 months and sows at 10 months.

Three immature pigs in the Murchison area, initially weighing 2.7, 3.6 and 11 kg, grew at 115, 107 and 98 g/day, respectively (a much lower rate than in domestic pigs); one sow gained 12 kg over 16 weeks between December and April (a weight gain of 106 g/day), during which time she weaned her litter (J. C. McIlroy, unpubl.).

Population Dynamics

Density. During the period of highest numbers of feral pigs in New Zealand (around 1947), local densities were estimated to vary from 123/km² down to 35/km² (Saywell, quoted by Wodzicki, 1950). Crude estimates of densities in the Murchison area during 1983-84 were 3–8/km² in a heavily hunted area and 12–43/km² in a less disturbed area with more abundant food and cover (J. C. McIlroy, unpubl.).

Mortality. Mortality among young feral piglets is generally high, ranging from 10-15% when food supplies and the weather are favourable to 90% when conditions are poor (Jezierski, 1977; Barrett, 1978; Giles, 1980). Autumn- and winter-born piglets are especially susceptible (Wodzicki, 1950). Challies (1975) estimated the mean mortality of piglets on Auckland I. at a minimum of 58% (litters lost *in toto* were not included). Pregnant and lactating sows are very susceptible to starvation during droughts or periods of food shortage; however, most adult sows will continue to breed even during adverse conditions, so that populations can increase rapidly when adequate high-protein food becomes available. In Australia, feral pigs can almost double their numbers every year even if mortality of piglets is 70-80% (Giles, 1980); the New Zealand population, which has a more stable food supply, more uniform climate and the opportunity to breed twice per year, probably has the same capacity. Saywell (in Wodzicki, 1950) commented that less than a year may suffice to replace a pig population depleted through extensive hunting in New Zealand.

PREDATORS, PARASITES AND DISEASES

Hunters and their dogs are the only predators of feral pigs in New Zealand. Ineson (1953) found three species of ectoparasites and eight species of endoparasites of feral pigs. Lice, *Haematopinus suis* were the most common (68% frequency), followed by the lungworm *Metastrongylus elongatus* (67%), the protozoan *Balantidium coli* (62%) and the roundworm *Ascaris suum* (43%). The hydatid tapeworm (*Echinococcus granulosus*) and liver fluke (*Fasciola hepatica*) are usually confined to domestic pigs. *Sarcocystis* spp. have been found in the diaphragmatic muscle of feral pigs examined at a game packing plant (Collins, Charleston & Wiens, 1980) and in muscle tissue of pigs from various areas of the North Island (Collins & Charleston 1979b). Ekdahl, Smith and Money (1970) reported cases of bovine tuberculosis, and Daniel (1967) an arbovirus, in New Zealand feral pigs. The incidence and intensity of parasitic infestations has no significant effect on feral pig numbers in New Zealand (Ineson, 1953).

ADAPTATION TO NEW ZEALAND CONDITIONS

There are few major differences between the feral pigs in New Zealand and those elsewhere in the world. Size, appearance and social organization are similar to those in other countries. Diet is similar, but supplies are probably more reliable in New Zealand than elsewhere; annual and seasonal changes in food abundance (particularly in relation to droughts, floods or differences in forest mast production) and pressure from predators can be much greater overseas, and these factors significantly affect breeding success and population dynamics. These may be the reasons why the densities of relatively undisturbed pig populations in New Zealand (e.g., in the late 1940s) were apparently much greater than most of those reported for feral pigs in different habitats in Australia (e.g., 1-80/km^2; Anon., 1978a; Boreham, 1981; Masters, 1981) and for feral pigs and wild boars in the United States and Europe (e.g., 0.4-79/km^2; Singer, 1981).

The present population of feral pigs in New Zealand is a good example of how feral animals can revert to a uniform type much more closely resembling their wild ancestors than their variable domesticated forebears (Clutton-Brock, 1987).

SIGNIFICANCE TO THE NEW ZEALAND ENVIRONMENT

DAMAGE. Pigs were among the first feral animals to become established in New Zealand. At first they were welcomed as a source of food by the Maori, early prospectors, whalers and settlers; but as they became more numerous and widespread, their habits of damaging crops and pastures and killing lambs and cast sheep changed their status to that of pests. Newborn lambs are probably rarely killed by pigs in New Zealand nowadays, except in back-country farms close to native forest; likewise, crop losses due to pigs are now only localized in New Zealand, although they can be much more widespread and significant in Australia (Pavlov, 1980; Tisdell, 1982). Damage to pastures, usually those adjacent to native forest, exotic plantations or large areas of gorse and bracken, is still common in many areas, especially during late autumn and winter. Damage to exotic plantations, especially in the early stages of growth, is caused by pigs rooting up young trees and eating their roots.

The consequences of the arrival of pigs for the mainland native flora and fauna have not been, and now cannot be, properly assessed; but they probably contributed to, for example, the reduction of the kakapo (*Strigops habroptilus*) and the constricted distribution of the orchid *Gastrodia cunninghamii* (Kirk, 1896). In modern times, feral pigs have the greatest effect of all the ungulates on remnant populations of the native land snail (*Powelliphanta* spp.) by rooting up and completely destroying their litter habitat and eating their eggs and up to 95% of the snails (Meads, Walker & Elliott, 1984). The impact of pigs on the vegetation of Auckland I., especially the endemic large-leaved subantarctic species of plants, and on nesting seabirds, was very destructive late last century. Challies (1975) suggested that the present reduced population may have established some sort of balance with the impoverished flora and fauna. However, more recently, Campbell and Rudge (1984) showed that the Auckland I. pigs were still reducing areas of *Chionochloa antarctica* tussock, by eating the tops and digging up the roots and, together with goats, were likely to eliminate the tussock from low altitude areas. On Aorangi I. (Poor Knights group), a low mixed forest regenerated, and many seabirds recolonized, after pigs were exterminated in 1938 (Harper, 1983).

CONTROL. In 1925 the government, in response to the largely unrestricted increase of pigs in New Zealand during World War I and their subsequent damage to farmlands, instigated a bounty scheme to encourage the control of pigs (Munro, 1927). This scheme, administered by the Department of Agriculture, involved a "bonus" payment of one shilling (or, later, three rounds of ammunition) for each snout and tail of a pig handed in to a representative of the NZFS. It continued, with interruptions due to financial reasons, until at least the late 1950s (Wodzicki, 1950; Martin, 1972; Holden, 1982). Some government departments also employed professional hunters. Large numbers of pigs were killed – e.g., over 40 000 in the Wellington district between 1925 and 1927 (Munro, 1927) and over 340 000 throughout New Zealand between 1926 and 1945 (Wodzicki, 1950) – but the density of pigs remained high. Consequently, poisoning trials began, initially with strychnine, phosphorus and arsenic, and later with the more effective 1080 poison. During the 1950s massive poisoning campaigns killed many pigs, and many more were killed under the national bounty scheme (e.g., 40 000 between 1955 and 1957; Martin, 1972). By the late 1950s pig numbers had been greatly reduced in most parts of New Zealand; this decline still continues, because of sport hunting, some limited control work by NZFS/DOC (Rammell & Fleming, 1978), and the clearing of their preferred habitats. Since feral pigs in New Zealand now have a negligible effect on the economy, and are not at present an important reservoir of any serious disease of economic stock, any more intensive attempt to control them is now unnecessary. On the other hand, traditional pig hunting is still a popular sport, and "wild" pork is a culinary speciality, both on local tables and in the game meat market in Europe. New Zealand is fortunate to have this resource; in parts of Australia, high rates of infection of melioidosis and sparganosis make wild-caught pigs unfit for human consumption.

J. C. M.

Family Bovidae

The bovids are a large family of 121 species in 47 genera. The family is divided into six subfamilies: the American pronghorn (a single species with a subfamily to itself), all cattle (23 species in eight genera), the duikers (17 species in 12 genera), the grazing antelopes (24 species in 11 genera), the gazelles and dwarf antelopes (30 species in 12 genera) and the goat antelopes (26 species in 13 genera). They are all ruminants; that is, they have evolved a multi-chambered stomach and the habit of chewing the cud in order to digest fibrous herbage. (The lagomorphs have evolved a different answer to the same problem; see p. 138.) They all have permanent horns, made of a solid bony core covered with a horny sheath, which grow continuously from the base throughout life, in contrast to the deer, which have naked bony antlers grown and shed every year.

There are five members of the family living wild in New Zealand. Three (cattle, sheep and goats) were brought in as domestic stock, and are still important to our pastoral economy, as well as having established in the wild as independent feral populations; and two (tahr and chamois) were brought in as game animals. The taxonomic relationships and distinguishing marks of the five are shown in Table 63. The only other members of the family known to have been liberated into the wild in New Zealand were three Himalayan bharal (p. 392), and a single gnu (*Connochaetes gnou*: Subfamily Hippotraginae, the grazing antelopes), brought from South Africa to Kawau I. by Sir George Grey in 1870 (Wodzicki, 1950).

Subfamily Bovinae

Genus *Bos*

The eight genera of wild cattle and spiral-horned antelopes belong to the subfamily Bovinae, which includes the buffalo, bison, bushbuck, kudu, nyala

Table 63: Distinguishing marks of bovids in New Zealand.

	Subfamily Bovinae	Subfamily Caprinae			
	Tribe Bovini	Tribe Rupicaprini	Tribe Caprini		
	Cattle *Bos taurus*	Chamois *Rupicapra rupicapra*	Tahr *Hemitragus jemlahicus*	Sheep *Ovis aries*	Goat *Capra hircus*
Scent glands	None	Behind horns	None	All feet, but no subcaudal	Front feet, plus subcaudal
Beard	None	None	None	None, either sex	Males, plus some feral females
Coat	Usually short	Short	Long	Crimped woolly fleece	Mostly short hair, or if longer, uncrimped, silky
Mane	None	Male, prominent	Male, very prominent	None	Male, moderate
Tail	Long, tufted	Short	Short, bare underneath	Long, pendulous, woolly	Short, tufted, bare underneath
Horns (both sexes)	Spread sideways, may be absent in either sex	Black, slender, erect and hooked	Short, stout, curve sharply back	Spiral backwards, tight curl	Sweep back, shallow curl
Sexual dimorphism in size	Pronounced	Slight	Pronounced	Moderate	Slight

Fig. 65 Feral cattle, Enderby I. (R. H. Taylor).

and eland. The true cattle, members of the genus *Bos*, comprise five species: the banteng and gaur of South-East Asia, the yak of the Tibetan plateau, the Cambodian forest ox, and the domestic cattle and their wild ancestors. All have horn cores which are circular in section, and most have a dark, short-haired coat; some have well-developed dewlaps and humps.

33. Feral cattle
Bos taurus Linnaeus, 1758

The alternative common name, "wild cattle", is best avoided, to save confusion with the true wild cattle of Asia. Some authorities refer domesticated cattle to the same species as their extinct wild ancestor, the aurochs (*B. primigenius*), but that policy is not followed here (p. 31).

DESCRIPTION (Fig. 65)
Distinguishing marks, Table 63 and skull key, p. 24.

Feral cattle can be distinguished from domestic stock only by their location and lack of ear marks or tags. Their size and conformation vary greatly depending on sex, age and breed. The male is heavier and larger, particularly around the head and neck. The hair is either straight or curly, and ranges from whitish to black with shades and blotches of red, roan, brown or buff. Both sexes can have horns, which are permanent and hollow, and grow throughout life over bony cores projecting from a prominent ridge on the skull. The horns of bulls are usually shorter and thicker than those of cows.

Cattle have no incisors or canine teeth in the upper jaw. In the lower jaw, eight sharp chisel-shaped teeth, six incisors and two canines, work against a cartilaginous dental pad in the upper jaw.

Dental formula $I^0/_3\ C^0/_1\ Pm^3/_3\ M^3/_3 = 32$.

FIELD SIGN
Cattle leave blunt two-toed footprints, almost as wide across as they are long, characterized by the distinctly concave inner margin of each half of the hoof.

The size of cattle tracks depends on the animal's breed and age, but those of adults can be up to 140 mm long by 120 mm wide. In thick scrub or bush country, well beaten trails can be found along main ridges, near river crossings, or between clearings. The droppings of feral cattle are very similar to those of domestic stock, although feral animals on coarse feed produce harder, layered droppings, about 60-100 mm across, rather than the familiar flat "cow pat". Smashed and twisted bushes, and "pawing pits" (see below) are a sure sign of the presence of feral bulls (Murie, 1954; R. H. Taylor, unpubl.).

MEASUREMENTS

There are few reliable measurements available. A mature bull (of Shorthorn type) shot in the Waihaha Valley, west of Lake Taupo, in 1950, measured: length of head and body, 2410 mm; tail, 1140 mm; spread across horns, 690 mm. A noticeably runty mature bull shot on Enderby I. in 1954 measured: shoulder height, 1140 mm; length of head and body, 2010 mm; tail, 660 mm; spread across horns, 360 mm (R. H. Taylor, unpubl.).

VARIATION

Feral cattle vary considerably in appearance and size. Crosses of Shorthorn, Ayrshire and other early dairy breeds are the most common in the North Island. Those on Enderby I. are reputed to be from pedigree Shorthorn stock (Falla, 1948), although they appear also to have some Friesian and Ayrshire blood. Herds previously present on Campbell I. and in the Catlins were also of dairy stock, but some South Island and Chatham I. animals show signs of Angus and Hereford ancestry. There are some exceptionally large feral bulls; those from long-established feral herds often appear stockier and to have proportionately higher withers and smaller hindquarters than modern farm animals (R. H. Taylor, unpubl.; NZFS, unpubl.).

HISTORY OF COLONIZATION

The first cattle seen in New Zealand were brought by Samuel Marsden to the Bay of Islands in 1814 (Nicholas, 1817). From then on, cattle were shipped regularly from Australia for the Bay of Islands Mission, for shore whaling stations, as items for trade with the Maori, and for European settlers in the North and South Islands. Ten were taken from Sydney to Mana I. in 1833. Cattle were first landed on the Chatham Is. in 1841, and on the Auckland Is. in 1849, by which time they could be seen as farm stock in most parts of mainland New Zealand (Thomson, 1922).

In 1819, within five years of the first introduction, John Butler reported that there were cattle in the bush at the Bay of Islands that had "been running wild for some years past" (Barton, 1927). Because much of the settlers' land was rough and unfenced, or still in heavy bush, cattle were very soon lost or escaped mustering, and went wild. Feral cattle were established in the Marlborough Sounds in 1839, and were abundant on Kapiti I. by 1840 (Thomson, 1922).

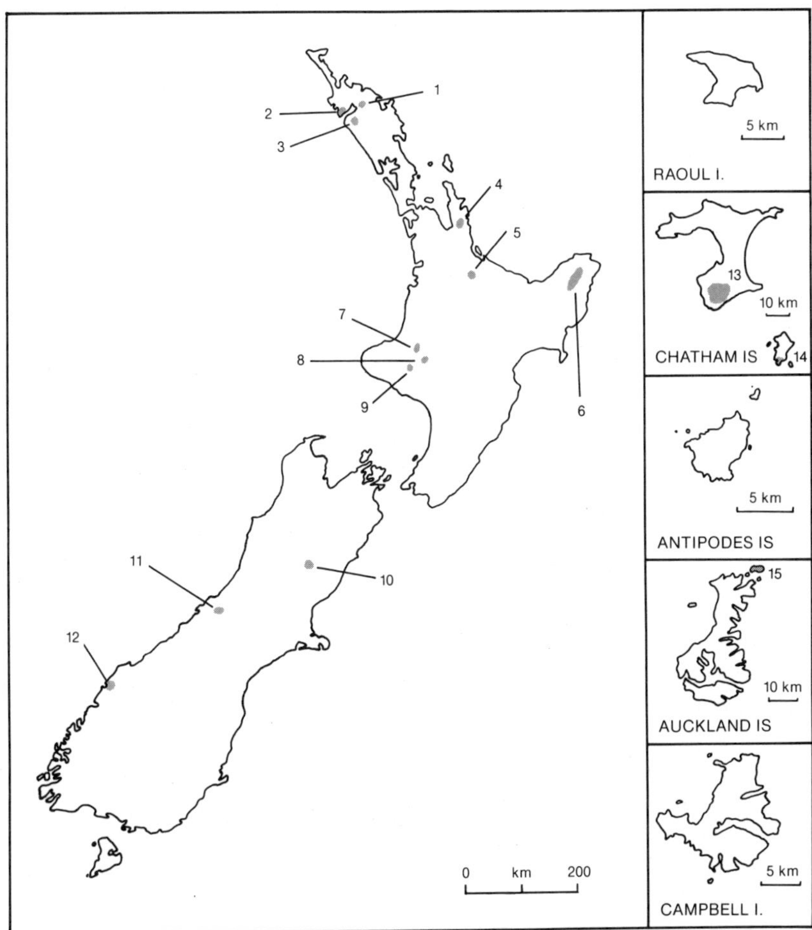

Map 32 Distribution of feral cattle in New Zealand. Herds numbered as in text (p. 376).

On the main islands of New Zealand, feral cattle were most widespread and in greatest numbers (sometimes in mobs of 100 or more) from the 1860s to the 1880s. In some areas hunting wild cattle became a sport (Thomson, 1922; Wodzicki, 1950; Davidson, 1965). As settlement progressed, feral cattle decreased and retreated rapidly; by 1916 they were "only found in the ranges distant from settlement" (Thomson, 1922).

In the late 1940s, feral cattle were still common in remote areas of forest and scrub throughout the three main islands, including many parts of Northland, Coromandel, the Volcanic Plateau, the Wanganui River region, valleys of the main North Island ranges from East Cape to the Ruahines, the South Island back country from Farewell Spit to Southland, and near the Ruggedy Range on Stewart I. They also lived on Chatham, Pitt, Campbell and Enderby Is. (Wodzicki, 1950; R. H. Taylor, unpubl.).

During the last 30 years feral cattle populations have steadily declined with the advance of land development, permanent settlement, more intensive farming and wild animal control campaigns. Nearly all of the remaining herds (including those on Chatham and Pitt Is.) are long established, although often reinforced by escapees from adjacent farms. The small herd on Enderby I., in the Auckland Is., is the only one that has remained completely isolated from domestic stock since last century (Taylor, 1971 and unpubl.; NZFS, unpubl.).

DISTRIBUTION

WORLD. The ancestors of today's Eurasian breeds of humpless cattle were the "wild aurochs" — large, formidable, long-legged and long-horned beasts — the last of which were hunted to extinction in Poland in 1627. Archaeological evidence suggests that cattle were first domesticated in the Middle East between 6000 and 5000 BC, and spread from there through Africa and Europe. Other early independent centres of domestication included Switzerland, Germany and Denmark. Hundreds of distinct breeds have been produced by artificial selection and transported throughout the world (Friend, 1978; Epstein & Mason, 1984).

NEW ZEALAND (Map 32). Feral cattle are no longer widespread or numerous, but at least 15 distinct populations exist (Johns & MacGibbon, 1986; NZFS, unpubl.). In the North Island there are 9 herds: (1) Puketi State Forest; (2) Warawara State Forest; (3) Waima and Mataraua State Forests and the eastern fringe of Waipoua State Forest; (4) the headwaters of the Hihi and Piraunui Streams and branches of the Tairua River in Coromandel State Forest Park; (5) Kaimai-Mamaku State Forest Park; (6) the eastern and northern flanks of the Raukumara Range from the Mangaotane to the Raukokore catchments; (7) the Tangarakau Valley in the headwaters of the Wanganui River; (8) the Mangatiti Valley, Wanganui catchment; and (9) the Upper Waitotara catchment. In the South Island there are now three herds: (10) Grantham State Forest and adjacent unoccupied Crown land near Hanmer; (11) the Wanganui River valley, Westland; and (12) Big Bay, South Westland. Two (Farewell Spit, Catlins State Forest Park) have recently (1975, 1979) been exterminated. Feral cattle formerly present on Stewart I. were shot out in the 1940s. On outlying islands there are still three herds; (13) on the Southern Tableland of Chatham I., (14) on Pitt I., and (15) on Enderby I.; one (Campbell I.) has recently (1984) been exterminated.

HABITAT

Most feral herds on the mainland now live in thickly bushed or scrub-covered areas, but this is an effect of hunting rather than a direct habitat preference. In earlier times large numbers were recorded out on open grassland in both the North and South Islands, though all populations would have had some bush or scrub in their range.

On scrub- and tussock-covered Campbell I. the feral cattle sheltered in the tall *Dracophyllum* scrub and fed mainly in the tussock grassland, but they restricted their range to about 440 ha on one corner of the 11 700 ha island. This area corresponded almost precisely to the only major area of limestone, which apparently provided essential minerals and a drier substrate (Taylor, 1976).

In the Raukumara Ranges feral cattle live mainly in forest and subalpine scrub, and at Grantham Forest, Hanmer, they inhabit large areas of thick manuka (*Leptospermum scoparium*) and broom (*Cytisus scoparius*) interspersed with clearings of grass, tussock and fern (NZFS, unpubl.). Without exception, feral cattle prefer easy slopes, and are most often encountered in valley bottoms, on flat terraces or on broad ridge tops.

FOOD

In mainland forests feral cattle browse on a very wide range of shrubs and young trees, including *Coprosma* spp., mahoe (*Melicytus ramiflorus*), broadleaf (*Griselinia* spp.), wineberry (*Aristotelia serrata*), kotukutuku (*Fuchsia excorticata*), five-finger (*Pseudopanax* spp.), pate (*Schefflera digitata*), rangiora (*Brachyglottis repanda*), and tawa (*Beilschmiedia tawa*). Ferns, particularly *Histiopteris incisa,* sedges, herbs and grasses are also staple foods (Aston, 1912; R. H. Taylor, unpubl.). The stomach of a feral bull shot in scrubland west of Lake Taupo contained mainly bracken (*Pteridium esculentum*) and grasses with a few leaves of tanekaha (*Phyllocladus trichomanoides*) (Bull, 1950).

The feral cattle on Enderby I. prefer to graze the taller herbs and grasses of the open sward. They are also commonly seen eating kelp (*Durvillaea antarctica*) washed up on the shore, and signs of their browsing have been found on southern rata (*Metrosideros umbellata*), *Myrsine divaricata, Bulbinella rossii, Poa litorosa* tussock, rushes, sedges and ferns (Taylor, 1971 and unpubl.).

Feral cattle are afoot at first light, feeding rapidly until the paunch is full, and then they alternate periods of chewing the cud with grazing throughout the day. Normally they ruminate lying down, but in wet weather they may stand with their backs to the wind. In bush country feral cattle will "walk down" tall saplings up to 6 m high, straddling the stem in order to bend the tops within reach, and then stripping off the leaves.

SOCIAL ORGANIZATION AND BEHAVIOUR

Feral cattle are diurnal, feeding during the day and seeking shelter and sleep at night. Those born in the wild are very shy, and quickly retreat into thick bush or scrub when alarmed.

The normal social units are: (1) family groups, comprising a mature breeding bull and usually up to five cows and their yearlings and calves; (2) small bachelor groups of mature males, and (3) solitary bulls. On Enderby I. there is considerable overlap in the home ranges of family and bachelor groups, but breeding bulls will tolerate the close proximity of other mature males only if none of the cows is in oestrus. The members of a family group maintain a loose association at all times and trot or gallop off together when disturbed, usually led by one of the adult cows (Taylor, 1971, 1976 and unpubl.).

During the breeding season, mature bulls roar, smash down low shrubs and "paw" the ground. At certain strategic sites, near ridge crests or in bush clearings, they make "pawing pits", often 2 m or more across, and use them year after year (Bull, 1950; R. H. Taylor, unpubl.). Otherwise feral cattle are normally silent.

Reproduction and Development

Males reach puberty at about 10 months of age, and thereafter are fecund throughout the year, but feral bulls do not mate until strong enough to compete for cows. Domestic cows can conceive at 6–10 months, but apparently very few do so in the wild, and no two-year-old heifers have been recorded with calves at foot on either Enderby I. or Campbell I. Cows may remain fertile for about 12 years and come in-season in spring or about 3 weeks after calving. The oestrous cycle is 3 weeks, and the gestation period about 9.5 months. Feral calves are most commonly born in late spring, and on Enderby and Campbell Is. they appear from early October to early January. Multiple births are unknown in feral herds. Calves are born with their eyes open, they stand and suckle almost at once, and within a few hours can follow their mother. They are usually weaned well before the next calf is born (Sandars, 1957; R. H. Taylor, unpubl.).

Population Dynamics

Density. By far the largest breeding population of feral cattle is that in the Raukumara Ranges, which is thought to number several hundred (Johns & MacGibbon, 1986). Other mainland populations are smaller, sometimes fewer than ten: all of them, and those in the Chatham Is., are now declining (often quite rapidly) because of shooting and other control measures (NZFS, unpubl.; R. H. Taylor, unpubl.). In the absence of human interference, the herd on Enderby I. (688 ha) has fluctuated between about 35 and 50 beasts over the last 20 years, with an average density of about 6 per 100 ha (Penniket, Garrick & Breese, 1986; Taylor, 1976 and unpubl.).

Sex ratio of adults varies widely between herds, probably because of the very small size of each. From 1963 to 1983 there were many more adult bulls than cows on Enderby I.; on Campbell I. the sex ratio was heavily in favour of cows.

Recruitment and mortality. Observations on Campbell I. over eight seasons between 1969 and 1983 suggested that a maximum of 25% of cows calved in the best year, with an overall average of only 10% calving per year. By comparison, on Enderby I. 47% of adult cows calved in 1965–66 and 67% in 1972–73. Survival of young was also better on Enderby (Taylor, 1971, 1975b, 1976 and unpubl.; Dilks & Wilson, 1979). Mortality of adults on Enderby may be accidental (through trying to reach choice food items near cliff tops or in bogs), or in calving (one case known); most of those found dead, both adults and yearlings, had died of no apparent cause (Penniket, Garrick & Breese, 1986; R. H. Taylor, unpubl.). Nothing is known of adult sex ratios, mortality, birth or recruitment rates in other feral herds in New Zealand.

Predators, Parasites and Diseases

In New Zealand, man is the only predator of feral cattle. Some are shot for skins and pet food, but recovery costs are high and most animals are simply killed as pests and left.

Apart from a record of the mite *Linguatula serrata,* nothing is known of the parasite burdens of feral cattle (Heath, 1976). A mature bull shot west of Lake Taupo in 1950 was examined for ectoparasites and tapeworms but none were found (Bull, 1950). In Northland, feral cattle are certain to be infested with cattle ticks.

In mainland areas where domestic herds are infected with or threatened by bovine tuberculosis, considerable numbers of feral cattle have been shot as potential carriers, but there appears to be no evidence that any were diseased (Hutton, 1976; NZFS, unpubl.). Large cornified growths have very occasionally been seen on the necks of old beasts on Campbell and Enderby Is. (Hutton, 1980; R. H. Taylor, unpubl.), but otherwise both these herds outwardly appeared healthy. However, there is circumstantial evidence that the Campbell I. cattle may have suffered from copper deficiency (Wilson, 1980; Hutton, 1980), which could explain their very low reproductive rate (p. 378).

SIGNIFICANCE TO THE NEW ZEALAND ENVIRONMENT

DAMAGE. Feral cattle can severely modify native vegetation by browsing, crushing and trampling (Aston, 1912; Wodzicki, 1950). In native forests they invariably lay bare the forest floor and eliminate nearly all young trees, shrubs and ferns, until only a few unpalatable or browse-resistant species (e.g., pepperwood, *Pseudowintera* spp., and *Histiopteris incisa*) remain. In subalpine environments feral cattle open up clearings by breaking down and browsing low-canopied vegetation. On Enderby I. they have not only altered the composition of the undergrowth in the southern rata forest, but prevent the regeneration of *Poa litorosa* tussock grassland and a variety of subantarctic endemic herbs, while encouraging the spread of *Bulbinella rossii* (Taylor, 1971 and unpubl.).

CONTROL. Feral cattle present very little danger to man, despite the many far-fetched stories told of the ferocity of "wild bulls": they rarely attempt to attack, except when unwittingly "spooked" at close range, or if wounded. In New Zealand feral cattle are generally thought to have no economic, scientific or aesthetic interest. Most remaining herds live in protection forests, forest parks or nature reserves, and so are in direct conflict with the management priorities and status of the land. Control campaigns are steadily removing them: for example, on Stewart I. (late 1940s), Farewell Spit (1975), Catlins State Forest Park (1979) and Campbell I. (1984). Objectively, the Shorthorns on Enderby I. have a considerable historic and scientific value, but little aesthetic appeal. Unfortunately, they live on a nature reserve, and so the current management plan calls for their complete removal. The growing attitude towards conserving feral farm animals in New Zealand, which is gaining special reserves for feral sheep and horses, is unlikely to influence the future of the remaining feral cattle herds, and they will probably completely disappear within the next 20 years.

R. H. T.

Subfamily Caprinae

The 13 genera and 26 species belonging to this subfamily are adapted to steep terrain in Eurasia and North America, and to climates ranging from hot deserts and steamy jungles to arctic barrens. They are divided into four tribes, of which two are represented in New Zealand. The chamois belongs to the relatively primitive tribe Rupicaprini, which have little sexual dimorphism and short, sharp horns; the tahr, goat and sheep belong to the more advanced tribe Caprini, which have pronounced sexual dimorphism and heavy horns specialized for frontal combat. The distinguishing marks of the four are shown in Table 63.

Tribe Rupicaprini

The four genera of goat-antelopes are *Capricornis* (two species), the serows of Asia and Japan; *Nemorhaedus*, the endangered Asian goral; *Oreamnos*, the American mountain goat; and *Rupicapra*, the chamois.

Genus *Rupicapra*

The single species *R. rupicapra* has about 10 subspecies, of which the nominate "alpine" form, *R. r. rupicapra*, was brought to New Zealand (recent revision of the genus suggests that the alpine form may be granted specific status: Nascetti *et al.*, 1985). The genus *Rupicapra*, probably of Asiatic origin, evolved in Asia Minor and Europe during the Pleistocene (Lovari, 1980).

34. Chamois

Rupicapra rupicapra rupicapra Couturier, 1938

The Latin name is usually abbreviated to *Rupicapra rupicapra* in New Zealand and Central Europe. The common name is of French origin.

DESCRIPTION (Fig. 66)

Distinguishing marks, Table 63, and skull key, p. 24.

The chamois is about the size of a goat, but has longer legs, a longer neck, and larger hooves. It is a graceful animal, with generally trim muscular body lines: males are 7-38% (average 16%) heavier than females. In summer the coat is short and mostly light fawn, though the colour varies individually from grey-brown through tan to a rich honey-gold tone. A dark brown or black band extends from the nose, around the eyes to the base of the horns; the frontal region between the eyes, the nasal ridge, cheeks, and lower jaw and throat are a contrasting white or pale fawn. A dark stripe extends along the nape of the neck and back to the tail. In winter the pelage is longer and thicker and dark brown, almost black. The winter coat of males is heavier than that of females, with a prominent mane and dorsal stripe. Males have long (100-150 mm) erectile hairs (the tips of which have a silver sheen) along the middle of the back. The inguinal region and rump of both sexes is white or pale fawn. Males have a dark, urine-stained pizzle area, a characteristic useful for distinguishing the sexes, which are otherwise not very different in appearance.

Fig. 66 Chamois (G. S. Roberts).

Table 64: Body measurements of chamois in New Zealand.

	Males	Females	n
Contemporary populations in general			
Live weight (kg)	25-45	19-35	
Eviscerated weight (kg)	17-25	13-21	
Total body length (nose to base of tail) (mm)	1 180-1 300	1 150-1 210	
Shoulder height (mm)	650-900	600-800	
Horn length (mm) average	205-235	180-220	
trophy class	250-330	250-330	
Local variation[1]			
Arawhata River (South Westland) 1970-71, late colonizing phase			
Eviscerated weight (kg)	21.4 ± 1.4	20.4 ± 0.4	67
Body length (mm)	1 180 ± 70	1 204 ± 60	97
Horn length (mm)	223.4 ± 3.5	214.0 ± 5.9	96
Avoca River (mid Canterbury) 1975-78, established population			
Eviscerated weight (kg)	16.4 ± 0.8	13.8 ± 0.4	36
Body length (mm)	1 246 ± 80	1 209 ± 50	159
Horn length (mm)	225.1 ± 6.0	184.3 ± 5.7	141

1. Westland and Canterbury data from Henderson and Clarke (1986).

The ears are pointed, internally covered with white hair, and with darker markings externally. The horns are black and slender, rising erect or nearly so from bony cores extending 50-80 mm from the skull, and curving backwards to form sharp hooks at the ends. Those of males are usually stouter and with more strongly developed hooks than those of females. The horns grow throughout an individual's life, though most rapidly during the first 3 years. Narrow annual growth rings, a few millimetres wide, may be seen near the horn base in chamois over 2 years old, although these rings are often indistinguishable in New Zealand. Paired postcorneal glands are positioned behind the horns and in front of the ears in both sexes. The skull is relatively long, with prominent sutures, especially between the various nasal bones. The orbits are large, and eyesight is good.

Dental formula $I^0/_3\ C^0/_1\ Pm^3/_3\ M^3/_3 = 32$.

Graceful, sure-footed and shy, chamois are renowned for their agility in steep terrain. The most common gait is a slow, measured walk associated with feeding, but they are capable of surprising speed. Agonistic encounters between males may result in chases descending and ascending up to 300 m in a few minutes. They are not afraid of water; they often swim rivers and may occasionally swim lakes.

FIELD SIGN

The pellets are about 20-25 mm long, black, oblong, and about 10 mm in diameter, seldom adhering. They are generally smaller and narrower than those of red deer, but similar in size to those of goats. Large accumulations of fresh and weathered pellets may be found in latrine areas, under sheltered bluffs and in alpine and subalpine shrublands.

Tracks may be observed in soil, streambed sand, or snow; the imprints of the digits measure about 25 × 70 mm and have sharply pointed toes with indentations made by the narrow outer rim. The toes of chamois running downhill at speed may be widely splayed. Worn pathways may be found in steep bluff terrain and subalpine shrublands. The most common alarm signal is a high-pitched whistle.

MEASUREMENTS

See Table 64.

VARIATION

Weight, physical size and horn length vary regionally, probably in relation to food resources (e.g., population phase and extent of food depletion), though the relationships between food supply and growth in post-peak populations are poorly understood. Chamois from colonizing populations in south Westland were larger and had significantly bigger horns than their contemporaries from central Westland (Lambert & Bathgate, 1973). The differences in body weight between central and peripheral populations may be considerable, up to 33% of eviscerated weight in favour of the peripheral animals (Bauer, 1982). Chamois from the Avoca River, Canterbury, were large framed but weighed less than Westland chamois (Table 64). The weight differences suggested better habitat and food resources for Westland chamois, but the large skeletal size of Avoca chamois was unexplained.

The colour of the summer coat varies to some extent, perhaps in relation to the condition or fatness of the animal. Pure white pelage (white phenotype) is rare and found only in the St. James and Amuri Ranges, North Canterbury (C. M. H. Clarke, unpubl.); genetic albinism is exceedingly rare in New Zealand. The possibility of genetic variation in New Zealand chamois, with reference to the population of origin, has not been investigated.

HISTORY OF COLONIZATION

The chamois was introduced into New Zealand from Austria in 1907 and 1914. The liberation stock was a gift to the New Zealand Government from the Emperor Franz Josef of Austria. There is some doubt as to the exact origin of the animals (C. M. H. Clarke, unpubl.): some authors report that they were captured near Mürsteg in the Austrian province of Styria, though others say they came from Ebensee, 150 km north-west of Mürsteg in upper Austria.

The foundation stock consisted of nine chamois released near Mt. Cook, eight (two males and six females) in 1907 (Fig. 67), and a further pair in 1914 (though only the female survived; the male was shot by a guide after it attacked some tourists: R. E. Lambert, unpubl.). Chamois then went through a classic irruptive colonization phase, similar to that described for other species (e.g., red deer, p. 448), although with some notable differences. Chamois were remarkably successful colonists: they dispersed sooner, further and faster (average 9.6 km/year for 1907-63; Christie, 1963) from their liberation point

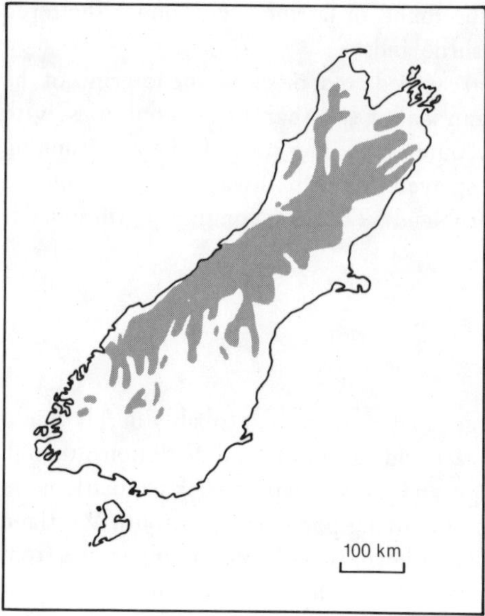

Map 33 Distribution of chamois in New Zealand in 1985.

than other introduced ungulates; and they dispersed in small groups, generally before resources became limiting. Between 1907 and 1920 chamois colonized an area of 2644 km²; by 1930, 7765 km²; by 1940, 16 175 km²; by 1950, 23 406 km²; by 1960, 31 118 km²; and by 1970, 36 136 km² (R. E. Lambert, unpubl.). At first they dispersed mainly north-east along the alpine chain, with subsidiary movements to the outlying ranges; extension to the south-west followed later. The linear rates of dispersal varied with direction of spread and from decade to decade. The maximum recorded rate for the northern population was 13.3 km/year from 1920 to 1930, though they probably reached 14-20 km/year between 1910 and 1920. By comparison, the maximum dispersal rate south from Mt. Cook was much slower (5.7 km/year).

Dispersal was initiated mainly by yearlings, though solitary adult males were also observed moving into new areas far ahead of the breeding population. Small numbers of chamois formed scattered breeding colonies, and later filled in the intervening areas as the population increased (R. E. Lambert, unpubl.). From the speed of dispersal and rapid rise in numbers, it is clear that the colonization phase of each local population was short, perhaps only about 10 years.

By the 1960s chamois had colonized most of the major mountain ranges along the southern axis of the South Island and were spreading into Fiordland and the mountains north of the Buller River. Although chamois have been sighted throughout Fiordland, their numbers have since shrunk because of intensive hunting pressure: the southernmost populations in the Kepler and Murchison mountains are small. Areas colonized during the last decade or so include the drier open ranges in South Canterbury and Otago (e.g., Hunter Hills, Kirkliston Range, the Remarkables, Garvie Mountains), the Tasman

Fig. 67 Preparing for the first liberation of chamois at Mt Cook in 1907. Crated chamois were conveyed via the Hooker cableway to the north bank of the Hooker River, where they were released close to the Tasman Valley. Tasman River and Mt Blackburn in background (Collett collection, Alexander Turnbull Library).

Mountains in the north, and the forests and rocky low altitude sites from the Wanganui to the Karangarua rivers in Westland.

DISTRIBUTION

WORLD. Chamois are native to most mountains, hills, gorges and high plateaux from Asia Minor (Turkey, USSR) across central and southern Europe to the Cantabrium Mountains in north-western Spain. The alpine form *R. r. rupicapra* is native to the central European Alps, from northern Yugoslavia and northern Italy through Austria, Switzerland and Germany to eastern France.

NEW ZEALAND (Map 33). Next to red deer, chamois are probably the most numerous and widespread wild ungulates in the South Island. They are found throughout the high country, mostly in alpine and subalpine areas but extending to low elevations in some areas (e.g., central Westland). They are mostly absent from the eastern and southern Otago ranges. Their distribution is related to the extent of favourable terrain and vegetation and the phase of colonization attained before aerial hunting hampered their unrestricted spread; in recently colonized districts subject to harrassment, there are only scattered colonies (e.g., in north-west Nelson, in Marlborough north of the Wairau River, and in Fiordland), but in areas colonized before 1970, chamois are more widespread. Solitary animals and occasional small groups may be seen far outside the usual range. In forested areas they are hard to observe, and their distribution is poorly defined.

HABITAT

Chamois occupy a wide range of mountain habitats, including alpine bluffs, alpine grasslands/herbfields, subalpine shrublands, and forest and rocky sites to sea level. Steep rugged areas are favoured, chosen largely for features such as slope, shelter and bedding sites (Christie, 1963; Espie, 1977; Clarke 1986). In alpine areas, chamois commonly concentrate on the zone within about 300 m of the timberline; in the Avoca River area they mainly use patchy grasslands and bluff grasslands, avoiding the larger areas of alpine vegetation on gentle slopes; within the eastern beech forests they use the zone immediately below the timberline, plus forest bluffs, scree margins, and also cool streambeds during summer; in the western rata/kamahi forests, they extend over a wide altitudinal range, living mainly on forest bluffs, ridges and spurs, and slips.

Where forests have been destroyed by fire (or were originally absent) they may extend to lower elevations, but chamois are increasingly preferring habitats affording cover and refuge from aerial hunting. Overall, the wily chamois is well adapted to the New Zealand high country, although its spread to low altitude forests and rocky places (e.g., river gorges) shows that it is not restricted by the extent of high country.

FOOD

Chamois are essentially grazers, but eat a wide range of plants, not necessarily in proportion to their abundance in the habitat. In the Avoca River region they feed mostly on herbaceous species and short sward grasses, e.g., *Poa colensoi, Rytidosperma setifolium* (Clarke, 1986). Local diet varies considerably in the summer, but in winter they eat more shrubs (*Dracophyllum* spp.). The three principal foods of chamois in the Cupola Basin, Nelson, are the short-sward grasses *Poa colensoi* and *Festuca* spp.; snowgrass (*Chionochloa* spp.); and shrubs (*Dracophyllum* spp.) (Batcheler & Christie, quoted in Gibb & Flux, 1973).

SOCIAL ORGANIZATION AND BEHAVIOUR

ACTIVITY. Chamois are mostly diurnal, though there may be extensive nocturnal activity between spring and summer (Clarke, 1986). In summer, chamois feed extensively from dawn for 3-4 hours, rest until 3-4 hours before dusk, and then resume intensive feeding. In winter, feeding is much less intensive and confined mostly to mid morning and mid afternoon (Christie, 1967).

DISPERSION. Home range size of females in Basin Creek (Avoca River) varied between 138 and 656 ha with a mean of 341 ha (Clarke & Henderson, 1984). Home ranges are largest in summer (mean 207 ha) and smallest in winter (mean 70 ha). Dominant females had significantly larger home ranges than subordinate animals. The mean daily distance travelled by chamois in Basin Creek varied between 0.2 and 0.5 km in adult females, and 0.2 and 0.7 km in adult males (C. M. H. Clarke, unpubl.). Males wandered about less predictably than females.

Over the whole Avoca River region (1000 km^2) there were several separate subpopulations of resident and migrant chamois (Clarke, 1986). Most were resident and tended to aggregate into distinct preferred areas varying in size from 1.7 to 7.1 km^2. The resident overwintering population in the Avoca study area comprised mainly females and subadults. Migrant chamois, more often males than females, ranged over a 6000 km^2 area of Canterbury and Westland, from summer or breeding ranges in Canterbury to wintering areas near the main divide or in Westland; they travelled along definite migration routes up to 60 km. The mean distances travelled by males (average 18.7 km), females (18.2 km) and unknowns (21.6 km) were not significantly different.

Chamois have a matriarchal social system. Females and young form loose unstable groups; males are mostly solitary. The only individual bonds are between mother and kid (Krämer, 1969). Chamois tend to be widely dispersed rather than gregarious. In Basin Creek (Avoca River, Canterbury), even though the home ranges of resident females overlapped, individuals spaced themselves widely (Clarke & Henderson, 1984). Only half of the animals sighted were in groups of two or more. The large herds reported in Europe and during the irruptive phase in New Zealand are not seen in the present, low-density New Zealand populations.

A loose social hierarchy exists among resident females (C. M. H. Clarke, unpubl.), among which the large-bodied females are dominant (Clarke & Henderson, 1984); but these relationships change frequently in a population influenced by seasonal migration and dispersal. Residents, regardless of age, have higher social status than migrants. Chamois use imposing behaviour to assert superiority, rather than territorial defence. Generally only a few males are territorial. Males are very intolerant of each other during the rut; broadside imposing displays are seen mainly between animals of similar rank and may lead to vigorous chases, although overt aggression, particularly frontal attacks with the horns, is uncommon in New Zealand. Scent marking, mainly of shrubs, may be observed throughout the year, although the glands of adult males are more fully developed during the breeding season. Marking is a form of imposing behaviour not specifically related to territoriality (Krämer, 1969).

REPRODUCTION AND DEVELOPMENT

Most chamois are sexually mature at 30 months, approximately half (48.5%) at 18 months (Henderson & Clarke, 1986), and a few at 6–7 months (Caughley, 1970d; Bauer, 1982).

The rut begins in early-mid May and peaks in late May to early June, occasionally continuing among young animals (<2 years) until mid July.

Gestation is variable, between 5½ and 6 months. The single kid is born during the 3 months from November to February (Christie, 1963; Clarke, in press). The median birth date in the Rangitata catchment (South Canterbury) was 23 November (Caughley, 1971b); in the Avoca Valley (mid Canterbury) it was in early December in the older adults, and late December to February in the 2-year-olds producing their first kid (Clarke, in press).

Birth weight averages 2.7 kg in the Avoca (Bauer, 1982) and elsewhere ranges between 2.4 kg and 3.2 kg.

Kids are extremely agile and are able to follow mothers within an hour of birth. Young kids are particularly playful.

The growth rate varies with food supplies. In colonizing populations and in exceptional cases individual chamois may attain 15 kg (60% adult weight) by 7 months (Bauer, 1982); most are fully grown by 3 years. Adult males are most fat in March-April and least fat after the rut in June-July, during which up to 80% of reserves are expended. Adult females are most fat in May-June and least during the birth season (Henderson & Clarke, 1986).

POPULATION DYNAMICS

The foetal sex ratio is 1:1 (Bauer, 1977; Henderson & Clarke, 1986), but the sex ratio of adults is generally imbalanced towards females (0.66–0.76 males:1 female), largely because mortality is higher among both juvenile and older males (R. J. Henderson, unpubl.). The sex ratios of local populations may vary considerably, because the sexes differ, not only in mortality rates but also in the proportion of residents to migrants; and the age structures within each sex are different in colonizing compared with long-established populations. The fecundity rate in New Zealand is high and has no significant local variation (Henderson & Clarke, 1986). After 2 years of age, 80–90% of females produce young in any given year; yearlings give birth most often in colonizing populations (Bauer, 1977) and in favourable years in long-established populations (e.g., Basin Creek) (Clarke, in press).

The size, structure and dynamics of the Basin Creek chamois population were studied in detail between 1973 and 1978 (Clarke & Henderson, 1981; Clarke, in press). Between 45 and 70 chamois occupied an area of 1400 ha, i.e., 3.2–5.0/100 ha. Numbers fluctuated within narrow limits, according to the seasonal availability of food and the spacing behaviour of individuals. Numbers were dynamically balanced by mortality, dispersal and outward migration offsetting high reproductive rates, inward migration, and recruitment of breeding females. The turnover rates of marked animals moving in and out of the Basin was high (about 25% of the marked population per month). The productivity rates of females ($\geq$2 years) varied from 77% to 100% (mean 91%). The mortality rates of kids up to 8 months old varied from 23% to 82% (mean 47%).

Annual growth layers in the tooth cementum are used for estimation of age in New Zealand chamois. Annual growth rings on the horns are used only by European workers, e.g., Bauer (1977). Horn growth rings are less distinct in New Zealand than in Europe.

The mean life expectancy of male and female chamois at birth was 2.81 years in Basin Creek and 3.47 years for female chamois in the Avoca River region (R. J. Henderson, unpubl.). Avoca females surviving to 2 years old had a further life expectancy of 5.61 years. Maximum age in New Zealand is about 19 years in females and 14 years in males.

PREDATORS, PARASITES AND DISEASES

Chamois have no predators in New Zealand other than man.

Damalinia longicornis is the sole ectoparasite recovered from chamois in New Zealand (Tenquist & Charleston, 1981). Gastrointestinal nematodes recorded in New Zealand include: *Ostertagia ostertagia, O. circumcincta, O. trifurcata, Oesophagostomum venulosum, Trichurus ovis, Trichostrongylus axei, T. vitrinus, Chabertia ovina, Spiculotragia spiculoptera, S. asymmetrica, Nematodirus spathiger, N. filicollis* and *N. abnormalis* (Andrews, 1973; Clark & Clarke, 1981). *Nematodirus abnormalis* does not appear to have been recovered from *R. r. rupicapra* before, and it was probably acquired from sheep grazing the same range. The only cestode recorded is *Moniezia expansa*. Lungworms (*Muellerius capillaris, M. tenuispiculatus* and *Dictyocaulus* sp.) are common, and the infection is often intensive, increasing with age, but lungworm infection alone is seldom fatal. Epidemics of European chamois blindness (keratoconjunctivitis) were common among high density populations in the Southern Alps between 1936 and 1962 (Daniel & Christie, 1963). Christie (1965) described localized high mortality of up to 50% of a population; chamois were seen dead and dying in hundreds. Epidemics of the viral disease facial eczema (*Ecthyma contagiosum*) were also common between 1940 and 1962. Neither disease appears to have had any effect on overall population density.

Periodontal disease may cause significant tooth loss in up to 20% of chamois in Canterbury/Westland, increasing in severity with age, though apparently not affecting the health of the animals (Pekelharing, 1974).

Bacterial pneumonia resulting from infection with *Pasteurella,* possibly linked with a virus, was a serious disease of juvenile chamois in the Avoca River region in 1976–78 (Henderson & Clarke, 1986). Occasional episodes of this disease were associated with late spring thaws and poor summer-autumn weather and caused high mortality, but may not be typical of chamois populations in New Zealand generally.

ADAPTATION TO NEW ZEALAND CONDITIONS

Environmental conditions encountered by chamois in New Zealand are distinctly different from those in Europe, and the New Zealand chamois may have developed a number of adaptations in response, including earlier maturation, greater productivity and a looser social structure which favours dispersal and migration.

Early breeding may be the result of the particular light conditions and extended growing season in New Zealand (Bauer, 1985), although early sexual maturation of the under 2-year-old females probably inhibits their growth rate. European chamois sometimes produce twins (never in New Zealand), but most do not begin breeding until aged 3 or 4 years. Consistently high productivity in both pre- and post-peak chamois populations in New Zealand may be the result of especially favourable environmental conditions.

Chamois from established New Zealand populations are considerably smaller than those in Europe. The average eviscerated weights of adult males (21.9 kg, n = 113) and adult females (18.1 kg, n = 154) from three post-peak populations were 21.3% and 17.4% lighter than adult European chamois from comparable populations (C. M. H. Clarke, unpubl.). Weight differences between the sexes are less marked in New Zealand than in Europe. On the other hand, weights of New Zealand chamois from pre-peak populations are similar to or slightly heavier than those from colonizing populations in Europe.

The local movements of chamois in New Zealand (often sufficient to be called migrations) are much more wide-ranging than in Europe and may be seen as an adaptation to the different climate and food resources available east and west of the Southern Alps (Clarke, 1986). Alpine areas east of the Main Divide support a generally sparse vegetation cover, consisting mostly of tall snowgrasses seldom eaten by chamois; food resources palatable to chamois are patchy and can be utilized only if chamois spread their activities thinly, which in turn entails a loose social organization. Western areas have more plentiful food resources, and a milder climate over winter. In summer, chamois favour drier eastern grassland/herbfield areas, but these cannot support summer numbers throughout the year, so many animals must migrate to western winter ranges. Most migratory chamois return to their birth places at least once a year.

Food resources for chamois are apparently more abundant in the European Alps; local densities of chamois are higher, and large herds gather on both summer and winter grounds (Schröder, 1971; Hamr, 1984).

The unpredictable New Zealand climate, particularly in summer, was mentioned by Christie (1967) as a cause of changes in activity and feeding intensity of chamois in Cupola Basin, Nelson. Chamois were sensitive to both hot and cold temperatures during summer and were less active and fed less under these conditions compared with settled or "average" weather; in winter they were seldom stressed by temperature extremes, but activity and feeding were much reduced compared with summer. These behavioural characteristics are believed to reflect adaptations to the climate of New Zealand, which is different (much less severe over winter, but less stable in summer) from that of their native Europe.

Niethammer (1971) compared the teeth of Austrian and New Zealand chamois and found greater dental variability and heavier wear in the New Zealand sample, suggesting a coarser diet for chamois in New Zealand. Mortality patterns in New Zealand are different from those in Europe, since (1) several serious diseases of chamois in Europe, including "sarcoptic mange", caused by the burrowing ectoparasite *Sarcoptes rupicaprae,* are not present in New Zealand; and (2) the winter climate is milder and over-winter survival is high. By contrast, mortality from disease and winter in the European Alps is generally severe, and has a major effect on populations.

Although naturally elusive animals, chamois in New Zealand have developed a number of behavioural patterns to avoid aerial hunting, including hiding in minimal cover, freezing motionless, dashing headlong for cover, moving nocturnally in open areas, and a preference for forested and shrubby areas. In alpine areas of the Avoca River region, chamois hid among rocks or under tall snowgrass tillers, remaining there for up to 3 hours (C. M. H. Clarke, unpubl.), even when only a distant helicopter was heard.

SIGNIFICANCE TO THE NEW ZEALAND ENVIRONMENT

Chamois were introduced for sport, but their number increased so rapidly that control measures became necessary within 25 years. Culling by government hunters on foot began in the 1930s; they removed about 90 000 chamois by 1982, but made little impact on the burgeoning population. Aerial hunting was much more effective, and during the 1970s this increase in pressure greatly reduced the then existing numbers (up to 200 were shot per day from a single helicopter), particularly in recently colonized areas, but it has not restricted the established distribution or prevented further spread (see p. 384). Together with the other introduced browsing and grazing animals, chamois have contributed to the vastly modified state of the mountain vegetation today. The areas that they harmed most were the alpine herbfields and shrublands, and sparsely vegetated rugged areas. Recently, some improvement and recovery of the vegetation has been noted, although there are probably some areas where chamois numbers are still high enough to prevent it.

Chamois have three economic values: (1) carcasses recovered by helicopter are sold by the "game meats industry", a very lucrative trade in the 1970s, which still continues though at a much reduced level (524 carcasses exported in 1985, average value NZ$45 each); (2) since the chamois is an internationally recognized and highly prized trophy animal, market prices for mounted trophies (>200 mm horn length) are substantial (1986 price from NZ$500 to $1000 each) and professional hunting guides are active in many areas (fees NZ$300–$500 per person per day plus trophy fee, total up to NZ$2500); and (3) sales to tourists of whole mounted chamois kids, winter skins, and mounted horns and heads below trophy specifications. Chamois also have a continuing attraction for recreational (non-commercial) hunters. However, chamois farming seems unlikely to develop, as those taken in the wild do not settle well into captivity, and even captive-bred chamois are liable to heavy losses. In contrast to red deer, therefore, chamois are seldom captured alive, except occasionally for export to zoological parks, or for release into enclosed hunting areas (e.g., Lilybank Station).

C. M. H. C.

Tribe Caprini

This group includes 17 species in five genera: *Hemitragus*, the three species of tahr; *Ammotragus*, the Barbary sheep; *Pseudois*, the blue sheep; *Capra*, the six species of goats, including the ibex and tur; and *Ovis*, the six species of sheep, including the urial, mouflon, argalis and American bighorn. They are advanced caprids with moderate to large differences between the sexes in size and appearance. The males have a well-developed social hierarchy, and the place of each individual in it is determined by confrontations, either ritualistic, aided in some species by long display hair, or actual, by frontal combat using the horns. Three genera are represented in New Zealand, by one species each. The distinguishing marks of the three are shown in Table 63. In addition, three

Himalayan bharal (*Pseudois nayaur*) were imported by the government and liberated at Mt. Cook in 1909, but died out (Thomson, 1922).

Genus *Hemitragus*

Hemitragus has three living species, of which two (*H. hylocrius*, the Nilgiri tahr from southern India, and *H. jayakari,* the Arabian tahr from Oman) survive only as isolated refugee populations and are classified as vulnerable or endangered. The third, *H. jemlahicus,* from the Himalayas, is somewhat threatened though not yet endangered in its homelands, and has also been introduced to several other countries. All tahr are well adapted to steep country; both sexes carry horns, and neither has malodorous skin glands; the females are about 60% of the size of the males.

35. Himalayan tahr

Hemitragus jemlahicus (Smith, 1826)
Synonym *Capra jemlahicus* Smith, 1826.

The English transliteration of the Nepali common name may be spelt tahr or thar; neither is "correct" (Caughley, 1971a; Levine, 1985), but tahr refers unambiguously to *Hemitragus*, and is more common in the international English-language scientific literature.

DESCRIPTION (Fig. 68)
Distinguishing marks, Table 63 and skull key, p. 24.

A mature bull in winter pelage has a dark face and muzzle; short, massive horns; and a huge shaggy mane of long (to >250 mm) hair hanging from the neck and shoulders. The mane is usually slate-grey to straw coloured, and the sides and hindquarters black to reddish-brown, the underside lighter; older animals tend to be darker, with a lighter band alongside the flanks, accentuating the darker mid-dorsal line. Bull tahr have a small red-brown rump patch. Younger bulls are more uniformly brown, and the mane is less conspicuous. Adult nannies are grey to rich brown in winter with a darker muzzle and legs, and light undersides. There is a vestige of a ruff on older nannies, and a darker mid-dorsal line on some. Younger nannies and kids are more uniformly brown, except for the black frontside of the legs; they are somewhat similar to brown goats (but these are found only outside tahr range) and chamois (but these have a slimmer build and longer, more curved horns).

In spring, both sexes shed the fine, dense, insulating wool-like underhair and the guardhairs progressively bleach to a straw colour. Tahr regain their winter coats between February and April.

The horns, present in both sexes, measure up to about 330 mm in mature bulls (maximum recorded 378 mm) and 190 mm in nannies. They are short, laterally flattened, curve sharply backwards, and have a prominent keel on the front edge. The horns grow by diminishing annual increments, so horn rings are laid down each winter, after the first, making individuals easy to age (Caughley, 1965b).

Fig. 68 Two mature bull tahr caught in courtship and agonistic displays in a disturbed group. The bull at far left assumes a typical courtship pose to a female (out of sight); the other bull is in a lateral aggressive display, directed at the first. Carneys Creek, Rangitata Valley, May 1963 (NZFS).

Both sexes have excellent eyesight, narrow pointed ears, a bare underside to the tail, no beard and no preorbital, inguinal or pedal glands. Mature males have a strong body odour over the rutting period, and exceptionally thick hides. Chromosome number 2n = 48.

Tahr are possibly the most accomplished climbers of all the ungulates and are particularly well adapted to life in precipitous and rocky terrain. The hooves have a hard rim of keratin surrounding a spongy-soft convex pad; there are well-developed dew claws; the sternum and the backs of the legs have pads of short dense hairs, which add traction on steep rock surfaces.

Nannies are much smaller than bulls (Table 65) and shorter-haired. They have four teats, although the anterior pair are rudimentary.

Dental formula $I^0/_3\ C^0/_1\ Pm^3/_3\ M^3/_3 = 32$.

FIELD SIGN

Tracks, evidence of feeding and pellets may be found in vegetated rock bluff systems, especially within the tahr's preferred altitude range of 1200–1800 m. Cloven hoofed imprints, typically about 45 × 35 mm, are blunt and squarish in shape. Tahr make clear pathways, especially around rock obstructions or throughout scrublands or bluffs; they leave well-used sidlings on scree slopes or horizontal tracks across snow slopes between rock outcrops. Browsing and grazing sign can be found on preferred food species (see below) in high alpine scrubland or tussock slopes. Pellets, about 15 × 7 mm, are typically brown to dark brown and uniformly cylindrical in shape. They are found in large piles,

Table 65: Body measurements of Himalayan tahr in New Zealand.

Age[1]	Males			Females		
	n	mean	SD	n	mean	SD
Body length (mm)[2]						
0-9	88	1 032	67.6	78	995	69.6
1-9	57	1 208	72.7	55	1 192	71.0
2-9	24	1 351	70.0	36	1 271	46.6
3-9	10	1 463	55.0	41	1 310	67.9
4-9	2	1 565	35.4	40	1 331	66.5
5-9	2	1 645	21.2	29	1 327	67.5
6-9	3	1 587	30.6	19	1 345	55.6
7-9+	4	1 702	96.1	53	1 355	62.2
Carcass weight (kg)[3]						
0-9	83	11.01	2.817	71	9.38	1.989
1-9	64	16.62	3.028	50	16.23	3.573
2-9	23	23.50	2.448	34	19.53	2.463
3-9	9	31.20	4.142	40	20.75	2.832
4-9	3	38.07	5.400	41	22.10	3.334
5-9	3	46.60	5.851	29	23.04	2.359
6-9	4	47.42	7.001	21	23.24	2.697
7-9+	3	48.67	1.155	51	23.77	3.294
Horn length (mm)[4]						
0-9	80	105.7	16.09	74	90.1	17.18
1-9	79	165.4	14.94	62	129.3	17.14
2-9	35	199.0	14.91	51	144.2	11.86
3-9	15	230.4	26.91	59	158.8	13.26
4-9	6	265.5	17.10	50	168.1	14.73
5-9	4	290.7	14.31	37	173.7	13.12
6-9	5	292.4	7.13	28	177.7	13.32
7-9	3	292.0	4.58	13	184.8	8.10
8-9	1	275.0	—	9	177.7	14.23
9-9	3	327.3	15.04	12	192.2	14.11
10-9+	1	295.0	—	36	193.2	16.89

1. Age in years and months.
2. Body length measured along the dorsal line from nose-tip to tail-tip.
3. Carcass weight from carcasses with skin retained but without viscera or heads, and with legs removed at the hock. Carcass weights are about two-thirds live weight. Adult males sampled here are somewhat lighter than average; the range of live weights often exceeds 125 kg.
4. Horn length measured from base of horn to horn-tip along the keel.

Data from K. G. Tustin (unpubl.).

especially in bedding places among rock bluffs or overhangs or on high ledges. The energetic observer may hear the rattle of dislodged stones and the shrill birdlike alarm whistle, and see the silhouette of inquisitive heads on a distant skyline.

MEASUREMENTS

See Table 65.

VARIATION

There is little variation in size or colour between individual tahr of the same age group. A number of tahr with white pelage have been seen in the Godley-Macauley Valley areas.

HISTORY OF COLONIZATION

Tahr were officially introduced by the New Zealand Government (under the then Tourist and Health Resorts Department), which wanted to create a hunting resource for sportsmen from New Zealand and, especially, overseas. The original 13 animals were gifted by the Duke of Bedford from his captive herd at Woburn Abbey, England, and released near the Hermitage at Mt. Cook: five (three females and two males) in 1904, and a further eight (two females and six males) in 1909 (Donne, 1924; Caughley, 1970a).

In 1919 four more (sex unknown) were added from the Wellington Zoo (Anderson & Henderson, 1961). The wild population established rapidly; groups of 13 were reported in 1915 (Thomson, 1922), 18-20 in 1916, and 50 in 1918; by the early 1920s the herd numbered over 100 animals (Donne, 1924).

Two other liberations failed to establish independent populations. One was in the North Island (three tahr) near Rotorua in 1909, and one at Franz Josef Glacier in Westland (four tahr) in 1913 (Thomson, 1922; Donne, 1924; Caughley 1970a).

The spread of the Mt. Cook population has been described in detail (Caughley, 1970a). By the mid 1930s breeding animals were dispersing northwards through the Murchison Valley, and southwards via the headwaters of the Dobson River. A decade later, tahr were present from the north bank of the Godley southwards to the headwaters of the Hopkins, and had begun colonizing west of the Main Divide of the Southern Alps into the Landsborough Valley, probably via the Elcho Pass.

By 1956 female tahr were established in the headwaters of both main tributaries of the Rangitata and were colonizing north Westland via the Perth, probably through the Sealey Pass from the Godley. To the south they extended their range in the Dobson, Hopkins and Huxley catchments, and had reached the Karangarua, Douglas and Copland valleys in South Westland.

By 1966 they were present in the Rakaia headwaters and in north Westland, including the Whataroa, Wanganui, Whitcombe and Hokitika valleys, from which they had colonized the Mathias tributary of the Rakaia. To the east they extended their range to include some of the Two Thumb and Ben McLeod ranges and most of the suitable terrain between the Godley and Tasman valleys; to the south the Ben Ohau Range and eastern bank of the Hunter almost to Lake Hawea. The south-western population meanwhile had expanded and included the Jacobs Valley.

By the mid 1970s further expansion had closed the "gap" between the north-western and south-western populations, and almost the total area of preferred tahr habitat available was occupied; beyond it, the rock bluff systems decline in size and scale, and tahr are more vulnerable to hunting. In recent times,

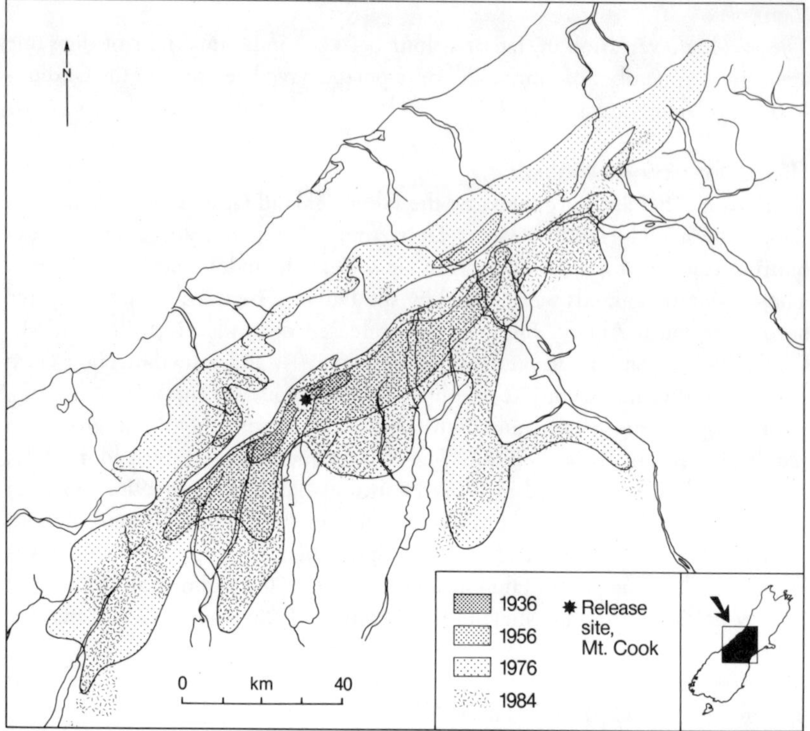

Map 34 Distribution of breeding tahr in New Zealand, at three stages of colonisation (maximum extent 1976) and since the recent retraction of range, last surveyed in 1984 (Parkes & Tustin, 1985).

aerial hunting has restricted the distribution of breeding populations to below their 1976 maximum (Parkes & Tustin, 1985).

From their liberation point at Mt. Cook, breeding populations of tahr expanded at between 0.64 and 3.84 km per year to the north, and 0.32 to 1.92 km per year to the south. Wandering males have frequently been recorded some distance outside the female breeding range, and precede them often by many years (Caughley, 1970a). As soon as the first female tahr arrived in an area the local population irrupted, and reached peak numbers after about 15-20 years (Caughley, 1970b; Tustin & Challies, 1978).

DISTRIBUTION

WORLD. The native range of the Himalayan tahr is a narrow zone along the southern flanks of the Himalayas, from the Pir Panjal Range in northern India eastwards through Nepal and Sikkim to Bhutan. Tahr were once continuously distributed in preferred habitat throughout this range, but are now reduced to isolated remnants by the increasing demands of settlement, hunting and land use (Schaller, 1977).

Tahr have also been introduced successfully in California and Ontario and on Table Mountain in South Africa (Schaller, 1977).

NEW ZEALAND (Map 34). The expansion of the Mt. Cook colony was steady from the time of liberation until about the mid-1970s, when some 6150 km^2 of subalpine terrain was inhabited by breeding populations (i.e., those including female tahr: Parkes & Tustin, 1985). At this time the northern limits bordered on the Waimakariri and Arahura catchments and included the suitable habitat of all major catchments with their origins against the Southern Alps Main Divide, both east and west, to a southern limit which included the Paringa Valley headwaters in South Westland, the upper and mid sections of the Landsborough Valley, and eastwards to the headwaters of the Wills and Makarora, most of the Hunter catchment, and the headwaters of the Dingle and Ahuriri rivers. To the east of the Main Divide they occupied the more precipitous country on the major lateral ranges, including the Ben Ohau, Two Thumb, Ben McLeod and Arrowsmith ranges.

However, ten years of intensive aerial hunting reduced the breeding range to about 4950 km^2 by the mid 1980s. The major boundary changes induced were in the more recently occupied, sparsely populated country, especially on the northern, north-western and southern limits, and a general contraction of the range in Westland (Parkes & Tustin, 1985). Whether the present breeding range will be sustained, or further reduced, depends to a large extent on the future politics of tahr management and control (p. 406). A possible biological limit to future dispersal is that female tahr tend to move into new areas only when density on their natal area is high: hence, the present populations are unlikely to extend to or beyond their former range while their densities generally remain low (Parkes & Tustin, 1985).

HABITAT

Massive rock bluff systems, the adjacent snow tussock basins, and the uppermost zone of the subalpine scrubland are the preferred habitat of tahr in New Zealand. Although they may range up to 2250 m and down to 750 m, they are usually found in the 1400–1700 m zone on mountain ranges where ridge crests exceed 1800 m. North and north-east facing slopes are preferred, because they are sunnier and less likely to accumulate snow in winter. Tahr tolerate rainfall ranging from over 7500 mm in the valley headwaters west of the divide to 1500 mm only 35 km to the east. The treeline on the western side of the Southern Alps is at about 900 m; then there is a dense band of subalpine scrubland up to 300 m wide; then alpine grassland and herbs, to a vegetation limit of about 1800 m, beyond which alpine barrens and permanent snow predominate. The country is typically very precipitous. On the eastern side the forest is patchy, with a short scrub zone. The central part of the range is mostly open; pockets of remnant podocarp-broadleaf forest are scattered through large areas of tussock grassland and scrublands, extending to the valley floors and broken by extensive screes and avalanche gullies.

Ecologically and climatically, tahr range in New Zealand is equivalent to that of their homelands in the Himalayas. There they prefer the subalpine and alpine zones at 3500–4500 m (Caughley, 1969), but also seasonally use a wide altitudinal range of habitat, from the mixed oak forests at 2500 m through the higher rhododendron and conifer zones to the alpine meadowlands around 5000 m (Schaller, 1977).

FOOD

Snow tussocks, alpine herbs and some subalpine scrubland plants make up the main diet of tahr, but the North Branch (Godley Valley) population used different forage types according to season. Within its home range four main vegetation communities were recognized: (1) matagouri scrub (*Discaria toumatou* dominant); (2) short tussock grassland (*Festuca novae-zelandiae, Agrostis tenuis* and *Anthoxanthum odoratum* dominant); (3) short podocarp scrub (*Podocarpus nivalis* dominant); and (4) tall snow tussock (*Chionochloa rigida* dominant).

Generally, tahr made most use of (3) and (4), although for over half the daylight time spent in them they were resting or moving, rather than feeding. In spring, about 25% of total feeding time was spent in each of these two communities, increasing progressively to over 60% in winter; short podocarp scrubland on the steeper sites was favoured in deep snow conditions in September, and the tall snow tussock communities earlier. The other two communities, (1) and (2), were used mostly for feeding, especially in early spring, since they were first to flush. Tahr spent 40% of their daylight time feeding in the matagouri in October and November, but then progressively less and less until April, when it was practically unused until the following spring. They spent about 20% of their time in the short tussock grassland in spring, and then suddenly much less after December, when dry conditions limit growth lower down and the higher altitude communities flush. In early autumn there is another period of growth, and tahr returned to the short tussock grassland for nearly 20% of their feeding time in March–April. The tahr did have some particular preferences: the flowerheads of *Aciphylla* and the seedheads from the montane flax, *Phormium cookianum*, were keenly sought in the summer months and, to a lesser extent, so were the seedheads from *Chionochloa* and the flowers of the mountain daisy, *Celmisia*.

In another study, plant fragments from the stomachs of 72 tahr shot in the adjacent Macaulay Valley in August 1975 were macroscopically identified (Tustin & Parkes, 1988). Ranked by frequency of occurrence, the most important items were *Chionochloa* spp. in 100% of the stomachs, *Podocarpus nivalis* (87%), *Celmisia* spp. including *C. coriacea, C. augustifolia, C. lyallii* (78%), *Dracophyllum uniflorum* and *D. pronum* (71%), *Gaultheria crassa* (37%), whipcord *Hebe* spp. including *H. lycopodioides, H. cheesemanii* (30%), small grasses including *Poa colensoi, Rytidosperma setifolia, Agrostis tenuis* (29%), *Helichrysum selago* (21%), *Coprosma pseudocuneata* (19%), *Hebe pinguifolia* (18%), *Cyathodes colensoi* (18%), *Raoulia eximia* (17%), *Hymenanthera alpina* (12%), *Olearia cymbifolia* (12%), *Aciphylla* spp. including *A. aurea, A. similis* (11%), *Coprosma parviflora* (10%) and *C. cheesemanii* (10%), *Myrsine nummularia* (7%), *Gaultheria depressa* (4%), *Pittosporum anomalum* (4%) and *Anisotome flexuosa* (4%).

Twenty-four rumens were subsampled and their contents separated and dried before weighing: 92.6% of the dried weight contents were *Chionochloa* spp., 3.8% *Podocarpus nivalis* and 1.7% *Dracophyllum* spp. (Tustin & Parkes, 1988).

Feeding occupies over 50% of the daylight hours between November and February (over 8 hours/day) but drops to about 30% from April to September (about 3 hours/day). Furthermore, tahr continue to feed more outside observable daylight hours in summer than in winter. The lowering of intake and activity over winter months helps to conserve energy, since foraging for low quality food among snowfall is expensive, and body fat reserves fall throughout late winter (Caughley, 1970c).

SOCIAL ORGANIZATION AND BEHAVIOUR

Tahr are in many ways intermediate between rupicaprids (e.g., chamois) and caprids (e.g., goats, sheep), and they resemble both in patterns of courtship and aggression (Geist, 1971; Schaller, 1977).

ACTIVITY. On a single rock bluff system in North Branch, a minor tributary of the Godley Valley, the activities of tahr were observed for 2½ years from mid 1978 (Tustin & Parkes, 1988). The bluff was about 2 km wide, ranged from 800 m to 2200 m in elevation, and about 30 tahr lived on the study area of 200 ha.

The routine of feeding and resting, the main daily activities of tahr, changes considerably between seasons. Total activity drops dramatically during winter; tahr spend less time feeding, and more time standing and resting, than in other seasons.

Typically, tahr rest during the middle part of the day, often well above the vegetation limit among rock bluffs. In mid to late afternoon they descend, feeding as they go, to a lower limit near dusk, and remain at lower levels until the following morning, when they return to their higher altitude resting places. Daily movement therefore tends to be vertical rather than horizontal. The daily descent is purposeful and often very rapid, and there seems to be remarkable unanimity in its timing and consistency in its amplitude, which changes with the seasons: at North Branch it averaged 450 m in midsummer, but decreased sharply in winter to less than half that distance. In late winter and early spring (August and September), when snowfall was greatest, the tahr kept to the steepest areas at around 1500 m, which offered the least opportunity for deep snow to accumulate. In late spring and early summer (October-December) they rested lower down, around 1320 m, well below the vegetation limit, and descended each day some 300-350 m to feed on the flush of spring growth which reaches lower-altitude plant communities first. As summer progressed and the scrubland and snow tussock flushed with new growth, and days became hotter, tahr rested higher and used vegetation at greater altitude. By autumn and early winter (May, June and July) they were resting at around 1600 m and were using only the upper limits of the vegetation at a considerably reduced intake. By this time the quality of the forage had declined with the heavy frosts after April (Tustin & Parkes, 1988).

DISPERSION. Tahr are gregarious animals and form easily recognizable social groupings. (1) Nanny-kid groups consist of females, yearlings, kids, and males up to the age of 2 years. They occupy particular rock bluff systems and appear to remain loyal to that area throughout their lives. (2) Mature bulls over 4 years old form discrete bachelor groups from the end of winter, after September, until just before the rut, around April-May. They occupy a summer range which is often exclusive, well separated and at higher elevations than the nanny-kid group range. (3) Young bulls, 2–3 years old, form separate and usually slightly smaller groups, and often range closer to the nanny-kid groups than the mature bulls. (4) Mixed-sex groups form in autumn before the onset of the rut, when the bull groups break up and bulls of all ages integrate with nanny-kid groups on the latter's range. Over the duration of the rut bulls compete for the females. The immature bulls often remain with a particular group and court all females unsuccessfully; mature bulls, on the other hand, remain with a group only if there is a female in oestrus, when they will court her and defend her against the attentions of other bulls. If no female is in oestrus, they move on. After the rut, the mature bulls become progressively more tolerant towards each other, and the mixed sex groups remain until bachelor groups reform when the snow retreats after September. (5) After parturition in November-December, the previous season's kids of both sexes form transient yearling groups, usually in the vicinity of nanny-kid groups. Similarly, within the nanny-kid groups, usually around February, temporary associations of kids may form for the duration of a feeding session, especially in the late afternoon to evening. These groups, sometimes containing eight or more kids, remain in one vicinity with one or two adults, while the rest of the group moves away feeding elsewhere.

The North Branch tahr were a loosely cohesive group (Tustin & Parkes, 1988). Generally, they would join up during an evening feeding session, not always as a single group, and split into smaller subgroups the following morning before resting. There were no apparent stable relationships among the subgroups other than between females and their offspring. They tended to group together when alarmed, usually climbing high into the rocks when disturbed from below, but readily hiding, crouching or "freezing" when disturbed by helicopters.

They were indifferent to, and did not mingle with, tahr from another group which occasionally used the far side of the bluffs. There was no evidence of any particular hierarchy among the North Branch group, although dominance was asserted from time to time, and no indication that some tahr assumed "leadership" or "sentinel" roles, as Anderson and Henderson (1961) assert.

REPRODUCTIVE BEHAVIOUR. In a courtship display, the bull faces the nanny, at right angles to her, with head and muzzle held high and mane erect. A bull can maintain this pose for more than 5 hours, adjusting his orientation towards the nanny to continue facing her as she moves, feeds or rests, and occasionally twisting his head through 90° to present his profile rather than his face (Fig. 68; K. G. Tustin, unpubl.). As courtship progresses his display intensifies and his pose becomes more exaggerated, until it is accompanied by a progressively more insistent "head nod", usually with the tongue extended or flicked, and a low grunt given with each "nod". Eventually the bull steps alongside and then behind the standing nanny; mounting and copulation are brief.

Young bulls often watch courtship displays, but keep a respectful distance and quickly retreat when a mature bull interrupts courtship to turn towards them; the more insistent intruders are repelled by a brief lunge, sometimes with a head-jerk threat.

An encounter between two breeding bulls of about the same size can lead to prolonged displays, sometimes lasting for an hour or more, but only very rarely settled by combat. Both bulls move with a stiff-legged, swaggering walk in a lateral display (Fig. 68), heads in the same direction, with manes and dorsal ridge erect, head forward and lowered, and horns exposed. Any movement by one will be matched by the other so that they remain parallel, and they may move some distance in this way. The confrontation usually ends when the larger bull slips in behind the other, which attempts to turn broadside on to block his opponent; if he lacks the room to manoeuvre and/or the confidence to maintain a static display, the encounter develops into a chase with the rearmost bull the victor. Only very rarely is such a dispute settled by fighting; it develops into a brief head-to-head wrestle, with violent pushing, until one loses balance. The victor will aim a lightning swipe with his horns to his opponent's body before the loser retreats, usually at a run.

Both courtship and agonistic behaviour among competing bulls involve striking visual displays: the long hairs of the mane are erected in ritualized presentations, making the animals appear twice their normal size (Schaller, 1973, 1977). However, the agonistic display involves presenting the largest possible profile, with horns exposed, whereas the courtship display presents only the frontal area, with the horns buried in the erected mane and minimized by the upraised muzzle. These are, respectively, the most and the least threatening postures. Furthermore, it may be one of the tahr's key adaptations to life on steep, unstable bluffs, to have small, rather than large horns, and to settle all but the most serious rivalry disputes by visual displays, rather than by combat.

Reproduction and Development

The rut lasts 6 weeks from late May to mid July; gestation is about 165 days, and nannies give birth to a single kid between the end of November and January (median date of birth 30 November, ± 18.5 days; Caughley, 1971b). Twins are known in captivity but are very rare (0.001% of all births) in the wild (K. G. Tustin, unpubl.).

Bulls are fecund by 2½ years, but those under 4½ years are not socially mature enough to obtain matings. Breeding adult males are in full winter pelage, at the peak of condition, and smell strongly from urinating on their undersides. They continually test for the onset of oestrus by mingling with groups of nannies, approaching each in a muzzle outstretched "low stretch" display, inducing them to urinate, then nuzzling the urine and responding with

Table 66: Life tables for female tahr from the Rangitata and Godley populations.

Age (years)	Rangitata (increasing)			Godley (stationary)		
	lx	*dx*	*qx*	*lx*	*dx*	*qx*
0	1.000	0.374	0.374	1.000	0.533	0.533
1	0.626	0.018	0.029	0.467	0.011	0.024
2	0.608	0.027	0.044	0.456	0.031	0.068
3	0.581	0.035	0.060	0.425	0.055	0.129
4	0.546	0.043	0.079	0.370	0.059	0.159
5	0.503	0.052	0.103	0.311	0.065	0.209
6	0.451	0.059	0.131	0.246	0.058	0.236
7	0.392	0.065	0.166	0.188	0.049	0.260
8	0.327	0.062	0.190	0.139	0.039	0.281
9	0.265	0.058	0.219	0.100	0.030	0.300
10	0.207	0.050	0.242	0.070	0.022	0.314
11	0.157	0.045	0.287	0.048	0.016	0.333
12	0.112			0.032		

Data from Caughley (1970b).

a head-raised "lip curl" or flehmen gesture.

Nannies separate from the group a day or two before giving birth in solitude, usually in cover of scrubland. After birth the female immediately consumes the placenta, and intermittently feeds while grooming the newborn kid. The kid will struggle to its feet within minutes, nurse within half an hour, walk actively, though somewhat unsteadily, within 2-3 hours, and practise "playing" in short, jerky movements within a day. Over the following few days the nanny leaves her kid for up to 7 hours while feeding and resting elsewhere and thereafter rejoins other nannies with newborn kids to form a small subgroup.

Young tahr grow fast. A pet male kid kept by G. Baker (unpubl.) weighed about 2 kg when found shortly after birth; after 2 days it could leap in 500-mm bounds; at 13 days it weighed 4.5 kg and could "jump a 100-cm fence".

Population Dynamics

During the dispersal of tahr into new country, the populations on the expanding fringes first rapidly increased, then levelled off, and eventually declined as food became limiting. Shot samples from populations at different distances from the point of liberation, representing the sequence of demographic statistics across a classic population irruption therefore describe 50-60 years of population change (Fig. 69, Table 66).

Life tables for female tahr showed that fecundity was 90% in adults from both increasing (Rangitata) and stationary peak (Godley) populations (Table 66), but less (75%) in post-peak or declining populations; 67% of young females in increasing populations had their first kid at 2 years of age, 27% at the peak. However, the rate of increase was governed mainly by variation in death rate, especially during the first year of life: 37% of juvenile tahr died before reaching one year of age during the increase phase, 53% at the peak, and over

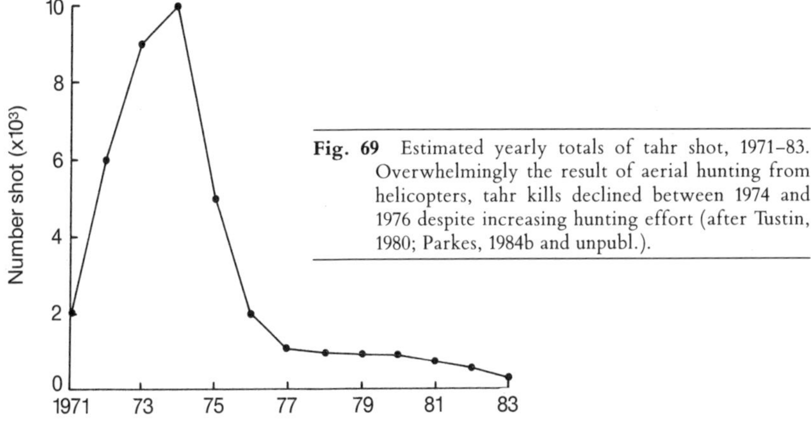

Fig. 69 Estimated yearly totals of tahr shot, 1971–83. Overwhelmingly the result of aerial hunting from helicopters, tahr kills declined between 1974 and 1976 despite increasing hunting effort (after Tustin, 1980; Parkes, 1984b and unpubl.).

59% during the decline. Fat reserves, a measure of condition, were, as expected, highest in increasing populations, lower at peak, and lower still during the decline, but recovered after the decline ceased (Caughley, 1970c).

The median age at death of adult female tahr in a peak population was 6.20 years, and 8.02 years in an increasing population (Caughley, 1970b). Females live longer (to 22 years) than males (to 14 years: n = 4000; K. G. Tustin, unpubl.). Accidental death or injury through avalanche, rockfall or loss of footing, especially on ice, is not uncommon.

The history of tahr in Carney's Creek, a small tributary valley (22 km^2) of the Rangitata River, has been observed from first occupation to the recent reductions resulting from hunting (Tustin & Challies, 1978), and may be taken to represent the history of tahr in most areas east of the Main Divide (areas west of the divide have had negligible attention from sports hunting or official control measures).

Bull tahr were first reported in Carney's Creek in 1939, but females did not establish for another 10 years. Despite some government and recreational hunting, the population increased to high densities by the mid 1960s. A census of what was probably the peak population in February 1965 counted 710 tahr in 45 groups. The mean group size was 15.8, although 50% of the tahr present were in groups of 31 or more, 37% in groups of 11–30, and 13% of 10 or fewer. The two largest groups comprised 69 and 62 animals. All but 6% were nanny-kid groups; there were five mature bulls, three of them solitary, plus six groups of from 2 to 10 immature bulls, numbering 35 in all. The density was one tahr to 3.0 ha (32.9 tahr/km^2) over the entire study area, or a biomass of 825 kg/km^2; these figures would be more than doubled if only the area of available vegetation were considered.

In 1965, an attempt to cut down the numbers of tahr in the area by aerial poisoning was unsuccessful. Recreational hunters accounted for about 1300 tahr over the next decade. After 1967 aerial hunting began as an official control measure, and Carney's Creek yielded 150 tahr on the first foray. Commercial game-meat recovery operations began in the area after 1972.

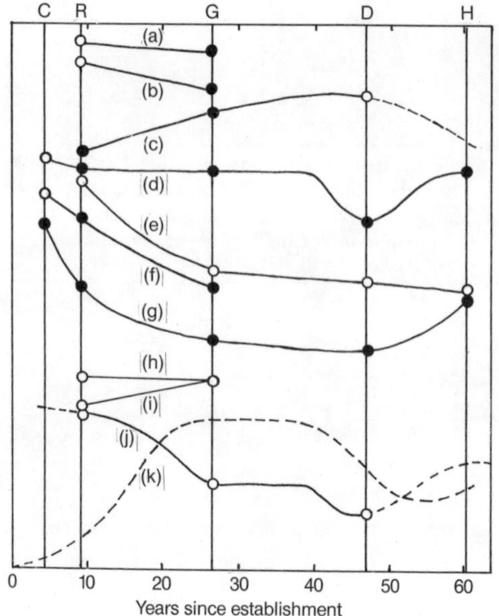

Fig. 70 Schematic representation of changes in a tahr population and its food supply during a population irruption. Vertical axis is a scale of magnitude, with separate scales and origins for each variable. Plots: (a) generation length; (b) median age of adult deaths; (c) juvenile mortality rate; (d) adult fecundity; (e) winter food supply; (f) juvenile fecundity; (g) fat reserves, females; (h) birth rate; (i) death rate; (j) rate of increase; (k) population density. Key: solid circles, reasonably accurate estimates; open circles, estimates presented with less confidence; broken lines, extrapolations inferred from trends of other statistics; vertical lines, estimated temporal position of five sampled populations – C, Copland; R, Rangitata; G, Godley; D, Dobson; H, Hooker (from Caughley, 1970b).

In February 1977 another census counted 48 tahr in 15 groups; the mean group size was 3.2, the largest had seven tahr. Except for two solitary immature males, most were nanny-kid groupings. The density of tahr in Carney's Creek had therefore been reduced by 93%, to one tahr per 45 ha (2.2 tahr/km^2), a biomass of 56 kg/km^2. Most of the survivors were found where the terrain offered most cover from aerial hunting, i.e., where relatively large areas of subalpine scrubland graded into steep dissected bluffs at high elevations. A further count in 1984 revealed 26 tahr in Carney's Creek (J. P. Parkes, unpubl.).

Because harvesting of tahr has been intense and evenly spread over the whole of the species' limited range, it is possible to make retrospective estimates of the changes in the total number of tahr in New Zealand since about 1970. Tustin reckoned there were 20 000–30 000 in 1975 (in Schaller, 1977) and 4000–8000 in 1980 (Tustin, 1980), although kill data now suggest that all these figures were too high; Parkes reckoned 1000–2000 in 1984 (Parkes, 1984b). The current estimate is that the total New Zealand tahr population probably numbered about 30 000–40 000 in 1970 and 4000 in 1987.

PREDATORS, PARASITES AND DISEASES

Man is the only predator of tahr in New Zealand. Mature bulls are a much prized sporting trophy; however, their wariness and rugged habitat make hunting on foot difficult and often dangerous, so sportsmen have had little effect on population numbers. By contrast, aerial hunting, for control and for commercial game-meat recovery, can devastate tahr populations within a short time.

Tahr have few diseases or parasites. Scabby mouth (contagious eczema) is not uncommon in young tahr from parts of the eastern ranges; it probably originates from domestic sheep, with which tahr sometimes share grazing in summer. Similarly, the nematode parasites *Oesophagastomum venulosum* and *Trichuris ovis*, both present in sheep, have been recorded from the caecum of tahr, and trichostrongylids from the abomasum and small intestine (Christie & Andrews, 1964). Pinkeye (keratoconjunctivitis) and mastitis (in lactating females) have been recorded. The only known host-specific parasite in New Zealand tahr is a mallophagan louse (*Damalinia hemitragi*) (Christie & Andrews, 1964).

ADAPTATION TO NEW ZEALAND CONDITIONS

None documented apart from the 6-month shift in breeding season.

SIGNIFICANCE TO THE NEW ZEALAND ENVIRONMENT

DAMAGE. The only objective study of vegetation changes induced by tahr (Caughley, 1970b) related the decreasing density of snow tussocks (*Chionochloa* spp.) and increasing turf grassland (*Poa colensoi*) in the Dobson/Hopkins, Hooker, Godley and Rangitata/Rakaia watersheds to the succession of stages during a population irruption of tahr, and in turn linked this with the population statistics of the resident tahr.

Other writers, e.g., Burrows (1974) and Douglas (1984), have made less systematic observations of the effects of browsing by tahr. The high densities that were attained locally, and especially the confinement of such numbers of browsing animals to narrow zones of high altitude scrubland during winter, also caused conspicuous changes in the composition and welfare of some subalpine scrublands, even eliminating them locally. Recently, the decrease in tahr density has allowed a substantial recovery of formerly depleted ranges, especially in the west where rainfall is high.

CONTROL. Tahr flourished under legal protection until 1930; then, rising concern about damage to mountainland vegetation forced the removal of protection, both from tahr and from all other introduced game animals. Shortly afterwards, teams of government hunters began control operations on the eastern side of the Southern Alps, and shot some 25 000 tahr within the first 25 years.

Ground shooting had only a local effect, however. About the early 1960s aerial poisoning was attempted (Douglas, 1967) but was largely unsuccessful. After 1967 aerial shooting by government-hired helicopter crews (normally engaged in red deer game-meat recovery) became the primary means of control, and had considerable effect in parts of Canterbury. These operations have continued until the present day, although in recent years they have mostly concentrated on the fringes of the species' range.

After 1971, markets for tahr as an export game meat were developed, and helicopters were used by commercial hunters for carcass recovery. The initial tallies were high, especially from areas where tahr were about at a population peak, and kills were limited mainly by the number of carcasses the helicopters and their crews could handle (Tustin, 1980). In these areas it was common for operators to recover over 100 tahr during a short winter's day — the record was 174. At first only the heavier bulls were shot, but as numbers decreased all tahr seen were shot. The result was a sudden and dramatic decrease in tahr numbers throughout their entire range, mostly over the four winters of 1972 to 1975 (Fig. 70). In all, some 40 000 tahr have been shot since 1970, 30 000 of these between 1972 and 1975. Since the late 1970s, commercial hunting has declined because of insufficient returns, but official control operations continue.

Tahr in New Zealand are not endangered by existing commercial or control operations, which are effective enough to permit close control of numbers, yet are still at the mercy of economics. On the other hand, there is a growing acknowledgement that it may be desirable to create a limited reserve especially for tahr, for sport hunting or for other recreational and aesthetic purposes.

K. G. T.

Genus *Capra*

The six species of wild goats all live in mountain country from the Pyrenees to China. One, *C. aegagrus*, the ancestor of the domestic goat, still lives in the Middle East. Linnaeus called the domestic goat *C. hircus*, a name which is still widely used (see p. 31), but the domestic goat is actually only a subspecies of the wild goat, and is called by some authorities *C. a. hircus*. The males of all goats have chin beards and a strong smell; both sexes possess subcaudal glands (but not preorbital, inguinal or pedal glands) and a flat tail which is bare on the underside.

36. Feral goat

Capra hircus Linnaeus, 1758

Breed names of domestic goats giving rise to feral populations in New Zealand are Toggenburg, Alpine, Saanen, Nubian, Old English, Angora, "unimproved", and "Swiss". "Cashmere" and "mohair" are fibre types, not breed names.

DESCRIPTION (Fig. 71)

Distinguishing marks, Table 63 and skull key, p. 24.

The coat is generally short haired with variable amounts of underfur (cashmere fleece, "pashmina"). It may be shaggy on the hindquarters of both sexes and on the neck and shoulders of males. Some populations of Angora origin have longer coats of wavy, silky, white hair. The main colour may be white, black, brown, or any combination of these. When males are predominantly brown, they tend to have a dark neck and shoulders and a dark line down the spine ("Bezoar" markings, so named from the Persian wild goat). Brown females commonly have just the dorsal black stripe. In both sexes the underbelly and legs are usually paler if the body is mainly brown or black.

Fig. 71 Feral goats, Hawke's Bay, 1966 (J. H. Johns).

Facial marks may be unstructured blotches of colour, or a symmetrical "reversed badger" pattern, i.e., a dark background with lighter eyebrows, and a lighter line running through the eye and down the muzzle.

The tail is short (100-130 mm), tufted, and can be carried erect; ears may be more or less erect (e.g., Rimutaka Range, Arapawa I.) or completely pendulous (Auckland I.).

All males, and some females, are bearded as adults. In males the beard continues as a mane down the throat and chest. Males also develop a neck and shoulder mane but females do not. Males have a prominent scrotum which may bear a small pair of false teats at the base. Lactating females have a conspicuous udder with two terminal teats. In some populations (see below), both sexes may have throat lappets (wattles).

The eyes are prominent with horizontal yellow pupils. Eyesight, hearing and smell are well developed.

In New Zealand both sexes have horns. In females they are slender and curve upwards and backwards, with a clear space between the bases. The tips usually diverge, although they sometimes curve in and may even point forwards. In males the horns are larger and may touch at the base. They sweep up and backwards, either in the more or less erect "scimitar" form or the "prisca" form, in which they grow outwards in an open spiral. The spiral is homonymous; that is, the horn on the right side of the head is right-winding (clockwise). Horns of true wild goats are heteronymous. Horns grow throughout life and bear rough ridges and growth lines.

Males are the larger sex, with clearly heavier forequarters, shaggier coats, and larger horns. They stand up to about 700 mm tall at the shoulder, compared with 660 mm for females (Tables 67, 68).

Chromosome number 2n = 60.

Dental formula $I^0/_3\ C^0/_1\ Pm^3/_3\ M^3/_3 = 32$.

Table 67: Body measurements of adult male goats in the New Zealand region

Locality	Horns (mm)				Weight (kg)		Measurements (mm)			Reference
	mean		max.							
	length	span	length	span	total	evisc.	length	shoulder height	girth	
Arapawa I.	360	423	405	485	40.8	27.9	1 322	694	886	Clark & Asher (1977)
Wairau	424	462	540	605	46.8	33.1	1 422	736	925	Clark & Asher (1977)
Tai Pari	400	465	—	—	29.7(36.3)	—	1 210	—	—	Williamson (1975)
Mt. Stokes	—	—	—	—	41.5	25.5	1 415	665	865	Clark & Asher (1977)
Dun Mountain	—	—	—	—	31.3	19	1 193	590	787	Clark & Asher (1977)
Oaro	400	320	483	580	—	—	1 321	—	—	Clark (1972)
							(1 460)			
Arapaoanui	504	566	590	710	32.7	18.4	1 298	658	815	Clark & Asher (1977)
Hawke's Bay	420	—	534	—	—	—	—	—	—	Rudge (1972)
King Country	—	—	—	—	—	20.2	1 269	680	—	Clark (1974)
Macauley I.	302	—	516	—	—	—	—	—	—	Williams & Rudge (1969)
Raoul I.	350	—	—	—	—	20.7	1 269	686	—	Rudge & Clark (1978)
Auckland I.	330	260	—	—	50[1]	—	1 300	680	920	Rudge & Campbell (1977)
Auckland I.[2]	325	370	—	—	58	—	1 333	780	—	Rudge (1986a)
Rimutaka	317	406	500	578	36.9	22.1	—	662	782	M. R. Rudge (unpubl.)
Range					(43.9)	(25.8)		(711)	(684)	

1. Estimated weight. 2. n=1

Numbers in parentheses = maximum weight; evisc. = eviscerated weight.

NZ Trophy record for greatest male horn length 1 009 mm, and for span, 1 263 mm.

Table 68: Body measurements of adult female goats in the New Zealand region.

Locality	Horns (mm)				Weight (kg)		Measurements (mm)			Reference
	mean		max.							
	length	span	length	span	total	evisc.	length	shoulder height	girth	
Arapawa I.[1]	150	150	—	—	22	10	1 230	635	700	Clark & Asher (1977)
Tai Pari	—	—	147	165	21.8 (25.9)	—	1 110 (1 260)	—	—	Williamson (1975)
Oaro	130	160	170	247	—	-	1 133 (1 270)	—	—	Clark (1972)
Hawke's Bay	150	—	178	—	—	—	—	—	—	Rudge (1972)
King Country	—	—	—	—	—	13.5	1 167	626	—	Clark (1974)
Macauley I.	157	—	241	—	—	—	—	—	—	Williams & Rudge (1969)
Raoul I.	135	—	—	—	—	13.3	1 170	638	—	Rudge & Clark (1978)
Auckland I.	55	70	—	—	25[2]	—	1 170	640	—	Rudge & Campbell (1977)
Auckland I.	140	140	165	190	33 (47)	—	1 225 (1 270)	670 (690)	—	Rudge (1986a)
Rimutaka Range	123	160	171	254	28.3 (35.4)	15.2 (19.9)	—	599 (647)	714 (813)	M.R. Rudge (unpubl.)

1. Estimated weight. 2. n=1

Numbers in parentheses = maximum weight; evisc. = eviscerated weight.

Field Sign

Tracks show paired hooves with pointed, slightly incurved tips. They are distinguishable from those of sheep (which are blunt at the tips) but may be confused with those of sika deer. Dimensions of each digit are about 55 × 25 mm in adults.

Pellets may be found singly or in groups. They are smooth and unfaceted, rarely adherent or pasty, with an elongated oval outline and smoothly rounded at each end, measuring up to 20 × 8 mm. They are easily confused with those of sika deer, and not reliably distinguishable from those of sheep or possums in some circumstances. They can also be confused with droppings of young red deer, which lack the characteristic dinge and nipple on the ends of the pellets of adult deer.

Their call is a "meh" rather than a "baa" (cf. sheep) and is easily confused with seabird cries. A pungent smell is produced from subcaudal glands by alarmed or rutting males.

When feeding, goats leave cleanly cut leaves and shoots, up to 1.5 m above ground, but they can reach to 2 m by standing on their hind legs. Their feeding sign is not absolutely distinguishable from that of possums, feral sheep, or deer, except that deer can browse higher and possums tend to shred leaves. Goats will bite bark, but so will deer and possums.

Measurements

See Tables 67, 68.

Variation

There is great variability in the colour, pattern and length of the coat, and in horn form and body size, related to the breed origins and nutritional condition of local populations.

The "scimitar" type of horn prevails in Orongorongo Valley, Raoul I., Arapawa I., and Marlborough populations, and the "prisca" type in Hawke's Bay/Gisborne, Northland and those parts of King Country where Angora strains were liberated.

The colour and pattern of the coat vary both between and within populations. All-white (frequent in Hawke's Bay/Gisborne, Hunua Range and Auckland I.), all-black (Raoul I. and King Country), all-brown (Arapawa I. and Rimutaka Range), and combinations of these main colours, are found in all populations. There are also "roans" of black/white, and black/brown.

Throat lappets (wattles) are considered to be a sign of "improved" ("Swiss") blood. They are found on goats in King Country (Clark, 1974), Hunua Range (Clark, 1976), Mt. Stokes, Marlborough, and Arapaoanui in Hawke's Bay (Clark & Asher, 1977). They are not found in goats of the Rimutaka Range or Auckland I. (Rudge & Campbell, 1977), Arapawa I. (Clark & Asher, 1977), Macauley I. (Williams & Rudge, 1969), Raoul I. (Rudge & Clark, 1978), Kaikoura (Clark, 1972), Tai Pari, Marlborough (Williamson, 1975), Wairau, Dun Mountain, Red Hills, or Takaka Valley (Clark & Asher, 1977). Genetic differences and relationships between populations have been revealed by the frequencies of transferrin genes (Asher, 1981).

HISTORY OF COLONIZATION

The first goats liberated in New Zealand, by James Cook in 1773 and 1777, were probably eaten by the Maori of the Marlborough Sounds (Beaglehole, 1961). Donne (1924) records that Cook also gave some goats to the Maori of Cape Kidnappers in Hawke's Bay; and Wodzicki (1950) refers to early liberations in North Auckland.

Throughout the 19th century, goats were brought in by sealers, whalers and settlers. They were liberated widely on the main islands and also on smaller islands to provide food for castaways. Their origins and spread have never been studied systematically. Some early records are listed in Table 69.

Goats were a convenient stock animal for settlers and were moved around the country by surveyors, gold-miners and farmers (Donne, 1924). As land was "developed", woody weeds (gorse, blackberry, briar) invaded it, and goats were brought in to eat them. The populations in Hawke's Bay, Taranaki and Marlborough owe their origin largely to this (Guthrie-Smith, 1929; Wodzicki, 1950). In the 1880s acclimatization societies introduced Angora goats to establish a commercial fibre industry (Thomson, 1922). This enterprise foundered, and the abandoned animals then added their genes to the feral populations (Thomson, 1922, quoting a letter received from B. C. Aston in 1916). Today their descendants are providing the breeding stock for a renascent goat fibre industry.

The early introductions soon gave rise to feral populations in forested mountainous areas. Warnings, by naturalists of the day, of their likely impact on the native vegetation were eventually vindicated (Kirk, 1896; Aston, 1912; Guthrie-Smith, 1969, in the preface to his 1926 second edition). Control campaigns have been waged since the 1930s, when goats were eventually recognized as a major forest pest, but that did not stop farmers moving them round the country to control weeds and then abandoning them to colonize the forests. Goats can still be found in most of the mainland localities they occupied 50 years ago, though in much lower numbers. Control work has been so successful on islands that only five populations now remain (Table 70).

DISTRIBUTION

WORLD. Many populations of varying origins and antiquity exist throughout the world (Lever, 1985). They are almost universally classed as pests, but some (e.g., in UK) are preserved for historical or local natural history value.

NEW ZEALAND (Map 35). Since organized observation began about 30 years ago, the mainland distribution has changed little, although numbers are now lower. Goats are scattered at low density with locally higher concentrations, in forested ranges and scrubby hill country in both main islands. In recent years they have been rediscovered as stock animals for scrub control and fibre production. As a result, populations which were mapped as feral by Wodzicki (1950) (e.g., Mahia Peninsula, Hawke's Bay) are now considered to be stock, and are not mapped here as feral.

Island populations now remain only on Auckland I. (about 120), Arapawa I. (<500), Forsyth I. (number unknown), D'Urville I. (very few), and Great

Table 69: Establishment of goats in some mainland areas of New Zealand.

Location	History	Reference
Nelson (Maitai Valley)	By 1850s large numbers escaped from farmed goats	Thomson (1922)
Kaikoura	Introduced in 1850, numerous by 1916	Sherrard (1966)
Southern Lakes/Skippers	Present by 1870s, immense numbers by 1916	Thomson (1922) Donne (1924)
SW South Island (Puysegur)	Liberated by Capt. Fairchild (?1890s), numerous by 1916	Thomson (1922) Begg & Begg (1973)
North Auckland (Hokianga)	Possibly introduced by Cook but not common in 1920s	Thomson (1922) Wodzicki (1950)
Hunua Range	Introduced 1870-90, throughout by 1912	Thomson (1922) Barton (1974)
Urewera	Donated to Maori in 1903, protected by tapu	Silvester (1964) Donne (1924)
King Country	Introduced for blackberry control 1910	Clark (1974)
Mt. Egmont	Introduced for blackberry control about 1910, became very numerous in wild by 1925	Atkinson (1954) Anon. (1966)
Rimutaka Range	Present by 1850s, abounded by 1888	Monrad (1968)
Ruahine Range	Introduced 1919 to south end, spread in large numbers to northern Ruahine by 1954	Greig (1946) Elder (1965)
Cape Kidnappers	Given by Cook to Maori 1777, abundant by 1920s	Donne (1924)
Hawke's Bay	Transported 1910 from Marlborough for weed control	Anon. (1978b)

Table 70: Distribution and history of goats on the offshore and outlying islands of New Zealand.

Island	Lat. (S)	Area (ha)	Introduction date	Fate of population	Reference
Adams	50°55′	9 896	*c.* 1885	Died out	Rudge & Campbell (1977)
Arapawa	41°15′	7 785	Introduced pre-1839	Still present	Wakefield (1845), M. R. Rudge (unpubl.)
Antipodes	49°40′	2 025	Introduced 1887-1903	Died out	Chilton (1909)
Auckland	50°45′	45 975	About 10 liberation sites in 19th century	One endured, at north end (about 120 remain)	Rudge & Campbell (1977)
Campbell	52°30′	11 216	Introduced 1888 and 1890	Died out	Cockayne (1903)
Cavalli	34°59′	799	Not known	Exterminated 1972	Miller (1976)
Cuvier	36°25′	799	Reported in 1905 and 1909	Exterminated 1961	Blackburn (1967)
D'Urville	40°50′	16 782	Not known	Still present	B. D. Bell (unpubl.)
East	37°40′	*c.* 8	Occasional, by fishermen	Exterminated 1968	B. D. Bell (unpubl.)
Enderby	50°30′	710	1850 onwards	Died out	Rudge & Campbell (1977)
Ernest	46°58′	75	Occasional, by fishermen	None now	B. D. Bell (unpubl.)
Ewing	50°32′	54	1850 onwards	Died out	Rudge & Campbell (1977)
Forsyth	40°58′	775	Not known	Still present	R. H. Taylor (unpubl.)
Great Barrier	36°11′	28 512	Not known	Still present	Atkinson (1973b), B. D. Bell (unpubl.)
Great King	34°06′	435	Introduced pre-1830 and 1889	Exterminated 1946	Baylis (1951), Holdsworth & Baylis (1967)
Great Mercury	36°40′	1 718	Introduced pre-1868	Still present	Wright (1976)
Kapiti	40°51′	1 970	Not known	Exterminated 1928	Esler (1967)
Macauley	30°14′	306	Introduced pre-1836	Exterminated 1967	Straubel (1954) Williams & Rudge (1969)
Mahurangi	36°50′	*c.* 32	Late 19th century	Removed 1915	Atkinson (1972)
Mangere	44°18′	113	Not known	Died out	Fleming (1939)
Maud	41°10′	309	Not known	Exterminated in 1970s	Meads (1976b)
Mokohinau (Burgess)	35°55′	52	Not known	Exterminated 1973	A. H. Whitaker (unpubl.)
Motuoruhi (Goat)	36°40′	70	Not known	None now	Newhook, Dickson & Bennett (1971)
Raoul	29°15′	2 938	Introduced pre-1836	Exterminated 1984	Straubel (1954), NZFS (unpubl.)
Snares	48°02′	*c.* 280	Introduced 1890-1900	Died out	Cockayne (1903)
South East	44°21′	219	Introduced pre-1900	Exterminated 1915-16	Ritchie (1970)
Steep-to	46°06′	*c.* 45	Not known	None now	B. D. Bell (unpubl.)
Whale	37°52	140	Occasional, by fishermen	None now	B. D. Bell (unpubl.)

Barrier I. (number unknown) (Table 70). Some tacit protection has been given to the goats on Arapawa I. (Dingwall & Rudge, 1984); and a programme is under way to assess the commercial value of goats from Auckland I. On the mainland, their status has now become confused as new commercial possibilities have been recognized. In forested land and in nature reserves they are still classed as pests.

HABITAT

Feral goats are descended from mountain dwellers of the Middle East, but in New Zealand they prosper in a wide range of conditions. They live in damp, peaty moorland on subantarctic Auckland I. (Rudge & Campbell, 1977), and in subtropical forest on Raoul I. (Rudge & Clark, 1978). Their altitudinal range extends from sea level to the alpine zone (Atkinson *et al.*, 1965). Habitats include farmland, scrub, native forest and exotic forest.

Because goats are essentially browsers, their characteristic habitat is forest- or scrub-covered upland, but they commute to rough grassland if they have the chance (Riney & Caughley, 1959; Clayton-Greene, 1976).

Goats prefer a rocky substrate, and because they are so agile on steep crags and narrow ledges, they can exploit places which deer and feral sheep cannot reach. Their choice of sunny rather than shady faces is often reflected in the relative damage to vegetation, and in erosion patterns. They will make intensive use of open places close to the shelter of forest or scrub (Riney & Caughley, 1959; Clark, 1974). Because their tastes in food are so wide, they are said to replace deer in depleted forest (Kean, 1959). In the Kaimai Range, goats apparently use the forest only for shelter and dietary variety (Dale & James, 1977).

FOOD

Goats are browsers rather than grazers and have catholic tastes. In the Rimutaka Range, 120 different plant items were eaten, 40 of which made up 98% of the diet (M. R. Rudge & D. J. Campbell, unpubl.); 47% of the diet was broadleaved plants, and in all seasons more than 50% of the diet consisted of just three woody plants, *Brachyglottis repanda, Griselinia lucida,* and *Melicytus ramiflorus*. An average rumen contained 12 different food items.

In summer the proportion of herbaceous monocotyledons increased; and in winter, plants typical of the forest interior (ferns) were more prominent. The diet did not vary with locality (coastal/interior), age, sex, or breeding status, nor from year to year.

On Raoul I. eight food items comprised 89% of the diet (total items 42), over half of which were broadleaved woody plants, with monocotyledons and ferns about equal (Parkes, 1984d). Jew's ear fungus (*Auricularia* sp.) seemed highly favoured.

On Mt. Taranaki individual rumens contained on average 19 species and sometimes over 30. Over the year, two species, *Asplenium bulbiferum* and *Ripogonum scandens*, made up 45% of the diet. Species which were eaten in significantly different proportions at different seasons were *Coprosma*

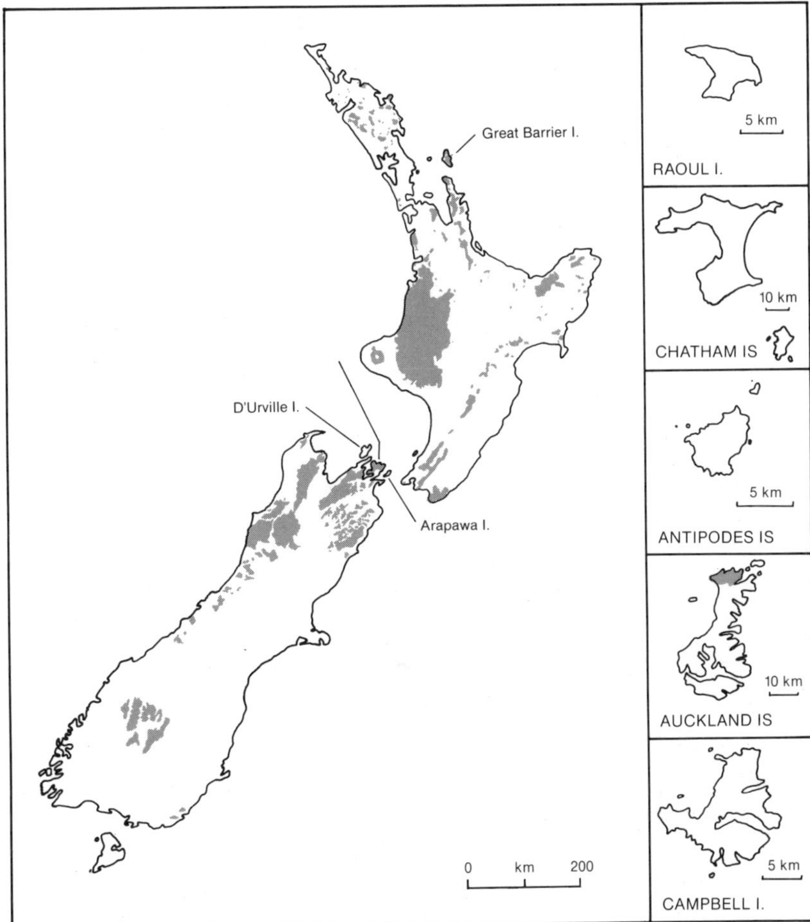

Map 35 Distribution of feral goats on the main islands of New Zealand in 1975 (M. R. Rudge, unpubl.). For further details, see Table 69; for distribution on offshore islands, see Table 70.

grandifolia, C. tenuifolia, Griselinia litoralis, Melicytus ramiflorus, Ripogonum scandens, Weinmannia racemosa and *Dicksonia squarrosa*. Clear patterns of preference and rejection were seen among the different plants within the browse zone (Mitchell, Fordham & John, 1987). Chemical analyses showed that goats received adequate N, P and minerals by virtue of eating a wide range of species (Mitchell, Grace & Fordham, 1987).

Despite their reputation, goats do find some plants unpalatable. In the Rimutaka Range these included upland pepperwood (*Pseudowintera colorata*), *Leptospermum* spp., harsh ferns of the genera *Cyathea, Dicksonia* and *Polystichum*, and monocotyledons such as *Microlaena, Uncinia, Gahnia* and *Cortaderia*. In the ground layer of a goat-infested forest, Moore and Cranwell (1934) described an induced dominance of the unpalatable rice grass *Microlaena avenacea*. Atkinson

Table 71: Breeding statistics for New Zealand goat populations.

Location	Age at 1st breeding	Frequency	Breeding season	Prenatal mort. (%)	Twin (%)	Reference
Tai Pari	F > 1 y	Twice/y at age 3	Dec-May	29	11	Williamson (1975)
Oaro	F < 1 y M > 1 y	Once/y	Mar	39	45	Clark (1972)
Hunua	F < 1 y	Once/y	Nov	—	50[1]	Clark (1976)
Arapawa I.	F > 1 y	—	Nov-Jun	—	—[1]	Clark & Asher (1977)
Arapawa I.	< 1 y	Twice/y	Dec-Feb & May-Jun	—	31	Challies & Christie (1984)
King Country	0.5 y both sexes	Twice/y	Jan	22-50	23	Clark (1974)
Macauley I.	F > 1 y	Twice/y	?Dec	33	39	Williams & Rudge (1969)
Rimutaka Range	F < 1 y	Twice/y	Nov	23	52	Rudge (1969)
Raoul I.	< 1 y both sexes	Once/y	Apr-Jun	—	14	Rudge & Clark (1978)
Raoul I.	F < 1 y	Twice/y	Jan-Sep	—	36[1]	Parkes (1984c)

1. Triplets recorded. y = year(s).

(1964b) lists unpalatable species of fern, monocotyledon herbs and broadleaved woody plants from various areas. Goats will, however, eat species poisonous to other animals, including ngaio, *Myoporum laetum* (Turbott, 1948) and tutu, *Coriaria* spp. (Parkes, 1984d).

The success of goats in eating such a wide range of plants rests on several characteristics: they have a great capacity to digest roughage, their lips are very mobile, they can climb sloping trunks and branches, and they are inquisitive and constantly "testing" the vegetation. Because they are so surefooted, and prefer open sites, they inhibit regeneration very effectively on steep, naturally eroding slopes.

The knowledge that goats eat harsh woody vegetation was put to use by the early settlers to combat invasion by blackberry (*Rubus fruticosus*) on newly cleared land (Guthrie-Smith, 1969). Goats were transported to Taranaki and Hawke's Bay for that purpose. Today they are again being used against sweet briar (Anon., 1981), gorse, and a wide range of pasture invaders (Batten, 1979; Krause, Beck & Dent, 1984; Radcliffe, 1985).

SOCIAL ORGANIZATION AND BEHAVIOUR

ACTIVITY is diurnal. Unsynchronized periods of feeding and resting alternate through the day, with peaks of activity in the morning and evening. In summer, rest includes loafing as well as cud-chewing. In subantarctic conditions almost the whole of daily activity is devoted to feeding, even in summer (Rudge &

Campbell, 1977). In wet, windy weather goats generally seek shelter rather than feed in the open.

DISPERSION. Group size and composition changes with population density, disturbance, time of day, and time of year. The basic social unit is a female and her offspring-of-the-year or previous year. These matriarchal groups aggregate to form larger groups to which males may become attached. Young males (1.5–3 years old) and adult males form distinct mixed-age bands, which continually form and dissolve. Grouping and parting seems to be a feature of social behaviour in all the populations studied (e.g., Riney & Caughley, 1959). Adult females are generally more alert than males in mixed groups and will direct the course of flight when disturbed. In both sexes, alarm signals are foot stamping, snorting, and an erect tail. The male's repertoire of communication includes pungent discharges from the subcaudal gland.

HOME RANGE. Females tend to occupy the home range of their mother, or an adjacent overlapping area, for life. Typically, this is about 20 ha, but it is very variable. Males do not defend territories, and they wander more widely than females. Their home ranges vary from 100 m to 20 km across. Riney and Caughley (1959) found that males had distinct winter (forest) and summer (grassland) ranges, whereas females lived at the forest-grassland interface all year round. Old males occupy small ranges alone or as pairs. Generally, goats are rather sedentary even when hunted (Riney & Caughley, 1959; Clark, 1974). Daily movements are erratic around the home range and may or may not finish at the same bedding site. Generally, goats feed along a meandering track, but occasionally move directly and quickly to a distant point.

REPRODUCTIVE BEHAVIOUR can be seen all year round, but it is at a peak in November (Rudge, 1969). During courtship, groups of males chase females relentlessly, making a gobbling sound and displaying flehmen (flared nostrils, extended tongue). At such time males, even juvenile, non-breeding males, spend long periods chasing each other and clashing horns together.

Play is a conspicuous feature of goat behaviour. Kids and their mothers, kids together, juveniles and mixed-age and mixed-sex groups will all run, jump, pirouette and slide on steep faces and screes, particularly at dusk, when juveniles will infect a whole group before it beds down (Rudge, 1969).

REPRODUCTION AND DEVELOPMENT

The goat is seasonally polyoestrous (Asdell, 1964). All studies in New Zealand show virtually year-round breeding, with one or sometimes two distinct peaks (e.g., on Arapawa I.) (Table 71). In most populations, females breed in their first year. They may bear twins then, but productivity (twinning and frequency of pregnancies) usually rises with age (Williams & Rudge 1969; Clark, 1974; Parkes, 1984c). Females can go into postpartum oestrus and so breed twice per year (Rudge, 1969). Up to half of all conceptions are of twins, and the frequency can vary by locality even within a population (Clark, 1974). Minimum estimates give prenatal mortality at between 22% and 50%. Males are fecund at 6 months, but seem to be excluded by larger males from mating until they are 3-4 years old. Females can breed at 6 months old.

Table 72: Growth of male and female feral goats in the Rimutaka Range.

	Males			Females		
	0–6 months	24–36 months	> 36 months	0–12 months	24–36 months	> 36 months
Horn length (mm)	59	261	317	46	109	122
Horn circumference (mm)	74	129	129	43	67	69
Chest girth (heart) (mm)	540	763	780	554	665	714
Shoulder height (mm)	498	643	663	498	576	599
Net body weight (kg)	8.3	19.9	22.1	8.0	13.1	15.1

Data from M. R. Rudge (unpubl.).

Table 73: Demographic statistics for New Zealand goat populations.

	Life span (y)	Sex ratio	Mortality (%)	Productivity (kids/100 FF)	Density/ha	Reference
Tai Pari	5	1:1.5	49 (annual)[1]	—	—	Williamson (1975)
Oaro	6	1:1	—	—	—	Clark (1972)
Hunua	10% of M and 17% of F were > 4 y	1:1.12	—	—	—	Clark (1976)
King Country	10	1:1.6	14 (to 1 y)	92 (per F > 8 mo)	—	Clark (1974)
Rimutaka Range	>10	1:1	15 (to 6 mo)	105 (per F > 12 mo)	—	Rudge (1969)
Arapawa I.	—	1:1.5	—	123 (per F > 12 mo)	—	Challies & Christie (1984)
Macauley I.	10[3]	1:1.65	34–57 (to 6 mo) 19.7 (annual)	170 (per F > 12[2] mo)	10	Williams & Rudge (1969)
Raoul I.	—	1:1	—	83–94	0.28–0.73	Rudge & Clark (1978)
Raoul I.	—	1:0.7 (birth) 1:7.3 (by 2 y)	—	170	< 0.1	Parkes (1984c)

1. But a hunted population, not stable. 2. At birth. 3. Exponential mean 4.5 y. y = year(s); mo = months.

Females about to give birth become solitary. Soon after birth, the kids are mobile but are hidden by the female while she forages. After about 4 days they move around together, and kids develop agility and a behavioural repertoire by playing or fighting with other animals. It can take up to a year for the mother-kid pair to integrate socially with other goats. After parturition, females may reunite with previous female offspring, but not with juvenile males. Young males thereafter develop bonds with others of similar age or older, which take them beyond the maternal home range.

At birth, kids weigh about 2 kg, and reach adult size by about 3 years old. Males are significantly larger in horn dimensions by 6 months, and in all other dimensions by 1 year. Table 72 shows the actual changes in various measurements observed in the Rimutaka population (M. R. Rudge, unpubl.). Information from a King Country population is almost identical (Clark, 1974).

The rate of growth varies from place to place and even within the same population, and is usually related to food supplies and intensity of control work (i.e., to density, and hence indirectly to food supplies). In the Wairau Valley, goat density was lower on the south bank. Age for age, jaws and horns were longer, and the animals were heavier and fatter than on the north bank (Bathgate, 1973b). On Raoul I., growth rate and ultimate size of horns, body length and body weight all increased significantly as density decreased as a result of control work (Rudge & Clark, 1978).

POPULATION DYNAMICS

AGE DETERMINATION. The most common method used has been the tooth eruption sequence, following the chronology given by Habermehl (1961) for domestic goats. In feral goats the sequence is similar, but varies within and between populations. In the Rimutaka Range population (M. R. Rudge, unpubl.) the sequence and corresponding ages in the lower jaws of known-age animals is:

tooth:	M_1	M_2	I_1	$P_3\ P_4\ (I_2P_2)$	M_3	I_3	C_1
months:	3-6	12-15	15	26	26-32	32	32-42

On Raoul I. the sequence is M_1; M_2; I_1 (P_3 P_4); P_2 $\pm$ (M_3 I_2); I_3 C_1 (J. M. Clark, unpubl.); and in the King Country M_1; M_2 I_1; P_3; P_4; I_2 $\pm$ P_2; M_3; I_3; C_1 (J. M. Clark, unpubl.). Annuli on horns do not correspond to years of age (Rudge, 1972).

Annuli in the cementum of the first incisor indicate year of age to within one year in 87% of cases (M. R. Rudge, unpubl.). Sometimes, "marbling" patterns in the cementum can obscure or prevent line counts. The failure of teeth and horns to form clear annual discontinuities in growth is attributed to the equable oceanic climate in much of New Zealand. In higher latitudes and more rigorous climates, annuli are clearer and are reliable indicators of age in both structures (Grieg, 1969; McTaggart, 1971; McDougall, 1975; Bullock & Pickering, 1984).

POPULATION STATISTICS have been obtained largely during control operations (Table 73). The only undisturbed population sampled lived on Macauley I.

Table 74: Parasites and diseases of goats in New Zealand (those marked * are common to sheep and goats).

Organism	Common name / disease
Ectoparasites	
Melophagus ovinus	* Sheep ked
Linognathus stenopsis	Sucking louse
Psoroptes cuniculi	Ear canker mite
Haemaphysalis longicornis	NZ cattle tick
Damalinia caprae	Goat biting louse
D. limbata (on Raoul I.)	Goat louse
Chorioptes bovis	* Chorioptic mange mite
Demodex caprae	Goat follicle mite
Lucilia sericata	* Green blow fly
Oestrus ovis	* Sheep nasal bot fly
Endoparasites	
Capillaria sp.	Lung worm (nematode)
Bunostomum trigonocephalum	Gut nematode
Dictyocaulus filaria	Lung nematode
Taenia hydatigena	* Tape worm (cestode)
Echinococcus granulosus	* Hydatids (cestode)
Fasciola hepatica	Liver fluke (trematode)
Haemonchus contortus	* Barber's pole worm (gut nematode)
Ostertagia spp.	* Gut nematodes
Trichostrongylus axei, T. capricola, T. colubriformis, T. vitrinus	* Gut nematodes
Trichuris ovis	* Whip worm (gut nematode)
Oesophagostomum venulosum	* Gut nematode
Sarcocystis spp.	* Muscle cyst
Stadelmannia circumcincta, S. trifurcata	Gut nematodes
Spiculopteragia spiculoptera	Gut nematode
Cooperia curticei	* Gut nematode
Skrjabinagia lyrata	Gut nematode
S. ovis	Gut nematode
Protozoa	
Eimeria spp.	Coccidosis
Giardia caprae	Giardiasis
Sarcocystis spp.	Sarcocystis
Toxoplasma gondi	Toxoplasmosis, abortion
Bacteria	
Bacteroides (Fusiformis) nodusus	Foot rot
Clostridium perfringens	Enterotoxaemia
C. tetani	Tetanus
Corynebacterium ovis	Caseous lymphadenitis
Escherichia coli	Mastitis
Leptospira hardjo, L. copenhageni, L. pomona	Leptospirosis
Listeria monocytogenes	Abortion, nervous disease
Mycobacterium paratuberculosis	Johne's disease
M. bovis, M. avium	Tuberculosis
Pasteurella haemolytica, P. multocida	Pneumonia
Salmonella spp.	Scours
Staphylococcus aureus	Mastitis
Streptococcus spp.	Mastitis
Dermatiphilus congolensis	Dermatitis
Yersinia pseudotuberculosis	Yersiniosis
Fungi	
Trichophyton sp.	Ringworm
Viruses	
CAE virus (a retrovirus)	Caprine arthritis-encephalitis
Contagious ecthyma virus	Scabby mouth
Herpes virus	Herpes vulvovaginitis
Corona virus	Scours
Rota virus	Scours

Data from Andrews (1973); Clarke (1972); Collins and Crawford (1978); Daniel (1967); Heath (1979b); Heath, Bishop and Tenquist (1983); Kettle (1982); Kettle and Wright (1985); Tenquist and Charleston (1981).

There, despite the highest density ever recorded in a feral population (10/ha), recruitment was high (170 kids/100 females/year). Kid mortality was also high (34-57%/year), but surviving animals could then live up to 10 years old.

In a population recovering from control in the Rimutaka Ranges, the net recruitment was sufficient to allow the population to double in about 2 years (Rudge & Smit, 1970). Similar recovery rates have since been confirmed in other populations (Clark, 1976; Challies & Christie, 1984).

On Raoul I., the population in areas where desultory control had been practised showed better growth rate and recruitment associated with richer forest vegetation (Rudge & Clark, 1978). This relationship led to the prediction that when systematic control was pursued, there would be increased twinning and more frequent parturition in response to reduced density and an increasing food supply as the forest regenerated. This was confirmed 10 years later, just before the population was finally exterminated. By then it had a potential doubling time of only 20 months (Parkes, 1984c).

PREDATORS, PARASITES AND DISEASES

Man is the only predator of goats in New Zealand. Pigs could kill kids, and have been seen feeding on goat carcasses, although there is no evidence of direct predation.

Rudge (1970) and Clark (1974) described abnormalities (disease and trauma) of teeth and supporting structures. These included misplacement and rotation, unequal wear and breakage of teeth, periodontal disease, loss of teeth, over-eruption, and vestigial teeth.

Goats can suffer severe accidental injuries from time to time and recover from them. Broken horns, legs, ribs and jaws, asymmetrical horns and detached horn sheaths have been described (Clark, 1974; M. R. Rudge, unpubl.). Footrot can range from mild (Clark, 1974) to crippling (M. R. Rudge, unpubl.). Goats on Auckland I. have grossly overgrown hooves but no true footrot (Rudge & Campbell, 1977).

Table 74 lists the parasites and diseases reported from New Zealand feral goats. Many are common to sheep, so when feral goats encroach onto farms, they create a problem of stock hygiene. Of particular concern is that goats can carry hydatids, a cestode which may be contracted by farm dogs when they eat infected goat meat. The dogs pass it on to sheep, and it greatly reduces their value at slaughter. To control this, hunters are instructed to bury carcasses or hang them in trees beyond the reach of dogs when they shoot goats on the fringe of farmland (Anon., 1972).

ADAPTATION TO NEW ZEALAND CONDITIONS

Goats have adapted well to New Zealand conditions and have become as much a pest here as they are in many other places (Darwin, 1845; Furon, 1958). Vegetation, fauna and land stability have been adversely affected, just as on other oceanic islands which had no native browsing mammals, such as Hawaii (Spatz & Mueller-Dombois, 1973), Galapagos (Perry, 1969) and Aldabra (Stoddart, 1981). By contrast, in man-modified lands like Britain, where there are no risks to pristine endemic biota, feral goats have been retained as one of the few large mammals remaining in the fauna (Milner, Goodier & Crook, 1968; Greig & Cooper, 1970).

Overseas studies largely corroborate New Zealand information on behaviour (Yocom, 1967 in Hawaii; Milner, Goodier & Crook, 1968 and Crook, 1969 in Wales; Greig, 1969 and Boyd, 1981 in Scotland; Coblentz, 1980a in California), feeding (Milner, Goodier & Crook, 1968; Coblentz, 1977) and performance in relation to food supplies (Yocom, 1967; Coblentz, 1980b).

All studies agree that breeding can extend over a large part of the year, though local populations differ in the intensity of seasonal peaks. An exception is the quadrimodal breeding cycle described on Santa Catalina I., California, by Coblentz (1980a). In northern latitudes breeding extends from August to February (Boyd-Watt, 1937; Marshall, 1937; Geist, 1960; Asdell, 1964; Milner, Goodier & Crook, 1968; Greig, 1969; McTaggart, 1971). There seems not to be a 6-monthly displacement in the Southern Hemisphere (Asdell, 1964; Caughley, 1971b; Holst, 1982).

Goats elsewhere are generally similar in body size to those in New Zealand, except for some substantially larger animals reported from British Columbia (Geist, 1960). Horns can be very much larger elsewhere, and in shape more like the scimitar horns of ancestral wild goats (e.g., see Milner, Goodier & Crook, 1968; Greig, 1969). Naturally polled feral goats have been reported in Canada (Geist, 1960) and Wales (Milner, Goodier & Crook, 1968) but not in New Zealand.

SIGNIFICANCE TO THE NEW ZEALAND ENVIRONMENT

DAMAGE. Goats were recognized as a threat to native vegetation from the 1890s on (Kirk, 1896; Aston, 1912; Guthrie-Smith, 1969). The effects of their browsing are especially clear on islands. The original subtropical forest of Macauley I. was reduced to rank grassland by goats after the island was burnt (Oliver, 1910). On nearby Raoul I., goats removed all palatable seedlings and epicormic shoots of pohutukawa (*Metrosideros excelsa*) (Parkes, 1984d).

On Great King I. (Turbott, 1948) and Cuvier I. (Atkinson, 1964b) goats induced a grassland/sedge ground layer in forest after 60–70 years of browsing. Any parts of the remaining broadleaf trees brought down by wind were rapidly eaten (Turbott, 1948). On Cuvier I., the Tararua Range and Mt. Egmont, salt was thought to attract goats to browse the seaward slopes (Atkinson, 1964b) but on Mt. Karioi the relationship between salt and preferred browse could not be confirmed (Clayton-Green, 1976).

On Auckland I., goats ate mainly introduced pasture grasses. Palatable forest plants survived if they grew in thickets of unpalatables, but the coastal association of the palatable shrub *Hebe elliptica* was destroyed. Goats ate only small amounts of the tussock grass *Chionochloa antarctica,* but their continuous browsing was sufficient to virtually eliminate it from the part of the island they inhabited (Campbell & Rudge, 1984). On Mt. Egmont, areas of subalpine scrub low enough for goats to climb onto and browse from above (<1.5 m) were destroyed within 4 years (Atkinson, 1964b).

CONTROL. The first control work was undertaken by local landowners in the pastoral areas around the southern lakes (Thomson, 1922). When official government action against goats began, it was mainly to appease pastoral runholders in Marlborough (Wodzicki, 1950). Private landowners have long exercised some control over goat numbers and encouraged hunters on their land (Donne, 1924). The most concentrated effort has been on islands, and many populations have now been eradicated (Table 70).

A constant problem with goats in the pastoral fringe has been their dual identity as both stock and pests. This, together with a dismissive attitude to native forest values, meant that for many years there was little official control work. Eventually various Acts of Parliament clarified the position and enabled DIA and NZFS to begin control campaigns in the 1930s. Even then, control of goats was seen as a sideline to campaigns against deer.

Over the past 10 years goats have been captured alive from helicopters instead of shot, and used to stock goat farms. However, in rugged areas such as the Wanganui and Motu River valleys and in the Paparoa Range, where live capture is impossible, control work to protect forests continues.

The traditional method of control was with foot hunters and dogs. Variations include hunters in helicopters, marking individuals with bells or radios to betray a group (Judas technique), and sticking poison gel on vegetation (Parkes, 1983).

MANAGEMENT. In 1976 it was suggested that remaining feral mammals, including goats, might have genetic and commercial value (Whitaker & Rudge, 1976). The conservation of rare breeds of farm stock emerged as a public issue around the goats of Arapawa I., which were seen as a remnant of the "Old English" breed (Rudge, 1978).

The commercial value of feral goats is now well established. Feral goats of Angora origin have been used to build up mohair-producing flocks, and common feral goats are now known to possess underfur of cashmere quality (Kettle & Wright, 1985; National Business Review, 1985). There is a lively trade in mustering feral animals for meat and for breeding up to higher fibre grades (Moorhouse & Batten, 1978; Kirton & Ritchie, 1979); as surrogate mothers to gestate transplanted high-quality ova in rapid breeding-up programmes for Angoras (Kettle & Wright, 1985); and for cheap control of woody weeds. Goats were taken from Auckland I. in 1986 and 1987 to be evaluated as producers of meat and fibre by DLS (Rudge, 1986a).

Thus, the wheel of fortune for goats has gone full circle, from farming resource to pest and back to farming resource. But the pest *potential* of goats is undiminished, and the increase in commercial prospects brings with it renewed dangers for the forests. Goats are notoriously hard to fence in, and on hobby farms and hill country farms bordering forests, goats will be a constant source of reinfestation.

The present time is therefore very significant. Diversification of livestock and primary products is being dynamically encouraged, but it will need to be matched by equally dynamic policing and control, if 50 years of effort to protect many native forests is not to be jeopardized.

M. R. R.

Genus *Ovis*

Wild sheep are native to Eurasia (four species) and North America (two species). The ancestor of the domestic sheep, the mouflon (*O. musimon*), comes from the Middle East and still survives there and in Europe, where it is widely introduced as a game animal. All six species have characteristic preorbital, pedal, and inguinal glands, but do not produce the offensive smell of goats. The males are larger than the females. All are highly gregarious, specialized grazers.

37. Feral sheep

Ovis aries Linnaeus, 1758

Breed names of domestic sheep giving rise to New Zealand feral populations are Merino, Leicester, Romney, Cheviot, Lincoln and Corriedale.

Description (Fig. 72)

Distinguishing marks, Table 63 and skull key, p. 24.

The coat is a crimped, woolly fleece with occasional hair fibres. It may be short and neat or long and untidy, depending on the population and the stage of the moult cycle. The predominant colour varies with the population and may be white, dark brown to black, or particoloured. The feral sheep of Campbell and main Chatham Is. and all the mainland populations are predominantly "white", with a small percentage (10–20%) pigmented. On Arapawa I. (Orwin & Whitaker, 1984) and Pitt I. (Rudge, 1983) more than 90% are predominantly "black", often with badger face marks and white-tipped tails. Self-shedding is common in pigmented populations (Arapawa and Pitt Is.) and is also observed elsewhere. The pendulous tail is up to 400 mm long.

The sexes are dimorphic, with males larger in all respects. Sheep of either sex may carry horns or scurs (small horny pads or cones, not visible above the fleece, which never develop into true horns). Horns bear alternating grooves and ridges. In females they are slender, more or less erect and curved backwards. In males they are more massive, with broad anterior surfaces, and after curving back they spiral outwards for up to two turns.

Chromosome number 2n = 54.

Dental formula $I^0/_3\ C^0/_1\ Pm^3/_3\ M^3/_3 = 32$.

Field Sign

Tracks and pellets left by feral sheep are indistinguishable from those of domestic sheep. They are readily confused with those of feral goats, and with sika deer and pigs where their ranges overlap. Tracks have blunt tips (cf. pointed in goats) and strongly convex outer margins, and show no dew claws in soft ground (cf. pig). Each digit measures up to 55 × 25 mm in adults.

Pellets may be adherent (rarely) or an amorphous faecal paste if the sheep have access to rich pasture. Single pellets have faceted sides, usually without a prominent nipple and dinge fore and aft (cf. deer), and with proportions more square than elongated (cf. goat). Dimensions are about 20 × 10 mm.

Fig. 72 Feral sheep (M. R. Rudge).

Fleece wool may be found on shrubs; the call is a "baa" rather than a "meh" (cf. goat).

MEASUREMENTS
See Table 75.

VARIATION
The proportion of individuals with horns, scurs, or neither varies in different populations. On Arapawa I., horns are found on 88% of males but no females (Orwin & Whitaker, 1984); on Pitt I., horns are found on 97% of males and 13% of females, and scurs on a further 54% of females (Rudge, 1983); on Campbell I., 67% of males are horned and 22% scurred, and 20% of females are horned and 29% scurred (M. R. Rudge, unpubl.); in the Waimakariri area, 95% of males and 2% of females are horned (Parsons, 1980). Other forms of variation are negligible, except colour (see above).

HISTORY OF COLONIZATION
Domestic sheep were brought to New Zealand throughout the 19th century. They came largely from Australia, and most were therefore merinos. Flocks ran on extensive, unfenced range, often including or bordering undeveloped country such as bush, scrub and tussock. Sheep rapidly became the basis of New Zealand's pastoral economy, and many new breeds were introduced. By the turn of the century crossbred sheep replaced merinos, except in the high country, and many merino flocks were abandoned.

Sheep were also taken to offshore and outlying islands for farming or as emergency food in case of shipwrecks. They died out naturally on Antipodes and Adams Is., were exterminated on Mangere, Southeast and Kapiti Is., and are still present on Arapawa, main Chatham, Pitt and Campbell Is.

DISTRIBUTION
WORLD. Sheep were domesticated in the Middle East from the mouflon, *O. musimon,* in the Neolithic era (about 6000 years BC) and have been continuously

Table 75: Body measurements of feral sheep in New Zealand.

	Body weight[1] (kg)		Shoulder height (mm)		Body length (mm)		Reference
	M	F	M	F	M	F	
Campbell I.	51	28	—	—	—	—	Wilson & Orwin (1964)
	65 (80)	35 (39)					Regnault (1980)
	46 (66)	41 (60)	680 (780)	641 (770)	686 (820)	651 (780)	Ballance (1985)
Arapawa I.	51 (45-60)	38 (36-39)	705	650	1260	1170	Orwin & Whitaker (1984)
Pitt I.	—	—	766	695	1240	1262	M. R. Rudge (unpubl.)
Omahaki	—	—	720	638	1317	1189	Orwin & Whitaker (1984)
Waimakariri	(43-62)	44 (34-56)	770	690	1250	1220	Parsons (1980)

1. Weights of farmed ewes average 33.4 kg (Coop & Clark, 1966).
Numbers in parentheses show maximum or range.

improved by artificial selection since then. Feral populations of primitive breeds exist in Britain, and possibly on Corsica and Sardinia. Elsewhere, feral populations are derived from various modern breeds (Lever, 1985).

NEW ZEALAND (Map 36). Feral sheep were formerly widely distributed in both main islands (Wodzicki, 1950), but the distinction between independent populations and mismustered stragglers was unclear. There are at present eight discrete herds on the mainland and four on islands, and these may now be accepted as self-sustaining, unmanaged, and truly feral. The herds are located and numbered on Map 36 as follows. (1) Mohaka River valley; (2) Ngaruroro River valley; (3) Wairau River valley; (4) Clarence River valley; (5) Waimakariri River valley; (6) Waianakarua River valley; (7) Waipori Gorge; (8) Hokonui Hills; (9) Arapawa I.; (10) main Chatham I.; (11) Pitt I.; (12) Campbell I.

HABITAT

Feral sheep occupy rough pasture in broken scrub and forest, from sea level to over 2000 m. Nowhere do they live exclusively in woody vegetation, but they do move through it and use it for shelter and some browse.

FOOD

Sheep are essentially grazers on pasture and herbaceous plants (cf. goats, which are browsers). On Pitt I., woody plants (introduced and native) comprise only 3% of the diet (Rudge, 1984). On Campbell I., sheep and fire together have almost eliminated the large endemic herbs and palatable tussock grasses. The

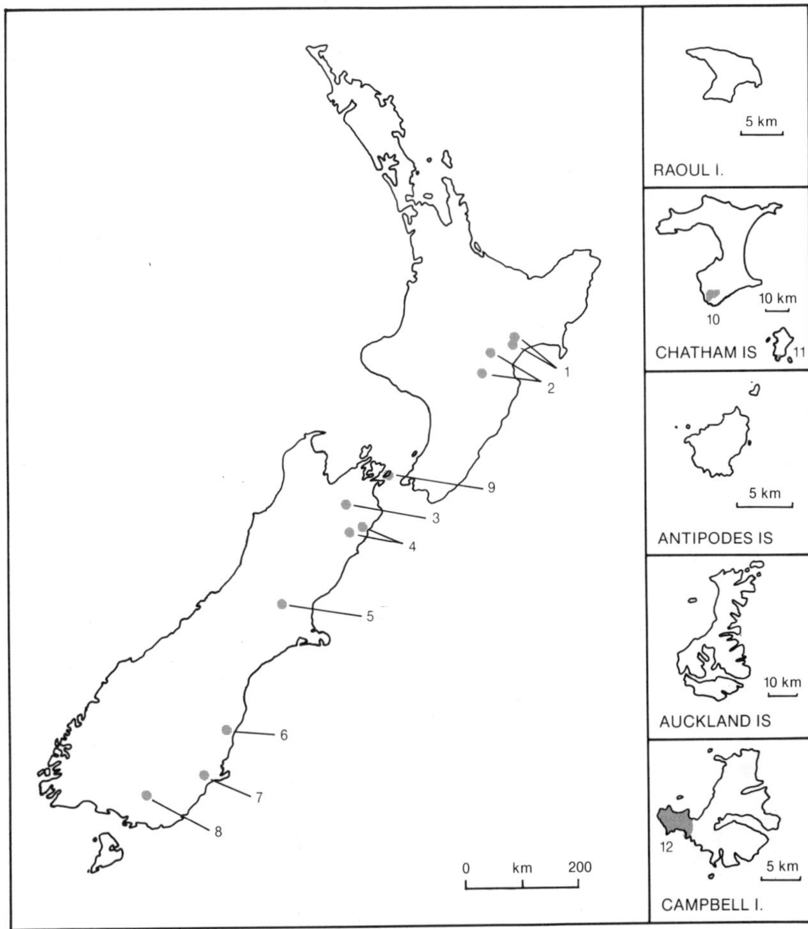

Map 36 Distribution of feral sheep in New Zealand. Herds numbered as in text (p. 426).

sheep now feed on induced swards (Wilson & Orwin, 1964).

Feral sheep feed in the open, but have penetrated the scrub on Campbell I. and opened it up (Wilson, 1980). They feed actively throughout the daylight hours on Campbell I. and in all weathers, and lie down to ruminate (Regnault, 1980). In winter 70% of their time is spent feeding and 17% ruminating (Ballance, 1985).

Social Organization and Behaviour

No populations have yet been studied throughout the year. All anecdotal records agree that the modal group size is two, comprising a female and her young of the year; that there is a strong home range attachment; and that up to nine rams may form a group during the non-breeding season. In the Waimakariri, males exhibited sentinel behaviour and led the escape (Parsons, 1980), but on Campbell I. females were the most alert and responsible for group cohesion (Regnault, 1980).

Table 76: Parasites and diseases of feral sheep on Campbell I. and Pitt I.

	Campbell I.	Pitt I.
Ectoparasites		
Melophagus ovinus (sheep ked)	+	+
Damalinia ovis (lice)	+	0
Chorioptes bovis (mange mite)	+	+
Ectoparasites found on mainland sheep but not in either island population were: *Psorergates ovis, Demodex* sp., *Psoroptes communis ovis, Linognathus pedalis, L. ovillus*		
Endoparasites		
Ostertagia sp.	+	+
Trichostrongylus axei	+	+
Nematodirus sp.	+	0
Bunostomum sp.	+	0
Taenia hydatigena	+	0
Cysticercus tenuicollis	+	0
Sarcocystis spp.	0	+
Muellerius capillaris	+	+
Endoparasites expected but not recorded in either population were: *Echinococcus granulosus, Taenia ovis*		
Diseases		
Toxoplasma gondii	-ve	+ve
Sarcocystis ovicanis	-ve	+ve
S. tenella	-ve	+ve
Brucella ovis	-ve	-ve
Leptospira pomona, L. icterohaemorrhagiae, L. hardjo, L. ballum	-ve	-ve
Johne's CFT	-ve	-ve

Parasites recorded as present (+) or absent (0). Results of tests for diseases shown as positive (+ve) or negative (-ve).
Data from Wilson and Orwin (1964); Heath (1979a, 1983); Hutton (1980).

Ballance (1985) studied behaviour on Campbell I. in winter. Home ranges were 41 ha (males) and 45 ha (females) and the mean distance moved each day was 400 m. Group size was 2–9 (mean 6, mode 2) with loose aggregations of up to 30. Regnault (1980) found that some areas were occupied exclusively by large groups of rams.

Alarm behaviour includes whistling (M. R. Rudge, unpubl.), foot stamping, coughing (Parsons, 1980) and snorting. When alarmed, feral sheep run for woody cover (observed on Pitt and Arapawa Is.), whereas domestic sheep generally run in the open. On Pitt I., and in the Waimakariri, feral males are known to mate with domestic females, but feral females will not accept domestic rams. In captivity, feral and domestic sheep keep apart of their own volition (D. F. G. Orwin, unpubl.; F. R. M. Cockrem, unpubl.).

Reproduction and Development

In all populations observed, breeding extends over a long period, with a peak in June and July (Parsons, 1980; Orwin & Whitaker, 1984; Ballance, 1985). In

Hokonui some may breed twice a year (Orwin & Whitaker, 1984). The gestation period is 150 days. Lambs appear at almost any time, but mostly from July to September, and also in March in the Waimakariri and on Campbell I. Campbell I. females come into oestrus at a younger age than domestic sheep (Bigham & Cockrem, 1982) and some breed before they are 12 months old (Rudge, 1986b).

Lambing rate (lambs/100 ewes) has been estimated at 68 on Pitt I. (Rudge, 1983), 23–81 on Campbell I. (Rudge, 1986b) and 20–55 on Arapawa I. (Orwin & Whitaker, 1984). Twins are rarely reported in the wild, but are common in feral stock returned to captivity (Bigham & Cockrem, 1982).

Lambs are born with 10 milk teeth in the lower jaw (3 incisors, 1 incisiform canine, 6 premolars), and 6 premolars in the upper jaw. Permanent teeth erupt in the sequence M_1; M_2; I_1; (M_3 $P_{2,3,4}$) $\gtreqless$ I_2; I_3; C_1 in Campbell I. sheep (Rudge, 1986b). Full body size is reached in 1.5–2 years (Parsons, 1980) or 2–3 years (Ballance, 1985).

POPULATION DYNAMICS

NUMBERS. There are at present about 1500 feral sheep in three protected flocks: Arapawa I. 100–120 (Orwin & Whitaker, 1984); Pitt I. about 500 (M. R. Rudge, unpubl.); and Campbell I. about 800 (Rudge, 1986b). All other flocks are hunted, and local informants indicate that numbers are less than 100 in Ngaruroro, Clarence, and Wairau; up to 500 in Hokonui and Waimakariri; and at least 10 000 in the Mohaka River Valley. On Campbell I., total numbers over the whole island declined from about 3000 in 1931 to about 1000 in 1961 (r = –0.05) (Wilson & Orwin, 1964), then increased to about 3000 in 1969 (r = +0.14). On the southern half of the island, the remnant left after control work in 1970 continued to grow at r = +0.053 till the next control operation in 1984 (Dilks & Wilson, 1979; Rudge, 1986b). Sheep may live up to 11 years on Campbell I. (M. R. Rudge, unpubl.), but none older than 5 years was recorded in the hunted Waimakariri population (Parsons, 1980).

SEX RATIOS of males to females are 1:4 on Arapawa I.; 1:3 in Waimakariri; 1:1 on Pitt I.; and 1:1.3 on Campbell I.

PREDATORS, PARASITES AND DISEASES

Predation has been reported only on Campbell I., where skuas (*Stercorarius skua lonnbergi*) prey on lambs (Spence, 1968) and harass adults (Regnault, 1980). Parasite burdens have been examined in detail on Campbell and Pitt Is. (Table 76).

Footrot (*Bacteroides nodosus*) and flystrike were not recorded in the island populations. On Campbell I., blood levels of selenium, zinc and iron were normal, but copper was appreciably higher in an area underlain by limestone (Hutton, 1980). Periodontal disease was reported from Campbell I. by Hutton (1980), but Suckling and Rudge (1977) interpreted it as a non-pathological growth form of the teeth developing in conditions of low abrasion.

In the Waimakariri population *Melophagus ovinus, Muellerius capillaris* and *Cysticercus tenuicollis* have been reported, but there were no gut nematodes, nor

Table 77: Comparison of fleece and fibre characters of feral and domestic sheep.

	Arapawa I.	Pitt I.	Waimak-ariri	Campbell I.	Raglan	Commercial merino
Mean ratio of secondary to primary follicles	6	4.9	10.5	5	3.7	11-46
Mean fibre diameter (μm)	23.1	24.4	20.1	27.3	40.6	27.1(5-50)
Fleece weight (kg)		1.1–3.6				3.5-5.0

any antibody reactions to *Brucella ovis, Leptospira pomona, L. hardjo* and Johne's disease (Parsons, 1980).

Horn and jaw abnormalities were scored in 1080 sheep shot on Campbell I. (M. R. Rudge, unpubl.). Of ten abnormal horns, four were growing into the soft tissues and causing sepsis, and others were variously broken off or split. Jaw and tooth abnormalities were found in 102 jaws. Periodontal disease was found in 41 jaws, often as a secondary effect of broken teeth, impaction of food, or malocclusion. Teeth were broken in 38 jaws. In nine jaws, infections had caused one mandible to inflate by between 1.2 and 1.8 times the thickness of its partner.

ADAPTATION TO NEW ZEALAND CONDITIONS

The Soay and other British feral sheep are of ancient lineage, so cannot be compared with recent feral populations derived from modern breeds (Jewell, Milner & Boyd, 1974; Lever, 1985). Feral merino sheep on Santa Cruz I. (California) and Mauna Kea (Hawaii) are about the same size as those in New Zealand: body weights of 50 kg male, 37 kg female are recorded on Santa Cruz (Griffin 1976); and 45 kg male, 34 kg female in Hawaii (Van Vuren, 1981). Males on Mauna Kea were usually horned and females were not. Behaviour and home range were similar to those of New Zealand feral sheep.

In several populations, a range of fleece, fibre and skin characteristics have been examined and compared (Orwin & Whitaker, 1984). Although feral flocks may produce wools of various mean fibre diameters, there is a common trend toward low greasy fleece weight, high bulk, low follicle density and a greater proportion of primary follicles (i.e., low secondary to primary S/P ratio) (Table 77) compared with commercial sheep of similar breed. For example, commercial merinos have a fibre diameter similar to those of the Arapawa, Pitt and Waimakariri flocks, but have a higher fleece weight and a higher S/P ratio. Each feral flock also has interesting wool characteristics of its own.

SIGNIFICANCE TO THE NEW ZEALAND ENVIRONMENT

DAMAGE. On the mainland, most feral sheep live in marginal scrub and bush habitats, where they cause little damage because they are essentially grazers.

They can create a local nuisance to farmers by mixing with domestic stock to disrupt breeding and spread ectoparasites.

In bush on Arapawa I., sheep (together with goats and pigs) prevented regeneration and damaged the forest floor with their hooves. Both effects are damaging to giant snails (*Powelliphanta hochstetteri*), which need deep humid litter for food and shelter (Dingwall & Rudge, 1984). The sheep and the snails have now been separated by fences.

On Pitt I., sheep (and cattle) also prevent regeneration and disturb the nesting areas of burrowing petrels (B. D. Bell, unpubl.).

Campbell I. is the best documented example of the impact of feral sheep on an island. After the fires and overstocking of the farming era, sheep maintained a highly modified vegetation almost devoid of large endemic herbs and tussock grasses. In 1970, the island was bisected by a fence, and all sheep were removed from the northern half. In the next 10 years there was a spectacular recovery in the vegetation north of the fence. Closely grazed sward was replaced in many places by large endemic herbs (*Anisotome* spp., *Stilbocarpa polaris, Pleurophyllum* spp.), and by the palatable tussock grasses *Poa foliosa* and *Chionochloa antarctica* (Meurk, 1982; Rudge, 1986b). The population of nesting royal albatrosses (*Diomedea e. epomophora*) did not decline with the increase in vegetation cover, as had been feared (Dilks & Wilson, 1979).

CONTROL. The status of feral sheep around the world varies. They are protected in Britain, where they have scientific and historic interest as a primitive breed, and hunted as pests in the Galapagos Is. and Hawaii, where they inhabit valuable nature reserves. In New Zealand, abandoned or strayed sheep have long been regarded as pests to be controlled by government and private shooters (Howard, 1965). Between 1951 and 1961 an average of 2000 were shot officially each year, but nowadays the number is about 100, mostly in special areas such as nature reserves (NZFS files). In recent years feral sheep have gradually been cleared from the Raglan peninsula and the Takitimu Range. They were exterminated from Mangere I. in 1968, Southeast I. in 1961 (Atkinson & Bell, 1973), and Kapiti I. in about 1928 (Esler, 1967). In 1970 all sheep to the north of the fence on Campbell I. were killed (about 1300); in 1984 the population on the southern side was exterminated (about 4000) except for about 800 fenced off on a peninsula (Ballance, 1985).

As populations became fewer and smaller, scientific interest in the remainder has increased. The case to preserve some *in situ* has been accepted as a contribution to the worldwide movement to conserve livestock breeds and varieties (Whitaker & Rudge, 1976; Rudge, 1982, 1984). Protected, representative populations have now been established on Pitt I. (Chatham group), Arapawa I. (Marlborough Sounds) and Campbell I. (subantarctic) (Rudge, 1983, 1986b; Dingwall & Rudge, 1984). They are being used for research on wool and fleece characters (Table 77), fertility, physiology and adaptation to feral life (Rudge, 1984).

M. R. R.

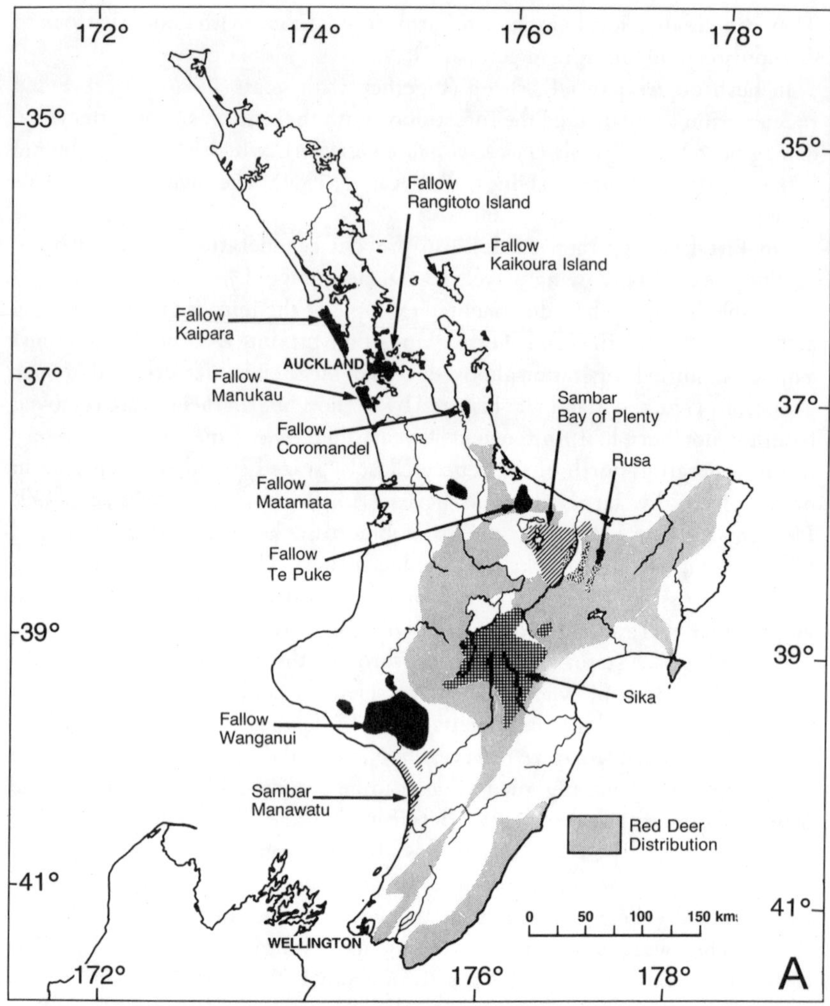

Map 37A General distribution of deer in the North Island (from Challies, 1985a). Challies accepted the presence of a herd of fallow deer on the Coromandel Peninsula, whereas Davidson & Nugent do not, so that herd is omitted from their chapter (p. 498) and Map 43.

Family Cervidae

The deer family contains about 36 species in 16 genera, grouped in four subfamilies. The Hydropotinae has one species (the Chinese water deer); the Muntiacinae has six species (the muntjacs and tufted deer of Asia); the Odocoilinae has 15 species (including the mule and white-tailed deer of North America, the roe deer of Eurasia, the moose and reindeer common to both, and the brockets and pudu of South America); and the Cervinae has 13 species (including red deer/wapiti, fallow, axis, sika, sambar, rusa and Père David's deer), all except wapiti native to Eurasia. Of these, only the Odocoilinae and

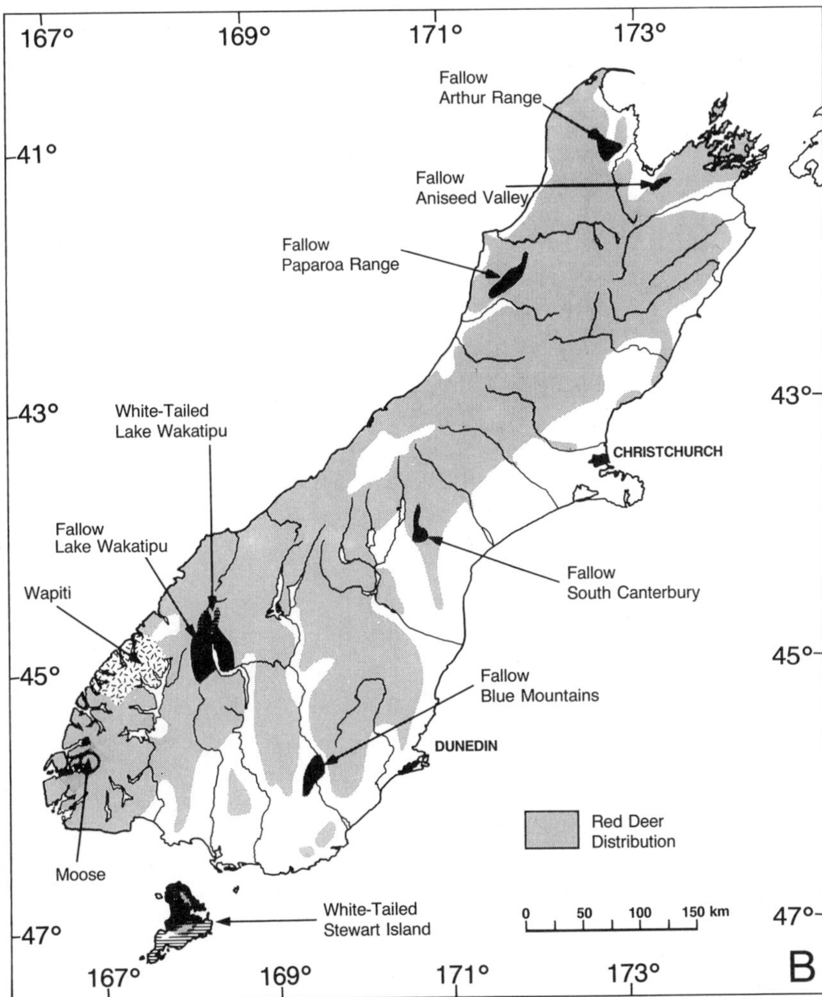

Map 37B General distribution of deer in the South Island (from Challies, 1985a).

Cervinae are represented in New Zealand, with a total of seven living species (one with two distinctly different subspecies); their distinguishing marks are listed in Table 78 and Fig. 73, and the relationship between their ranges in New Zealand is shown in Map 37. Red deer are sympatric with all other species over parts of their range, but none of the others is sympatric one with another except possibly in the Bay of Plenty (p. 487). The antlers of deer, worn by males only, are made of naked bone and are grown and shed every year; by contrast, the horns of bovids, usually found in both sexes, are made of a bony core covered by a horny sheath, and they grow throughout life.

Table 78: Distinguishing marks of cervids in New Zealand.

	Order of size in NZ	Distribution	Antlers (mature stags)	Coat			Tail
				Adults (S: Summer; W: Winter)	Young calves	Rump patch (size & colour)	
Subfamily Cervinae							
Red deer *Cervus elaphus scoticus*	4	Widespread	Round, erect, 10+ tines	Plain, red-brown (S), grey-brown (W)	Spotted	Medium, creamy-white	Short, light red-brown
Wapiti *C. e. nelsoni*	2	Fiordland	Round, erect, 12+ tines	Plain, pale grey-brown, dark head & neck (W)	Spotted	Large, extending on to lower back, creamy-white	Short, creamy-white
Sika deer *C. nippon*	6	Central North Island	Round, erect, 8 tines	Spotted, red-brown (S), plain brown (W)	Spotted	Medium, white bordered with black	Long, white with black stripe
Sambar *C. unicolor*	3	Manawatu, Bay of Plenty	Round, erect, 6 tines	Plain dark brown (S), darker-brown (W)	Plain	Small, creamy-white	Long, dark brown
Rusa deer *C. timorensis*	5	Galatea district	Round, erect, 6 tines	Plain brown (S), grey-brown (W)	Plain	Small, light brown	Long, brown
Axis deer *Axis axis*	6	Extinct?	Round, erect, 6 tines	Spotted, red-brown	Spotted	Medium, white bordered with brown	Long, brown with white fringe
Fallow deer *Dama dama*	6	Localized	Palmate, erect, medium size	Spotted or plain, several colour phases	Spotted	Medium, white or brown, depending on phase	Long, black and/or white
Subfamily Odocoilinae							
White-tailed deer *Odocoileus virginianus*	6	Otago Stewart I.	Round, curved forward, 8+ tines	Plain, red-brown (S), grey-brown (W)	Spotted	Medium, white	Long, brown with white fringe
Moose *Alces alces*	1	Fiordland	Palmate, extend sideways, very large	Plain, dark brown	Plain	None	Short, dark brown

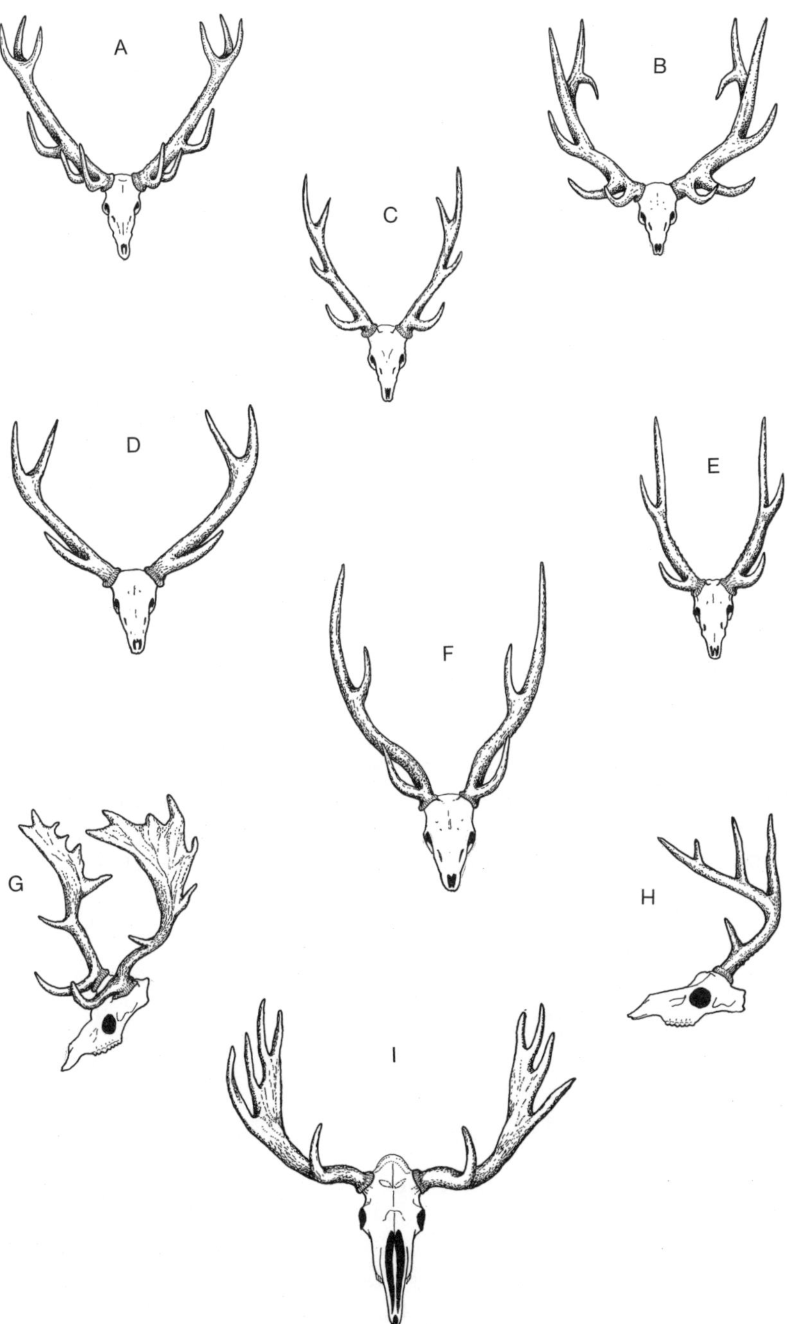

Fig. 73 Antlers of deer species introduced to New Zealand (not to scale). The axis deer is probably extinct, and the moose nearly so, but their antlers might still be found (J. Lavas). A, red deer; B, wapiti; C, sika; D, sambar; E, rusa; F, axis; G, fallow; H, whitetailed; I, moose.

Subfamily Cervinae

The subfamily Cervinae contains one large genus (*Cervus*) and three smaller genera: *Axis,* with four species, all Asiatic, and *Elaphurus* and *Dama* with one species each. Five forms of *Cervus* and one each of *Dama* and *Axis* are or have been established in the wild in New Zealand. *Elaphurus davidiensis*, Père David's deer, was formerly native to China but is now extinct there. From 1983, small numbers of these deer have been imported to New Zealand from Britain for diversification and improvement of stock on deer farms. At first the new deer bred and did well, but they proved very susceptible to malignant catarrhal fever, and by 1987 few of the pure-bred stock had survived. Some Père David-red deer hybrids remain, but their future is uncertain (Dearlove, 1987).

Genus *Cervus*

Of the seven species of *Cervus*, four have been brought to New Zealand, but one, *C. elaphus* (= *C. canadensis*) has been brought both from Europe (red deer, *C.e. scoticus*) and from North America (wapiti, *C.e. nelsoni*). Variation within the *C. elaphus/C. canadensis* group (total 23 subspecies; Whitehead, 1972) includes a cline of increasing body size and of lightening body colour eastward from Britain and Europe, across Asia to North America. The British red deer and the wapiti represent opposite ends of the cline, but are still fully interfertile; when they met in New Zealand, they freely hybridized, proving that wapiti cannot, as formerly (Macdonald, 1984), be regarded as a distinct species. The other three species of this genus brought here are all native to Asia: *C. nippon* (eastern Asia and Japan), *C. timorensis* (Indonesia) and *C. unicolor* (India and South-East Asia). The temperate species of *Cervus* have a distinct, short autumn breeding season and are often larger; the tropical species have a longer, more diffuse season and most, except *C. unicolor*, are relatively small.

38. Red deer

Cervus elaphus scoticus Lönnberg, 1906

DESCRIPTION (Fig. 74)

Distinguishing marks, Table 78 and skull key, p. 24.

A medium sized, round-antlered deer with a uniform, plain brown body, lighter below; creamy-white rump patch; short tail, light reddish brown above; and adult antlers with 10 or more tines, the uppermost pointing upwards in a cluster.

Both sexes and all ages (except young calves) are similar in colour. The winter pelage consists of a thick woolly undercoat and long, coarse, finely crimped guard hairs, mostly brown or grey-brown but with a faint dorsal stripe. Adults have no spots, except a few which have a single row of light brown spots along either side of the back. The throat and underside are light grey grading to creamy-white between the hind legs. Stags have a prominent brown mane and are often dark grey to black on the belly. The rump patch is creamy-white, grading into light reddish brown above the tail and onto the

Fig. 74 Red deer in velvet, Otago, 1967 (J. H. Johns).

lower back, bordered on each side by a dark brown to black line. The tail is short and matches the colour of the upper rump. The ears are pointed, as long as or longer than half the length of the head, and dark brown outside with long, creamy-white hair inside. The winter coat is moulted over several weeks in late spring (from late October through December), earlier in the older and best conditioned deer. The summer pelage, consisting of short, stiff guard hairs with little or no undercoat and no mane, is bright red-brown when new, but becomes progressively lighter in colour during the summer as it wears; it often has a dark dorsal stripe, most pronounced on the neck and lower back; elsewhere it is similar to the winter coat. In late summer (from mid February through April) the new winter guard hairs grow through the summer coat, which is gradually shed. The muffle is blackish and hairless, and the hooves are grey to black.

There are three pairs of glands. Two, the facial or preorbital glands (with narrow openings directly in front of the eyes) and the metatarsal glands (on the outer sides of the tarsi), are inconspicuous, and the third, in the tail, is much larger.

Newborn calves are brown or reddish brown with a dark dorsal stripe and a creamy to light brown rump patch; white spots are scattered on the back and flanks, with a double row along the middle of the back from the neck to the base of the tail, and a single row in an arc on each side of the rump. As the calves grow and their hair lengthens, the spots gradually blur and disappear by about 2 months of age. The first coat is shed during late summer and replaced by a winter coat of thick under-wool and long guard hairs similar to that of adults but softer and fluffier.

Antlers are grown and cast annually by males from their second year. The growth of velvet antler starts between early September and December, soon after the previous antlers have cast, and is completed when the dried velvet is frayed from the hard antler between mid January and mid March. Older stags and those in good condition tend to cast and start growing their new antlers earlier than do younger stags and those in poorer condition. The velvet antler is soft and flexible at first, bulbous in appearance, grey or brownish, and grows by elongation of the main beams and budding of the tines from the apex at their respective heights. Newly-frayed antlers are cream coloured and bloody, but soon become stained brown or blackish with light tips. Yearlings typically grow a pair of unbranched spikes. The antlers of older stags (Fig. 73) have a lightly pearled main beam with a number of unbranched tines. The number of tines and size of antlers depends on the age and condition of the stag. Two-year-olds commonly have 4 to 8 points, rarely spikes, and occasionally 10 points; mature stags have 10 to 12 or more points.

Dental formula $I^0/_3$ $C^1/_1$ $Pm^3/_3$ $M^3/_3$ = 34.

The upper canines are smooth, rounded and loosely set alone on either side of the mouth; the lower canines are incisoriform and set in the front of the mouth as nominal fourth incisors. The molars are bicuspid except for the mandibular third molar, which is tricuspid (tooth nomenclature follows Riney, 1951).

FIELD SIGN

The pellets of red deer are hard and well shaped, glossy black or dark brown when fresh, fading in time to dull light brown; they are usually elongated, with a slight depression at one end and pointed at the other, but can also have a variety of other forms, and are adherent when moist. They differ in size with the age and sex of the animal; those of adult females average about 12 mm in diameter and 20 mm in length. Individual defaecations produce from about 30 to 150 pellets. They are either aggregated or scattered depending on whether the deer was standing or moving.

The footprints show two sharply defined pointed toes, with rounded, poorly defined heels, normally touching along the whole length but splayed and imprinting the dew claws in soft ground and at speed. The size of the prints depends on age and sex; those of adult females average about 65 mm long and 50 mm wide and those of younger deer are proportionately shorter and more pointed. When walking and trotting the hind feet overprint the forefeet, but when galloping the feet imprint separately in unevenly spaced groups of four. Well defined trails are formed along commonly used routes, which may become trenched in areas of high use.

Shrubs and saplings are frayed by stags preparing for the rut. Wet wallow holes are formed by stags before and during the rut, and both wet and dry wallows are used to some extent by both sexes at other times of the year.

MEASUREMENTS

See Table 79. Red deer differ greatly in size depending on the quality of their nutrition during their formative years (see below). The statistics in Table 79 are from a composite sample of mature deer (6-7 years and older) shot in South Westland between 1968 and 1974, which included animals from populations at different densities and levels of nutrition. The growth rates of red deer in all but the most recently colonized areas of New Zealand have increased over the last 20 years, as deer numbers have been reduced by commercial hunting (p. 456; Challies, 1985a); hence most mature animals are now larger than the means given in Table 79.

VARIATION

GENETIC. Several genetically different strains of red deer were introduced into New Zealand, some of which established local herds distinguishable by stature, pelage colour, antler form, or antler tine configuration, e.g., the presence or absence of bez tines (Clarke, 1973). As the colonizing herds coalesced and interbred, their distinct characters became blurred and are now largely lost.

SEASONAL. There are two pronounced seasonal changes in conformation: the temporary enlargement of the neck of adult stags before the rut, and the seasonal cycle of fat deposition and removal which causes corresponding fluctuations in weight. During summer, average mature stags increase in body weight by 35% to 40%, and breeding hinds up to 10%, compared with late winter.

ENVIRONMENTAL. Until the 1970s, there were substantial differences between local populations in different parts of the country, in density, growth rate,

Table 79: Body measurements of red deer in South Westland 1968-74.

	Sex	Mean	SD	95% range[1]	n
Carcass weight (kg)[2]	M	71.4	18.9	34.3-108.5	31
	F	45.2	9.1	27.3-63.1	109
Body length (mm)[3]	M	1 936	143	1 655-2 217	84
	F	1 768	96	1 579-1 957	150
Foot length (mm)[4]	M	526	27	473-579	57
	F	497	19	460-534	93
Jaw length (mm)[5]	M	293.9	15.7	263.1-324.7	150
	F	273.5	12.3	249.3-297.7	150

1. The 95% ranges approximate the limits of growth of wild red deer in New Zealand high country.
2. Body cleaned of viscera including entire gut, trachea, reproductive tract and udder, and with head and hocks removed (i.e., German market carcass weight). These weights were of animals taken in November and December when most red deer are relatively lean. By March the carcass weights of mature stags and breeding hinds are on average 35–40% and up to 10% heavier, respectively. Carcass weight is equivalent to 55–65% of live weight, depending on the condition of the animal (Mitchell, 1970; Smith, 1974).
3. From tip of nose to tip of tail, excluding hair, over the curve of the back.
4. From proximal end of calcaneum to tip of extended hoof.
5. From distal end of dentary to proximal edge of mandibular condyle, diagonally across jaw.

Data adapted from Challies (1978).

antler size, fecundity and general health, depending mainly on the phase of colonization they were in (p. 448) and on the types, timing and intensity of hunting they had sustained. Deer from high density populations were lighter (averaging 55% to 65% of body weight at low density), with smaller antlers; fewer females bred as yearlings, and more females of all ages failed to breed; and the mortality of young animals was higher (Challies, 1978, 1985a). These differences greatly exceeded any variation due to the origin of the local founding stock. During the last 20 years, commercial hunting has everywhere greatly reduced the numbers of deer. The surviving populations are now generally well fed, in good condition, and highly productive throughout their range.

HISTORY OF COLONIZATION

ORIGIN OF NEW ZEALAND RED DEER. More than 250 red deer of British origin were imported into New Zealand between 1851 and about 1919. About one-third were shipped direct from Britain, and the rest were bred from British stock previously imported into Victoria, Australia. The direct imports came from the Scottish highlands (from Black Mount and Glenavon Forests via Invermark Forest) and from various English deer parks, particularly Thorndon Hall, Windsor Great Park, Warnham Court and Stoke Park; the Australian stock were park-bred, primarily from the Windsor Great Park strain (Donne, 1924; Logan & Harris, 1967).

Only the Scottish deer were of a pure wild strain; the English park herds were interbred mixtures of local and other park stocks, with additions from wild herds in other parts of the British Isles and the Continent. German and Danish red deer were introduced into both the Thorndon Hall and Windsor Great Park herds in the 17th century, and their diluted bloodlines were later passed on to the Warnham Court herd with the introduction of Thorndon Hall animals (Whitehead, 1950). Many of these park herds developed, or were selected to produce, their own characteristics, such as a distinctive antler form, which have at times been noticeable in some of New Zealand's wild herds (Clarke, 1973).

About 1000 red deer had been liberated into the wild in New Zealand by 1923, including most of the imported animals, plus some translocated from earlier established wild herds or local game parks. All but a few of the original herds were derived from deer of two or more strains, the result of interbreeding of different stocks before or after release, or the addition of stags of different strains to existing herds (Logan & Harris, 1967).

ESTABLISHMENT AND SPREAD. Most liberations were made in or adjacent to forested and mountainous areas where the resulting herds had space and seclusion to multiply and spread. Over 50 release sites are known, spanning the lengths of the North and South Islands, and on Stewart I. Most liberations were multiple, comprising several releases of mixed groups of stags and hinds, often supplemented in later years with additional stags. Typically only small numbers of deer were liberated at each site; single releases were generally of fewer than 10 animals, and the total numbers liberated at one site rarely exceeded 30 animals (Logan & Harris, 1967). Most groups successfully adapted to their new habitats and thrived. Although their early histories are largely unrecorded, the initial growth of many of the resulting herds must have been rapid (Wodzicki, 1950), probably in the order of 25% to 30% per year.

Stags colonized new areas first, preceding hinds by 5–10 years, and occasionally by as much as 20 years. The long-term sustained dispersal rate of the local breeding populations averaged 1.6 km per year (Caughley, 1963), depending on the terrain, vegetation cover and expansion phase of the herd. Red deer spread fastest along valleys and ridges and through areas of native grasslands; slowest across mountain ranges, developed farmland, and through unbroken expanses of forest. Well-established herds spreading through favourable habitat have achieved short-term, local dispersal rates of up to 11 km per year (Clarke, 1971).

By the late 1940s red deer had colonized most suitable ranges in both the North and South Islands, and the original discrete herds had coalesced (Wodzicki, 1950). Since then they have spread also into the Raukumara, Kaimai, and Rangitoto ranges in the central North Island, and most of the remaining deer-free areas of central and south Westland, and northern Fiordland. Red deer are still colonizing parts of the Raukumara Range, and the

Catlins area of eastern Southland, and seem likely to eventually reach the Coromandel Peninsula, and also the Milford catchment in northern Fiordland, unless prevented by man or by interspecific competition (p. 451).

DISTRIBUTION

WORLD. The British subspecies, *Cervus elaphus scoticus,* is found throughout the central and western highlands of Scotland, in south-west Scotland, on many of the islands in the Inner and Outer Hebrides, and in parts of England and Ireland. In England there are also localized wild herds of various park strains, derived from liberations or escapes of stock bred on deer parks.

NEW ZEALAND (Map 38). Red deer have established throughout the high country and forests of the two main islands of New Zealand; the northern half of Stewart I.; on Secretary and Resolution Is.; and on D'Urville I. At present (1987) they are absent from the Coromandel and Northland peninsulas and Taranaki in the North Island, and Banks Peninsula in the South Island.

HABITAT

Red deer have adapted to, and thrived in, habitats ranging from high mountainous areas and steep hill country to river flats and coastal lowlands; all the major indigenous and exotic forest types; and native scrublands and grasslands at all altitudes. They avoid the vegetation associations over ultrabasic rocks, e.g., in the Red Hills area of South Westland. When undisturbed they favour areas with a mixture of forest and short scrub or grassland, typical of the timberline and lower forest margin zones in predominantly forested high country, but in the past have also readily colonized unbroken expanses of both forest and native grassland. Local habitat use has depended mainly on the distribution of preferred food plants, and the need for shelter from inclement weather and when hunted.

The main restrictions on their use of otherwise suitable habitat have been human occupation and, to a lesser extent, competition from other ungulates (p. 451). Red deer will thrive on improved pastures and are positively attracted to some crops such as turnips and other brassicas; but they have been excluded from most agricultural lands by uncontrolled hunting, persisting only in areas with adjoining cover and where hunting is restricted. They also thrive in coniferous plantations and have managed to survive in most plantation forests despite intensive hunting. Hunting has also affected the local distribution of deer elsewhere, mainly by limiting their use of scrub and grassland. Sustained hunting from helicopters since the mid 1960s has virtually eliminated all deer from the high country above the timberline, and those that survive in the adjacent forests are now concentrated in the areas furthest from the main forest margins (Challies, 1985a).

FOOD

Red deer are opportunistic and highly adaptable feeders that both browse and graze. The composition of their diet is largely determined by what is locally available, and this in turn depends on the vegetation type and its past use. On farmland deer can live entirely on pasture, but in the wild they take varying

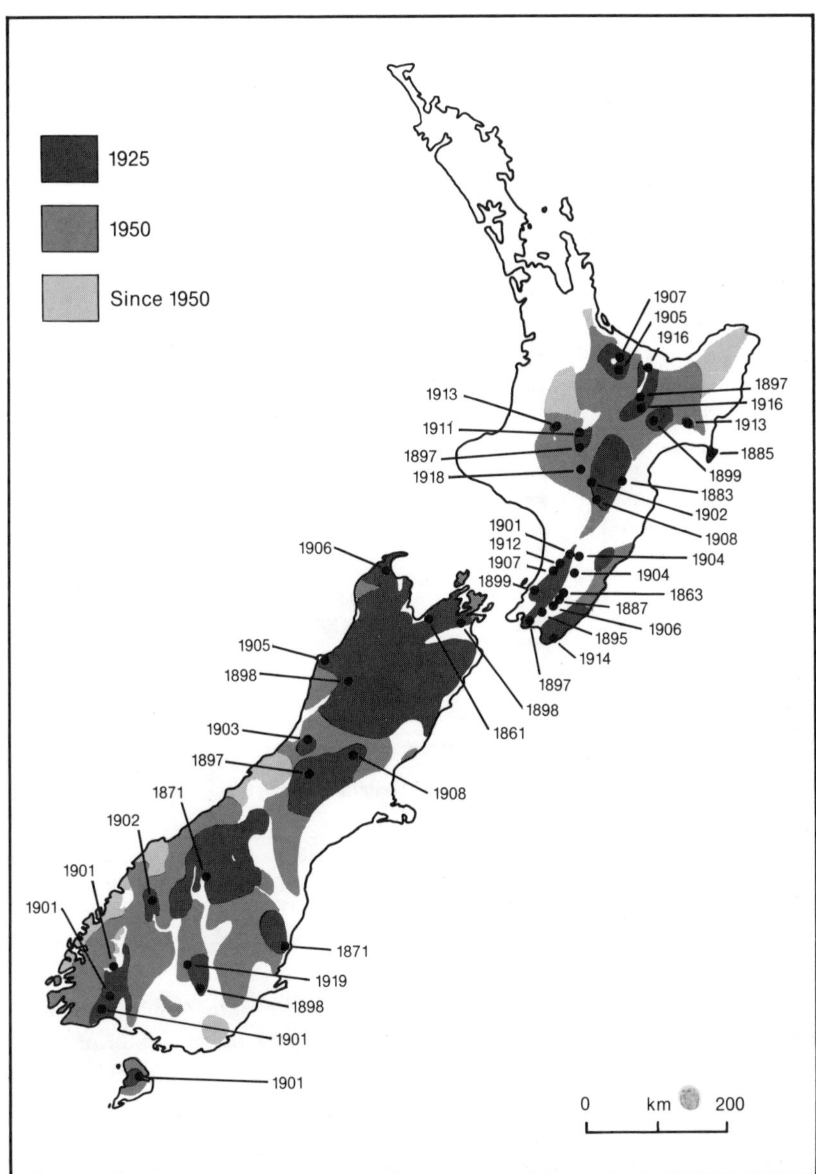

Map 38 Distribution of breeding red deer in New Zealand at three stages of colonisation, with dates and sites of the more significant primary introductions. Many other known liberations failed to establish new populations or only added to existing ones (from Challies, 1978, 1985a; Clarke, 1971; Forbes, 1924; Logan & Harris, 1967; Perham, 1922; Wodzicki, 1950). Red deer are also present on D'Urville, Secretary and Resolution Is.

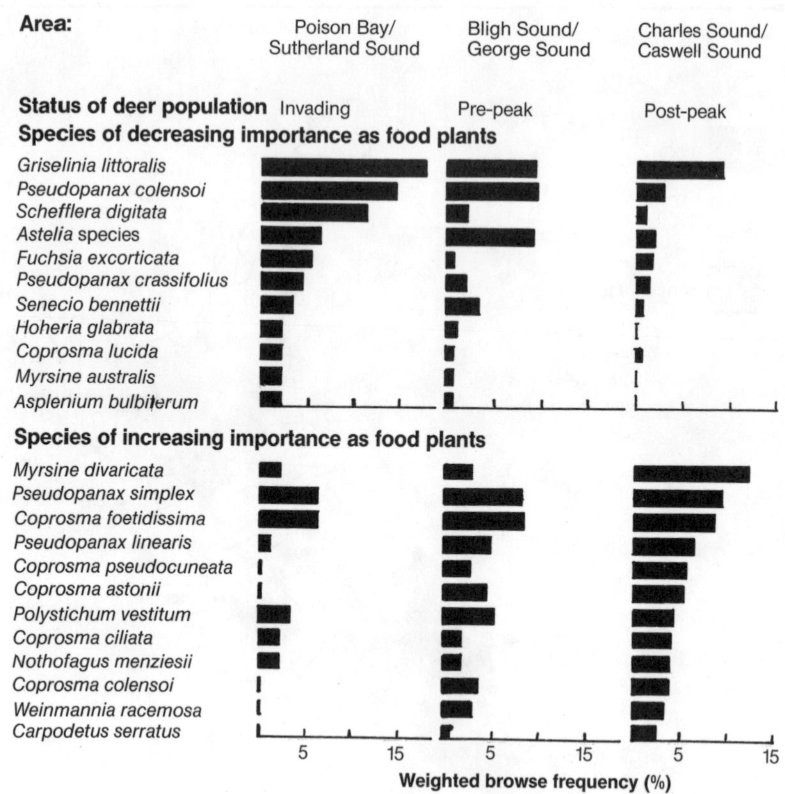

Fig. 75 Relative importance of food plants eaten by red deer and wapiti in an invading population compared with those in a pre-peak and a post-peak population. Weighted estimates of browse frequency are expressed as a percentage of the total browse record (adapted from Wardle, 1984).

amounts of browse. In subalpine areas and around forest margins browse may comprise only 20% to 30% of deer diet, whereas in the forested interior it can be as high as 70% to 80% of the diet (Gibb & Flux, 1973; Lavers, 1978; C. N. Challies, unpubl.). The remainder of their food is mainly herbs and grasses as available, depending on area and season.

Red deer also have strong food preferences. Deer colonizing new range usually had the choice of many abundant native food plants, and at first they fed very selectively, taking only the most preferred species, and then often only on selected sites. In forested areas these included *Pseudopanax* spp., large-leaved *Coprosma* species, *Schefflera digitata, Griselinia littoralis* and the fern *Asplenium bulbiferum.* The leaves and bark of the araliads *Pseudopanax colensoi* and *P. arboreus* were particularly attractive, as was the bark of *Coprosma lucida, C. robusta, Schefflera digitata* and sometimes *Griselinia littoralis,* though not to the same extent (Fig. 75; Wardle, 1984). In the grasslands they sought out a range of large-leaved herbs, e.g., *Ranunculus* spp. and Apiaceae such as *Anisotome* spp., and some grasses, e.g., the tall tussocks *Chionochloa flavescens* and *C. pallens,* especially those growing on fertile sites.

As the numbers of deer increased, the most favoured food plants disappeared from accessible areas. The deer were then forced to take alternative and progressively less attractive species, which, in forested areas, resulted in greater browsing of the shrub tier, including *Myrsine divaricata,* the smaller species of *Pseudopanax, Coprosma foetidissima,* small-leaved *Coprosma* species, and the fern *Polystichum vestitum.* Where red deer have remained in high numbers, their browsing pattern has continued to change. Next in order of preference were the saplings of the main canopy trees; and eventually the deer became increasingly dependent on crops of ephemeral seedlings, epicormic shoots of trees, e.g., *Weinmannia racemosa* and *Griselinia littoralis,* herbs and grasses of the forest floor, lichens and fungi, and cast leaves from the forest canopy (Wardle, 1984). The diet of deer living in native scrub and grasslands has undoubtedly changed in a similar way, but this is largely unrecorded.

SOCIAL ORGANIZATION AND BEHAVIOUR

DISPERSION. Red deer are sociable animals, forming single-sex groups for most of the year. The female groups include adult hinds, their calves, and previous female offspring, and are led by one or more older and presumably experienced hinds. Groups range from two or three animals up to loose herds of 50 to 150, depending mainly on the local population density and intensity of hunting. The male groups include stags of 2–3 years old and older, and range in size from one or two animals to perhaps 20 to 30. They have a linear hierarchy dominated by the older and larger stags with the largest antlers, but the social order becomes unstable while the stags are shedding and growing their antlers, when changes in rank are common. The composition of the separate-sex groups is not rigid; occasionally hinds are found with groups of stags, and more often there are young stags associating with hind groups.

In the weeks before the birth of a new calf, a hind becomes increasingly antagonistic towards her previous calf and repeatedly chases it away, eventually driving it from the hind group. These yearlings (both sexes) tend then to congregate, especially in November and December. When the hind groups reform again after parturition, the female yearlings are accepted back into their family group. The male yearlings normally stay together, although they can become loosely attached to either male or female groups.

HOME RANGE. Hinds and their female offspring are fairly sedentary, remaining in the same general area, probably throughout life. By comparison, stags are only loosely attached to their natal area; yearlings and 2-year-olds, particularly, are inclined to wander away and have been recorded moving up to 32 km (Gibb & Flux, 1973). Adult stags move seasonally between their stag group range and the rutting area, which may be several kilometres apart. In colonizing populations, stags tended to spend most of the year outside the range of hinds, moving back into hind areas for the rut (Banwell, 1968).

Red deer may also shift their ranges in response to seasonal changes in climate and in the availability of food. During winter, deer usually remain within forest cover, mainly on the mid valley slopes; when plants flush in spring, they tend to move to the forest edge and out into the open to feed. If undisturbed, red deer remain in the general area of the forest edge throughout summer and autumn, moving between the adjacent forests, scrub and grasslands. Where winter conditions are severe, some deer have separate summer and winter ranges, moving from one to the other in early winter and again in spring.

VOICE. Red deer are normally silent outside of the rut (or "roar", see below). Both sexes will give a gruff bark as a warning when they sense danger. Hinds use a nasal bleating call to maintain contact with their calves, which have a high-pitched bleat, and may scream if alarmed.

REPRODUCTIVE BEHAVIOUR. Male groups disband before the rut, and the adult stags move away and establish separate rutting areas on which they attempt to gather a harem of hinds. Successful stags continuously herd their hinds in a group away from other stags, an exhausting task which leaves them little time to rest or feed. The hinds are served when they come into oestrus, and then the groups drift apart as the "master" stags lose condition and energy; the hinds are then often pursued by the younger and unsuccessful stags, but the sexes soon segregate again.

Throughout the rut stags are antagonistic towards one another as they compete for hinds; they roar periodically, especially in the early morning and evening, partly as a warning of their presence and partly as a threat. The roar is typically a rumbling, guttural bellow similar to the bellow of cattle but with a deeper intonation, followed usually by several grunts. Some at times give only a single resonant groan. Rutting stags will roar singly, or if there are several within earshot, they roar in response to one another. Young stags approaching a stag with hinds are chased off with grunts and visual threats. Stags that are physically well matched follow their vocal challenges with a visual assessment at close range; if this does not convince one or other that he is inferior, a fight ensues. Fights are relatively uncommon and are largely a ritualized test of strength. The stags face each other with their antlers interlocked, and furiously push and twist, each trying to force the other back or throw him off balance. Once one decides that his opponent is the stronger he breaks off the fight and flees, often chased by the victor; the vanquished stag is more likely to be struck in the flank or rear and injured at this stage than during the fight itself. The winning stag takes possession of all the hinds involved.

During the rut stags wallow in muddy pools, into which they urinate and defecate. The covering of mud accentuates the smell given off by a rutting stag and can give it a larger, darker appearance. Wet and dry wallows are used occasionally by both sexes at other times of the year, particularly during late spring when they are shedding their winter coat.

REPRODUCTION AND DEVELOPMENT

The reproductive cycle is highly synchronized. Stags become fertile about February, when their antlers reach full development, and remain fertile for up to 8 months (Mitchell, 1973). The rut begins in late March and extends throughout April; most conceptions are in early to mid April. Poor-conditioned and first-breeding hinds tend to conceive later than do good-conditioned and mature hinds. Most hinds conceive on their first oestrus; those that do not, come into oestrus again at about 18-day intervals, if necessary up to 8–9 times (Lincoln, Youngson & Short, 1970; Guinness, Lincoln & Short, 1971). Gestation ranges from about 221 to 252 days, averaging 234 days (Guinness, Gibson & Clutton-Brock, 1978). Most calves are single and born in late November and December; the peak is in early December (median date 9 December; Caughley, 1971b), but a few calves are born in January, and occasionally in February. Twin calves are rare, normally about one pair in 500 to 1000 births. The sex ratio is near equality; one sample of foetuses from the mixed red deer-wapiti herd in northern Fiordland comprised 121 males and 115 females (Smith, 1974).

Shortly before parturition, hinds leave their group and find a sheltered place in which to give birth. The birth itself is generally quick and uncomplicated; the calf is immediately licked clean by its mother and is soon on its feet and suckling. After a normal birth the hind consumes the placenta. The calf is bedded down for the first 5–10 days of life, spending most of its time either sleeping or quietly resting; it is active only when joined by its mother. Hinds return to feed and groom their calves several times a day. The calves start to run with their mothers as soon as they are strong enough, and the two remain in close company until shortly before the hind's next calf is born. Hinds with calves start to reform their family groups 2–4 weeks after parturition.

Calves are born with a complete deciduous dentition of 22 teeth ($I^0/_3$ $C^1/_1$ $Pm^3/_3$ $M^0/_0$), which are systematically replaced with permanent teeth between 12 and 30 months of age. A further three teeth (permanent molars) are added to the back of each molariform tooth row, completing the set by 24 months of age (Challies, 1978).

Calves are suckled throughout summer and autumn. A few are weaned in April-May at 5 to 6 months old (Challies, 1978), but the majority are weaned during the winter and subsequent spring. In South Westland in the mid 1970s, 47% (n = 17) of adult hinds (3 years old or more) were lactating during late July and early August, and 16% (n = 43) were lactating during late September and October. Of those, 87% (n = 15) were also pregnant (C. N. Challies, unpubl.). In northern Fiordland in the late 1960s, 11% (n = 100) of pregnant adult hinds and 27% (n = 33) of non-pregnant hinds were still lactating during late October and early November (Smith, 1974).

Hinds first breed as yearlings or 2-year-olds (i.e., produce their first calf at 2 or 3 years old) and only rarely as 3-year-olds. Although some calves have well-developed ovarian follicles at 4 to 5 months of age (Daniel, 1963), they do not breed. The proportion of hinds breeding as yearlings ranges from about 10% to 60–70%, and the proportion of older hinds breeding from 65% to 95% (Challies, 1978, 1985a). Breeding success is directly related to the well-being of the herd and is limited especially by small body size and/or poor condition during the rut (Hamilton & Baxter, 1980). Hinds maintain their fertility throughout life, at least in good-conditioned herds. In South Westland in the mid 1970s, 91.7% (n = 12) of hinds 15 years old and older were fecund, compared with 94.3% (n = 298) of 5–9-year-olds, and 92.2% (n = 77) of 10–14-year-olds (C. N. Challies, unpubl.). Stags are fertile seasonally from the time they fray their first antlers, usually as yearlings (Mitchell, 1973).

Red deer increase in body size rapidly during their first 2 years of life, and more slowly thereafter until at least 7 years of age. By 2 years stags have reached 85% to 90%, and hinds 90% to 95%, of their mature skeletal size. Males grow faster than females from birth, and are obviously larger from about 12 months of age. Skeletal growth tends to be seasonal, slowest in winter, when calorific intake is usually low. This seasonality is more pronounced in the proximal parts of the body than in the head and legs, which have priority for growth in young animals (Challies, 1978).

Body weight follows the same general growth pattern as does the skeleton, but is also affected by changes in body conformation and condition. The musculature of both sexes grows and matures during the first 2 to 3 years of life; the consequent increase in body mass relative to skeletal size gives a progressively more robust appearance. Sexual dimorphism in weight is very pronounced; adult hinds average about 65% the weight of lean stags of a comparable age (Challies, 1978).

POPULATION DYNAMICS

NATURAL FLUCTUATIONS IN NUMBERS. Many of the local red deer populations in New Zealand have passed through a classic population irruption (Holloway, 1950; Riney *et al.,* 1959; Caughley, 1970b; Clarke, 1976a), typically comprising three phases: an initial, usually rapid, increase in numbers; a period of sustained high density; and then a natural decline in numbers, caused by the progressive reduction in the quantity and quality of food available per capita (Fig. 76).

The general characteristics of the successive phases are as follows (adapted from Challies, 1985a). (1) The *pre-peak* phase begins with the establishment of a new breeding population, either by liberation or the extension of existing range, and lasts until the population reaches peak density in 10–30 years. The rate of increase in numbers is exponential at first, but declines as density increases and food shortages start to limit population growth. (2) The *peak* population phase spans the short (3–13 years) period of highest, comparatively stable numbers, when the deer are temporarily in balance with their habitat, but at a level exceeding its long-term carrying capacity. (3) The *post-peak* phase has two stages: first, a short-term natural decline in numbers, leading to, second, a relatively stable lower-density post-decline population, more or less adjusted to the sustainable carrying capacity of the habitat.

The duration of the first two phases varies for reasons which are not clear but are probably related to climate, topography, vegetation and hunting pressure (Clarke, 1976a). Light to moderate levels of hunting have tended to prolong the irruption, and intensive hunting has tended to hold the population in the pre-peak phase.

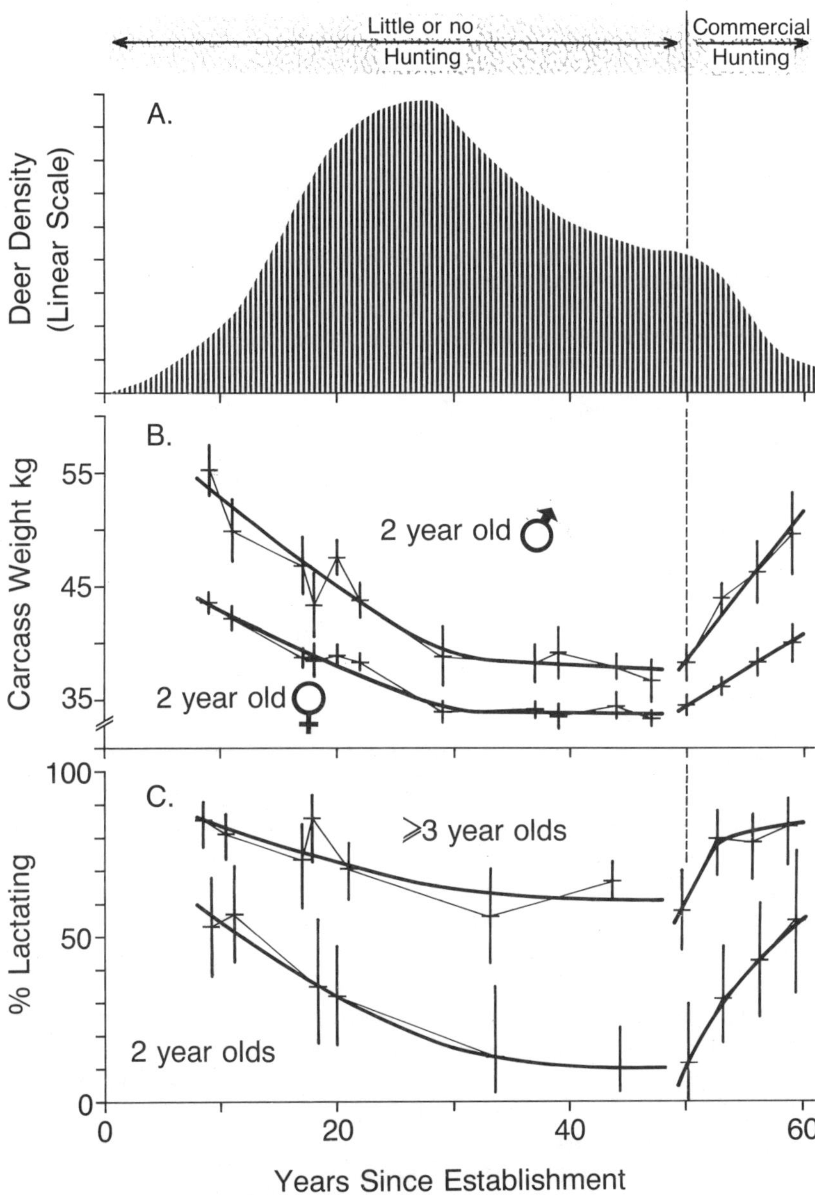

Fig. 76 Changes in population density, body size and productivity of red deer during an irruption and subsequent period of intensive commercial hunting. A, relative density; B, standard carcass weights (equivalent to lean German market carcass weight, estimated from skeletal measurements); C, proportion of females lactating in mid to late summer. In B and C for years 0–50, estimates are from separate populations established for different periods; for years 50–60, estimates are from one population sampled sequentially. Means and percentages shown with 95% confidence limits (adapted from Challies, 1985a).

DEMOGRAPHY. The demography of colonizing populations of red deer has not been described in detail, but is probably similar to that in other ungulates (Caughley, 1970b): the rapid population increase is the result of high fecundity and early breeding of females, and reduced mortality, especially of calves. The slowing of the increase, and eventually the stasis and following decrease of the population, are caused by a simultaneous decrease in fecundity and increase in mortality. Fecundity starts to decline midway through the pre-peak phase, and reaches its minimum level around the end of the peak phase. The decline is especially pronounced in 2-year-olds, and older hinds seem able to maintain relatively high birth rates under most conditions (Challies, 1978, 1985a). Mortality, especially of calves during winter and early spring, is probably the most important factor limiting deer numbers (Caughley, 1970b) and is consistent with the prolonged decline in numbers characteristic of the post-peak phase (Clarke, 1976a). Catastrophic die-offs of poor-conditioned deer of all ages have been reported occasionally, mainly early in the post-peak phase, but these are the exception (Challies, 1985a).

Changes in the rate of increase in a population are certainly reflected in its age structure, though that effect has not been documented in red deer. The only mortality and survivorship statistics published to date are for a post-peak population; they are based on a shot sample of female red deer taken in 1966–67 from a long-established population with an apparent zero rate of increase (Caughley, 1971c). The age frequencies in that sample showed that 41% of the female calves born survived to 2 years, 38% to 5 years, and only 5% to 10 years of age. Mortality rates were high in both calves and yearlings, negligible for 2–5-year-olds, and rapidly increased again after 5 years of age. The mean age of hinds 1 year old and older during the season of births was 4.3 years. Stags generally have a higher rate of natural mortality than hinds; long-established populations hunted from helicopters for the first time typically yielded about 80 stags per 100 hinds 1 year old and older. The longevity of red deer in the wild is about 12 to 14 years for stags and 15 to 20 years for hinds (C. N. Challies, unpubl.).

Most red deer populations in New Zealand have been intensively hunted for at least the last 15–20 years. This has (1) reduced life expectancy, since deer are killed before they would otherwise have died, especially the stags; and (2) reduced deer densities, and thereby increased the food available per head, enabling the surviving females to maximize their productivity (Fig. 76). These populations therefore typically have a low mean age, and a sex ratio biased in favour of females. In one population in South Westland that had been intensively hunted from helicopters continuously for 10 to 13 years, 31% of the male and 47% of the female calves born survived to 2 years, 5% and 14% to 5 years, and 1% and <1% to 10 years of age respectively. The mean ages of stags and hinds 1 year old and older during the season of births was 2.1 and 2.8 years respectively; and the all-age sex ratio from 1 year of age was 62 stags to 100 hinds (data based on the age and sex composition of the kill weighted seasonally and adjusted for the on-going reduction in deer numbers: C. N. Challies, unpubl.).

CONDITION AND FATNESS. Body fat is concentrated mainly in three depots: the marrow of the long bones; the abdominal cavity, especially along the dorsal wall, around the kidneys, and in the mesenteries; and the subcutaneous connective tissue, especially over the rump. During favourable conditions, fat is deposited first in the long-bone marrow, then in the viscera, and lastly under the skin; this order is reversed during inanition (Riney, 1955a).

The fatness of deer varies according to age, sex, reproductive status, and the season of the year. Calves and yearlings usually remain lean throughout the year, since growth has a greater priority than storage at that age. The fatness of hinds 2 years and older tends to increase in summer and autumn, and decrease over winter to a seasonal minimum in late winter and spring. This cycle is usually quite pronounced in non-lactating females; those that are suckling calves do not normally have much fat at any season. Stags of 2 years and older accumulate fat rapidly during the summer to a maximum just before the rut, lose most of it during the rut, and then remain lean for the rest of the year (Mitchell, McCowan & Nicholson, 1976; Challies, 1978).

INTERACTIONS WITH OTHER WILD UNGULATES. Red deer are now locally sympatric with all the other wild deer in New Zealand, and they freely hybridize with wapiti, and to a lesser extent also with sika deer (pp. 465, 468). Some degree of competition between red deer and several other deer species is also recognizable but largely unrecorded. The best-documented example is that sika deer, when in high numbers, appear able to displace red deer in time, especially in lower altitude forests (McKelvey, 1959; Kiddie, 1962). Circumstantial evidence also suggests some form of competitive exclusion of red deer by both fallow and white-tailed deer (Kean, 1959). Where fallow deer are already established and in high numbers before meeting with red deer, they have remained the dominant species on at least part of their range, especially at lower altitudes. In the Blue Mountains in east Otago, where fallow deer have been present in high numbers since about 1900, red deer are absent. White-tailed deer have thrived in the low altitude forests of Stewart I. to the exclusion of red deer, which are now present only in low numbers in less favoured habitats. All three (sika, fallow and white-tailed deer) are smaller bodied than red deer.

Red deer are also slow to spread into areas where there are already large numbers of feral goats; this is probably the main reason why they have not already colonized the Coromandel Peninsula and the remaining parts of the Raukumara Range.

PREDATORS, PARASITES AND DISEASES

Red deer have no predators in New Zealand except man.

Most wild deer carry small ectoparasite burdens which cause few problems to healthy animals. Records include the lice *Damalina longicornis* and *Solenopotes burmeisteri*; a hair follicle mite *Demodex* sp. (Nutting *et al.*, 1975); the sheep ked *Melophagus ovinus* (P. R. Wilson, quoted in Mason, 1988); and the cattle tick *Haemaphysalis longicornis*. The cattle tick is restricted to the northern North Island and is common in some areas of the Urewera ranges. It attaches to the ears, neck, shoulders and inside flanks of deer, and heavy infestations result in the loss of hair from these regions (Andrews, 1964).

By comparison, the known fauna of endoparasites of red deer is large; Mason (1988) lists 25 species from wild and/or farmed stock in New Zealand, comprising three species of protozoa, two trematodes, two cestodes and 18 nematodes. The following list concerns only those recorded in wild deer. Two protozoa, *Sarcocystis* sp. and *Toxoplasma gondii*, are both recorded near Rotorua; their definitive hosts are dogs and cats, respectively (Collins & Charleston, 1979b; Collins, 1981b). The common liver fluke *Fasciola hepatica* is limited in distribution to parts of Hawke's Bay-Poverty Bay, Nelson, central Otago and central Westland, the range of its intermediate hosts, particular species of aquatic snails (Andrews, 1964). Deer can be infected with false hydatids, *Taenia hydatigena*, or cysticercus tennuicollis, its larval stage, but not, as far as is known, with either true hydatids, *Echinococcus granulosus*, or sheep measles, *Taenia ovis* (Sweatman & Williams, 1962; Andrews, 1964). The nematode *Dictyocaulus viviparus* (cattle lungworm) is common in wild deer, especially in calves (Charleston, 1980). The tissue worm *Elaphostrongylus cervi*, first identified from wild red deer and wapiti from Fiordland, is now found in farmed deer throughout the country. The gastrointestinal species of nematodes are distributed as follows: nine in the abomasum, five in the small intestine and two in the large intestine and caecum. Five of these worms have been found only in cervids; the others are common also to domestic stock (Mason, 1988). The significance of these worms to wild populations is not known.

There have been reports of wild deer infected with bacterial diseases common in domestic animals, but no clinical cases of viral diseases have been confirmed (Mackintosh, 1988). The first diagnosis of bovine tuberculosis (caused by *Mycobacterium bovis*) from a New Zealand red deer was in a yearling stag shot in 1970 near Inangahua in Westland, an area with a history of tuberculosis in both cattle and brushtail possums (p. 92). Since then, further sporadic cases of *M. bovis* in wild red deer have been diagnosed in both the North and South Islands, mainly around and west of Lake Taupo, in the Wairarapa, in west Nelson-north Westland, and occasionally in Southland. In all these areas tuberculosis is present in brushtail possums, which are probably the main source of infection for deer (Beatson, 1985a; Mackintosh & Beatson, 1985).

Red deer are susceptible to leptospirosis (*Leptospira pomona*) and brucellosis (*Brucella* sp.), although no clinical cases of either have been diagnosed in the wild (Mackintosh, 1988). A serological survey of 109 red deer shot in Kaingaroa Forest in the central North Island during 1963-65 found one reactor to each, both near the sites of recent outbreaks of brucellosis and *L. pomona* in cattle on farms. The deer were probably exposed to infection while grazing on farmland adjacent to the forest, and therefore should be regarded as accidental hosts. During this survey faecal samples from 30 red deer were cultured for *Salmonella*, but all were negative (Daniel, 1966a).

Renal calculi were found in the kidneys of 8 of 325 red deer examined in South Westland. The numbers of calculi found ranged from one to at least 28, with a combined weight of up to 1.6 g. Larger numbers, and larger calculi, found incidentally in three other deer, ranged up to at least 140 renal calculi and 850 urethral calculi in one kidney, with a total weight of 76 g. The reason for this relatively high prevalence of urolithiasis is not known (Reynolds, 1982).

Other diseases diagnosed in farmed red deer in New Zealand include malignant catarrhal fever, yersiniosis, facial eczema and a number of clostridial diseases (Mackintosh, 1988); some of these may also be present, but probably rare, in wild populations.

ADAPTATION TO NEW ZEALAND CONDITIONS

The obvious success of red deer in New Zealand has been attributed to a combination of favourable conditions including the moist, temperate climate with mild winters; luxuriant, varied, evergreen vegetation; and a lack of predators and native competitors.

New Zealand red deer have traditionally been considered larger-bodied and more fecund than their parent stocks — as a consequence, it was assumed, of the better quality habitat in New Zealand for deer compared with the British Isles. This is largely true, although populations of deer, both in New Zealand and in the British Isles, have varied greatly in character. During the period of their establishment in New Zealand, deer near the colonizing fronts had virtually unlimited supplies of high quality foods; their growth rates, antler sizes and productivity matched those in the very best of the English deer parks and far exceeded those in many wild herds in the British Isles. At the other extreme, the growth rates and productivity of long-established unhunted populations in New Zealand were similar to those of the average red deer herds in the highlands and islands of Scotland.

Wild red deer in New Zealand are relatively free from clinical disease. Many of the serious conditions found in deer overseas are unknown here. This is probably partly due to the natural quarantine effect of the long sea voyages endured by the original stock introduced into Australasia (see also pp. 112, 158). The breeding season is 6 months offset from that on the deer's native range.

SIGNIFICANCE TO THE NEW ZEALAND ENVIRONMENT

DAMAGE. Colonizing populations of red deer undoubtedly made a big impact on most habitats during their initial irruptions. At first they fed very selectively, concentrating on the few most palatable plant species (p. 444), sometimes causing spectacular damage locally (Mark & Baylis, 1975). As these species were depleted, the deer progressively shifted on to less preferred plants (Fig. 75); eventually the density and complexity of the forest understorey were greatly reduced by the loss of palatable herbs, shrubs, understorey woody species and seedlings through over-browsing, and of some subcanopy trees by bark-stripping (Holloway, 1950). Average reductions of 50% to 60% in the densities of woody stems (James & Wallis, 1969), and of nearly 90% in the densities of seedlings and saplings of shrub-hardwood species (Wardle, 1984), have been attributed to red deer browsing. The reduction in palatables was often accompanied by large increases in some unpalatable and browse-resistant species such as *Pseudowintera colorata*. It is not clear to what degree the vegetation attains a new level of stability under continuous browsing and grazing. There is evidence that deer numbers continue to decline naturally for several decades after reaching peak density (Clark, 1976a), so it seems likely that the reduction in the amount of browsable biomass is continuing.

Exclosure plots have shown that forests modified by deer will readily recover if protected from browsing. At first the response is limited to growth of epicormic shoots on trees and new sprouts on ferns, and it is usually several years before seedlings of palatable species become conspicuous. Succession back to a predominantly woody understorey usually depends on a few species suited to the local conditions. Most of these (characterized by rapid growth and often large leaves) are typically preferred food plants (Jane & Pracy, 1974). They form a new understorey, which is not necessarily of the same composition as the original. Deer control measures have generally not been intensive enough to allow anywhere near the amount of regeneration in the open forest that has been observed in exclosures. Once the understorey has been depleted, it probably takes relatively few deer to keep it in that state. Where control has been really effective, the remaining deer have concentrated on the most preferred plants, allowing some response in the less-preferred ones (Wardle, 1984).

Assessment of the significance of the impact of deer browsing on the native vegetation of New Zealand depends on the objectives of local land management plus various other considerations, including changes in human perception of the problem (Caughley, 1983). In national parks and similar mainland protected natural areas, where the native biota is supposed to be kept as near intact as practical, red deer have already caused irreversible changes in the vegetation. On lands managed primarily for the protection of soil and water, the case is less clear-cut. Damage to the vegetative cover by browsing, grazing and trampling may increase the *potential* for soil erosion, but it is difficult to isolate this effect from natural erosion (Grant, 1985) and the contributions made by other wild animals, domestic stock and repeated burning.

CONTROL. The primary consideration in the management of deer in New Zealand has long been the need to control their numbers artificially. As early as 1906 it was apparent to some acclimatization societies (who then had responsibility for deer management) that the well-being and trophy value of their herds were deteriorating as deer increased in numbers, and that to prevent this they would have to be culled. Many thousands of the poorest quality animals were shot for that reason. By the early 1920s, deer at locally high densities were displacing farm stock and damaging pasture, crops and pine plantations as well as modifying native vegetation (Perham, 1922). Protection was removed from red deer in the acclimatization society districts between 1923 and 1930, and bounties were paid by some societies and by the DIA for all animals killed, regardless of quality. In 1927 the then State Forest Service began its own operations to prevent damage in State Forest plantations. The problem persisted, and in 1930 remaining protection was lifted from all deer species, and the responsibility for their control passed to the government (Wodzicki, 1950).

From 1931 DIA had responsibility for deer control under the Animals Protection Game Act 1921-22 (and later under the Wildlife Act 1953). Hunters shot deer for a basic wage plus a bonus for each skin or tail returned. At first they concentrated on reducing deer numbers on pastoral lands, but soon recognized that the principal threat was to soil and water conservation in the high country catchments of the major rivers. The main effort was then redirected to those areas, leaving the more accessible parts of the deer range to private hunters where possible. The numbers of deer killed in these operations increased steadily from about 8000 a year in the early 1930s to 41 000 in 1940, dropped to an average of 13 000 a year during the Second World War, then recovered by the mid 1950s to 40 000-60 000 animals a year. Skins were recovered from 38% of the deer shot up to 1952, when this part of the scheme was discontinued. It now seems unlikely that many of the control operations before the late 1950s were of sufficient intensity or duration to have reduced deer numbers significantly over large areas. Although the tallies obtained at those times seemed large by contemporary standards, they were small compared with those obtained subsequently by commercial hunters in the same areas.

Responsibility for deer control was transferred to NZFS under the Noxious Animals Act 1956 (later replaced by the Wild Animal Control Act 1977). Their policy was to concentrate all available resources into a few priority areas selected on the basis of the downstream values at risk (Riney, 1956); control operations ceased in the non-priority areas. This policy appeared to have the desired effect, resulting in reduced browsing on preferred plant species, lower densities of deer pellets, and larger-bodied animals (p. 449). The number of deer killed in these operations declined sharply from the late 1950s to about 25 000 animals a year in the mid 1960s, mainly because deer became harder to find in the more intensively hunted areas. The downward trend continued as the developing game meats industry expanded, and commercial hunting operations were extended to include the initially less productive priority areas. So effective and sustained was this hunting that it quickly negated the need for control efforts in most areas. The few remaining non-commercial deer control operations are now restricted to areas where lower animal numbers are required than can be obtained by commercial or recreational hunting; they accounted for only 1133 deer in the year 1981-82.

COMMERCIAL USE. Commercial interest in wild deer in New Zealand started with skins and meat for export and has more recently shifted to live animals for local farm stock. Wild deer have also provided a variety of by-products, including antlers in velvet, tails, testicles and sinews for quasi-medicinal use in eastern Asia, and maxillary canines for use in hunting jewellery in Europe. These enterprises have tended to be opportunistic, and the numbers of animals taken have fluctuated in response to changes in market conditions and the availability of deer.

Skins were the main saleable product obtained from deer until the early 1960s. A total of 34 000 deer skins was exported in the 5 years to 1935; over 70% of these were obtained in deer control operations. Between the early 1940s and late 1950s, skin exports ranged from 50 000 to 103 000 per year. Hunting of deer specifically for skins ceased during the 1960s with the development of the more lucrative market for whole carcasses. Overall, hunting for skins accounted for twice as many animals as official control operations had up to that time.

Commercial hunting for venison originated about 1958-59, mainly for export to Europe, especially West Germany. Carcasses were at first supplied by foot hunters working in areas accessible by road, jetboat or fixed-wing aircraft. Helicopters were used at first to service foot hunters; shooting deer from helicopters began about 1965 in west Otago and South Westland, where red deer were then in high numbers (Fig. 77). By the late 1960s aerial hunting was commonly practised throughout the high country of the South Island, but application of the method in the North Island was hampered by public opposition, and some State lands have only recently become available for this purpose.

There have been no limits on the numbers, age or sex of deer that could be taken and only a few local restrictions on the season in which they could be hunted. Hunting methods have been governed mainly by the statutory provisions for the use of firearms and aircraft. Conditions for the handling of carcasses are more stringent; the Game Regulations 1975 set maximum time periods for the delivery of carcasses to chillers and factories, and require that hearts, lungs, livers and kidneys remain attached to carcasses for inspection before processing. So far over 1.5 million deer carcasses have been processed and exported, along with large quantities of by-products.

The intensive and unrestricted nature of commercial hunting has induced changes in distribution and substantial reductions in deer numbers on most parts of their range. For example, in the Arawata Valley, South Westland, where intensive aerial hunting started in 1965, deer pellet counts declined by 75% between 1970 and 1980 (the equivalent of an approximately 90% decrease since 1965); the remaining deer are concentrated on the mid and lower slopes furthest from the timberline, the area most intensively hunted (Challies, 1977, 1985b). Commercial hunting has also affected the well-being (p. 440), and age and sex composition of the residual populations. Males are taken at much younger age on average than are females, resulting in an overall sex ratio from 1 year of age of about 62 stags to 100 hinds. This, when combined with high fecundity and low natural mortality, gives these populations a harvestable increment equivalent to about 35% of the numbers existing before the new season's calves are added.

In recent years, attention has shifted from killing to live capture for farm stock. The farming of deer behind fences became legal in 1969 with the introduction of regulations under the Animals Act 1967 and Noxious Animals Act 1956. After a slow start, interest in deer farming increased rapidly from the mid 1970s, and by 1985 there were nearly 400 000 deer on farms, mostly red

Fig. 77 Recovering carcasses of red deer for the game meat industry from near the head of Caswell Sound, 1970 (C. N. Challies).

deer, plus a few fallow, sika and wapiti. Of these, at least 60 000 had been caught in the wild and the remainder had been bred from this stock.

The boom in the live capture of deer started about 1977 and reached a peak in 1979–80 when about 25 000 deer were taken in 12 months. Since then the annual take has decreased. During the period of highest demand for live deer, all available animals were captured regardless of their sex or age; more recently, females have been selected and any males encountered shot for carcasses.

By substantially increasing the value of wild deer, the demand for farm stock has prolonged the viability of commercial hunting. All the same, many helicopter operators have withdrawn from the industry during the last few years, primarily because it became uneconomic to continue hunting. It is now clear that wild deer have been over-exploited in a boom-and-bust fashion for the last 20 years. This has, of course, had the beneficial side-effect of eliminating the need to otherwise control deer numbers except in a few special areas.

The future of commercial hunting, especially from helicopters, is open to speculation. Its viability at present deer densities depends primarily on the prices obtained for female deer caught live, and secondarily on the prices for deer carcasses. However, the demand for, and therefore the value of, captured deer is likely to decline as deer farming becomes more self-sufficient in stock and less speculative. This decline has been hastened by the government's recent changes in the ways livestock farmers are taxed. If commercial hunters have to revert to taking deer carcasses for most of their income, the industry is unlikely to survive in its present form.

The decline in the demand for deer is of concern to managers of national and forest parks and other reserved lands and high country forests. Deer numbers have been reduced to lower levels than were believed possible, even in the areas previously given highest priority for deer control. Any reduction now in hunting effort will result in a rapid increase in deer numbers. In the event that all commercial hunting operations were to cease, deer numbers would probably double within three years, and double again every four years until vegetation condition again limited population growth.

C. N. C.

39. Wapiti

Cervus elaphus nelsoni (Bailey, 1935)

Synonym *Cervus canadensis nelsoni* Bailey 1935 (see p. 436).

Also called elk or Rocky Mountain elk in North America, but this name should be avoided in New Zealand (see p. 514).

DESCRIPTION (Fig. 78)

Distinguishing marks, Table 78 and skull key, p. 24.

Wapiti are the largest round-antlered deer in the world. Both sexes have a uniformly light-coloured body, contrasting strongly with the dark head, neck and legs, especially in males and in winter. The winter pelage consists of a thick woolly undercoat with long guard hairs, further lengthened on the neck to a distinct mane. The head, neck and legs of adults are dark chestnut-brown to almost black; the sides and back are yellowish to brownish grey; the underside is blackish, with a white patch between the hind legs. The rump patch is large and a uniform cream to tawny, bordered with dark brown to black on the sides, extending dorsally above the tail onto the lower back, and grading into the body colouring. The winter coat is moulted over a period of several weeks in late spring, during which the animals have a patchy and unkempt appearance. The summer pelage consists of short, stiff guard hairs with little or no undercoat and no obvious mane, and similar in colour to the winter coat except that the body tends to be more tawny, even reddish or light bay. In late summer the new winter guard hairs grow through the summer hair, which is gradually shed. The muffle is blackish and hairless; the ears are about half the length of the head and dark brown to blackish outside with light hairs inside; the hooves are blackish; and the tail is short and the same uniform colour as the rump patch.

Fig. 78 Wapiti (G. Roberts).

Newborn calves are tawny brown, some with a dark dorsal stripe. They have a pronounced yellowish brown rump patch, and a speckling of light spots over the body, arranged in the same pattern as in red deer but more diffuse, and disappearing at about 2 months of age. Their first coat is shed during late summer and replaced by winter pelage with thick under-wool and long guard hairs similar to that of adults but browner. The contrasting colour pattern of adults is developed progressively through subsequent moults.

Antlers (Fig. 73) are grown and cast annually by males from their second year. The growth of velvet antler starts between mid August and November, depending on the age and condition of the animal, and is completed between late January and mid March, when the dried velvet is frayed from the hard antler. Yearlings grow a pair of unbranched spikes. Two-year-olds commonly have a total of 4 to 8 antler points, whereas mature wapiti have at least 10 and usually 12 or more points (Smith, 1974), and reach maximum antler size at about 7 years of age (Flook, 1970). Some New Zealand wapiti lack the bez tine (Smith, 1974).

Dental formula $I^0/_3$ $C^1/_1$ $Pm^3/_3$ $M^3/_3$ = 34.

FIELD SIGN

Wapiti sign is similar to that of red deer, but proportionately larger. Pellets are typically hard, well-shaped, black or dark brown, and aggregated or scattered; they are adherent if moist. The footprints show two moderately pointed toes, in outline more like the footprints of cattle than those of red deer. Well developed trails are found in areas of high use. Shrubs and tree saplings are frayed by bulls with their antlers during the few months before the rut, and wet wallow holes are also formed at this time.

Table 80: Body measurements of adult (6+ years) female[1] wapiti in Fiordland, early 1970s.

	Mean	SD	95% range	n
Carcass weight (kg)[2]	82.5	11.0	60.9-104.1	14
Body length (mm)	2 088	114	1 864-2 312	16
Foot length (mm)	567	18	532-602	16
Jaw length (mm)	314.7	6.5	301.9-327.5	14

1. Equivalent figures for wapiti bulls not available.
2. For definitions of measurements, see Table 79.
Data from C.N. Challies (unpubl.).

MEASUREMENTS

See Table 80 (females only). Smith (1974) gives total body weights of 215, 224 and 260 kg, and carcass weights of 151, 146 and 184 kg, for three bull wapiti shot in Fiordland during the rut; these animals were 5, 6, and 8 years old respectively.

VARIATION

The variation in physical growth rates, condition and productivity, usually associated with a population irruption, has not been described in wapiti in New Zealand, but undoubtedly happened much as in red deer.

HISTORY OF COLONIZATION

The present wapiti population originates from a single liberation of 18 animals at the head of George Sound in March 1905. These were the survivors of 20 wapiti obtained from two captive herds in the eastern United States; 10 from the National Zoological Park in Washington, DC, as part of an exchange of animals, and 10 purchased from H. E. Richardson's Indian Game Reserve in Brookfield, Massachusetts (Donne, 1924). These wapiti were "almost certainly" *C. e. nelsoni* bred from stock originally obtained from Wyoming (Miers, 1966; O. J. Murie, unpubl.). Banwell's (1966) suggestion that the group obtained from Richardson's Game Reserve may have included eastern wapiti, *C. e. canadensis*, seems unlikely, as the eastern subspecies disappeared from its native range early in European settlement and was probably extinct well before the New Zealand animals were acquired (Murie, 1951; Miers, 1966).

Two earlier introductions of wapiti, of unknown origin, to Kawau I. some time in the 1870s, and later in Dunedin, were unsuccessful (Thompson, 1922; Banwell, 1966).

After release the wapiti were left relatively undisturbed, so little is known of their early acclimatization and dispersal. Visitors to the liberation site in 1909 reported seeing "numerous footprints of young wapiti and calves", showing that the original animals had bred (Donne, 1924). In 1921, a small government expedition confirmed that wapiti were well established in the Lake Alice and Lake Katherine areas at the head of George Sound, and had dispersed as far as Lake Marchant at the head of Caswell Sound. In the same year Southland Acclimatisation Society rangers found about 20 wapiti in the Wapiti River valley on the eastern side of the Main Divide (Donne, 1924; Banwell, 1966). In 1924 hunting parties reported wapiti sign along the full length of the Stillwater Valley (Donald, 1927) and also in the Edith Valley. Wapiti clearly thrived and increased rapidly in the 15-20 years following their release; by 1925 they had colonized an area of about 100 km^2.

Wapiti continued dispersing at an average rate of about 0.64 km per year (Caughley, 1963), more slowly than red deer (p. 441), probably because of the difficult mountainous terrain. By 1950 female wapiti had reached the head of Bligh Sound to the north, the Glaisnock and Worsley Valleys to the east, the Doon and Irene Valleys to the south, and the west coast between George and Caswell Sounds. Since then they have colonized most of the remaining country between the west coast and Lake Te Anau, except the Murchison Mountains, and further extended their range both north and southwards. Males tended to colonize new range before females; mature males were shot as far afield as the Eglinton Valley in 1924 (about 35 km from the nearest known female wapiti), the Transit River near Milford Sound in 1964, and Bradshaw Sound to the south in 1965 (Banwell, 1966).

DISTRIBUTION

WORLD. Rocky Mountain elk are indigenous to the Rocky Mountains and intermontane areas of North America between latitudes 60° and 35°N. They are still present throughout much of that area, and, over the last 75 years, have been introduced to other parts of the United States and Canada (Bryant & Maser, 1982).

NEW ZEALAND (Map 39). The only wild herd established in New Zealand occupies a discrete area in northern Fiordland west of Lake Te Anau.

HABITAT

Northern Fiordland is mountainous, rising from sea level in the west, and Lake Te Anau at 200 m asl in the east, up to 1000 to 1700 m on the major ridges, and to >1800 m on several peaks. It is a typical, recently glaciated landscape with deep, steep-sided, U-shaped valleys and narrow dividing ridges. The basement rocks are mainly Paleozoic metamorphics with some Tertiary sedimentary deposits in the east, and the soils are shallow and predominantly podzolized and infertile. The climate is cloudy, mild and wet, with prevailing westerly winds. Average annual rainfall is high, about 5600 to 6350 mm along the western coastline, increasing with altitude to 7000 to 7350 mm at 1000 m, decreasing eastwards to about 1500 mm to 2000 mm on the western side of Lake Te Anau (Smith, 1974). Rain falls throughout the year, although the summer months are a little wetter on average than the winter months. In winter much of the precipitation above 1000 m falls as snow. Cloud is common, especially on the ridges exposed to the west, and sunshine hours are correspondingly low. Temperatures are generally moderate because of the maritime influence. Mean daily maximum temperatures in summer at low altitudes range from about 17° to 22°C, and mean minimum temperatures in winter from about –1° to +2°C (Smith, 1974).

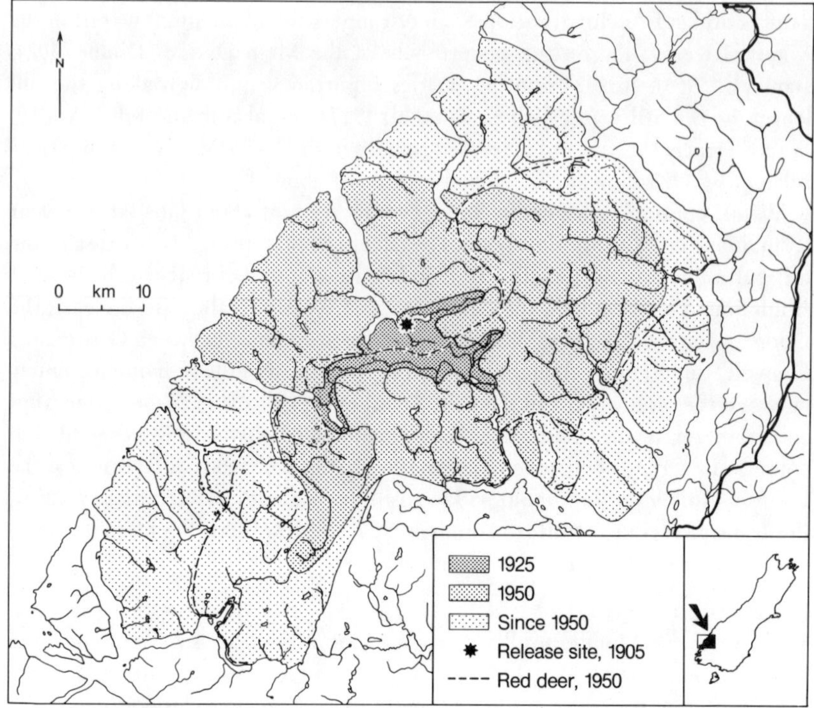

Map 39 Distribution of breeding wapiti in New Zealand in 1925 and 1950, and the additional areas colonised by wapiti and wapiti-red deer hybrids since 1950. Central area enclosed by dotted line shows extension into wapiti range by red deer as at 1950.

The area is forested up to about 900 m asl, except for localized valley bogs and recent slip scars. Silver beech (*Nothofagus menziesii*) is the predominant forest canopy species throughout most of the area from sea level to the timberline. Other species forming the canopy locally include mountain beech (*N. solandri* var. *cliffortioides*), rimu (*Dacrydium cupressinum*) at low elevations near the sea or lakes, and southern rata (*Metrosideros umbellata*) on steep valley sides. These forests have, or originally had, a well-developed small tree tier of, for example, kamahi (*Weinmannia racemosa*), *Pseudopanax simplex* and broadleaf (*Griselinia littoralis*); a dense understorey of shrubs, tree ferns and ferns; and a forest floor covered with mosses and liverworts. Above the forest is a discontinuous belt of subalpine scrub and extensive *Chionochloa* grasslands, grading into herbfields at the higher elevations. Detailed descriptions of the vegetation of this area are given by Poole (1951a,b) and Wardle, Hayward and Herbert (1971).

FOOD

On their native range wapiti eat a wide variety of woody plants, herbs and grasses, the proportions of each food type varying considerably between habitat and seasonally (Nelson & Leege, 1982). In New Zealand their diet appears also to be quite variable, and much the same as that of red deer.

The foods of Fiordland wapiti have been described from samples of the rumen contents of 39 animals shot from the mixed red deer-wapiti herd during the 1949 New Zealand-American Fiordland Expedition (R. Mason, in Poole, 1951a). About 80% of their diet comprised leaves and twigs of woody plants, and 15% sedges and grasses. Ferns were found in the majority of samples but only in small quantities; a few samples also contained some moss and bark, and one sample some lichens. The number of different plants identified in each sample ranged from 4 to 21.

In the deer shot in forest, the most frequent and abundant item was leaves of broadleaf (*Griselinia littoralis*), which comprised 25% of their food. Other plants frequently found were silver beech (*Nothofagus menziesii*), kamahi (*Weinmannia racemosa*), lancewood (*Pseudopanax crassifolium*) and the sedges *Carex* and *Uncinia* spp. combined. Although found in 85% of the samples, silver beech was generally present only in small quantities. The ferns eaten most often were *Blechnum capense* and *Polystichum vestitum,* which occurred in 44% and 26% of the samples, respectively. Other common species were *Neopanax colensoi,* mountain ribbonwood (*Hoheria glabrata*), *Carmichaelia* sp., mountain beech, tree fuchsia (*Fuchsia excorticata*) and *Hebe salicifolia*. In the deer shot above timberline, the most frequently found species were the snow tussocks *Chionochloa* spp., the shrubs *Gaultheria* sp., *Hebe* sp., and *Coprosma serrulata*, the herbs *Celmisia* spp. and *Astelia nervosa*, and mountain flax (*Phormium cookianum*) (Poole, 1951a; Mason, 1966). See also Fig. 75.

SOCIAL ORGANIZATION AND BEHAVIOUR

Wapiti in Fiordland are similar in social organization and in patterns of behaviour and habitat use to red deer (p. 442), and mix and interbreed freely with them.

The rutting call of bull wapiti differs markedly from the "roar" of red deer (p. 446). Typically it begins on a low note, glides upward to a prolonged, high, bugle-like note, then drops quickly to a series of grunts. The "bugles" of individual wapiti vary considerably in character and pitch. Calls intermediate between the "roar" of red deer and "bugle" of wapiti have been attributed to hybrids.

REPRODUCTION AND DEVELOPMENT

The rut begins in mid March (about two weeks earlier than red deer), and most cows are served by early April. Those not conceiving during their first oestrus recycle at about 21-day intervals, if necessary three or four times. Gestation takes 247 to 262 days (Bubenik, 1982), and most calves are born between late November and mid December. Ten percent (n = 303) of the breeding adults, and a greater proportion of 2-year-olds, are still pregnant at the end of December (Smith, 1974). A single calf is born (twins are rare). The sex ratio at birth is near equality, although it may slightly favour males (Smith, 1974). Calves are suckled throughout the summer and autumn and are weaned during the following winter and spring. About 20% (n = 54) of all adult cows (i.e., about 28% of those that were lactating during the previous summer) are still lactating in late October (Smith, 1974).

Cows first breed as yearlings or 2-year-olds. In the late 1960s about 23% (n = 91) of cows produced a calf at 2 years of age, compared with an average of 73% (n = 270) of cows 3 years and older (Smith, 1974).

The growth and development of young wapiti is the same as in red deer (p. 447).

POPULATION DYNAMICS

NATURAL FLUCTUATIONS IN NUMBERS. What little information there is on the growth of the wapiti herd suggests that there were local irruptions similar to those described in red deer (p. 448). Wapiti appear generally to have increased in numbers rapidly in areas recently colonized; they were reported to be "numerous" in the southern tributaries of the Wapiti River catchment in 1932 (A. Sutherland, quoted in Anon., 1932), in "considerable numbers" in the Glaisnock Valley in 1934 (Sutherland, 1934) and "plentiful" in the headwaters of the Lugar Burn in 1937 (K. H. Miers & P. C. Logan, unpubl.). But by 1949 there were few wapiti in the earliest colonized areas around the head of George Sound and the Stillwater Valley, where previously they had been more common. Members of the New Zealand-American Fiordland Expedition concluded from their observations that the apparent decline in numbers there was part of a natural cycle resulting from the depletion of available browse (Poole, 1951a).

DEMOGRAPHY. The sex and age composition in a recently established population of wapiti in the Stuart Mountains was observed during the mid 1950s (Miers, 1962). The sex ratio (excluding young under 1 year old) was 72 bulls to 100 cows (n = 399). Calves comprised 21% of the population, 54 per 100 cows 2 years old and older, and yearlings comprised 16% and 43 per 100 cows. Deer culled from the long-established, little-hunted mixed red deer-wapiti population around Caswell Sound in May 1970 had a similar sex and age composition (Challies, 1970). The overall sex ratio in animals 1 year old and older was in the order of 80 bulls to 100 cows. Calves comprised about 25% and yearlings 15% of the population. The age frequencies in the sample showed that about 43% of the female calves born survived to 2 years, 30% to 5 years, and 10% to 10 years of age. The mean age of cows 1 year old and older was 5.0 years (n = 291) (C. N. Challies, unpubl.).

CONDITION AND FATNESS. Wapiti have the same seasonal pattern of fat deposition and inanition as red deer (p. 451), except that rutting bulls lose condition about two weeks before red deer stags, since they breed earlier (Smith, 1974). Similar levels of fatness have been recorded in red deer, wapiti and their hybrids living in the same areas. During the late 1960s and early 1970s, the longer-established populations of wapiti were in relatively poor condition compared with the average for red deer populations elsewhere in New Zealand (Challies, 1970; Smith, 1974).

HYBRIDISM BETWEEN WAPITI AND RED DEER. Red deer dispersing north and south-westward through Fiordland (p. 441) eventually reached the expanding wapiti range. Males of the two subspecies probably met during the 1920s, and females about 1940. By 1950 wapiti retained exclusive use of only the Edith Valley and the areas between Bligh, George and Caswell Sounds; by 1970 they were sympatric with red deer throughout their range.

Supposed "hybrids", intermediate in size and colouring between wapiti and red deer, have been observed in increasing numbers in the mixed population since 1932 (the first reported was a male from the headwaters of the Wapiti River; Anon., 1949b). In 1938, government-employed hunters shot five "hybrids" in the Stillwater Valley, and one in the Glaisnock Valley (E. B. Davison, unpubl.). In 1948 government hunters reckoned that 37% of 268 deer seen in the eastern catchments were hybrids, 31% wapiti and 32% red deer (Banwell, 1966). Subsequent reports have consistently mentioned "hybrids" as comprising a substantial proportion of the mixed population. By the early 1970s the deer in the wapiti area ranged in physical characteristics from typical red deer through a continuous series of intermediate forms to typical wapiti, with a preponderance of red deer and intermediate forms (C. N. Challies, unpubl.). Skull measurements from 177 adult female deer shot in the eastern catchments in 1966-67 (Caughley, 1971d) showed that the mixed population had a continuous range of skull forms, closely correlated with a similar range of morphological descriptions of type based mainly on size and pelage. That sample was estimated to comprise 52% hybrids, 40% wapiti and only 8% red deer (Batcheler & McLennan, 1977).

The mixed herd was intensively hunted from helicopters between 1973 and 1981, selecting for red deer and obvious hybrids. This did not have the expected effect of increasing the proportion of "wapiti-like" deer in the surviving population. During 1982 and 1983 the herd was hunted non-selectively, primarily for live animals. Of 618 female deer (other than calves) captured in these operations, only 17.8% were considered by experienced observers to be wapiti or "wapiti-like" hybrids. Over 90% of these came from areas colonized by wapiti before 1950, mostly from the western side of their range, the part most recently colonized by red deer. Few wapiti were encountered outside those areas; only 2 of the 76 deer caught south of Caswell Sound, and 4 of 43 caught north of Bligh Sound, were classified as wapiti or "wapiti-like" (NZFS, unpubl.). These trends suggest that the pure wapiti type will not survive in the wild in the presence of red deer.

PREDATORS, PARASITES AND DISEASES

Wapiti have no predators in New Zealand except man.

Ectoparasites are rare. The biting louse *Damalina longicornis* was found on an adult bull shot in the Worsley Valley in 1965 (Andrews, 1969). No other ectoparasites were obtained from that animal, nor were any found on three wapiti inspected in 1949 (Poole, 1951a).

Five species of endoparasites have been identified from pure type wild wapiti in New Zealand, four nematodes and one cestode. Two of the nematodes, *Spiculopteragia spiculoptera* and *Oesophagostomum venulosum*, were found in the abomasum and caecum, respectively, of an adult bull shot in the Large Burn in 1949 (Andrews, 1973). The cattle lung worm *Dictyocaulus viviparus* and tissue worm *Elaphostrongylus cervi* have both been found in live wapiti taken from Fiordland for farm stock (Mason & McAllum, 1976; Mason, 1979). One of six wapiti examined specifically for tapeworms contained an omental cyst of false hydatids, *Taenia hydatigena* (Sweatman & Williams, 1962). In addition, a red deer-wapiti hybrid bull, shot near Lake Marchant in 1962, had a further three nematodes (*Skrjabinagia kolchida, Rinadia mathevossiani* and *Spiculapteragia asymmetrica*) in the abomasum (Andrews, 1973). All of these parasites are also known from red deer.

No clinical cases of disease have been reported in wild wapiti.

Adaptation to New Zealand Conditions

Wapiti readily adapted to the wet, forested, steep-sided mountain country of Fiordland and, once established, steadily increased and spread. They reached high numbers and subsequently decreased naturally in at least the earlier colonized parts of their range (in the manner described for red deer; see p. 448 and Fig. 76), with a consequent decline in physical condition. In the 1960s and early 1970s Fiordland wapiti were substantially smaller in body size compared both with wapiti from North American populations and with those taken from the Fiordland herd in the past, and they were generally in poor condition and had low fertility (Smith, 1974). Body weights recorded during the 1960s were about 65-75% those of comparable North American animals. As deer numbers in Fiordland have since been substantially reduced, it is probable that this trend has now reversed (see Fig. 76 for red deer). The breeding season is 6 months offset from that on their native range.

Significance to the New Zealand Environment

DAMAGE. Wapiti seem to have modified their habitat in a similar manner to red deer (p. 453); the only apparent difference is that wapiti browse to a greater height and are more inclined to bark-chew than red deer (Poole, 1951a).

CONTROL. Wapiti were introduced to provide trophy hunting, and that, along with the need to control their numbers, has been the main consideration in their management. After a period of complete protection, the herd was opened to hunting in 1923 under the control of the Southland Acclimatisation Society. For the next 12 years a limited number of licences was issued annually to hunt bull wapiti, during which time many fine trophies were taken (Banwell, 1966). The remaining protection was removed from wapiti in 1935, and in 1938 shooters employed by DIA began "culling" the herd. Those operations continued intermittently through to the early 1950s, along with a little trophy hunting and some commercial hunting for skins (Smith, 1974). In 1954 the New Zealand Deerstalkers' Association took over management of the herd and introduced programmes of selective culling and controlled trophy hunting. Between 1954 and 1962, 112 trophy wapiti bulls were taken, and 3952 red deer and hybrids were culled (Henderson, 1965). These programmes were maintained for a further 5 years under the control of the Fiordland National Park Board, with the co-operation of the NZ Deerstalkers' Association.

Concern about the propriety of managing deer for hunting in a national park (Fiordland was reserved as a national park in 1904, the year before wapiti were liberated), and about the adequacy of current animal control measures in Fiordland, increased during the 1960s. As a result, in 1969-70 the Forest Research Institute (NZFS) undertook a survey of the deer populations and habitat conditions in northern Fiordland. They concluded that deer numbers were too high over the whole of the wapiti range, and that about 6000 animals needed to be removed immediately to prevent their habitat being further modified (Wardle *et al.*, 1970). Trial culls using helicopters were made in the Caswell and Charles Sounds area in 1970 and 1971, during which over 1200 red deer and hybrids were killed. The whole of the wapiti range was hunted commercially from helicopters for the first time in 1973, when about 3200 deer were taken, and these operations have continued intermittently since. All of the legitimate commercial hunting in northern Fiordland has been selective against red deer and recognizable hybrids, except in 1982 and 1983, when all deer were permitted to be taken. The net result of this hunting has been an overall reduction of about 80% in the size of the mixed red deer-wapiti population between 1969 and 1984 (Nugent, Parkes & Tustin, 1987).

The future management of the wapiti in Fiordland has been controversial. Hunters have argued that the herd is unique and should have special status as a hunting resource, while conservationists have argued that they should be treated in the same way as other introduced animals in the park, and reduced to the lowest possible numbers. In 1980 the then Minister of Lands, in an attempt to resolve the conflict, asked for an investigation into the feasibility of establishing a herd of wapiti outside Fiordland. Several areas in the South Island were identified as potentially suitable, and one, in west Nelson, was surveyed in detail (Davis & Orwin, 1985). No area met all the criteria set, and the proposal lapsed. The State retained ownership of 110 of the more wapiti-like animals taken from the park alive during 1982 and 1983 as a nucleus from which to breed wapiti for relocation. The cows have subsequently been bred to bulls imported from Canada so as to upgrade the quality of the captive herd. It now seems unlikely that any of these animals or their offspring will be released into the wild, either in Fiordland or elsewhere. At present (1988), the herd is being managed by Land Corporation Ltd. near Te Anau, and any income accruing to the State from it is to be used for the benefit of recreational hunting generally.

C. N. C.

40. Sika deer

Cervus nippon Temminck, 1836

Also called Japanese deer, spotted deer.

DESCRIPTION (Fig. 79)

Distinguishing marks, Table 78 and skull key, p. 24.

The overhair of the summer coat of sika deer is a sleek bright chestnut, with white spots variable in size and extent, grading through paler brown towards the white belly hair and down the grey-brown legs; the underhair is not readily identifiable. A black dorsal stripe runs from the head and expands into a patch at the base of the tail. The hair on the head is a paler extension of the bright coat, fading into a distinct, off-white chevron across the frontal area and around the eyes, accentuating the preorbital glands. There is a patch of white on each side of the muffle, and along the upper lips; the larger oval patches of off-white hair continue back from the chin, and pale brown hair extends down the throat. There are stiff, black vibrissae and softer, finer hairs on the muffle. An occasional sika has a white throat patch. The ears are smaller and more rounded than in red deer, with conspicuous, whitish, sparse inner hair (thick and matted in winter). The rump patch is a large, round, black-margined area of coarse, crimped hair, divided into two by the central black stripe down the long, white tail; a few longer hairs may extend the main tail hair another 100 mm, to a recorded 330 mm. The rump patch can be flared into a prominent, fan-like area of white when the sika is alarmed or disturbed. The metatarsal glands on the outer rear legs have a conspicuous tuft of hair, from cream to tan in colour, and varying in shape but usually oval with the lower edge narrowing and sweeping forward. No tarsal gland is evident, but there are interdigital (pedal) glands, and tail glands at the root of the tail.

The overhair of the winter coat is uniform brownish black with faint spots, which looks quite black when wet or distant; it is much thicker than the summer coat, and the male has a mane. The overhairs are pale grey until near their ends, where there is a black section tipped with gold, giving the brownish black look; there is fine, white underhair forming an insulating mat. Towards the belly the black pigment of the overhair is lost, and the hairs are gold tipped, shading into darker grey-brown on the legs. The white belly hair is thick and tipped with grey, but towards the rear becomes very sparse. The rump patch and metatarsal tuft of the newborn sika fawns are light brown for 2-3 months. Fawns lose their spots as they grow their winter coats, at about 4 months. The skull and antlers (Fig. 73) of sika are similar to those of red deer but smaller, and the antlers lack the bez tine. Adults normally have brow, trez, and two top tines (total eight points) (Davidson, 1973a).

Dental formula $I^0/_3\ C^1/_1\ Pm^3/_3\ M^3/_3 = 34$.

FIELD SIGN

Sika make many trails for travelling and feeding. The hoofprints are tapered in front and rounded at the rear, sometimes showing the imprints of dew claws. The pellets are usually smaller and less adherent than those of red deer of

Fig. 79 Sika deer (J. H. Johns/NZFS).

comparable age. Stags approaching the rut thrash shrubs with their antlers and make rutting "scrapes" (bare areas about 0.5 m across, sometimes much larger, near shrub edges on river flats, or in the forest; Kiddie, 1962). On the Oamaru and Mohaka flats, scrapes were 6 to 25 m apart, averaging 12 m. Species identification of deer sign in the field is unreliable without positive observation, or the certain evidence of cast antlers.

MEASUREMENTS

See Table 81.

VARIATION

The original sika brought to New Zealand were supposed to be of the Manchurian type (Donne, 1924). Later evidence suggests that in fact they were of very mixed origin, and hence are classified simply as *Cervus nippon* (Davidson, 1973a). New Zealand sika have a 2:1 ratio of blackish to reddish velvet, and there are also variations in spotting patterns and in the size and extent of the metatarsal tuft; some have a white throat patch. Discriminant function analysis of measurements suggests both variation within the population and a hybrid gradient between at least two subspecies (Davidson, 1973a).

HISTORY OF COLONIZATION

The first attempt to establish sika in New Zealand was unsuccessful. Three individuals (called by Thomson (1922) and Donne (1924) *Cervus sika*) were supplied to the Otago Acclimatisation Society by John Bathgate of Dunedin in 1885 and liberated on the Otekaiki Estate, near Oamaru. They were probably shot out by the settlers; there is no suggestion that they could not tolerate the conditions, since sika are known to be adaptable and hardy (Crandall, 1964).

A second group of six deer (three males and three females) was gifted to New Zealand by the Duke of Bedford from his Woburn Abbey Park herd. They were quarantined for six months on Somes Island (Wellington Harbour)

Table 81: Body measurements of adult[1] sika deer in New Zealand.

	Males			Females		
	mean	SD	n	mean	SD	n
Linear measurements (mm)						
Tail length (skeletal)	128	16.3		112	19.8	
Head length[2]	326	20.5		295	18.3	
Ear length	136	7.0		128	6.2	
Rear foot length	415	17.2		391	13.6	
Hoof/metatarsal length	306	16.3		286	12.2	
Height at shoulder	892	48.5		817	42.0	
Chest girth	919	55.5		822	48.0	
Body length	1 398	84.7		1 313	66.0	
			47			60
		range			range	
Total weight (kg)	62.5	41.0–82.5	54	49.5	36.5–70	64
Eviscerated weight (kg)	41.0	26.5–57.0	54	31.0	20.0–50.0	64

1. Adults: animals over 3 years old.
2. Over the pelt.
Measurements from Davidson (1973a); weights from M. Davidson (unpubl.).

and then, with a fawn born on the ship, taken in early 1905 to the property of W. N. Smith, Taharua (now Poronui) Station (about 39°S) on the eastern side of the northern Kaimanawa Mountains, now Kaimanawa State Forest Park (Thomson, 1922; Donne, 1924). The release site was probably Merrylees Clearing, at the edge of silver beech (*Nothofagus menziesii*) and red beech (*N. fusca*) forest between the upper and lower homesteads. The remains of crates said to have been used to transport them were still there in the early 1930s (K. East, unpubl.). An unpublished Department of Tourist and Publicity report (1900-07) states that two of the males were killed at the time of release, but also that, by 1906, there were two fawns at foot. Hence the four adult survivors and their offspring established the present herd.

The rate of natural dispersal between 1905 and 1930 is unknown, although since the first stalking season was opened in 1925 (McKinnon & Coughlan, 1960), there must by then have been a substantial buildup of the herd. The estimated dispersal rate of 0.6 km/year (Table 82) for this period, therefore, is probably an underestimate. The total range occupied extended relatively slowly, compared with that of red deer, and the limits shown in Map 40 do not appear to have changed much since the early 1970s. Sika were expected to move west and north from the west Taupo forests into indigenous protection forests, but apparently have not; indeed, since shrublands are rapidly being converted to other uses, sika range may in fact be contracting (Davidson, 1973a and unpubl.).

There have been several illegal transplantations of sika to areas beyond the range of their natural dispersal. For example, a male was shot at Te Aroha in

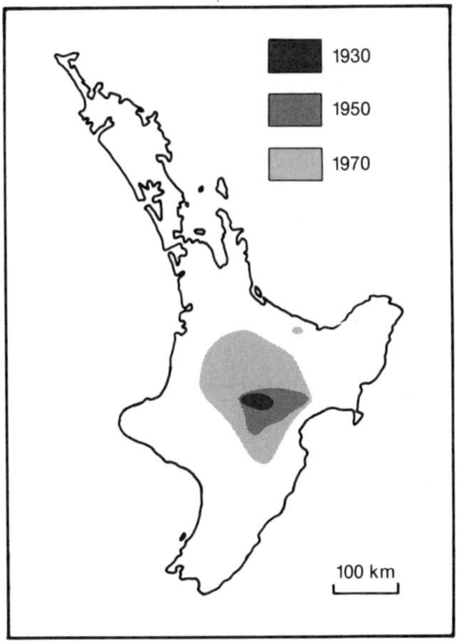

Map 40 Distribution of sika deer (including non-breeding stragglers) at three stages of colonisation (from Davidson, 1973a).

1953, and another at Alfredton in the Wairarapa in 1964. An illegal liberation of sika was made in the Awakeri Hills, 15 km south-west of Whakatane, about 1944 (M. J. Daniel, unpubl.), and by 1974 these had built up to a good hunting population (E. Green, unpubl.).

DISTRIBUTION

WORLD. Sika are native to eastern Asia. Some authorities recognize seven distinct regional subspecies, all in Asia, Japan, China, Manchuria and Formosa (Dansie & Wince, 1968-70); others claim 13 subspecies (Whitehead, 1972).

There has been a very long history of introductions of sika deer to Europe from different parts of their native range. For example, there were five subspecies of sika at Woburn Abbey at the turn of the century, identified as Japanese, Manchurian, Formosan, Pekin and a melanistic form (Kerama); and even these sika were probably of mixed blood before they reached Britain (Lydekker, 1901; Glover, 1956; Ratcliffe, 1987). Introduced populations are established in Britain, Ireland, Czechoslovakia, Denmark, France, Germany, Austria, Poland, USSR, and North America (Lever, 1985).

NEW ZEALAND (Map 40). There is one herd in the central North Island (38°-40°S), occupying a single near-continuous range concentrated in the Kaimanawa and Kaweka State Forest Parks and the Ahimanawa Range, but also straggling to the southern Urewera National Park, northern Ruahine State Forest Park, and the west Taupo forests. There is also a small herd in the Awakeri Hills in the Bay of Plenty. The extreme limits for sightings of sika (including animals probably illegally translocated out of the natural range) are

Table 82: Dispersal of the national herd of sika deer, 1905-70, from the liberation point in eastern Kaimanawa State Forest Park.

Years	Number of sightings		First sightings		Breeding groups	
	Lone stags	Breeding groups	Area/year (km²)	Rate/year (km)	Area/year (km²)	Rate/year (km)
1905-30	0	2	803	0.6	803	0.6
1931-40	2	4	2512	1.2	1917	0.9
1941-50	2	2	4273	0.9	3186	0.7
1951-60	3	12	10412	1.1	6915	1.5
1961-70	9	7	20202	2.1	11836	1.4

The presumed disperal route was down the Mohaka Valley, between the Ahimanawa Range and the Kaweka State Forest Park, to the southern Kawekas (Davidson, 1973a).

172 km north, 125 km west, 185 km south, 60 km east, and 130 km north-east (Awakeri) of the 1905 liberation point (Davidson, 1973a and unpubl.). The breeding range is much smaller than this.

HABITAT

The northern Kaimanawa and Kaweka State Forest Parks and the Ahimanawa Range are upfolds of greywacke rock, covered to various depths with pumice and ash from eruptions of the active volcanoes of Tongariro National Park. The vegetation of these areas has been well described (McKelvey & Nicholls, 1957; Elder, 1959, 1962, 1965). Sika range throughout the dominant forest types of red beech with kamahi (*Weinmannia racemosa*) and shrub hardwoods; red and silver beech with kamahi; mountain beech (*Nothofagus solandri* var. *cliffortioides*) and red beech; and black beech (*N. solandri* var. *solandri*), silver and red beech with shrub hardwoods (McKelvey & Nicholls, 1957). With the exception of mountain totara (*Podocarpus hallii*), podocarp forest trees such as miro (*P. ferrugineus*), matai (*P. spicatus*), rimu (*Dacrydium cupressinum*) and maire (*Nestegis gymnelaea* and *N. lanceolata*) are almost confined to the western side of the watershed and are sparse there. Many palatable species have been replaced by *Neomyrtus pedunculata*, pepperwood (*Pseudowintera colorata* and *P. axillaris*) and mingimingi (*Cyathodes juniperina* and *C. fasciculata*). Over considerable areas the crown fern (*Blechnum discolor*) has taken the place of palatable understorey shrubs and forest floor plants; locally, so has the fern *Dixonia lanata*. In the eastern Kaweka Range, the Ahimanawa Range and the northern Ruahine State Forest Park, forest was cleared for farming, abandoned about 1905, and has since been reverting to shrubland (Elder, 1965). Since about mid-century, however, some of these shrublands have been converted to commercial pine forest, which is not preferred sika habitat (Davidson, 1979). In the Oamaru Valley, the rainfalls for 1965 and 1966 were 2134 mm and 2184 mm (Davidson & Gannaway, 1967); a provisional mean annual rainfall of 1829 mm was estimated from these and from longer-term records from nearby rain stations (Grant, 1969). In the eastern Kaweka, sika range onto the open tops above the bushline in summer, but move downward in winter (Q. Roberts, unpubl.). In this eastern area manuka shrubland registers higher temperatures than forest in summer, but in winter it is the reverse, with great extremes of temperature (Cunningham, 1964).

FOOD

Sika are primarily grazers, taking both indigenous and introduced grasses and herbs of the forest and shrubland for most of the year. They also browse understorey trees and shrubs such as broadleaf (*Griselinia littoralis*), putaputaweta (*Carpodetus serratus*), all *Pseudopanax* species, fuchsia (*Fuchsia excorticata*), wineberry (*Aristotelia serrata*), pate (*Schefflera digitata*) and many species of *Coprosma*. On the Oamaru River flats, fescue (*Festuca novae-zelandiae*) grows in wet places, and silver tussock (*Poa caespitosa*) where it is drier. Associated introduced grasses are Yorkshire fog (*Holcus lanatus*), catsear (*Hypochaeris radicata*) and sweet vernal (*Anthoxanthum odoratum*), but most prominent is the ragwort (*Senecio jacobaeus*), which flowers from January to April. Sika have been observed browsing on *Dracophyllum subulatum* through most of the year, but more particularly in April and September. In winter, sika leave the frosty flats and move up into the beech forest, living off beech and other seedlings, sparse understorey browse, fallen leaves, mosses, lichens and fungi, especially the violet beech tuft fungus (*Rhizopogon violaceus*) in autumn (Davidson, 1973a,b and unpubl.). When plentiful, fungi may comprise some 70% of the gut content (D. H. Vowles, unpubl.).

SOCIAL ORGANIZATION AND BEHAVIOUR

ACTIVITY. If undisturbed, sika of both sexes and all ages will graze together on valley flats from spring to autumn (Davidson, 1973a,b); under severe hunting pressure, however, they emerge from cover only at night.

VOICE AND COMMUNICATION. Sika are more wary and more vocal than red deer. They usually give a shrill whistle when alarmed, and flare the rump patch as they bound away. At other times they may stand in full view out on the open flats, giving their high-pitched whistle while looking directly at the disturbance. They may whistle in the night for no apparent reason. Sika also have a great variety of other calls. A hind will scream, yelp and snort for hours, when camps are occupied within her nightly feeding area; she may even gallop around the camp giving calls ranging from a squeak to a harsh bark. In daylight, calls made by one may alert others or may be ignored; of two deer disturbed in a forest, one may whistle and the other sneak away in a crouching gait without making a sound. As well as visual and auditory communication, sika also make use of olfactory signals subserved by the skin glands. Human footprints left during a morning check on the Oamaru Valley flats obviously put up an invisible but strong olfactory barrier. Hinds venturing out of the forest in the evening and encountering the scent would whirl around and disappear, not usually returning in daylight (M. M. Davidson, unpubl.). Stags were bolder, and after retreating a little, perhaps several times, would "work" along the scent line until reaching a "gap" and then cross hurriedly and run down the flat before settling down to feed.

Table 83: Dispersal of tagged sika deer, 1964–74.

	Distance (km)	Time (months)
All sika deer	2.2 (0.41-17.71)	6.9 (1-97)
Yearling males	2.5	6.5 (3-17)
Yearling females	1.9	4.5 (1-11)
Adult males	5.9	9.0 (1-15)
Adult females	1.7	24.5 (3-97)

Data given are means (and ranges) of linear distances travelled by 54 sika deer from tagging to kill site, and time taken, determined from 6-monthly checks over 10 years, 1964–74 (Davidson, 1979)

SOCIAL RELATIONSHIPS. Domination by the stag during the rut is temporary and does not constitute leadership; the leader of the herd is an experienced hind who maintains a strict social hierarchy in the herd. Sika juveniles are more independent than those of red deer, and even fawns may lead on occasion; if suddenly alarmed, hind, fawn and yearling will scatter in different directions, whereas in red deer, the hind, fawn and yearling will run away in single file (M. M. Davidson, unpubl.).

Sika stags which have cast their antlers may stand on their hind legs and spar by "boxing". Hinds likewise box with other hinds, and hinds with rebellious juveniles, including fawns. Also, hinds and fawns, and hind, fawn and yearling groups, race and prance and box, and even fawn pairs play chasing and boxing. A sika fawn which tries to suckle its dam, and is repulsed, is quite liable to stand up on its hind legs and strike her sharply on the rump with its forefeet. Some sika males feed in bachelor groups of all ages; at the time to return to the forest, the junior male (generally very reluctantly) is forced to take the lead, the others following in age order with the largest, often a stag with trophy antlers, bringing up the rear (M. M. Davidson, unpubl.).

REPRODUCTIVE BEHAVIOUR. The main rutting period is late March to early May. From early March, stags make scrapes in the ground with forefeet and antlers, and fling the turf aloft. They rub the neck and sides of the head, or the antlers, into the scrape several times, and may lower the whole body onto it. A group of juvenile males may take turns behaving in this way on a single scrape. Scraping behaviour is most common in April, tapering off in May, but is also occasionally observed outside the rutting season (recorded in August and November).

The sika stag roar is a shrill whistle/scream, rising and falling several times, the number of "signals" varying from one to seven. Of 318 roars recorded from 26 April 1963 to 28 November 1966, 5% comprised two signals, 20% had three signals, 42% had four, 22% had five, 9% had six and 2% had seven; the modal number was four (M. M. Davidson, unpubl.). The peak of the roar is mid April, but roars have been heard in every month except February. Stags may round up hinds during the rut (Kiddie, 1962) or follow one hind only (Davidson,

1973b); they occasionally fight bitterly between themselves.

Adult sika stags cast their antlers in November-December (younger males progressively later) and new ones start growing at once. Antlers are stripped of dead velvet and hardened by mid March.

REPRODUCTION AND DEVELOPMENT

The main birth period is November to January, and the calculated median birth date is 13 December (Davidson, 1976). One fawn is usual, but there have been three reports of twin embryos (K. East, unpubl.; D. H. Vowles, unpubl.). After the fawn is born, the hind eats the afterbirth, cleans the fawn and herself, and lies down to rest. In 2-3 hours she resumes feeding, and the fawn can run around. The spread of foetal ages, determined from autopsy and observation, suggests the possibility of out-of-season breeding, or minor ruts. Sika hinds still bearing foetuses were shot on 27 February 1970 (foetus 2.5 kg), 29 February 1970 (2.3 kg) and 23 April 1971; a sika fawn about a fortnight old was seen in the Oamaru Valley on 10 June 1970 (D. H. Vowles, unpubl.). Some sika first mate as yearlings (about 16 months old; Davidson, 1976). A sample of 49 foetuses (29 males, 20 female) collected from outside the Oamaru Valley in 1964-66 was recovered from hinds of age groups distributed as follows: 1-2 years, 10 (4 male, 6 female); 2-3 years, 13 (6 male, 7 female); 3-4 years, 6 (2 male, 4 female); 4-5 years, 5 (4 male, 1 female); 5-6 years, 8 (7 male, 1 female); 6-7 years, 2 (male); 7-8 years, 3 (2 male, 1 female); 8-9 years, 2 (male). The younger hinds (2-3 years and 1-2 years) apparently had a higher reproductive success, but this population was outside the research area and was subject to considerable hunting pressure (M. M. Davidson, unpubl.). Foetal resorption has been recorded twice in sika. In early 1963, a hind carrying a dead foetus was shot in the Tauranga-Taupo Valley, northern Kaimanawa (G. R. Hampton, unpubl.); in January 1965 an apparently healthy hind, able to run, was shot and found to be carrying a rotting foetus, the head partly formed, but without legs (W. Wilhelmson, unpubl.). Sika appear to mature faster than other deer, both in the field (Davidson, 1973b) and in captivity (Crandall, 1964); they also accumulate fat reserves more rapidly than other deer (Crandall, 1964), which perhaps explains why they are often reported to be in superior condition (Davidson, 1973b).

POPULATION DYNAMICS

No census of sika is feasible. Marked sika seldom travel further than 6 km over 1-2 years (Table 83). Sex ratio is 1:1 in fawns, but very variable thereafter (see Table 84). Maximum longevity of sika in the wild is usually about 12 years (Table 84), although a hind skeleton found in the bush on the Oamaru River flats was aged at >15 years (M. M. Davidson, unpubl.). Sika and red deer may hybridize where their ranges overlap (Davidson 1973a,b; Ratcliffe, 1987).

PREDATORS, PARASITES AND DISEASES

Man is the only predator of sika deer in New Zealand.

Tests for *Brucella abortus, Leptospira* spp. and *Salmonella* spp. showed only one reactor to *B. abortus*, a sika stag in contact with cattle (Daniel, 1967). In

Table 84: Age structure and sex ratio of sika deer collected in 1964-66.

Age[1] (years)	No. of males	No. of females	Ratio M:F
0-1	20	20	100:100 (1:1)
1-2	22	18	122:100 (1:0.8)
2-3	12	22	55:100 (1:1.8)
3-4	10	10	100:100 (1:1)
4-5	10	9	111:100 (1:0.9)
5-6[2]	5	13	38:100 (1:2.6)
6-7[2]	1	3	33:100 (1:3.0)
7-8	4	5	80:100 (1:1.25)
8-9	3	4	75:100 (1:1.3)
9-10	2	2	100:100 (1:1)
10-11	1	3	33:100 (1:3.0)
11-12	0	1	—
Total	90	110	82:100

1. Ages determined from cementum layering.
2. Number of sika in the 5-6 and 6-7 year age classes may have been reduced by poison trials in 1960-61 (Daniel, 1966b).

Data from M. Davidson (unpubl.).

captivity, however, many sika die from malignant catarrhal fever (Yerex, 1982). Endoparasites recorded from sika include the lungworm *Dictyocaulus viviparus* (P. C. Mason, unpubl.), and two nematodes of the abomasum, *Spiculopteragia assymmetrica* and *S. spiculoptera* (Andrews, 1973).

Malformed antlers are quite common in sika, but malformed skulls, dentition and feet less so. A case of polydactylism was found in the right front leg of a sika hind shot in the Otupua Valley, Ahimanawa Range, in 1969; a small extra metacarpal carried a digit with three phalanges, a hoof and a dew claw (Davidson, 1971). A sika five-point stag, shot in the same area, seemingly a healthy and well-grown animal with small but sturdy antlers, had external genitalia in aberrant form and position (J. Erceg, unpubl.).

ADAPTATION TO NEW ZEALAND CONDITIONS

Sika in New Zealand tend to be of more uniform size than in Asia, where they increase in size from south to north. In Japan sika prefer the open grasslands (Ito, 1967), and Bromley (1956) suggests they originated in the steppe. Elsewhere, sika are said to favour woodland edges. Like other introduced deer species, sika have adjusted to annual cycles displaced by six months from those of the Northern Hemisphere, in rutting, antler fall (older males first) and regrowth, and birth of young.

SIGNIFICANCE TO THE NEW ZEALAND ENVIRONMENT

DAMAGE. Considerable modification to indigenous forest followed the introduction of sika deer to the northern Kaimanawas (McKelvey, 1959), but it is indistinguishable from that due to red deer (p. 453).

CONTROL. Small numbers of sika were shot during official control operations for red deer, which began in the 1930s and continued to the late 1950s. Experienced men were sought to hunt sika, as they were considered harder to shoot than red deer. Since then, control of sika has been left to recreational hunters, who greatly prize the species as a trophy animal (Thornton, 1933; Douglas, 1959). Recreational hunters on foot have had to compete with commercial recovery operations both on the ground and from helicopters hunting for venison, velvet and live deer for farms and wildlife parks. In the eastern Kawekas, there is now a Recreational Hunting Area, in which sport hunters are protected from competition with commercial operators. The sport hunters pay for this privilege by accepting the responsibility for keeping sika numbers low; since soil erosion in the eastern Kawekas is severe, only minimum numbers of deer are now acceptable.

Sika do not adapt well to captivity and are vulnerable to malignant catarrhal fever, an important disease of farmed deer throughout New Zealand (Beatson, 1985b). It caused high mortality among farmed sika at first, but recently some farms have been more successful with sika, e.g., Lochinver Station in the central North Island. Sika provide both velvet and superior quality venison.

M. M. D.

41. Sambar deer

Cervus unicolor unicolor (Kerr, 1792)
Synonym *Cervus axis unicolor* Kerr, 1792.
Also called sambur, sambhur.

DESCRIPTION (Fig. 80)
Distinguishing marks, Table 78 and skull key, p. 24.

Sambar are large-bodied deer of solid build. The pelage is a uniform brown, varying in shade from light brown to almost black in older animals, grading to brown edged with tan to rust red on the rump. Hinds are generally lighter coloured. The body hair is coarse textured, stiff and straight with no underfur, and stands erect when the animal is excited; mature stags have a conspicuous ruff of unkempt hair around the neck. The hair on the groin, belly and inner sides of the legs is finer, and grey to mid brown. The tail is long, bushy and black at the tip. The bare muzzle extends noticeably below the nostrils to form a broad rim. Above the nostrils the nasal arch is prominent and slightly convex, and the forehead is broad and flat. The antler pedicles are short and stout. The ears are large and rounded.

The heavy, three-tined antlers (Fig. 73) are little advanced from the primitive single spike. The structure of the antlers is solid, with little interior cellular formation and a rough, deeply corrugated surface.

The prominent preorbital scent glands secrete a strong-smelling substance stored in large gland sacs. These can be everted at will, but are normally hidden in large deep pits below the eye sockets. The supraorbital glands, on the forehead above the eyes, are poorly developed; the metatarsal glands are marked by tufts of hair on the outer side of the hind legs above the feet; there are interdigital glands between the toes of both front and hind feet.

The permanent incisiform teeth are relatively small.

Dental formula $I^0/_3\ C^1/_1\ Pm^3/_3\ M^3/_3 = 34$.

Although alert and wary, undisturbed sambar have been known to bed down within sight and sound of farm and forestry workers. If suspicious, sambar will slip quietly away or retreat slowly in a stiff-legged posture, with head and tail erect. When caught unawares, a stag will explode from cover with a sudden loud "pook" sound, perhaps an intimidation action originally used as a defence against predators.

FIELD SIGN

Sambar, although slightly larger than red deer, have smaller and narrower hooves. The hoofprint of a large stag will measure, at most, 70 × 45 mm; those of a hind are smaller, narrower and leave a shallower impression. Prints are frequently found along stream edges, around ponds and swamps, and in the soft mud of drying watercourses during summer.

Signs of grazing and browsing are scattered, because sambar feed selectively, generally while on the move. Where flax (*Phormium tenax*) is common, sambar feed on the soft base of the leaf blades and leave much-fragmented remains of the upper stem material (Kelton, 1981).

Pellets are black, pointed at one end, average 20 × 10 mm, and are often deposited in piles within thick cover where sambar like to pause on their trails, to observe before venturing into the open. Sambar habitually rub their antlers on trees and shrub stems, not only to clean off dried velvet but for long afterwards as well, often leaving a long bare slash on the tree and a scattering of shredded bark at the base.

Stags mark their territories by rubbing secretion from their preorbital glands on to shrubs and overhanging branches, if necessary by rearing up on their hind legs to do so. This behaviour is the cause of much trampling around the base of the tree, and of mud and hair adhering to the tips of the branches above.

Both stags and hinds wallow, particularly during summer. They form a network of trails within thick cover throughout their home range, or make use of old cattle trails and timber extraction drag trails.

MEASUREMENTS

Few measurements of New Zealand sambar are available. A 7-day-old male fawn weighed 7 kg, stood 545 mm high at the shoulder and measured 820 mm total length (over the curve), including the tail length of 120 mm. A juvenile male (spiker) stood 1270 mm at the shoulder and measured 2060 mm, including the tail length of 250 mm.

Generally, stags may be expected to measure on average 1370 mm high at the shoulder, with a total body length of 2130 mm. Hinds are smaller by about 220 mm at the shoulder. The whole body weight of stags is usually up to 227 kg (Roberts, 1968) and may reach 245 kg (Farmer, 1965). Hinds weigh between 113 and 157 kg (Roberts, 1968).

Fig. 80 Sambar deer, Rotorua district, 1966 (J. H. Johns).

VARIATION

The two main sambar populations in New Zealand show little physical variation because they are founded from the same basic stock. There is some variation in the antler formation; for example, within Kaingaroa Forest stags carrying untypical antlers with up to 10 points have been recorded (Farmer, 1965). Similar untypical antlers with 7, 9, and 12 points are possible on the native range of the sambar (Field, 1907; Cuffe, 1914; Burton, 1930), but variation in antler formation of stags from the Whakatane region (M. J. W. Douglas, unpubl.) suggests that the New Zealand examples may be due to hybridization with rusa deer.

HISTORY OF COLONIZATION

A single pair of sambar was imported from Sri Lanka and released on the Carnarvon Estate, Rangitikei, by Mr Falconer Larkworthy in 1875. They thrived, increased in numbers and spread throughout the then swampy and shrub-covered coastal dune country. In 1893, after illicit hunting had reduced the herd to not more than 30 animals (Harris, 1971a), they were given protection under the Animals Protection Act of 1880. By 1900 they numbered 100 (Shailer, 1957) and in 1906 the Wellington Acclimatisation Society issued a limited number of licences for the hunting of sambar. All protection was removed in 1930, but they continued to thrive, and reached an estimated 400-

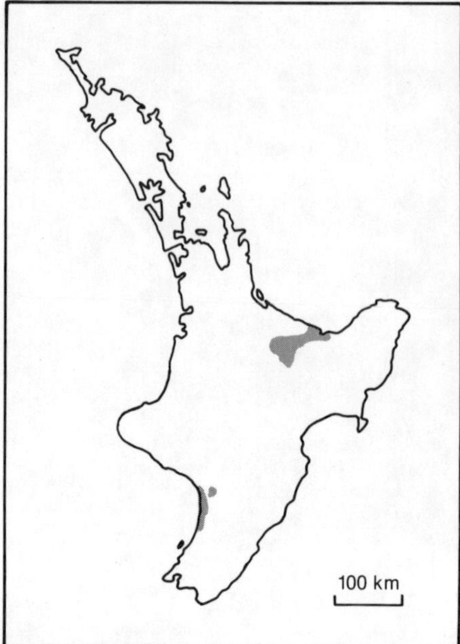

Map 41 Distribution of sambar deer in New Zealand.

500 animals by 1947 (Wodzicki, 1950). By 1957 sambar were present along a 100 km coastal strip, up to 10 km wide, between the Waitotara and Hokio rivers (Riney, 1957), possibly extending south to Paraparaumu (Harris, 1966). In about 1950 the herd also extended inland along the Turakina and Whangaehu river valleys.

Several translocations from the Manawatu stock were made to the Bay of Plenty district. In 1876 a stag from the Manawatu herd and a hind from Sri Lanka were liberated near Morrinsville, but the resulting herd was shot out by local hunters (Harris, 1966). Later, more translocations from the Manawatu herd were made between 1914 and 1921 to the Whakatane and Waimana valleys; on Mt. Edgecumbe; and near Lake Rerewhakaitu (Harris, 1971a). Probably there were also other unrecorded liberations; for example, two sambar stags and a hind were released unofficially in the Kaharoa area north of Lake Ngongotaha about 1924 (D. B. Seager, unpubl.). Both the Manawatu and the Bay of Plenty populations have declined in range and density since about 1950, and in February 1982 all sambar deer were given full protection by Ministerial order for a period of five years (Douglas, 1983).

DISTRIBUTION

WORLD. The sambar deer has been described as the typical forest deer of South-East Asia (Prater, 1948), widely distributed and subdivided into 16 subspecies (Whitehead, 1972) differing in body and antler size. Shoulder heights range from 1525 mm in the west of the range to 610 mm in the east, and antler size also decreases eastwards. The nominate subspecies from Sri Lanka (*Cervus unicolor*

unicolor), which is the source of the New Zealand stock, is one of the large-bodied forms. The wild sambar now established in Australia also originated from Sri Lanka (Allison, 1980).

NEW ZEALAND (Map 41). There are two discrete herds of sambar in New Zealand, separated by 120 km direct distance. The Manawatu herd is scattered along the whole coastal belt from Levin to Harakeke just south of Wanganui, and inland along the Turakina and Whangaehu river valleys (Douglas, 1983). The Bay of Plenty herd is concentrated south-east of Rotorua and extending north-east to Whakatane, but also with small outlying populations well outside the range of the main herd (NZFS, unpubl.).

HABITAT

Sambar are normally forest-dwelling animals, but they occupy a wide variety of habitats in New Zealand. At the time of liberation much of lowland Manawatu consisted of swamp, low shrub-clad ridges, and remnant indigenous forest, rich in food and cover. With progressive development of land for agriculture, sambar became restricted to remaining undeveloped areas, from which they foraged onto farmland. The extension of pine plantations along the coast and on adjacent farmland from the early 1940s has provided cover but little food during summer when the plantations are dry. Sambar therefore spend much of the summer feeding on farmland and lying-up in small areas of remnant cover and shelterbelts, returning to the pine forests in winter (Douglas, 1983). The inland foothills of the Turakina/Whangaehu catchments still support remnant patches of indigenous forest and shrubs, particularly within steep gullies, which provide sufficient cover adjacent to pasture and croplands.

The Bay of Plenty region has extensive tracts of original and regenerating indigenous forests, very large pine forests and many lakes, as well as agricultural land. Sambar frequent a wide variety of habitats within this region, ranging from lake-edge, overgrown gullies, isolated scrub patches and raupo (*Typha orientalis*) swamps within farmland to steep shrub-filled gullies and flax-bordered streams within exotic pine forests, and the indigenous forest of the Urewera ranges.

Sambar leave cover after dark, and range widely if necessary, according to the distribution of available food sources, before returning to cover at dawn. They form complex trail systems within and between their cover and adjacent open feeding sites. The main trails, frequently used, generally follow natural hollows and gullies with side trails leading off to lying-up areas on ridge systems. Windfalls or fences are no obstacle; a standing jump of 1.5 m apparently requires little effort for a sambar.

FOOD

A small population of sambar living in a wetland area of the Manawatu were observed to forage on some 16 plant species (Kelton, 1981). Flax (*Phormium tenax*) and tall fescue (*Festuca arundinacea*) were the main winter foods (March-October), and floating sweet grass (*Glyceria declinata*) and reed canary grass

(*Phalaris arundinacea*) in summer (November-February). Analysis of pellets supported these findings. Sambar living within pine forests graze on various introduced grasses and browse the emerging shoots of briar (*Rosa canina*), blackberry (*Rubus fruticosus*) and buds of young pines. In spring and autumn the bark of coniferous trees is sometimes torn off and eaten. Improved pasture, maize and root crops such as swedes and chou moellier are frequently included in their diet.

SOCIAL ORGANIZATION AND BEHAVIOUR

Adult stags remain solitary throughout the year except during the rut, whereas juvenile males and females form small social or family groups of 2–5 animals. Stags do not round up a harem of hinds during the rut, but mark out and defend a territory against intruding stags. Hinds living within or close to the area are attracted to the stag by his roaring and scent, but do not remain with him for long. After the rut, the stag resumes his solitary existence (Whitehead, 1972). Most stags cast their old antlers between November and December, and grow new ones from January to March (Riney, 1957); these are hard between June and November (Rudd, 1978; Kelton, 1982).

REPRODUCTION AND DEVELOPMENT

The breeding cycle of sambar is somewhat indefinite and debatable. They rut at any time from late May until December, most often between June and August (Rudd, 1978; Kelton, 1981). Both stags and hinds may first breed at 15 months (Draisma, 1979), depending on condition and body size. A hind observed mating in captivity produced a fawn 264 days later (Douglas, 1983).

Newborn fawns are pale brown with indistinct spots which fade within several days. A narrow band of black hair runs the full length of the back. The rump patch and the insides of the ears and legs are creamy white. Shedding of the deciduous teeth starts at about 15–16 months of age, and the full complement of permanent teeth is acquired by 30–32 months (Van Bemmel, 1949).

POPULATION DYNAMICS

The density and distribution of both herds in New Zealand have been limited by overhunting from about 1950 to 1982, and also to some extent by habitat changes. Agricultural and forestry development have destroyed large areas of swamp and shrub habitat formerly used by sambar, although they have also provided new habitat on farmland and within production forest plantations, particularly in the Manawatu region (Douglas, 1983).

PREDATORS, PARASITES AND DISEASES

There are no natural predators of sambar in New Zealand, except man, and possibly stray dogs that may take the occasional young fawn. Little is known about the parasites and diseases of sambar even in their countries of origin; here they may accidentally acquire the cattle tick *Haemaphysalis longicornis* (Tenquist & Charleston, 1981) and the gastrointestinal nematode *Spiculopteragia spiculoptera* (Andrews, 1973).

Adaptation to New Zealand Conditions

The social system of sambar in New Zealand has not changed. In India, the average size of 73 groups containing two or more individuals, but excluding an adult stag, was 2.3 (Schaller, 1967). The season of rut is still extended, as in tropical deer, rather than synchronized, as in the temperate deer introduced into New Zealand.

Within their Asian range sambar inhabit mountain country up to 2285 m in Sri Lanka, and forests at up to 3656 m in Formosa (Whitehead, 1972); yet the Manawatu herd has never colonized the adjacent and extensive tracts of forested mountains of the Tararua or Ruahine ranges.

The New Zealand record trophy head, with length of 1029 mm and spread of 1022 mm, was obtained in the Manawatu during the early 1920s (Forbes, 1924). This is not much smaller than the world record head obtained in India, which measured 1273 mm in length, and larger than the Sri Lankan record head of 825 mm (Whitehead, 1972).

Significance to the New Zealand Environment

Sambar have had little detrimental effect on the New Zealand environment. They may damage root and maize crops, browse young trees and strip bark in forest plantations, but generally only seasonally and locally. Although the sambar is widely considered to be the most challenging of all deer species to hunt on foot, the continued existence of both herds as a resource for recreational hunters will depend on the tolerance of forest and land owners and on sound research-based management. The Ministerial order of 1982 prohibiting all sambar hunting expired on 31 January 1987 (Government Gazette, 4 March 1982), but has been extended indefinitely.

M. J. W. D.

42. Rusa deer

Cervus timorensis Blainville, 1822

The common name "rusa" is the Malay word for "deer" and is also applied by the Malays to sambar.

Description (Fig. 81)

Distinguishing marks, Table 78 and skull key, p. 24.

Although generally similar to sambar deer, rusa are smaller, with pointed ears and a long narrow tail, slightly bushy towards the tip. The summer coat of stags is generally a dark reddish-brown changing to a dark greyish brown in autumn, and of hinds a pale yellowish-red in summer, becoming greyish red in winter. The chin, throat and underparts are cream-coloured. During the rut stags develop a heavy and dark-coloured mane. The texture of the coat is coarse, except on the underparts; the individual hairs are flat, crimped and banded towards the tip, whereas in sambar the hair is round, straight and unbanded. Fawns are pale tan, unspotted, with white chin, throat and underparts. The preorbital gland pits are small. The hooves are heart-shaped. Typical antlers are slender and three-tined (Fig. 73).

Dental formula $I^0/_3\ C^1/_1\ Pm^3/_3\ M^3/_3 = 34$.

Rusa are remarkably agile even over rough terrain, and when suddenly disturbed, flee at high speed holding the head low and neck outstretched. This curious running posture is displayed not only when escaping through cover, but also while in the open.

FIELD SIGN

The hoofprints of a mature rusa stag can measure up to 60 × 38 mm and are frequently found along a network of regularly used trails connecting feeding and lying-up areas. Bedding sites and tunnel trails may be found in clearings covered with bracken fern (*Pteridium aquilinum*) within dense cover (Allen, 1976). Rusa are sedentary by nature, and the evidence of their browsing and grazing is often clear. Wallows and signs of antler-thrashing are similar to those made by other deer.

MEASUREMENTS

Few measurements are available for the New Zealand rusa, but generally a mature stag measures about 1060 mm high at the shoulder, with a total body length of 1560 mm, and tail length of 180–250 mm. Live stags weigh about 122 kg. Hinds measure up to 810 mm at the shoulder and can weigh up to 70 kg. New Zealand rusa do not produce antlers of world trophy quality, and heads measuring 760 mm length by 660 mm spread are considered top class for the area (Allison, 1980).

VARIATION

Stags in the Whakatane River area are larger than those from around Galatea, and those frequenting the Hanimahihi clearings often develop extra points on one side (Rudd, 1980). Photographs taken by hunters of animals shot within the Whakatane River catchment support Rudd's observation (M. J. W. Douglas, unpubl.). These atypical antler formations may result from hybridization between rusa and sambar, whose ranges overlap in that area.

HISTORY OF COLONIZATION

Rusa deer were introduced into New Zealand by accident. In November 1907, eight "sambar" deer were obtained from New Caledonia, two for the then Department of Tourist and Health Resorts, and six for Mr H. R. Benn of Lake Okareka near Rotorua. Benn subsequently liberated all eight animals at Galatea in 1908 (Donne, 1924), where they established a herd of wild deer known to local people as "little sambar". These deer were eventually identified as rusa deer (Riney, 1955b).

The early establishment of the herd is scarcely known, except that at first it spread very slowly. Even after 40 years it apparently remained restricted to the low country around Galatea near the liberation site (Wodzicki, 1950; Riney, 1955b). In the late 1950s rusa were shot in the scrublands on the western bank of the Rangitaiki River adjacent to forest plantations, and on the eastern bank between Galatea and Waiohau (Allen, 1976).

After 1960, the dispersal of rusa suddenly accelerated. Development of farmland in the Galatea and Waiohau areas, and the regeneration of scrub

Fig. 81 Rusa deer, mature stag with hinds and fawns (M. J. Douglas).

hardwood stands in adjacent areas previously burned, increased the food supply available to rusa. Local hunters considered that numbers were highest in the early 1960s (Allen, 1976). By the mid 1960s a small herd established itself around the junction of the Whakatane River and the Ohane Stream (Holden, 1971). By 1973 the known distribution of the herd, as mapped then by NZFS, was the largest ever and apparently continuing to expand (Allen, 1976), although the density was unknown. Allen listed several related reasons for this increased dispersal, including higher social intolerance in the peak population of the early 1960s; increased hunting pressure as the game-meat industry, which started buying venison for export in the early 1960s, gained momentum; and decreased competition as culling operations greatly reduced the numbers of red deer throughout the Urewera ranges.

DISTRIBUTION

WORLD. Rusa deer are native to the Indonesian archipelago, from the western tip of Java to the Moluccas. Much of the "original" distribution is due to introductions, since deer have been liberated onto islands outside their indigenous range for centuries (Van Bemmel, 1949). Recent known introductions in the area include West Irian, Papua New Guinea and New Britain. There are six subspecies (Whitehead, 1972).

Rusa from Java, *Cervus timorensis russa,* were introduced to the Indian Ocean island of Mauritius in 1639 (Owadally & Butzler, 1972). Deer from this stock were subsequently taken to New Caledonia and thence to New Zealand (Harris, 1971a) and Australia (Allison, 1980). The New Zealand rusa could have been of mixed blood (p. 487); so their exact subspecific status is uncertain (Whitehead, 1972). Other rusa populations in Australia resulted from liberations of Moluccan rusa, *C. t. moluccensis,* onto islands within the Torres Strait; on the western tip of Cape York; and on North-East I., just off Groote I., in the Gulf of Carpentaria (Bentley, 1967; Allison, 1980).

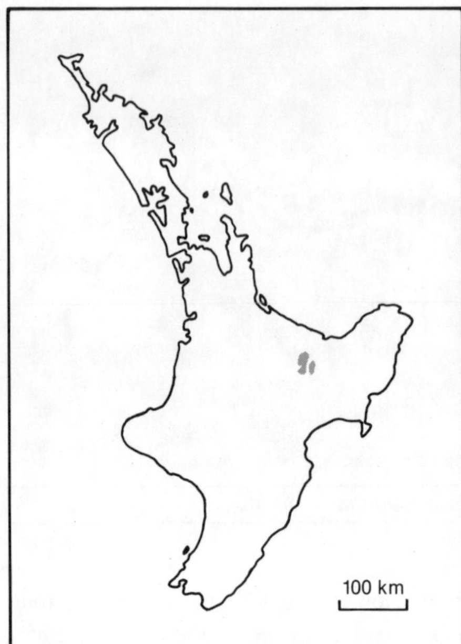

Map 42 Distribution of rusa deer in New Zealand.

NEW ZEALAND (Map 42). The main herd is found south-west of Rotorua within the Galatea and Waiohau valleys, the foothills of the Ikawhenua Range, and the headwaters of the Whakatane River. Several small herds have become established around Ruatoki North and Kutarere, in the Waimana catchment, and around Minginui and Ruatahuna (Allen, 1976; NZFS, unpubl.).

HABITAT

The main habitat now used by rusa deer provides thick cover on a warm aspect adjacent to open grazings (e.g., swamps, river flats or open forest clearings). There are three major habitat types in the range of the main Galatea herd (Allen, 1976). (1) A scrub hardwood forest type on moderately steep slopes, where the predominant cover is rewarewa (*Knightia excelsa*), manuka (*Leptospermum scoparium*), three-finger (*Pseudopanax colensoi*), and *Gaultheria antipoda,* interspersed with bracken fern (*Pteridium aquilinum* var. *esculentum*), and adjacent to farm pasture. (2) Stands of bracken fern and manuka, backing on to the edge of exotic forest plantations on the western side of the Rangitaiki River, and fronting onto farmlands and grassed firebreaks. (3) A three-tiered forest habitat, common throughout the Urewera ranges, consisting of red beech (*Nothofagus fusca*), northern rata (*Metrosideros robusta*) and tawa (*Beilschmiedia tawa*) in the upper tier; mahoe (*Melicytus ramiflorus*) and tawa in the second tier; stinging nettle (*Urtica ferox*), rewarewa and small-leaved coprosmas (*Coprosma* spp.) in the understorey; and hook grass (*Uncinia uncinata*) and crown fern (*Blechnum discolor*) covering much of the forest floor.

Food

Rusa prefer mixed grass sward communities for grazing (Van Bemmel, 1949; Mahood, 1978; Douglas, 1982). The Galatea herd makes heavy use of farm pasture (Allen, 1976) and root crops such as carrots and swedes (Roberts, 1968). Within scrub hardwood forest they browse on three-finger, bracken tips, flax tips and heads of young manuka; in clearings, they graze on clover (*Lotus major*) and new growth of stinging nettle and hook grass (Reiher, 1963; Allen, 1976).

Social Organization and Behaviour

Rusa are wary, semi-nocturnal animals, and spend much of the day lying-up within thick cover or sunning themselves on ridge clearings. For most of the year they live in family groups or small herds, on relatively small home ranges, depending on the distribution of suitable cover and feeding areas.

During the rut, dominant stags collect a harem of hinds and roar mainly during the night (Allen, 1976). The roar is similar to that of a red deer (p. 446) but short, low and deep in tone. Both stags and hinds have been observed to wander in small groups at this season (Allen, 1976). Stags also wallow more frequently during the rut, and often entangle bundles of vegetation on their antlers while wallowing or antler-thrashing trees and scrub. This behaviour seems to be deliberate in rusa in New Zealand (Roberts, 1968), in Australia (Mahood, 1980), and on Mauritius (Douglas, 1982), and it may function as an intimidation display. A confrontation between evenly matched stags, not resolved by deterrence, develops into a vicious fight in which one or both may be severely wounded. Stags cast their antlers in December and January; the new antler growth is complete and cleaned of velvet by May (Allen, 1976).

Reproduction and Development

The rut usually starts in mid July and continues into August, and fawns are born during March and April (Roberts, 1968). The gestation period is about 240 days (Douglas, 1982), and twin fawns are rare.

Population Dynamics

The density, age structure, survival, mortality, sex ratio, etc. of rusa in New Zealand are unknown. For historical changes in dispersal rate, see p. 484.

There is some possibility that the New Caledonian rusa brought to New Zealand had already been exposed to hybridization with sambar. Both rusa and sambar were liberated on New Caledonia (Harris, 1971a) before the rusa that eventually arrived in New Zealand were bought and transported, and hybridization between these two species has been recorded in captivity (Slee, 1984; van Mourik & Schurig, 1985). At present there is no reliable information on the possible hybridization between sambar and rusa either within the colonizing stock or between the established New Zealand populations where their ranges overlap.

Predators, Parasites and Diseases

Rusa have no predators in New Zealand except man, and no known diseases, but they are liable to pick up the cattle tick *Haemaphysalis longicornis* (Tenquist

& Charleston, 1981) and the cattle lungworm *Dictyocaulus viviparus* (Wilson & Collier, 1981), presumably from farm pasture.

ADAPTATION TO NEW ZEALAND CONDITIONS

The body size and antler quality in the Galatea herd are lower than in the rusa of Australia, Mauritius and particularly New Caledonia, but basically the New Zealand population does not differ greatly from either the original or the other introduced populations.

Hunting of rusa in New Zealand is intensive and unrestricted, and rusa here are relatively silent. In Mauritius, where hunting is restricted to a short open season, rutting stags roar more frequently, and both hinds and fawns are extremely vocal, particularly while on the move and feeding.

SIGNIFICANCE TO THE NEW ZEALAND ENVIRONMENT

The detrimental effects of rusa browsing within the first two habitat types listed on p. 486 are minimal. Recreational and venison hunters became so successful at maintaining control of the Galatea population that eventually the local farmers attempted to protect at least those animals living within the bounds of individual farms. However, the present extension of the distribution of rusa has taken them into the interior forest (the third habitat type listed), which is more susceptible to overbrowsing (Allen, 1976). Furthermore, much of this forest habitat type is within the Urewera National Park, and under the National Parks Act 1980, all introduced fauna is, on principle, unwelcome in all protected natural areas. Some reassessment of the biology of rusa deer, and their relationship to the habitat they occupy in New Zealand, is now required.

M. J. W. D.

Genus *Axis*

The four species of *Axis* are all Asiatic; *A. axis* and *A. porcinus* live in India and Sri Lanka (*porcinus* extending into Burma and Thailand), and the other two are confined to islands.

43. Axis deer

Axis axis (Erxleben, 1777)

Synonym *Cervus axis* Erxleben, 1777.

Also called chital or spotted deer.

DESCRIPTION (Fig. 82)

Distinguishing marks, Table 78 and skull key, p. 24.

The coat is reddish brown, spotted all over with white in both sexes, at all seasons, and all ages. A dark stripe runs from the nape to the tip of the long pointed tail and is bordered on each side by one or two rows of white spots along the back. These spots often merge into a continuous streak on the lower flanks. The chin, upper throat, insides of the ears, inner sides of the legs, and underside of the tail are all white. The head is darker brown on the face, and with a black band across the broad muffle. The male has no mane, and usually no tusks. The preorbital gland is small relative to that of the red deer; the hind

Fig. 82 Axis deer (C. Andrew Henley/ NPIAW).

feet have a well-developed pedal gland; the metatarsal gland is concealed by long hairs. The antlers are three-tined and slender (Fig. 73).

FIELD SIGN

Indistinguishable from that of other deer.

MEASUREMENTS

Bucks stand about 915 to 965 mm at the shoulder and weigh about 68-82 kg, exceptionally 110 kg (Wodzicki, 1950). The antlers reach a length of about 760 mm; the record is 1016 mm (Riney, 1955b).

HISTORY OF COLONIZATION

In 1867, seven axis deer from Melbourne (originally imported to Australia from Calcutta) were liberated by the Otago Acclimatisation Society in the Goodwood Bush, between Oamaru and Palmerston. Within 10 years they were believed to number around 100 and were beginning to cause damage to crops. By 1890-95, they had been exterminated by settlers (Donne, 1924).

Around the turn of the century, when the then Tourist Department was actively encouraging the liberation of new game animals, further attempts were made to establish axis deer (of unknown origin) on Kapiti I. (one pair, 1893); in Tongariro National Park south of Lake Taupo (five deer, 1907); and on the shores of Dusky Sound (five or six deer, 1908). The Kapiti group had increased to only three when, in 1906, the first fulltime caretaker was appointed to the island (declared a specially protected sanctuary for native wildlife in 1897), and all were shot. The Tongariro group died out. The Dusky Sound group was the only one to establish, but the animals did not thrive and did not spread far from the liberation site. They did survive at least 40 years, since an axis stag was shot in the Dusky Sound area in April 1948 by K. W. Dalrymple (Wodzicki, 1950); but there have been no confirmed records since. It seems likely that the axis deer disappeared from New Zealand some time during the 1950s or 1960s (Tustin, 1973).

DISTRIBUTION

WORLD. Axis deer are native to India, Nepal and Sri Lanka. They have been introduced to USSR, Yugoslavia, USA, Argentina, Brazil, Uruguay, Australia, the Andaman Is., and Hawaii, as well as New Zealand.

NEW ZEALAND. The longest-established herd occupied a small area around the head of Dusky Sound, but is now probably extinct. Their exact location is unknown, but the general area is shown in Map 45.

SIGNIFICANCE TO THE NEW ZEALAND ENVIRONMENT

Practically nothing is known about the biology of the axis deer in Fiordland, or anywhere else in New Zealand. Although they are regarded overseas as among the most beautiful of deer, they seem to have attracted little interest here — certainly less than the moose, whose history in New Zealand has been similar (p. 515). Their impact during their brief sojourn here was presumably slight, except only in the Goodwood Bush, the sole location where, temporarily, they reached sufficient numbers to cause any damage.

C. M. K.

Genus *Dama*

Dama is a medium sized deer inhabiting woodland and woodland edges, native to southern Europe, Asia Minor, and possibly northern Africa. It has only one species, with two subspecies.

44. Fallow deer

Dama dama dama (Linnaeus, 1758)

Synonyms *Cervus dama* Linnaeus, 1758; *Cervus* (*Dama*) *mesopotamicus* Brooke, 1875.

All individuals established in the wild in New Zealand are *D. dama dama,* the west European fallow. However, in November 1985, two male and one female Mesopotamian fallow (*D. d. mesopotamica*) were imported, and a number of *dama* × *mesopotamica* hybrids have since been successfully reared on an Auckland deer farm (Otway, 1986; G. W. Asher, unpubl.).

DESCRIPTION (Fig. 83)

Distinguishing marks, Table 78 and skull key, p. 24. Also, see front right-hand endpaper for colour plates.

The fallow is one of the smaller deer species in New Zealand, reaching (in adult males) a shoulder height of 900-1000 mm. The forelegs are slightly shorter than the hind legs, so the line of the back slopes forwards. Yearling males have unbranched "spike" antlers up to 250 mm long, while those of older males are branched, up to about 700-800 mm long and with the upper half of the main beam flattened into a palm up to 200 mm wide (Fig. 73). Males have no mane, but a prominent larynx, and a conspicuous tuft of hair on the penis sheath, which also has a scent gland. Both sexes have long pointed ears, long tail hair, and preorbital, metatarsal and rear interdigital scent glands.

Fig. 83 Fallow deer, black phase (NZFS).

Table 85: Body measurements of fallow deer in New Zealand.

Body measurements of adult Blue Mountains fallow in 1972 (from Baker 1973)

Measurement	Male			Female		
	mean	SD	n	mean	SD	n
Tail length (skeletal) (mm)	188.5	19.5	22	157.5	15.6	46
Ear length (mm)	142.9	10.3	23	131.5	4.0	41
Hind leg length (mm)	415.4	19.4	16	370.0	17.6	40
Shoulder height (mm)	888.8	41.0	20	778.0	30.1	42
Girth (mm)	923.6	60.4	18	792.3	50.0	40
Body length (including tail vertebrae) (mm)	1 573.1	90.0	23	1 386.8	99.3	42
Carcass weight (minus viscera & head) (kg)	35.9	7.3	28	21.7	2.7	56

Liveweight of farmed fallow (kg)

		Mean	SD	n	Reference
Females	at birth	3.5	—	59	Asher & Adams (1985)
Males	at birth	3.9	—	62	Asher & Adams (1985)
	at 1.2 years	49.8	1.5	8	Asher (1985a)
	at 2.2 years	57.5	4.5	14	Asher (1985a)
	at 3.2 years	63.3	5.1	12	Asher (1985a)
	older	79.6	7.0	30	Asher (1985a)

Lower jaw lengths (mm) of adult fallow from several populations

Location	Male			Female			Reference
	mean	SD	n	mean	SD	n	
England (8 localities)	200	—	70	191	—	50	Chapman & Chapman (1975)
Blue Mts (1971)	200	7.2	41	183	4.5	75	Baker (1973)
Blue Mts (1982–86)	203	7.1	43	188	4.5	104	G. Nugent (unpubl.)
Mt. Arthur (1985–86)	197	4.8	15	—	—	—	G. Nugent (unpubl.)
N. Island farms (1984)	204	1.3	4	192	4.1	69	G. Nugent (unpubl.)

Chapman and Chapman (1975) recognized four main colour phases (black, common, menil and white). (1) In New Zealand, black phase fallow are the most common (Egan, 1969, 1970). They have a browny-black back, rump and tail, and a paler grey-brown underside and neck. The dark colouring readily distinguishes black phase fallow from other deer species in New Zealand, although the duller and browner winter coat could conceivably be confused with that of sika deer (Taylor-Page, 1959). (2) The so-called common colour phase animals, the characteristic type of English fallow, are relatively rare in

New Zealand (Egan, 1969, 1970; Baker, 1973). In summer, the sides and back are a light red-brown with conspicuous pale or white spots and a black dorsal stripe extending down the tail. The underside is pale brown fading into white, and the rump patch is white with a partial black border. In New Zealand, these red-brown animals are frequently referred to as "Spanish" fallow. In winter, however, the red-brown colour is lost and the sides and back become a grey-black colour, similar to that of the black phase except that the white underside and rump patch are retained. (3) The menil phase is basically a paler version of the summer common-coloured fallow without the dark winter coat. The "yellow" animals referred to by Egan (1969) were probably menil phase fallow. (4) The white phase is, in fact, usually a very pale brown or cream, at least in young animals; older ones may be almost pure white. Any or all of the colour phases may coexist indefinitely within the same herd.

Fallow are agile and fast, capable of speeds up to 65 km/hour when pursued (Chapman & Chapman, 1975) and can clear obstacles over 2 m high. When hunted, they are extremely wary and secretive. In the Blue Mountains, for example, only one deer was seen by hunters for every seven hours hunted in 1984, even though the average density of deer exceeded one/15 ha, and seven deer were seen briefly for each one actually shot (Nugent, 1985). Fallow spend much of the day lying-up in heavy cover, and feed mainly at dawn and dusk. When disturbed, they frequently travel only a few hundred metres before stopping, or circling around behind the disturbance.

Dental formula $I^0/_3$ $C^0/_1$ $Pm^3/_3$ $M^3/_3$ = 32.

FIELD SIGN

The hoofprints of adult fallow does measure about 500-600 mm × 300-400 mm. When walking, the print of the hind hoof usually overlaps that of the fore, and the two halves of each hoof come neatly together at the tip. In prints made at speed, the tips are splayed, and, in soft ground, may show the impression of the dew claws. The pellets are smaller than those of red deer and usually separate, although all the pellets in a group may be adherent; there were an average of 52 ± 22.8 (SD) pellets in 144 pellet groups deposited by an Otago herd of eight farmed fallow of mixed age and sex (G. Nugent, unpubl.). Pellets, tracks and browsing sign of fallow cannot readily be distinguished from those of other similar sized deer, but cast palmated antlers are a reliable sign. In April and May, fresh "scrapes" in the earth, and shrubs severely battered by antler rubbing or thrashing, are commonly seen where mature stags have established rutting territories (Taylor-Page, 1959). Well worn "play rings", usually around an old stump, are also evidence that fallow are present (Cadman, 1966).

MEASUREMENTS

See Table 85.

VARIATION

Black, red-brown and white fallow have been reported in varying proportions in wild herds in New Zealand, though the white and menil phases are always rare (Egan, 1969, 1970; Baker, 1973). Smith (1980) ascribes most of the colour variation to three genetic loci; the white variety has recessive alleles at all three. There is also considerable additional variation within each of the four major colour phases.

Apart from the obvious variability in coat colour and, to a lesser extent, in antler form (e.g., Egan, 1969, 1970; Clarke, 1976b; Smith, 1980), fallow appear to have less natural genetic variation than other deer species. Pemberton and Smith (1985) did not detect any polymorphism at 30 loci in 794 fallow from 37 locations in England and Wales. They suggested that the lack of variation in most characters was due to population bottlenecks, as herds were partially domesticated during their transfer to new areas by man. By contrast, coat colour and antler form are the type of characters on which artificial selection operates (at the expense of other variability) when species are domesticated (Zeuner, 1968).

Most, if not all, wild fallow in New Zealand are descended from English stock. The lower variability in the live weights and oestrous cycle lengths of farmed fallow deer compared with red deer (Asher & Adams, 1985), and the lower coefficients of variation in adult jaw lengths of Blue Mountains fallow (males 3.5% and females 2.4%; Baker, 1973) compared with well-conditioned red deer in Westland (males 4.1% and females 2.9%; Challies, 1978) also suggest that New Zealand fallow have a limited genetic base. There is a legend that the small Te Puke herd includes stock from a private liberation of fallow from the Spanish Pyrenees (Egan, 1969). This has never been authenticated, but, if true, makes that herd the most likely source of any genetic variation.

The apparently limited genetic variability makes it likely that most variation in body size, both within and between herds, is due to nutrition. For example, the increase in average jaw length of adult Blue Mountains fallow between 1971 and the 1980s (Table 85) followed a decrease in deer density and an increase in food supply. Body size may also vary seasonally, as adult males lose up to 25% of their live weight during the rut (Asher & Kilgour, 1985).

HISTORY OF COLONIZATION

There were at least 24 successful liberations of fallow in New Zealand between 1860 and 1910 (Challies, 1985a). Some initially successful populations later died out, or merged. In the 1980s, 13 discrete wild populations remain (Table 86); the histories of the nine largest surviving herds are as follows.

(1) *South Kaipara herd.* The owner of the South Kaipara Head Station in 1900, Alfred Buckland, was a foundation member of the Auckland Acclimatisation Society, and introduced fallow for sport and meat; in the early years venison was sold to station employees (Lane, 1982). In the 1920s hunting by invitation was instituted under licence, and there was also some culling because of damage to crops. All protection was removed in 1930. After 1939, the South Kaipara Head estate was subdivided into small farms. About 1955, the owner of the largest block, Waioneke (880 ha), started a conservation programme for fallow. The herd flourished, and by 1969 up to 1700 deer were present on private farms. The herd produced the top New Zealand fallow deer trophies for five of the six years to 1971 (Lane, 1982). Since the development of

commercial hunting and of deer farming during the 1970s, the numbers of wild fallow have fallen (Spiro, 1979). Most of those left are found in Woodhill State Forest.

(2) *Matamata herd.* This herd was established from English stock in 1877. Thomson (1922) stated that the "herd has increased very largely, and is noted for the fine heads of the stags", and reckoned that over 1000 deer had been harvested. Since then, gradual habitat loss through land clearance has greatly reduced the herd; the small residual population lives mainly on private land.

(3) *Te Puke herd.* Established by the transfer of stock from Motutapu I. in 1884 and augmented by additions from Otago in 1908, this population was also initially successful, but has decreased because of habitat loss and intensified hunting pressure. Live trapping for farm stock has further reduced numbers in recent years, and only small numbers remain. Unlike the predominantly black phase fallow populations elsewhere in New Zealand, about half the Te Puke herd are of the "Spanish" strain (Egan, 1969).

(4) *Wanganui herd.* This herd developed from two liberations, one well up the Wanganui River in 1877 (Thomson, 1922; Donne 1924), and the other at Featherstone Stream, Kaiwhaiki, later in the 1880s (Walton, 1986). The growth of the herd was apparently slow to start with, although the first fallow shooting licences were issued before 1889. Further attempts to add to the herd quality were not successful until 1935, when a "Spanish" strain doe from the Te Puke herd was introduced (Walton, 1986). By the 1920s, the herd was "too numerous", and several hundred animals were removed (Donne, 1924). All protection was lifted in 1930. However, the herd continued to spread, and at its peak occupied an area of 60 × 40 km. An isolated group of fallow to the north suggests that the spread of this herd was probably assisted by man (W. J. Simmons, unpubl.). From the 1960s onward numbers declined to very low levels on public land and on many farms. On other farms, however, the deer are protected, and some high density pockets remain (P. Butler, unpubl.).

(5) *Mount Arthur herd.* This herd was established by translocation of an unknown number of deer, probably from Otago. There were two liberations, one at Barrons flat near Takaka, and the other at Flora Saddle. It reached peak densities in the 1940s, but declined soon after (Clarke, 1976b).

(6) *Aniseed Valley herd.* This herd is derived from an original liberation of English stock in 1864, plus additional animals transferred from the Wakatipu herd in 1906. It may have reached peak densities as early as the 1920s, after which intensified hunting kept numbers in check, although perhaps also stimulating dispersal (Clarke, 1976b). The range of the herd has not changed in recent years.

(7) *Paparoa herd.* Red deer were absent from the Paparoa Range in the late 1880s, when fallow were translocated there from Motutapu I.; so this herd increased quickly and expanded north. Peak numbers were reached between 1925 and 1935, and persisted till the 1950s (Clarke, 1976b). By the 1980s, fallow remained only in pockets along the eastern flank of the Paparoa Range, and densities were low and static (Morse, 1981).

Table 86: History, location, and status of the 13 herds of wild fallow deer in New Zealand.

Herd	Liberation date(s)	Liberation site(s)	Origin	No. liberated	Present habitat and status	Reference
(1) South Kaipara	1900	Lake Otoatoa	Motutapu I.	7	Sand dune/pine forest/ shrubland; population estimated at *c.*800 in 1985	Lane (1982); Broome (1985); Deuss (1983a)
(2) Kaikoura I.	?	?	?	?	Perhaps up to 300 wild fallow in grassland, manuka, and gorse in the early 1980s.	NZFS (unpubl.)
(3) Rangitoto I. and Motutapu I.[1]	1860	Motutapu I.	Wales (?)	4 (?)	Although about 1000 present before 1920, the population is now near extinction, and confined to lowland coastal forest on Rangitoto.	Donne (1924); Thomson (1922); J. W. A. Dyer (unpubl.); K. Purdon (unpubl.); A. H. Leigh (unpubl.)
(4) S. Manukau Head	*c.*1910	Rangiriri Point	Onehunga Zoo	2	Rare, in scattered shrubland around private farmland.	N. Douglas (unpubl.)
(5) Matamata	1877	Matamata Range Maungakawa Range	England	18	Low density in lowland podocarp hardwood forest; 10–20 kills p.a. reported from Te Miro Reserve.	Donne (1924); Ashby (1967)
(6) Te Puke	1884 1908	Tauranga ?	Motutapu I. Otago	?	Moderate to low densities in cut-over lowland forest, and on farmland; probably <100 killed or captured annually.	J. W. A. Dyer (unpubl.); M. C. Llewellyn (unpubl.)

(7) Wanganui	1877	Wanganui	England	10	Low to moderate densities scattered in pockets over a wide area, including farmland, lowland forest, scrubland and pine forest; probably the largest remaining population.	Donne (1924); Walton (1986)
	1880s	Featherstone Stream	?	7–8		
(8) Mt. Arthur Range	*c.*1908	Barrons Flat	Otago, or Motutapu I.	?	Low densities in beech forest, and forest margins; 73 kills reported in 1982–83.	Clarke (1976b)
	1910	Flora Saddle	Otago	?		
(9) Aniseed Valley	1864	Aniseed Valley	England	3	Scrubby vegetation backing on to beech forest, some on private farmland; 30–40 reported kills p.a. in 1980s.	Clarke (1976b)
	1906	Teal Valley	Greenstone	5		
(10) Paparoa Range	*c.*1887	Doolan Creek	Motutapu I.	7	Low density, in dense rain forest and farm margins.	Donne (1924); Clarke (1976b)
(11) South Canterbury	*c.*1878	Albury Park	Tasmania(?)	?	Never abundant, but moderate densities in some small protected areas, mainly in scrubby gullies on farmland.	J. A. Anderson (unpubl.); Davidson (1965); Christie and Andrews (1966)
(12) Wakatipu	1887	Caples, Greenstone Valleys	Tasmania	18	Severely culled since 1970s, in beech forest and grassy valley flats; now <50 kills p.a. reported.	?
	*c.*1900	?	Otago (?)	?		D. Flook (unpubl.)
(13) Blue Mountains	1869	Tapanui	Tasmania	12	Total about 2000. Beech and pine forest, and scrubby farm margins.	Baker (1972)
	1871	Tapanui	England (?)	1		

1. Joined at low tide.

(8) *Wakatipu herd.* The origins of this herd are not certain, but the foundation stock was probably supplied by the Otago Acclimatisation Society from the Blue Mountains. By the 1930s, a forest ranger, D. Dunn, reported extensive damage to the indigenous forest, and that "not a vestige of feed remained" (Rose, 1977). Subsequent culling (including poisoning trials; Daniel, 1966b) may have reduced the herd slightly, but the greatest reduction was achieved after the commercialization of hunting.

(9) *Blue Mountains herd.* The original animals were imported from Tasmania in 1869, with the intention of establishing an English-style deer park at Tapanui. The deer became very numerous, and by 1899 had dispersed into the adjacent Blue Mountains. By 1903, over 2000 fallow bucks had been shot by licensed stalkers, including many which yielded large trophies (Donne, 1924). The herd increased further, causing damage to farm crops and eating out the understorey of the indigenous forest. In 1923 all protection was removed from the herd, and in 1926, 3890 deer tails were collected by NZFS from the Blue Mountains area (Baker, 1972). Despite continued heavy hunting pressure and poisoning trials, the population remained high until the 1950s and was generally in poor condition (Wodzicki, 1950; Shearing, 1965; Daniel, 1967; Baker, 1973). After 1960 hunting pressure increased, and the reported kill from the state forests making up the core of the range declined from 2000 per year in the early 1960s, to around 400 in 1984 (Nugent, 1985).

There were also some unsuccessful liberations. Fallow disappeared after release at Cairn Bush (Otago) in 1899 (Wodzicki, 1950); in the Morven Hills (Otago) in 1867 (Baker, 1972); and at Hawarden (Canterbury) in 1882 (Lamb, 1964). Two fallow released on Kapiti I. in 1900 either died out or were shot when the island was made a sanctuary (Wodzicki, 1950). Fallow deer liberated near Whangamoa (Nelson) were exterminated about 1912 because they damaged crops (Clarke, 1976b). A herd established on Kawau I. by Sir George Grey in 1862-64 became extinct about 1966 (K. Purdon, unpubl.). A small herd near Coromandel, derived from releases around 1900 near Thames, is reported to have died out after 1963 (K. Purdon, unpubl.), but there are persistent rumours that wild fallow still survive there (G. W. Asher, unpubl. and Map 37). Christie and Andrews (1966) refer to a northern Auckland population, and Wodzicki (1950) states that fallow were present in Herbert State Forest, north of Dunedin, but there are no recent reports of fallow in these areas.

The total number of fallow imported before 1910 was probably only 50-60 animals, so many of the liberations involved translocations of stock from elsewhere in New Zealand. All the documented importations came from England, Wales or Tasmania, and the Tasmanian fallow herds were themselves probably all derived from English stock (Bentley, 1967; Chapman & Chapman, 1980). Since the advent of deer farming in the 1970s, farmers interested in livestock improvement have begun a second wave of importations, including stock from outside Britain (p. 490; Otway, 1986; Massey, 1987).

DISTRIBUTION

WORLD. Fallow are indigenous to countries bordering the Mediterranean

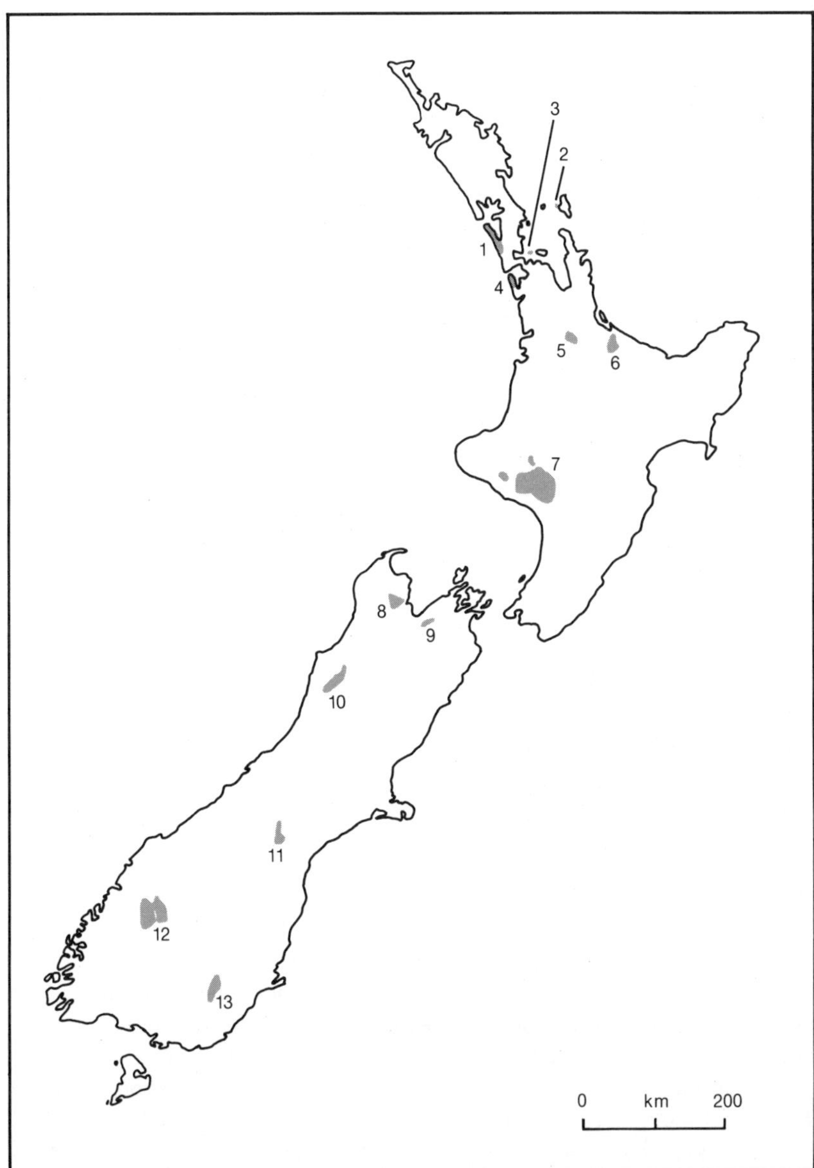

Map 43 Distribution of fallow deer in New Zealand. Herds numbered as in Table 86.

(Lister, 1984). They have been introduced to all continents (38 countries) and are now among the most widespread of ungulates. The most complete summaries of their present world distribution are provided by Chapman and Chapman (1980) and Lever (1985).

NEW ZEALAND (Map 43, Table 86). Fallow are the second most widespread deer species in the wild in New Zealand. The 13 wild populations are scattered from north of Auckland to Southland (latitudes 36° to 46°S). The herd on

Rangitoto-Motutapu Is. (the two islands are joined at low tide) is near extinction, and the South Manukau, Kaikoura I. and South Canterbury (Burkes Pass–Two Thumb region) herds are small, and, although wild, almost totally confined to private farmland. The remaining nine herds are truly wild and well established on both public and private lands as follows.

(1) *South Kaipara*: Wild fallow are virtually restricted to Woodhill State Forest (14 700 ha), although there are thousands of fallow on nearby farms (Spiro, 1979).

(2) *Matamata*: Fallow are present in the Maungakawa and Matamata ranges (Te Miro and Te Tapui Scenic Reserves) and on private land around Whitehall (S. Kelton, unpubl.).

(3) *Te Puke*: Oripi and Otanewainuku State Forests, plus Otawa and other Scenic Reserves, in patches of Crown land and on private farmland (M. C. Llewellyn, unpubl.).

(4) *Wanganui*: From the Waitotara River to the Whangaehu River, and northwards on both sides of the Wanganui River beyond Jerusalem, including Lismore State Forest; also south-east of the Waitotara Forest in Taranaki, in Waitotara, Tarere, and Kapara State Forests, and Patokino Reserve (S. D. Kelton & W. J. Simmons, unpubl; J. Barkla, unpubl.).

(5) *Mount Arthur*: Most of the Mount Arthur Tablelands region, the Grecian and Flora catchments, and the lower Cobb valley (Hayward, 1983).

(6) *Aniseed Valley*: Bryant Range, Pelorus River valley, Mount Richmond and extending into Marlborough (Clarke, 1976b).

(7) *Paparoa Range*: From Blackball to the Inangahua Landing, mainly on the eastern flank of the Paparoa Range (Morse, 1981).

(8) *Wakatipu*: Greenstone and Caples valleys, and Mount Creighton; plus a few in the Moonlight and adjacent catchments in the Shotover region, on the eastern foothills of the Humboldt Mountains, and occasionally in the Earnslaw, Rees and Dart catchments (Cuddihy & Ross, 1979).

(9) *Blue Mountains*: Beaumont, Rankleburn and Tapanui State Forests, and on private farmland around the south-east and northern margins of these. Also in pockets of bush on the eastern bank of the Clutha River, and in Glendhu State Forest, 15 km to the north-east (N. Hills, unpubl.).

In addition, fallow may legally be farmed in most rural areas of the North and South Islands, and many farms are well outside the present range of wild fallow. It is inevitable that escapes from farms or illegal liberations will extend the distribution of wild fallow deer; such incidents have already been reported from near Oxford (10 deer) and on Banks Peninsula (2 deer) in Canterbury.

HABITAT

Fallow occupy a wide range of habitats, ranging from improved pasture on farms to dense rain forest on the west coast of the South Island (Table 86). Most wild fallow live in indigenous forest or shrubland, especially around the scrubby margins of farms whose owners provide some protection from hunting. They also occupy commercial plantations (Lismore, Woodhill, Blue Mountains). Fallow in the South Island formerly made some use of the subalpine tussock zone in the Mount Arthur, Blue Mountains, Paparoa and Wakatipu areas, but the species always appeared to be more at home in the low altitude forests, and the low density populations of the 1980s seldom venture above timberline (e.g., Buckingham, 1985).

FOOD

Fallow eat a wide range of plant species, including most, if not all, of those eaten by domestic stock. In the wild in New Zealand they prefer the same indigenous plants as other deer (see red deer, p. 444, and Wodzicki, 1950), their diet depending on which of these is locally available. They have a high ratio of rumen volume to body weight compared with other cervids (Hoffman, 1985). Hoffman suggests that this reflects a greater adaptation to non-selective grazing of short sward grasses and herbs than in other cervids, which tend to concentrate more on selective browsing of trees and shrubs. Around farm margins and on deer farms, fallow readily graze introduced grass and herb species, but they can also thrive in indigenous forests even where grasses are scarce.

A study of fallow deer diet in the Blue Mountains between 1982 and 1984 (G. Nugent, unpubl.) showed marked differences in diet between habitat types. In commercial pine forest, where deer densities were low and introduced plant species common, the five most important summer foods (in decreasing order of importance) were grasses, herbs, fungi (mainly mushrooms), broadleaf (*Griselinia littoralis*) and small-leaved *Coprosma* species. In winter, broadleaf became more important and fungi less important, but grass was still by far the most common food. Fallow also ate small quantities of introduced weeds such as gorse (*Ulex europaeus*), broom (*Cytisus scoparius*), blackberry (*Rubus fruticosa*) and Himalayan honeysuckle (*Leycesteria formosa*). Woody browse appeared to be an important dietary requirement; even on lush pasture fallow still occasionally ate gorse and broom. None of the planted conifers was touched.

In silver beech forests (*Nothofagus menziesii*), the important summer foods were fungi, broadleaf, lichen (mainly *Usnea* spp.), *Coprosma* spp. and grasses; the main winter foods were broadleaf, *Coprosma* spp., *Carpodetus serratus* and *Pseudopanax* spp. The heavy reliance on fungi and lichen reflected the scarcity of palatable foliage within the reach of deer, as most developing seedlings of preferred species were eliminated at an early age. Most of the leaves eaten had fallen from the canopy, especially in winter when they were knocked down by snow and wind.

SOCIAL ORGANIZATION AND BEHAVIOUR

Undisturbed fallow tend to be gregarious, with a highly developed social tolerance (Massey, 1985). The Wakatipu herd, in particular, was well known for the large groups (100+), containing all age and sex classes, that grazed the valley flats in the 1950s. However, heavy hunting during the 1960s and 1970s reduced the herd so much that groups of more than three or four became rare by the 1980s (R. Martin, unpubl.).

Home range size depends to some extent on the distribution of resources, but is often <100 ha (Chapman & Chapman, 1975). Ranges are not exclusive; those of males tend to be larger than those of females, and divided into separate summer and rutting ranges.

Fallow deer have a variety of vocalizations. During the rut, males groan repeatedly and loudly to advertise their location. When startled, females, in particular, often bark loudly several times, at about 10–20-second intervals. Does with young fawns may produce a medium high-pitched bleat, and deer of any age may accompany submission postures with a higher-pitched mewing sound. Fawns make a different high-pitched peeping sound when distressed, or to contact their dams. Younger deer, in particular, may scream when wounded or captured.

Fallow also communicate with each other visually. When mildly disturbed, one member of a group may adopt an "alert" posture, with the neck extended vertically and the body held rigid, and the rest of the group will often follow suit even if unaware of the danger. They may also stamp their hooves, and walk stiff-legged towards the potential danger, usually holding the tail upright. Fallow may flee a mild disturbance with a slightly exaggerated rhythmic trot, or a bouncing gait in which all four feet leave the ground simultaneously, but they usually flee a sudden disturbance at the gallop, often with a low "woof" of alarm.

The rut begins in April. Males lose interest in food and their necks swell. They establish rutting territories by making several scrapes in the ground with their forefeet, and urinating in them. Surrounding bushes are thrashed and scent-marked with secretions from the preorbital glands. Rutting males on farms may groan for several weeks, but those in hunted populations often restrict groaning to one or two days at the peak of the rut (R. Webb, D. Harrison, unpubl.). In the Blue Mountains, Baker (1973) heard groaning periodically between 31 March and 17 May, mostly between 22 April and 6 May.

Antlers are cast in October or November, and the development of new ones is usually complete by February (Riney, 1954).

It is generally accepted that females are attracted to the rutting stand by the presence of the male. They are not usually herded up into a harem as in other deer (Chapman & Chapman, 1975), except on farms (Anon., 1984c). In high density undisturbed populations, males do not usually breed until at least four years of age, because subordinate bucks are excluded from mating by the older bucks. However, most of the heavily hunted wild populations in New Zealand now contain few such fully mature bucks (Baker, 1973), so even yearling males may get the chance to breed. Mating behaviour is described in detail by Asher and Kilgour (1985).

Reproduction and Development

Fallow reproduction in New Zealand is described by Asher (1985b) and Baker (1973). Does usually breed for the first time at 16 months of age, although it is possible for fawns to conceive in their first year of life (Ueckerman & Hansen, 1968). Fallow are polyoestrous. Does may have up to six oestrous cycles during the breeding season, with a mean cycle length of 22.4 $\pm$ 1.3 (SD) days. The first oestrus of the season is synchronized within a 2-week period in late April, although young does breeding for the first time usually come into oestrus a week later than older does. Most breeding does conceive during the first cycle. The average gestation period of 88 farm does was 234.2 ± 2.7 (SD) days (Asher, 1985b), and the main peak of births was in mid December. There was a second, much smaller peak in births about 21–22 days later, and some births (2-3%) as late as March. Twinning is rare.

Sterba and Klusak (1984) and Chapman and Chapman (1975) describe the development of fallow. Newborn fawns tend to remain hidden for several weeks, except when the doe returns to feed them. They begin to take solid food almost immediately, although rumination does not begin until the second or third week after birth. Lactation may continue for up to 9 months, but many fawns are weaned at 4-5 months.

The full adult dentition is usually in place by 2.5 years of age (Chapman & Chapman, 1975). Males begin to develop antlers at 7-9 months of age and reach sexual maturity in their second year (Baker, 1973). Females appear to reach full body size by 6 years of age, whereas males may not be fully grown until their ninth year.

Population Dynamics

There are no accurate estimates of the total number of wild fallow in New Zealand, but it is unlikely there were many more than 5000 in 1985 (some details on individual herds, past and present, are given in Table 86). By contrast, there were probably about 30 000 fallow deer on farms at that time (G. W. Asher, unpubl.). Pemberton and Smith (1985) cite a 1984 world population estimate of 164 000.

Baker (1973) reports a pregnancy rate of 91.4% for 70 wild does more than 1 year old. The pregnancy rate of does in their second year is generally lower than that of older does. The sex ratio at birth is probably close to 1:1 (Baker, 1973; Sterba & Klusak, 1984; Asher & Adams, 1985).

Although fallow may live for up to 32 years (Ueckerman & Hansen, 1968), the oldest male among 351 deer shot in the Blue Mountains in 1971 was 8 years old and the oldest female 12 (Baker, 1973). In the 1980s mortality not due to hunting appeared to be rare in the Blue Mountains, as no natural mortality was observed among an average of 30 young deer during a 2-year radio-tracking study (G. Nugent, unpubl.). In unprotected herds in New Zealand most deer are shot within the first two years of life, males sooner, on average, than females (Baker, 1973; Nugent, 1985). Such mortality patterns produce a predominantly female adult population with a high annual productivity. Nugent (1985) estimated that 40 fawns were born for every 100 deer present in the Blue Mountains before the 1984 fawning season.

Fallow tend to remain close to where they were born, which explains why they were slow to disperse from liberation points; the Wanganui herd spread at 0.8 km/year (Caughley, 1963), the South Kaipara herd at 0.6 km/year (from Lane, 1982) and the Paparoa herd at 0.8 km/year (Clarke, 1976b). Most dispersal is by young males nearing the end of their first year, and typically, these move less than 3 km from their natal range (G. Nugent, unpubl. telemetry data, Blue Mountains). By contrast, shifts of up to 32 km have been recorded in red deer in the Cupola basin (Gibb & Flux, 1973).

Fallow may have a competitive advantage over red deer, as Challies (1985a) noted that red deer have tended not to establish in areas, such as the Blue Mountains and Wakatipu, where fallow had already reached high density. Kean (1959) considers that fallow can displace red deer except at high altitudes. Batcheler (1960) suggested that fallow in Scotland were better adapted to a mature forest environment than red deer.

PREDATORS, PARASITES AND DISEASES

In New Zealand the only predators of fallow are man and dogs. Although dogs are seldom used specifically to catch deer, the dogs used to hunt pigs occasionally catch them, especially fawns up to three months old (S. Cook, unpubl.).

On farms fallow are prone to facial eczema, pneumonia, ryegrass staggers and possibly leptospirosis, but yersiniosis and malignant catarrhal fever are unheard of (Massey, 1985). Free-ranging fallow are generally regarded as disease resistant and healthy (Spiro, 1979). However, Christie and Andrews (1966), Daniel (1967), Andrews (1973) and others report the incidence of a number of parasites and diseases in fallow in New Zealand, as follows. Protozoa: *Eimeria* sp. (Collins, 1981b). Trematoda: Paramphistomatidae (P. C. Mason, unpubl.). Cestoda: *Taenia hydatigena* (Sweatman & Williams, 1962) and *Moniezia* sp. (P. C. Mason & N. Gladden, unpubl.). Nematoda: *Dictyocaulus viviparus* in the lungs (Wilson & Collier, 1981), *Apteragia odocoilei, Spiculopteragia asymmetria* and *Trichostrongylus axei* in the abomasum (Andrews, 1973; P. C. Mason, unpubl.), *Capilleria* sp. in the small intestine (P. C. Mason, unpubl.), *Oesophagostomum venulosum* and *Trichuris* sp. in the large intestine (Andrews, 1973; P. C. Mason & N. Gladden, unpubl.). Acari: *Haemophysalis longicornis* (cattle tick) on the skin (P. R. Wilson, unpubl.).

ADAPTATION TO NEW ZEALAND CONDITIONS

Despite an apparently limited genetic base, fallow deer have adapted well to New Zealand conditions. When provided with adequate cover or protection, they have occupied a wide range of habitats and reached high densities in many areas. Although they are normally animals of the forest margin, fallow in the Blue Mountains and elsewhere have thrived in forested areas with few clearings. Their patchy distribution (relative to red deer) appears to be a consequence of their site tenacity and slow dispersal rate. As with red deer, the period of highest metabolic demand (fawning and lactation) starts in December, after the main spring flush of plant growth. Because of this, fallow are, at this stage, less suited to farming than sheep or cattle, particularly in areas prone to summer drought. However, crossbreeding experiments with Persian fallow now in progress (Otway, 1986) may produce an earlier-calving strain more likely to do well on New Zealand farms in future.

At the low densities prevailing in the 1980s, wild fallow in New Zealand appear to be about the same size as both farmed fallow in New Zealand and

English park fallow, since the lower jaw lengths of adults are much the same regardless of origin (Table 85). However, overseas the most frequent coat colour phase is red-brown with conspicuous spots, whereas in New Zealand (except in the Te Puke herd) it is the black phase with faint spots or none.

SIGNIFICANCE TO THE NEW ZEALAND ENVIRONMENT

DAMAGE AND CONTROL. Fallow deer have caused severe modification to indigenous vegetation in some areas (Wodzicki, 1950; Christie & Andrews, 1966; Rose, 1977; Bathgate, 1977; Hayward, 1983), although most populations live in areas already massively modified by man. Major control campaigns to alleviate damage were conducted in the Blue Mountains and Wakatipu areas (Wodzicki, 1950; Daniel, 1966b; Baker, 1972), and there has been smaller-scale control work in other areas (Donne, 1924; Harris, 1967; Lane, 1982). The general reductions in density since the 1960s have removed the need for active control.

CURRENT MANAGEMENT. The flighty nature and small size of fallow deer, and the low demand for their antler velvet, have made them less commercially desirable than red deer (Yerex, 1982). One consequence of this is that fallow have been substantially involved in recent developments relating to recreational hunting.

Much of the range of the Blue Mountains herd was gazetted as the first Recreational Hunting Area (RHA) in 1980 (under the 1977 Wild Animals Control Act), and is now managed with the needs of sport hunters in mind (Nugent & Mawhinney, 1987). Since then, recreational hunting has further decreased the deer population within the RHA, and in late 1986 restrictions on hunting were imposed (NZFS, 1986). RHAs have also been established in the Wakatipu (1981) and Mount Arthur (1981) areas, and similar hunting restrictions imposed (Hayward & Hosking, 1986).

In the North Island, the fallow population in Woodhill has been managed for recreational hunting (Lane, 1982; Deuss, 1983b; Broome, 1985), although Woodhill is not gazetted as an RHA. In Wanganui, positive management intended to build up the herd in Lismore State Forest began in April 1985 with the liberation of 21 fallow does from Blue Mountains and South Kaipara stock (Simmons, 1985; Benes, 1986). This was the first officially sanctioned release of farm deer into the wild since deer were declared noxious animals in the 1930s, and reflects the recent fundamental change in official attitudes to wild animal management in New Zealand.

The remaining herds of fallow on public land are also at low density, and few animals are harvested annually (Table 86). Legal access to these herds is usually difficult to obtain.

Farming of fallow began slowly, because of the limited availability of stock, and because fallow were more difficult to handle using conventional livestock techniques (Fitzi & Monk, 1977; Spiro, 1979; Massey, 1985, 1986). However Waioneke Park, one of the first large deer farms, demonstrated that the species could be successfully farmed (Spiro, 1979), and substantial numbers of wild fallow have been caught using pen traps, tranquillizer darts, drive nets, or helicopters (Spiro, 1979; Yerex, 1982). As with other deer, fallow have a lower fat content and higher muscle to bone ratio than sheep and cattle (Gregson & Purchas, 1985). Many farmers now believe that fallow provide a better quality venison than red deer, and fallow are becoming increasingly common on farms (Massey, 1986; Walton, 1986).

M. M. D. & G. N.

Subfamily Odocoilinae

The second of the two largest subfamilies of cervids, the Odocoilinae, has nine genera. Six are found only in the New World (five confined to Central and South America, and one, *Odocoileus*, centred in North America); one (*Capreolus*) only in Eurasia, and two (*Alces* and *Rangifer*) right across the northern Holarctic.

Representatives of two genera, *Odocoileus* and *Alces*, have established local populations in New Zealand. In addition, three individuals of a South American species (?*Hippocamelus bisulcus*) are reported to have been released at some unspecified location by the Auckland Acclimatisation Society in 1870, but died out (Thomson, 1922; Wodzicki, 1950).

Genus *Odocoileus*

The two species of *Odocoileus* are highly regarded as game animals by North American hunters, which is probably the reason that both were brought to New Zealand. Only *O. virginianus* became established. Five *O. hemionus*, the black-tailed or mule deer, were liberated near Tarawera, Hawke's Bay (not Bay of Plenty: Johns & MacGibbon, 1986:75) in 1905. The Hawke's Bay Acclimatisation Society reported them as "increasing in March 1915" (Thomson, 1922), but, for unknown reasons, the group did not survive. Another nine *O. hemionus* were reported to have been liberated in the Rees Valley, north of Lake Wakatipu, also in 1905 (Southland Acclimatisation Society, n.d.), but it is likely that these animals were really misidentified *O. virginianus* (C. N. Challies, unpubl.). There is no other record of this release, but the present Wakatipu herd of white-tailed deer originated from nine deer released in the Rees Valley in 1905. About 1876 the Auckland Acclimatisation Society imported some "Californian deer", probably *O. hemionus*, and liberated them near Mercer and in the Piako and Thames districts, but they were killed by Maori (Ashby, 1967).

45. White-tailed deer

Odocoileus virginianus borealis (Miller, 1900)
Synonym *Odocoileus americanus borealis* Miller, 1900.
Also called Virginia or Virginian deer, whitetail.

DESCRIPTION (Fig. 84)
Distinguishing marks, Table 78 and skull key, p. 24.

A small to medium sized, round-antlered deer with a plain, uniform coat. Both sexes and all ages, except very young calves, are similar in colour. The

Fig. 84 White-tailed deer in velvet, lower Dart Valley, 1969 (J. H. Johns).

winter coat is grey-brown, with long, coarse, crimped guard hairs overlying a woolly underfur; the light brown summer coat has short guard hairs and little or no underwool. Males have no mane. The underside, rump patch, chin and upper throat, and around the muzzle and eyes, are white. The tail is long, red or grey-brown fringed with white above, wholly white below. The ears are pointed, dark grey-brown outside and on the inner tip, and white inside. The muffle and hooves are dark grey-black. The young are spotted on the back and flanks and have a double row of spots extending from the neck to the base of the tail.

There are four sets of scent glands: tarsal glands on the inner hind legs, with a prominent, erectile tarsal tuft of white hair; metatarsal glands on the outer

Table 87: Body measurements of adult[1] white-tailed deer on Stewart Island, 1981-1987.

		Mean	SD	95% range	n
Live weight (kg)	M	54.3	9.5	35.6-73.0	33
	F	39.7	6.1	27.7-51.7	38
Body length (mm)[2]	M	1 598	97	1 407-1 789	36
	F	1 501	71	1 362-1 640	44
Foot length (mm)[2]	M	465	20	426-504	37
	F	442	16	411-473	50
Jaw length (mm)[2]	M	212.6	7.7	197.5-227.7	57
	F	201.9	7.5	187.2-216.6	69

1. Adult: animals of 3 years old or over .
2. Measurements described in footnotes to Table 79.
Data from C. N. Challies (unpubl.).

hind legs; interdigital glands between the hooves; and preorbital glands in the corners of the eyes. Usually the tail is carried close to the rump when the deer is standing, walking or trotting, but is thrown upright into a conspicuous white "flag" when it flees at a bounding gallop. The antlers curve forwards, with vertical tines (Fig. 73). Yearlings have spikes, two-year-olds usually four points, and mature bucks eight or more points.

Dental formula $I^0/_3\ C^0/_1\ Pm^3/_3\ M^3/_3 = 32$.

FIELD SIGN

Travelling and feeding trails of white-tailed deer have a criss-cross pattern of major and subsidiary pathways. Hoofprints and pellets are similar to those of red deer (p. 439) but proportionately smaller. Near rutting time, small trees and shrubs are rubbed and thrashed by bucks cleaning their antlers, and rutting "scrapes" (p. 469) are formed. Moulted hair is shed in bedding areas and rubbed off on vegetation; cast antlers are unmistakable.

MEASUREMENTS

See Table 87. The statistics given are from a composite sample of adult deer, three or more years old, taken on the east coast of Stewart I. in 1981, and at the southern end of the island in 1985 and 1987. Those from the south weighed on average 15% less than those from the east. All the deer were killed in mid to late summer. There are no comparable data from the Wakatipu herd.

The antlers of 20 bucks with six or more points, taken from the east coast of Stewart I. in 1981, measured: length, mean 412 mm, 95% range 306–518 mm; spread, mean 344 mm, range 228–460 mm (C. N. Challies, unpubl.).

VARIATION

Both the existing herds of white-tailed deer in New Zealand came from the same stock, but the colonizing animals responded differently to the new conditions. Those on Stewart I. bred and dispersed rapidly; those at Lake Wakatipu increased in numbers and spread only slowly.

HISTORY OF COLONIZATION

At the first attempt to establish white-tailed deer in New Zealand, in 1901, two males and two females were released in the Takaka Valley, Nelson, but this group did not survive (Thomson, 1922; Donne, 1924).

In 1905, the government imported 19 white-tailed deer from New Hampshire, USA. Nine (two males and seven females) were released at Cook's Arm, Port Pegasus (Stewart I.), and nine, sexes unknown, in the Rees Valley at the head of Lake Wakatipu, west Otago; the remaining one was added to the Takaka group (Thomson, 1922; Donne, 1924). The Lake Wakatipu herd slowly extended its range up the Rees Valley, into the lower Dart Valley, and onto the Forbes Mountains. The long-term dispersal rate of this herd was about 0.34 km/year (Harris, 1981). The Stewart I. population increased and spread northwards, reaching the northern tip of the island about 1926 (R. B. Trail, unpubl.). That required a dispersal rate of at least 2.9 km/year (Harris, 1981); Caughley (1963) gives 1.0 km/year for Stewart I. generally. By 1949 they were reported in all coastal areas of the island, and numerous in some places, e.g., Mason Bay, but less so in the Lords River area and absent from the alpine regions (Corboy, 1949; C. M. Schofield, unpubl.). They reached Ulva I., in Paterson Inlet, by 1930 (M. McArthur, unpubl.).

DISTRIBUTION

WORLD. The native range is North, Central and northern South America, and a few islands in the Caribbean (30 subspecies recognized by Baker, 1984). There is a general trend of increasing body size from south to north, and a lightening of coat colour from wetter to drier areas. White-tailed deer have been naturalized in the wild in Finland, Czechoslovakia, Yugoslavia and some previously unoccupied Caribbean islands (Baker, 1984). The subspecies introduced into New Zealand is native to the north-eastern states of USA and eastern provinces of Canada, around the Great Lakes.

NEW ZEALAND (Map 44). There are two established herds of white-tailed deer in New Zealand, separated by about 220 km. One is at the head of Lake Wakatipu, including the Rees and lower Dart Valleys and the lower slopes of parts of the Humboldt, Forbes, and Richardson Mountains. The other occupies the coastal forests of Stewart I., declining in numbers inland and scarce above 300 m asl (Harris, 1981).

HABITAT

The Lake Wakatipu herd occupies low altitude beech (*Nothofagus* spp.) forest below 600 m, and river flats, particularly those containing a mixture of tussock and introduced grasses. These forests have been substantially modified by browsing, and preferred species of shrubs and trees are rare or absent. The area, at a latitude of some 45°S, is a typical glaciated landscape with steep-sided valleys. It lies on the eastern (rain-shadow) side of the Southern Alps, and the climate varies between extremes of temperature and of dry and wet periods; the annual rainfall is about 1000 mm, and there are heavy snowfalls in winter (Bathgate, 1976).

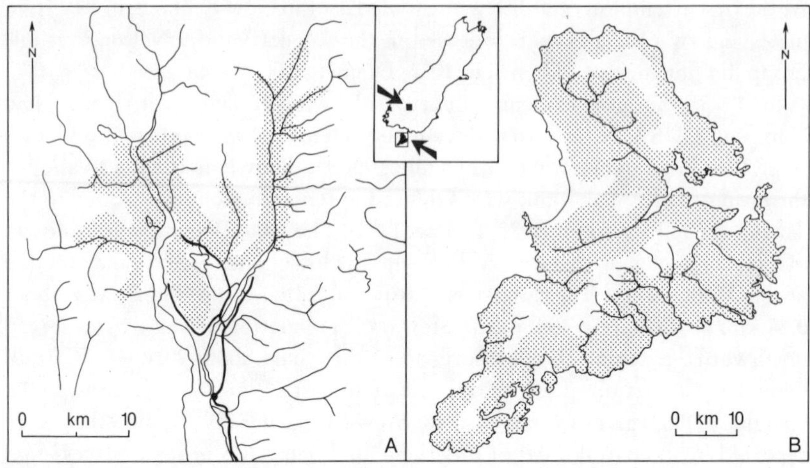

Map 44 Distribution of white-tailed deer in New Zealand in 1982, (A) at head of Lake Wakatipu, (B) on Stewart I. Adapted from Harris (1981).

On Stewart I. there is no beech forest, and the white-tailed deer there inhabit lowland mixed podocarp-hardwood forests, especially those near the coast. The canopy is predominately rimu (*Dacrydium cupressinum*) and miro (*Podocarpus ferrugineus*), often as emergents, plus rata (*Metrosideros umbellata*), and kamahi (*Weinmannia racemosa*). Beneath this is an extensive subcanopy of hardwoods such as broadleaf (*Griselinia littoralis*), *Pseudopanax simplex, P. crassifolius, Carpodetus serratus* and *Coprosma lucida,* and the tree fern *Dicksonia squarrosa.* These species often form the main canopy on wetter sites where supplejack (*Ripogonum scandens*) is also common. The only coastal areas not forested are small patches of dunes and swamps, and where the forest near the shoreline has suffered recent dieback. In those places the invading ground cover is mainly of unpalatable plants such as *Histiopteris incisa* and *Carex solandri* (Veblen & Stewart, 1980).

Stewart I. straddles the 47°S latitude line, but has a maritime climate lacking the extremes of temperature and precipitation of the Lake Wakatipu area. The annual rainfall of 1500 mm is evenly spread throughout the year, and it seldom snows.

Food

White-tailed deer are mainly browsers. They sample a great variety of trees, shrubs and, to a lesser extent, ferns, fungi and algae, but the bulk of their food is taken from only a few species. On Stewart I. broadleaf and the leaves, stems and fruit of supplejack together comprise >50% of their annual diet. Another 102 food items have been recorded, but only 18 of these formed >0.5% of the diet (Table 88; Nugent & Challies, 1988).

The Lake Wakatipu herd feeds on the understorey trees and shrubs of the

Table 88: Main foods of white-tailed deer on the north and east coasts of Stewart Island.

	% of diet		% of diet
Trees and shrubs		**Climbers**	
Griselinia littoralis	34.6	*Ripogonum scandens* — fruit	11.3
Carpodetus serratus	4.4	— leaves	4.6
Coprosma foetidissima	4.0	— stems	2.7
Weinmannia racemosa	3.3	**Seaweeds**	1.3
Metrosideros umbellata	3.1	**Fungi**	1.1
Pseudopanax crassifolius	2.4	**Ferns**	
Pittosporum spp.	1.1	*Dicksonia squarrosa*	2.1
Fuchsia excorticata	0.8	*Blechnum fluviatile*	1.4
Podocarpus ferrugineus	0.7	Others	3.3
Coprosma lucida	0.6	**Grasses and sedges**	2.6
Pseudopanax simplex	0.6	**Herbs**	1.2
P. edgerleyi	0.6	**Minor plant foods**	5.8
Other species	2.1	**Unidentified material**	5.0

Data obtained from samples of rumen contents from 160 deer shot between 1979 and 1985. The samples were sorted into items, dried to constant weight, and combined by weighting the samples from each season equally to give an estimate of annual diet (Nugent & Challies, 1988). Only items comprising > 0.5% of diet have been listed. A total of 104 foods were identified to species (80) or genus (24). Included in the category "minor plant foods" were 22 ferns, 27 herbs, and 26 trees, shrubs and climbers.

beech forest, mainly mountain beech (*Nothofagus solandri* var. *cliffortioides*), but also silver beech (*N. menziesii*) and red beech (*N. fusca*). The highly preferred broadleaf, *Pseudopanax* spp. and *Coprosma* spp. are scarce. River flats provide indigenous and introduced grasses and herbs, with occasional matagouri (*Discaria toumatou*); the deer also graze the improved grasslands on farms (Bathgate, 1976).

White-tailed deer tend to feed throughout the day in winter, morning and evening in summer. In adverse weather, they may remain in shelter for several days, and come out to feed only after a storm is past (K. Purdon, unpubl.). They move freely when feeding, nipping foliage rather than concentrating on individual shrubs, and will stand on their hind legs to reach preferred browse.

SOCIAL ORGANIZATION AND BEHAVIOUR

White-tailed deer live in small family groups, usually of does and their offspring, while the bucks live separately. Their territories are probably small, although on Stewart I. they seem to shift around in spring. When disturbed they move, soundlessly and swiftly, in a circle back to the point of departure (Schofield, 1948). White-tailed deer do not wallow, but take readily to water and will swim long distances. They are quick to learn, and can live close to human activities; they will let a person or dog pass them unseen, before departing silently. They are generally quiet, but will snort or stamp when disturbed. Calves and their dams bleat to each other. Bucks do not call during the rut but may bleat when chasing a doe.

Old antlers are cast during July and August, and the new ones are hardened by March. The rut extends from late April to the end of May, with the peak in the middle of May. Males do not round up females into a harem, but follow individual females around, searching for those that are receptive.

Reproduction and Development

Gestation is 187–222 days (Haugen & Davenport, 1950; Haugen, 1959), and most calves are born in December. Multiple births are rare. In the Lake Wakatipu herd there is only one recent record of twin foetuses (Bathgate, 1976); on Stewart I. at present, about 10% of pregnancies of adult does produce twins (C. N. Challies, unpubl.).

Most does give birth to their first calf at two years of age. Calves weigh about 2 kg at birth. They begin to browse at a few weeks old; they become functional ruminants at about two months and are weaned soon after (Verme & Ullrey, 1984). The deciduous teeth comprise $I^0/_3$ $C^0/_1$ $Pm^3/_3$. These are replaced by permanent teeth over the period between 5 and 20 months of age, and another three molars in each jaw are added at 3–18 months (Severinghaus, 1949).

Population Dynamics

The Stewart I. herd went through a typical population irruption at first, which was apparently quite substantial and associated with pronounced modification of the vegetation. Despite that, and continuous hunting, they are still at relatively high densities in preferred areas of coastal forest (one deer per 2–4 ha). Inland and at higher altitudes, densities are much lower (<1/10 ha) (C. N. Challies, unpubl.). They are generally in poor condition; 20 adult deer shot at the southern end of the island in February 1987 had kidney fat indices ranging from 3.8 to 11.9% (C. N. Challies, unpubl.). Poor condition is also reflected in their relatively small body size, small antlers and low reproductive rate.

In 1974 there were about 840 white-tailed deer in the Lake Wakatipu herd, or one deer per 12.8 ha. The proportion of white-tailed deer in the mixed red deer/white-tailed deer population was about 30% in the Dart Valley, 40% in the Rees Valley, and 50% in the Earnslaw Burn (Bathgate, 1976). The Lake Wakatipu herd has long been in poor condition, with high juvenile mortality (Daniel, 1967; Bathgate, 1976). Possible explanations for this include unsuitable or degraded habitat, competitive exclusion by red deer, or the high incidence of parasites and disease acquired from domestic stock (Bathgate, 1976; Harris, 1981). None of these suggestions has been tested.

Predators, Parasites and Diseases

Man is the only predator of white-tailed deer in New Zealand. The only disease to have been diagnosed was typical pleuropneumonia (Daniel, 1966b), which in 1963 killed many white-tailed deer, particularly juveniles in the Lake Wakatipu herd. Tests for *Brucella* and *Leptospira* in 1963–65 proved negative in both herds (Daniel, 1967).

Endoparasites recorded from white-tailed deer in New Zealand are as follows. Cestodes: cysts of *Taenia hydatigena* (Christie & Andrews, 1965).

Nematodes: *Dictyocaulus viviparus* in the lungs, *Apteragia odocoilei, Haemonchus contortus, Ostertagia circumcincta, O. ostertagi, Skrjabinagia kolchida, Spiculopteragia asymmetrica,* and *S. spiculoptera* in the abomasum, *Capillaria* sp. and *Cooperia curticei* in the small intestine, *Oesophagostomum venulosum* in the large intestine (P. C. Mason, unpubl.). Ectoparasites: *Damalinia lipeuroides* and *D. parallela* (Andrews, 1973).

ADAPTATION TO NEW ZEALAND CONDITIONS

The white-tailed deer on Stewart I. thrived at first, but they now have smaller bodies and antlers, and a lower reproductive rate, than those of the same subspecies in their native country. In New Hampshire yearling (1.5 years) white-tailed deer weigh on average 64 kg in males and 57 kg in females (Halls, 1984), substantially more than adult deer on Stewart I. (Table 87). These differences indicate that the Stewart I. herd is at present in poor condition. The same is probably also true of the Wakatipu herd, even though they live at lower density in a very different habitat, but there are no data available for comparison.

SIGNIFICANCE TO THE NEW ZEALAND ENVIRONMENT

DAMAGE. The Lake Wakatipu herd is too small to present any danger to the extensive local protection forests. On Stewart I., white-tailed deer had become locally very abundant by 1920, and concern was being expressed about the severe modification of the island vegetation (Thomson, 1922). By sustained browsing of the palatable shrubs, ferns and regenerating saplings of the hardwood subcanopy trees, they opened up the lower tiers of most coastal forests and limited the replacement of canopy trees. In the 1970s, the rata-kamahi forests of the north and east coasts suffered a severe dieback, attributed mainly to climatic or sea conditions. In these forests, deer (and possums) have restricted the regeneration of woody species (Veblen & Stewart, 1980). In 1981 a trial poisoning programme temporarily removed deer from a block of coastal forest and was followed by a vigorous response in the vegetation, mainly the regeneration of less preferred species (Challies & Burrows, 1984).

CONTROL. White-tailed deer were protected until 1919, when shooting under licence began, but in 1926 all protection was removed. Hunters were employed to cull deer on Stewart I. periodically between 1927 and 1958. A total of 6380 deer, both red and white-tailed, was culled between 1937 and 1952 (Harris, 1981). Since 1958 the control of white-tailed deer on Stewart I. has been left mainly to recreational hunters, and, during the last 15-20 years, also to commercial hunters operating both on foot and from helicopters. Their combined pressure does not appear to have been intensive enough to reduce the number of white-tailed deer much below carrying capacity. A trial natural-bait poisoning operation on the east coast in 1981 killed >90% of the local population (Challies & Burrows, 1984). If more intensive control of white-tailed deer on Stewart I. is required in the future, poisoning is an obvious alternative to hunting.

The white-tailed deer on Stewart I. are popular with hunters because of their novelty value. Hunting is allowed all year round, but is concentrated between March and June, when the deer, in their winter pelts and with the bucks in hard antler, make attractive trophies. However, hunting success is low, averaging about 1-2 deer per hunter per trip; the total kill is about 1000 deer per year.

M. M. D. & C. N. C.

Genus *Alces*

Most of the nine genera in this subfamily are confined to North and South America, but two range right across the Holarctic, from Alaska and Canada into northern Eurasia and Siberia. Both have only one species, large animals known by well-established common names, different on either side of the North Atlantic; *Rangifer tarandus* is called "reindeer" in Eurasia and "caribou" in North America, and *Alces alces* is called "elk" in Eurasia and "moose" in North America.

By chance, the wapiti (p. 458) is also called "elk" in North America. Both kinds of "elk" have been brought to New Zealand, the only country they have in common, but neither is known by that name here (although, unfortunately, the deer farming industry is beginning to call wapiti "American elk"). *Alces alces* came from North American stock, and retains its local common name "moose".

46. Moose

Alces alces andersoni Peterson, 1952

Also called elk in Europe, but this name should be avoided in New Zealand (see above).

DESCRIPTION (Fig. 85)

Distinguishing marks, Table 78 and skull key, p. 24.

The moose is the largest member of the deer family. The short coarse-haired summer coat is blackish brown, lighter on the legs, lightish grey on the belly, and with no rump patch and no underfur. The winter coat is smoky grey, with long, coarse guard hairs and a fine, woolly underfur. The newborn moose calf has a reddish brown, unspotted coat, which later turns dark brown. The head of the moose is long, with a broad, prehensile upper lip, long ears, and small, deep-set eyes. The neck is short, and in both sexes a growth of skin, the dewlap or "bell", hangs from the throat. There is a downward slope from the massive shoulders to the rump, and the body is short in relation to height; the legs are long and slim, and the tail is so short it may appear to be absent. Antlers in the male project laterally from the head with brow, trez, and guard tines and horizontal palms (Fig. 73).

Dental formula $I^0/_3\ C^0/_1\ Pm^3/_3\ M^3/_3 = 32$.

FIELD SIGN

Moose make large, tapered hoofprints, fore larger than rear (the opposite of other deer). The rear impression usually overlaps that of the forefoot. Dew claw prints nearly always show, whereas in red deer they show only on soft

Fig. 85 Moose, shot in Seaforth Valley, Fiordland, April 1929 (see Table 89) (E. J. Herrick, courtesy L. Harris).

surfaces. Moose make a high browse line (some 2.7 m, cf. around 2 m in red deer) on some food species, especially *Fuchsia excorticata,* and leave signs of their feeding activities characteristically different from those of red deer. Whereas red deer only nip terminal shoots, moose wrench down branches and, using their prehensile upper lip and a sideways movement of the head, strip them of leaves. They may leave the ground littered with stripped branches of up to 50 mm diameter at the base. They eat large-diameter stems, up to 20 mm on some plants, e.g., *Pseudopanax* spp., and occasionally "walk down" saplings to feed on foliage out of normal reach. High-level bark-biting is also characteristic of moose. The pellets are larger than those of red deer. Cast antlers are unmistakable.

Measurements

The only measurements available are from six moose shot in Fiordland (Table 89).

Variation

Unknown.

History of Colonization

Four moose (two bulls and two cows), survivors of the 14 imported from Saskatchewan, Canada (59°N) by the New Zealand Government in 1900, were liberated in the Hokitika Valley, Westland, but died out. A second group was brought from Saskatchewan in 1909, and these, four bulls and six cows, were landed at Supper Cove, Dusky Sound, on 6 April 1910 (Donne, 1924).

Table 89: Body measurements of six moose from Fiordland, 1929-1952.

Date, hunter	Sex & estimated age	Shoulder height (mm)	Total length (mm)	Skull length (mm)	Whole weight (kg)	Antlers (mm)		
						Length		Spread
						L	R	
April 1929 (E. J. Herrick)	M Unknown but old	1 829	2 743	—	—	835	867	959
April 1934 (E. J. Herrick)	M Unknown	—	—	—	—	740	721	854
April 1951 (R. V. Francis-Smith)	F 5-6 years	1 829	2 438+	—	>455	—	—	—
August 1951 (J. Macintosh)	F 4 years	1 727	2 438	660	450	—	—	—
March 1952 (P. J. Lyes)	M 5-6 years	1 981	2 718	—	—	727	749	1 118
February 1956 (unknown)	M Unknown	1 900	2 400	—	—	—	—	—

Data from Thornton (1933, 1949); Curtis (1951); Harris (1971b); Tustin (1973); Tinsley (1983).

By the early 1920s, surveys reported that the Fiordland moose were established and breeding. A limited number of hunting licences were issued by the Southland Acclimatisation Society each year from 1923 until all protection was removed from wild ungulates in 1930. During that period only two adult bulls were shot under licence, one in the Seaforth Valley in 1929 and the other in a tributary valley of Wet Jacket Arm in 1934, both by E. J. Herrick.

The establishment of a herd of moose at this latitude (45°S) was considered a major achievement at the time, and the progress of the herd, regularly reported in the *Southland Times* and the *New Zealand Fishing and Shooting Gazette* from the 1920s to 1950s, was followed with great interest by the sportsmen of the time.

There is little information on the dispersal of the herd, since few people visited southern Fiordland in the 1920s to 1940s, and what few records exist are often contradictory, difficult to confirm and sometimes puzzling. However, in 1923 moose or signs of moose were reliably reported from Long Sound, Fanny Bay, Cascade Cove and Wet Jacket Arm. Within 10 years there were other reports of moose from Resolution I., Anchor I. and even from near Tuatapere. It seems likely that individual moose eventually reached most parts of the Dusky Sound area (there are extensive calm waterways, and moose are excellent swimmers), but the nucleus of the herd remained in the Seaforth Valley for several decades. Around the 1950s there were more reports of moose in the tributary valleys of Wet Jacket Arm. This was the time that red deer were invading the Dusky Sound area from the south and east; since then reports of moose have progressively declined, despite a considerable increase in human activities and opportunities for sightings (Tustin, 1972, 1973, 1974).

Probably about 30 moose have been shot over the years, mostly when numbers were highest from the late 1930s to 1950s. Many were shot by fishermen seeing moose standing around the edges of the sounds or swimming in the water; most were undocumented. Of the bulls shot, the largest recorded specimen was taken in Wet Jacket Arm in 1952 by P. J. Lyes. There are occasional reliable reports of moose sign to the present day, but no confirmed sightings or carcasses over the last 30 years (Tinsley, 1983; K. G. Tustin, unpubl.).

DISTRIBUTION

WORLD. The seven subspecies of moose have a circumpolar range in the Northern Hemisphere from 45° to 70°N, i.e., throughout the boreal coniferous forests of the Holarctic (Dansie & Wince, 1968-70).

NEW ZEALAND (Map 45). The one small population of moose is confined to a small area of Fiordland National Park (45°S). The last extensive survey, from February to mid April 1972, confirmed that moose still existed in the Dusky Sound-Wet Jacket Arm area, but estimated that there were probably fewer than 25 individuals altogether. Moose apparently preferred to stay around the heads of the two sounds and in their tributaries, where the forest is more diverse than further west (Tustin, 1972, 1973, 1974). Subsequent inspections by searching parties still report occasional moose sign (browsing, hoofprints,

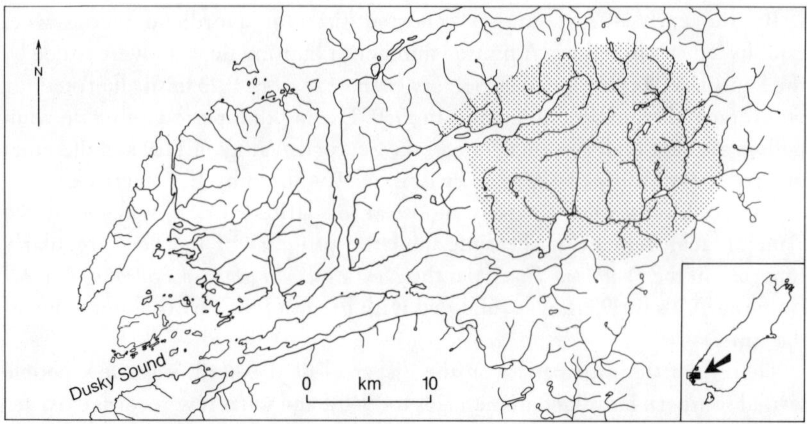

Map 45 Approximate distribution of moose in New Zealand in early 1970s (from Tustin, 1973). Axis deer were also once present in this area, but exactly where is unknown.

pellets) within these areas (Tinsley, 1983; K. G. Tustin, unpubl.).

HABITAT

The Dusky Sound area is a typical glaciated landscape in the east, with very steep valley sides and catchment headwaters, but grades towards more gentle country near the western coastline. The fiords are deep, narrow waterways with many forested islands. The forest tends to be mostly a beech-podocarp mixture in the west, with large areas of stunted scrubland carpeted by sphagnum and swampy ground cover. Towards the heads of the sounds the forest is more diverse, dominated by tall beech forest and with many patches of seral species on recently disturbed sites or around clearings. Swamps are common on the larger valley floors. Above the treeline at about 900 m asl is alpine grassland. The climate is mild and very wet (estimated annual rainfall >5000 mm; Tustin, 1974).

FOOD

Moose are almost wholly browsers and eat grasses only occasionally, unlike red deer, which are grazers as well as browsers. This is partly because moose have relatively long legs and short necks, so cannot conveniently reach low-growing herbage (Kean, 1959). However, they can browse vegetation up to a height of 2.7 m (far above the reach of red deer), and in summer they favour the seral species, especially fuchsia (*Fuchsia excorticata*), pate (*Schefflera digitata*), wineberry (*Aristotelia serrata*), broadleaf (*Griselinia littoralis*), ribbonwood (*Hoheria glabrata*), five-finger (*Pseudopanax simplex* and *P. colensoi*), mahoe (*Melicytus ramiflorus*), bramble (*Rubus* sp.) and tutu (*Coriaria arborea*); and to a lesser extent *Pseudopanax arboreum, Coprosma rotundifolia, C. colensoi* and *Myrsine divaricata* (Tustin, 1973, 1974 and unpubl.). The food requirements of moose are massive, averaging 10-20 kg fresh weight (4 kg dry weight) of vegetation per day, depending on season, bulk and availability (Anon., 1967).

SOCIAL ORGANIZATION AND BEHAVIOUR

Moose are solitary rather than gregarious animals, although a cow with a calf, or two or more subadults, may associate together for some time. The antlers are shed in August-September, and the new growth is hard by March-April, the period of the rut (Donne, 1924). The male associates with and mates with one female at a time (unlike the harem mating of the red deer), and the equivalent of his rutting call is usually a grunt or "throaty gulp", less often a low, rising moo. The female, however, does most of the calling, described variously as a long, low, bleating note (Donne, 1924), or a sort of braying call with a bell-like echo (Tinsley, 1983). The call of the calf is moo-like but with a harsher ending, and its dam calls to it with frequent grunts. In forested country moose are very secretive and elusive, remarkably so for such large animals.

REPRODUCTION AND DEVELOPMENT

There are no data on the reproduction of moose in Fiordland, although they did breed at first.

POPULATION DYNAMICS

Unknown.

PREDATORS, PARASITES AND DISEASES

Man is the only potential predator of moose in Fiordland. There is no information on their parasites or diseases.

ADAPTATION TO NEW ZEALAND CONDITIONS

Unlike many of the ungulates introduced into New Zealand, moose have never been numerous, but they have survived in the wild for at least 75 years so far. However, they have been declining over the last 30 years, and may well become extinct. All indications suggest that Fiordland is only a marginally suitable environment for moose, and that they are being excluded from it by competition from red deer.

The forests and extensive swamplands of the Seaforth Valley only superficially resemble the moose's favoured habitats in its homelands. New Zealand has no equivalents of the aquatic plants which are seasonally important to moose in the Northern Hemisphere, and the small-leaved shrubland plants, sedges and sphagnum mosses which surround the Fiordland swamps are unpalatable. Red deer share the moose's preference for seral vegetation but are present in vastly greater numbers, and they readily graze the grasses on the adjacent clearings, and (at least before the advent of aerial hunting) the alpine grasslands, which are not used by moose. Seral species form the bulk of the moose's diet, but unlike the aspens, alders, willows and poplars of the Northern Hemisphere, the equivalent New Zealand species do not readily repair browsing damage by sprouting from the base. The Fiordland moose are therefore progressively removing their own main food supply by their destructive method of feeding. Their ability to reach higher than red deer is only a temporary advantage. Most of the important seral species are deciduous, so winter forage is very limited.

On the other hand, aerial hunting has greatly reduced the numbers of red deer throughout Fiordland (p. 456), and a general and dramatic recovery of the vegetation has followed. History will show whether this increase in food supply will be sufficient to check the decline of moose.

From the scant data available (Table 89), it appears that moose in New Zealand have not attained the size of animals of the same subspecies in their homeland, and the few trophy heads secured from Fiordland are greatly inferior. For example, Canadian moose aged 3+ years already have a mean shoulder height of 1980 mm in males, 1850 mm in females, average total lengths of 2690 mm and 2450 mm, and whole weights of 414 kg and 417 kg (Blood, McGillis & Lovaas, 1967).

SIGNIFICANCE TO THE NEW ZEALAND ENVIRONMENT

None.

M. M. D. & K. G. T.

Glossary and Abbreviations

Abomasum The fourth chamber of the stomach of a ruminant (after the rumen, reticulum, and omasum), corresponding to the stomach proper of other mammals.

Adaptation The features of an animal which adjust it to its environment. Adaptations may be genetic, produced by evolution and not alterable by the individual within its lifetime, or phenetic, the result of responses by individuals to environmental conditions during their lifetime, particularly while they are actively growing, or, in adults, during the breeding season. Natural selection favours the survival and breeding success of individuals whose adaptations best suit them to average contemporary conditions, at the expense of those with less suitable adaptations.

Adult A fully developed, sexually mature animal.

Aerial hunting The practice of shooting game or pest animals (usually deer, goats, chamois, or tahr) directly from a helicopter in flight.

Agonistic Competitive or aggressive (behaviour).

Agouti A grizzled colouration produced by alternate light and dark banding on individual hairs.

Alveolus Tooth socket.

Anoestrus State of sexual quiescence in females, between seasons or oestrous cycles.

Antarctic Convergence The region between 50° and 55°S, where cold, dense Antarctic seawater meets and slides beneath lighter, warmer subantarctic water.

Antitragus A prominent flap of skin, continuous with the outer margin of the ear in bats.

Antlers (cf. Horns) Deciduous bony outgrowths from the pedicel (a projection from the skull) on the head of a male deer, produced each year before the rut and shed afterwards. Covered in soft skin (velvet) while growing, but naked bone when mature. *Length:* From the outside bottom edge of the burr (coronet flange) to the tip of the furthest point, measured along the outside curve of the main beam. *Spread:* The widest horizontal outside width across the main beams, as measured between parallel perpendiculars.

APDC/B Agricultural Pest Destruction Council/Board.

Arboreal Living in trees.

asl Above sea level.

Baculum *Os penis* or penis bone, characteristic of some male mammals.

Bergmann's Rule A generalization claiming that mammals become larger in higher latitudes, by comparison with related animals living nearer the equator, since conservation of body heat is relatively less expensive in larger animals.

Bifid Forked.

Biomass The total mass of living animal matter in a given unit area, calculated from the number of individuals multiplied by their average body weight.

Biota The total community of all life forms living in a given place.

Blastocyst A very early embryo, still at the stage of a hollow ball of cells floating free in the uterus.

Bovid A member of the family Bovidae (in New Zealand, cattle, sheep, goats, chamois, and tahr).

Browsing Feeding on woody plants and ferns (browse), cf. **grazing**, feeding on grass.

C/100TN Captures per hundred trap-nights. A measure of relative density widely used in New Zealand to follow changes in populations of rodents and mustelids. See Fig. 42 and Cunningham and Moors (1983), Nelson and Clark (1973).

Caecum A blind diverticulum of the hindgut, usually immediately before the colon.

Cache (noun) Stored food. Also (verb) to store food for future use.

Caprinid A member of the subfamily Caprinae (family Bovidae) — in New Zealand, chamois, tahr, sheep, and goats.

Carnassial teeth Opposing pair of teeth (Pm^4 and M_1) in carnivores, especially shaped to shear past each other like scissor blades.

Carnivore A member of the order Carnivora (in New Zealand, the cat, dog, stoat, weasel, ferret, sea lion, fur seal, or any of the five true seals); hence a **carnivorous diet**, one that includes much fresh meat.

Caudal Pertaining to the tail.

Cervid A member of the family Cervidae (deer).

Chromosomes Thread-like strands within the nucleus of a cell, carrying the genes. In body cells the chromosomes are paired, and their number is expressed as 2n (the diploid number); in the process of producing the reproductive cells or gametes, each pair splits up, and the diploid number is restored only at fertilization.

Cline A gradual and progressive change in some attribute of a species (e.g., body weight), measured over a substantial geographic distance.

cm Centimetre(s).

Cohort A group of animals all recruited into a population in the same breeding season or birth pulse.

Commensal Two species of animals closely dependent on each other for food. The term is also used of rodents making use of man-made buildings or other resources, but still wild; cf. domesticated.

Condylobasal length Measure of the total length of the skull, from the occipital condyles to the anterior edge of the premaxillary bone (excluding the teeth).

Coprophagy (see also refection). The habit of eating faeces, often those containing incompletely digested material.

Corpora lutea Small yellow bodies in the ovaries, each one formed in the space vacated by a released ovum (hence, counts of corpora lutea indicate the ovulation rate; see fecundity).

Crepuscular Active at dawn and dusk.

Cusp A prominence on a premolar or molar tooth.

Delayed implantation A period ranging from a few hours to 9–10 months, during which development of the blastocyst (qv) is interrupted, and it remains floating free in the uterus.

Dental formula A conventional shorthand means of summarizing the number and arrangement of the teeth. The numbers of each of the four types of teeth, in each half of the upper and lower jaws, are given in order: I (incisors), C (canines), Pm (premolars), and M (molars), plus the total number of teeth. Individual teeth may be referred to by letter and number (denoting position), using superscripts for teeth in the upper jaw and subscripts for teeth in the lower jaw: e.g., Pm^4, the fourth upper premolar.

DIA Department of Internal Affairs (parent department to the National Museum and the former NZWS).

Diastema A natural gap in a row of teeth, especially between the incisors and premolars in herbivorous mammals (rodents, lagomorphs, horses) in which the canines are absent.

Digital glands Scent-producing glands located between the toes.

Digitigrade Walking on the toes (cf. plantigrade).

Dimorphism The permanent co-existence of two distinctly different body forms in the same species: e.g., sexual dimorphism (marked differences between male and female), common in carnivores and deer. **Polymorphism** Co-existence of three or more distinct forms: e.g., the colour phases of ship rats and possums.

Dioestrous Having two oestrous cycles in a season.

Distal Furthest from the point of origin or centre of mass (cf. proximal).

Diurnal Active in daylight.

DLS Department of Lands and Survey (disbanded 31 March 1987, see DOC). Parent department to the former National Parks Authority.

DOC Department of Conservation, established 1 April 1987 to take over the conservation-related functions of parts of DLS, NZWS, and NZFS.

Domestic Individual animals of a domesticated species living under direct human influence (cf. feral).

Domesticated Species which have been tamed and bred under human control for many centuries, often now distinctly different in form and behaviour from their wild ancestors. (cf. domestic, feral).

Dorsal On or pertaining to the back of an animal.

DSIR Department of Scientific and Industrial Research.

Endemic Found only in a given country or area (cf. indigenous, native).

Epiphyses Terminal sections of long bones and vertebrae which ossify separately from the main bone, in order to leave room for growth, and become permanently fused to it only when growth is complete. The extent of fusion can sometimes be used to indicate relative age.

Eruption Explosive activity of volcanoes, and (metaphorically) the emergence of teeth from the gums of young mammals, often in a regular order (the **eruptive sequence**) useful in age-determination. Not correct when applied to populations (cf. irruption).

Eruptive fluctuation Phrase used by some authors to describe the explosive increase in numbers and range of a successful colonizing species in favourable new habitat, and the subsequent decline in numbers, with or without decline in range. Better to use irruption (qv).

Exotic (of species) Not native (in New Zealand, includes man and all species introduced by man, either deliberately or accidentally).

Fecund Capable of producing ova or sperm. **Fecundity** of females may be measured by counting the number of ova released (cf. corpora lutea) or embryos *in utero*.

Feral Describes individuals or populations of domesticated species living in the wild, free of human control.

Fertility The number of young actually produced and reared to a specified age (cf. fecundity, which measures only *potential* fertility).

Flehmen A characteristic facial expression in which the head is raised, with lips pulled back, nose wrinkled and teeth exposed. Often seen in animals sniffing scent marks or socially important odours.

Founder effect The tendency for new populations established from small numbers of colonists to have a slightly different genetic constitution from the parent stock, since a small colonizing group cannot adequately sample the full genetic variability of the species.

FP Forest Park.

FRI Forest Research Institute.

Gestation Pregnancy.

Guard hair The long contour hairs which give colour and texture to the pelage; cf. **underfur**, the softer, denser, finer hairs which provide insulation.

ha hectare(s).

Hauling ground Beach or rocks used by seals for hauling out, but not for breeding (cf. rookery).

Hibernaculum Winter nest of hibernating species, e.g. hedgehog.

Holarctic The Northern Hemisphere temperate and boreal zones, in both Eurasia and North America.

Home range The area traversed by an animal in the course of its daily activities, whether or not it is defended from intrusions by other animals (cf. territory).

Horns Permanent bony outgrowths, covered with a keratin (qv) sheath, from the head of a male or female bovid (cf. antlers).

Hybrid The offspring of parents of different species or subspecies.

Indigenous Native to, but not confined to, a given country or area.

Induced ovulation Ovulation that can be stimulated only by copulation, as opposed to **spontaneous** or **cyclic ovulation**, stimulated by environmental or internal cues regardless of whether there are any males present.

Inguinal Pertaining to the groin.

Insectivore A member of the order Insectivora (in New Zealand, only the hedgehog); an **insectivorous diet**, one that includes a lot of insects.

Introduced Brought to New Zealand with human help, deliberate or accidental. If the intended meaning includes man as well, exotic (qv) may be better.

Irruption The explosive increase in numbers, with or without extension of range, characteristic of opportunistic species when first colonizing new habitat or during particularly favourable seasons (cf. eruption). The distinction between an irruption and an eruptive fluctuation (qv) is unnecessarily precise and not always known; "irruption" is recommended for both.

IUCN International Union for the Conservation of Nature and Natural Resources.

Karyotype The set of chromosomes in a cell, especially when arranged in conventional order for description.

Keratin A strong flexible substance, produced by the skin, from which claws, hooves and horns are formed.

Kiore (pronounced "kee-aw-reh") Maori name for *Rattus exulans*.

kg Kilogram(s).

km Kilometre(s).

Kuri (pronounced "koo-ree") Maori name for *Canis familiaris*.

Lagomorph A member of the order Lagomorpha (in New Zealand, a rabbit or a brown hare).

Lifespan Maximum age to which an animal can live in ideal conditions (e.g., in captivity), cf. longevity.

Longevity The age to which animals live in natural conditions.

m Metre(s).

MAF Ministry of Agriculture and Fisheries.

Man The human species, including both male and female individuals.

Marsupial A member of the order Marsupialia (in New Zealand, the five species of wallabies and the brushtail possum).

Melanism Darkness of colour due to the pigment melanin.

Metabolic rate The rate at which the chemical processes of the body proceed.

mm Millimetre(s).

Molar Posterior chewing tooth, or cheek tooth.

Monoestrous Having only one oestrous cycle per season.

Monogamic Having only one mate per season, or for life.

Muffle Area of thick roughened skin between the upper lip and nostrils of ruminants.

Mustelid A member of the family Mustelidae (in New Zealand, only the weasel, stoat, and ferret).

Muttonbirds Several species of sea birds, mostly burrowing shearwaters, traditionally exploited by the Maori for food.

Native A species having colonized New Zealand without human help, at any time from Tertiary to contemporary times, and now maintaining a self-sustaining population here (cf. indigenous, endemic).

Nearctic Temperate and boreal North America.

Niche A complex idea, variously defined, analogous to the "address" or the "profession" of an animal in its world.

Nocturnal Active at night.

NP National Park.

NZFS New Zealand Forest Service (disbanded 31 March 1987, see DOC).

NZWS New Zealand Wildlife Service (disbanded 31 March 1987, see DOC).

Opportunist A species adapted to take swift, short-term advantage of temporary resources.

Pakihi A type of native vegetation found on bogs or heathland.

Palaearctic Temperate and boreal Eurasia.

Parturition The process of giving birth.

Pelage The hairy or furry coat of a mammal.

Pellets Hard, dry faeces produced by herbivorous mammals (cf. scats).

Pheromone An aromatic secretion that carries a message to another animal of the same species.

Placenta A structure that connects the foetus to the wall of the mother's uterus, and acts to ensure the transfer of nutrients to, and removal of waste products from, the foetal bloodstream.

Placental mammals Those having a well-developed placenta, i.e., all mammals except marsupials and monotremes.

Placental scars Traces of former placental attachment sites, visible as grey spots of varying intensity on the uterine walls of, for example, rodents. Useful as indicators of past pregnancies.

Plantigrade Walking on the whole foot, including the heel (cf. digitigrade).

Polygamous Having more than one mate per season. **Polygynous** males mate with more than one female; **polyandrous** females mate with more than one male.

Polymorphism See dimorphism.

Polyoestrous Having two or more oestrous cycles per season.

Population A more or less distinct, interbreeding group of animals of the same species.

Postpartum Immediately after birth.

Prehensile Capable of grasping.

Productivity The number of offspring produced in a stated time per breeding female or per population.

Proximal Closest to the point of origin or centre of mass (cf. distal).

Refection A process (e.g., in lagomorphs) in which an animal reingests its own faeces and re-digests them as a normal means of increasing the rate of recovery of food value from fibrous vegetation.

RHA Recreational Hunting Area, managed for the benefit of sport hunters; commercial hunting excluded.

Rookery Beach used by seals for pupping and mating (cf. hauling ground).

Rostrum The forward (nose) area of the skull, anterior to the orbits.

Ruminant A mammal with a multichambered stomach specialized for storing, regurgitating, and later remasticating vegetation, "chewing the cud".

Rupicaprid A member of the tribe Rupicaprini (family Bovidae) — in New Zealand, only the chamois.

Scats Faeces, usually of a carnivore (cf. pellets).

Scent mark A site at which chemical messages are received and deposited by animals of the same species, in the form of urine, faeces or the secretions of specialized scent glands. Also the action of depositing such marks.

SE Standard error.

SF State Forest.

Suid A member of the family Suidae (pigs).

Sympatric (of two species) Living in the same place, or having overlapping distributions.

Tapetum lucidum A layer of reflective material behind the retina, which increases the capacity of the eye to gather light at night; results in the green "eyeshine" of, for example, cats when caught in a beam of bright light. The eyes of animals with no tapetum lucidum shine red.

Tapu (Maori) Sacred.

Taxonomy The science of naming and describing animals.

TB Tuberculosis.

Territory A defended home range (qv).

Tines The points of a deer's antlers. The total is the sum of points on both sides.

Track The impression made by an individual footprint (cf. **trail,** the line of footprints made by an animal moving across a soft surface).

Tragus An upright flap of skin inside the opening of the ear in bats.

TVNZ Television New Zealand.

Ungulate A member of either of the two orders Perissodactyla (odd-toed ungulates) and Artiodactyla (even-toed ungulates).

Velvet See antlers.

Ventral Pertaining to the underside of an animal.

Vibrissae Whiskers.

Weka A New Zealand endemic flightless bird of the genus *Gallirallus*.

Weta A New Zealand endemic flightless insect of the order Orthoptera.

Withers The shoulders of a horse.

References

Abildgård F, Andersen J, Barndorff-Nielsen O, 1972. The hare population (*Lepus europaeus* Pallas) of Illumo Island, Denmark. A report on the analysis of the data from 1957-1970. Danish Review of Game Biology 6: 1-32.

Adams M, 1851. The Journal of Martha Adams. MS., Alexander Turnbull Library, Wellington.

Advani R, 1984. Ecology, biology and control of black rat *Rattus rattus* in Minicoy Island. Journal of Plantation Crops 12: 11-16.

Advani R, Idris M, 1982. Neophobic behaviour of the house rat *Rattus rattus rufescens*. Zeitschrift für Angewandte Zoologie 69: 139-144.

Agricultural Pests Destruction Council, 1982. Fourteenth Annual Report. Wellington.

Aitken V, Kroef P, Pearson A, Ricketts W, 1979. Report on observations of feral horses in the southern Kaimanawas (February 1979) to the Kaimanawa Wild Horse Committee. Unpublished report, Faculty of Veterinary Science, Massey University, Palmerston North.

Allen RB, 1976. The significance of rusa deer (*Cervus timoriensis*). Unpublished B.Sc. thesis, University of Canterbury, Christchurch.

Allen RB, Rose AB, 1983. Regeneration of southern rata (*Metrosideros umbellata*) and kamahi (*Weinmannia racemosa*) in areas of dieback. Pacific Science 37: 433-442.

Allison C, 1980. The Hunter's Manual of Australia and New Zealand. AH and AW Reed, Wellington.

Allo J, 1970. The Maori dog: a study of the Polynesian dog in New Zealand. Unpublished M.A. thesis, University of Auckland.

Allo J, 1971. The dentition of the Maori dog in New Zealand. Records of the Auckland Institute and Museum 8: 29-45.

Allo Bay-Petersen J, 1979. The role of the dog in the economy of the New Zealand Maori. *In* Anderson A (Ed.) Birds of a Feather, pp. 165-181. British Archaeological Reports, Oxford, International Series 62.

Almǎsan H, Cazacu I, 1976. Der Hase in der Sozialistischen Republik Rumänien. *In* Pielowski Z and Pucek Z (Eds) Ecology and Management of European Hare Populations, pp. 29-31. Polish Hunting Association, Warsaw.

Amaya JN, Alsina MG, Brandani AA, 1979. Ecology of the European hare (*Lepus europaeus* P.). II. Reproduction and bodyweight of a population near San Carlos de Bariloche, Argentina. Serie: Ecologia y Control de la Fauna Silvestre, Informe Technico No. 9, 36 pp. INTA Bariloche.

Anderson A, 1980. The archaeology of Raoul Island (Kermadecs) and its place in the settlement history of Polynesia. Archaeology and Physical Anthropology in Oceania 15: 131-141.

Anderson A, 1981. Pre-European hunting dogs in the South Island, New Zealand. NZ Journal of Archaeology 3: 15-20.

Anderson A, 1984. The extinction of moa in southern New Zealand. *In* Martin PS and Klein RG (Eds) Quaternary Extinctions, pp. 728-740. University of Arizona Press, Tucson.

Anderson JA, Henderson JB, 1961. Himalayan thar in New Zealand. NZ Deerstalkers Association Special Publication No. 2, 37 pp.

Anderson JW, 1954. The production of ultrasonic sounds by laboratory rats and other mammals. Science 118: 808-809.

Anderson PK, 1970. Ecological structure and gene flow in small mammals. Symposia of the Zoological Society of London 26: 299-325.

Andrews JRH, 1964. The arthropod and helminth parasites of red deer (*Cervus elaphus* L.) in New Zealand. Transactions of the Royal Society of NZ 5: 97-121.

Andrews JRH, 1969. The parasitology of some wild ruminants introduced into New Zealand. Unpublished Ph.D. thesis, Victoria University of Wellington.

Andrews JRH, 1973. A host-parasite check list of helminths of wild ruminants in New Zealand. NZ Veterinary Journal 21: 43-47.

Andrews JRH, Daniel MJ, 1974. A new species of *Hymenolepis* (Cestoda: Hymenolepididae) from the New Zealand long-tailed bat, *Chalinolobus tuberculatus*. NZ Journal of Zoology 1: 333-336.

Angas GF, 1847. Savage Life and Scenes in Australia and New Zealand (2 vols). Smith Elder, London.

Angehr G, 1984. A bird in the hand: Andreas Reischek and the stitchbird. Notornis 31: 300-311.

Angermann R, 1983. The taxonomy of Old World *Lepus*. Acta Zoologica Fennica 174: 17-21.

Angot M, 1954. Observation sur les mammifères marins de L'Archipel de Kerguelen avec une étude detaillée de l'elephant de mer, *Mirounga leonina* (L.). Mammalia 18: 1-111.

Anon., 1842. 'Nelson Examiner' and 'NZ Chronicle', 18 June.

Anon., 1932. The wapiti stalkers. NZ Fishing and Shooting Gazette 5: 8.

Anon., 1949a. Stoat essays long swim. NZ Outdoor 12: 6.

Anon., 1949b. The stalking of wapiti (*Cervus canadensis*) in New Zealand. NZ Fishing and Shooting Gazette 17(2): 5-20.

Anon., 1951. Wild horses hunted, 22 more 'Brumbies' taken in Marlborough back country. Marlborough Express, 29 March.

Anon., 1966. Goats in Egmont area. Forestry News 16: 1.

Anon., 1967. Moose in British Columbia, Canada. Wildlife Review 4(6): 15-18.

Anon., 1972. Goat-shooting creates hydatids problem. NZ Outdoor 37(5): 6.

Anon., 1975. Tuberculosis in opossums. What's New in Forest Research 27: 1-4.

Anon., 1978a. Feral pigs in New South Wales. Australian Meat Research Committee Review No. 35, pp. 1-11.

Anon., 1978b. A portfolio of oldies. NZ Wildlife 7(55 & 56): 79.

Anon., 1980. Feral cat parasitism. Surveillance 7(3): 7.

Anon., 1981. Beating weeds without chemicals. 1. Sweet briar. Aspect 2: 4-5.

Anon., 1982. Wild horses have lived in forest park for 100 years. Taupo Times, 27 July.

Anon., 1984a. The black rat's home in Roman Britain. New Scientist 20: 22.

Anon., 1984b. Wild Horses of the Kaimanawa Ranges. NZ Wildlife Service, Wellington.

Anon., 1984c. Mating performance among fallow. The Deer Farmer 18: 29-30.

Anon., 1986. Tenders for capture of wild horses. The Press, 24 November.

Archer M, 1975. Abnormal dental development and its significance in dasyurids and other marsupials. Memoirs of the Queensland Museum 17: 251-265.

Arnold HR (Ed.), 1978. Provisional atlas of the mammals of the British Isles. The Institute of Terrestrial Ecology, Huntingdon, England.

Asdell SA, 1964. Patterns of Mammalian Reproduction (2nd edition). Cornell University Press.

Ashby CR, 1967. The centenary history of the Auckland Acclimatisation Society 1867-1967. Auckland Star Printers, Auckland, 143 pp.

Asher GW, 1981. Transferrin gene frequencies of feral and domestic goats in New Zealand. NZ Journal of Zoology 8: 447-452.

Asher GW, 1985a. Meat production from fallow deer. *In* Fennessey PF and Drew

KR (Eds) Biology of Deer Production, pp. 299-302. Royal Society of NZ Bulletin 22.

Asher GW, 1985b. Oestrus cycle and breeding season of farmed fallow deer, *Dama dama.* Journal of Reproduction and Fertility 75: 521-529.

Asher GW, Adams JL, 1985. Reproduction of farmed red and fallow deer in northern New Zealand. *In* Fennessey PF and Drew KR (Eds) Biology of Deer Production, pp. 217-224. Royal Society of NZ Bulletin 22.

Asher GW, Kilgour R, 1985. The fallow deer rut. The Deer Farmer 25: 40-44.

Ashton EH, Thomson APD, 1955. Some characteristics of the skulls and skins of the European polecat, Asiatic polecat and domestic ferret. Proceedings of the Zoological Society of London 125: 317-335.

Aspisov OI, Popov VA, 1940. Factors determining fluctuation in the numbers of ermines. *In* King CM (Ed.) Biology of Mustelids: some Soviet Research, pp. 109-131. DSIR Bulletin 227 (1980).

Aston BC, 1912. Some effects of imported animals on the indigenous vegetation. Transactions and Proceedings of the NZ Institute 44: 19-24.

Atkinson GG, 1954. Aliens in the forest. Forest and Bird 112, May.

Atkinson GG, Logan JR, Cookson T, Fokerd SE, 1965. Preliminary report and plan for control of noxious animals, Mount Egmont National Park. Unpublished report, NZ Forest Service, pp. 1-70.

Atkinson IAE, 1964a. The flora, vegetation and soils of Middle and Green Islands, Mercury Islands group. NZ Journal of Botany 2: 385-402.

Atkinson IAE, 1964b. Relations between feral goats and vegetation in New Zealand. Proceedings of the NZ Ecological Society 11: 39-44.

Atkinson IAE, 1972. Report on Mahurangi Island Mercury Bay. Unpublished report, Botany Division, DSIR.

Atkinson IAE, 1973a. Spread of the ship rat (*Rattus r. rattus* L.) in New Zealand. Journal of the Royal Society of NZ 3: 457-472.

Atkinson IAE, 1973b. Protection and use of the islands in the Hauraki Gulf Maritime Park. Proceedings of the NZ Ecological Society 20: 103-114.

Atkinson IAE, 1977. A reassessment of factors, particularly *Rattus rattus* L., that influenced the decline of endemic forest birds in the Hawaiian Islands. Pacific Science 31: 109-133.

Atkinson IAE, 1978. Evidence for effects of rodents on the vertebrate wildlife of New Zealand islands. *In* Dingwall PR, Atkinson IAE and Hay C (Eds) The Ecology and Control of Rodents in New Zealand Nature Reserves, pp. 7-30. Department of Lands and Survey Information Series 4.

Atkinson IAE, 1985. The spread of commensal species of *Rattus* to oceanic islands and their effects on island avifaunas. *In* Moors PJ (Ed.) Conservation of Island Birds, pp. 35-81. International Council for Bird Preservation Technical Publication No. 3.

Atkinson IAE, 1986. Rodents on New Zealand's northern offshore islands: distribution, effects and precautions against further spread. *In* Wright AE and Beever RE (Eds): The Offshore Islands of Northern New Zealand, pp. 13-40. Department of Lands and Survey Information Series No. 14.

Atkinson IAE, Bell BD, 1973. Offshore and outlying islands. *In* Williams GR (Ed.) The Natural History of New Zealand, pp. 372-392. AH and AW Reed, Wellington.

Atkinson IAE, Campbell DJ, 1966. Habitat factors affecting saddlebacks on Hen Island. Proceedings of the NZ Ecological Society 13: 35-40.

Axel HE, 1956. Predation and protection at Dungeness Bird Reserve. British Birds 49: 193-312.

Badan D, 1979. The ecology of mice (*Mus musculus* L.) in two forests near Auckland. Unpublished M.Sc. thesis, University of Auckland.

Badan D, 1986. Diet of the house mouse (*Mus musculus* L.) in two pine forests and a kauri forest. NZ Journal of Ecology 9: 137-141.

Bailey AM, Sorensen JH, 1962. Subantarctic Campbell Island. Proceedings of the Denver Museum of Natural History 10: 1-305.

Baker AN, 1983. Whales and Dolphins of New Zealand and Australia. Victoria University Press, Wellington.

Baker K, 1972. The history of the Blue Mountain fallow (*Dama dama*) herd. NZ Wildlife 30: 27-31.

Baker K, 1973. Reproductive biology of fallow deer (*Dama dama*) in the Blue Mountains of New Zealand. Unpublished M.Sc. thesis, University of Otago, Dunedin.

Baker RH, 1984. Origin, classification and distribution. *In* Halls LK (Ed.) White-tailed Deer — Ecology and Management, pp. 1-18. Stackpole Books, Harrisburg, Pennsylvania.

Baker-Gabb DJ, 1981. The diet of the Australasian harrier (*Circus approximans*) in the Manawatu-Rangitikei sand country, New Zealand. Notornis 28: 241-254.

Balham RW, 1949. Some ecological studies of the waterfowl in the Manawatu district of New Zealand. Unpublished M.Sc. thesis, Victoria University of Wellington.

Ballance AP, 1985. Aspects of the biology of Campbell Island feral sheep (*Ovis aries* L.). Unpublished M.Sc. thesis, Massey University, Palmerston North.

Bamford JM, 1970. Estimating fat reserves in the brush-tailed possum, *Trichosurus vulpecula* Kerr (Marsupialia: Phalangeridae). Australian Journal of Zoology 18: 415-425.

Bamford JM, 1972. The dynamics of possum (*Trichosurus vulpecula* Kerr) populations controlled by aerial poisoning. Unpublished Ph.D. thesis, University of Canterbury, Christchurch.

Bamford J, Hill J, 1985. Environmental impact report on a proposal to introduce myxomatosis as another means of rabbit control in New Zealand. Commissioned by Agricultural Pests Destruction Council, Wellington.

Bamford JM, Martin JT, 1971. A method for predicting success of aerial poisoning campaigns against opossums. NZ Journal of Science 14: 313-321.

Banwell DB, 1966. Wapiti in New Zealand. AH and AW Reed, Wellington.

Banwell DB, 1968. The highland stags of Otago. AH and AW Reed, Wellington.

Barbehenn KR, 1974. Estimating density and home range size with removal grids: the rodents and shrews of Guam. Acta Theriologica 19: 191-234.

Barbour RA, 1977. Anatomy of marsupials. *In* Stonehouse B and Gilmore D (Eds) The Biology of Marsupials, pp. 237-272. MacMillan Press, London.

Barlow ND, Clout MN, 1983. A comparison of 3-parameter, single species population models in relation to the management of brushtail possums in New Zealand. Oecologia (Berlin) 60: 250-258.

Barnes RFW, Tapper SC, Williams J, 1983. Use of pastures by brown hares. Journal of Applied Ecology 20: 179-185.

Barnett JL, How RA, Humphreys WF, 1976. Possum damage to pine plantations in north-eastern New South Wales. Australian Forest Research 7: 185-195.

Barnett SA, 1975. The Rat: a Study in Behaviour (revised edition). University of Chicago Press, Chicago.

Barnett SA, Spencer MM, 1951. Feeding, social behaviour and interspecific competition in wild rats. Behaviour 3: 229-242.

Barrett RH, 1971. Ecology of the feral hog in Tehama County California. Unpublished Ph.D. thesis, University of California.

Barrett RH, 1978. The feral hog on the Dye Creek Ranch, California. Hilgardia 46: 283-355.

Barrett-Hamilton GEH, Hinton MAC, 1910-21. A History of British Mammals. Gurney and Jackson, London.

Barton IL, 1974. *In* Clark JM, 1976, Tane 22: 129-134.

Barton RJ (Comp.), 1927. Earliest New Zealand. Palamontain and Petherick, Masterton.

Batcheler CL, 1960. A study of the relations between roe, red, and fallow deer, with special reference to Drummond hill forest, Scotland. Journal of Animal Ecology 29: 375-384.

Batcheler CL, 1975. Probable limit of error of the point distance-neighbour distance estimate of density. Proceedings of the NZ Ecological Society 22: 28-33.

Batcheler CL, 1982. Quantifying "bait quality" from number of random encounters required to kill a pest. NZ Journal of Ecology 5: 129-139.

Batcheler CL, 1983. The possum and rata-kamahi dieback in New Zealand: a review. Pacific Science 37: 415-426.

Batcheler CL, McLennan MJ, 1977. Cranimetric study of allometry, adaptation and hybridism of red deer (*Cervus elaphus scoticus*, L.) and wapiti (*C. e. nelsoni*, Bailey) in Fiordland, New Zealand. Proceedings of the NZ Ecological Society 24: 57-75.

Batcheler CL, Darwin JA, Pracy LT, 1967. Estimation of opossum (*Trichosurus vulpecula*) populations and results of poison trials from trapping data. NZ Journal of Science 10: 97-114.

Batchelor TA, 1980. The social organisation of the brush-tailed rock wallaby (*Petrogale penicillata penicillata*) on Motutapu Island. Unpublished M.Sc. thesis, University of Auckland.

Bathgate JL, 1973a. Summary of questionnaire returns. *In* Assessment and Management of Introduced Animals in New Zealand Forests, pp. 102-116. NZ Forest Service, Forest Research Institute Symposium No. 14.

Bathgate JL, 1973b. Report of a resurvey of the introduced mammals of the Wairau catchment. NZ Forest Service, Protection Forestry Report 126: 1-40.

Bathgate JL, 1976. Wakatipu whitetailed deer — a population census. Unpublished report, NZ Forest Service.

Bathgate JL, 1977. Blue Mountains fallow. A population census. Unpublished report, NZ Forest Service (Southland Conservancy), 12 pp.

Batten GJ, 1979. Controlling scrubweeds with goats. NZ Journal of Agriculture 139(4): 31-32.

Bauer JJ, 1977. Population dynamics of chamois at Lake Wanaka. Unpublished M.Sc thesis, University of Otago, Dunedin.

Bauer JJ, 1982. Untersuchungen zur Dynamik von stabilen und kolonisierenden Gemsenpopulationen (*Rupicapra rupicapra* L.) Neuseelands. Unpublished Ph.D. thesis, Albert-Ludwigs-Universität, Freiburg, West Germany.

Bauer JJ, 1985. Fecundity patterns of stable and colonising chamois populations of New Zealand and Europe. *In* Lovari S (Ed.) The Biology and Management of Mountain Ungulates, pp. 154-165. Croom Helm, London.

Baverstock PR, Adams M, Maxson LR, Yosida TH, 1983. Genetic differentiation among karyotypic forms of the black rat, *Rattus rattus*. Genetics 105: 969-983.

Bay-Petersen J, 1984. Competition for resources: the role of pig and dog in the Polynesian agricultural economy. Journal de la Société des Oceanistés 77: 121-129.

Baylis GTS, 1951. Incipient forest regeneration on Great Island, Three Kings Group. Records of the Auckland Institute and Museum 4(2): 103-109.

Beaglehole JC (Ed.), 1961. The Journals of Captain James Cook on His Voyages of Discovery. Vol. 2. The Voyage of the Resolution and Adventure 1772-1775. Cambridge University Press for the Hakluyt Society.

Beaglehole JC (Ed.), 1967. The Journals of Captain James Cook on His Voyages of Discovery. Vol. 3. The Voyage of the Resolution and Discovery, 1776-1780. Cambridge University Press for the Hakluyt Society.

Beatson NS, 1985a. Tuberculosis in red deer in New Zealand. *In* Fennessy PF and

Drew KR (Eds) Biology of Deer Production, pp. 147-150. Royal Society of NZ Bulletin 22.

Beatson NS, 1985b. Field observations of malignant catarrhal fever in red deer in New Zealand. *In* Fennessy PF and Drew KR (Eds) Biology of Deer Production, pp. 135-137. Royal Society of NZ Bulletin 22.

Begg AC, Begg NC, 1966. Dusky Bay. In the Steps of Captain Cook. Whitcombe and Tombs, Christchurch.

Begg AC, Begg NC, 1973. Port Preservation. Whitcombe and Tombs, Christchurch.

Bell BD, 1978. The Big South Cape Islands rat irruption. *In* Dingwall PR, Atkinson IAE and Hay C (Eds) The Ecology and Control of Rodents in New Zealand Nature Reserves, pp. 33-40. Department of Lands and Survey, Wellington.

Bell BD, 1981. Breeding and condition of possums *Trichosurus vulpecula* in the Orongorongo Valley, near Wellington, New Zealand 1966-1975. *In* Bell BD (Ed.) Proceedings of the first symposium on marsupials in New Zealand, pp. 87-139. Zoology Publications of Victoria University of Wellington No. 74.

Bell BD, 1983. The challenge of the "stoat invasion" on Maud Island. Forest and Bird 227: 12-14.

Bell BD, 1986. The conservation status of New Zealand wildlife. NZ Wildlife Service Occasional Publication No. 12: 1-103.

Bell CJE, Bell I, 1971. Subfossil moa and other remains near Mt Owen, Nelson. NZ Journal of Science 14: 749-758.

Bell DJ, 1981. Estimating the density of possums *Trichosurus vulpecula*. *In* Bell BD (Ed.) Proceedings of the first symposium on marsupials in New Zealand, pp. 177-182. Zoology Publications of Victoria University of Wellington No. 74.

Bell J, 1977. Breeding season and fertility of the wild rabbit, *Oryctolagus cuniculus* (L.) in North Canterbury, New Zealand. Proceedings of the NZ Ecological Society 24: 79-83.

Bellingham PJ, 1979. The birds of Ponui (Chamberlain's) Island, Hauraki Gulf, August 1978. Tane 25: 17-21.

Bellingham PJ, Hay JR, Hitchmough RA, McCallum J, 1982. Birds of Rakitu (Arid) Island. Tane 28: 141-147.

Benes J, 1986. Preliminary report: Fallow deer programme Lismore State Forest 136. Unpublished report, NZ Forest Service (Wellington Conservancy), 9 pp.

Bentley A, 1967. An Introduction to the Deer of Australia. John Gartner, Hawthorn Press, Melbourne, 224 pp.

Bentley EW, 1964. A further loss of ground by *Rattus rattus* L. in the United Kingdom during 1956-61. Journal of Animal Ecology 33: 371-373.

Bentley EW, Taylor EJ, 1965. Growth of laboratory-reared ship rats (*Rattus rattus* L.). Annals of Applied Biology 55: 193-205.

Berger NM, 1944. Acclimatization of fur-animals in west Siberia. Zoologicheskii Zhurnal 23: 267-274.

Berry RJ, 1968. The ecology of an island population of the house mouse. Journal of Animal Ecology 37: 445-470.

Berry RJ, Jakobson ME, 1974. Vagility in an island population of the house mouse. Journal of Zoology (London) 173: 341-354.

Berry RH, Searle AH, 1963. Epigenetic polymorphism of the rodent skeleton. Proceedings of the Zoological Society of London 140: 577-615.

Berry RJ, Tricker BJK, 1969. Competition and extinction: The mice of Foula, with notes on those of Fair Isle and St. Kilda. Journal of Zoology (London) 158: 247-265.

Bertram GCL, 1940. The biology of the Weddell and crabeater seals. Scientific Reports of the British Graham Land Expedition, 1934-37, 1: 1-139.

Best E, 1942. Forest Lore of the Maori. Dominion Museum Bulletin 14.

Best HA, 1974. A preliminary report on the natural history and behaviour of Hooker's sea lion at Enderby Island, Auckland Islands, New Zealand, December 1972 to March 1973. Ministry of Agriculture and Fisheries, Fisheries Technical Report No. 132.

Best LW, 1968. The ecology of *Rattus rattus rattus* L. in selected areas of the South Island, NZ. Unpublished M.Sc. thesis, University of Canterbury, Christchurch.

Best LW, 1969. Food of the roof-rat, *Rattus rattus rattus* (L.), in two forest areas of New Zealand. NZ Journal of Science 12: 258-267.

Best LW, 1973. Breeding season and fertility of the roof rat, *Rattus rattus rattus*, in two forest areas of New Zealand. NZ Journal of Science 16: 161-170.

Bettesworth DJ, 1972a. Aspects of the ecology of *Rattus norvegicus* on Whale Island, Bay of Plenty, New Zealand. Unpublished M.Sc. thesis, University of Auckland.

Bettesworth DJ, 1972b. *Rattus exulans* on Red Mercury Island. Tane 18: 117-118.

Beveridge AE, 1964. Dispersal and destruction of seed in central North Island podocarp forests. Proceedings of the NZ Ecological Society 11: 48-55.

Beveridge AE, Daniel MJ, 1965. Observations of a high population of brown rats (*Rattus norvegicus* Berkenhout, 1767) on Mokoia Island, Lake Rotorua. *NZ Journal of Science* 8: 174-189.

Beveridge AE, Daniel MJ, 1966. A field trial of new rat poison, compound S-6999, on brown rats. Proceedings of the NZ Ecological Society 13: 40-43.

Bhardwaj D, Khan JA, 1974. Food requirements of "black rat", *Rattus rattus* L. Journal of the Bombay Natural History Society 71: 605-608.

Bickham JW, Daniel MJ, Haiduk MW 1980. Karyotype of *Mystacina tuberculata* (Chiroptera: Mystacinidae). Journal of Mammalogy 61: 322-324.

Biggins JG, 1979. Olfactory communication in the brushtailed possum, *Trichosurus vulpecula* Kerr, 1792 (Marsupialia, Phalangeridae). Unpublished Ph.D. thesis, Monash University, Melbourne.

Biggins JG, 1984. Communication in possums: a review. *In* Smith A and Hume I (Eds.) Possums and Gliders, pp. 35-57. Surrey Beatty and Sons, Chipping Norton, New South Wales.

Biggins JG, Overstreet DH, 1978. Aggressive and non-aggressive interactions among captive populations of the brush-tail possum, *Trichosurus vulpecula* (Marsupialia, Phalangeridae). Journal of Mammalogy 59: 149-159.

Bigham ML, Cockrem FR, 1982. Feral sheep, have they any value? NZ Journal of Agriculture 144(3): 3-4.

Bishop JA, Hartley DJ, 1976. The size and age structure of rural populations of *Rattus norvegicus* containing individuals resistant to the anticoagulant poison warfarin. Journal of Animal Ecology 45: 623-646.

Blackburn A, 1967. A brief survey of Cuvier Island. Notornis 14(1): 3-8.

Blackwell JM, 1981. The role of the house mouse in disease and zoonoses. Symposia of the Zoological Society of London 47: 591-616.

Blood D, McGillis JR, Lovaaš AL, 1967. Weights and measurements of moose in Elk Island National Park, Alberta. Canadian Field Naturalist 81(4): 263-269.

Boersma A, 1974. Opossums in the Hokitika River catchment. NZ Journal of Forestry Science 4: 64-75.

Bolliger A, Whitten WK, 1940. Observations on the urine of *Trichosurus vulpecula*. Australian Journal of Science 2: 178.

Bolliger A, Whitten WK, 1948. The paracloacal (anal) glands of *Trichosurus vulpecula*. Journal and Proceedings of the Royal Society of New South Wales 82: 36-43.

Bokonyi S, 1974. The Przevalsky Horse. Souvenir Press, London.

Bokonyi S, 1984. Horse. *In* Mason IL (Ed.) Evolution of Domesticated Animals, pp. 162-173. Longman, New York.

Bomford M, Redhead T, 1987. A field experiment to examine the effects of food quality and population density on reproduction of wild house mice. Oikos 48: 304-311.

Bonner WN, 1968. The fur seal of South Georgia. British Antarctic Survey Scientific Reports No. 56.

Boreham P, 1981. Some aspects of the ecology and control of feral pigs in the Gudgenby Nature Reserve. Australian Capital Territory Conservation Service, Conservation Memorandum 10.

Bowie JY, 1980. A study of the nematode fauna of *Macropus rufogriseus fruticus* and *Macropus eugenii*. Unpublished M.Sc. thesis, University of Canterbury, Christchurch.

Bowie J, Bennett W, 1983. Parasites from possums in Waitahuna Forest. NZ Veterinary Journal 31: 163.

Bowring LD, Stonehouse B, 1968. Birth of elephant seal (*Mirounga leonina*) in New Zealand. NZ Journal of Marine and Freshwater Research 2: 574-575.

Boyd IL, 1981. Population changes and the distribution of a herd of feral goats (*Capra* sp.) on Rhum, Inner Hebrides, 1960-1978. Journal of Zoology (London) 193(3): 287-304.

Boyd-Watt H, 1937. On the wild goat in Scotland. Journal of Animal Ecology 6: 15-20.

Bradbury JW, 1977. Lek mating behaviour in the hammer-headed bat. Zeitschrift für Tierpsychologie 45: 225-255.

Bradbury JW, 1981. The evolution of leks. *In* Alexander AD and Tinkle DW (Eds) Natural Selection and Social Behaviour, pp. 138-169. Chiron Press, New York.

Braestrup FW, Thorsonn G, Wesenberg-Lund E, 1949. Vort lands dyreliv, Pt 1 pattedyr, fugle, krybdyr, padder. Gylendalske Boghandel, Nordisk Forlag, Copenhagen.

Bremner AG, Butcher CF, Patterson GB, 1984. The density of indigenous invertebrates on three islands in Breaksea Sound, Fiordland, in relation to the distribution of introduced animals. Journal of the Royal Society of NZ 14: 379-386.

Briese LA, Smith MH, 1973. Competition between *Mus musculus* and *Peromyscus polionotus*. Journal of Mammalogy 54: 968-969.

Brinck C, Erlinge S, Sandell M, 1983. Anal sac secretion in mustelids: a comparison. Journal of Chemical Ecology 9: 727-745.

Brisbin IL, Smith MV, Smith MH, 1977. Feral swine studies at the Savannah River Ecology Laboratory: an overview of program goals and designs. *In* Wood GW (Ed.) Research and Management of Wild Hog Populations, pp. 71-90. Belle W. Baruch Forest Science Institute, Georgetown.

Brockie RE, 1958. The ecology of the hedgehog, (*Erinaceus europaeus* L.) in Wellington, New Zealand. Unpublished M.Sc. thesis, Victoria University of Wellington.

Brockie RE, 1959. Observations of the food of the hedgehog in New Zealand. NZ Journal of Science 2: 121-136.

Brockie RE, 1960. Road mortality of the hedgehog (*Erinaceus europaeus* L.) in New Zealand. Proceedings of the Zoological Society of London 134: 505-508.

Brockie RE, 1964. Dental abnormalities in European and New Zealand hedgehogs. Nature 202: 1355-1356.

Brockie RE, 1974a. Studies on the hedgehog, *Erinaceus europaeus* L., in New Zealand. Unpublished Ph.D. thesis, Victoria University of Wellington.

Brockie RE, 1974b. The hedgehog mange mite *Caparinia tripilis* in New Zealand. NZ Veterinary Journal 22: 243-247.

Brockie RE, 1975. Distribution and abundance of the hedgehog (*Erinaceus europaeus* L.) in New Zealand, 1869-1973. NZ Journal of Zoology 2: 445-462.

Brockie RE, 1976. Self-anointing by wild hedgehogs, *Erinaceus europaeus*, in New Zealand. Animal Behaviour 24: 68-71.

Brockie RE, 1977. Leptospiral infections of rodents in the North Island. NZ Veterinary Journal 25: 89-91.

Brockie RE, 1982. Effects of commercial hunters on the number of possums, *Trichosurus vulpecula*, in the Orongorongo Valley, Wellington. NZ Journal of Ecology 5: 21-28.

Brockie RE, Bell BD, White AJ, 1981. Age structure and mortality of possum *Trichosurus vulpecula* populations from New Zealand. *In* Bell BD (Ed.) Proceedings of the first symposium on marsupials in New Zealand, pp. 63-86. Zoology Publications of Victoria University of Wellington No. 74.

Brockie RE, Cowan PE, Efford MG, White AJ, Bell BD, 1979. The demography of *Trichosurus vulpecula* in Australia and New Zealand. Unpublished internal report, Ecology Division, DSIR.

Brockie RE, Fitzgerald AE, Green WQ, Morris JY, Pearson AJ, 1984. Research on possums in New Zealand. Wildlife Research Liaison Group Review 5.

Brockie RE, Moeed A, 1986. Animal biomass in a New Zealand forest compared with other parts of the world. Oecologia 70: 24-34.

Brockie RE, Till DG, 1977. *Leptospira ballum* isolated from hedgehogs. NZ Veterinary Journal 25: 28-29.

Broekhuizen S, 1976. The situation of hare populations in the Netherlands. *In* Pielowski Z and Pucek Z (Eds) Ecology and Management of European Hare Populations, pp. 23-24. Polish Hunting Association, Warsaw.

Broekhuizen S, 1979. Survival in adult European hares. Acta Theriologica 24: 465-473.

Broekhuizen S, Kalsbeek F, 1969. Weisses Handgelenkband beim Feldhasen auch in den Niederlanden festgestellt. Zeitschrift für Jagdwissenschaft 15: 164-165.

Broekhuizen S, Maaskamp F, 1980. Behaviour of does and leverets of the European hare (*Lepus europaeus*) whilst nursing. Journal of Zoology (London) 191: 487-501.

Broekhuizen S, Maaskamp F, 1981. Annual production of young in European hares (*Lepus europaeus*) in the Netherlands. Journal of Zoology (London) 193: 499-516.

Broekhuizen S, Maaskamp F, 1982. Movement, home range and clustering in the European hare (*Lepus europaeus* Pallas) in the Netherlands. Zeitschrift für Säugetierkunde 47: 22-32.

Bromley GF, 1956. Ecology of the wild spotted deer in the Maritime Territory. *In* Jurgenson PB (Ed.) Studies of Mammals in Government Preserves. Ministry of Agriculture of the USSR, Moscow. (English translation by A Biron and LS Cole 1961; Israel Program for Science Translation.)

Bronson FH, 1979. The reproductive ecology of the house mouse. Quarterly Review of Biology 54: 265-299.

Brooker MG, Ridpath MG, 1980. The diet of the wedge-tailed eagle, *Aquila audax*, in Western Australia. Australian Wildlife Research 7: 433-452.

Brooks JE, Rowe FP, 1979. Commensal rodent control. World Health Organization Unpublished Document WHO/VBC/79.726.

Broome K, 1985. Fallow deer in Woodhill resurvey 1985. Unpublished report, NZ Forestry Service (Auckland Conservancy), 21 pp.

Brosset A, 1963. Statut actuel des mammifères de îles Galapagos. Mammalia 27: 323-338.

Brown JE, Lasiewski RC, 1972. Metabolism of weasels: the cost of being long and thin. Ecology 53: 939-943.

Brown KG, 1952. Observations on the newly born leopard seal. Nature (London) 170: 982-983.

Brown KG, 1957. The leopard seal at Heard Island, 1951-54. Australian National Antarctic Research Expeditions Internal Reports 16.

Bryant LD, Maser C, 1982. Classification and distribution. *In* Thomas JW and Toweill DE (Eds) Elk of North America — Ecology and Management, pp. 1-59. Stackpole Books, Harrisburg, Pennsylvania.

Bryden MM, Lim GHK, 1969. Blood parameters of the southern elephant seal (*Mirounga leonina*), in relation to diving. Comparative Biochemistry and Physiology 28: 139-148.

Bubenik AB, 1982. Physiology. *In* Thomas JW and Toweill DE (Eds) Elk of North America — Ecology and Management, pp. 125-179. Stackpole Books, Harrisburg, Pennsylvania.

Buchanan J, 1876. Botany of Kawau. Transactions of the NZ Institute 9: 503-527.

Buckingham RP, 1985. Wildlife of Wakatipu State Forest 1982-84. Unpublished report, NZ Forest Service (Southland Conservancy), 38 pp.

Buckingham RP, Elliott G, 1979. An ecological survey of the fauna of D'Urville Island. Unpublished internal report, Ecology Division, DSIR.

Bull PC, 1950. Birds and mammals observed in the country between Lake Taupo and the Hauhungaroa Range, January 1-13, 1950. Animal Ecology Section, DSIR, Report No. 9.

Bull PC, 1953a. Hedgehogs and ground-nesting birds. Annual Report of the Ornithological Society of NZ 1: 32.

Bull PC, 1953b. Parasites of the wild rabbit, *Oryctolagus cuniculus* (L.) in New Zealand. NZ Journal of Science and Technology B 34: 341-372.

Bull PC, 1964. Ecology of helminth parasites of the wild rabbit *Oryctolagus cuniculus* (L.) in New Zealand. DSIR Bulletin 158.

Bull PC, 1969. The smaller placental mammals of Canterbury. *In* Knox GA (Ed.) The Natural History of Canterbury, pp. 400-417. Reed, Wellington.

Bull RM, 1967. A study of the large New Zealand weevil, *Lyperobius huttoni* Pascoe, 1876 (Coleoptera: Curculionidae Molytinae). Unpublished M.Sc. thesis, Victoria University of Wellington.

Buller WL, 1877. On the proposed introduction of the polecat into New Zealand. Transactions and Proceedings of the NZ Institute 9: 634-635.

Buller WL, 1893. Notes on the bats of New Zealand. Transactions and Proceedings of the NZ Institute 25: 50-52.

Buller, WL, 1895. Some curiosities of bird life. Transactions and Proceedings of the NZ Institute 27: 134-142.

Buller WL, 1896. Notes on New Zealand ornithology, with an exhibition of specimens. Transactions and Proceedings of the NZ Institute 28: 326-358.

Buller WL, 1905. Supplement to the 'Birds of New Zealand' (2 vols). London, published by the author.

Bullock DJ, Pickering SP, 1984. The validity of horn ring counts to determine the age of Scottish feral goats (*Capra* (domestic)). Journal of Zoology (London) 202(4): 561-564.

Bunn TJ, 1979. The effects of food supply on the breeding behaviour and population ecology of kiore on Tiritiri Matangi Island. Unpublished M.Sc. thesis, University of Auckland.

Burnett TD, 1927. The Mackenzie Country. *In* Speight R, Wall A and Laing RM (Eds) Natural History of Canterbury. Philosophical Institute of Canterbury, Christchurch.

Burrows CJ, 1974. A botanist's view on the thar problem. Tussock Grasslands and Mountainlands Institute Review 28: 5-18.

Burton RW, 1930. Abnormal horns of sambar (*Cervus unicolor*). Journal of the Bombay Natural History Society 34: 1058-1059.

Busser J, Zweep A, van Oortmerssen GA, 1974. Variability in the aggressive behaviour of *Mus musculus domesticus*, its possible role in population structure. *In* van Abeelan JHF (Ed.) The Genetics of Behaviour, pp. 185-199. North Holland, Amsterdam.

Butler, S, 1863. A First Year in Canterbury Settlement. Brassington AC and Maling PB (Eds). Reprinted Blackwood & Janet Paul, Auckland, 1964.

Byrne D, 1973. Prehistoric coprolites: a study of human and dog coprolites from prehistoric archaeological sites in the North Island of New Zealand. Unpublished M.A. thesis, University of Auckland.

Cadman WA, 1966. The fallow deer. Forestry Commission Leaflet No. 52. Her Majesty's Printing Office, London, 39 pp.

Calaby JH, 1971. The current status of Australian Macropodidae. Australian Zoologist 16: 17-29.

Calaby JH, 1983. Red-necked wallaby *Macropus rufogriseus*. *In* Strahan R (Ed.) The Australian Museum Complete Book of Australian Mammals, pp. 239-241. Angus and Robertson, Sydney.

Calhoun JB, 1963. The ecology and sociology of the Norway rat. United States Department of Health, Education and Welfare, Public Health Service Publication 1008.

Campbell DJ, 1978. The effects of rats on vegetation. *In* Dingwall PR, Atkinson IAE and Hay C (Eds) The Ecology and Control of Rodents in New Zealand Nature Reserves, pp. 99-120. Department of Lands and Survey Information Series 4.

Campbell DJ, 1984. The vascular flora of the DSIR study area, lower Orongorongo Valley, Wellington, New Zealand. NZ Journal of Botany 22: 223-270

Campbell DJ, Rudge MR, 1984. Vegetation changes induced over ten years by goats and pigs at Port Ross, Auckland Islands (Subantarctic). NZ Journal of Ecology 7: 103-118.

Campbell DJ, Moller H, Ramsay GW and Watt JC, 1984. Observations on foods of kiore (*Rattus exulans*) found in husking stations on northern offshore islands of New Zealand. NZ Journal of Ecology 7: 131-138

Campbell PA, 1973. The feeding behaviour of the hedgehog (*Erinaceus europaeus* L.) in pasture land in New Zealand. Proceedings of the NZ Ecological Society 20: 35-40.

Carrick R, Csordas SE, Ingham SE, 1962. Studies on the southern elephant seal, *Mirounga leonina* (L.). IV. Breeding and development. Commonwealth Scientific and Industrial Research Organisation, Wildlife Research 7: 161-197.

Carrick R, Csordas SE, Ingham SE, Keith K, 1962. Studies on the southern elephant seal, *Mirounga leonina* (L.). III. The annual cycle in relation to age and sex. Commonwealth Scientific and Industrial Research Organisation, Wildlife Research 7: 119-160.

Carrick R, Ingham SE, 1962a. Studies on the southern elephant seal, *Mirounga leonina* (L.). I. Introduction to the series. Commonwealth Scientific and Industrial Research Organisation, Wildlife Research 7: 89-101.

Carrick R, Ingham SE, 1962b. Studies on the southern elephant seal, *Mirounga leonina* (L.) II. Canine tooth structure in relation to function and age determination. Commonwealth Scientific and Industrial Research Organisation, Wildlife Research 7: 102-118.

Carrick R, Ingham SE, 1962c. Studies on the southern elephant seal, *Mirounga leonina* (L.). V. Population dynamics and utilisation. Commonwealth Scientific and Industrial Research Organisation, Wildlife Research 7: 198-206.

Carroll ALK, 1968. Foods of the harrier. Notornis 15: 23-28.

Carter ME, Cordes DO, 1980. Leptospirosis and other infections of *Rattus rattus* and *Rattus norvegicus*. NZ Veterinary Journal 28: 45-50.

Cassels R, 1983. Prehistoric man and animals in Australia and Oceania. *In* Peel LJ and Tribe DE (Eds) Domestication, Conservation and the Use of Animal Resources, pp. 41-62. Elsevier Science Publishers B.V., Amsterdam.

Cassels R, 1984. The role of prehistoric man in the faunal extinctions of New Zealand and other Pacific Islands. *In* Martin PS and Klein RG (Eds) Quaternary Extinctions, pp. 741-767. University of Arizona Press, Tucson.

Catt DC, 1975. Growth, reproduction and mortality in Bennett's wallaby (*Macropus rufogriseus fruticus*) in South Canterbury, New Zealand. Unpublished M.Sc. thesis, University of Canterbury, Christchurch.

Catt DC, 1977. The breeding biology of Bennett's wallaby (*Macropus rufogriseus fruticus*) in South Canterbury, New Zealand. NZ Journal of Zoology 4: 401-411.

Catt DC, 1979. Age determination in Bennett's wallaby, *Macropus rufogriseus fruticus* (Marsupalia), in South Canterbury, New Zealand. Australian Wildlife Research 6: 13-18.

Catt DC, 1981. Growth and condition of Bennett's wallaby (*Macropus rufogriseus fruticus*) in South Canterbury, New Zealand. NZ Journal of Zoology 8: 295-300.

Caughley G, 1963. Dispersal rates of several ungulates introduced into New Zealand. Nature 200: 280-281.

Caughley G, 1965a. Standardizing the common name "possum" for *Trichosurus vulpecula*. Tuatara 13: 30.

Caughley G, 1965b. Horn rings and tooth eruption as criteria of age in the Himalayan thar, *Hemitragus jemlahicus*. NZ Journal of Science 8: 333-351.

Caughley G, 1966. Mortality patterns in mammals. Ecology 47: 906-918.

Caughley G, 1969. Habitat of the Himalayan thar *Hemitragus jemlahicus* (H. Smith). Journal of the Bombay Natural History Society 67: 103-105.

Caughley G, 1970a. Liberation, dispersal and distribution of Himalayan thar (*Hemitragus jemlahicus*) in New Zealand. NZ Journal of Science 13: 220-239.

Caughley G, 1970b. Eruption of ungulate populations, with emphasis on Himalayan thar in New Zealand. Ecology 51: 53-72.

Caughley G, 1970c. Fat reserves of Himalayan thar in New Zealand by season, sex, area and age. NZ Journal of Science 13: 209-219.

Caughley G, 1970d. Population statistics of chamois. Mammalia 24: 194-199.

Caughley G, 1971a. The name of the Himalayan ****. NZ Wildlife 32: 20-21.

Caughley G, 1971b. The season of births for Northern Hemisphere ungulates in New Zealand. Mammalia 35: 204-219.

Caughley G, 1971c. Demography, fat reserves and body size of a population of red deer *Cervus elaphus* in New Zealand. Mammalia 35: 367-383.

Caughley G, 1971d. An investigation of hybridisation between free-ranging wapiti and red deer in New Zealand. NZ Journal of Science 14: 993-1008.

Caughley G, 1983. The Deer Wars: the Story of Deer in New Zealand. Heinemann, Auckland.

Cawthorn MW, 1975. Preliminary report on the feeding habits and behaviour of the New Zealand sea-lion, *Phocarctos hookeri*, at the Auckland Islands, January 1975. Unpublished report to the Director-General, Department of Lands and Survey.

Cawthorn MW, Crawley MC, Mattlin RH, Wilson GJ, 1985. Research on pinnipeds in New Zealand. Wildlife Research Liaison Group Research Review No.7.

Challies CN, 1970. Fatness, size, fecundity, and survival as indicators of the planes of nutrition for deer in south-west Fiordland. NZ Forest Service, Forest Research Institute, Protection Forestry Branch Report 90: 1-24.

Challies CN, 1975. Feral pigs (*Sus scrofa*) on Auckland Island: status, and effects on vegetation and nesting sea birds. NZ Journal of Zoology 2: 479-490.

Challies CN, 1976. Feral pigs in New Zealand. *In* Whitaker AH and Rudge MR (Eds), The Value of Feral Farm Mammals in New Zealand, pp. 23-25. NZ Department of Lands and Survey Information Series 1.

Challies CN, 1977. Effects of commercial hunting on red deer densities in the Arawata Valley, South Westland 1972-76. NZ Journal of Forestry Science 7(3): 263-273.

Challies CN, 1978. Assessment of the physical well-being of red deer (*Cervus elaphus* L.) populations in South Westland, New Zealand. Unpublished Ph.D. thesis, University of Canterbury, Christchurch.

Challies CN, 1985a. Establishment, control, and commercial exploitation of wild deer in New Zealand. *In* Fennessy PF and Drew KR (Eds) Biology of Deer Production, pp. 23-36. Royal Society of NZ Bulletin 22.

Challies CN, 1985b. Commercial hunting of wild red deer in New Zealand. *In* Beason SL and Robertson SF (Eds) Game Harvest Management, pp. 279-287. Caesar Kleberg Wildlife Research Institute, Kingsville, Texas.

Challies CN, Burrows L, 1984. Deer control and vegetation response on Stewart Island. NZ Forest Service, Forest Research Institute, What's New in Forest Research 126: 1-4.

Challies CN, Christie AHC, 1984. Breeding season and productivity of feral goats on Arapawa Island. *In* Dingwall PR and Rudge MR (Eds) Biological and Ecological Values of Arapawa Island Scenic Reserve, pp. 59-61. NZ Department of Lands and Survey Information Series 13.

Chapman DI, Chapman NG, 1975. Fallow Deer: Their History, Distribution and Biology. Terence Dalton Ltd, Lavenham, Suffolk, 271 pp.

Chapman DI, Chapman NG, 1980. The distribution of fallow: a worldwide review. Mammal Review 10: 60-138.

Charleston WAG, 1980. Lungworm and lice of the red deer (*Cervus elaphus*) and the fallow deer (*Dama dama*) – a review. NZ Veterinary Journal 28: 150-152.

Charleston WAG, Innes JG, 1980. Seasonal trends in the prevalence and intensity of spiruroid nematode infections of *Rattus r. rattus*. NZ Journal of Zoology 7: 141-145.

Cheeseman TF, 1894. Notes on the New Zealand bat. Transactions and Proceedings of the NZ Institute 26: 218-222.

Childs JE, 1986. Size-dependent predation on rats (*Rattus norvegicus*) by house cats (*Felis catus*) in an urban setting. Journal of Mammalogy 67: 196-199.

Chilton C (Ed.) 1909. The Subantarctic Islands of New Zealand. Government Printer, Wellington.

Chitty D, 1954. The study of the brown rat and its control by poison. *In* Chitty D and Southern HN (Eds) Control of Rats and Mice, pp. 160-291. Clarendon Press, Oxford.

Choate T, 1965. Mammal and bird studies at Jackson's Bay. Science Record 15: 59-61.

Choate TS, 1967. Terrestrial vertebrates of the Monowai-Grebe River region. Science Record 17: 51-53.

Christie AHC, 1963. The ecology of chamois (*Rupicapra rupicapra* L.) in an alpine basin in Southern Nelson. Unpublished M.Sc. thesis, Victoria University of Wellington.

Christie AHC, 1965. Blindness in chamois. Protection Forestry Newsletter 3: 8.

Christie AHC, 1967. The sensitivity of chamois and red deer to temperature fluctuations. Proceedings of the NZ Ecological Society 14: 34-39.

Christie AHC, Andrews JRH, 1964. Introduced ungulates in New Zealand: (a) Himalayan thar. Tuatara 12: 69-77.

Christie AHC, Andrews JRH, 1965. Introduced ungulates in New Zealand: (b) Virginia deer. Tuatara 13: 1-8.

Christie AHC, Andrews JRH, 1966. Introduced ungulates in New Zealand: (d) Fallow. Tuatara 14: 82-88.

Clabby J, 1976. The Natural History of the Horse. Weidenfeld and Nicolson, London.

Clapperton BK, 1985. Olfactory communication in the ferret (*Mustela furo* L.) and its application in wildlife management. Unpublished Ph.D. thesis, Massey University, Palmerston North.

Clapperton BK, in press. Scent-marking behaviour of the ferret (*Mustela furo* L.). Animal Behaviour.

Clapperton BK, Fordham RA, Sparksman RI, 1987. Preputial glands of the ferret *Mustela furo* L. (Carnivora: Mustelidae). Journal of Zoology (London) 212: 356-361.

Clapperton BK, Minot EO, Crump DR, 1988. An olfactory recognition system in the ferret *Mustela furo* L. (Carnivora: Mustelidae). Animal Behaviour 36: 541-553.

Clapperton BK, Minot EO, Crump DR, 1989. Scent lures from the anal sac secretions of the ferret *Mustela furo* L. Journal of Chemical Ecology 15: 291-308.

Clark DA, 1980. Age and sex-dependent foraging strategies of a small mammalian omnivore. Journal of Animal Ecology 49: 549-563.

Clark DA, 1981. Foraging patterns of black rats across a desert-montane forest gradient in the Galapagos Islands. Biotropica 13: 182-194.

Clark DA, 1982. Foraging behaviour of a vertebrate omnivore (*Rattus rattus*): meal structure, sampling and diet breadth. Ecology 63: 763-772.

Clark DB, 1980. Population ecology of *Rattus rattus* across a desert-montane forest gradient in the Galapagos Islands. Ecology 61: 1422-1433.

Clark JM, 1972. Ecology of feral goats (*Capra hircus* L.) near Kaikoura, New Zealand. Unpublished B.Sc. thesis, University of Canterbury, Christchurch.

Clark JM, 1974. Ecology of feral goats (*Capra hircus* L.) in the southern King Country, New Zealand. Unpublished M.Sc. thesis, University of Canterbury, Christchurch.

Clark JM, 1976. A note on the feral goat (*Capra hircus* L.) in the Hunua ranges, Auckland. Tane 22: 129-134.

Clark JM, 1977. *Bertiella trichosuri* Khalil (Cestoda: Anoplocephalidae) in opossums (*Trichosurus vulpecula*) from northern Taranaki. NZ Journal of Zoology 4: 95.

Clark JM, Asher, GW, 1977. The feral goats of Arapawa Island. Unpublished report, Ecology Division, DSIR.

Clark MR, Dingwall PR, 1985. Conservation of islands in the Southern Ocean: a review of the protected areas of Insulantarctica. IUCN, Cambridge.

Clark WC, 1980. *Ancylostoma brasiliense* in the Kermadec Islands. NZ Veterinary Journal 28: 193.

Clark WC, 1981. *Cylicospirura advena* n.sp. (Nematoda: Spirocercidae) a stomach parasite from a cat in New Zealand, with observations on related species. Systematic Parasitology 3: 185-191.

Clark WC, Clarke CMH, 1981. Parasites of chamois in New Zealand. NZ Veterinary Journal 29: 144.

Clark WC, McKenzie JC, 1982. North Island kiwi, *Apteryx australis mantelli* (Apterygiformes: Aves): A new host for *Toxocara cati* (Nematoda: Ascaridoidea) in New Zealand. Journal of Parasitology 68: 175-176.

Clarke CMH, 1971. Liberations and dispersal of red deer in northern South Island districts. NZ Journal of Forestry Science 1: 194-207.

Clarke CMH, 1973. Dispersal of four strains of red deer in northern South Island districts. NZ Journal of Forestry Science 3: 342-354.

Clarke CMH, 1976a. Eruption, deterioration and decline of the Nelson red deer herd. NZ Journal of Forestry Science 5: 235-249.

Clarke CMH, 1976b. Fallow deer in the north of the South Island. NZ Wildlife 52: 5-13.

Clarke CMH, 1986. Chamois movements and habitat use in the Avoca River area, Canterbury, New Zealand. NZ Journal of Zoology 13: 175-198.

Clarke CMH, in press. Size, structure, and dynamics of a chamois population in Basin Creek, Canterbury, New Zealand. NZ Journal of Zoology.

Clarke CMH, Henderson RJ, 1981. Natural regulation of a non-hunted chamois population. NZ Journal of Ecology 4: 126-127.

Clarke CMH, Henderson RJ, 1984. Home range size and utilization by female chamois (*Rupicapra rupicapra* L.) in the Southern Alps, New Zealand. Acta Zoologica Fennica 171: 287-291.

Clayton-Greene KA, 1976. The vegetation of Mt Karioi and of forested areas in the Waikato, North Island, New Zealand. Unpublished M.Sc. thesis, University of Waikato, Hamilton.

Clemens WA, 1977. Phylogeny of the marsupials. *In* Stonehouse B and Gilmore D (Eds) The Biology of Marsupials, pp. 51-68. MacMillan Press, London.

Clout MN, 1977. The ecology of the possum (*Trichosurus vulpecula* Kerr) in *Pinus radiata* plantations. Unpublished Ph.D. thesis, University of Auckland.

Clout MN, 1980. Ship rats (*Rattus rattus* L.) in a *Pinus radiata* plantation. NZ Journal of Ecology 3: 141-145.

Clout MN, 1982. Determination of age in the brushtail possum using sections from decalcified molar teeth. NZ Journal of Zoology 9: 405-408.

Clout MN, Barlow ND, 1982. Exploitation of brushtail possum populations in theory and practice. NZ Journal of Ecology 5: 29-35.

Clout MN, Efford MG, 1984. Sex differences in the dispersal and settlement of brushtail possums (*Trichosurus vulpecula*). Journal of Animal Ecology 53: 737-749.

Clout MN, Gaze PD, 1984. Brushtail possums (*Trichosurus vulpecula* Kerr) in a New Zealand beech (*Nothofagus*) forest. NZ Journal of Ecology 7: 147-155.

Clout MN, Hay JR, 1981. South Island kokako (*Callaeas cinerea cinerea*) in *Nothofagus* forest. Notornis 28: 256-259.

Clutton-Brock J, 1981. Domesticated Animals from Early Times. British Museum (Natural History) and William Heinemann, London.

Clutton-Brock J, 1987. A Natural History of Domesticated Mammals. Cambridge University Press and British Museum (Natural History), London.

Clutton-Brock J, Corbet GB, Hills M, 1976. A review of the family Canidae, with a classification by numerical methods. Bulletin of the British Museum (Natural History), Zoology 29(3).

Coblentz BE, 1977. Some range relationships of goats on Santa Catalina Island, California. Journal of Range Management 30(6): 415-419.

Coblentz BE, 1980a. A unique ungulate breeding pattern. Journal of Wildlife Management 44(4): 929-933.

Coblentz BE, 1980b. Effects of feral goats on the Santa Catalina island ecosystem. *In* Power D (Ed.) the Californian Islands, pp. 167-170. Santa Barbara Museum of Natural History.

Cockayne L, 1903. A botanical excursion during mid winter to the southern islands of New Zealand. Transactions and Proceedings of the NZ Institute 36 (19 n.s.): 225-333.

Cockayne L, 1919. An economic investigation of the montane tussock-grassland of New Zealand. III. Notes on depletion of the grassland. NZ Journal of Agriculture 19: 129-138.

Cody A, 1981. New Zealand bats. NZ Speleological Bulletin 6(116): 361-366.

Coleman JD, 1981. Tuberculosis and the control of possums *Trichosurus vulpecula* — an expensive business. *In* Bell BD (Ed.) Proceedings of the first symposium on marsupials in New Zealand, pp. 211-220. Zoology Publications of Victoria University of Wellington No. 74.

Coleman JD, Green WQ, 1984. Variations in the sex and age distributions of brush-tailed possum populations. NZ Journal of Zoology 11: 313-318.

Coleman JD, Gillman A, Green WQ, 1980. Forest patterns and possum densities within podocarp/mixed hardwood forests on Mt Bryan O'Lynn, Westland. NZ Journal of Ecology 3: 69-84.

Coleman JD, Green WQ, Polson JG, 1985. Diet of brushtail possums over a pasture-alpine gradient in Westland, New Zealand. NZ Journal of Ecology 8: 21-35.

Colenso W, 1890. Bush notes. Transactions and Proceedings of the NZ Institute 23: 477-491.

Collier GE, O'Brien SJ, 1985. A molecular phylogeny of the Felidae: immunological distance. Evolution 39: 473-487.

Collins GH, 1973. A limited survey of gastro-intestinal helminths of dogs and cats. NZ Veterinary Journal 21: 175-176.

Collins GH, 1981a. Sarcocystis in rats from Stewart Island. NZ Journal of Zoology 8: 129.

Collins GH, 1981b. Studies in *Sarcocystis* species VIII: *Sarcocystis* and *Toxoplasma* in red deer (*Cervus elaphus*). NZ Veterinary Journal 29: 126-127.

Collins GH, Charleston WAG, 1979a. Studies on *Sarcocystis* species: 1. Feral cats as definitive hosts for sporozoa. NZ Veterinary Journal 27: 80-84.

Collins GH, Charleston WAG, 1979b. Studies on *Sarcocystis* species II: Infection in wild and feral animals — prevalance and transmission. NZ Veterinary Journal 27: 134-135.

Collins GH, Charleston WAG, Wiens BG, 1980. Studies on *Sarcocystis* species. VI. A comparison of three methods for the detection of *Sarcocystis* species in muscle. NZ Veterinary Journal 28: 173.

Collins GH, Crawford SJS, 1978. *Sarcocystis* in goats: prevalence and transmission. NZ Veterinary Journal 26: 288.

Coman BJ, 1972. A survey of the gastro-intestinal parasites of the feral cat in Victoria. Australian Veterinary Journal 48: 133-136.

Condy PR, 1977. Ross seal, *Ommatophoca rossi* (Gray, 1850), with notes from the results of surveys conducted in the King Haakon VII Sea, Jan/Feb 1974-76. Proceedings of a symposium on endangered wildlife in southern Africa, Pretoria, 1976, pp. 1-16.

Cook BR, 1975. Tuberculosis in possums, Hohonu Mountain. MAF/NZ Forest Service Project 117. Ministry of Agriculture and Fisheries Animal Health Division Technical Report.

Cooke BD, 1982. A shortage of water in natural pastures as a factor limiting a population of rabbits, *Oryctolagus cuniculus* (L.), in arid northeastern South Australia. Australian Wildlife Research 9: 456-476.

Coop IE, Clark IR, 1966. The influence of live weight on wool production and reproduction in high country flocks. NZ Journal of Agricultural Research 9: 165-181.

Copson GR, 1986. The diet of the introduced rodents *Mus musculus* L. and *Rattus rattus* L. on subantarctic Macquarie Island. Australian Wildlife Research 13: 441-445.

Corbet GB, 1964. The Identification of British Mammals. British Museum (Natural History), London.

Corbet GB, 1978. The Mammals of the Palaearctic Region: A Taxonomic Review. British Museum (Natural History), London.

Corbet GB, 1984. The Mammals of the Palaearctic Region: A Taxonomic Review. Supplement. British Museum (Natural History), London.

Corbet GB, Harris S (Eds), 1990. The Handbook of British Mammals, 3rd edition. Blackwell Scientific Publications, Oxford.

Corbet GB, Hill JE, 1980. A World List of Mammalian Species. British Museum (Natural History), London.

Corbet GB, Hill JE, 1986. A World List of Mammalian Species, 2nd edition. British Museum (Natural History), London.

Corbet GB, Southern HN (Eds), 1977. The Handbook of British Mammals, 2nd edition. Blackwell Scientific Publications, Oxford. (For 3rd edition, see Corbet & Harris, 1990.)

Corbett L, Newsome A, 1975. Dingo society and its maintenance: a preliminary analysis. *In* Fox MW (Ed.) The Wild Canids, pp. 369-379. Van Nostrand Reinhold, New York.

Corboy JD, 1949. Stewart Island 1948-49 summer season. Unpublished report, Department of Internal Affairs.

Corner LA, Presidente PA, 1981. *Mycobacterium bovis* infection in the brushtailed possum (*Trichosurus vulpecula*): II. Comparison of experimental infections with an Australian cattle strain and a New Zealand possum strain. Veterinary Microbiology 6: 351-366.

Courtenay-Latimer M, 1961. Two rare seal records for South Africa. Annals of the Cape Province Museum 1: 102.

Couturier MAJ, 1938. Le Chamois. Published by the author (B. Arthaud, Ed.), Grenoble.

Cowan J, 1907. Kapiti Island. Report to Dept of Tourism and Health Resorts. C.8A. Appendix to the Journals of the House of Representatives, New Zealand, Vol. II, pp. 1-4. Wellington.

Cowan PE, 1981. Early growth and development of roof rats, *Rattus rattus* L. Mammalia 45: 239-250.

Cowan PE, Atkinson I, Bell B, 1985. Kapiti Island — the last possum? Forest and Bird 16: 12-13.

Cowan PE, Barnett SA, 1975. The new-object and new-place reactions of *Rattus rattus* L. Zoological Journal of the Linnaean Society 56: 219-234.

Cowan PE, Moeed A, 1987. Invertebrates in the diet of brushtail possums, *Trichosurus vulpecula*, in lowland podocarp/broadleaf forest, Orongorongo Valley, Wellington, New Zealand. NZ Journal of Zoology 14: 163-177.

Cowan PE, Waddington DC, Daniel MJ, Bell BD, 1985. Aspects of litter production in a New Zealand lowland podocarp/broadleaf forest. NZ Journal of Botany 23: 191-199.

Cox JE, Taylor RH, Mason R, 1967. Motunau Island, Canterbury, New Zealand. An ecological survey. DSIR Bulletin 178.

Craig JL, 1986. The effects of kiore on other fauna. *In* Wright AE and Beever RE (Eds) The Offshore Islands of Northern New Zealand, pp. 75-83. Department of Lands and Survey Information Series No. 16.

Crandall LS, 1964. The Management of Wild Animals in Captivity. University of Chicago Press.

Crawley MC, 1972. Distribution and abundance of New Zealand fur seals on the Snares Islands, New Zealand. NZ Journal of Marine and Freshwater Research 6: 115-126.

Crawley MC, 1973. A live-trapping study of Australian brush-tailed possums, *Trichosurus vulpecula* Kerr, in the Orongorongo Valley, Wellington, New Zealand. Australian Journal of Zoology 21: 75-90.

Crawley MC, 1975. Growth of New Zealand fur seal pups. NZ Journal of Marine and Freshwater Research 9: 539-545.

Crawley MC, 1978. Weddell seal harvesting at Scott Base, McMurdo Sound, Antarctica, 1970-76. NZ Journal of Ecology 1: 132-137.

Crawley MC, Brown DL, 1971. Measurements of tagged pups and a population estimate of New Zealand fur seals on Taumaka, Open Bay Islands, Westland. NZ Journal of Marine and Freshwater Research 5: 389-395.

Crawley MC, Cameron DB, 1972. New Zealand sea lions, *Phocarctos hookeri*, on the Snares Islands. NZ Journal of Marine and Freshwater Research 6: 127-132.

Crawley MC, Stark JD, Dodgshun PS, 1977. Activity budgets of New Zealand fur seals, *Arctocephalus forsteri*, during the breeding season. NZ Journal of Marine and Freshwater Research 11: 777-788.

Crawley MC, Warneke R, 1979. New Zealand fur seal. *In* Mammals in the Seas, Volume II: Pinniped Species Summaries and Report on Sirenians, pp. 45-48. Food and Agriculture Organization Fisheries Series No. 5, Rome.

Crawley MC, Wilson GJ, 1976. The natural history and behaviour of the New Zealand fur seal (*Arctocephalus forsteri*). Tuatara 22: 1-29.

Croft JD, Hone LJ, 1978. The stomach contents of foxes, *Vulpes vulpes*, collected in New South Wales. Australian Wildlife Research 5: 85-92.

Crook IG, 1969. Feral goats of North Wales. Animals 12(1): 13-15.

Crook IG, 1973. The tuatara, *Sphenodon punctatus* Gray, on islands with and without populations of the Polynesian rat, *Rattus exulans* (Peale). Proceedings of the NZ Ecology Society 20: 115-120.

Crosby A, 1986. Ecological Imperialism: the Biological Expansion of Europe, 900-1900. Cambridge University Press, Cambridge.

Crowcroft P, 1959. Spatial distribution of feeding activity in the wild house mouse (*Mus musculus* L.). Annals of Applied Biology 47: 150-155.

Crowcroft P, 1966. Mice All Over. Foulis, London.

Crowcroft P, Rowe FP, 1963. Social organisation and territorial behaviour in the house mouse (*Mus musculus* L.). Proceedings of the Zoological Society of London 140(3): 44-62.

Croxall JP, Evans PGH, Schreiber RW (Eds) 1984. Status and conservation of the world's seabirds. ICBP Technical Publication No. 2.

Crozet JM, 1891. Voyage to Tasmania, New Zealand, the Ladrone Islands, and the Philippines in the years 1771-1772. H Ling Roth (Transl.). Truslove and Shirley, London.

Cruise RA, 1823. Journal of a Ten Months' Residence in New Zealand [1820]. Republished 1957 by Pegasus Press, Christchurch, AG Bagnall (Ed.).

Crump DR, 1980a. Thietanes and dilthiolanes from the anal gland of the stoat (*Mustela erminea*). Journal of Chemical Ecology 6: 341-347.

Crump DR, 1980b. Anal gland secretion of the ferret (*Mustela putorius* forma *furo*). Journal of Chemical Ecology 6: 837-844.

Crump DR, Moors PJ, 1985. Anal gland secretions of the stoat (*Mustela erminea*) and the ferret (*Mustela putorius* forma *furo*). Some additional thietane compounds. Journal of Chemical Ecology 11: 1037-1043.

Csordas SE, 1958. Breeding of the fur seal, *Arctocephalus forsteri* Lesson, at Macquarie Island. Australian Journal of Science 21: 87-88.

Csordas SE, 1963. Sea lions on Macquarie Island. Victorian Naturalist 80: 32-35.

Csordas SE, Ingham SE, 1965. The New Zealand fur seal, *Arctocephalus forsteri* (Lesson), at Macquarie Island, 1949-64. Commonwealth Scientific and Industrial Research Organisation, Wildlife Research 10: 83-99.

Cuddihy MJ, Ross AD, 1979. Forests, grasslands and animals of Mount Aspiring National Park and Dart State Forest. Unpublished report, NZ Forest Service (Southland Conservancy), 53 pp.

Cuffe CTW, 1914. Abnormal sambhar horns. Journal of the Bombay Natural History Society 23: 356.

Cumming DHM, 1984. Wild pigs and boars. *In* Macdonald D (Ed.) The Encyclopaedia of Mammals 2, pp. 500-503. G. Allen and Unwin, London.

Cunningham A, 1964. Reports on the headwaters of the Tutaekuri River, Hawke's Bay. Unpublished report, NZ Forest Service.

Cunningham A, 1979. A century of change in the forests of the Ruahine Range, North Island, New Zealand. NZ Journal of Ecology 2: 11-21.

Cunningham DM, Moors PJ, 1983. A guide to the identification and collection of New Zealand rodents. NZ Wildlife Service Occasional Publication 4.

Curtis OM, 1951. Back to the moose country. NZ Fishing and Shooting Gazette 18(11): 19-22.

Cuthbertson K, 1974. Pighunting in New Zealand. AH and AW Reed, Wellington.

Dagg AI, Windsor DE, 1972. Swimming in northern terrestrial mammals. Canadian Journal of Zoology 50: 117-130.

Dailey MD, Brownell RL, 1972. A checklist of marine mammal parasites. *In* Ridgway SH (Ed.) Mammals of the Sea: Biology and Medicine, pp. 528-589. Charles C. Thomas, Illinois.

Dale RW, James JL, 1977. Frost and environment in the Kaimai Range. NZ Forest Service Technical Paper 65.

Daniel MJ, 1963. Early fertility of red deer hinds in New Zealand. Nature 200: 380.

Daniel MJ, 1966a. A preliminary survey of the incidence of brucellosis, leptospirosis, and salmonellosis in red deer in New Zealand. NZ Journal of Science 9: 399-408.

Daniel MJ, 1966b. Early trials with sodium monofluoroacetate (compound 1080) for the control of introduced deer in New Zealand. NZ Forest Service Technical Paper 51.

Daniel MJ, 1967. A survey of diseases in fallow, Virginia and Japanese deer, chamois, tahr, and feral goats and pigs in New Zealand. NZ Journal of Science 10: 949-963.

Daniel MJ, 1969. A survey of rats on Kapiti Island, New Zealand. NZ Journal of Science 12: 363-372.

Daniel MJ, 1970. Bat sightings on Kapiti Island, New Zealand, 1906-1969. Proceedings of the NZ Ecological Society 17: 136-138.

Daniel MJ, 1972. Bionomics of the ship rat (*Rattus r. rattus*) in a New Zealand indigenous forest. NZ Journal of Science 15: 313-341.

Daniel MJ, 1973. Seasonal diet of the ship rat (*Rattus r. rattus*) in lowland forest in New Zealand. Proceedings of the NZ Ecological Society 20: 21-30.

Daniel MJ, 1975. First record of an Australian fruit bat (Megachiroptera: Pteropodidae) reaching New Zealand. NZ Journal of Zoology 2: 227-231.

Daniel MJ, 1976. Feeding by the short-tailed bat (*Mystacina tuberculata*) on fruit and possibly nectar. NZ Journal of Zoology 3: 391-398.

Daniel MJ, 1978. Population ecology of ship and Norway rats in New Zealand. *In* Dingwall PR, Atkinson IAE, and Hay C (Eds) The Ecology and Control of Rodents in New Zealand Nature Reserves, pp. 145-152. Department of Lands and Survey Information Series 4.

Daniel MJ, 1979. The New Zealand short-tailed bat, *Mystacina tuberculata*: a review of present knowledge. NZ Journal of Zoology 6: 357-370.

Daniel MJ, 1981. First record of a colony of long-tailed bats in a *Pinus radiata* forest. NZ Journal of Forestry 26: 108-111.

Daniel M, Baker A, 1986. Collins Guide to the Mammals of New Zealand. W. Collins Publishers, Auckland.

Daniel MJ, Christie AHC, 1963. Untersuchungen über Krankheiten der Gemse (*Rupicapra rupicapra* L.) und des Thars (*Hemitragus jemlaicus* Smith) in den Sudalpen von Neuseeland. Schweizer Archiv für Tierheilkunde 105: 399-411.

Daniel MJ, Williams GR, 1981. Long-tailed bats (*Chalinolobus tuberculatus*) hibernating in farm buildings near Geraldine, South Canterbury. NZ Journal of Zoology 8: 425-430.

Daniel MJ, Williams GR, 1983. Observations of a cave colony of the long-tailed bat (*Chalinolobus tuberculatus*) in North Island, New Zealand. Mammalia 47: 71-80.

Daniel MJ, Williams GR, 1984. A survey of the distribution, seasonal activity and roost sites of New Zealand bats. NZ Journal of Ecology 7: 9-25.

Daniel MJ, Yoshiyuki M, 1982. Accidental importation of a Japanese bat into New Zealand. NZ Journal of Zoology 9: 461-462.

Dansie O, Wince W, 1968-70. Deer of the world. Supplements to Deer, Vol. 1, No. 7, 8 and 10; and Vol. 2, No. 1 and 2.

Dards JL, 1983. The behaviour of dockyard cats: interactions of adult males. Applied Animal Ethology 10: 133-153.

Darwin C, 1845. Journal of Researches into the Natural History and Geology of the Countries Visited during the Voyage Round the World of HMS Beagle under the command of Captain Fitzroy RN. John Murray, London.

Davidson J, 1984. The Prehistory of New Zealand. Longman Paul, Auckland.

Davidson MM, 1965. Changes in wild mammal populations in Canterbury. NZ Forest Service Information Series 52.

Davidson MM, 1971. A case of polydactylism in sika deer in New Zealand. Journal of Wildlife Diseases 7: 109-110.

Davidson MM, 1973a. Characteristics, liberation and dispersal of sika deer (*Cervus nippon*) in New Zealand. NZ Journal of Forestry Science 3(2): 153-180.

Davidson MM, 1973b. Use of habitat by sika deer. *In* Orwin J (Ed.) Assessment and management of introduced animals in New Zealand forests. NZ Forest Service, Forest Research Institute Symposium 14: 55-67.

Davidson MM, 1976. Season of parturition and fawning percentages of sika deer (*Cervus nippon*) in New Zealand. NZ Journal of Forestry Science 5(3): 355-357.

Davidson MM, 1979. Movement of marked sika (*Cervus nippon*) and red deer (*Cervus elaphus*) in Central North Island. NZ Journal of Forestry Science 9(1): 77-88.

Davidson MM, Gannaway PA, 1967. Climate observations in the Kaweka Range made in conjunction with a sika deer research project. Unpublished report No. 39, NZ Forest Service, Protection Forestry Branch.

Davies FJ, 1980. The prehistoric environment of the Dunedin area: the approach of salvage prehistory. Unpublished M.A. thesis, University of Otago, Dunedin.

Davies JL, 1957. A hedgehog road mortality index. Proceedings of the Zoological Society of London 128: 606-608.

Davis DE, 1953. The characteristics of rat populations. Quarterly Review of Biology 28: 373-401.

Davis DE, Emlen JT, Stokes AW, 1948. Studies on home range in the brown rat. Journal of Mammalogy 29: 207-225.

Davis LS, 1979. Social rank behaviour in a captive colony of Polynesian rats (*Rattus exulans*). NZ Journal of Zoology 6: 371-380.

Davis MR, Orwin J (Eds), 1985. Report on a survey of the proposed wapiti area, West Nelson. NZ Forest Service, Forest Research Institute Bulletin 84: 1-245.

Davis RA, 1979. Unusual behaviour by *Rattus norvegicus*. Journal of Zoology 188: 298.

Dawson TJ, 1983. Monotremes and Marsupials: the Other Mammals. Studies in Biology No. 150. Edward Arnold, London.

Dawson TJ, Hulbert AJ, 1970. Standard metabolism, body temperature, and surface areas of Australian marsupials. American Journal of Physiology 218: 1233-1238.

Day MG, 1968. Food habits of British stoats (*Mustela erminea*) and weasels (*Mustela nivalis*). Journal of Zoology (London) 155: 485-497.

Deanesly R, 1934. The reproductive processes in certain mammals. VI. The reproductive cycle of the female hedgehog. Philosophical Transactions of the Royal Society of London. B 223: 239-276.

Deanesly R, 1935. The reproductive processes of certain mammals. IX. Growth and reproduction in the stoat (*Mustela erminea*). Philosophical Transactions of the Royal Society of London. B 225: 459-492.

Deanesly R, 1944. The reproductive cycle of the female weasel (*Mustela nivalis*). Proceedings of the Zoological Society of London 114: 339-349.

Dearborn JH, 1965. Food of Weddell seals at McMurdo Sound, Antarctica. Journal of Mammalogy 46: 37-43.

Dearlove A, 1987. Hybrid survivors keep Père David hopes alive. NZ Journal of Agriculture, June 28-29.

Debrot S, 1981. Trophic relations between the stoat (*Mustela erminea*) and its prey, mainly the water vole (*Arvicola terrestis* Sherman). *In* Chapman JA and Purseley D (Eds) Proceedings of the First Worldwide Furbearer Conference, Frostburg MD, pp. 1259-1289.

Debrot S, 1983. Fluctuations de populations chez l'hermine (*Mustela erminea* L.). Mammalia 47: 323-332.

Debrot S, 1984. The structure and dynamics of a stoat (*Mustela erminea*) population. Revue Ecologie (Terre et Vie) 39: 77-88.

Debrot S, Mermod C, 1983. The spatial and temporal distribution pattern of the stoat (*Mustela erminea* L.). Oecologia 59: 69-73.

Debrot S, Weber JM, Marchesi P, Mermod C, 1985. The day and night activity pattern of the stoat (*Mustela erminea* L.). Mammalia 49: 13-17.

Deem JM, Jenkinson GM, 1914. The rabbit pest: control by effective fencing. NZ Journal of Agriculture, July 20: 62-66.

Delaney MJ, 1982. Mammal Ecology. Blackie & Son, Glasgow.

Delattre P, 1984. Influence de la pression de prédation exercée par une population de belettes (*Mustela nivalis*) sur un peuplement de Microtidae. Acta Oecologica, Oecologia Generalis 5: 285-300.

Delong KT, 1967. Population ecology of feral house mice. Ecology 48: 611-634.

De Master DP, 1979. Weddell seal. *In* Mammals in the Seas. Volume II: Pinniped Species Summaries and Report on Sirenians, pp. 130-134. Food and Agriculture Organisation Fisheries Series No. 5, Rome.

Deuss F, 1983a. Fallow deer in Woodhill. Unpublished report, NZ Forest Service (Auckland Conservancy), 39 pp.

Deuss F, 1983b. The deerhunter in Woodhill. Unpublished report, NZ Forest Service (Auckland Conservancy), 20 pp.

Dewsbury DA, Lanier DL, Miglietta A, 1980. A laboratory study of climbing behaviour in 11 species of muroid rodents. American Midland Naturalist 130: 66-72.

Diaz de la Guardia R, Camacho JPM, Ladron de Guevara RG, 1981. Autosomal and sex-chromosomal polymorphism in a wild population of the Norway rat, *Rattus norvegicus*. Genetica 56: 93-97.

Dick AMP, 1985. Rats on Kapiti Island, New Zealand: co-existence and diet of *Rattus norvegicus* Berkenhout and *Rattus exulans* Peale. Unpublished M.Sc. thesis, Massey University, Palmerston North.

Dickinson CD, Scott PP, 1956. Nutrition of the cat. 1. A practical stock diet supporting growth and reproduction. British Journal of Nutrition 10: 304-311.

Dieffenbach E, 1841. An Account of the Chatham Islands. Journal of the Royal Geographical Society 11: 195-215.

Dieffenbach E, 1843. Travels in New Zealand, 2 volumes. John Murray, London.

Dilks PJ, 1979. Observations on the food of feral cats on Campbell Island. NZ Journal of Ecology 2: 64-66.

Dilks PJ, Wilson PR, 1979. Feral sheep and cattle and royal albatrosses on Campbell Island; population trends and habitat changes. NZ Journal of Zoology 6: 127-139.

Dingwall PR, Rudge MR, 1984. Biological and ecological values of Arapawa Island Scenic Reserve. Department of Lands and Survey Information Series 13.

Dingwall PR, Atkinson IAE, Hay C, 1978. The ecology and control of rodents in New Zealand nature reserves. Department of Lands and Survey Information Series 4.

Donald VE, 1927. Deer stalking in New Zealand. NZ Fishing and Shooting Gazette 1: 4-6.

Donne TE, 1924. The Game Animals of New Zealand. John Murray, London.

Doré AB, 1918. Rat trypanosomes from New Zealand. NZ Journal of Science and Technology 1: 200.

Douglas CE, 1894. On the Westland Alps. Appendices to the Journal of the NZ House of Representatives, 1894 C1: 71-75.

Douglas, MH 1967. Control of thar (*Hemitragus jemlahicus*): evaluation of a poisoning technique. NZ Journal of Science 10: 511-526.

Douglas MH, 1984. The warning whistle for the Himalayan thar. Forest and Bird 15: 2-6.

Douglas MJW, 1970a. Movements of hares, *Lepus europaeus* Pallas, in high country in New Zealand. NZ Journal of Science 13: 287-305.

Douglas MJW, 1970b. Foods of harriers in a high country habitat. Notornis 17: 92-95.

Douglas MJW, 1982. Biology and management of rusa deer on Mauritius. Tigerpaper IX (3): 1-10. Food and Agriculture Organization of the United Nations, Bangkok.

Douglas MJW, 1983. Status and future management of the Manawatu sambar deer herd. NZ Forest Service, Forest Research Institute Bulletin 30.

Douglas N, 1959. The Douglas Score, a Handbook on the Measuring of Antlers, Horns and Tusks. New Zealand Deerstalkers' Association, Wellington.

Downes TW, 1926. Maori rat-trapping devices. Journal of the Polynesian Society 25: 228-234.

Draisma M, 1979. Some aspects of the biology of wild sambar in Victoria, Australia. NZ Deerstalkers' Association International Wildlife Forum.

Druett J, 1983. Exotic Intruders. The Introduction of Plants and Animals into New Zealand. Heinemann, Auckland.

Dunn E, 1977. Predation by weasels (*Mustela nivalis*) on breeding tits (*Parus* sp.) in relation to the density of tits and rodents. Journal of Animal Ecology 46: 633-652.

Dunnet GM, 1964. A field study of local populations of the brush-tailed possum, *Trichosurus vulpecula*, in eastern Australia. Proceedings of the Zoological Society of London 142: 665-695.

Dunson WA, Lazell JD, 1982. Urinary concentrating capacity of *Rattus rattus* and other mammals from the lower Florida Key. Comparative Biochemistry and Physiology 71: 17-21.

Dwyer PD, 1960a. New Zealand bats. Tuatara 8(2): 61-71.

Dwyer PD, 1960b. Studies on New Zealand Chiroptera. Unpublished M.Sc. thesis, Victoria University of Wellington.

Dwyer PD, 1962. Studies on the two New Zealand bats. Zoology Publications from Victoria University of Wellington 28: 1-28.

Dwyer PD, 1970. Size variation in the New Zealand short-tailed bat. Transactions of the Royal Society of NZ (Biological Sciences) 12: 239-243.

Dwyer PD, 1978. A study of *Rattus exulans* (Peale) (Rodentia: Muridae) in the New Guinea Highlands. Australian Journal of Wildlife Research 5: 221-248.

Dymond JR, 1922. The European hare in Ontario. Canadian Field-Naturalist 36: 142-143.

Earle A, 1832. Narrative of a Residence in New Zealand. Journal of a Residence in Tristan da Cunha. McCormick EH (Ed.), 1966. Clarendon Press, Oxford.

East K, Lockie JD, 1965. Further observations on weasels (*Mustela nivalis*) and stoats (*Mustela erminea*) born in captivity. Journal of Zoology (London) 147: 234-238.

East R, 1972. Starling (*Sturnus vulgaris* L.) predation on grass grub (*Costelytra zealandica* (White) (Melolonthinae) populations in Canterbury. Unpublished Ph.D. thesis, Lincoln College, Canterbury.

Ecke DH, 1958. Analysis of populations of the roof rat in southwest Georgia. United States Public Health Monograph 27: 1-18.

Edwards GP, Ealey EHM, 1975. Aspects of the ecology of the swamp wallaby *Wallabia bicolor* (Marsupialia: Macropodidae). Australian Mammalogy 1: 307-317.

Efford MG, 1976. *Rattus exulans* in Polynesia — a case of morphometric divergence. Unpublished B.Sc. (Hons) thesis, Victoria University of Wellington.

Efford MG, 1981. Growth rates of possum pouch young in the Orongorongo Valley 1980-81. Internal report, Ecology Division, DSIR.

Efford MG, Karl BJ, Moller H, 1988. Population ecology of *Mus musculus* on Mana Island. Journal of Zoology (London) 216: 539-563.

Egan H, 1969. Some notes on the colour phases of fallow deer in New Zealand. NZ Wildlife 27: 14-17.

Egan H, 1970. Further notes on the colour phases of fallow deer. NZ Wildlife 31:9.

Egoscue HJ, 1970. A laboratory colony of the Polynesian rat, *Rattus exulans*. Journal of Mammalogy 51: 261-266.

Ekdahl MO, Smith BL, Money DFL, 1970. Tuberculosis in some wild and feral animals in New Zealand. NZ Veterinary Journal 18: 44-45.

Elder NL, 1959. Vegetation of the Kaweka Range. Transactions of the Royal Society of NZ 87(1-2): 9-26.

Elder NL, 1962. Vegetation of the Kaimanawa Ranges. Transactions of the Royal Society of NZ (Botany) 2: 1-37.

Elder NL, 1965. Vegetation of the Ruahine Range: an introduction. Transactions of the Royal Society of NZ (Botany) 3: 13-66.

Elton CS, 1953. The use of cats in farm rat control. British Journal of Animal Behaviour 1: 151-155.

Epstein H, Mason IL, 1984. Cattle. *In* Mason IL (Ed.) Evolution of Domesticated Animals, pp. 6-27. Longman, New York.

Erickson AW, Hofman RJ, 1974. Antarctic seals. *In* Bushnell VC (Ed.) Antarctic Mammals, p. 8. Antarctic Map Folio Series 18, American Geographical Society, New York.

Erickson AW, Siniff DB, Cline DR, Hofman RJ, 1971. Distributional ecology of Antarctic seals. *In* Deacon, Sir George (Ed.) Procedings of a Symposium on Antarctic Ice and Water Masses, pp. 55-76. Scientific Community Antarctic Research, Brussels.

Erickson AW, Denney RN, Brueggeman JJ, Sinha AA, Bryden MM, Otis J, 1974. Seal and bird populations of Adelie, Clarie and Banzare coasts. Antarctic Journal 9: 292-296.

Erlinge S, 1974. Distribution, territoriality and numbers of the weasel *Mustela nivalis* in relation to prey abundance. Oikos 25: 308-314.

Erlinge S, 1977a. Spacing strategy in stoat *Mustela erminea*. Oikos 28: 32-42.

Erlinge S, 1977b. Agonistic behaviour and dominance in stoats (*Mustela erminea* L.). Zeitschrift für Tierpsychologie 44: 375-388.

Erlinge S, 1979a. Movement and daily activity pattern of radio-tracked male stoats, *Mustela erminea*. *In* Amlaner CJ Jr. and Macdonald DW (Eds) A Handbook on Biotelemetry and Radio Tracking, pp. 703-710. Pergamon Press, Oxford.

Erlinge S, 1979b. Adaptive significance of sexual dimorphism in weasels. Oikos 33: 233-245.

Erlinge S, 1981. Food preference, optimal diet and reproductive output in stoats *Mustela erminea* in Sweden. Oikos 36: 303-315.

Erlinge S, 1983. Demography and dynamics of a stoat *Mustela erminea* population in a diverse community of vertebrates. Journal of Animal Ecology 52: 705-726.

Erlinge S, Frylestam B, Göransson G, Högstedt G, Liberg O, Loman J, Nilsson IN, von Schantz T, Sylvén M, 1984. Predation on brown hare and ring-necked pheasant populations in southern Sweden. Holarctic Ecology 7: 300-304.

Erlinge S, Sandell M, 1986. Seasonal changes in the social organisation of male stoats *Mustela erminea*: an effect of shifts between two decisive resources. Oikos 47: 57-72.

Erlinge S, Sandell M, 1988. Coexistence of stoat *Mustela erminea* and weasel *Mustela nivalis*: social dominance, scent communication and reciprocal distribution. Oikos 53: 242-246.

Erlinge S, Sandell M, Brinck C, 1982. Scent marking and its territorial significance in stoats *Mustela erminea*. Animal Behaviour 30: 811-818.

Esler AE, 1967. The vegetation of Kapiti Island. NZ Journal of Botany 5: 394.

Esler AE, 1971. Inner Islands of Hauraki Gulf. Challenger Island. Unpublished report, Botany Division, DSIR, 13 pp.

Espie PR, 1977. Classification of the habitat and niche of the chamois in the Nina catchment, Lewis Pass. Unpublished B.Ag.Sc. thesis, Lincoln College, University of Canterbury.

Evans RL, Katz EM, Olson NL, Dewsbury DA, 1978. A comparative study of swimming behaviour in eight species of muroid rodents. Bulletin of the Psychonomic Society 11: 168-70.

Ewer RF, 1971. The biology and behaviour of a free-living population of black rats (*Ruttus rattus*). Animal Behaviour Monographs 4: 127-174.

Ewer RF, 1973. The Carnivores. Cornell University Press, Ithaca.

Fain A, 1963. Les Acariens psoriques parasites des chauves-souris. XXV *Chirophagoides mystacopis* N.G., n.sp. (Sarcoptidae: Sarcoptiformes). Bulletin et Annales de la Société Royale d'Entomologie de Belgique 99(11): 159-167.

Fain A, 1968. Étude de la variabilité de *Sarcoptes scabei* avec une revision des Sarcoptidae. Acta Zoologica et Pathologica 47: 1-196.

Fairweather AAC, Brockie RE, Ward GD, 1987. Brushtail possums (*Trichosurus vulpecula*) sharing dens: a potential infection route for bovine tuberculosis. NZ Veterinary Journal 35: 15-16.

Fall MW, Medina AB, Jackson WB, 1971. Feeding patterns of *Rattus rattus* and *Rattus exulans* on Eniwetok Atoll, Marshall Islands. Journal of Mammalogy 52: 69-76.

Falla RA, 1948. The outlying islands of New Zealand. NZ Geographer 4: 127-154.

Falla RA, 1965. Birds and mammals of the subantarctic islands. Proceedings of the NZ Ecological Society 12: 63-68.

Falla RA, Taylor RH, Black C, 1979. Survey of Dundas Island, Auckland Islands, with particular reference to Hooker's sea lion (*Phocarctos hookeri*). NZ Journal of Zoology 6: 347-355.

Farmer A, 1965. Sambur deer in Rotorua. Protection Forestry Newsletter 3(2): 5-7.

Feilden HW, 1890. Notes on the terrestrial mammals of Barbados. Zoologist 48: 52-55.

Feist JD, McCullough DR, 1975. Reproduction in feral horses. Journal of Reproduction and Fertility, Supplement 23: 13-18.

Feist JD, McCullough DR, 1976. Behaviour patterns and communication in feral horses. Zeitschrift für Tierpsychologie 41: 337-371.

Feldman HW, 1926. Unit character inheritance of color in the black rat, *Mus rattus* L. Genetics 11: 456-465.

Fertig DS, Edmonds VW, 1969. The physiology of the house mouse. Scientific American 221(4): 103-110.

Field JA, 1907. Abnormal sambar horns. Journal of the Bombay Natural History Society 17: 1020.

Filmer JF, 1953. Disappointing tests of myxomatosis as rabbit control. NZ Journal of Agriculture 87: 402-404.

Fineran BA, 1973. A botanical survey of seven mutton-bird islands, south-west of Stewart Island. Journal of the Royal Society of NZ 3(4): 475-526.

Fisher HI, Baldwin PH, 1946. War and the birds on Midway Atoll. Condor 48: 3-15.

Fitzgerald AE, 1976. Diet of the opossum *Trichosurus vulpecula* (Kerr) in the Orongorongo Valley, Wellington, New Zealand, in relation to food-plant availability. NZ Journal of Zoology 3: 399-419.

Fitzgerald AE, 1978. Aspects of the food and nutrition of the brushtailed possum *Trichosurus vulpecula* (Kerr, 1792), Marsupialia: Phalangeridae, in New Zealand. *In* Montgomery GG (Ed.) The Ecology of Arboreal Folivores, pp. 289-303. Smithsonian Institution Press, Washington, D.C.

Fitzgerald AE, 1981. Some effects of the feeding habits of the possum *Trichosurus vulpecula*. *In* Bell BD (Ed.) Proceedings of the first symposium on marsupials in New Zealand, pp. 41-50. Zoology Publications of Victoria University of Wellington No. 74.

Fitzgerald AE, 1984a. Diet of the possum (*Trichosurus vulpecula*) in three Tasmanian forest types and its relevance to the diet of possums in New Zealand forests. *In* Smith A and Hume I (Eds) Possums and Gliders, pp. 137-143. Surrey Beatty and Sons, Chipping Norton, New South Wales.

Fitzgerald AE, 1984b. Diet overlap between kokako and the common brushtail possum in central North Island, New Zealand. *In* Smith A and Hume I (Eds) Possums and Gliders, pp. 569-573. Surrey Beatty and Sons, Chipping Norton, New South Wales.

Fitzgerald AE, Clarke RTJ, Reid CSW, Charleston WAG, Tarttelin MF, Wyburn RS, 1981. Physical and nutritional characteristics of the possum (*Trichosurus vulpecula*) in captivity. NZ Journal of Zoology 8: 551-562.

Fitzgerald AE, Wardle P, 1979. Food of the opossum *Trichosurus vulpecula* (Kerr) in the Waiho Valley, South Westland. NZ Journal of Zoology 6: 339-345.

Fitzgerald BM, 1964. Ecology of mustelids in New Zealand. Unpublished M.Sc. thesis, University of Canterbury, Christchurch.

Fitzgerald BM, 1977. Weasel predation on a cyclic population of the montane vole (*Microtus montanus*) in California. Journal of Animal Ecology 46: 367-397.

Fitzgerald BM, 1978. Population ecology of mice in New Zealand. *In* The ecology and control of rodents in New Zealand nature reserves. Department of Lands and Survey Information Series 4, pp. 163-171.

Fitzgerald BM, 1988. Diet of domestic cats and their impact on prey populations. *In* Turner DC and Bateson P (Eds) The Domestic Cat: the Biology of its Behaviour, pp. 123-144. Cambridge University Press, Cambridge.

Fitzgerald BM, Karl BJ, 1979. Foods of feral house cats (*Felis catus* L.) in forest of the Orongorongo Valley, Wellington. NZ Journal of Zoology 6: 107-126.

Fitzgerald BM, Karl BJ, 1986. Home range of feral house cats (*Felis catus* L.) in forest of the Orongorongo Valley, Wellington, New Zealand. NZ Journal of Ecology 9: 71-81.

Fitzgerald BM, Veitch CR, 1985. The cats of Herekopare Island, New Zealand; their history, ecology and effects on birdlife. NZ Journal of Zoology 12: 319-330.

Fitzgerald BM, Karl BJ, Moller H, 1981. Spatial organisation and ecology of a sparse population of house mice (*Mus musculus*) in a New Zealand forest. Journal of Animal Ecology 50: 489-518.

Fitzgerald BM, Meads MJ, Whitaker AH, 1986. Food of the kingfisher (*Halcyon sancta*) during nesting. Notornis 33: 23-32.

Fitzgerald BM, Johnson WB, King CM, Moors PJ, 1984. Research on mustelids and cats in New Zealand. Wildlife Research Liaison Group Research Review No. 3.

Fitzi H, Monk DR, 1977. Farming fallow deer. NZ Agricultural Science 11(4): 170-171.

Flack JAD, Lloyd BD, 1978. The effect of rodents on the breeding success of the South Island robin. *In* The ecology and control of rodents in New Zealand nature reserves. Department of Lands and Survey Information Series 4, pp. 59-66.

Fleming CA, 1939. Birds of the Chatham Islands. Emu 38: 380-413.

Fleming CA, 1962. The extinction of moas and other animals during the Holocene period. Notornis 10: 113-117.

Fleming CA, 1979. The Geological History of New Zealand and its Life. Auckland University Press/Oxford University Press.

Flook DR, 1970. Causes and implications of an observed sex differential in the survival of wapiti. Canadian Wildlife Service, Report Series 11: 1-71.

Flux JEC, 1965a. Timing of the breeding season in the hare, *Lepus europaeus* Pallas, and rabbit, *Oryctolagus cuniculus* (L.). Mammalia 29: 557-562.

Flux JEC, 1965b. Incidence of ovarian tumors in hares in New Zealand. Journal of Wildlife Management 29: 622-624.

Flux JEC, 1966. Occurrence of a white wrist band on hares in New Zealand. Journal of Zoology (London) 148: 582-583.

Flux JEC, 1967a. Hare numbers and diet in an alpine basin in New Zealand. Proceedings of the NZ Ecological Society 14: 27-33.

Flux JEC, 1967b. Reproduction and body weights of the hare *Lepus europaeus* Pallas, in New Zealand. NZ Journal of Science 10: 357-401.

Flux JEC, 1970. Life history of the mountain hare (*Lepus timidus scoticus*) in north-east Scotland. Journal of Zoology (London) 161: 75-123.

Flux JEC, 1980. High incidence of missing posterior upper molars in hares (*Lepus europaeus*) in New Zealand. NZ Journal of Zoology 7: 257-259.

Flux JEC, 1981a. Field observations of behaviour in the genus *Lepus*. *In* Myers K and MacInnes CD (Eds) Proceedings of the World Lagomorph Conference held in Guelph, Ontario, August 1979, pp. 377-394. University of Guelph, Guelph.

Flux JEC, 1981b. Prospects for hare farming in New Zealand. NZ Agricultural Science 15: 24-29.

Flux JEC, Fullagar PJ, 1983. World distribution of the rabbit *Oryctolagus cuniculus*. Acta Zoologica Fennica 174: 75-77.

Forbes HO, 1893. A list of the birds inhabiting the Chatham Islands. Ibis 6 (5): 521-546.

Forbes J, 1924. New Zealand Deer Heads. Country Life, London.

Ford-Robertson J de C, Bull PC, 1966. Some parasites of the kiore, *Rattus exulans*, on Little Barrier and Hen Islands, New Zealand. NZ Journal of Science 9: 221-224.

Forster JGA, 1777. A Voyage Round the World in His Britannic Majesty's Sloop Resolution (2 vols). London.

France NW, 1969. A record wild boar of 1905 vintage. NZ Wildlife 25: 28-30.

Fraser KW, 1979. Dynamics and condition of opossum (*Trichosurus vulpecula* Kerr) populations in the Copland Valley, Westland, New Zealand. Mauri Ora 7: 117-137.

Fraser KW, 1985. Biology of the rabbit (*Oryctolagus cuniculus* (L.)) in Central Otago, New Zealand, with emphasis on behaviour and its relevance to poison control operations. Unpublished Ph.D. thesis, University of Canterbury, Christchurch.

Freeland WJ, Winter JW, 1975. Evolutionary consequences of eating: *Trichosurus vulpecula* (Marsupialia) and the genus *Eucalyptus*. Journal of Chemical Ecology 1: 439-455.

Friend JB, 1978. Cattle of the World. Blandford Press, Poole.

Frylestam B, 1976. The European hare in Sweden. *In* Pielowski Z and Pucek Z (Eds) Ecology and Management of European Hare Populations, p. 33. Polish Hunting Association, Warsaw.

Furon R, 1958. The gentle little goat, archdespoiler of the earth. UNESCO Courier 1958(1): 30-32.

Galef BG, Clark MM, 1971. Social factors in poison avoidance and feeding behaviour. Journal of Comparative and Physiological Psychology 75: 341-357.

Gales RP, 1982. Age- and sex-related differences in diet selection by *Rattus rattus* on Stewart Island, New Zealand. NZ Journal of Zoology 9: 463-466.

Gaskin DE, 1972. Whales, Dolphins and Seals: with Special Reference to the New Zealand Region. Heinemann, London.

Geist V, 1960. Feral goats in British Columbia. The Murrelet 41(3): 34-40.

Geist V, 1971. Mountain Sheep. University of Chicago Press, Chicago.

Getty R (Ed.), 1975. Sisson and Grossman's The Anatomy of the Domestic Animals, 5th edition. W. B. Saunders Company, Philadelphia.

Gibb JA, 1967. What is efficient rabbit destruction? Tussock Grasslands and Mountain Lands Institute Review 12: 9-14.

Gibb JA, Flux JEC, 1973. Mammals. *In* Williams GR (Ed.) The Natural History of New Zealand, pp. 334-371. AH and AW Reed, Wellington.

Gibb JA, Flux JEC, 1983. Why New Zealand should not use myxomatosis in rabbit control operations. Search 14(1-2): 41-43.

Gibb JA, Ward GD, Ward CP, 1969. An experiment in the control of a sparse population of wild rabbits (*Oryctolagus c. cuniculus* L.) in New Zealand. NZ Journal of Science 12: 509-534.

Gibb JA, Ward CP, Ward GD, 1978. Natural control of a population of rabbits, *Oryctolagus cuniculus* (L.), for ten years in the Kourarau enclosure. DSIR Bulletin 223.

Gibb JA, White AJ, Ward CP, 1985. Population ecology of rabbits in the Wairarapa, New Zealand. NZ Journal of Ecology 8: 55-82.

Gibbons J, 1984. Iguanas of the South Pacific. Oryx 18: 82-91.

Gibson RN, 1972. Metazoan parasites of rodents in New Zealand. Unpublished M.Sc. thesis, University of Canterbury, Christchurch.

Gibson RN, 1973a. Notes on the burrow system of a colony of *Rattus norvegicus* (Berkenhout, 1767) near Christchurch. Mauri Ora 1: 49-53.

Gibson RN, 1973b. Reproduction of the house mouse, *Mus musculus* L., in the Christchurch area. Mauri Ora 1: 43-48.

Gibson RN, 1986. Some ectoparasites on rodents in New Zealand I. Fleas (Insecta: Siphonaptera). Mauri Ora 13: 81-92.

Gibson RN, Pilgrim RLC, 1986. Some ectoparasites on rodents in New Zealand II. Sucking lice (Insecta: Anoplura). Mauri Ora 13: 93-102.

Giffin JG, 1976. Ecology of the feral sheep on Mauna Kea. Pittmann-Robertson project W-15-5 study no. 11. Final report. State of Hawaii, Department of Lands and Resources, Division of Fish and Game.

Gilbert JR, Erickson AW, 1977. Distribution and abundance of seals in the pack ice of the Pacific sector of the Southern Ocean. *In* Llano GA (Ed.) Adaptations within Antarctic Ecosystems, pp. 703-740. Proceedings of the third SCAR symposium on Antarctic biology. Smithsonian Institution, Washington, D.C.

Gilbert N, Myers K, Cooke BD, Dunsmore JD, Fullagar PJ, Gibb JA, King DR, Parer I, Wheeler SH, Wood DH, 1987. Comparative dynamics of Australasian rabbit populations. Australian Wildlife Research 14: 491-503.

Giles JR, 1980. The ecology of feral pigs in western New South Wales. Unpublished Ph.D. thesis, University of Sydney.

Gillies R, 1877. Notes on some changes in the fauna of Otago. Transactions of the NZ Institute 10: 306-319.

Gilmore DP, 1966. Studies on the biology of *Trichosurus vulpecula* Kerr. Unpublished Ph.D. thesis, University of Canterbury, Christchurch.

Gilmore DP, 1969. Seasonal reproductive periodicity in the male Australian brush-tailed possum (*Trichosurus vulpecula*). Journal of Zoology (London) 157: 75-98.

Glover R, 1956. Notes on the sika deer. Journal of Mammalogy 37(1): 99-104.

Godley C, 1951. Letters from early New Zealand 1850-1853. Godley JR (Ed.). Whitcombe and Tombs, Wellington.

Goldfinch AJ, Molnar RE, 1978. Gait of the brush-tailed possum (*Trichosurus vulpecula*). Australian Zoologist 19: 277-289.

Gomez JC, 1960. Correlation of a population of roof rats in Venezuela with seasonal changes in habitat. American Midland Naturalist 63: 177-193.

Goodall DM, 1977. A History of Horse Breeding. Robert Hale, London.

Grant PJ, 1956. Opossum damage in beech forests, Ruahine Range, Hawke's Bay. NZ Journal of Forestry 7: 111-113.

Grant PJ, 1969. Rainfall patterns in the Kaweka Range. Journal of Hydrology, NZ 8(1): 17-34.

Grant PJ, 1985. Major periods of erosion and alluvial sedimentation in New Zealand during the Late Holocene. Journal of the Royal Society of NZ 15: 67-121.

Grant PJ, 1989. A hydrologist's contribution to the debate on wild animal management. NZ Journal of Ecology 12 (Suppl.): 165-169.

Green LMA, 1963. Distribution and comparative histology of cutaneous glands in certain marsupials. Australian Journal of Zoology 11: 250-272.

Green WQ, 1984. A review of ecological studies relevant to management of the common brushtail possum. *In* Smith A and Hume I (Eds) Possums and Gliders, pp. 483-499. Surrey Beatty and Sons, Chipping Norton, New South Wales.

Green WQ, Coleman JD, 1981. A progress report on the movements of possums *Trichosurus vulpecula* between native forest and pasture. *In* Bell BD (Ed.) Proceedings of the first symposium on marsupials in New Zealand, pp. 51-62. Zoology Publications of Victoria University of Wellington No. 74.

Green WQ, Coleman JD, 1984. Response of a brush-tailed possum population to intensive trapping. NZ Journal of Zoology 11: 319-328.

Green WQ, Coleman JD, 1986. Movements of possums (*Trichosurus vulpecula*) between forest and pasture in Westland, New Zealand: implications for bovine tuberculosis transmission. NZ Journal of Ecology 9: 57-69.

Green WQ, Coleman JD, 1987. Den sites of possums, *Trichosurus vulpecula*, and frequency of use in mixed hardwood forest in Westland, New Zealand. Australian Wildlife Research 14: 285-292.

Gregson JE, Purchas RW, 1985. The carcass composition of male fallow deer. *In* Fennessey PF and Drew KR (Eds) Biology of deer production. Royal Society of NZ Bulletin 22, pp. 295-298.

Greig BDA, 1946. Tararua story. Tararua Tramping Club.

Greig JC, 1969. The ecology of feral goats in Scotland. Unpublished M.Sc. thesis, University of Edinburgh.

Greig JC, Cooper B, 1970. The 'wild' sheep of Britain. Oryx 10(6): 383-388.

Griffiths M, Davies D, 1963. The role of the soft pellets in the production of lactic acid in the rabbit stomach. Journal of Nutrition 80: 171-180.

Grigera DE, Rapoport EH, 1983. Status and distribution of the European hare in South America. Journal of Mammalogy 64: 163-166.

Grue H, King CM, 1984. Evaluation of age criteria in New Zealand stoats (*Mustela erminea*) of known age. NZ Journal of Zoology 11: 437-443.

Guiler ER, 1968. The fauna of Tasmania. Tasmanian Yearbook 2: 55-60.

Guiler ER, Banks DM, 1958. A further examination of the distribution of the brush possum, *Trichosurus vulpecula*, in Tasmania. Ecology 39: 89-97.

Guinness FE, Gibson RM, Clutton-Brock TH, 1978. Calving times of red deer (*Cervus elaphus*) on Rhum. Journal of Zoology (London) 185: 105-114.

Guinness F, Lincoln GA, Short RV, 1971. The reproductive cycle of the female red deer, *Cervus elaphus* L. Journal of Reproduction and Fertility 27: 427-438.

Gulamhusein AP, Tam WH, 1974. Reproduction in the male stoat, *Mustela erminea*. Journal of Reproduction and Fertility 41: 303-312.

Gulamhusein AP, Thawley AR, 1974. Plasma progesterone levels in the stoat. Journal of Reproduction and Fertility 36: 405-408.

Guthrie-Smith H, 1929. Blackberry control by goats. NZ Journal of Agriculture 38: 16-19.

Guthrie-Smith H, 1969. Tutira. The Story of a New Zealand Sheep Station (4th edition). AH and AW Reed, Wellington.

Guy PA, 1984. *Ollulanus tricuspis* in domestic cats — prevalance and methods of post-mortem diagnosis. NZ Veterinary Journal 32(6): 81-84.

Guzman RF, 1982. *Cheyletiella blakei* (Acari: Cheyletiellidae) hyperparasite on the cat flea *Ctenocephalides felis felis* (Siphonaptera: Pulicidae) in New Zealand. NZ Entomologist 7: 322-323.

Guzman RF, 1984. A survey of cats and dogs for fleas: with particular reference to their role as intermediate hosts of *Dipylidium caninum*. NZ Veterinary Journal 32(5): 71-73.

Gwynn AM, 1953a. The status of the leopard seal at Heard Island and Macquarie Island, 1948-1950. Australian National Antarctic Research Expeditions Internal Report No. 3.

Gwynn AM, 1953b. Notes on the fur seals at Macquarie Island and Heard Island. Australian National Research Expeditions Internal Report No. 4.

Habermehl KH, 1961. Die Altersbestimmung bei Haustieren, Peltztieren und beim Jagdbaren Wild. Paul Parey, Berlin.

Hall LS, Richards GC, 1979. Bats of eastern Australia. Queensland Museum Booklet No. 12, pp. 1-66.

Hall-Martin AJ, 1974. Observations on population density and species composition of seals in the King Haakon VII Sea, Antarctica. South African Journal of Antarctic Research No. 4: 34-39.

Halls LK (Ed.), 1984. White-tailed Deer: Ecology and Management. Stackpole Books, Harrisburg, Pennsylvania.

Hamilton JE, 1939. The leopard seal *Hydrurga leptonyx* (De Blainville). Discovery Report 18: 239-264.

Hamilton WJ, 1933. The weasels of New York. American Midland Naturalist 14: 289-344.

Hamilton WJ, Baxter KL 1980. Reproduction in farmed red deer. 1. Hind and stag fertility. Journal of Agricultural Science (Cambridge) 95: 261-273.

Hamr J, 1984. Home range sizes and determinant factors in habitat use and activity of the chamois (*Rupicapra rupicapra* L.) in Northern Tyrol, Austria. Unpublished Ph.D. thesis, Leopold-Franzens Universität, Innsbruck, Austria.

Hanson RP, Karstad L, 1959. Feral swine in the southeastern United States. Journal of Wildlife Management 23: 64-74.

Hardy AR, Taylor KD, 1979. Radio tracking of *Rattus norvegicus* on farms. *In* Amlaner CJ and Macdonald DW (Eds) A Handbook on Biotelemetry and Radiotracking, pp. 657-665. Pergamon Press, Oxford.

Harper PC, 1983. Biology of the Buller's shearwater (*Puffinus bulleri*) at the Poor Knights Islands, New Zealand. Notornis 30(4): 299-318.

Harris J, 1982. Tracing the early Falkland Islands history. NZ Dominion, 4 May, p. 6.

Harris LH, 1966. Hunting Sambar Deer. NZ Forest Service, Wellington.

Harris LH, 1967. Hunting Fallow Deer. NZ Forest Service, Wellington.

Harris LH, 1971a. Notes on the introduction and history of sambar deer in New Zealand. NZ Wildlife 35: 33-42.

Harris LH, 1971b. Notes on the introduction and history of moose in New Zealand. NZ Wildlife 32: 29-37.

Harris LH, 1981. White-tailed deer in New Zealand. NZ Wildlife 64, Supplement, 12 pp.

Harris PM, Dellow DW, Broadhurst RB, 1985. Protein and energy requirements and deposition in the growing brushtail possum and Rex rabbit. Australian Journal of Zoology 33: 425-436.

Harrison JL, 1954. The natural food of some rats and other mammals. Bulletin of the Raffles Museum, Singapore 25: 157-165.

Harrison JL, 1957. Habitat of some Malayan rats. Proceedings of the Zoological Society of London 128: 1-21.

Hart K, 1979. Feral pig problems on the South Coast. Agricultural Gazette of New South Wales 90: 18-21.

Hartman L, 1964. The behaviour and breeding of captive weasels (*Mustela nivalis* L.) NZ Journal of Science 7: 147-156.

Hathaway SC, Blackmore DK, 1981. Ecological aspects of the epidemiology of infection with leptospires of the Ballum serogroup in the black rat (*Rattus rattus*) and the brown rat (*Rattus norvegicus*) in New Zealand. Journal of Hygiene (Cambridge) 87: 427-436.

Hathaway SC, Blackmore DK, Marshall RB, 1981. Leptospirosis in free-living species in New Zealand. Journal of Wildlife Diseases 17: 489-496.

Haugen AO, 1959. Breeding records of captive white-tailed deer in Alabama. Journal of Mammalogy 56(1): 151-159.

Haugen AO, Davenport LA, 1950. Breeding records of white-tailed deer in the Upper Peninsula of Michigan. Journal of Wildlife Management 14(3): 290-295.

Hay JR, 1981. The kokako. Unpublished report, Forest Bird Research Group, Rotorua.

Hayman DL, 1977. Chromosome number — constancy and variation. *In* Stonehouse B and Gilmore D (Eds) The Biology of Marsupials, pp. 27-48. MacMillan Press, London.

Hayward JD, 1983. The forests and animals of the North-West Nelson Recreational Hunting Area. Unpublished report, NZ Forest Service (Nelson Conservancy), 46 pp.

Hayward J, Hosking T, 1986. North West Nelson Recreational Hunting Area wild animal control plan. Unpublished report, NZ Forest Service (Nelson Conservancy), 49 pp.

Heath ACG, 1976. The parasitology of some feral Artiodactyla in New Zealand. *In* Whitaker AH and Rudge MR (Eds) The Value of Feral Farm Mammals in New Zealand, pp. 44-50. Department of Lands and Survey Information Series 1.

Heath ACG, 1979a. Ectoparasites of feral sheep on Campbell Island. NZ Journal of Zoology 6: 141-144.

Heath ACG, 1979b. Arthropod parasites of goats. NZ Journal of Zoology 6: 655.

Heath ACG, 1983. Parasites of the feral sheep of Pitt Island, Chatham group. NZ Journal of Zoology 10: 365-370.

Heath ACG, Bishop DM, Daniel MJ, 1987b. A new laelapine genus (Acari: Laelapidae) from the bat *Mystacina tuberculata* in New Zealand. Journal of the Royal Society of NZ 17: 31-39.

Heath ACG, Bishop DM, Tenquist JD, 1983. The prevalance and pathogenecity of *Chorioptes bovis* (Hering, 1845) and *Psoroptes cuniculi* (Delafond, 1859) (Acari: Psoroptidae) infestations in feral goats in New Zealand. Veterinary Parasitology 13: 159-169.

Heath ACG, Julian AF, Daniel MJ, Bishop DM, 1987a. Mite infestation (Acari: Laelapidae) of New Zealand short-tailed bats, *Mystacina tuberculata*, in captivity. Journal of the Royal Society of NZ 17: 41-47.

Heath ACG, Rush-Munro RM, Rutherford DM, 1971. The hedgehog — a new host record for *Notoedres muris* (Acari: Sarcoptidae). NZ Entomologist 5: 100-103.

Henderson JB, 1965. Case for the association's retention of wapiti management. NZ Wildlife 11: 22-30.

Henderson RJ, Clarke CMH, 1986. Physical size, condition, and demography of chamois (*Rupicapra rupicapra* L.) in the Avoca River region, Canterbury, New Zealand. NZ Journal of Zoology 13: 65-73.

Hermans R, 1984. Feral Horses in Aupouri Forest. Internal report, NZ Forest Service, Auckland.

Hermans RFG, Cribb SD, Farquhar CA, Lyttle NG, 1982. Behaviour and population characteristics of the feral horses of the Kaimanawas. Unpublished report, Massey University, Palmerston North.

Herter K, 1938. Die Biologie der Europaïschen Igel. Schöps Verlag, Leipzig.

Hewson R, 1963. Moults and pelages in the brown hare *Lepus europaeus occidentalis* de Winton. Proceedings of the Zoological Society of London 141: 677-687.

Hewson R, Healing TD, 1971. The stoat, *Mustela erminea*, and its prey. Journal of Zoology (London) 164: 239-244.

Hewson R, Taylor M, 1968. Movements of European hares in an upland area of Scotland. Acta Theriologica 13: 31-34.

Hickson REH, Moller H, Garrick AS, 1986. Poisoning rats on Stewart Island. NZ Journal of Ecology 9: 111-121.

Hill DA, Robertson HA, Sutherland WJ, 1983. Brown rats (*Rattus norvegicus*) climbing to obtain sloes and blackberries. Journal of Zoology (London) 200: 302.

Hill JE, Daniel MJ, 1985. Taxonomy of the New Zealand short-tailed bat *Mystacina*

Gray, 1843 (Chiroptera, Mystacinidae). Bulletin of the British Museum (Natural History) Zoology, 48(4): 279-300.

Hill JE, Smith JD, 1984. Bats, a Natural History. British Museum (Natural History), London.

Hindwood KA, 1940. The birds of Lord Howe Island. Emu 40: 1-86.

Hirata DN, Nass RD, 1974. Growth and sexual maturation of laboratory-reared, wild *Rattus norvegicus, R. rattus,* and *R. exulans* in Hawaii. Journal of Mammalogy 55: 472-474.

Hitchmough RA, 1980. Kiore (*Rattus exulans*) on Motukawanui Island, Cavalli group, northern New Zealand. Tane 26: 161-168.

Hocken TM, 1879. Reminiscences of Edwin Palmer. Thomas Papers, Hocken Library, Dunedin (unpublished).

Hocking GJ, 1981. The population ecology of the brush-tailed possum, *Trichosurus vulpecula* (Kerr), in Tasmania. Unpublished M.Sc. thesis, University of Tasmania, Hobart.

Hoffman RR, 1985. Digestive physiology of the deer — their morphophysiological specialisation and adaption. *In* Fennessey PF and Drew KR (Eds) Biology of deer production. Royal Society of NZ Bulletin 22, pp. 393-408.

Hofman RJ, 1975. Distribution patterns and population structure of Antarctic seals. Unpublished Ph.D. thesis, University of Minnesota.

Hofman RJ, Erickson A, Siniff D, 1973. The Ross seal (*Ommatophoca rossi*). *In* Seals, IUCN Publications New Series, Supplementary Paper No. 39, pp. 129-139. International Union for the Conservation of Nature and Natural Resources, Morges, Switzerland.

Hogg MJ, Skegg PDG, 1961. Moreporks in a nesting box. Notornis 9: 133-134.

Holden P, 1971. Pack and Rifle. AH and AW Reed, Wellington.

Holden P, 1982. The Wild Pig in New Zealand. Hodder and Stoughton, Auckland.

Holdsworth M, Baylis GTS, 1967. Vegetation of Great King Island, Three Kings group, in 1963. Records of the Auckland Institute and Museum 6(3): 175-184.

Holford GH, 1927. The acclimatisation of stock in Canterbury. *In* Speight R, Wall A and Laing RM (Eds) Natural History of Canterbury, pp. 273-299. Philosophical Institute of Canterbury, Christchurch.

Holislova V, Obrtel R, 1986. Vertebrate casualties on a Moravian road. Acta Scientiarum Naturalium Brno 20: 1-44.

Hollis CJ, Robertshaw JD, Harden RH, 1986. Ecology of the swamp wallaby (*Wallabia bicolor*) in north-eastern New South Wales. 1. Diet. Australian Wildlife Research 13: 355-365.

Holloway BA, 1976. A new bat-fly family from New Zealand (Diptera: Mystacinobiidae). NZ Journal of Zoology 3: 279-301.

Holloway BA, 1984. Larvae of New Zealand Fanniidae (Diptera: Calyptrata). NZ Journal of Zoology 11: 239-257.

Holloway JT, 1950. Deer and the forests of western Southland. NZ Journal of Forestry 6: 123-137.

Holloway JT, 1959. Noxious-animal problems of the South Island alpine watersheds. NZ Science Review 17: 21-28.

Holloway JT, 1973. The status quo in animal control and management: a research assessment. *In* Assessment and management of introduced animals in New Zealand forests, pp. 125-130. New Zealand Forest Service, Forest Research Institute Symposium No. 14.

Holloway JT, Wendelken WJ, Morris JY, Wraight MJ, Wardle P, Franklin DA, 1963. Report on the condition of the forests, subalpine scrublands and alpine grasslands of the Tararua Range. New Zealand Forest Service, Forest Research Institute Technical Paper No. 41.

Holst PJ, 1982. Age, hair colour, live weight, and fertility of two samples of Australian feral goat, *Capra hircus*. Australian Wildlife Research 8(3): 549-554.

Honacki JH, Kinman KE, Koeppl JW, 1982. Mammal Species of the World: a Taxonomic and Geographic Reference. Allen Press and the Association of Systematics Collections, Lawrence, Kansas.

Hope RM, 1970. Genetic variation in natural and laboratory populations of *Trichosurus vulpecula* and *Sminthopsis crassicaudata*. Unpublished Ph.D. thesis, University of Adelaide.

Hope RM, 1972. Observations on the sex ratio and the position of the lactating mammary gland in the brush-tailed possum, *Trichosurus vulpecula* (Kerr) (Marsupialia). Australian Journal of Zoology 20: 131-137.

Hopkins M, 1953. Distance perception in *Mus musculus*. Journal of Mammalogy 34: 393.

Horne RSC, 1979. Seasonal and altitudinal variations in diet and abundance of the European hare (*Lepus europaeus* Pallas) in Tongariro National Park, New Zealand. Unpublished M.Sc thesis, Massey University, Palmerston North.

Houtcooper WC, 1978. Food habits of rodents in a cultivated ecosystem. Journal of Mammalogy 59: 427-430.

Howard WE, 1959. The rabbit problem in New Zealand. DSIR Information Series 16.

Howard WE, 1964. Modification of New Zealand's flora by introduced mammals. Proceedings of the NZ Ecological Society 11: 59-62.

Howard WE, 1965. Control of introduced mammals in New Zealand. DSIR Information Series 45: 1-96.

Hughes RL, Hall LS, 1984. Embryonic development in the common brushtail possum *Trichosurus vulpecula*. *In* Smith A and Hume I (Eds) Possums and Gliders, pp. 197-212. Surrey Beatty and Sons, Chipping Norton, New South Wales.

Humphreys WF, Bradley AJ, How RA, Barnett JL, 1984. Indices of condition of phalanger populations: a review. *In* Smith A and Hume I (Eds) Possums and Gliders, pp. 59-77. Surrey Beatty and Sons, Chipping Norton, New South Wales.

Hurst R, 1984. Identification and description of larval *Anisakis simplex* and *Pseudoterranova decipiens* (Anisakidae: Nematoda) from New Zealand waters. NZ Journal of Marine and Freshwater Research 18: 177-186.

Huson LW, Davis RA, 1980. Discriminant functions to aid identification of faecal pellets of *Rattus norvegicus* and *Rattus rattus*. Journal of Stored Products Research 16: 103-104.

Huson LW, Rennison BS, 1981. Seasonal variability of Norway rat (*Rattus norvegicus*) infestation of agricultural premises. Journal of Zoology (London) 194: 257-260.

Hutton FW, 1868. Notes on the birds of the Little Barrier Island. Transactions and Proceedings of the NZ Institute 1: 106. (Second edition 1875.)

Hutton FW, 1877. Note on the Maori rat. Transactions of the NZ Institute 9: 348.

Hutton HB, 1976. Animal health. *In* Whitaker AH and Rudge MR (Eds) The Value of Feral Farm Mammals in New Zealand, pp. 51-53. Department of Lands and Survey, Wellington.

Hutton HB, 1980. The disease status of the sheep population of Campbell Island in 1975. *In* Preliminary Reports of the Campbell Island Expedition 1975-76. Department of Lands and Survey Reserves Series 7: 130-143.

Hyett J, Shaw N, 1980. Australian Mammals. A Field Guide for New South Wales, South Australia, Victoria and Tasmania. Thomas Nelson, Melbourne.

Il'in V, 1962. The European hare in Irkutsk Oblast. Sel'skoe Khoz Sibiri 12: 71.

Imber MJ, 1975. Petrels and predators. XII Bulletin of the International Council for Bird Preservation: 260-263.

Imber MJ, 1985. Exploitation by rats *Rattus* of eggs neglected by gadfly petrels *Pterodroma*. Cormorant 12: 82-93.

Imber MJ, 1987. Breeding ecology and conservation of the black petrel (*Procellaria parkinsoni*). Notornis 34: 19-39.

Ineson MJ, 1953. A comparison of the parasites of wild and domestic pigs in New Zealand. Transactions of the Royal Society of NZ 82: 579-609.

Ingham SE, 1960. The status of seals (Pinnipedia) at Australian Antarctic stations. Mammalia 24: 422-430.

Innes JG, 1977. Biology and ecology of the ship rat *Rattus rattus rattus* (L.) in Manawatu (NZ) forests. Unpublished M.Sc. thesis, Massey University, Palmerston North.

Innes, JG, 1979. Diet and reproduction of ship rats in the northern Tararuas. NZ Journal of Ecology 2: 85-86.

Innes JG, Skipworth JP, 1983. Home ranges of ship rats in a small New Zealand forest as revealed by trapping and tracking. NZ Journal of Zoology 10: 99-110.

Inns RW, 1980. Ecology of the Kangaroo Island wallaby, *Macropus eugenii* (Desmarest) in Flinders Chase National Park, Kangaroo Island. Unpublished Ph.D. thesis, University of Adelaide.

Ito T, 1967. Ecological studies on the Japanese deer, *Cervus nippon centralis* Kishida on Kinkazon Island. I. The distribution and population structure. Bulletin of the Marine Biological Station, Asamushi Tohoku University 13(1): 57-62.

Izawa M, Doi T, Ono Y, 1982. Grouping patterns of feral cats (*Felis catus*) living on a small island in Japan. Japanese Journal of Ecology 32: 373-382.

Jackson WB, 1962. Reproduction. *In* Storer TI (Ed.) Pacific Island Rat Ecology, pp. 92-107. Bernice P. Bishop Museum Bulletin 225.

Jackson WB, 1965. Litter size in relation to latitude in two murid rodents. American Midland Naturalist 73: 245-247.

Jackson WB, Strecker RL, 1962a. Home range studies. *In* Storer TI (Ed.) Pacific Island Rat Ecology, pp. 113-123. Bernice P. Bishop Museum Bulletin 225.

Jackson WB, Strecker RL, 1962b. Ecological distribution and relative numbers. *In* Storer TI (Ed.) Pacific Island Rat Ecology, pp. 45-63. Bernice P. Bishop Museum Bulletin 225.

James IL, 1974. Mammals and beech (*Nothofagus*) forests. Proceedings of the NZ Ecological Society 21: 41-44.

James IL, Wallis FP, 1969. A comparative study of the effects of introduced mammals on *Nothofagus* forest at Lake Waikareiti. Proceedings of the NZ Ecological Society 16: 1-6.

Jane GT, 1981. Application of the poisson model to the bait-interference method of possum *Trichosurus vulpecula* assessment. *In* Bell BD (Ed.) Proceedings of the first symposium on marsupials in New Zealand, pp. 185-195. Zoology Publications of Victoria University of Wellington No. 74.

Jane GT, Green TGA, 1983. Vegetation mortality in the Kaimai Range, North Island, New Zealand. Pacific Science 37: 385-389.

Jane GT, Pracy LT, 1974. Observations on two animal exclosures in Haurangi forest over a period of twenty years (1951-1971). NZ Journal of Forestry 19: 102-113.

Jarman PJ, Johnson KA, 1977. Exotic mammals, indigenous mammals, and land-use. Proceedings of the Ecological Society of Australia 10: 147-166.

Jeffrey SM, 1977. Rodent ecology and land use in Western Ghana. Journal of Applied Ecology 14: 741-755.

Jemmett JE, Evans JM, 1977. A survey of sexual behaviour and reproduction of female cats. Journal of Small Animal Practice 18: 31-37.

Jensen B, 1978. Resultater af fangst med kassefaelder. Natura Jutlandica 20: 129-136.

Jewell PA, Fullagar PJ, 1966. Body measurements of small mammals; sources of error and anatomical changes. Journal of Zoology 150: 501-509.

Jewell PA, Milner C, Boyd JM, 1974. Island Survivors: the Ecology of the Soay Sheep of St Kilda. Athlone Press, London University.

Jezierski W, 1977. Longevity and mortality rate in a population of wild boar. Acta Theriologica 22(24): 337-348.

Johns JH, MacGibbon RJ (Eds), 1986. Wild Animals in New Zealand. Reed Methuen, Auckland.

Johnson DH, 1962. Rodents and other Micronesian mammals collected. *In* Storer TI (Ed.) Pacific Island Rat Ecology, pp. 21-38. Bernice P. Bishop Museum Bulletin 225.

Johnson KA, 1977. Methods for the census of wallaby and possum in Tasmania. Tasmanian National Parks and Wildlife Service, Wildlife Division Technical Report 77/2.

Johnston TH, Mawson PM, 1940. New and known nematodes from Australian marsupials. Proceedings of the Linnean Society, New South Wales LXV: 468-476.

Johnston PG, Sharman GB, 1979. Electrophoretic, chromosomal and morphometric studies on the red-necked wallaby, *Macropus rufogriseus* (Desmarest). Australian Journal of Zoology 27: 433-441.

Johnstone GW, 1985. Threats to birds on subantarctic islands. *In* Moors PJ (Ed.) Conservation of island birds, pp. 101-121. International Council for Bird Preservation Technical Publication No. 3.

Jolly JN, 1976. Movements, habitat use and social behaviour of the opossum, *Trichosurus vulpecula*, in a pastoral habitat. Unpublished M.Sc. thesis, University of Canterbury, Christchurch.

Jolly JN, 1983. Little-spotted kiwi research on Kapiti Island 1980-1982. Wildlife, A Review 12: 33-39.

Jolly JN, Spurr EB, 1981. Damage by possums to erosion-control plantings. *In* Bell BD (Ed.) Proceedings of the first symposium on marsupials in New Zealand, pp. 205-210. Zoology Publications of Victoria University of Wellington No. 74.

Jones E, 1977. Ecology of the feral cat, *Felis catus* (L.), (Carnivora: Felidae) on Macquarie Island. Australian Wildlife Research 4: 249-262.

Jones E, Coman BJ, 1981. Ecology of the feral cat, *Felis catus* (L.), in south-eastern Australia. I. Diet. Australian Wildlife Research 8: 537-547.

Jones E, Coman BJ, 1982a. Ecology of the feral cat, *Felis catus* (L.), in south-eastern Australia. II. Reproduction. Australian Wildlife Research 9: 111-119.

Jones E, Coman BJ, 1982b. Ecology of the feral cat, *Felis catus* (L.), in south-eastern Australia. III. Home ranges and population ecology in semi-arid north-west Victoria. Australian Wildlife Research 9: 409-420.

Jordan KHE, 1947. On a new genus and species of bat-fleas from the Pelorus Islands and New Zealand. Transactions of the Royal Society of NZ 76: 208-210.

Joyce JP, Rattray DV, Parker JP, 1971. The utilization of pasture and barley by rabbits. I. Feed intakes and liveweight gains. NZ Journal of Agricultural Research 14: 173-179.

Julian AF, 1981. Tuberculosis in the possum *Trichosurus vulpecula*. *In* Bell BD (Ed.) Proceedings of the first symposium on marsupials in New Zealand, pp. 163-174. Zoology Publications of Victoria University of Wellington No. 74.

Julian AF, 1984. Possum distribution and recent spread in northern Northland: with observations on the vegetation. Unpublished internal report, Ecology Division, DSIR.

Kalkowski W, 1967. Olfactory bases of social orientation in the white mouse. Folia Biologica (Cracow) 15: 69-87.

Kaluziński J, Bresiński W, 1976. The effect of the European hare and roe deer populations on the yields of cultivated plants. *In* Pielowski Z and Pucek Z (Eds) Ecology and Management of European Hare Populations, pp. 247-253. Polish Hunting Association, Warsaw.

Kami HT, 1966. Foods of rodents in the Hamakua district, Hawaii. Pacific Science 20: 367-373.

Karl BJ, Best HA, 1982. Feral cats on Stewart Island; their foods and their effects on kakapo. NZ Journal of Zoology 9: 287-294.

Karnoukhova NG, 1971. Age determination of brown and black rats. Soviet Journal of Ecology 2: 144-147.

Kean RI, 1959. Ecology of the larger wildlife mammals of New Zealand. NZ Science Review 17: 35-37.

Kean RI, 1964. "Opossum" or "possum". Tuatara 12: 155-156.

Kean RI, 1965. "Possum" or "opossum". Tuatara 13: 192.

Kean RI, 1966. Marsupials (Part 2). Tuatara 15: 25-45.

Kean RI, 1967. Behaviour and territorialism in *Trichosurus vulpecula* (Marsupialia). Proceedings of the NZ Ecological Society 14: 71-78.

Kean RI, 1971. Selection for melanism and low reproductive rate in *Trichosurus vulpecula* (Marsupialia). Proceedings of the NZ Ecological Society 18: 42-47.

Kean RI, 1975. Growth of the opossum (*Trichosurus vulpecula*) in the Orongorongo Valley, Wellington, New Zealand 1956-1961. NZ Journal of Zoology 2: 435-444.

Kean RI, Pracy LT, 1953. Effects of the Australian opossum (*Trichosurus vulpecula* Kerr) on indigenous vegetation in New Zealand. Proceedings of the 7th Pacific Science Congress 4: 1-8.

Keiper RR, Keenan MA, 1980. Nocturnal activity patterns of feral ponies. Journal of Mammalogy 61: 116-118.

Kelly GC, 1980. Landscape and nature conservation. *In* Molloy LF et al. (Eds) Land alone endures, pp. 63-87. DSIR Discussion Paper 3.

Kelton SD, 1981. Biology of sambar deer (*Cervus unicolor* Kerr, 1792) in New Zealand with particular reference to diet in a Manawatu flax swamp. Unpublished M.Sc. thesis, Massey University, Palmerston North.

Kelton SD, 1982. Sambar deer. NZ Wildlife 20: 8-14.

Kepler CB, 1967. Polynesian rat predation on nesting Laysan albatrosses and other Pacific seabirds. Auk 84: 426-430.

Kerle JA, 1983. The population biology of the northern brushtail possum. Unpublished Ph.D. thesis, Macquarie University, Sydney.

Kerle JA, 1984. Variation in the ecology of *Trichosurus:* its adaptive significance. *In* Smith A and Hume I (Eds) Possums and Gliders, pp. 115-128. Surrey Beatty and Sons, Chipping Norton, New South Wales.

Kerr IGC, Williams JM, Ross WD, Pollard JM, 1986. The classification of land according to degree of rabbit infestation in Central Otago. Proceedings of the NZ Grasslands Association 48: 65-70.

Kettle PR, 1982. Ectoparasites of goats: their importance and control. DGH Seminar, Whangarei, pp. 26-39.

Kettle PR, Wright DE, 1985. The New Zealand goat industry. The Agricultural Research Division perspective. Ministry of Agriculture and Fisheries, Wellington, 51 pp.

Khalil LF, 1970. *Bertiella trichosuri* n.sp. from the brush-tailed opossum *Trichosurus vulpecula* (Kerr) from New Zealand. Zoologischer Anzeiger 185: 442-450.

Kiddie DG, 1962. The sika deer (*Cervus nippon*) in New Zealand. NZ Forest Service Information Series 44: 1-35.

Kildemoes A, 1985. The impact of introduced stoats (*Mustela erminea*) on an island population of the water vole, *Arvicola terrestris*. Acta Zoologica Fennica 173: 193-195.

King CM, 1974. The nematode *Skrjabingylus nasicola* (Metastrongyloidea) in mustelids: a new record for New Zealand. NZ Journal of Zoology 1: 501-502.

King CM, 1975a. Report on the mustelid live-trapping project in the Orongorongo Valley, 1972-75. Unpublished report, Ecology Division, DSIR.

King CM, 1975b. The sex ratio of trapped weasels (*Mustela nivalis*). Mammal Review 5: 1-8.

King CM, 1975c. The home range of the weasel (*Mustela nivalis*) in an English woodland. Journal of Animal Ecology 44: 639-668.

King CM, 1980a. Field experiments on the trapping of stoats (*Mustela erminea*). NZ Journal of Zoology 7: 261-266.

King CM, 1980b. Population ecology of the weasel *Mustela nivalis*, on British game estates. Holarctic Ecology 3: 160-168.

King CM, 1981a. The reproductive tactics of the stoat, *Mustela erminea*, in New Zealand forests. *In* Chapman JA and Purseley D (Eds) Proceedings of the First Worldwide Furbearer Conference, pp. 443-468. Frostburg MD.

King CM, 1981b. The effects of two types of steel traps upon captured stoats (*Mustela erminea*). Journal of Zoology (London) 195: 553-554.

King CM, 1982a. Age structure and reproduction in feral New Zealand populations of the house mouse (*Mus musculus*) in relation to seed fall of southern beech. NZ Journal of Zoology 9: 467-480.

King CM, 1982b. Stoat observations. Landscape 12: 12-15.

King CM, 1983a. The relationships between beech (*Nothofagus* sp.) seedfall and populations of mice (*Mus musculus*), and the demographic and dietary responses of stoats (*Mustela erminea*), in three New Zealand forests. Journal of Animal Ecology 52: 141-166.

King CM, 1983b. The life history strategies of *Mustela nivalis* and *Mustela erminea.* Acta Zoologica Fennica 174: 183-184.

King CM, 1984a. Immigrant Killers. Introduced Predators and the Conservation of Birds in New Zealand. Oxford University Press, Auckland.

King CM, 1984b. The orgin and adaptive advantages of delayed implantation in *Mustela erminea.* Oikos 42: 126-128.

King CM, 1985a. Interactions between woodland rodents and their predators. Symposium of the Zoological Society of London 55: 219-247.

King CM, 1985b. Stoat in the dock. Forest and Bird 16(3): 7-9.

King CM, 1989. The advantages and disadvantages of small size to weasels, *Mustela* spp. *In* Gittleman JL (Ed.) Carnivore Behaviour, Ecology and Evolution, pp. 302-334. Cornell University Press, Ithaca, New York.

King CM, Edgar RL, 1977. Techniques for trapping and tracking stoats (*Mustela erminea*): a review and a new system. NZ Journal of Zoology 4: 193-212.

King CM, McMillan CD, 1982. Population structure and dispersal of peak year cohorts of young stoats (*Mustela erminea*) in two New Zealand forests, with especial reference to control. NZ Journal of Ecology 5: 59-66.

King CM, Moody JE, 1982. The biology of the stoat (*Mustela erminea*) in the national parks of New Zealand. NZ Journal of Zoology 9: 49-144.

King CM, Moors PJ, 1979a. The life history tactics of mustelids, and their significance for predator control and conservation in New Zealand. NZ Journal of Zoology 6: 619-622.

King CM, Moors PJ, 1979b. On co-existence, foraging strategy and the biogeography of weasels and stoats (*Mustela nivalis* and *M. erminea*) in Britain. Oecologia 39: 129-150.

King DR, Wheeler SH, Schmidt GL, 1983. Population fluctuations and reproduction of rabbits in a pastoral area on the coast north of Carnarvon, Western Australia. Australian Wildlife Research 10: 97-104.

King JE, 1959. The northern and southern populations of *Arctocephalus gazella.* Mammalia 23: 19-40.

King JE, 1964a. Seals of the World. British Museum (Natural History), London.

King JE, 1964b. A note on the increasing specialisation of the seal fore-flipper. Journal of Anatomy (London) 98: 476-477.

King JE, 1968. The Ross and other Antarctic seals. Australian Natural History 16: 29-32.

King JE, 1969. The identity of the fur seals of Australia. Australian Journal of Zoology 17: 841-853.

King JE, 1972. Observations on phocid skulls. *In* Harrison RJ (Ed.) Functional Anatomy of Marine Mammals, vol. I, pp. 81-115. Academic Press, London and New York.

King JE, 1983. Seals of the World 2nd edition. Oxford University Press, Oxford.

King WB, 1973. Conservation status of birds of central Pacific islands. Wilson Bulletin 85: 89-103.

Kingsmill E, 1962. An investigation of criteria for estimating age in the marsupials *Trichosurus vulpecula* Kerr and *Perameles nasuta* Geoffrey. Australian Journal of Zoology 10: 597-617.

Kinloch DI, 1973. Ecology of the parma wallaby, *Macropus parma* Waterhouse, 1846, and other wallabies on Kawau Island New Zealand. Unpublished M.Sc. thesis, University of Auckland.

Kirk HB, 1920. Opossums in New Zealand. Appendices to the Journal of the House of Representatives, NZ, Session 1, H-28: 1-12.

Kirk T, 1896. The displacement of species in New Zealand. Transactions of the NZ Institute 28: 1-27.

Kirkpatrick TH, 1964. Molar progression and macropod age. Queensland Journal of Agricultural Science 21: 163-165.

Kirkpatrick TH, 1970. The swamp wallaby in Queensland. Queensland Agricultural Journal 96: 335-336.

Kirkpatrick TH, 1983. Black-striped wallaby *Macropus dorsalis. In* Strahan R (Ed.) The Australian Museum Complete Book of Australian Mammals, p. 238. Angus and Robertson, Sydney.

Kirsch JAW, Calaby JH, 1977. The species of living marsupials: an annotated list. *In* Stonehouse B and Gilmore D (Eds) The Biology of Marsupials, pp. 9-26. MacMillan Press, London.

Kirton AH, Ritchie JMW, 1979. Goat farming. NZ Journal of Agricultural Science 13(3): 134-139.

Kitchener DJ, 1975. Reproduction in female Gould's wattled bat, *Chalinolobus gouldii* (Gray) (Vespertilionidae), in Western Australia. Australian Journal of Zoology 23: 29-42.

Kitchener DJ, Coster P, 1981. Reproduction in female *Chalinolobus morio* (Gray) (Vespertilionidae) in south-western Australia. Australian Journal of Zoology 29: 305-320.

Knowlton JE, 1986. A sighting device for estimating molar index to determine age from macroped skulls. Australian Wildlife Research 11: 451-454.

Knowlton JE, Panapa N, 1982. Dama wallaby survey, Okataina Scenic Reserve. Unpublished report, NZ Forest Service.

Koeppl JW, Slade NA, Turner RW, 1979. Spatial associations of *Rattus exulans* in Central Java, Indonesia. Journal of Mammalogy 60: 795-802.

Koford CB, 1968. Peruvian desert mice: water independence, competition, and breeding cycle near the equator. Science 160: 552-553.

Kooyman GL, 1965. Leopard seals of Cape Crozier. Animals 6: 59-63.

Kooyman GL, 1966. Maximum diving capacities of the Weddell seal, *Leptonychotes weddelli*. Science 151: 1553-1554.

Kooyman GL, 1968. An analysis of some behavioural and physiological characteristics related to diving in the Weddell seal. Antarctic Research Series 11: 227-261.

Kooyman GL, 1975. A comparison between day and night diving in the Weddell seal. Journal of Mammalogy 56: 563-574.

Kooyman GL, 1981a. Weddell seal: *Leptonychotes weddelli* Lesson, 1826. *In* Ridgway SH and Harrison RJ (Eds) Handbook of Marine Mammals, Vol. 2: Seals, pp. 275-296. Academic Press, London.

Kooyman GL, 1981b. Weddell Seal: Consummate Diver. Cambridge University Press, Cambridge.

Kooyman GL, 1981c. Leopard seal: *Hydrurga leptonyx* Blainville, 1820. *In* Ridgway

SH and Harrison RJ (Eds) Handbook of Marine Mammals, Vol. 2: Seals, pp. 261-274. Academic Press, London.

Kooyman GL, 1981d. Crabeater seal: *Lobodon carcinophagus* (Hombron and Jacquinot, 1842). *In* Ridgway SH and Harrison RJ (Eds) Handbook of Marine Mammals, Vol. 2: Seals, pp. 221-235. Academic Press, London.

Kooyman GL, Kerem DH, Campbell WB, Wright JJ, 1973. Pulmonary gas exchange in freely diving Weddell seals, *Leptonychotes weddelli*. Respiratory Physiology 12: 271-282.

Kovács G, 1983. Survival pattern in adult European hares. Acta Zoologica Fennica 174: 69-70.

Krämer A, 1969. Soziale Organisation und Sozialverhalten einer Gemspopulation (*Rupicapra rupicapra* L.) der Alpen. Zeitschrift für Tierpsychologie 26: 889-964.

Kratochvil J, Kratochvil Z, 1976. The origin of the domesticated forms of the genus *Felis* (Mammalia). Zoologicki Listy 25(3): 193-208.

Krause MA, Beck AC, Dent JB, 1984. The economics of controlling gorse in hill country: goats versus chemicals. Agricultural Economics Unit, Lincoln College, Research Report 149.

Kristiansson H, 1984. Ecology of a hedgehog, *Erinaceus europaeus* population in southern Sweden. Unpublished Ph.D. thesis, University of Lund, Sweden.

Kristoffersson R, 1964. An apparatus for recording general activity of hedgehogs. Annales Akademiae Scientiarum Fennica A IV 79: 1-7.

Kristoffersson R, Soivio A, 1964. Hibernation in the hedgehog (*Erinaceus europaeus* L.) Annales Akademiae Scientiarum Fennica A IV 82: 1-17.

Kristoffersson R, Suomalainen P, 1964. Studies on the physiology of the hibernating hedgehog: 2. Changes of body weight of hibernating animals. Annales Akademiae Scientiarum Fennica (A) 76: 1-11.

Kruuk H, 1964. Predators and anti-predator behaviour of the black-headed gull (*Larus ridibundus* L.). Behaviour, Supplement 11.

Kurtén B, 1971. The Age of Mammals. Weidenfeld and Nicholson, London.

Kurz JC, Marchinton RL, 1972. Radiotelemetry studies of feral hogs in South Carolina. Journal of Wildlife Management 36: 1240-1248.

Kuschel G, 1975. Biogeography and Ecology in New Zealand. W. Junk Publishers, The Hague.

Kutzer E, Frey H, 1976. Die Parasiten der Feldhasen (*Lepus europaeus*) in Österreich. Berliner und Münchener Tierarztliche Wochenschrift 89: 480-483.

Lamb RC, 1964. Birds, Beasts and Fishes. Caxton Press, Christchurch.

Lambert RE, Bathgate JL, 1973. Determination of the plane of nutrition of chamois. Proceedings of the NZ Ecological Society 24: 48-56.

Lane R, 1982. The history of the South Kaipara fallow deer herd. NZ Wildlife 9 (66/67): 22-30.

Lattanzio RM, Chapman JA, 1980. Reproductive and physiological cycles in an island population of Norway rats. Bulletin of the Chicago Academy of Science 12: 1-68.

Lavers RB, 1973a. Aspects of the ecology of the ferret, *Mustela putorius* forma *furo* L., at Pukepuke Lagoon. Proceedings of the NZ Ecological Society 20: 7-12.

Lavers RB, 1973b. Results of a live-trapping study on a population of ferrets at Pukepuke Lagoon. Unpublished report to the Controller of Wildlife, Wellington.

Lavers RB, 1978. The diet of red deer (*Cervus elaphus*) in the Murchison Mountains: a preliminary report. *In* Seminar on the Takahe and its Habitat, pp. 187-198. Fiordland National Park Board, Invercargill.

Lavers RB, Mills JA, 1978. Stoat studies in the Murchison Mountains, Fiordland. *In* Seminar on the Takahe and its Habitat, pp. 222-233. Fiordland National Park Board, Invercargill.

Lavrov NP, 1944. Effect of helminth invasions and infectious diseases on variations in numbers of the ermine, *Mustela erminea*. *In* King CM (Ed.) Biology of Mustelids: Some Soviet Research, pp. 170-187. British Library Lending Division, Boston Spa, Yorkshire, UK, 1975.

Law PG, Burstall T, 1956. Macquarie Island. Australian National Antarctic Research Expeditions, Internal Report No. 14.

Lawrence MJ, Brown RW, 1973. Mammals of Britain, their tracks, trails and signs (revised edition). Blandford Press, London.

Laws RM, 1953. The elephant seal (*Mirounga leonina* Linn). I. Growth and age. Falkland Islands Dependencies Survey, Scientific Report No. 8.

Laws RM, 1956. The elephant seal (*Mirounga leonina* Linn). II. General, social and reproductive behaviour. Falkland Islands Dependencies Survey, Scientific Report No. 13.

Laws RM, 1958. Growth rates and ages of crabeater seals, *Lobodon carcinophagus* (Jacquinot and Pucheran). Proceedings of the Zoological Society of London 130: 275-288.

Laws RM, 1960. The southern elephant seal (*Mirounga leonina* Linn.) at South Georgia. Norsk Hvalfangst-tid 49: 466-476, 520-542.

Laws RM, 1973. The current status of seals in the Southern Hemisphere. *In* Seals. IUCN Publications, New Series, Supplementary Paper No. 39: 144-161.

Laws RM, 1977a. The significance of vertebrates in the Antarctic marine ecosystem. *In* Llano GA (Ed.) Adaptations within Antarctic ecosystems, pp. 411-438. Proceedings of the third SCAR symposium on Antarctic biology. Smithsonian Institution, Washington, D.C.

Laws RM, 1977b. Seals and whales of the Southern Ocean. Philosophical Transactions of the Royal Society of London, B 279: 81-96.

Laws RM, 1979. Southern elephant seal. *In* Mammals in the seas. Volume II: Pinniped Species Summaries and Report on Sirenians, pp. 106-109. FAO Fisheries Series No. 5, Rome.

Laws RM, Christie EC, 1976. Seals and birds killed or captured in the Antarctic Treaty area, 1970-73. Polar Record 18: 318-320.

Lazell JD, Sutterfield TW, Giezentanner WD, 1984. The population of rock wallabies (genus *Petrogale*) on Oahu, Hawaii. Biological Conservation 30: 99-108.

Leach BF, 1979. Excavations in the Washpool Valley, Palliser Bay. *In* Leach B and Leach M (Eds) Prehistoric Man in Palliser Bay, pp. 67-136. Bulletin of the National Museum of NZ 21.

Leathwick JR, 1984. Phenology of some common trees, shrubs and lianes in four central North Island forests. NZ Forest Service, Forest Research Institute Bulletin No. 72.

Leathwick JR, Hay JR, Fitzgerald AE, 1983. The influence of browsing by introduced mammals on the decline of the North Island kokako. NZ Journal of Ecology 6: 55-70.

Leslie PH, Ranson RM, 1954. The amount of wheat consumed by the brown rat. *In* Chitty D and Southern HN (Eds) Control of Rats and Mice, pp. 335-349. Clarendon Press, Oxford.

Leslie PH, Venables VM, Venables LSV, 1952. The fertility and population structure of the brown rat (*Rattus norvegicus*) in corn-ricks and some other habitats. Proceedings of the Zoological Society of London 122: 187-238.

Le Souef AS, 1930. Occasional notes 2. Alteration in character of wallabies acclimatised on Kawau Island, New Zealand. Australian Zoologist 6(2): 110-111.

Lever C, 1985. Naturalized Mammals of the World. Longman, London.

Levine MJ, 1985. Himalayan thar in New Zealand: issues in management of an introduced mammal. Unpublished M.Sc. thesis, Canterbury University, Christchurch. (See also NZ Wildlife 79: 20-21.)

Leyhausen P, 1979. Cat Behavior: the Predatory and Social Behavior of Domestic and Wild Cats. Garland Series in Ethology. Garland STPM Press, New York.

Liberg O, 1980. Spacing patterns in a population of rural free roaming domestic cats. Oikos 35: 336-349.

Liberg O, 1984. Social behaviour in free-ranging domestic and feral cats. *In* Anderson RS (Ed.) Nutrition and Behaviour in Dogs and Cats, pp. 175-181. Pergamon Press, Oxford.

Lidicker WZ Jr, 1966. Ecological observations on a feral house mouse population declining to extinction. Ecological Monographs 36: 27-50.

Lidicker WZ Jr, 1976. Social behaviour and density regulation in house mice living in large enclosures. Journal of Animal Ecology 45: 677-697.

Lincoln GA, Youngson RW, Short RV, 1970. The social and sexual behaviour of the red deer stag. Journal of Reproduction and Fertility, Supplement 11: 71-103.

Lindsay CJ, Ordish RC, 1964. The food of the morepork. Notornis 11: 154-158.

Lindsey AA, 1937. The Weddell seal in the Bay of Whales, Antarctica. Journal of Mammalogy 18: 127-144.

Lindsey GD, Nass RD, Hood GA, Hirata DN, 1973. Movement patterns of Polynesian rats (*Rattus exulans*) in sugarcane. Pacific Science 27: 239-246.

Ling JK, 1968. The skin and hair of the southern elephant seal, *Mirounga leonina* (L.) III. Morphology of the adult integument. Australian Journal of Zoology 16: 629-645.

Ling JK, Bryden MM, 1981. Southern elephant seal: *Mirounga leonina* Linnaeus, 1758. *In* Ridgway SH and Harrison RJ (Eds) Handbook of Marine Mammals, Vol. 2: Seals, pp. 297-327. Academic Press, London.

Ling JK, Button CH, Ebsary BA, 1974. A preliminary account of gray seals and harbor seals at Saint-Pierre and Miquelon. Canadian Field Naturalist 88: 461-468.

Lister AM, 1984. Evolutionary and ecological origins of British deer. Proceedings of the Royal Society of Edinburgh 82B: 205-229.

Lockie JD, 1966. Territory in small carnivores. Symposia of the Zoological Society of London 18: 143-165.

Logan PC, Harris LH, 1967. Introduction and establishment of red deer in New Zealand. NZ Forest Service, Information Series 55: 1-36.

Lovari S, 1980. Revision of *Rupicapra* genus. I. A statistical re-evaluation of Coutourier's data on the morphometry of six chamois subspecies. Bollettino di Zoologia 47: 113-124.

Lovegrove T, 1985. Saddlebacks on Kapiti Island: Fourth Annual Report, 1984-85 Season. Unpublished report, Department of Lands and Survey, NZ.

Luomala K, 1960. A history of the binomial classification of the Polynesian native dog. Pacific Science 14: 193-223.

Luomala K, 1962. Additional eighteenth-century sketches of the Polynesian native dog, including the Maori. Pacific Science 16: 170-180.

Lydekker R, 1901. The Great and Small Game of Europe, Western and Northern Asia and America. Their Distribution, Habits, and Structure. Rowland Ward, London.

Lyne AG, Verhagen AMW, 1957. Growth of the marsupial *Trichosurus vulpecula* and a comparison with some higher mammals. Growth 21: 167-195.

McCallum J, 1980. Reptiles of the northern Mokohinau group. Tane 26: 53-59.

McCann C, 1955. Observations on the polecat (*Putorius putorius* Linn.) in New Zealand. Records of the Dominion Museum 2: 151-165.

McCartney WC, 1970. Arboreal behaviour of the Polynesian rat (*Rattus exulans*). Bioscience 20: 1061-1062.

McCartney WC, Marks J, 1973. Inter- and intraspecific aggression in two species of the genus *Rattus*: evolutionary and competitive implications. Proceedings of the Pennsylvania Academy of Science 47: 145-148.

McClintock MK, Adler NT, 1978. The role of the female during copulation in wild and domestic Norway rats (*Rattus norvegicus*). Behaviour 67: 67-96.

Macdonald D (Ed.), 1984. The Encyclopaedia of Mammals; 2 vols. George Allen and Unwin, London.

Macdonald DW, Apps PJ, 1978. The social behaviour of a group of semi-dependent farm cats, *Felis catus*: a progress report. Carnivore Genetics Newsletter 3(7): 256-268.

McDougall P, 1975. The feral goats of Kielderhead moor. Journal of Zoology (London) 176(2): 215-246.

McFadden I, 1984. Composition and presentation of baits, and their acceptance by kiore (*Rattus exulans*). NZ Wildlife Service Technical Report No. 7.

McGlone MS, 1983. Polynesian deforestation of New Zealand: a preliminary synthesis. Archaeology in Oceania 18: 11-25.

McIlroy JC, 1983. The sensitivity of the brushtail possum (*Trichosurus vulpecula*) to 1080 poison. NZ Journal of Ecology 6: 125-132.

McIntosh IG, Adams TW, 1955. The control of rats and mice. NZ Journal of Agriculture 90: 229-235.

McKean J, 1975. The bats of Lord Howe Island with a description of a new nyctophiline bat. Australian Mammalogy 1: 329-332.

McKean JL, Hamilton-Smith E, 1967. Litter size and maternity sites in Australian bats (Chiroptera). Victorian Naturalist 84: 203-205.

McKelvey PJ, 1959. Animal damage in North Island protection forests. NZ Science Review 17: 28-34.

McKelvey PJ, Nicholls JL, 1957. A provisional classification of North Island forests. NZ Journal of Forestry 7(4): 84-101.

McKenna PB, [n.d.] Parasites of domestic animals in New Zealand — checklist. Technical report, Animal Health Division, Ministry of Agriculture and Fisheries.

McKenna PB, Charleston WAG, 1980a. Coccidia (Protozoa: Sporozoasida) of cats and dogs. I. Identity and prevalence in cats. NZ Veterinary Journal 28: 86-88.

McKenna PB, Charleston WAG, 1980b. Coccidia (Protozoa: Sporozoasida) of cats and dogs. III. The occurrence of a species of *Besnoitia* in cats. NZ Veterinary Journal 28: 120-122.

McKinnon AD, Coughlan L, 1960. Data on the establishment of some introduced animals in New Zealand forests (2 vols). Unpublished report, NZ Forest Service, Wellington.

Mackintosh CG, 1988. Diseases of deer. Wildlife Society, NZ Veterinary Association Publication 2: 68-73.

Mackintosh CG, Beatson NS, 1985. Relationships between diseases of deer and those of other animals. *In* Fennessy PF and Drew KR (Eds) Biology of Deer Production, pp. 77-82. Royal Society of NZ Bulletin 22.

Mackintosh CG, Henderson T, 1984. Potential wildlife sources of *Yersinia pseudotuberculosis* for farmed red deer (*Cervus elaphus*). NZ Veterinary Journal 32: 208-210.

Mackintosh GR, 1944. Wild pig poisoning. NZ Journal of Agriculture 68(4): 261-263.

Mackintosh JH, 1973. Factors affecting the recognition of territory boundaries by mice (*Mus musculus*). Animal Behaviour 21: 464-470.

Mackintosh JH, 1981. Behaviour of the house mouse. Symposia of the Zoological Society of London 47: 337-365.

MacLean FS, 1955. The history of plague in New Zealand. NZ Medical Journal 54: 131-143.

MacLennan DG, 1984. The feeding behaviour and activity patterns of the brushtail possum, *Trichosurus vulpecula*, in an open eucalypt woodland in southeast Queensland. *In* Smith A and Hume I (Eds) Possums and Gliders, pp. 155-161. Surrey Beatty and Sons, Chipping Norton, New South Wales.

McLeod S, 1986. The feeding and behavioural interaction of Bennett's wallaby (*Macropus rufogriseus rufogriseus*) with domestic sheep (*Ovis aries*) in South Canterbury, New Zealand. Unpublished M.Sc. thesis, University of Canterbury, Christchurch.

McNab AG, Crawley MC, 1975. Mother and pup behaviour of the New Zealand fur seal, *Arctocephalus forsteri* (Lesson). Mauri Ora 3: 77-88.

McNab R, 1907. Murihiku and the Southern Islands. Wilson and Horton, Auckland.

McNab R, 1913. The Old Whaling Days — a History of Southern New Zealand 1830-1840. Whitcombe and Tombs, Christchurch.

McKnight T, 1976. Friendly Vermin. University of California Press, Berkeley.

McTaggart HS, 1971. Observations on the behaviour of an island community of feral goats. British Veterinary Journal 2127(8): 399-340.

Mahood IT, 1978. Royal deer herds. *In* Parks and Wildlife 2(2): 52-53. New South Wales National Parks and Wildlife Service, Sydney.

Mahood IT, 1980. A report on rusa deer (*Cervus timoriensis*) in Royal National Park. Unpublished report, New South Wales National Parks and Wildlife Service, Sydney, pp. 1-14.

Maling PB (Ed.), 1958. The Torlesse Papers: the Journals and Letters of Charles Obins Torlesse Concerning the Foundation of the Canterbury Settlement in New Zealand 1848-51. Pegasus Press, Christchurch.

Mandahl N, Fredga K, 1980. A comparative chromosome study by means of G., C. and Nor-bandings of the weasel, the pygmy weasel and the stoat (*Mustela*, Carnivora, Mustelidae). Hereditas 93: 75-83.

Mansfield AW, 1958. The breeding behaviour and reproductive cycle of the Weddell seal (*Leptonychotes weddelli* Lesson). Falkland Islands Dependencies Survey, Scientific Report 18: 1-41.

Manson DCM, 1972. A new species of *Ornithonyssus* (Acarina: Dermanyssidae) from a New Zealand bat. NZ Journal of Science 15: 465-472.

Mantalenakis SJ, Ketchel MM, 1966. Frequency and extent of delayed implantation in lactating rats and mice. Journal of Reproduction and Fertility 12: 391-394.

Marjeed SK, Cooper JE, 1984. Lesions associated with *Capillaria* infestation in the European hedgehog (*Erinaceus europaeus*). Journal of Comparative Pathology 94: 625-628.

Mark AF, Baylis GTS, 1975. Impact of deer on Secretary Island, Fiordland, New Zealand. Proceedings of the NZ Ecological Society 22: 19-24.

Marlow BJ, 1975. The comparative behaviour of the Australasian sea lions *Neophoca cinerea* and *Phocarctos hookeri* (Pinnipedia: Otariidae). Mammalia 39: 159-230.

Marples BJ, 1942. A study of the little owl, *Athene noctua*, in New Zealand. Transactions and Proceedings of the Royal Society of NZ 72: 237-252.

Marples BJ, 1967. Notes on the phenotypes of cats observed in New Zealand and in Thailand. Carnivore Genetics Newsletter 1(3): 43-44.

Marples MJ, Smith JMB, 1960. The hedgehog as a source of human ringworm. Nature 188: 867-868.

Marples MJ, Smith JMB, 1962. *Trichophyton terrestre* as a resident in hedgehog skin. Sabouraudia 2: 100-107.

Marples RR, 1955. *Rattus exulans* in Western Samoa. Pacific Science 9: 171-176.

Marshall FHA, 1937. On the changeover in the oestrus cycle in animals after transference across the equator, with further observations on the incidence of the breeding seasons and the factors controlling sexual periodicity. Proceedings of the Royal Society of London B 122: 413-428.

Marshall JT, 1977. Roof rat. *In* Lekagul B and McNeely JA (Eds) Mammals of Thailand, pp. 475-478. Association for the Conservation of Wildlife.

Marshall JT, Sage RD, 1981. Taxonomy of the house mouse. Symposia of the Zoological Society of London 47: 15-25.

Marshall WH, 1961. A note on the food habits of feral cats on Little Barrier Island, New Zealand. NZ Journal of Science 4: 822-824.

Marshall WH, 1963. The ecology of mustelids in New Zealand. DSIR Information Series 38.
Martin HB, 1885. Objections to the introduction of beasts of prey to destroy the rabbit. Transactions and Proceedings of the NZ Institute 17: 179-182.
Martin JT, 1972. Wild pigs. Potential vectors of foot and mouth disease. NZ Journal of Agriculture I 25(6): 18-23.
Martin JT, 1975. Movement of feral pigs in North Canterbury, New Zealand. Journal of Mammalogy 56: 914-915.
Martinet L, 1977. Reproduction et fertilité du lièvre en captivité. *In* Pesson P (Ed.) Écologie du petit Gibier et Aménagement des Chasses, pp. 265-272. Gauthier-Villars, Paris.
Martinet L, Raynaud F, 1972. Mecanisme possible de la superfoetation chez la hase. Compte Rendu Hebdomadaire des Séances de l'Academie des Sciences, Paris 274: 2683-2686.
Maser C, Mate BR, Franklin JF, Dyrness CT, 1981. Natural history of Oregon coast mammals. US Department of Agriculture, Forest Service General Technical Report, PNW-133: 220.
Mason PC, 1975. New parasite records from the South Island. NZ Veterinary Journal 23: 69.
Mason PC, 1979. Tissue worm in red deer, biology and significance. Aglink FPP 249, Ministry of Agriculture and Fisheries, Wellington.
Mason PC, 1988. Parasitism in deer in New Zealand: a review. Wildlife Society, NZ Veterinary Association Publication 2: 75-91.
Mason PC, McAllum HJF, 1976. *Dictyocaulus viviparus* and *Elaphostrongylus cervi* in wapiti. NZ Veterinary Journal 24: 23.
Mason R, 1966. Foods of the Fiordland wapiti. *In* Banwell DB (Ed.) Wapiti in New Zealand, pp. 160-163. AH and AW Reed, Wellington.
Massey W, 1985. Fallow. A scientific approach to the breed. The Deer Farmer 23: 18-20.
Massey W, 1986. The Waikato formula is working. The Deer Farmer 33: 36-40.
Massey W, 1987. Fallow in the cattle heartland. The Deer Farmer 36: 14-15.
Masters KB, 1981. Private life of the wild pig. Journal of Agriculture, Department of Agriculture Western Australia 22(3): 103-105.
Matthews LH, 1929. The natural history of the elephant seal. Discovery Reports 1: 234-255.
Matthews LH, 1952. British Mammals. Collins, London.
Mattlin RH, 1978a. Pup mortality of the New Zealand fur seal (*Arctocephalus forsteri* Lesson). NZ Journal of Ecology 1: 138-144.
Mattlin RH, 1978b. Population biology, thermoregulation and site preference of the New Zealand fur seal, *Arctocephalus forsteri* (Lesson, 1828), on the Open Bay Islands, New Zealand. Unpublished Ph.D. thesis, University of Canterbury, Christchurch.
Mattlin RH, 1981. Pup growth of the New Zealand fur seal *Arctocephalus forsteri* on the Open Bay Islands, New Zealand. Journal of Zoology (London) 193: 305-314.
Maynes GM, 1973. Reproduction in the parma wallaby, *Macropus parma* Waterhouse. Australian Journal of Zoology 21: 331-351.
Maynes GM, 1974. Occurrence and field recognition of *Macropus parma.* Australian Zoologist 18: 72-87.
Maynes GM, 1977a. Distribution and aspects of the biology of the parma wallaby, *Macropus parma*, in New South Wales. Australian Wildlife Research 4: 109-125.
Maynes GM, 1977b. Breeding and age structure of the population of *Macropus parma* on Kawau Island, New Zealand. Australian Journal of Ecology 2: 207-214.
Maynes GM, 1983. Parma wallaby *Macropus parma. In* Strahan R (Ed.) The Australian Museum Complete Book of Australian Mammals, pp. 230-231. Angus and Robertson, Sydney.

Maynes GM, Sharman GB, 1983. Brush-tailed rock-wallaby *Petrogale penicillata*. *In* Strahan R (Ed.) The Australian Museum Complete Book of Australian Mammals, pp. 211-212. Angus and Robertson, Sydney.

Meads MJ, 1976a. Effects of opossum browsing on northern rata trees in the Orongorongo Valley, Wellington, New Zealand. NZ Journal of Zoology 3: 127-139.

Meads MJ, 1976b. Visit to Chetwode Is., Middle Trio Is., Cook Strait: Maud Is., Marlborough Sounds from 23 September to 7 October 1976. Unpublished report, Ecology Division, DSIR.

Meads MJ, Walker KJ, Elliott GP, 1984. Status, conservation and management of the land snails of the genus *Powelliphanta* (Mollusca: Pulmonata). NZ Journal of Zoology 11: 277-306.

Medway, Lord, 1969. The Wild Mammals of Malaya and Offshore Islands Including Singapore. Oxford University Press, Oxford.

Meeson J, 1885. The plague of rats in Nelson and Marlborough. Transactions of the New Zealand Institute 17: 199-207.

Melland E, 1890. Notes on a paper entitled 'The Takahe in Western Otago' by Mr James Park, FGS. Transactions and Proceedings of the NZ Institute 22: 295-300.

Menkhorst PA, 1984. The application of nestboxes in research and management of possums and gliders. *In* Smith A and Hume I (Eds) Possums and Gliders, pp. 517-525. Surrey Beatty and Sons, Chipping Norton, New South Wales.

Merchant JC, 1983. Swamp wallaby *Wallabia bicolor*. *In* Strahan R (Ed.) The Australian Museum Complete Book of Australian Mammals, pp. 261-262. Angus and Robertson, Sydney.

Merchant JC, Calaby JH, 1981. Reproductive biology of the red-necked wallaby (*Macropus rufogriseus banksianus*) and Bennett's wallaby (*M. r. rufogriseus*) in captivity. Journal of Zoology (London) 194: 203-217.

Merton DV, 1970. Kermadec Islands expedition reports: a general account of birdlife. Nortornis 17(3): 147-199.

Merton DV, 1972. Cuvier — an island restored. Forest and Bird 184: 7-9.

Merton DV, 1978. Controlling introduced predators and competitors on islands. *In* Temple SA (Ed.) Endangered Birds: Management Techniques for Preserving Threatened Species, pp. 121-128. University of Wisconsin Press, Madison.

Metge J, 1976. The Maoris of New Zealand, revised edition. Routledge and Kegan Paul, London.

Meurk CD, 1982. Regeneration of subantarctic plants on Campbell Island following exclusion of sheep. NZ Journal of Ecology 5: 51-58.

Meylan A, 1967. Les chromosomes de *Mustela erminea cicognanii* Bonaparte (Mammalia, Carnivora). Canadian Journal of Genetics and Cytology 9: 569-574.

Miers KH, 1962. Herd composition and effective reproduction of wapiti (*Cervus canadensis*) of eastern Fiordland. Proceedings of the NZ Ecological Society 9: 31-33.

Miers KH, 1966. Origins of genus *Cervus* and relationship of wapiti and red deer. *In* Banwell DB (Ed.) Wapiti in New Zealand, pp. 148-150. AH and AW Reed, Wellington.

Miers KH, 1973. Animals in forests: present policy and position in animal control and management. *In* Assessment and management of introduced animals in New Zealand forests, pp. 89-96. NZ Forest Service, Forest Research Institute Symposium No. 14.

Miers KH, 1985. Wild animal control: changing emphasis towards the 1990s. Forest and Bird 16(3): 5-6.

Millener PR, 1981. The Quaternary avifauna of the North Island, New Zealand. Unpublished Ph.D. thesis, University of Auckland.

Millener PR, 1988. Contributions to New Zealand's Late Quaternary avifauna. 1: *Pachyplicas*, a new genus of wren (Aves: Acanthisittidae), with two new species. Journal of the Royal Society of NZ 18: 383-406.

Miller EH, 1971. Social and thermo-regulatory behaviour of the New Zealand fur seal, *Arctocephalus forsteri* (Lesson, 1828). Unpublished M.Sc. thesis, University of Canterbury, Christchurch.

Miller EH, 1974. Social behaviour between adult male and female New Zealand fur seals, *Arctocephalus forsteri* (Lesson), during the breeding season. Australian Journal of Zoology 22: 155-173.

Miller EH, 1975. Annual cycle of fur seals, *Arctocephalus forsteri* (Lesson), on the Open Bay Islands, New Zealand. Pacific Science 29: 139-152.

Miller PJ, 1976. The lizards of the Cavalli Islands, northeastern New Zealand. Unpublished report to Commissioner of Crown Lands, Whangarei.

Mills JA, Mark AF, 1977. Food preferences of takahe in Fiordland National Park, New Zealand, and the effect of competition from introduced red deer. Journal of Animal Ecology 46: 939-958.

Mills JA, Ryder JP, Shaw PW, McLay R, 1977. Further birth records of elephant seals *Mirounga leonina* in New Zealand (note). NZ Journal of Marine and Freshwater Research 11: 789-791.

Milner C, Goodier R, Crook IG, 1968. Feral goats in Wales. Nature in Wales 11(1): 3-11.

Ministry of Agriculture and Fisheries, 1986. Cattle tuberculosis. Surveillance 13(3): 1-37.

Mitchell B, 1970. The potential output of meat as estimated from the vital statistics of natural and park populations of red deer. Deer 2: 453-456.

Mitchell B, 1973. The reproductive performance of wild Scottish red deer, *Cervus elaphus*. Journal of Reproduction and Fertility, Supplement 19: 271-285.

Mitchell B, McCowan D, Nicholson IA, 1976. Annual cycles of body weight and condition in Scottish red deer, *Cervus elaphus*. Journal of Zoology (London) 180: 107-127.

Mitchell RJ, Fordham RA, John A, 1987. The annual diet of feral goats (*Capra hircus* L.) in lowland rimu-rata-kamahi forest on eastern Mt Taranaki (Mt Egmont). NZ Journal of Zoology 14: 179-192.

Mitchell RJ, Grace ND, Fordham RA, 1987. The nitrogen and mineral content of seven native plant species preferred by feral goats (*Capra hircus* L.) in lowland rimu-rata-kamahi forest on eastern Mt Taranaki (Mt Egmont). NZ Journal of Zoology 14: 193-196.

Moller H, 1977. Ecology of *Rattus exulans* on Tiritiri Matangi Island. Unpublished M.Sc. thesis, University of Auckland.

Moller H, 1978. A weta and rodent study on Arapawa Island. Unpublished report, Ecology Division, DSIR.

Moller H, 1985a. Tree wetas (*Hemideina crassicruris*) (Orthoptera: Stenopelmatidae) of Stephens Island, Cook Strait. NZ Journal of Zoology 12: 55-69.

Moller H, 1985b. Killing of Weddell seals at McMurdo Sound, Antarctica. Unpublished report, Ecology Division, DSIR.

Moller H, Craig JL, 1986. The population ecology of *Rattus exulans* on Tiritiri Island, and a model of comparative population dynamics in New Zealand. NZ Journal of Zoology 14: 305-328.

Moller H, Tilley JAV, 1986. Rodents and their predators in the eastern Bay of Islands. NZ Journal of Zoology 13: 563-572.

Monrad G, 1968. Memoirs of Alfred Matthews, 'Wairongomai', Featherston. Alexander Turnbull Library MS 920/1968.

Moore LB, 1976. The changing vegetation of Molesworth Station, New Zealand, 1944 to 1971. DSIR Bulletin 217.

Moore LB, Cranwell LM, 1934. Induced dominance of *Microlaena avenacea* (Raoul) Hook. f. in a New Zealand rain-forest area. Records of the Auckland Institute and Museum 1: 219-230.

Moorehouse SR, Batten GJ, 1978. Goat meat for export 1. The market. NZ Journal of Agriculture 137(12): 22-25.

Moors PJ, 1975. Introduced predators and the South Island robin. Wildlife — A Review 6: 26-31.

Moors PJ, 1977. Studies of the metabolism, food consumption and assimilation efficiency of a small carnivore, the weasel (*Mustela nivalis* L.) Oecologia 27: 185-202.

Moors PJ, 1978. Methods for studying predators and their effects on forest birds. *In* Dingwall PR, Atkinson IAE and Hay C (Eds) The Ecology and Control of Rodents in New Zealand Nature Reserves, pp. 47-56. Department of Lands and Survey Information Series 4.

Moors PJ, 1979. Observations on the nesting habits of the European hedgehog in the Manawatu sand country, New Zealand. NZ Journal of Zoology 6: 489-492.

Moors PJ, 1980. Sexual dimorphism in the body size of mustelids (Carnivora): the roles of food habits and breeding systems. Oikos 34: 147-158.

Moors PJ, 1981. Report on studies of kiore on Macauley Island, November 1980. Unpublished report, NZ Wildlife Service.

Moors PJ, 1983a. Predation by stoats (*Mustela erminea*) and weasels (*Mustela nivalis*) on nests of New Zealand forest birds. Acta Zoologica Fennica 174: 193-196.

Moors PJ, 1983b. Predation by mustelids and rodents on the eggs and chicks of native and introduced birds in Kowhai Bush, New Zealand. Ibis 125: 137-154.

Moors PJ, 1985a. Norway rats (*Rattus norvegicus*) on the Noises and Motukawao Islands, Hauraki Gulf, New Zealand. NZ Journal of Ecology 8: 37-54.

Moors PJ, 1985b. Eradication campaigns against *Rattus norvegicus* on the Noises Islands, New Zealand, using brodifacoum and 1080. *In* Moors PJ (Ed.) Conservation of island birds, pp. 143-155. International Council for Bird Preservation Technical Publication 3.

Moors PJ, Atkinson IAE, 1984. Predation on seabirds by introduced animals, and factors affecting its severity. *In* Croxall JP, Evans PGH and Schreiber RW (Eds) Status and conservation of the world's seabirds, pp. 667-690. International Council for Bird Preservation Technical Publication 2.

Moors PJ, Lavers RB, 1981. Movements and home range of ferrets (*Mustela furo*) at Pukepuke Lagoon, New Zealand. NZ Journal of Zoology 8: 413-423.

Morgan DR, 1981a. Current research and development of baits for aerial control of the possum *Trichosurus vulpecula*. *In* Bell BD (Ed.) Proceedings of the first symposium on marsupials in New Zealand, p. 221. Zoology Publications of Victoria University of Wellington No. 74.

Morgan DR, 1981b. Development of a tracer technique for monitoring bait acceptance in brush-tailed possum (*Trichosurus vulpecula* Kerr) populations. NZ Journal of Forestry Science 11: 271-277.

Morgan DR, 1982. Field acceptance of toxic and non-toxic baits by brush-tailed possum (*Trichosurus vulpecula* Kerr) populations. NZ Journal of Ecology 5: 36-43.

Morgan DR, Copland AJ, 1985. Geographical distribution of possums (*Trichosurus vulpecula*) and rock wallabies (*Petrogale penicillata*) on Rangitoto Island. Unpublished report, NZ Forest Service, Forest Research Institute.

Morgan DR, Sinclair MJ, 1983. A bibliography of the brush-tailed possum (*Trichosurus vulpecula* Kerr). NZ Forest Service, Forest Research Institute Bulletin No. 25. (See also Brockie et al., 1984.)

Morgan DR, Batcheler CL, Peters JR, 1986. Why do possums survive aerial poisoning operations? *In* Salmon TP (Ed.) Proceedings of the Twelfth Vertebrate Pest Conference, pp. 210-214. University of California, Davis.

Morris B, 1961. Some observations on the breeding season of the hedgehog and the rearing and handling of the young. Proceedings of the Zoological Society of London 136: 201-206.

Morris P, 1964. The hedgehog in captivity. Bulletin of the Mammal Society of the British Isles 22: 13-14.

Morris P, 1969. Some aspects of the ecology of the hedgehog (*Erinaceus europaeus* L.). Unpublished Ph.D. thesis, University of London.

Morris P, 1973. Winter nests of the hedgehog (*Erinaceus europaeus* L.). Oecologia (Berlin) 11: 299-313.

Morris P, 1977. Pre-weaning mortality in the hedgehog (*Erinaceus europaeus* L.). Journal of Zoology (London) 182: 162-167.

Morris P, 1984. An estimate of the minimum body weight necessary for hedgehogs (*Erinaceus europaeus*) to survive hibernation. Journal of Zoology (London) 203: 291-293.

Morris P, 1985. The effects of supplementary feeding on the movement of hedgehogs (*Erinaceus europaeus*). Mammal Review 15: 23-32.

Morris P, English MP, 1973. Transmission and course of *Trichophyton erinacei* infections in British hedgehogs. Sabouraudia 11: 42-47.

Morrison K, 1980. Bird and stoat encounters in Fiordland. Notornis 27: 324.

Morrow H, 1975. New Zealand Wild Horses. Millwood Press, Wellington.

Morse PM, 1981. Wildlife values and wildlife conservation of Buller and North Westland. NZ Wildlife Service Fauna Survey Unit report No. 29, 185 pp.

Morton EK, 1957. Crusoes of Sunday Island. G. Bell and Sons, London.

Mosby JM, Wodzicki K, Thompson HR, 1973. Food of the kimoa (*Rattus exulans*) in the Tokelau Islands and other habitats in the Pacific. NZ Journal of Science 16: 799-810.

Mosley MP, 1978. Erosion in the south-eastern Ruahine Range: its implications for downstream river control. NZ Journal of Forestry 23: 21-48.

Müller G, 1890. Quoted in Thomson (1922: 74).

Müller H, 1970. Beiträge zur Biologie des Hermelins, *Mustela erminea* Linné, 1758. Säugetiere Mitteilungen 18: 293-380.

Munro D, 1927. The wild pig nuisance — poisoning trials in Wellington land district. NZ Journal of Agriculture 35(6): 364-368.

Munro D, Wright R, 1933. The rabbit pest and its control. NZ Journal of Agriculture, 20 Jan.: 26-37.

Munro H, 1917. The rabbit pest. NZ Journal of Agriculture, 20 Oct.: 206-337.

Murie OJ, 1951. The Elk of North America. Stackpole Co., Harrisburg, Pennsylvania.

Murie OJ, 1954. A Field Guide to Animal Tracks. Houghton Mifflin, Boston.

Murison WD, 1877. Note on the wild dog. Appendix to R. Gillies, Notes on some changes in the fauna of Otago. Transactions and Proceedings of the NZ Institute 10: 306-324.

Murphy CR, Smith JR, 1970. Age determination of pouch young and juvenile Kangaroo Island wallabies. Transactions of the Royal Society of South Australia 94: 15-20.

Murphy EC, 1986. A comparison of an island and mainland population of mice in the Marlborough Sounds. NZ Journal of Ecology 9: 161.

Murphy EC, 1988. The cestode *Vampirolepis straminea* in mice: a new record for New Zealand. NZ Journal of Zoology 15: 423-424.

Murray MD, Nicholls DG, 1965. Studies on the ectoparasites of seals and penguins. 1. The ecology of the louse *Lepidophthirus macrorhini* Enderlein on the southern elephant seal, *Mirounga leonina* (L.). Australian Journal of Zoology 13: 437-454.

Musser GG, 1977. *Epimys benguetensis*, a composite, and one zoogeographic view of rat and mouse faunas in the Philippines and Celebes. American Museum Novitates 2624: 1-15.

Musser GG, Califia D, 1982. Results of the Archbold expeditions. No. 106. Identities of rats from Pulau Maratua and other islands off East Borneo. American Museum Novitates 2726: 1-30.

Myers JH, 1974. Genetic and social structure of feral house mouse populations on Grizzly Island, California. Ecology 55: 747-759.

Myers K, 1971. The rabbit in Australia. *In* den Boer PJ and Gradwell GR (Eds)

Proceedings of the Advanced Study Institute on 'Dynamics of Numbers in Populations' (Oosterbeek, 1970), pp. 478-506. Pudoc, Wageningen.

Myers K, Gilbert N, 1968. Determination of age of wild rabbits in Australia. Journal of Wildlife Management 32: 841-849.

Myers K, Poole WE, 1962. A study of the biology of the wild rabbit, *Oryctolagus cuniculus* (L.), in confined populations. III. Reproduction. Australian Journal of Zoology 10: 225-267.

Myers K, Poole WE, 1963. A study of the biology of the wild rabbit, *Oryctolagus cuniculus* (L.) in confined populations V. Population dynamics. Australian Wildlife Research 8: 166-203.

Mykytowycz R, 1965. Further observations on the territorial function and history of the submandibular cutaneous (chin) glands in the rabbit, *Oryctolagus cuniculus* (L.). Animal Behaviour 13: 400-412.

Mykytowycz R, 1966a. Observations on odoriferous and other glands in the Australian wild rabbit, *Oryctolagus cuniculus* (L.), and the hare *Lepus europaeus* P. 1. The anal gland. CSIRO Wildlife Research 11: 11-29.

Mykytowycz R, 1966b. Observations on odoriferous and other glands in the Australian wild rabbit, *Oryctolagus cuniculus* (L.), and the hare *Lepus europaeus* P. II. The inguinal glands. CSIRO Wildlife Research 11: 49-64.

Mykytowycz R, 1966c. Observations on odoriferous and other glands in the Australian wild rabbit, *Oryctolagus cuniculus* (L.), and the hare *Lepos europaeus* P. III. Harder's lachrymal and submandibular glands. CSIRO Wildlife Research 11: 65-90.

Mykytowycz R, Dudzinski ML, 1966. A study of the weight of odoriferous and other glands in relation to social status and degree of sexual activity in the rabbit, *Oryctolagus cuniculus* (L.). CSIRO Wildlife Research 11: 31-47.

Nascetti G, Lovari S, Lonfranchie P, Berducou C, Mattiucci S, Rossi L, Bullini L, 1985. Revision of *Rupicapra* genus III. Electrophoretic studies demonstrating species distinction of chamois population of the Alps from those of the Apennines and Pyrenees. *In* Lovari S (Ed.) The Biology and Management of Mountain Ungulates, pp. 56-82. Croom Helm, London.

Nasimovich AA, 1949. The biology of the weasel in Kola Peninsula, in connection with its competitive relations with the ermine. Zoologicheskii Zhurnal 28: 177-182 (in Russian: translation available from the Elton Library, Oxford, UK).

National Business Review, 1985. Goats: investment special feature. National Business Review 16(34): 33.

Nelson JR, Leege TA, 1982. Nutritional requirements and food habits. *In* Thomas JW and Toweill DE (Eds) Elk of North America, Ecology and Management, pp. 323-367. Stackpole Books, Harrisburg, Pennsylvania.

Nelson L Jr, Clark FW, 1973. Correction for sprung traps in catch/effort calculations of trapping results. Journal of Mammalogy 54: 295-298.

Newhook FJ, Dickson EM, Bennett KJ, 1971. A botanical survey of some offshore islands of the Coromandel Peninsula. Tane 17: 97-117.

Newman DG, 1986. Can tuatara and mice co-exist? The status of the tuatara, *Sphenodon punctatus* (Reptilia: Rynchocephalia), on the Whangamata Islands. *In* Wright AE and Beever RE (Eds) The offshore islands of northern New Zealand, pp. 179-185. Department of Lands and Survey Information Series 14.

Newsome AE, 1969. A population of house mice temporarily inhabiting a South Australia wheatfield. Journal of Animal Ecology 38: 341-359.

Newsome AE, Catling RC, Parer I, 1983. The effect of predator removal on rabbit populations. Seventh Australian Vertebrate Pest Control Conference, Dubbo, pp. 156-160.

Newsome AE, Corbett LK, 1975. Outbreaks of rodents in Central Australia: causes, preventions and evolutionary considerations. *In* Prakash I and Ghosh P (Eds) Rodents in Desert Environments, pp. 117-153. Junk, the Hague.

Newsome AE, Corbett LK, Catling PC, Burt RJ, 1983. The feeding ecology of the dingo. I. Stomach contents from trapping in south-eastern Australia, and the non-target wildlife also caught in dingo traps. Australian Wildlife Research 10: 477-486.

Newsome AE, Cowan PE, Ives PM, 1982. Homing by wild house-mice displaced with or without the opportunity to see. Australian Wildlife Research 9: 421-426.

New Zealand Forest Service, 1986. Blue Mountains Recreational Hunting Area wild animal control plan revision 1986. Unpublished report, NZ Forest Service (Southland Conservancy), 13 pp.

Nichol R, 1981. Preliminary report on excavations at the Sunde Site, N38/24, Motutapu Island. Newsletter of the NZ Archaeological Association 24: 237-256.

Nicholas JL, 1817. Narrative of a Voyage to New Zealand, vol. 1. James Black, London.

Nicholas M, 1982. Spatial organisation of *Rattus exulans* on Tiritiri Matangi Island. Unpublished M.Sc. thesis, University of Auckland.

Nicholson AJ, Warner DW, 1953. The rodents of New Caledonia. Journal of Mammalogy 34: 168-179.

Niethammer G, 1971. Die gemsen Neuseelands. Zeitschrift für Säugetierkunde 36: 228-238.

Noirot E, 1968. Ultrasounds in young rodents. 2. Changes with age. Animal Behaviour 16: 129-134.

Norman FI, 1970. Food preferences of an insular population of *Rattus rattus*. Journal of Zoology (London) 162: 493-503.

Nowak RM, Paradiso JL, 1983. Walker's Mammals of the World (4th edition). 2 vols. Johns Hopkins University Press, Baltimore.

Nugent G, 1985. Blue Mountains deer research. Preliminary results 1982-85. Unpublished report, NZ Forest Service, Forest Research Institute, 37 pp.

Nugent G, Challies CN, 1988. Diet and food preferences of white-tailed deer in north-eastern Stewart Island. NZ Journal of Ecology 11: 61-71.

Nugent G, Mawhinney K, 1987. Recreational hunters' views on fallow deer management in the Blue Mountains, Otago. NZ Forestry 32(1): 32-35.

Nugent G, Parkes JP, Tustin KG, 1987. Changes in the density and distribution of red deer and wapiti in northern Fiordland. NZ Journal of Ecology 10: 11-21.

Nutting WB, Kettle PR, Tenquist JD, Whitten LK, 1975. Hair follicle mites (*Demodex* spp.) in New Zealand. NZ Journal of Zoology 2: 219-222.

Obara Y, 1982. Comparative analysis of karyotypes in the Japanese mustelids, *Mustela nivalis namiyeii* and *M. erminea nippon*. Journal of the Mammalogical Society of Japan 9: 59-69.

Obrtel R, Holisova V, 1980. The numbers of owned domestic cats in the urban environment of Brno. Folia Zoologica 29(2): 97-106.

Ogle CC, 1981. Great Barrier Island wildlife survey. Tane 27: 177-200.

O'Gorman F, 1963. Observations on terrestrial locomotion in Antarctic seals. Proceedings of the Zoological Society of London 141: 837-850.

Oliver WRB, 1910. The vegetation of the Kermadec Islands. Transactions of the NZ Institute 42: 118.

Oliver WRB, 1955. New Zealand Birds (2nd edition). AH and AW Reed, Wellington.

Opotiki Minute Book No. 1, 1878. Held in the archives of the Maori Land Court, Rotorua, New Zealand.

Øritsland T, 1970a. Biology and population dynamics of Antarctic seals. *In* Holdgate MW (Ed.) Antarctic Ecology, Vol. I, pp. 361-366. Academic Press, London and New York.

Øritsland T, 1970b. Sealing and seal research in the Southwest Atlantic pack ice, Sept-Oct 1964. *In* Holdgate MW (Ed.) Antarctic Ecology, Vol. I, pp. 367-376. Academic Press, London and New York.

Øritsland T, 1977. Food consumption of seals in the Antarctic pack ice. *In* Llano GA (Ed.) Adaptation within Antarctic ecosystems, pp. 749-768. Proceedings of the third SCAR symposium on Antarctic ecology. Smithsonian Institution, Washington, D.C.

O'Rourke F, 1970. The Fauna of Ireland. An Introduction to the Land Vertebrates. Mercier Press, Cork.

Orwin DFG, Whitaker AH, 1984. Feral sheep (*Ovis aries* L.) of Arapawa Island, Marlborough Sounds, and a comparison of their wool characteristics with those of four other feral flocks in New Zealand. NZ Journal of Zoology 11: 201-224.

Osborn TAB, 1933. Effect of introductions of exotic plants and animals into Australia. Proceedings of the 5th Pacific Science Congress 1: 809-810.

Otway W, 1986. Mesopotamian fallow. A new breed in New Zealand. The Deer Farmer 32: 13-17.

Owadally AW, Butzler W, 1972. The Deer in Mauritius. Alpha Printing, Port Louis.

Owens JMR, 1981. New Zealand before Annexation. *In* Oliver WH and Williams BR (Eds) The Oxford History of New Zealand, pp. 28-53. Oxford University Press, Wellington.

Palen GF, Goddard GV, 1966. Catnip and oestrus behaviour in the cat. Animal Behaviour 14: 372-377.

Parer I, 1977. The population ecology of the wild rabbit, *Oryctolagus cuniculus* (L.) in a Mediterranean-type climate in New South Wales. Australian Wildlife Research 4: 171-205.

Parkes JP, 1975. Some aspects of the biology of the hedgehog (*Erinaceus europaeus* L.) in the Manawatu, New Zealand. NZ Journal of Zoology 2: 463-472.

Parkes JP, 1983. Control of feral goats by poisoning with compound 1080 on natural vegetation baits, and by shooting. NZ Journal of Forestry Science 13: 266-274.

Parkes JP, 1984a. Home ranges of radio-telemetered hares (*Lepus capensis*) in a sub-alpine population in New Zealand: implications for control. Acta Zoologica Fennica 171: 279-281.

Parkes JP, 1984b. Distribution and density of thar in New Zealand. Unpublished report, NZ Forest Service, Forest Research Institute, Protection Forestry Branch.

Parkes JP, 1984c. Feral goats on Raoul Island. I. Effects of control methods on their density, distribution and productivity. NZ Journal of Ecology 7: 85-93.

Parkes JP, 1984d. Feral goats on Raoul Island. II. Diet and notes on the flora. NZ Journal of Ecology 7: 95-101.

Parkes JP, Brockie RE, 1977. Sexual differences in hibernation of hedgehogs in New Zealand. Acta Theriologica 22: 384-386.

Parkes JP, Tustin KG, 1985. A reappraisal of the distribution and dispersal of female Himalayan thar in New Zealand. NZ Journal of Ecology 8: 5-10.

Parkes JP, Tustin K, Stanley L, 1978. The history and control of red deer in the takahe area, Murchison Mountains, Fiordland National Park. NZ Journal of Ecology 1: 145-152.

Parsons SR, 1980. The feral sheep (*Ovis aries* L.) of Woodstock Station, Canterbury, New Zealand. Unpublished B.Sc. (Hons) project, University of Canterbury, Christchurch.

Pascal M, 1980. Structure et dynamique de la population de chats harets de l'archipel des Kerguelen. Mammalia 44(2): 161-182.

Patton JL, Yang SY, Myers P, 1975. Genetic and morphologic divergence among introduced rat populations (*Rattus rattus*) of the Galapagos Archipelago, Ecuador. Systematic Zoology 24: 296-310.

Paulian P, 1953. Pinnipèdes, cétaces, oiseaux des Iles Kerguelen et Amsterdam. Mission Kerguelen, 1951. Mémoires de l'Institut Scientifique de Madagascar 8: 111-234.

Pavlov PN, 1980. The diet and ecology of the feral pig (*Sus scrofa*) at Girilambone, New South Wales. Unpublished Ph.D. thesis, Monash University, Melbourne.

Payton IJ, 1985. Southern rata (*Metrosideros umbellata* Cav.) mortality in Westland, New Zealand. *In* Turner H and Tranquillini W (Eds) Establishment and tending of subalpine forest: research and management, pp. 207-214. Third International Workshop, IUFRO, Riederalp, Switzerland, 1984. Eidgenossiche Anstalt für das forstliche Versuchswessen, Berlichte 270.

Peirson HM, Owtram IH, 1945. The origin of pig breeds. NZ Journal of Agriculture 71(2): 391-394.

Pekelharing CJ, 1970. Cementum deposition as an age indicator in the brush-tailed possum, *Trichosurus vulpecula* Kerr (Marsupialia). Australian Journal of Zoology 18: 71-76.

Pekelharing CJ, 1974. Paradontal disease as a cause of tooth loss in a population of chamois (*Rupicapra rupicapra* L.) in New Zealand. Zeitschrift für Säugetierkunde 39: 250-255.

Pekelharing CJ, 1979. Fluctuations in opossum population along the north bank of the Taramakau catchment and its effects on the forest canopy. NZ Journal of Forestry Science 9: 212-224.

Pekelharing CJ, Reynolds RN, 1983. Distribution and abundance of browsing mammals in Westland National Park in 1978, and some observations on their impact on the vegetation. NZ Journal of Forestry Science 13: 247-265.

Pelikan J, 1981. Patterns of reproduction in the house mouse. Symposia of the Zoological Society of London 47: 205-229.

Pemberton J, Smith RH, 1985. Lack of biochemical polymorphism in British fallow deer. Heredity 55: 199-207.

Penney RL, Lowry G, 1967. Leopard seal predation on Adelie penguins. Ecology 48: 878-882.

Penniket A, Garrick A, Breese E (Eds), 1986. Preliminary reports of Expeditions to the Auckland Islands Nature Reserve 1973-1984. Department of Lands and Survey, Wellington.

Perham AN, 1922. Deer in New Zealand. Appendix to the Journals of the House of Representatives, New Zealand C-3A.

Perry JS, 1945. The reproduction of the wild brown rat (*Rattus norvegicus* Erxleben). Proceedings of the Zoological Society of London 115: 19-46.

Perry R, 1969. Conservation problems in the Galapagos. Micronesica 5(2): 275-281.

Peters JA, 1972. Vertebrate pests in New Zealand: research on control. *In* Dana RH (Ed.) Proceedings of the 5th Vertebrate Pest Control Conference, pp. 101-103. University of California, Davis.

Philpott A, 1919. Notes on the birds of South-Western Otago. Transactions of the NZ Institute 51: 216-224.

Pickard CR, 1984. The population ecology of the house mouse (*Mus musculus*) on Mana Island. Unpublished M.Sc. thesis, Victoria University of Wellington.

Pielowski Z, 1971. Length of life of the hare. Acta Theriologica 16: 89-94.

Pielowski Z, 1972. Home range and degree of residence of the European hare. Acta Theriologica 17: 93-103.

Pielowski Z, 1976. Number of young born and dynamics of the European hare population. *In* Pielowski Z and Pucek Z (Eds) Ecology and Management of European Hare Populations, pp. 75-78. Polish Hunting Association, Warsaw.

Pierce RJ, 1982. A comparative ecological study of pied and black stilts in South Canterbury. Unpublished Ph.D. thesis, University of Otago, Dunedin.

Pierce RJ, 1986. Differences in susceptibility to predation during nesting between pied and black stilts (*Himantopus* spp.) Auk 103: 273-280.

Pierson ED, Sarich VM, Lowenstein JM, Daniel MJ, 1982. *Mystacina* is a phyllostomoid bat. Bat Research News 23: 78.

Pierson ED, Sarich VM, Lowenstein JM, Daniel MJ, Rainey WE, 1986. A molecular link between the bats of New Zealand and South America. Nature (London) 324: 60-63.

Pilton PE, Sharman GB, 1962. Reproduction in the marsupial *Trichosurus vulpecula*. Journal of Endocrinology 25: 119-136.

Pitt-Turner J, 1974. Beagles and hares. Evening Post, Wellington. 13 Nov.

Pocock RI, 1936. The polecats of the genera *Putorius* and *Vormella* in the British Museum. Proceedings of the Zoological Society of London 1936: 691-723.

Polack JS, 1838. New Zealand: Being a Narrative of Travels and Adventures during a Residence in that Country between the Years 1831 and 1837 (2 vols.) Richard Bentley, London.

Pollock GA, 1970. The South Island otter — a reassessment. Proceedings of the NZ Ecological Society 17: 129-135.

Pollock, GA, 1974. The South Island otter — an addendum. Proceedings of the NZ Ecological Society 21: 57-61.

Poole AL, 1951a. Preliminary reports of the New Zealand-American Fiordland Expedition investigations in Fiordland, New Zealand, in 1949. DSIR Bulletin 103.

Poole AL, 1951b. Flora and vegetation of the Caswell and George Sounds District. Transactions of the Royal Society of NZ 79: 62-83.

Poole AL, Johns JH, 1970. Wild Animals in New Zealand. AH and AW Reed, Wellington.

Poole TB, 1972. Some behavioural differences between the European polecat, *Mustela putorius*, the ferret, *M. furo* and their hybrids. Journal of Zoology (London) 166: 25-35.

Poole TB, 1973. The aggressive behaviour of individual male polecats (*Mustela putorius, M. furo* and hybrids) towards familiar and umfamiliar opponents. Journal of Zoology (London) 170: 395-414.

Poulter TC, 1968. Underwater vocalisation and behaviour of pinnipeds. *In* Harrison RJ, Hubbard RC, Peterson RS, Rice CE and Schusterman RJ (Eds) The Behaviour and Physiology of Pinnipeds, pp. 69-84. Appleton-Century-Crofts, New York.

Pounds CJ, 1981. Niche overlap in sympatric populations of stoats (*Mustela erminea*) and weasels (*Mustela nivalis*) in northeast Scotland. Unpublished Ph.D. thesis, University of Aberdeen.

Powell RA, 1979. Mustelid spacing patterns: variations on a theme by *Mustela*. Zeitschrift für Tierpsychologie 50: 153-165.

Powell RA, 1985. Possible pathways for the evolution of reproductive strategies in weasels and stoats. Oikos 44: 506-508.

Pracy LT, 1974. Introduction and liberation of the opossum (*Trichosurus vulpecula*) into New Zealand. NZ Forest Service Information Series No. 45.

Pracy LT, 1975. Opossums (1). New Zealand's Nature Heritage 3: 873-882.

Pracy LT, 1980. Opossum Survey Report. Agricultural Pests Destruction Council, NZ.

Pracy LT, Kean RI, 1968. Tapeworms found in opossums. NZ Journal of Agriculture 117: 21-22.

Pracy LT, Kean RI, 1969. The opossum in New Zealand (habits and trapping). NZ Forest Service Information Series No. 40.

Prater SH, 1948. The Book of Indian Animals. Bombay Natural History Society, Bombay.

Prentiss PG, Wolf AV, Eddy HA, 1959. Hydropenia in cat and dog. Ability of the cat to meet its water requirements solely from a diet of fish or meat. American Journal of Physiology 196: 625-632.

Presidente PJA, 1978. Diseases seen in free ranging marsupials and those held in captivity. *In* Fauna Part B, pp. 457-471. Post-graduate Committee on Veterinary Science, Proceedings No. 36, University of Sydney.

Presidente PJA, 1984. Parasites and diseases of brushtail possums (*Trichosurus* spp.): occurrence and significance. *In* Smith A and Hume I (Eds) Possums and Gliders, pp. 171-187. Surrey Beatty and Sons, Chipping Norton, New South Wales.

Proctor-Gray E, 1984. Dietary ecology of the coppery brushtail possum, green ringtail possum and Lumholtz's tree kangaroo in north Queensland. *In* Smith A and Hume I (Eds) Possums and Gliders, pp. 129-135. Surrey Beatty and Sons, Chipping Norton, New South Wales.

Pullar EM, 1950. The wild (feral) pigs of Australia and their role in the spread of infectious diseases. Australian Veterinary Journal 26: 99-109.

Purchas TPG, 1981. Predation on starling (*Sturnus vulgaris*) in nest boxes in Hawke's Bay. Notornis 28: 39-40.

Purey-Cusp JR, McClymont RB, 1979. Stewart Island Land Management Study, 1978. NZ Forest Service and Department of Lands and Survey, Wellington.

Quay WB, Tomich PQ, 1963. A specialised midventral sebacious glandular area in *Rattus exulans*. Journal of Mammalogy 44: 537-542.

Racovitza EG, 1900. La vie des animaux et des plantes dans l'Antarctique. Bulletin de la Société Royale Belge de Geographie, Brussels 24: 177-230.

Radcliffe JE, 1985. Grazing management of goats and sheep for gorse control. NZ Journal of Experimental Agriculture 13: 181-190.

Rammell EG, Fleming PA, 1978. Compound 1080. Properties and use of sodium monofluoroacetate in New Zealand. Animal Health Division, Ministry of Agriculture and Fisheries, New Zealand.

Ramsay GW, 1978. A review of the effects of rodents on the New Zealand invertebrate fauna. *In* Dingwall PR, Atkinson IAE and Hay C (Eds) The Ecology and Control of Rodents in New Zealand Nature Reserves, pp. 89-95. Department of Lands and Survey Information Series 4.

Rand RW, 1956. Notes on the Marion Island fur seal. Proceedings of the Zoological Society of London 126: 65-82.

Ratcliffe F, 1931. The flying fox (*Pteropus*) of Australia. Commonwealth Council for Scientific and Industrial Research Bulletin 53: 1-81.

Ratcliffe PR, 1987. Distribution and current status of sika deer, *Cervus nippon*, in Great Britain. Mammal Review 17: 39-58.

Ratumaitavuki MG, Kirk EJ, 1982. Report on bones of Kaimanawa horses: material received 22 February 1982. Unpublished report, Department of Physiology and Anatomy, Massey University, Palmerston North.

Ray C, 1970. Population ecology of Antarctic seals. *In* Holdgate MW (Ed.) Antarctic Ecology, Vol. I, pp. 398-414. Academic Press, London and New York.

Ray C, 1981. Ross seal: *Ommatophoca rossi* Gray, 1844. *In* Ridgway SH and Harrison RJ (Eds) Handbook of Marine Mammals, Vol. 2: Seals, pp. 221-235. Academic Press, London.

Raymond M, Bergeron JM, 1982. Réponse numerique de l'hermine aux fluctuations d'abondance de *Microtus pennsylvaticus*. Canadian Journal of Zoology 60: 542-549.

Redhead RE, 1968. An analysis of pellets cast by harrier hawks. Notornis 15: 244-247.

Redhead RE, 1969. Some aspects of the feeding of the harrier. Notornis 16: 262-284.

Redhead TD, 1982. Reproduction, growth and population dynamics of house mice in irrigated and non-irrigated cereal farms in New South Wales. Unpublished Ph.D. thesis, Australian National University, Canberra.

Redhead TD, Enright N, Newsome AE, 1985. Causes and predictions of outbreaks of *Mus musculus* in irrigated and non-irrigated cereal farms. Acta Zoologica Fennica 173: 123-127.

Regnault WR, 1980. Report on sheep and wool observations. *In* Preliminary reports of the Campbell Island Expedition 1975-76, pp. 144-148. Department of Lands and Survey Reserves Series 7.

Reid B, 1986. Kiwis, opossums and vermin. A survey of opossum hunting; and of target and non-target tallies. Fur Facts 7: 37-49.

Reiher AH, 1963. Some notes on the Javan rusa deer (*Cervus timoriensis*). Protection Forestry Newsletter 1(5): 25-26.

Reischek A, 1886. Notes on New Zealand ornithology: observations on *Pogonornis cincta* (Dubus); stitch-bird (Tiora). Transactions and Proceedings of the NZ Institute 18: 84-87.

Reischek A, 1887. Ornithological notes. Transactions and Proceedings of the NZ Institute 19: 188-193.

Reischek A, 1888. Notes on rats. Transactions of the NZ Institute 20: 125-126.

Reischek A, 1930. Yesterdays in Maoriland (translated by HEL Priday). Jonathan Cape, London.

Rempe U, 1970. Morphometrische Untersuchungen an Iltisschadeln zur Klarung der Verwandtschaft von Steppeniltis, Waldiltis und Frettchen. Analyse eines "Grenzfalles" zwischen Unterart und Art. Zeitschrift für wissenschaftliche Zoologie 180: 185-367.

Reynolds RN, 1982. Urolithiasis in a wild red deer (*Cervus elaphus*) population. NZ Veterinary Journal 30: 25-26.

Richards EC, 1952. The Chatham Islands, their Plants, Birds and People. Simpson and Williams, Christchurch.

Richardson BJ, Wood DH, 1982. Experimental ecological studies on a subalpine rabbit population. I. Mortality factors acting on emergent kittens. Australian Wildlife Research 9: 443-450.

Ricketts W, 1979. The surviving wild horses of the Kaimanawas. Newsletter 26, The Anti-Cruelty Society, Palmerston North.

Rieck W, 1956. Der Junghasenanteil auf den Strecken 1953/54 und 1954/55. Zeitschrift für Jagdwissenschaft 2: 160-164.

Riney T, 1951. Standard terminology for deer teeth. Journal of Wildlife Management 15: 99-101.

Riney T, 1954. Antler growth and shedding in a captive group of fallow deer (*Dama dama*) in New Zealand. Transactions of the Royal Society of NZ 82: 569-578.

Riney T, 1955a. Evaluating condition of free-ranging red deer (*Cervus elaphus*) with special reference to New Zealand. NZ Journal of Science and Technology (B) 36: 430-463.

Riney T, 1955b. Identification of big game animals in New Zealand. Dominion Museum Handbook 4.

Riney T, 1956. Comparison of occurrence of introduced animals with critical conservation areas to determine priorities for control. NZ Journal of Science and Technology (B)38: 1-18.

Riney T, 1957. Sambar (*Cervus unicolor*) in sand hill country. Proceedings of the NZ Ecological Society 5: 26-27.

Riney T, Caughley G, 1959. A study of home range in a feral goat herd. NZ Journal of Science 2: 157-170.

Riney T, Watson JS, Bassett C, Turbott EG, Howard WE, 1959. Lake Monk expedition, an ecological study in Southern Fiordland. DSIR Bulletin 135: 1-75.

Ritchie IM, 1970. A preliminary report on a recent botanical survey of the Chatham Islands. Proceedings of the NZ Ecological Society 17: 52-56.

Roach RW, Turbott EG, 1953. Notes on the long-tailed bat (*Chalinolobus morio*). NZ Science Review 11: 161.

Roberts G, 1968. Game Animals in New Zealand. AH and AW Reed, Wellington.

Robertson CJR, Law E de H, Wakelin DJ, Courtney SP, 1984. Habitat requirements of wetland birds in the Lower Waitaki River catchment New Zealand. NZ Wildlife Service Occasional Publications 2(6).

Robertson HA, Meads MJ, 1979. An ecological survey of the Haast-Arawata region of South Westland, March-April 1979. Unpublished report, Ecology Division, DSIR.

Robertson LN, 1976. Mustelids on the Otago peninsula. Unpublished Dip. Sci. thesis, University of Otago, Dunedin.

Robinson R, 1982. Evolution of the domestic cat. Carnivore 5: 4-13.

Robinson R, Cox HW, 1970. Reproductive performance in a cat colony over a 10-year period. Laboratory Animals (London) 4(1): 99-112.

Robinson RA, Daniel MJ, 1968. The significance of salmonella isolations from wild birds and rats in New Zealand. NZ Veterinary Journal 16: 53-55.

Rogers LM (Ed.), 1961. The Early Journals of Henry Williams. Pegasus Press, Christchurch.

Rogers PM, 1979. Ecology of the European wild rabbit, *Oryctolagus cuniculus* (L.) in the Camargue, Southern France. Unpublished Ph.D. thesis, University of Guelph, Canada.

Rolls EC, 1969. They All Ran Wild. The Story of Pests on the Land in Australia. Angus and Robertson, Sydney.

Rose A, 1977. Forests, deer and opossums in the Greenstone and Caples Valleys. Unpublished report, NZ Forest Service (Southland Conservancy), 31 pp.

Roser RJ, Lavers RB, 1976. Food habits of the ferret (*Mustela putorius furo* L.) at Pukepuke Lagoon, New Zealand. NZ Journal of Zoology 3: 269-275.

Ross, Sir James 1847. A Voyage of Discovery and Research in the Southern and Antarctic Regions During the Years 1839-43. (Reprinted 1969, Augustus M. Kelley, New York).

Rounsevell D, Eberhard I, 1980. Leopard seals, *Hydrurga leptonyx* (Pinnipedia), at Macquarie Island from 1949 to 1979. Australian Wildlife Research 7: 403-415.

Rowe FP, 1967. Notes on rats in the Solomon and Gilbert Islands. Journal of Mammalogy 48: 649-650.

Rowe FP, 1981. Wild house mouse biology and control. Symposia of the Zoological Society of London 47: 575-589.

Rowe FP, Bradfield A, Redfern R, 1974. Food preferences of wild house-mice (*Mus musculus* L.). Journal of Hygiene 73: 473-478.

Rowe FP, Bradfield A, Quy RJ, Swinney T, 1985. Relationship between eye lens weight and age in the wild house mouse (*Mus musculus*). Journal of Applied Ecology 22: 55-61.

Rowett HGQ, 1974. The Rat as a Small Mammal. John Murray, London.

Rowlands IW, 1972. Reproductive studies in the stoat. Journal of Zoology (London) 166: 574-576.

Rudd J, 1978. The antler growth pattern in the Bay of Plenty sambar herd. NZ Wildlife 7(55/56): 33-35.

Rudd J, 1980. Rusa through the eyes of a South Islander. NZ Wildlife 60: 41-44.

Rudge MR, 1969. Reproduction of feral goats (*Capra hircus*) near Wellington, New Zealand. NZ Journal of Science 12: 817-827.

Rudge MR, 1970. Dental and periodontal abnormalities in two populations of feral goats *Capra hircus* L. in New Zealand. NZ Journal of Science 13: 260-267.

Rudge MR, 1972. Horns as indicators of age in goats (*Capra hircus* L.). NZ Journal of Science 15: 255-263.

Rudge MR, 1978. First come, first preserved. NZ Listener, 23 September.

Rudge MR, 1982. Conserving feral farm mammals in New Zealand. NZ Agricultural Science 16: 157-160.

Rudge MR, 1983. A reserve for feral sheep on Pitt Island, Chatham group, New Zealand. NZ Journal of Zoology 10: 349-364.

Rudge MR, 1984. Distribution, status and preservation of feral sheep in New Zealand. *In* Blair HT (Ed.) Coloured Sheep and their Products, pp. 295-301. Black and Coloured Sheep Breeders Association, Masterton, NZ.

Rudge MR, 1986a. Capturing goats on Auckland Island for commercial evaluation. DSIR Ecology Division Report 3.

Rudge MR, 1986b. The decline and increase of feral sheep (*Ovis aries* L.) on Campbell Island (subantarctic). NZ Journal of Ecology 9: 89-100.

Rudge MR, Campbell DJ, 1977. The history and present status of goats on the Auckland Islands (New Zealand Subantarctic) in relation to vegetation changes induced by man. NZ Journal of Botany 15: 221-253.

Rudge MR, Clark JM, 1978. The feral goats (*Capra hircus* L.) of Raoul Island and some effects of hunting on their density and body size. NZ Journal of Zoology 5: 581-589.

Rudge MR, Smit T, 1970. Expected rate of increase of hunted populations of feral goats (*Capra hircus* L.) in New Zealand. NZ Journal of Science 13: 256-259.

Russell EM, 1974. The biology of kangaroos (Marsupalia-Macropodidae) Mammal Review 4: 1-59.

Russell EM, 1982. Patterns of parental care and parental investment in marsupials. Biological Reviews 57: 423-486.

Russell EM, 1985. The metatherians: order Marsupialia. *In* Brown RE and MacDonald DW (Eds) Social Odours in Mammals, Vol. 1, pp. 45-104. Clarendon Press, Oxford.

Rutland J, 1890. On the habits of the New Zealand bush rat. Transactions of the NZ Institute 22: 300-307.

Sadleir RMFS, Tyndale-Biscoe CH, 1977. Photoperiod and the termination of embryonic diapause in the marsupial *Macropus eugenii.* Biology of Reproduction 16: 605-608.

Sagar P, 1979. Harrier attacks hare. OSNZ News No. 13: 2.

Sage RD, 1981. Wild mice. *In* Foster HL, Small JD and Fox JD (Eds) The Mouse in Biomedical Research, Vol. I, pp. 39-90. Academic Press, New York.

Sage RD, Whitney JB III, Wilson AC, 1986. Genetic analysis of a hybrid zone between Domesticus and Musculus mice (*Mus musculus* complex): Hemoglobin polymorphisms. Current Topics in Microbiology and Immunology 127: 75-85.

Sage RD, Heyneman D, Lim Kee-Chong, Wilson AC, 1986. Wormy mice in a hybrid zone. Nature 324: 60-62.

St Paul R, 1977. A bushman's seventeen years of noting birds. Part F. Notornis 24: 65-74.

Sales GD, 1972. Ultrasound and mating behaviour in rodents with some observations on other behavioural situations. Journal of Zoology (London) 168: 149-164.

Sandars E, 1957. A Beast Book for the Pocket. Oxford University Press, London.

Sandell M, 1984. To have or not to have delayed implantation: the example of the weasel and stoat. Oikos 42: 123-126.

Sandell M, 1986. Movement patterns of male stoats *Mustela erminea* during the mating season: differences in relation to social status. Oikos 47: 63-70.

Saxton JW, 1849. Diary. Nelson Provincial Museum.

Schafer M, 1975. The Language of the Horse. Kaye and Ward, London.

Schaller GB, 1967. The Deer and the Tiger — a Study of Wildlife in India. University of Chicago Press, Chicago.

Schaller G, 1973. Observations of Himalayan thar (*Hemitragus jemlahicus*). Journal of the Bombay Natural History Society 70: 1-24.

Schaller G, 1977. Mountain Monarchs. University of Chicago Press, Chicago.

Schauenberg P, 1969. L'identification du chat forestier d'Europe *Felis s. silvestris* Schreber, 1777 par une méthode ostéométrique. Revue Suisse de Zoologie 76(2): 433-441.

Schauenberg P, 1977. Longueur de l'intestin du chat forestier *Felis silvestris* Schreber. Mammalia 41(3): 357-360.

Scheffer VB, 1958. Seals, Sea Lions and Walruses. Stanford University Press, Stanford.

Schein MW, 1950. The relation of sex ratio to physiological age in the wild brown rat. American Naturalist 84: 489-496.

Schevill WE, Watkins WA, 1965. Underwater calls of *Leptonychotes* (Weddell seal). Zoologica 50: 45-46.

Schevill WE, Watkins WA, 1971. Directionality of the sound beam in *Leptonychotes weddelli* (Mammalia: Pinnipedia). Antarctic Research Series 18: 163-168.

Schneider E, 1976. Der zeitige Kenntnisse über das Paarungsverhalten des Feldhasen. *In* Pielowski Z and Pucek Z (Eds) Ecology and Management of European Hare Populations, pp. 41-53. Polish Hunting Association, Warsaw.

Schneider E, 1978. Der Feldhase. Biologie Verhalten, Hege und Jagd. BLV Verlagsgesellschaft, Munich.

Schneider E, Leipoldt M, 1983. DNA relationship within the genus *Lepus* in S.W. Europe. Acta Zoologica Fennica 174: 31-33.

Schofield CM, 1948. Virginian deer (*Cariacus virginianus*). Unpublished Field Investigation Report No. 8, Department of Internal Affairs.

Scholander PF, 1940. Experimental investigations on the respiratory function in diving mammals and birds. Hvalradets Skrifter Norske Videnskaps-Akademie Oslo 22: 1-131.

Schröder W, 1971. Untersuchungen zur Ökologie des Gemswildes (*Rupicapra rupicapra* L.) in einem Vorkommen der Alpen. Zeitschrift für Jagdwissenschaft 17: 113-167.

Schwabe HW von, 1979. Eco-ethological studies on the ability to dispersal of the black rat (*Rattus rattus* L.). Zoologische Jahrbücher, Abteilung für Systematik, Ökologie und Geographie der Tiere 106: 124-168.

Schwarz E, Schwarz HK, 1943. The wild and commensal house mouse, *Mus musculus* Linnaeus. Journal of Mammalogy 24: 59-72.

Searle AG, 1968. Comparative Genetics of Coat Colour in Mammals. Logos Press, London.

Searle AG, Dhaliwal SS, 1957. The rats of Singapore Island. Proceedings of the 9th Pacific Science Congress 19: 12-14.

Segal AN, 1975. Postnatal growth, metabolism and thermoregulation in the stoat. Soviet Journal of Ecology 6: 28-32.

Senzota RBM, 1982. The house rat enters the Serengeti. African Journal of Ecology 20: 211-212.

Severinghaus CA, 1949. Tooth development and wear as criteria of age in white-tailed deer. Journal of Wildlife Management 13(2): 195-216.

Shailer LC, 1957. The sambar deer. The Roar 3: 19. NZ Deerstalkers Association Inc., Wellington.

Sharman GB, Maynes GM, 1983. Rock-wallabies *Petrogale, Peradorcas*. *In* Strahan R (Ed.) The Australian Museum Complete Book of Australian Mammals, pp. 207-208. Angus and Robertson, Sydney.

Sharp A (Ed.), 1971. Duperrey's Visit to New Zealand in 1824. Alexander Turnbull Library, Wellington.

Shaughnessy PD, 1970. Serum protein variation in southern fur seals, *Arctocephalus* spp., in relation to their taxonomy. Australian Journal of Zoology 18: 331-343.

Shaw WB, 1983. Tropical cyclones: determinants of pattern and structure in New Zealand's indigenous forests. Pacific Science 37: 405-414.

Shearing R, 1965. Some reminiscences of an early hunter. NZ Wildlife 10: 5-11.

Shepherd T, 1940. Journal of T. Shepherd, Port Pegasus, 1826. *In* Howard B (Ed.) Rakiura. A History of Stewart Island, New Zealand, pp. 355-365. AH and AW Reed, Dunedin.

Sherrard JM, 1966. Kaikoura, a History of the District. Caxton Press, Christchurch.

Short J, 1982. Habitat requirements of the brush-tailed rock-wallaby, *Petrogale penicillata*, in New South Wales. Australian Wildlife Research 9: 239-246.

Short J, 1985. The functional response of kangaroos, sheep, and rabbits in an arid grazing system. Journal of Applied Ecology 22: 435-447.

Shorten M, 1954. The reaction of the brown rat towards changes in its environment.

In Chitty D and Southern HN (Eds) Control of Rats and Mice, pp. 307-334. Clarendon Press, Oxford.

Sidorowicz J, 1975. Some observations on the ecology of rodents in the Orongorongo Valley (New Zealand). Mammalia 39: 643-647.

Silvester WB, 1964. Forest regeneration problems in the Hunua Range, Auckland. NZ Journal of Ecology 11: 1-15.

Simmons WJ, 1985. Deer liberation at Lismore Forest near Wanganui. Unpublished report, NZ Forest Service.

Simms DA, 1979. Studies of an ermine population in southern Ontario. Canadian Journal of Zoology 57: 824-832.

Singer FJ, 1981. Wild pig populations in the national parks. Environmental Management 5: 263-270.

Singleton GR, 1983. The social and genetic structure of a natural colony of house mice, *Mus musculus*, at Healeville Wildlife Sanctuary. Australian Journal of Zoology 31: 155-166.

Singleton GR, Hay DA, 1983. The effect of social organisation on reproductive success and gene flow in colonies of wild house mice, *Mus musculus*. Behavioural Ecology and Sociobiology 12: 49-56.

Singleton GR, Redhead TD, 1989. House mouse plagues in the Victorian Mallee Region. *In* Noble J and Bradstock R (Eds) Mediterranean Landscapes in Australia: Mallee Ecosystems and their Management [in press].

Singleton RJ, 1972. New Zealand fur seals at Three Kings Islands. NZ Journal of Marine and Freshwater Research 6: 649-650.

Siniff DB, Bengston JL, 1977. Observations and hypotheses concerning the interactions among crabeater seals, leopard seals, and killer whales. Journal of Mammalogy 58: 414-416.

Siniff DB, Reichle RA, 1976. Biota of Antarctic pack ice: R/V Hero cruise, 75-76. Antarctic Journal 11: 61.

Siniff DB, Cline DR, Erickson AW, 1970. Population densities of seals in the Weddell Sea, Antarctica. *In* Holdgate MW (Ed.) Antarctic Ecology, Vol. I, pp. 377-394. Academic Press, London and New York.

Siniff DB, Tester JR, Kuechle VG, 1971. Some observations on the activity patterns of Weddell seals as recorded by telemetry. Antarctic Research Series 18: 173-180.

Siniff DB, Stirling I, Hofman R, De Master D, Reichle R, Kirby R, 1975. Population studies of Weddell seals in eastern McMurdo Sound. Antarctic Journal 10: 120.

Siniff DB, De Master DP, Hofman RJ, Eberhardt LL, 1977. An analysis of the dynamics of a Weddell seal population. Ecological Monographs 47: 319-335.

Siniff DB, Stirling I, Bengston JL, Reichle RA, 1979. Social and reproductive behaviour of crabeater seals (*Lobodon carcinophagus*) during the austral spring. Canadian Journal of Zoology 57: 2243-2255.

Skira IJ, 1978. Reproduction of the rabbit, *Oryctolagus cuniculus* (L.), on Macquarie Island, Subantarctica. Australian Wildlife Research 5: 317-326.

Sladden B, 1926. Tuhua, or Mayor Island. NZ Journal of Science and Technology 8: 193-210.

Slee K, 1984. The Sambar deer in Victoria. *In* Deer Refresher Course. Postgraduate Committee on Veterinary Science, Proceedings 72: 559-572.

Smit FGAM, 1979. The fleas of New Zealand (Siphonaptera). Journal of the Royal Society of NZ 9: 143-232.

Smith A, Hume I (Eds), 1984. Possums and Gliders. Surrey Beatty and Sons, Chipping Norton, New South Wales.

Smith IWG, 1981a. Prehistoric mammalian fauna from the Coromandel Peninsula. Records of the Auckland Institute and Museum 18: 107-125.

Smith IWG, 1981b. Mammalian fauna from an Archaic site on Motutapu Island, New Zealand. Records of the Auckland Institute and Museum 18: 95-105.

Smith JC, 1981. Senses and communication. Symposia of the Zoological Society of London 47: 367-393.

Smith JMB, 1964. Some microbiological aspects of the short-eared European hedgehog, *Erinaceus europaeus*, in New Zealand. Unpublished Ph.D. thesis, University of Otago, Dunedin.

Smith JMB, 1968. Diseases of hedgehogs. Veterinary Bulletin 38: 425-430.

Smith JMB, Marples MJ, 1963. *Trichophyton mentagrophytes* var. *erinacei*. Sabouraudia 3: 1-10.

Smith JMB, Robinson RA, 1964. *Salmonella typhimurium* in New Zealand hedgehogs. NZ Veterinary Journal 12: 111-112.

Smith JMB, Rush-Munro FM, McCarthy M, 1969. Animals as a reservoir of human ringworm in New Zealand. Australasian Journal of Dermatology 10: 169-182.

Smith K, 1986. The diet of the ship rat *Rattus rattus rattus* L. in a kauri forest in Northland, New Zealand. Unpublished M.Sc. thesis, University of Canterbury, Christchurch.

Smith MCT, 1974. Biology and Management of the Wapiti (*Cervus elaphus nelsoni*) of Fiordland, New Zealand. NZ Deerstalkers Association Inc., Wellington.

Smith MJ, 1983. Tammar wallaby *Macropus eugenii*. *In* Strahan R (Ed.) The Australian Museum Complete Book of Australian Mammals, pp. 232-233. Angus and Robertson, Sydney.

Smith MJ, Brown BK, Frith HJ, 1969. Breeding of the brush-tailed possum, *Trichosurus vulpecula* (Kerr), in New South Wales. CSIRO Wildlife Research 14: 181-193.

Smith RH, 1980. The genetics of fallow deer and their implications for management. Deer 5: 79-83.

Snell GD (Ed.), 1941. Biology of the Laboratory Mouse. Jackson Laboratory, Blakiston, Philadelphia.

Sorensen JH, 1950. Elephant seals of Campbell Island. Cape Expedition Series Bulletin 6: 1-31.

Sorensen JH, 1969. New Zealand fur seals with special reference to the 1946 open season. NZ Marine Department Fisheries Technical Report No. 42.

Southern HN (Ed.), 1954. Control of Rats and Mice, Vol. 3. Clarendon Press, Oxford.

Southern HN (Ed.), 1964. The Handbook of British Mammals (1st edition). Blackwell Scientific Publications, Oxford.

Southland Acclimatisation Society, [n.d.] History of the Southland Acclimatisation Society. The Southland News Co. Ltd, Invercargill.

Southwick CH, 1966. Reproduction, mortality and growth of murid rodent populations. Indian Rodent Conference Proceedings, pp. 152-175.

Sowman WCR, 1981. Meadow, mountain, forest and stream. The provincial history of the Nelson Acclimatisation Society. Nelson Acclimatisation Society, Nelson.

Sparrman A, 1953. A Voyage Round the World with Captain James Cook in H.M.V. Resolution. Robert Hale Ltd, London.

Spatz G, Mueller-Dombois D, 1973. The influence of feral goats on koa tree reproduction in Hawaii Volcanoes National Park. Ecology 54(4): 870-876.

Spence A, 1968. A story of the Campbell Islands. Tussock Grasslands and Mountain Lands Institute Review 15: 63-75.

Spencer HJ, Davis DE, 1950. Movements and survival of rats in Hawaii. Journal of Mammalogy 31: 154-157.

Spiro JM, 1979. Aspects of biology important in the farming management of fallow deer (*Dama dama*) at South Kaipara Head. Unpublished M.Sc. thesis, University of Auckland.

Spurr EB, 1979. A theoretical assessment of the ability of bird species to recover from an imposed reduction in numbers, with particular reference to 1080 poisoning. NZ Journal of Ecology 2: 46-63.

Spurr EB, 1981. Modelling the effects of control operations on possum *Trichosurus vulpecula* populations. *In* Bell BD (Ed.) Proceedings of the first symposium on marsupials in New Zealand, pp. 223-233. Zoology Publications of Victoria University of Wellington No. 74.

Spurr EB, Jolly JN, 1981. Damage by possums *Trichosurus vulpecula* to farm crops and pasture. *In* Bell BD (Ed.) Proceedings of the first symposium on marsupials in New Zealand, pp. 197-203. Zoology Publications of Victoria University of Wellington No. 74.

Stack JW, 1874. On the disappearance of the larger kinds of lizard from North Canterbury. Transactions of the NZ Institute 7: 295-297.

Stains HJ, 1975. Distribution and taxonomy of the Canidae. *In* Fox MW (Ed.) The Wild Canids, pp. 3-26. Von Nostrand Reinhold, New York.

Statham HL, 1984. The diet of *Trichosurus vulpecula* (Kerr) in four Tasmanian forests. *In* Smith A and Hume I (Eds) Possums and Gliders, pp. 213-219. Surrey Beatty and Sons, Chipping Norton, New South Wales.

Stead EF, 1936. Notes on the short-tailed bat (*Mystacops tuberculatus*). Transactions of the Royal Society of NZ 66: 188-191.

Stead EF, 1937. The Maori rat. Transactions of the Royal Society of NZ 66: 178-181.

Stebbings RE, 1977. Order Chiroptera: Bats. *In* Corbet GB and Southern HN (Eds) The Handbook of British Mammals (2nd edition). Blackwell Scientific Publications, Oxford.

Steel S, 1975. Tally-ho — a'hunting we will go. Evening Post, 20 Sept.

Stenseth NC, 1985. Optimal size and frequency of litters in predators of cyclic prey: comments on the reproductive biology of stoats and weasels. Oikos 45: 293-296.

Sterba O, Klusak K, 1984. Reproductive biology of fallow deer *Dama dama.* Acta Scientarum Naturalium Academiae Scientarum Bohemoslovacae (Brno) 18(6): 1-46.

Stevens GR, 1980. New Zealand Adrift. The Theory of Continental Drift in a New Zealand Setting. AH and AW Reed, Wellington.

Stevens GR, 1985. Lands in Collision. Discovering New Zealand's Past Geography. DSIR Science Information Publishing Centre, Wellington.

Stewart GH, Veblen TT, 1982a. Regeneration patterns in southern rata (*Metrosideros umbellata*) — kamahi (*Weinmannia racemosa*) forest in central Westland, New Zealand. NZ Journal of Botany 20: 55-72.

Stewart GH, Veblen TT, 1982b. A commentary on canopy tree mortality in Westland rata-kamahi protection forest. NZ Journal of Forestry 27: 168-188.

Stewart GH, Veblen TT, 1983. Forest instability and canopy tree mortality in Westland, New Zealand. Pacific Science 37: 427-431.

Stirling I, 1969a. Ecology of the Weddell seal in McMurdo Sound, Antarctica. Ecology 50: 573-586.

Stirling I, 1969b. Tooth wear as a mortality factor in the Weddell seal, *Leptonychotes weddelli.* Journal of Mammalogy 50: 559-565.

Stirling I, 1970. Observations on the behaviour of the New Zealand fur seal (*Arctocephalus forsteri*). Journal of Mammalogy 51: 766-778.

Stirling I, 1971a. Studies on the behaviour of the South Australian fur seal, *Arctocephalus forsteri* (Lesson). I. Annual cycle, postures and calls, and adult males during the breeding season. Australian Journal of Zoology 19: 243-266.

Stirling I, 1971b. Studies on the behaviour of the South Australian fur seal, *Arctocephalus forsteri* (Lesson). II. Adult females and pups. Australian Journal of Zoology 19: 267-273.

Stirling I, 1971c. *Leptonychotes weddelli.* Mammalian Species No. 6. American Society of Mammalogists.

Stirling I, 1971d. Population dynamics of the Weddell seal (*Leptonychotes weddelli*) in McMurdo Sound, Antarctica, 1966-68. *In* Burt WH (Ed.) Antarctic Pinnipedia, pp. 141-161. Antarctic Research Series Vol. 18. American Geophysical Union, Washington, D.C.

Stirling I, 1971e. Population aspects of Weddell seal harvesting at McMurdo Sound, Antarctica. Polar Record 15: 653-667.

Stirling I, 1972. Regulation of numbers of an apparently isolated population of Weddell seals (*Leptonychotes weddelli*). Journal of Mammalogy 53: 107-115.
Stirling I, Greenwood DR, 1972. Observations on a stabilising population of Weddell seals. Australian Journal of Zoology 20: 23-25.
Stirling I, Warneke RM, 1971. Implications of a comparison of the airborne vocalisations and some aspects of the behaviour of the two Australian fur seals *Arctocephalus* spp., on the evolution and present taxonomy of the genus. Australian Journal of Zoology 19: 227-241.
Stock A, 1875. Notice of the existence of a large bat in New Zealand. Transactions of the NZ Institute 8: 180.
Stoddart DR, 1981. History of goats in the Aldabra Archipelago. Atoll Research Bulletin 255: 23-26.
Stolt B-O, 1979. Colour pattern and size variation of the weasel, *Mustela nivalis*, in Sweden. Zoon 7: 55-62.
Strahan R (Ed.), 1980. Recommended common names of Australian mammals. Bulletin of the Australian Mammal Society 6: 13-23.
Strahan R (Ed.), 1983. The Complete Book of Australian Mammals. Angus and Robertson, Sydney.
Strange F, 1850. The Canterbury Plains. Canterbury Papers 3: 77-80.
Straubel CR (Ed.), 1954. The whaling journal of Captain WN Rhodes. Barque "Australian" of Sydney 1836-1838. Whitcombe and Tombs, Christchurch.
Strecker RL, Jackson WB, 1962. Habitats and habits. *In* Storer TI (Ed.) Pacific Island Rat Ecology, pp. 64-73. Bernice P. Bishop Museum Bulletin 225.
Street RJ, 1964. Feeding habits of the New Zealand fur seal. NZ Marine Department Fisheries Technical Report No. 9.
Stubbe M, 1973. Das Hermelin (*Mustela erminea* L.). *In* Stubbe H (Ed.) Buch der Hege, Vol. 1: Haarwild, pp. 288-303. VEB Deutscher Landwirtschaftsverlag, Berlin.
Studholme EC, 1954. Te Waimate. Early Station Life in New Zealand. AH Reed, Dunedin.
Suckling GW, Rudge MR, 1977. Changes with age in the length of central incisor teeth and their clinical crowns in sheep. NZ Journal of Agricultural Research 20: 145-149.
Sumption KJ, Flowerdew JR, 1985. The ecological effects of the decline in rabbits (*Oryctolagus cuniculus* L.) due to myxomatosis. Mammal Review 15: 151-186.
Sutherland A, 1934. Into the Glaisnock Valley after wapiti. NZ Fishing and Shooting Gazette 8: 11-12.
Sweatman GK, 1962. Parasitic mites of non-domesticated animals in New Zealand. NZ Entomologist 3: 15-23.
Sweatman GK, Williams RJ, 1962. Wild animals in New Zealand as hosts of *Echinococcus granulosus* and other taeniid tapeworms. Transactions of the Royal Society of NZ, Zoology 2: 221-250.
Sweeney JM, Sweeney JR, 1982. Feral hog (*Sus scrofa*). *In* Chapman JA and Feldhamer GA (Eds) Wild Mammals of North America, pp. 1099-1113. Johns Hopkins University Press, Baltimore.

Tamarin RH, Malecha SR, 1971. The population biology of Hawaiian rodents: demographic parameters. Ecology 52: 383-394.
Tancred T, 1856. Notes on the natural history of the province of Canterbury, in the middle island of New Zealand. Edinburgh New Philosophical Journal 3: 5-38.
Tapper SC, 1979. The effect of fluctuating vole numbers (*Microtus agrestis*) on a population of weasels (*Mustela nivalis*) on farmland. Journal of Animal Ecology 48: 603-617.
Tapper SC, 1982. Using estate records to monitor population trends in game and

predator species, particularly weasels and stoats. Proceedings of the 14th International Congress of Game Biology, pp. 115-120.

Tapper SC, Green RE, Rands MRW, 1982. Effects of mammalian predators on partridge populations. Mammal Review 12: 159-167.

Tate ML, 1981. The autumn-winter diet of the New Zealand fur seal *Arctocephalus forsteri* (Lesson) with special reference to its cephalopod prey. Unpublished Diploma in Wildlife Management thesis, Otago University, Dunedin.

Taylor JM, Horner BE, 1973. Results of the Archbold Expeditions. No. 98. Systematics of native Australian *Rattus* (Rodentia: Muridae). Bulletin of the American Museum of Natural History 150: 1-130.

Taylor KD, Quy RJ, 1978. Long distance movements of a common rat (*Rattus norvegicus*) revealed by radio-tracking. Mammalia 42: 63-71.

Taylor M, 1984. Bone refuse from Twilight Beach. Unpublished M.A. thesis, University of Auckland.

Taylor NM (Ed.), 1966. The Journal of Ensign Best 1837-1843. Government Printer, Wellington.

Taylor R, 1848. A Leaf from the Natural History of New Zealand; or a Vocabulary of its Different Productions with their Native Names. Robert Stokes, Wellington.

Taylor, Rev. R, 1870. Te Ika a Maui; or New Zealand and its Inhabitants (2nd edition). H. Ireson Jones, Wanganui.

Taylor RH, 1971. Influence of man on vegetation and wildlife of Enderby and Rose Islands, Auckland Islands. NZ Journal of Botany 9: 225-268.

Taylor RH, 1975a. What limits kiore (*Rattus exulans*) distribution in New Zealand? NZ Journal of Zoology 2: 473-477.

Taylor RH, 1975b. The distribution and status of introduced mammals on the Auckland Islands, 1972-73. *In* Yaldwyn JC (Ed.) Preliminary results of the Auckland Islands expedition 1972-1973, pp. 233-243. Department of Lands and Survey, Wellington.

Taylor RH, 1976. Feral cattle in New Zealand. *In* Whitaker AH and Rudge MR (Eds) The Value of Feral Farm Mammals in New Zealand, pp. 13-14. Department of Lands and Survey, Wellington.

Taylor RH, 1978a. Distribution and interactions of rodent species in New Zealand. *In* Dingwall PR, Atkinson IAE and Hay C (Eds) Ecology and control of rodents in New Zealand nature reserves, pp. 135-141. Department of Lands and Survey Information Series No. 4.

Taylor RH, 1978b. *In* Dingwall PR, Atkinson IAE and Hay C (Eds) Ecology and control of rodents in New Zealand nature reserves, p. 173. Department of Lands and Survey Information Series No. 4.

Taylor RH, 1979. Predation on sooty terns at Raoul Island by rats and cats. Notornis 26(2): 199-202.

Taylor RH, 1982. New Zealand fur seals at the Bounty Islands. NZ Journal of Marine and Freshwater Research 16: 1-9.

Taylor RH, 1984. Distribution and interactions of introduced rodents and carnivores in New Zealand. Acta Zoologica Fennica 172: 103-105.

Taylor RH, Magnussen WB, 1965. Immobilising live-trapped opossums with succinylcholine chloride. NZ Journal of Science 8: 531-536.

Taylor RH, Tilly JAV, 1984. Stoats (*Mustela erminea*) on Adele and Fisherman Islands, Abel Tasman National Park, and other offshore islands in New Zealand. NZ Journal of Ecology 7: 139-145.

Taylor-Page FJ, 1959. Field Guide to British Deer. The Mammal Society of the British Isles, London. 86 pp.

Taylor-Page FJ, 1970. The Sussex Mammal Report 1969. Sussex Naturalist Trust, Henfield.

Teal FJ, 1975. Pleasant River Excavations 1959-1962. Unpublished research essay, University of Otago, Dunedin.

Telle HJ, 1966. Beitrag zur Kenntnis der Verhaltensweise von Ratten. Zeitschrift für Angewandte Zoologie 53: 129-196.

Temme M, 1982. Feeding pattern of the Polynesian rat *Rattus exulans* in the Northern Marshall Islands. Zeitschrift für Angewandte Zoologie 69: 463-479.

Temple-Smith PD, 1984. Reproductive structures and strategies in male possums and gliders. *In* Smith A and Hume I (Eds) Possums and Gliders, pp. 89-106. Surrey Beatty and Sons, Chipping Norton, New South Wales.

Tenquist JD, Charleston WAG, 1981. An annotated checklist of ectoparasites of terrestrial mammals in New Zealand. Journal of the Royal Society of NZ 11: 257-285.

Ternovsky DV, 1983. The biology of reproduction and development of the stoat *Mustela erminea* (Carnivora, Mustelidae). Zoologicheskii Zhurnal 62: 1097-1105. (In Russian: translation available from Department of Internal Affairs, Wellington.)

Tetley H, 1945. Notes on British polecats and ferrets. Proceedings of the Zoological Society of London 115: 212-217.

Thomas MD, Warburton B, Coleman JD, 1984. Brush-tailed possum (*Trichosurus vulpecula*) movements about an erosion control planting of poplars. NZ Journal of Zoology 11: 429-436.

Thomas O, 1888. Catalogue of the Marsupialia and Monotremata in the Collection of the British Museum (Natural History). British Museum, London.

Thomson C, Challies CN, 1988. Diet of feral pigs in the podocarp-tawa forests of the Urewera Ranges. NZ Journal of Ecology 11: 73-78.

Thomson GM, 1922. The Naturalisation of Animals and Plants in New Zealand. Cambridge University Press, Cambridge.

Thomson JA, Pears FN, 1962. The functions of the anal glands of the brushtail possum. Victorian Naturalist 78: 306-308.

Thornton FE, 1933. Records and notable heads secured in New Zealand. NZ Fishing and Shooting Gazette 6(9): 3-5; 6(10): 4.

Thornton FE, 1949. The stalking of moose (*Alces malchis*) in New Zealand. NZ Fishing and Shooting Gazette 17(2): 16-18.

Thouars A du Petit, 1841. Voyage autour le monde sur la Fregate la Venus: . . . pendant les années 1836-1839. Relation Tome III, pp. 115-116. Paris.

Tikhomirov EA, 1975. Biology of the ice forms of seals in the Pacific sector of the Antarctic. *In* Ronald K and Mansfield AW (Eds) Biology of the seal. Proceedings of a symposium held in Guelph, 14-17 August 1972, pp. 409-412. Rapports et Procès-verbaux des Réunions du Conseil International pour l'Exploration de la Mer, Vol. 109.

Tinsley R, 1983. Call of the Moose and Other Fiordland Hunting Adventures. AH and AW Reed, Wellington.

Tipene P, 1980. A stranger among us. NZ Journal of Agriculture 141(2): 71-73.

Tisdell CA, 1982. Wild Pigs: Environmental Pest or Economic Resource? Pergamon Press, Sydney.

Titcomb M, 1969. Dog and man in the ancient Pacific. Bernice P. Bishop Museum Special Publication 59, Honolulu.

Todd NB, 1962. Inheritance of the catnip response in domestic cats. Journal of Heredity 53: 54-56.

Tomich PQ, 1968. Coat colour in wild populations of the roof rat in Hawaii. Journal of Mammalogy 49: 74-82.

Tomich PQ, 1970. Movement patterns of field rodents in Hawaii. Pacific Science 24: 195-234.

Tomich PQ, 1979. Studies of leptospirosis in natural host populations. 1. Small mammals of Waipio Valley, Island of Hawaii. Pacific Science 33: 257-279.

Tomich PQ, 1981. Community structure of introduced rodents and carnivores. *In* Mueller-Dombois D, Bridges KW and Carson HL (Eds) Island Ecosystems.

Biological Organisation in Selected Hawaiian Ecosystems, pp. 301-309. Hutchinson Ross, Pennsylvania.

Tomich PQ, Kami HT, 1966. Coat colour inheritance of the roof rat in Hawaii. Journal of Mammalogy 47: 423-430.

Travers HH, 1868. On the Chatham Islands. Transactions and Proceedings of the NZ Institute 1: 119-127. (2nd edition 1875.)

Triggs SJ, 1982. Comparative ecology of the possum, *Trichosurus vulpecula*, in three pastoral habitats. Unpublished M.Sc. thesis, University of Auckland.

Triggs SJ, 1987. Population and ecological genetics of the brush-tailed possum (*Trichosurus vulpecula*) in New Zealand. Unpublished Ph.D. thesis, Victoria University of Wellington.

Trotter MM, 1972. A moa-hunter site near the mouth of the Rakaia River, South Island. Records of the Canterbury Museum 9: 129-150.

Trotter MM, McCulloch B, 1984. Moas, men and middens. *In* Martin PS and Klein RG (Eds) Quaternary Extinctions, pp. 708-727. University of Arizona Press, Tucson.

Troughton E, 1951. Furred Animals of Australia. Angus and Robertson, Sydney.

Troughton E, 1965. Notes from an Australian mammalogist on the usage of the common name possum. Tuatara 13: 192-193.

Tumanor IL, Levin VG, 1974. Age and seasonal changes in some physiological indices of *Mustela nivalis* L. and *Mustela erminea* L. *In* King CM (Ed.) Biology of Mustelids: Some Soviet Research, Vol II, pp. 192-196. DSIR Bulletin 227 (1980).

Turbott EG, 1948. Effects of goats on Great Island, Three Kings, with descriptions of vegetation quadrats. Records of the Auckland Institute and Museum 3(4-5): 253-272.

Turbott EG, 1949. Observations on the occurrence of the Weddell seal in New Zealand. Records of the Auckland Institute and Museum 3: 377-379.

Turbott EG, 1952. Seals of the Southern Ocean. *In* Simpson FA (Ed.) The Antarctic Today, pp. 195-215. AH and AW Reed, Wellington.

Turbott EG, 1961. Birds. *In* Hamilton WM (Comp.) Little Barrier Island (Hauturu) 2nd edition, pp. 136-175. DSIR Bulletin 137.

Turner RW, Padmowirjono S, Martoprawiro S, 1975. Dynamics of plague transmission cycle in Central Java (ecology of mammalian hosts with special reference to *Rattus exulans*). Bulletin Penelitian Kesehatan Health Studies in Indonesia 3: 41-71.

Tustin KG, 1972. Moose in Fiordland. Unpublished report No. 113, NZ Forest Service, Forest Research Institute.

Tustin K, 1973. Moose in Fiordland. An account of a wild moose chase. NZ Wildlife 6(41): 5-15.

Tustin KG, 1974. Status of moose in New Zealand. Journal of Mammalogy 55(1): 199-200.

Tustin KG, 1980. Recent changes in Himalayan thar populations and their effect on recreational hunting. NZ Wildlife 61: 40-48.

Tustin KG, Challies CN, 1978. The effects of hunting on the numbers and group sizes of Himalayan thar (*Hemitragus jemlahicus*) in Carneys Creek, Rangitata Catchment. NZ Journal of Ecology 1: 153-157.

Tustin KG, Parkes JP, 1988. Daily movement and activity of female and juvenile Himalayan thar (*Hemitragus jemlahicus*) in the eastern Southern Alps. NZ Journal of Ecology 11: 51-59.

Tyndale-Biscoe CH, 1955. Observations on the reproduction and ecology of the brush-tailed possum, *Trichosurus vulpecula* Kerr (Marsupialia), in New Zealand. Australian Journal of Zoology 3: 162-184.

Tyndale-Biscoe CH, 1973. Life of Marsupials. Edward Arnold, London.

Tyndale-Biscoe CH, 1984. Reproductive physiology of possums and gliders. *In* Smith A and Hume I (Eds) Possums and Gliders, pp. 79-87. Surrey Beatty and Sons, Chipping Norton, New South Wales.

Tyndale-Biscoe CH, Williams RM, 1955. A study of natural mortality in a wild population of the rabbit, *Oryctolagus cuniculus* (L.). NZ Journal of Science and Technology B 36: 561-580.

Uekermann E, Hansen P, 1968. Das Damwild. Verlag Paul Parey, Hamburg and Berlin.

Vaisfeld MA, 1972. A contribution to the ecology of ermine during the cold season in the European moult. *In* King CM (Ed.) Biology of Mustelids: Some Soviet Research, Vol. II, pp. 1-10. DSIR Bulletin 227 (1980).

Van Aarde RJ, 1983. Demographic parameters of the feral cat *Felis catus* population at Marion Island. South African Journal of Wildlife Research 13: 12-16.

Van Bemmel ACV, 1949. Revision of the Rusine deer in the Indo-Australian Archipelago. Treubia 20(2): 191-262.

Van Mourik S, Schurig V, 1985. Hybridization between sambar (*Cervus* (*Rusa*) *unicolor*) and Rusa (*Cervus* (*Rusa*) *timorensis*) deer. Zoologische Anzeiger (Jena) 214 (3/4, S): 177-184.

van Soest RWM, Van Bree PJH, 1969. On the moult in the stoat, *Mustela erminea* Linnaeus, 1758, from the Netherlands. Bijdragen tot de Dierkunde 39: 63-68.

Van Vuren D, 1981. The feral sheep on Santa Cruz Island: status and impacts. Arlington, Virginia, Nature Conservancy.

Van Wijngaarden A, Bruijns MFM, 1961. De hermelijnen, *Mustela erminea* L., van Terschelling. Lutra 3: 35-42.

Vaz Ferreira R, 1956. Caracteristics generales de las islas Uruguayas habitadas por lobos marinos. Trabajos Sobre Islas de Lobos y Marinos, No. 1, Service Oceanografica Pesca, Montevideo.

Veblen TT, Stewart GH, 1980. Comparison of forest structure and regeneration on Bench and Stewart Islands, New Zealand. NZ Journal of Ecology 3: 50-68.

Veblen TT, Stewart GH, 1982. The effects of introduced wild animals on New Zealand forests. Annals of the Association of American Geographers 72: 372-397.

Veitch CR, 1980. Feral cats on Little Barrier Island. Wildlife — A Review No. 11: 62-64.

Veitch CR, 1982. Eradication of cats from Little Barrier Island. Landscape 11: 27-29.

Veitch CR, 1983. A cat problem removed. Wildlife — A Review No. 12: 47-49.

Veitch CR, 1985. Methods of eradicating feral cats from offshore islands in New Zealand. *In* Moors PJ (Ed.) Conservation of island birds, pp. 125-141. International Council for Bird Preservation Technical Publication No. 3.

Venables LSV, Venables UM, 1955. Birds and Mammals of Shetland. Oliver and Boyd, Edinburgh.

Verbenne G, Leyhausen P, 1976. Marking behaviour of some Viverridae and Felidae. Time-interval analysis of the marking pattern. Behaviour 58: 192-253.

Verme LJ, Ullrey DE, 1984. Physiology and nutrition. *In* Halls LK (Ed.) White-tailed deer — ecology and management, pp. 91-118. Stackpole Books, Harrisburg, Pennsylvania.

Vestjens WJM, Hall LS, 1977. Stomach contents of forty-two species of bats from the Australasian region. Australian Wildlife Research 4: 25-35.

Voller R, 1969. Some statistics from a population of brush-tailed opossums sampled within the breeding season. Unpublished M.Sc. thesis, University of Canterbury, Christchurch.

Vujcich MV, 1979. Aspects of the biology of the parma (*Macropus parma* Waterhouse) and dama (*Macropus eugenii* Desmarest) wallabies with particular emphasis on social organisation. Unpublished M.Sc. thesis, University of Auckland.

Vujcich VC, 1979. Feeding ecology of the parma, *Macropus parma* (Waterhouse) and tammar *Macropus eugenii* (Desmarest) wallabies on Kawau Island. Unpublished M.Sc. thesis, University of Auckland.

Wakefield E, 1842. Letter dated 1842 from Emily Wakefield to Mrs Allom. Alexander Turnbull Library, Wellington.

Wakefield EJ, 1845. Adventure in New Zealand. John Murray, London.

Wakefield E, 1848. The Handbook of New Zealand. Parker, London.

Walker EP, 1964. Mammals of the World. Johns Hopkins Press, Baltimore.

Wallis FP, James IL, 1972. Introduced animal effects and erosion phenomena in the northern Ruahine forests. NZ Journal of Forestry Science 17: 21-36.

Walton T, 1986. Wanganui. The hot-shot hope from the hard hills. The Deer Farmer 33: 28-35.

Warburton B, 1977. Ecology of the Australian brush-tailed possum (*Trichosurus vulpecula* Kerr) in an exotic forest. Unpublished M.Sc. thesis, University of Canterbury, Christchurch.

Warburton B, 1983. Infestations of *Bertiella trichosuri* (Cestoda: Anoplocephalidae) in possums from Claverly, north Canterbury. NZ Journal of Zoology 10: 221-224.

Warburton B, 1986. Wallabies in New Zealand: history, current status, research, and management needs. Forest Research Bulletin No. 114.

Ward GD, 1978. Habitat use and home-range of radio-tagged opossums *Trichosurus vulpecula* (Kerr) in New Zealand lowland forest. *In* Montgomery GG (Ed.) The Ecology of Arboreal Folivores, pp. 267-287. Smithsonian Institution Press, Washington, D.C.

Ward GD, 1984. Comparison of trap- and radio-revealed home ranges of the brush-tailed possum (*Trichosurus vulpecula* Kerr) in New Zealand lowland forest. NZ Journal of Zoology 11: 85-92.

Ward GD, 1985. The fate of young radio-tagged common brushtail possums, *Trichosurus vulpecula*, in New Zealand lowland forest. Australian Wildlife Research 12: 145-150.

Wardle J, 1974. Influence of introduced mammals on the forest and shrublands of the Grey River headwaters. NZ Journal of Forestry Science 4: 459-486.

Wardle JA, 1984. The New Zealand beeches, ecology, utilisation and management. NZ Forest Service, Wellington.

Wardle J, Hayward J, Herbert J, 1971. Forests and scrublands of Northern Fiordland. NZ Journal of Forestry Science 1: 80-115.

Wardle JA, Evans GR, Tustin KG, Challies CN, 1970. Summary report on the vegetation and introduced mammals of northern Fiordland. NZ Forest Service, Forest Research Institute, Protection Forestry Branch Report 91, pp. 1-20.

Wards I, 1976. New Zealand Atlas. Government Printer, Wellington.

Waring GH, 1983. Horse Behaviour. Noyes Publications, New Jersey.

Warneke RM, 1982. Seals in the Australasian region. *In* Mammals in the Seas, Vol. IX: small cetaceans, seals, sirenians and others, pp. 431-475. FAO, Rome.

Warner RE, 1985. Demography and movements of free-ranging domestic cats in rural Illinois. Journal of Wildlife Management 49(2): 340-346.

Watson JS, 1951. The rat problem in Cyprus. Colonial Research Publications 9.

Watson JS, 1954. Reingestion in the wild rabbit, *Oryctolagus cuniculus* (L.). Proceedings of the Zoological Society of London 124: 615-624.

Watson JS, 1956. The present distribution of *Rattus exulans* (Peale) in New Zealand. NZ Journal of Science and Technology 37: 560-570.

Watson JS, 1957. Reproduction of the wild rabbit, *Oryctolagus cuniculus* (L.), in Hawke's Bay, New Zealand. NZ Journal of Science and Technology B 38: 451-482.

Watson JS, 1960. The New Zealand "otter". Records of the Canterbury Museum 7: 175-183.

Watson JS, 1961. Rats in New Zealand: a problem of interspecific competition. Proceedings of the Ninth Pacific Science Congress 19 (Zoology): 15-16.

Watson JS, Taylor RH, 1955. Reingestion in the hare *Lepus europaeus* Pal. Science 121: 314.

Watson JS, Tyndale-Biscoe CH, 1953. The apophyseal line as an age indicator for the wild rabbit, *Oryctolagus cuniculus* (L.). NZ Journal of Science and Technology (B) 34: 427-435.

Watt JC, 1986. Beetles (Coleoptera) of the offshore islands of northern New Zealand. *In* Wright AE and Beever RE (Eds) The offshore islands of northern New Zealand, pp. 221-228. Department of Lands and Survey Information Series No. 16.

Watt RJ, 1975. Notes on some morphological characteristics of kuri crania. Newsletter of the NZ Archaeological Association 18: 140-144.

Watts CHS, 1970. The foods eaten by some Australian desert rodents. South Australian Naturalist 44: 71-74.

Watts CHS, Aslin HJ, 1981. The Rodents of Australia. Angus and Robertson, Sydney.

Watts CHS, Braithwaite RW, 1978. The diet of *Rattus lutreolus* and five other rodents in southern Victoria. Australian Wildlife Research 5: 47-57.

Weber JM, Mermod C, 1983. Experimental transmission of *Skrjabingylus nasicola*, parasitic nematode of mustelids. Acta Zoologica Fennica 174: 237-238.

Weinbren MP, Weinbren BM, Jackson WB, Villella JB, 1970. Studies on the roof rat (*Rattus rattus*) in the El Verde forest. *In* Odum H and Pigeon RF (Eds) A tropical rainforest. A study of irradiation and ecology at El Verde, Puerto Rico, pp. 169-181. US Atomic Energy Commission, Washington D.C.

Wellington Acclimatisation Society, 1892. Annual Report, Wellington.

Wellington Acclimatisation Society, 1893. Annual Report, Wellington.

Wemmer C, Collins L, 1978. Communication patterns in two phalangerid marsupials; the grey cuscus (*Phalanger gymnotis*) and the brush opossum (*Trichosurus vulpecula*). Säugetierkundliche Mitteilungen 26: 161-172.

Westerskov K, 1955. The pheasant in New Zealand. Department of Internal Affairs Wildlife Publications 40.

Westlin LM, Jeffsson B, Meurling P, 1982. The nasal pad of the European hare, *Lepus europaeus* — a histologic and scanning electron microscopic study. Säugetierkundliche Mitteilungen 30: 221-226.

Weston RJ, Repenning CA, Fleming CA, 1973. Modern age of supposed Pliocene seal, *Arctocephalus caninus* Berry (= *Phocarctos hookeri* Gray) from New Zealand. NZ Journal of Science 16: 591-598.

Whitaker AH, 1973. Lizard populations on islands with and without Polynesian rats, *Rattus exulans* Peale. Proceedings of the NZ Ecological Society 20: 121-130.

Whitaker AH, 1978. The effects of rodents on reptiles and amphibians. *In* Dingwall PR, Atkinson IAE and Hay C (Eds) The ecology and control of rodents in New Zealand nature reserves, pp. 75-86. Department of Lands and Survey Information Series 4.

Whitaker AH, Rudge MR (Eds) 1976. The value of feral farm mammals in New Zealand. Department of Lands and Survey Information Series 1.

Whitaker JO Jr, 1966. Food of *Mus musculus, Peromyscus maniculatus bairdi* and *Peromyscus leucopus* in Vigo County, Indiana. Journal of Mammalogy 47: 473-486.

White T, 1890. On rats and mice. Transactions of the NZ Institute 23: 194-201.

White T, 1897. On rats, and their nesting in small branches of trees. Transactions and Proceedings of the NZ Institute 30: 303-309.

Whitehead GK, 1950. Deer and their Management. Country Life, London.

Whitehead GK, 1972. Deer of the World. Constable, London.

Whittle P, 1955. An investigation of periodic fluctuations in the New Zealand rabbit population. NZ Journal of Science and Technology B 37: 54-58.

Wildhaber CA, 1984. Chin rubbing behaviour in ferrets (*Mustela putorius furo*). Unpublished B.Sc. (Hons). dissertation, Massey University, Palmerston North.

Williams CK, Turnbull HL, 1983. Variations in seasonal nutrition, thermoregulation and water balance in two New Zealand populations of the common brushtail possum, *Trichosurus vulpecula* (Phalangeridae). Australian Journal of Zoology 31: 333-343.

Williams GR (Ed.), 1973. The Natural History of New Zealand: an Ecological Survey. AH and AW Reed, Wellington.

Williams GR, 1976. The New Zealand wattlebirds (Callaeatidae). *In* Frith HJ and Calaby JH (Eds) Proceedings of the 16th International Ornithological Congress, pp. 161-170. Australian Academy of Science, Canberra.

Williams GR, Given DR, 1981. The Red Data Book of New Zealand. Rare and endangered species of endemic terrestrial vertebrates and vascular plants. Nature Conservation Council, Wellington.

Williams GR, Harrison M, 1972. The laughing owl *Sceloglaux albifacies* (Gray, 1844). A general survey of a near-extinct species. Notornis 19: 4-19.

Williams GR, Rudge MR, 1969. A population study of feral goats (*Capra hircus* L.) from Macauley Island, New Zealand. Proceedings of the NZ Ecological Society 16: 17-28.

Williams H, Williams M, 1822. Letters and papers of Henry and Marianne Williams, 1822-24. MS., Alexander Turnbull Library, Wellington.

Williams JM, 1974a. The ecology and behaviour of *Rattus* species in relation to the yield of coconuts and cocoa in Fiji. Unpublished Ph.D. thesis, University of Bath, England.

Williams JM, 1974b. The effect of artificial rat damage on coconut yields in Fiji. PANS 20: 275-282.

Williams JM, 1974c. Rat damage to coconuts in Fiji. Part 1. Assessment of damage. PANS 20: 379-391.

Williams JM, 1975. Rat damage to coconuts in Fiji. Part 2. Efficiency and economics of damage reduction methods. PANS 21: 19-26.

Williams JM, 1983. The impact of biological research on rabbit control policies in New Zealand. Acta Zoologica Fennica 174: 79-83.

Williams JM, Robson DL, 1985. Rabbit ecology and management in the western Pest Destruction Board. Ministry of Agriculture and Fisheries, Agricultural Research Special Publication.

Williams JM, Bell J, Ross WD, Broad TM, 1986. Rabbit (*Oryctolagus cuniculus*) control with a single application of 50 ppm brodifacoum cereal baits. NZ Journal of Ecology 9: 123-136.

Williams L, 1980. Kohika coprolites. Unpublished M.A. research essay, University of Auckland.

Williamson GM, 1986. The ecology of the dama wallaby (*Macropus eugenii* Desmarest) in forests at Rotorua, with special reference to diet. Unpublished M.Sc. thesis, Massey University, Palmerston North.

Williamson MJ, 1975. A study of feral goats (*Capra hircus* L.) from Tai Pari, French Pass area. Unpublished B.For.Sc. thesis, School of Forestry, University of Canterbury, Christchurch.

Willis M, 1982. Some of My Best Friends are Animals. Whitcoulls, Christchurch.

Wilson CJN, Ambraseys NN, Bradley J, Walker GPL, 1980. A new date for the Taupo eruption, New Zealand. Nature (London) 288: 252-253.

Wilson DC, 1878. Disappearance of the small birds of New Zealand. Transactions of the NZ Institute 10: 239-242.

Wilson EA, 1907. Weddell's seal, Mammalia. National Antarctic Expeditions, 1901-04, Natural History 2: 1-66. British Museum (Natural History), London.

Wilson E, 1976. Hakora ki te Iwi. The Story of Captain Howell and his Family. Times Printing Service, Invercargill.

Wilson GJ, 1974. Distribution, abundance and population characteristics of the New Zealand fur seal (*Arctocephalus forsteri*). Unpublished M.Sc. thesis, University of Canterbury, Christchurch.

Wilson GJ, 1975. A second survey of seals in the King Haakon VII Sea, Antarctica. South African Journal of Antarctic Research 2105: 31-36.

Wilson GJ, 1979. Hooker's sea lions in southern New Zealand. NZ Journal of Marine and Freshwater Research 13: 373-375.

Wilson GJ, 1981. Distribution and abundance of the New Zealand fur seal, *Arctocephalus forsteri*. Fisheries Research Division Occasional Publication No. 20.

Wilson H, 1986. Long-term study of regeneration following the fire of March 1970 on the Liebig Range, Mount Cook National Park. Botany Division Newsletter No. 109: 12-14.

Wilson PR, 1980. Report on short term visit to Campbell Island. *In* Preliminary Reports of the Campbell Island Expedition 1975-76. Department of Lands and Survey Reserves Series 7: 153.

Wilson PR, Collier AJ, 1981. Lungworm in deer: a survey of veterinary practices. *In* Proceedings of a deer seminar for veterinarians, Queenstown, pp. 85-93. Deer Advisory Panel of the NZ Veterinary Association.

Wilson PR, Orwin DFG, 1964. The sheep population of Campbell Island. NZ Journal of Science 7: 460-490.

Winter JW, 1963. Observations on a population of the brush-tailed opossum (*Trichosurus vulpecula* Kerr). Unpublished M.Sc. thesis, University of Otago, Dunedin.

Winter JW, 1976. The behaviour and social organization of the brushtailed possum (*Trichosurus vulpecula* Kerr). Unpublished Ph.D. thesis, University of Queensland.

Winter JW, 1980. Tooth wear as an age index in a population of the brushtailed possum, *Trichosurus vulpecula* (Kerr). Australian Wildlife Research 7: 359-369.

Wirtz WO, 1972. Population ecology of the Polynesian rat, *Rattus exulans*, on Kure Atoll, Hawaii. Pacific Science 16: 433-463.

Wirtz WO, 1973. Growth and development of *Rattus exulans*. Journal of Mammalogy 54: 189-202.

Wodzicki KA, 1950. Introduced mammals of New Zealand: an ecological and economic survey. DSIR Bulletin 98.

Wodzicki KA, 1969. Preliminary report on damage to coconuts and on the ecology of the Polynesian rat (*R. exulans*) in the Tokelau Islands. Proceedings of the NZ Ecological Society 16: 7-12.

Wodzicki KA, 1978–79. Relationships between rats and man in the Central Pacific. Ethnomedizine 3/4: 433-446.

Wodzicki KA, Bull PC, 1951. The small mammals of the Caswell and George Sounds areas. *In* Preliminary reports of the New Zealand-American Fiordland Expedition, pp. 62-69. DSIR Bulletin 103.

Wodzicki KA, Darwin JH, 1962. Observations on the reproduction of the wild rabbit (*Oryctolagus cuniculus* L.) at varying latitudes in New Zealand. NZ Journal of Science 5: 463-474.

Wodzicki KA, Flux JEC, 1967a. Guide to introduced wallabies in New Zealand. Tuatara 15: 47-59.

Wodzicki KA, Flux JEC, 1967b. Re-discovery of the white-throated wallaby, *Macropus parma* Waterhouse 1846, on Kawau Island, New Zealand. Australian Journal of Science 29: 429-430.

Wodzicki KA, Flux JEC, 1971. The parma wallaby and its future. Oryx 11: 40-47.

Wodzicki KA, Robertson FH, 1959. Birds, with a note on the mammal *Rattus exulans* (Peale). *In* Hamilton WM and Baumgart IL (Comps) White Island, pp. 70-82. DSIR Bulletin 127.

Wodzicki KA, Taylor RH, 1984. Distribution and status of the Polynesian rat *Rattus exulans*. Acta Zoologica Fennica 172: 99-101.

Wodzicki KA, Wright S, 1984. Introduced birds and mammals in New Zealand and their effect on the environment. Tuatara 27: 77-104.

Wood DH, 1980. The demography of a rabbit population in an arid region of New South Wales, Australia. Journal of Animal Ecology 49: 55-79.

Wood GW, Brenneman RE, 1977. Research and management of feral hogs on Hobcaw

Barony. *In* Wood GW (Ed.) Research and Management of Wild Hog Populations, pp. 23-35. Belle W. Baruch Forest Science Institute, Georgetown.

Wood-Jones F, 1931. The cranial characters of the Hawaiian dog. Journal of Mammalogy 12: 39-41.

Woodward PW, 1972. The natural history of Kure Atoll, northwestern Hawaiian Islands. Atoll Research Bulletin 164: 1-318.

Worth CB, 1950. Field and laboratory observations on roof rats, *Rattus rattus* (Linnaeus), in Florida. Journal of Mammalogy 31: 293-304.

Worthy TH, 1984. Faunal and floral remains from F1, a cave near Waitomo. Journal of the Royal Society of NZ 14: 367-377.

Worthy TH, 1987. Osteology of *Leiopelma* (Amphibia; Leiopelmatidae) and descriptions of three new subfossil species. Journal of the Royal Society of NZ 17: 201-251.

Wright AE, 1976. Auckland University field club scientific trip to Great Mercury Island, May 1975. Tane 22: 1-3.

Wright PL, 1950. Development of the baculum of the long tailed weasel. Proceedings of the Society for Experimental Biology and Medicine (New York) 75: 820-822.

Wroot AJ, 1985. Foraging in the European hedgehog *Erinaceus europaeus*. Mammal Review 15: 2.

Yabe T, 1982. Habitats and habits of the roof rat *Rattus rattus* on Torishima, the Izu Islands. Journal of the Mammalogical Society of Japan 9: 20-24.

Yalden DW, Hosey GR, 1971. Feral wallabies in the Peak district. Journal of Zoology (London) 165: 513-520.

Yerex D (Ed.), 1982. The Farming of Deer. World Trends and Modern Techniques. Agricultural Promotional Associates, Wellington, 176 pp.

Yocom CF, 1967. Ecology of feral goats in Haleakala National Park, Maui, Hawaii. American Midland Naturalist 77: 418-451.

Yom-Tov Y, Green WQ, Coleman JD, 1986. Morphological trends in the common brushtail possum, *Trichosurus vulpecula*, in New Zealand. Journal of Zoology (London) 208: 583-593.

Yosida TH, 1973. Evolution of karyotypes and differentiation in 13 *Rattus* species. Chromosoma (Berlin) 40: 285-297.

Yosida TH, 1980. Cytogenetics of the Black Rat. University of Tokyo Press, Tokyo.

Yosida TH, Kato H, Tsuchiya K, Sagai T, Moriwaki K, 1974. Cytogenetical survey of black rats, *Rattus rattus*, in southwest and central Asia, with special regard to the evolutional relationship between three geographical types. Chromosoma (Berlin) 45: 99-109.

Young RA, 1979. Observations on parturition, litter size, and foetal development at birth in the chocolate wattled bat, *Chalinolobus morio* (Vespertilionidae). Victorian Naturalist 96: 90-91.

Zeuner FE, 1968. A History of Domesticated Animals. Hutchinson, London.

Zima J, Král B, 1984. Karyotypes of European mammals III. Acta Scientiarum Naturalium (Brno) 18(9): 1-51.

Zörner H, 1977. Ergebnisse der Untersuchung über die Ernährung des Feldhasen — *Lepus europaeus* (Pallas, 1778) — im Wildforschungsgebiet Hakel. Beiträge zur Jagd- und Wildforschung 10: 255-266.

Zörner H, 1981. Der Feldhase. Die Neue Brehm-Bucherei. Ziemsen, Wittenberg Lutherstadt.

Zotov VD, 1949. Forest deterioration in the Tararuas due to deer and opossums. Transactions and Proceedings of the Royal Society of NZ 77: 162-165.

Index

Alces alces 514
Alpaca 358
Arctocephalus forsteri 246
aries, Ovis 424
Axis axis 488

Bandicoot, southern brown 34
Bat, greater New Zealand short-tailed 131
lesser New Zealand short-tailed 123
New Zealand long-tailed 117
Bharal 392
bicolor, Wallabia 64
Bos taurus 373

caballus, Equus 349
Canis familiaris 281
Capra hircus 406
Cat, house 330
marsupial 34
Cattle, feral 373
catus, Felis 330
Cavia porcellus 173
Cervus elaphus nelsoni 458
e. scoticus 436
nippon 468
timorensis 483
u. unicolor 477
Cetaceans 13
Chalinolobus tuberculatus 117
Chamois 380
Chipmunk, gray 173
Connochaetes gnou 371
cuniculus cuniculus, Oryctolagus 138

Dama d. dama 490
Dasyurus sp. 34
Deer, axis 488
fallow 490
mule 506
red 436
rusa 483
sambar 477
sika 468
white-tailed 507
unidentified South American 506
Dog, European 281
Polynesian 281
Dolphins 13
dorsalis, Macropus 57

edwardsi, Herpestes 280
elaphus nelsoni, Cervus 458
elaphus scoticus, Cervus 436
Elk 514
Equus caballus 349
zebra 349
Erinaceus europaeus occidentalis 99
erminea, Mustela 288
europaeus occidentalis, Erinaceus 99
europaeus occidentalis, Lepus 161
exulans, Rattus 175

familiaris, Canis 281
Felis catus 330
Ferret 320
forsteri, Arctocephalus 246
furo, Mustela 320

glama, Lama 358
gnou, Connochaetes 371
Gnu 371
Goat, feral 406

Hare, brown 161
Hedgehog, west European 99
hemionus, Odocoileus 506
Hemitragus jemlahicus 392
Herpestes edwardsi 280
hircus, Capra 406
hookeri, Phocarctos 256
Horse, feral 349
Hydrurga leptonyx 268

Isoodon obesulus 34

jemlahicus, Hemitragus 392

Kangaroo, unidentified 35
Kiore 175
Kuri 281

Lama glama 358
pacos 358
leonina, Mirounga 262
Leptonychotes weddelli 268
leptonyx, Hydrurga 268
Lepus europaeus occidentalis 161
Little red flying fox 136
Llama 358
Lobodon carcinophagus 274
lotor, Procyon 280

Macropus dorsalis 57
eugenii 35
parma 51
robustus 35
r. rufogriseus 44
Mirounga leonina 262
Mongoose, Indian grey 280
Moose 514
Mouse, house 225
Mus musculus 225
Mustela erminea 288
furo 320
nivalis vulgaris 313
Mystacina robusta 131
tuberculata 123

nayaur, Pseudois 392
New Zealand bats 115
New Zealand fur seal 246
New Zealand 'otter' 287
New Zealand sea lion 256
nippon, Cervus 468
nivalis vulgaris, Mustela 313
norvegicus, Rattus 192

obesulus, Isoodon 34
Odocoileus virginianus borealis 507
hemionus 506
Ommatophoca rossi 277
Oryctolagus c. cuniculus 138
Ovis aries 424

pacos, Lama 358
parma, Macropus 51
penicillata penicillata, Petrogale 58
peregrinus, Pseudocheirus 34
Phocarctos hookeri 256
Pig, feral 358
guinea 173
porcellus, Cavia 173
Possum, brushtail 68
ringtail 34
Potoroo, long-nosed 34
Potorous tridactylus 34
Procyon lotor 280
Pseudocheirus peregrinus 34
Pseudois nayaur 392
Pteropus scapulatus 136

Rabbit, European 138
Raccoon 280
Rat, Norway 192
Polynesian 175
ship 206
Rattus exulans 175
norvegicus 192
rattus 206
robusta, Mystacina 131
robustus, Macropus 35
rossi, Ommatophoca 277
rufogriseus rufogriseus, Macropus 44
Rupicapra r. rupicapra 380

scrofa, Sus 358
scapulatus, Pteropus 136
Seal, crabeater 274
leopard 271
New Zealand fur 246
Ross 277
southern elephant 262
Weddell 268
Sheep, blue 392
feral 424
Squirrel, brown California 173
Stoat 288
striatus, Tamias 173
Sus scrofa 358

Tahr, Himalayan 392
Tamias striatus 173
taurus, Bos 373
timorensis, Cervus 483
Trichosurus vulpecula 68
tridactylus, Potorous 34
tuberculata, Mystacina 123
tuberculatus, Chalinolobus 117

unicolor unicolor, Cervus 477

virginianus borealis, Odocoileus 507
vulpecula, Trichosurus 68

Wallabia bicolor 64
Wallaby, Bennett's 44
black-striped 57
brushtailed rock 58
dama 35
parma 51
roan 35

swamp 64
unidentified 35
Wapiti 458
weddelli, Leptonychotes 268
Weasel, common 313
Whale 13

Zebra 349
zebra, Equus 349

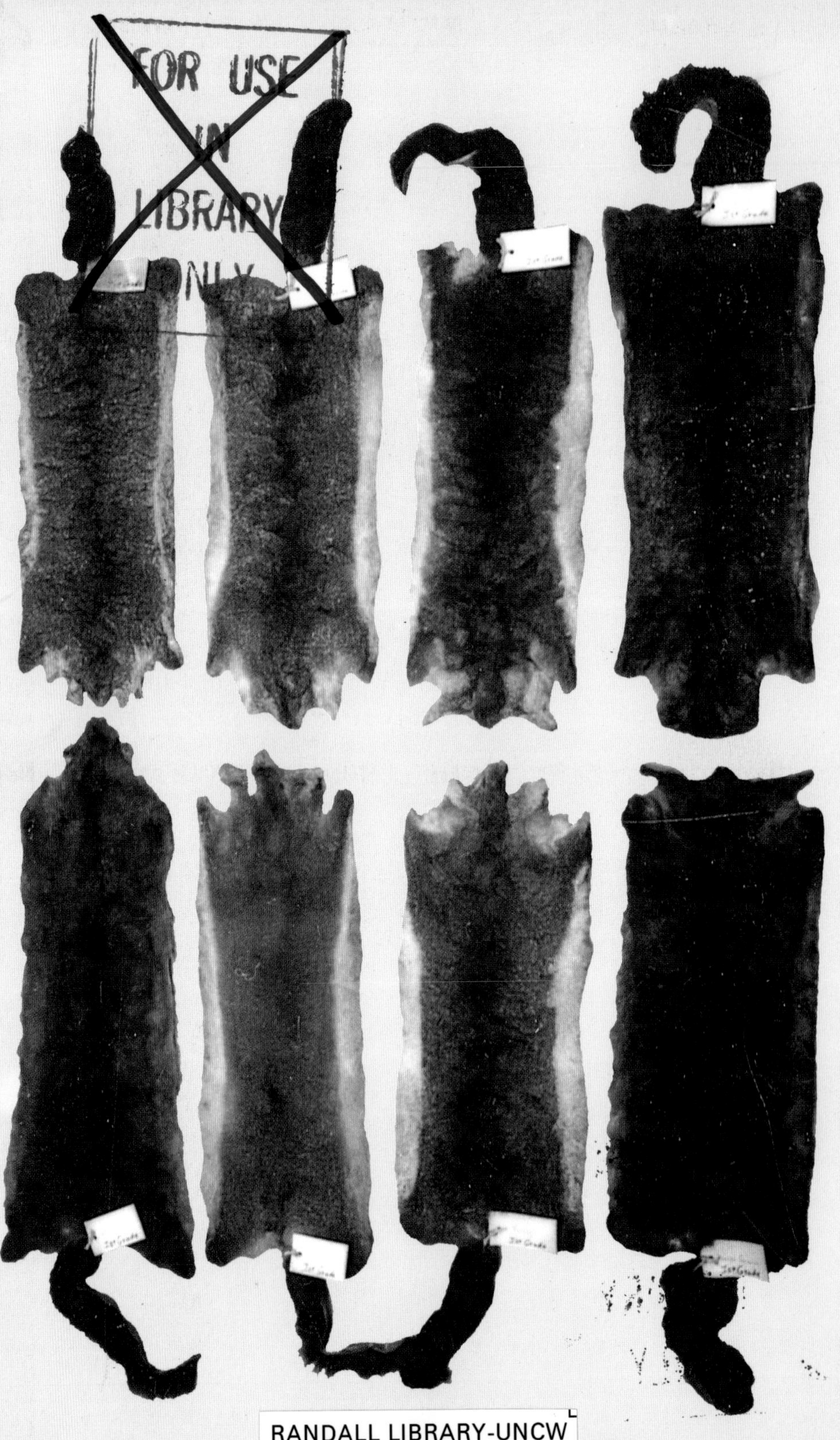

FOR USE IN LIBRARY ONLY

RANDALL LIBRARY-UNCW

3 0490 0425608 +